Biomedical Engineering and its Applications in Healthcare

Sudip Paul
Editor

Biomedical Engineering and its Applications in Healthcare

Springer

Editor
Sudip Paul
Department of Biomedical Engineering
North-Eastern Hill University
Shillong, Meghalaya, India

ISBN 978-981-13-3704-8 ISBN 978-981-13-3705-5 (eBook)
https://doi.org/10.1007/978-981-13-3705-5

This Springer imprint is published by the registered company Springer Nature Singapore Pte Ltd.
The registered company address is: 152 Beach Road, #21-01/04 Gateway East, Singapore 189721, Singapore

Preface

This book *Biomedical Engineering and its Applications in Healthcare* illustrates the importance and significance of biomedical engineering in the modern healthcare system. Biomedical engineering plays an important role from diagnosis and analysis to treatment and recovery and has entered the public conscience through the proliferation of implantable medical devices, such as pacemakers and artificial hips, to the more futuristic technologies, such as stem cell engineering and 3D printing of biological organs. The book encompasses the logical sequence of the chapters starting from the basic introduction of biomedical engineering and different tools and techniques for medical diagnostics and treatment to recent advancements. It provides comprehensive and integrated information on rehabilitation engineering that is involved in the designing of artificial body parts, its underlying principles, and standards. It also provides the conceptual framework to understand the relationship of ethical policies in medical practice and philosophical moral reasoning. Finally, the book highlights a number of challenges associated with modern healthcare technologies.

The last two decades have seen a drastic shift in the healthcare technology. The only and the most validating backgrounds are being the research applications across all relevant fields came together to bridge the gap between the engineering applications to the medical science. There lies the success of the name "biomedical engineering" which is the application of engineering disciplines, technology, principles, and design concepts to medicine and biology. With the different domains spread across the biomedical engineering divisions, the utmost objective in research and education is to improve the quality of living, reduce the impact of disease in our daily life, and promote for a sophisticated infrastructural facility to enhance the collaboration of biomedical engineering researches across the globe.

Biomedical engineering is spread across several sub-disciplines (but not limited to) such as bioinstrumentation, biostatistics, biomechanics, biomaterials, biosignal processing, rehabilitation, and the safety aspects, keeping in mind the core and emerging branches. This book, which is a collaborative effort of various contributors coming hand in hand, consists of 12 parts with 30 chapters.

Part I gives the basic overview of human physiology. Understanding the subject on which we would apply our engineering skills is an essential part of research and design. ***Part II*** establishes the application physics into medical sciences. ***Part III***

gives the overview of biosensors and transducers on biosignal detection and acquisition of different biophysical parameters. ***Part IV*** gives an overview of the use of biological materials and biological architectures for information processing systems and new devices especially for physically disabled. The sensitive biological element may be a biologically derived material that recognizes, interacts, and binds the subjects under study. ***Part V*** deals with the overview and design aspects of the most important components of biomedical engineering, i.e., bioinstrumentation. Over the last few decades, the advances in this field have revolutionized the healthcare world to newer heights. Later part emphasizes on a very recent technique for telemonitoring applications.

Part VI presents an introduction to biomaterials and its characteristics and application in therapeutic and diagnostic fields. Biomaterials have now established a strong footing in history with large number of multinational companies that have joined hands investing substantial financial capital in developing new products. Detailed discussion on analytical devices used for detection of chemical substances, which combines a biological component with a physiochemical detector, is also presented. ***Part VII*** covers the different forms of biological signals which are generated in our body at different levels like molecular level, cell level, or organ level. These signals can be in the form of EEG, EMG, EOG, EKG, etc. The second part shows the significance of these biological signals in designing of different assistive devices in rehabilitation. ***Parts VIII and IX*** deal with the two very important aspects of bioengineering, i.e., biosignal processing and bio-image processing. Substantial insight has been given in these two fields of study, keeping in mind the latest updates in the two fields like application of artificial neural network (ANN) and noninvasive techniques. ***Part X*** deals with the convergence of pathology with physiology – the pathophysiology of diseases causing physical disability. Calibration, repair, and safety are crucial aspects of a bioinstrumentation that come under the service perspective of a biomedical engineer. Since a biomedical engineer has to deal with the most complicated machinery on Earth, regular calibration of the instruments becomes the need of the hour. Safety aspects also need to be handled with utmost priority. ***Part XI*** focuses on a very important aspect of bioengineering which is rehabilitation engineering. This part specifically focuses on biomechanical perspectives of locomotion in the lower limb joints. ***Part XII*** on medical ethics and policies is also a crucial aspect for a biomedical engineer in designing, calibration, and implementation of prosthetic and biomaterials. All these aspects are covered under this part.

As an editor, I have tried to touch almost all aspects related to biomedical engineering. My heartful thanks to all the chapter contributors of the book.

Shillong, India Sudip Paul

Acknowledgment

I would like to extend my gratitude to all the chapter authors for their sincere and timely support to make this book a grand success.

I am equally thankful to all the executive board members of Springer Nature for their kind approval for me as an editor of this book.

I would also like to extend my sincere thanks to Dr. Bhavik Sawhney, Associate Editor, Biomedicine, Springer Nature, and Ms. RaagaiPriya ChandraSekaran, Production Editor (Books), Springer Nature, for their valuable suggestions and encouragement throughout the project.

It is with immense pleasure to express my thankfulness to my research scholars and colleagues for their support, love, and motivation in all my efforts during this project.

I am grateful to all the reviewers for their timely review and consent which helped me a lot to improve the quality of the book.

I am speechless! I can barely find words to express all the wisdom, love, and support given to me by my beloved parents and my beloved wife and sweet son for their unconditional love, fidelity, endurance, and encouragement. I express my deep gratitude for their love, without which, this book project would not have been completed.

There are so many others whom I may have inadvertently left out, but I sincerely thank all of them for their help.

Contents

Part I Basic Overview of Human Physiology

1 **Human Anatomy and Physiology** 3
Dhiviyalakshmi Lakshmipathy

2 **Anatomically Real Microwave Tissue Phantoms** 43
Gyanendra Sheoran and Vineeta Kumari

3 **Biophotonics in Disease Diagnosis and Therapy** 65
Shrutidhara Biswas, Vlad Bogdan Gavra, Anand Kant Das, and Umakanta Tripathy

Part II Basics of Bioelectronics

4 **Overview of Medical Electronics for Physically Disabled** 89
Vinay Kumar Pandey and Sudip Paul

5 **Advances in Diagnostic Techniques for Therapeutic Intervention** . . 105
Geetesh Verma, Radhika Kesharwani, Pabbala Veeresh, Harpreet Kaur, Deepaneeta Sarmah, Vignesh Kotian, Leela Mounica, Anupom Borah, Kiran Kalia, and Pallab Bhattacharya

Part III Biosensors and Transducers

6 **Biosensors and its Transducers** 125
Janani Viswananthan and Gopu Govindasamy

7 **Advance Biomedical Sensors and Transducers** 153
Harishchandra Digambar Jirimali

Part IV Biomaterials and Its Medical Applications

8 **Introduction to Ideal Characteristics and Advanced Biomedical Applications of Biomaterials** 171
Govinda Kapusetti, Namdev More, and Mounika Choppadandi

9 **Nanostructured Materials and Their Biomedical Application** 205
Sudip Mondal and Junghwan Oh

10 **Ceramics for Biomedical Applications** 229
Shyamal Mandal

11 **Biopolymeric Scaffolds for Tissue Engineering Application** 249
Nalini Ranganathan, A. Mugeshwaran, R. Joseph Bensingh, M. Abdul Kader, and Sanjay K. Nayak

12 **Surface Modifications in Ti-Based Orthopaedic Implants** 275
Sudip K. Sinha

Part V Bioinstrumentation and Its Design Aspects

13 **Biomedical Instrumentation: Focus Toward Point-of-Care Devices** 297
Sandeep Choudhary, Gaurav Pandey, Rupsha Mukherjee, and Abhijeet Joshi

14 **Techniques/Tools to Study Epigenetic Biomarkers in Human Cancer Detection** 327
Vivek Kumar, Alka Singh, Priyanka Gautam, and Manisha Sachan

Part VI Techniques Related to Disease Diagnosis and Therapeutic

15 **Paper-Based Sensors for Biomedical Applications** 355
Mohd Aurif Shergujri, Rabeuj Jaman, Arup Jyoti Baruah, Mrityunjoy Mahato, Davidson Pyngrope, L. Robindro Singh, and Manashjit Gogoi

16 **Emerging Point-of-Care Diagnostic Methods for Disease Detection** 377
Smriti Singh, Pranav Tripathi, and Seema Nara

17 **Computer Aided Design and Synthesis for Marker Proteins of HT Carcinoma Cells: A Study** 399
Shruti Jain and Durg Singh Chauhan

Part VII Biosignals and Its Significance

18 **Biomedical Signal Analysis and Its Usage in Healthcare** 423
Abdulhamit Subasi

19 **Importance of Bio-signal for Rehabilitative Engineering** 453
Uvanesh Kasiviswanathan and Neeraj Sharma

Part VIII Medical Imaging and Image Processing

20 **Elements of Medical Image Processing** 473
T. Emami, S. S. Janney, and S. Chakravarty

21 **"Overview of X-Ray Tube Technology"** 519
M. Anburajan and Jitendar Kumar Sharma

22 **Prospects of Avalanche Transit Time Terahertz Radiation Source in Biomedical Imaging: Application Feasibility in Health Engineering** 549
Moumita Mukherjee

Part IX Pathophysiology of Diseases Causing Physical Disability

23 **Pathophysiology of the Disease Causing Physical Disability** 573
Sachchida Nand Rai, Hareram Birla, Saumitra Sen Singh, Walia Zahra, Aaina Singh Rathore, Hagera Dilnashin, and Surya Pratap Singh

24 **Physical Impairments Associated with Diseases: A Pathophysiological Approach** 597
Vignesh Kotian, Leela Mounica, Deepaneeta Sarmah, Harpreet Kaur, Geetesh Verma, Radhika Kesharwani, Pabbala Veeresh, Anupom Borah, Kiran Kalia, and Pallab Bhattacharya

Part X Calibration, Repair and Safety Aspects

25 **The Technical Support: Repair, Preventive Maintenance, and Inspection** 621
Mana Sezdi

Part XI Advances in Biomedical Engineering

26 **Recent Advances on Polymer Nanocomposite-Based Radiation Shielding Materials for Medical Science** 639
Abhijit Nath, Aunggat Shah, Sanjeev Bhandari, Manashjit Gogoi, and Mrityunjoy Mahato

27 **Biodegradable Composite Scaffold for Bone Tissue Regeneration** 657
Sandip Bag

28 **Advancements and New Technologies in Drug Delivery System** . . . 681
Ajay Kumar Sahi, Pooja Verma, Pallawi, Kameshwarnath Singh, and Sanjeev Kumar Mahto

29 Recent Advances in the au NP Treatment Strategies of Lung Cancers 701
Parth Malik and Rakesh Kumar Ameta

Part XII Medical Ethics and Policies

30 Medical Ethics and Policies Related to Biomedical Equipment 733
Ramkrishna Mondal

About the Editor

Dr. Sudip Paul is currently an Assistant Professor at the Department of Biomedical Engineering, School of Technology, North-Eastern Hill University (NEHU), Shillong, India. He received his Ph.D. from the Indian Institute of Technology (Banaras Hindu University), Varanasi, with specialization in electrophysiology and brain signal analysis. He was selected as a Postdoc Fellow in 2017–2018 under the Biotechnology Overseas Associateship for scientists working in the north-eastern states of India, supported by the Department of Biotechnology, Government of India. He has organized several workshops and conferences, most significantly the 29th Annual Meeting of the Society for Neurochemistry, India, and IRBO/APRC Associate School in 2017. Dr. Sudip has published more than 90 international journal and conference papers and has also filed 4 patents. Recently, he completed three book projects and is currently serving as the editor for a further two. Dr. Sudip is a member of several societies and professional bodies, including APSN, ISN, IBRO, SNCI, SfN, and IEEE. He was awarded the First Prize in the Sushruta Innovation Award 2011, sponsored by the Department of Science and Technology, Govt. of India, and numerous other awards, including the World Federation of Neurology (WFN) travelling fellowship, Young Investigator Award, and IBRO and ISN Travel Awards. Dr. Sudip has served as an editorial board member for a variety of international journals. He has presented his research in the USA, Greece, France, South Africa, and Australia.

Part I

Basic Overview of Human Physiology

Human Anatomy and Physiology

1

Dhiviyalakshmi Lakshmipathy

Abstract

Biomedical engineering is the discipline which deals with all the devices and instruments that will interact with the human system. This chapter will facilitate the learner to understand the basic underlying concepts of body planes, movements and directions which is quite important in imaging devices like ultrasound, CT and MRI. Biosignal is the source for many types of equipment like ECG machine where biosignal's generation and conduction are the key mechanisms to know. Hence, it is very essential for anyone who comes in close interaction with the equipments to have a wise knowledge on anatomy and physiology of human system.

Keywords

Body planes · Movements · Directions · Biosignals

1.1 Introduction

Man's cognitive expression of an object is descriptive of the following: How does the object look? What is the colour? What is the shape? What is the size? The same sense of appreciating structure of the organs is human anatomy. In ancient medicine, where technology did not make a footprint in medical sciences, the only choice left with the physician to diagnose a condition or disease is simply cut open and explore. Thus, early medical diagnosis was made by surgical operation. In general, most of the medical terms are derived from greek language, the word "anatome" means "to cut open". Hence, anatomy is defined as the study of the structure of the organs of the

Dhiviyalakshmi Lakshmipathy (✉)
Department of Biomedical Engineering, Sri Ramakrishna Engineering College, Coimbatore, Tamilnadu, India

S. Paul (ed.), *Biomedical Engineering and its Applications in Healthcare*,
https://doi.org/10.1007/978-981-13-3705-5_1

Table 1.1 Sub-specializations of human anatomy

Sub-class	Specialization
Gross anatomy (or) macroscopic anatomy	Study of large structures of the body
Regional anatomy	Study of the structures that contribute to specific body regions
Systemic anatomy	Study of the structures that contribute to specific body systems
Microscopic anatomy (or) histology	Study of very small structures of the body using magnification
Developmental anatomy	Study of the structural changes that occur in the body throughout the lifespan
Embryonic anatomy	Study of the stages of the growth of an embryo

body and their interactions. Human physiology is defined as the study of the functions of the organs and their interactions with other organs and system of the human body. Physiology is a multifaceted process of physical, chemical, electrical, electro-chemical, mechanical interaction of the living system (Waugh and Grant 2014).

To relate our discussion with the definition stated earlier, let us discuss about heart. The heart is a conical organ, comprising four chambers: two auricles and two ventricles. The heart pumps the blood into the vascular system. The electrical stimulation from the nucleus solitarius located in the medulla oblongata is conducted to the heart, which excites the heart muscles to execute contraction and relaxation as a result of which the blood is pumped. In this, describing the shape, chambers and structure is an anatomical approach towards the organ, and describing about the electrical stimulation, conduction and contraction is physiology of the system. There are many sub-specializations in human anatomy, which are mentioned in Table 1.1.

1.2 Medical Terminologies

The naming system followed in anatomy and physiology will have a very close relation to anyone of the following or to all of location, structure and function. Medical terminology is the language of mentioning the body parts and its organization and functions (Marieb 2007; Hall 2013). It is generally made up of certain keywords, like root word or base word, which are the soul of the terminology, and it also consists of prefixes and suffixes which add meaning to the root word such as the condition of the system.

For example:

Phonocardiography: In this if we split the words *phono*, *cardio* and *graphy*, *phono* means sound, *cardio* means heart, and *graphy* means to record. Thus, the medical terminology phonocardiography means recording of heart sounds.

Stethoscope: In this medical terminology, *stetho* means chest and *scope* means instrument. Hence, stethoscope means an instrument used to examine the chest.

1.3 Body Planes

When we know how to label a part or a system with the use of medical terminology, then the guide for directing the appropriate labelling of the parts, organs and systems is with the help of body planes. Body planes help in mapping the body parts. By definition, body planes are imaginary planes that intersect with the body dividing it into different regions.

Anatomical position: In order to classify the body planes, we need to orient our body in a definite position to make a standardized description of body plane. To define the anatomical position the person should stand erect with hands on the sides with palms facing forward. Thus, the anatomical positioning is important while defining body planes. As shown in Fig. 1.1, there are two such positions:

1. Prone position – where the body is lying facing the ground
2. Supine position – where the body is lying on the ground with face pointing upwards

The various body planes used in medical coding are shown in Fig. 1.2.

Transverse plane: It is the plane that divides the body into two halves: upper and lower part. The upper part above the imaginary line is denoted as superior or cranial. The lower part below the imaginary plane is denoted as inferior or caudal region.

Frontal plane: This plane divides the body into two regions which is front and back. Front is also known as ventral or anterior and the back is known as dorsal or posterior.

Fig. 1.1 Body planes

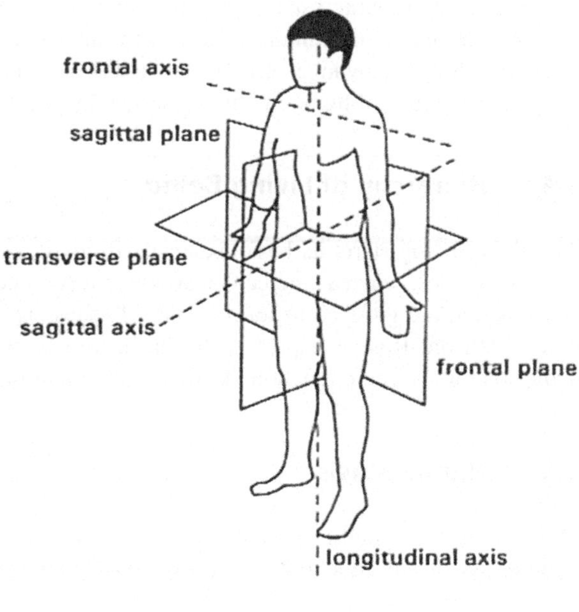

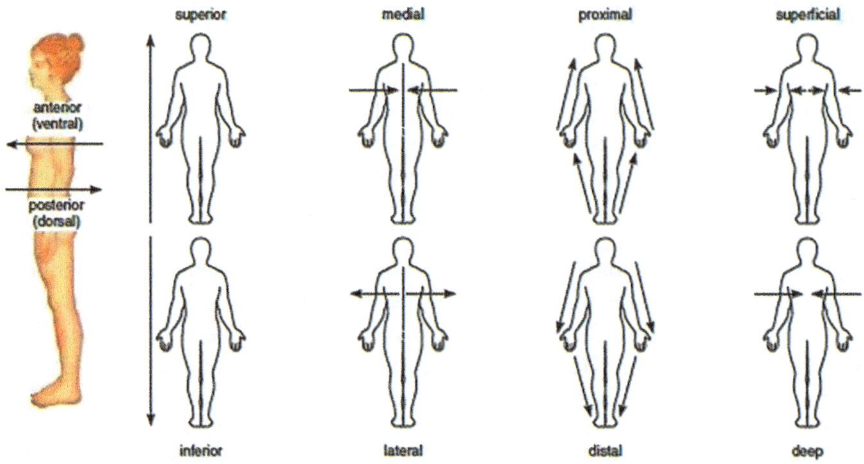

Fig. 1.2 Labelling of parts using body planes

Median plane: This plane divides the body into right and left halves. The region or part near the plane is denoted as proximal and away is distal.

Positions, directions and movements: Terms of position and direction describe the position of one body part relative to another, usually along one of the three major body planes. Table 1.2 shows the various positions and their description.

Table 1.3 shows the various terminologies used to mention the movements of body parts.

Body cavity: The various systems of the body such as digestive system, respiratory system and excretory system need to be accommodated within the body. Body cavities are the natural space within the body which help in housing these organs. For example, mediastinum is the cavity in which the heart is accommodated [Sarada Subramanyam et al.]. The cavity is surrounded by lungs and ribcage. The various cavities of human body are shown in Fig. 1.3.

1.4 Hierarchy of Living Being

Any living being starts its basic existence from an atom; atoms group together to form molecules, whereas molecules group to form a cell. Cells of similar structure and function are grouped to form tissues. Tissues are grouped together to form an organ [Allison, Rose, Chapter 1], which in turn is grouped as a system which is dedicated for a specific function like digestion, respiration, etc., as shown in Fig. 1.4.

1.5 Homeostasis

Maintaining a constant internal environment within a body refers to homeostasis. It requires synchronized activity of various systems within the human body.

Table 1.2 Various body positions

Position	Reference
Superior	Refers to a structure being higher than another structure in the body
Inferior	Refers to a structure being lower than another structure in the body
Anterior	Refers to a structure being more in front than another structure in the body
Posterior	Refers to a structure being more in back than another structure in the body
Medial	Refers to a structure being closer to the midline or median plane of the body than another structure of the body
Lateral	Refers to a structure being farther away from the midline than another structure of the body
Distal	Refers to a structure being farther away from the root of the limb than another structure in the limb
Proximal	Refers to a structure being closer to the root of the limb than another structure in that limb
Superficial	Refers to a structure being closer to the surface of the body than another structure
Deep	Refers to a structure being closer to the core of the body than another structure
Ventral	Towards the front or belly
Dorsal	Towards the back
Prone	Lying face down
Supine	Lying face up
Unilateral	Pertaining to one side of the body
Bilateral	Pertaining to both sides of the body

1.6 Cell the Basic Unit of Living System

A cell is the fundamental structural and functional unit of any living being. A typical cell can be viewed using microscope. Each type of cell is designed to perform a specific task. Cells also differ from each other in their function, but the basic fundamental characteristics of cell remain the same, for example, the mechanism of conversion of nutrients into energy is common across all types of cells. There are about 75 trillion types of cells in our body. Almost all types of cells are reproductive in nature, that is, if a cell is destroyed, new cells are produced to replace the lost cell. The physiological behaviour of a cell is determined by not only the internal components and factors but also the external factors of the surrounding environment of the cell. Because about one third of the cellular fluid is available in the surrounding spaces of the cell known as extracellular fluids, this fluid is circulated throughout the body along with the blood, in the spaces between blood vessel and tissues where material exchange is by diffusion. The body fluid is mainly a water solution with ions and other substances. The composition of the extracellular fluid is the same as that of the intracellular fluid but the difference is in the concentration (Marieb 2007) (Table 1.4).

Table 1.3 Body movements

Flexion	Bending a joint or decreasing the angle between two bones	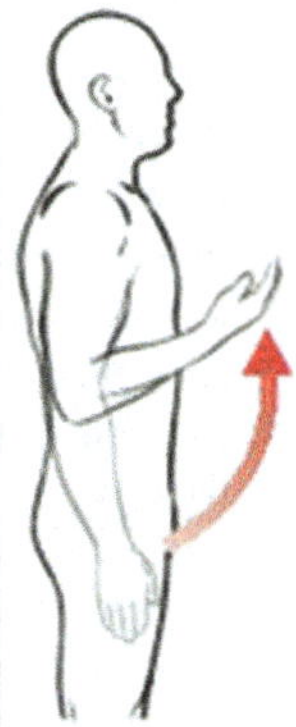 Flexion
Extension	Straightening a joint or increasing the angle between two bones	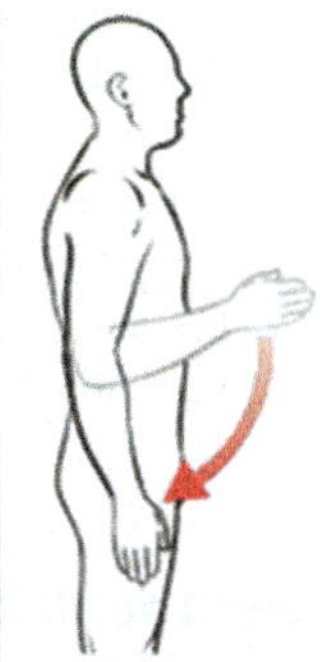 Extension
Hyperextension	Excessive extension of the parts at a joint beyond anatomical position	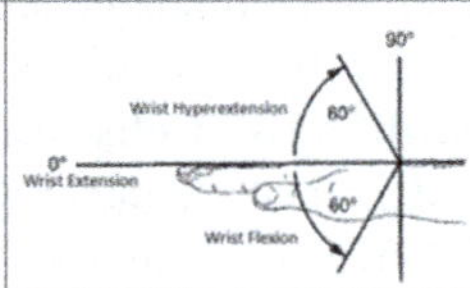
Adduction	Moving a body part towards the midline of the body	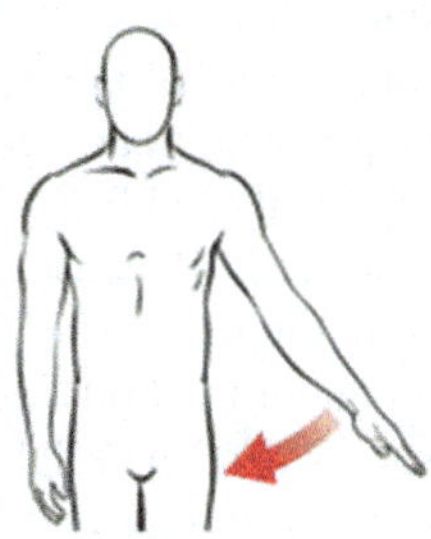

(continued)

Table 1.3 (continued)

Abduction	Moving a body part away from the midline of the body	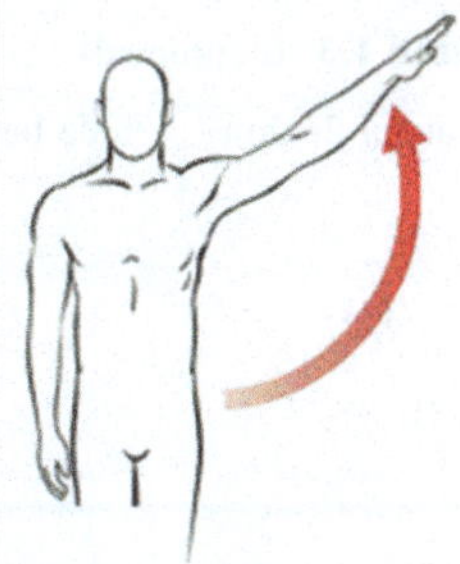
Pronation	Turning the arm or foot downward	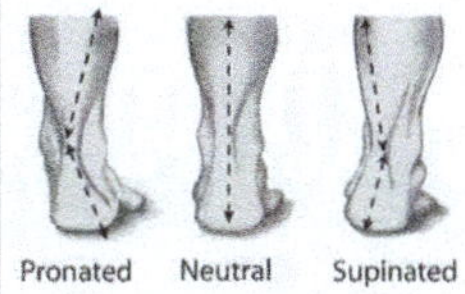
Supination	Turning the arm or foot upward	
Retraction	Moving a part backward	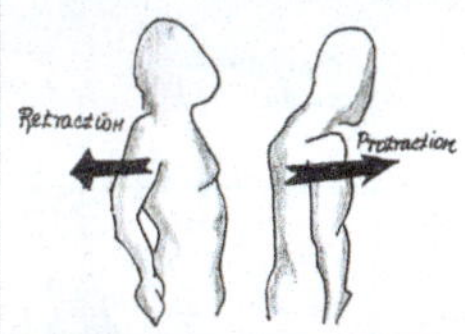
Protraction	Moving a part forward	
Elevation	Raising a part	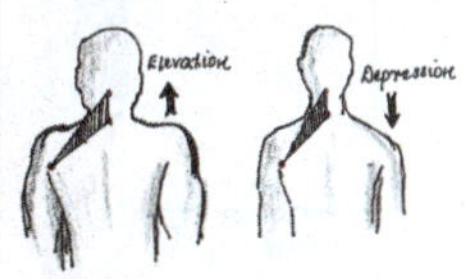
Depression	Lowering a part	
Rotation	Turning on a single axis	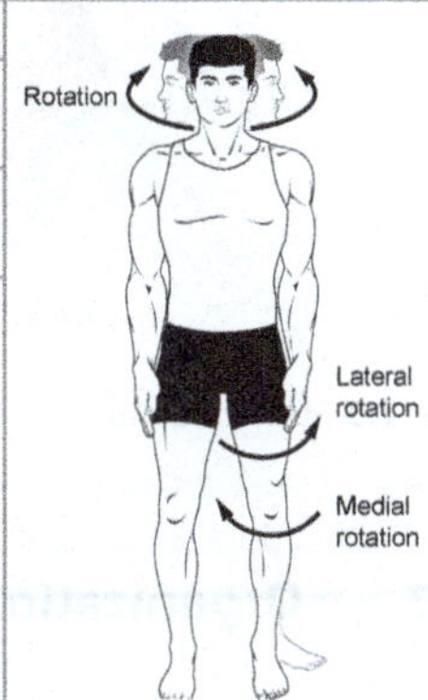
Circumduction	Tri-planar, circular motion at the hip or shoulder	
External rotation	Rotation of the hip or shoulder away from the midline	
Internal rotation	Rotation of the hip or shoulder towards the midline	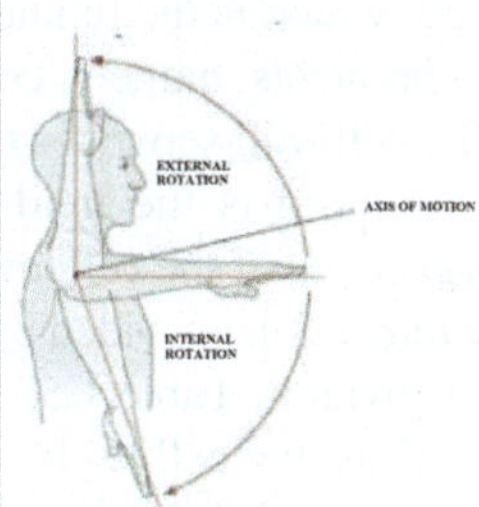

(continued)

Table 1.3 (continued)

Lateral flexion	Side-bending left or right	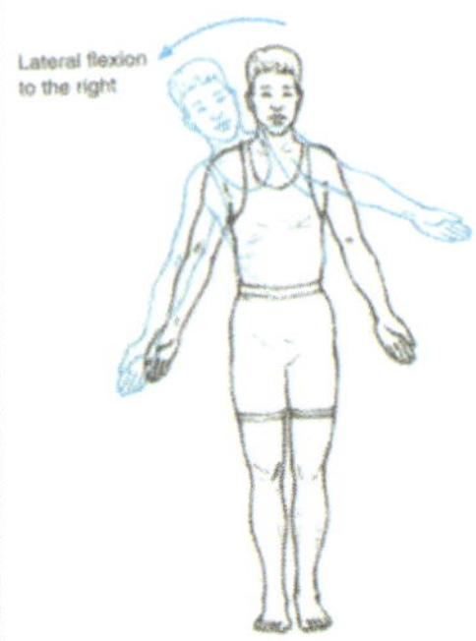

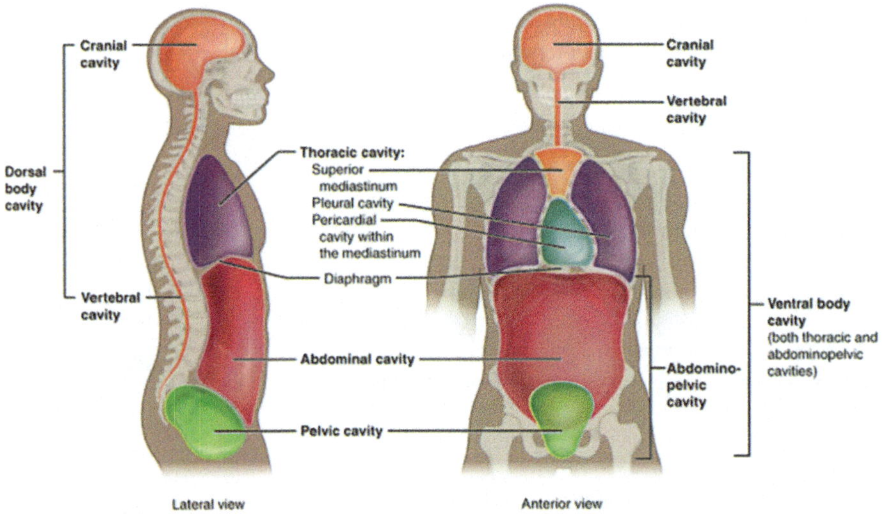

Fig. 1.3 Body cavities

1.7 Organization of a Cell

Cell, which is the fundamental unit of any living system, consists of three important components, namely, cell wall, cytoplasm and nucleus, which are shown in Fig. 1.5. The cell wall separates the internal environment of a cell from the extracellular fluid. Cytoplasm is the fluidic region contained within the cell wall and in which the organelles of the cell lay suspended. Nucleus is separated from the cytoplasm by the demarcation provided by nuclear membrane. The functions of various cell organelles are given in Table 1.5.

Thus, the cell itself is a self-sustained living organism provided the nutrients and oxygen supply are available continuously. When one think about how a cell is communicating with its environment and look into the cellular mechanism, cell is

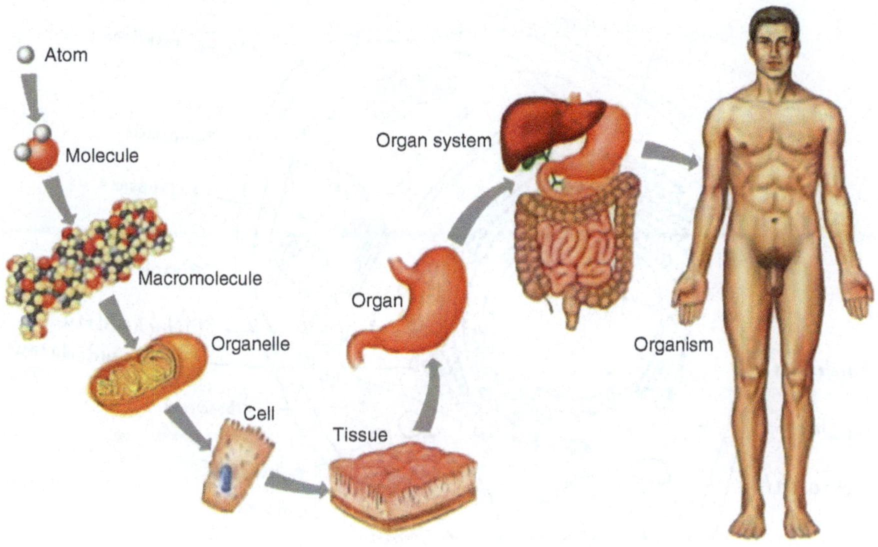

Fig. 1.4 Levels of organization of human being

Table 1.4 Difference between extracellular and intracellular fluid composition

Component	ECF	ICF
Sodium, chloride, bicarbonate ions	High	Low
Oxygen, glucose, fatty acids and amino acids	High	Low
Potassium, magnesium and phosphate ions	Low	High
Carbon dioxide, cellular waste products	High	Low

continuously engaged in the process of energy production, transportation, etc. Thus, there is a continuous need for movement of material across the cell membrane and within the cell.

Cell membrane is so designed to effectively transport materials across them. The structural organization of cell membrane and its function is clearly expressed in fluid mosaic model (shown in Fig. 1.6). According to the theory, the membrane is a fluid mosaic of lipids, proteins and carbohydrates embedded in a phospholipid bilayer.

Thus the cell membrane is made up of double layer of phospholipids where the phosphate forms the head and lipid the tail. Phosphate is hydrophilic and fatty acid tail is hydrophobic. Proteins are embedded in the membrane as shown in Fig. 1.7.

Cell membranes have pores (holes) in it which allows molecules to move across selectively. Molecules commonly move laterally and rarely **flip transversely** (flip-flop) across the double layer due to its selectiveness across the membrane as shown in Fig. 1.8.

The proteins embedded in the membrane shown in Fig. 1.9 are arranged in two different arrangements.

- Integral proteins
- Peripheral proteins

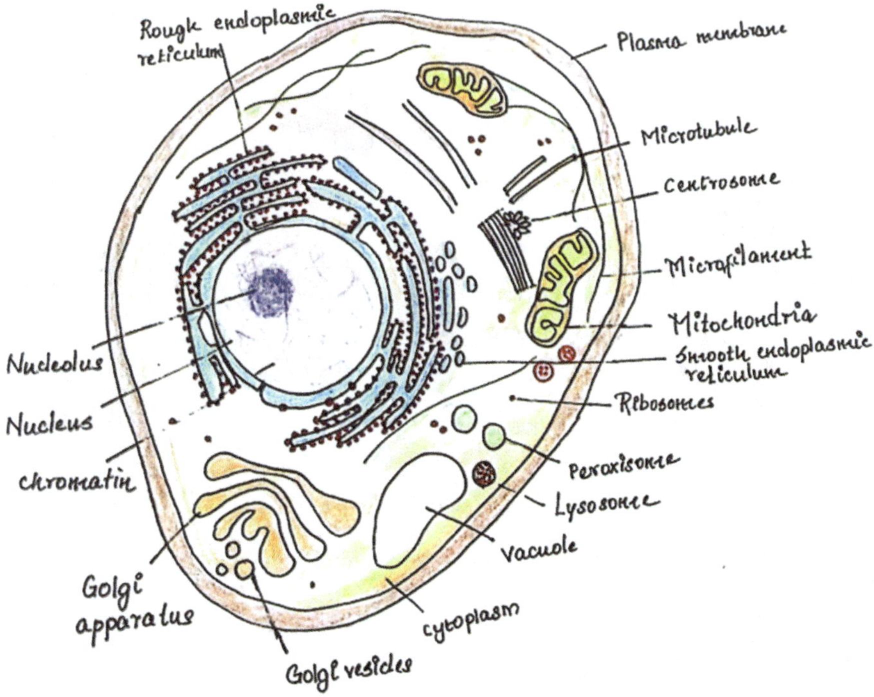

Fig. 1.5 Structure of cell

Table 1.5 Structure and function of cell organelles

Organelle	Structure	Function
Mitochondria	Double membrane with inner layer folding	Energy house of the cell. Converts glucose into energy
Vacuole	Large double-layered structure	Storage unit of the cell. Stores nutrients, water, waste
Lysosome	Small circular structure filled with enzymes	Digestive unit of the cell. Involves in active waste removal process of cell
Endoplasmic reticulum	Double-layered membrane. Two types: rough endoplasmic reticulum and smooth endoplasmic reticulum	Involves in transduction of protein
Ribosomes	Small round structures	Involves in protein synthesis
Golgi apparatus	Curved sac-like structure	Helps in transportation of protein between endoplasmic reticulum and cell membrane
Cytoskeleton	Fibre-like strands	Provides structural stability to the cell

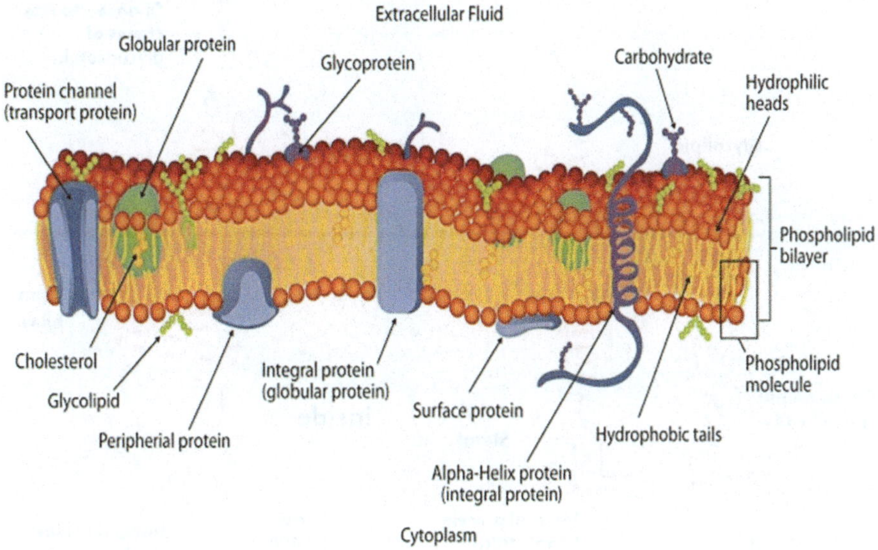

Fig. 1.6 Structure of cell membrane

Fig. 1.7 Polarity of the cell membrane

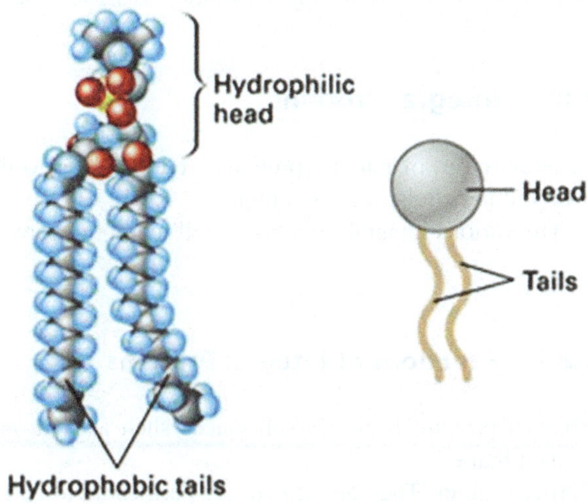

Fig. 1.8 Movement of phospholipids to allow substance movement

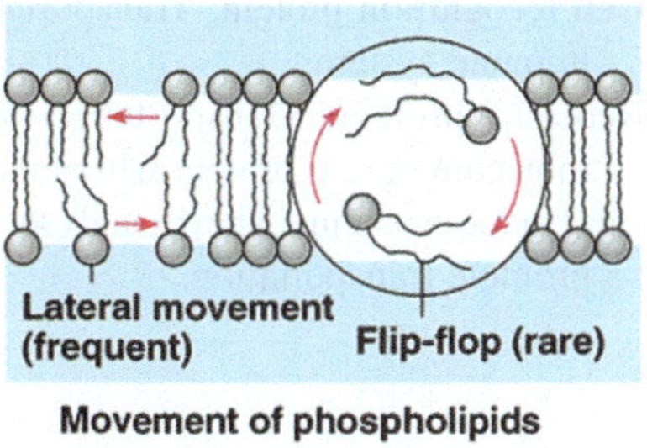

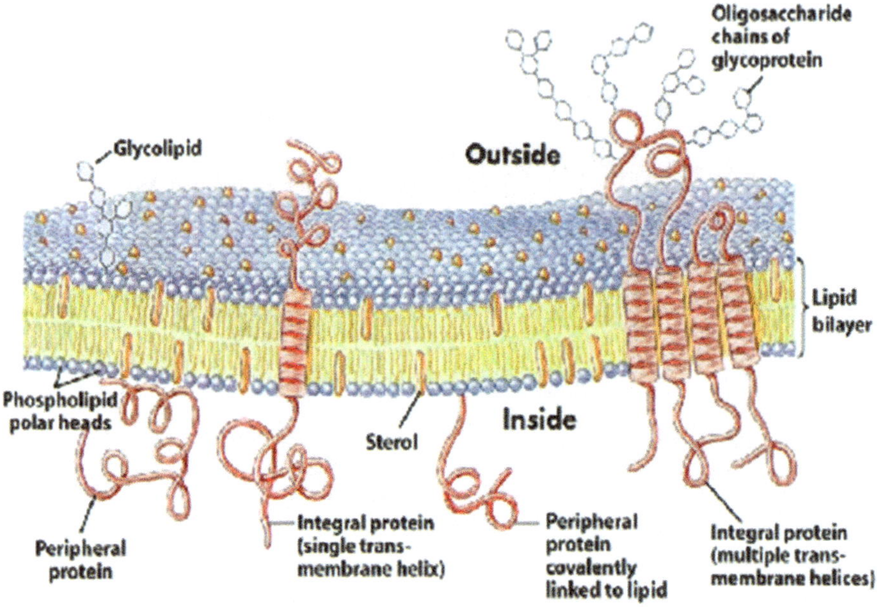

Fig. 1.9 Cell membrane showing embedded protein and cholesterol

1.8 Integral Proteins

Different types of integral proteins are embedded to the cell membrane which are involved in cellular transportation

The various integral proteins of cell membrane are shown in Fig. 1.10.

1.8.1 Functions of Integral Proteins

Channel protein: It allows substances such as hydrogen ions to move across the membrane.

Carrier protein: They are selectively permeable to certain molecules. For example, sodium pumps allow the movement of sodium ions only.

Cell recognition protein: Transportation can happen on reception of stimulus from immune system.

Receptor protein: Transportation of molecules is executed by the adhesion of molecule on the protein where the shape is modified allowing transportation.

Enzymatic protein: They catalyse specific reactions in the protein which will promote transportation.

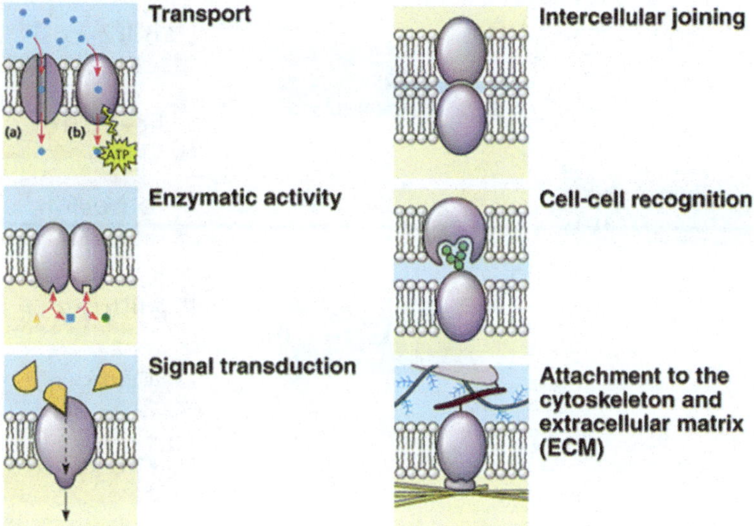

Fig. 1.10 Integral proteins

1.9 Peripheral Proteins

They are embedded in the cell membrane providing strength and stability to the cell membrane along with the cytoskeleton which supports the cytoplasm.

The above are the means by which transportation of molecules and materials takes place across cell membrane with the aid of integral proteins and this process is called facilitated diffusion. As shown in Fig. 1.11, the various cellular transportation mechanisms are as follows.

1.9.1 Active Transportation

It is a means of transportation where the membrane conducts on receiving an external stimulus.

Endocytosis: It is called cell eating. It is the process by which the materials needed within a cell is carried into the cell. The molecules are surrounded by a false finger-like projections formed at the membrane called pseudopodium which engulfs the molecules and are transported into the cell as phagosome.

Exocytosis: It is also called cell excretion. The materials needed to be transported out of the cell cannot diffuse through the cell membrane and are removed by expelling the materials across the membrane. For example, proteins synthesized need to be exported out of the cell.

The mechanism by which this process is carried out is shown in Fig. 1.12.

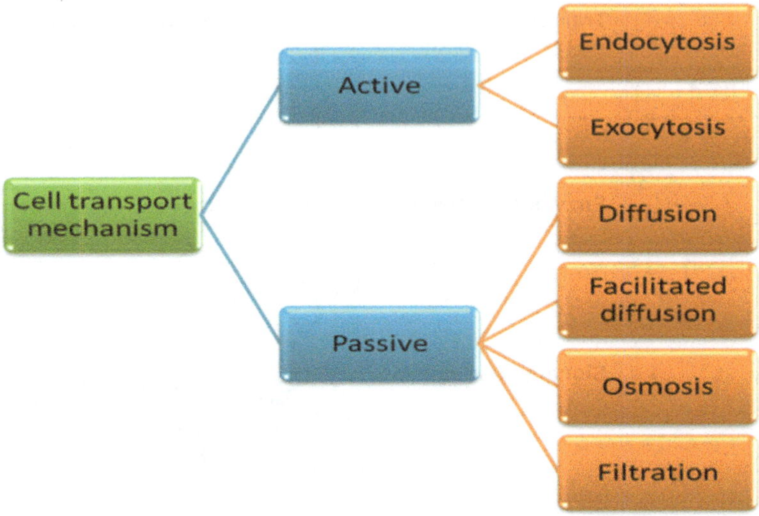

Fig. 1.11 Modes of cellular transportation

1.9.2 Passive Transport

1.9.2.1 Diffusion

It is a process of movement of substance across a concentration gradient and it always occurs from higher concentration towards lower concentration region until an equilibrium is attained. Rate of movement depends on the concentration difference, size of the material and the thermal energy of the solution.

1.9.2.2 Facilitated Diffusion

It is also a type of diffusion which is aided by proteins of the cell membrane; hence it is called a carrier-mediated transportation. Specific or selective transportation of materials happens by this process. Transportation needs an external stimulation or structural modification of the gate protein.

1.9.2.3 Osmosis

It is also a special type of diffusion as it is process of movement of substance from higher concentration to lower concentration through a selectively permeable membrane. In Fig. 1.13, it is shown that when two different solutions such as water and sodium chloride are separated by a selectively permeable membrane, water molecules which are highly permeable can move across the membrane with ease. In contrast, sodium and chlorine which are non-permeable molecules cannot pass through the membrane. In this situation if the pressure gradient across the membrane is adjusted in such a way that the mobility of molecules is stopped, then the pressure involved is known as osmotic pressure.

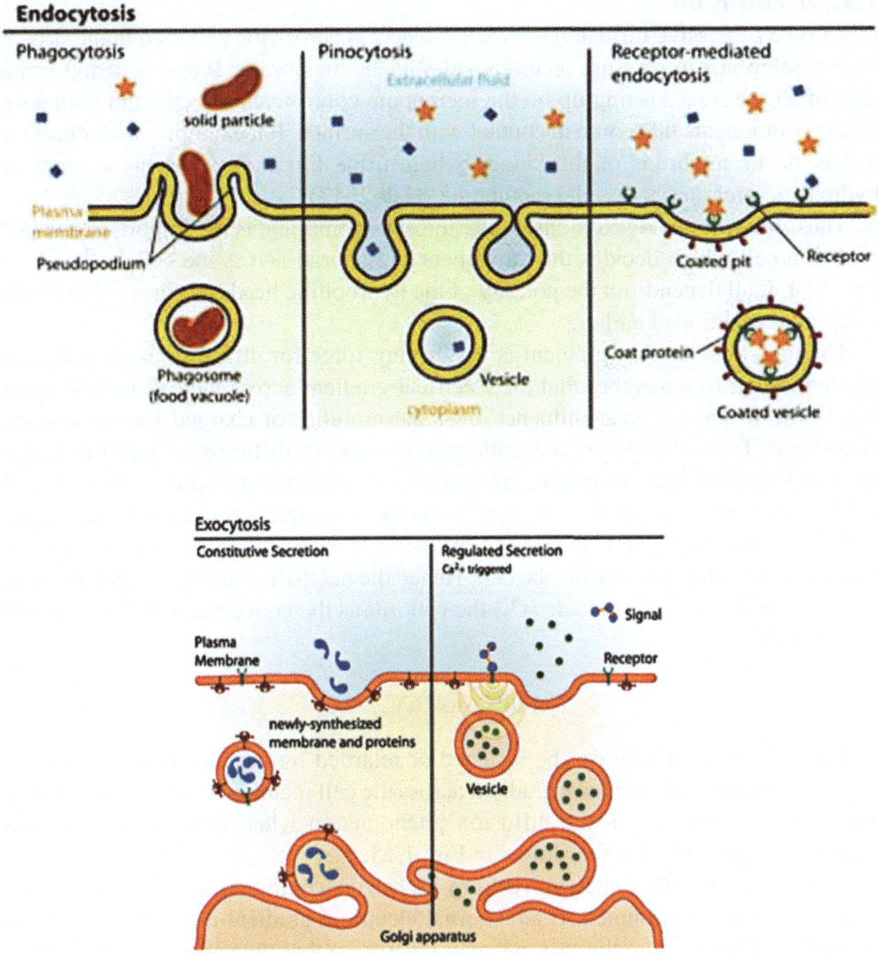

Fig. 1.12 Active transportation

Fig. 1.13 Osmosis

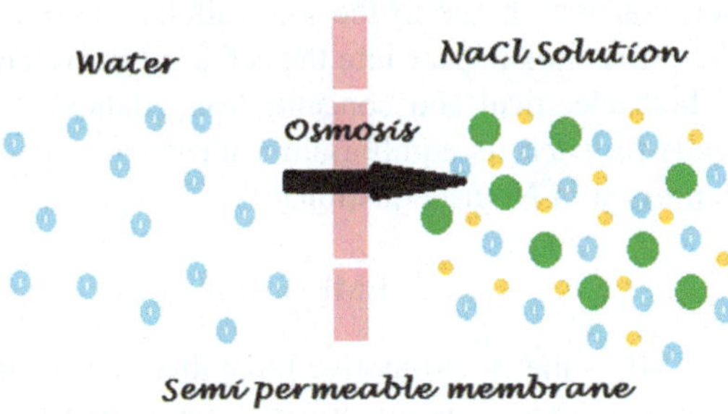

1.9.2.4 Filtration

It is a type of passive diffusion of small molecules across the cell membrane driven by the difference in pressure on either sides of the membrane. Pressure acting is the sum of all the forces acting up on the membrane contributed directly and indirectly by the components that come in contact with the surface. For example, such filtration occurs in the nephrons of the kidney where urine filtration occurs as a result of hydrostatic pressure across the membrane (Hall 2013).

Thus, from all the discussions made, the cell membrane is the important component of a cell which decides the movement of material across the cell as shown in Fig. 1.14. It all depends on the polarity of the hydrophilic heads of the phospholipids bilayers as mentioned earlier.

Though concentration gradient is the driving force for diffusion to happen, it is also important to remember that the electrical gradient across the cell membrane is also going to have a great influence over the mobility of charged ions across the membrane. Thus, the factors that influence the rate of diffusion of material in the desired direction include electric potential and pressure difference. The rate of diffusion of substances into a cell depends on the concentration of ions in the outside of the cell, and conversely, the rate of diffusion of ions outside the cell depends on the concentration of ions inside the cell. Hence the net diffusion rate of a cell is given by concentration of ions outside (C_o) the cell minus the concentration of ions inside (C_i) the cell.

$$\text{Net diffusion } \alpha \ C_o - C_i \tag{1}$$

This net diffusion rate can be fastened or retarded by introducing or modifying electrical gradient and pressure gradient across the cell membrane (Decoursey 2003). There is a controversy in the diffusion phenomenon when factors like electrical gradient are applied which is shown in Fig. 1.15.

As per theory, substances move from higher concentration to lower concentration. In such an environment if an external electrical gradient is maintained across the cell membrane, then the process may reverse so that the substance will start to move from lower to higher concentration. This can happen only if the electrical gradient is greater or higher than the concentration gradient across the membrane. In Figure 1.15 the concentration gradient is assumed to be at constant, and if an electrical gradient is introduced like negative energy to the extracellular matrix and positive charge to the intracellular matrix, then the negative ions move from the extracellular space into the cell as negative charge repels the negative ions away. If both electrical and concentration gradient strength is equal, then the diffusion tends to reach its equilibrium with no or very less substance movement which is expressed in Nernst equation.

$$\text{EMF (in millivolts)} = \pm 61 \log \text{Ci/Co} \tag{2}$$

EMF is the electromotive force that induces the movement of charges across the cell membrane. It is important to know that the EMF of a cell as all nerve-to-cell

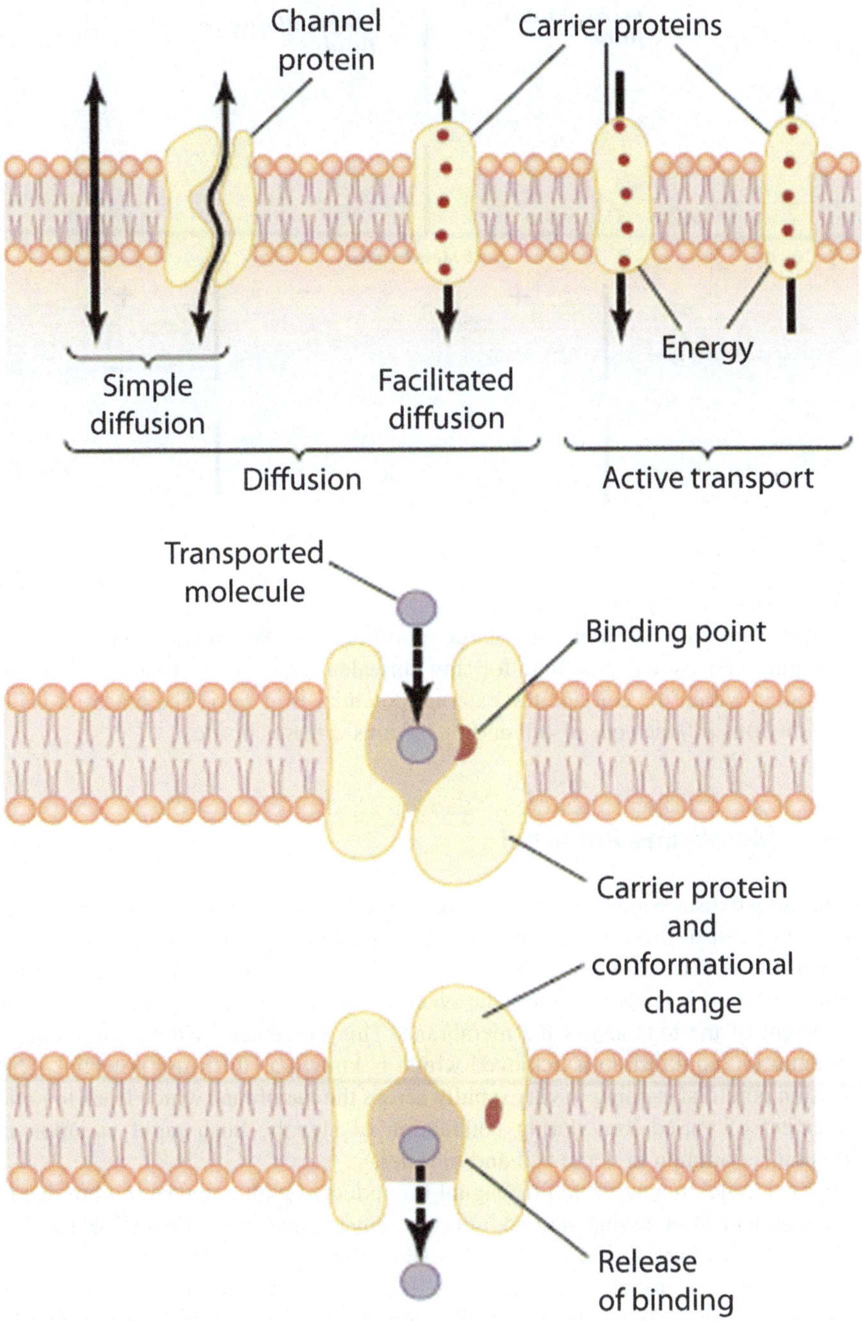

Fig. 1.14 Cellular transportation mechanism

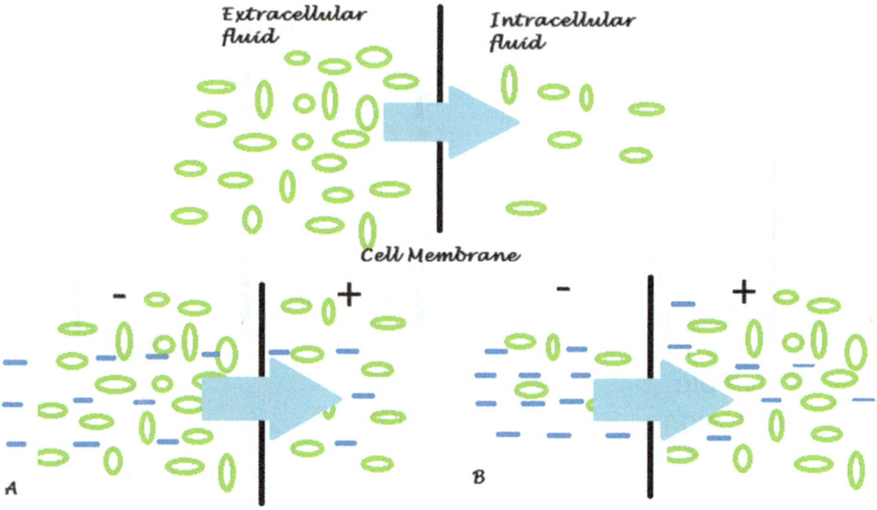

Fig. 1.15 Effect of electric gradient on diffusion

communication are based on this charge mobility. This Nernst equation is used for calculating the Nernst potential for any univalent ions. It is assumed that this potential is within the cell and the external potential is at zero. The potential is (+) when the ions diffuses out of cell and (−) if ions diffuses inside.

1.10 Membrane Potential

All the above discussions define that there exists electrical potential across almost all the type of cell in their cell membrane. This membrane potential is responsible for most of the cellular signalling. When there is concentration difference across the membrane then it leads to membrane electrical potential that in turn leads to the movement of the ions across the membrane. This movement will continue until a concentration equilibrium is achieved which is known as diffusion potential. This diffusion potential is going to vary rapidly across the membrane which leads to rapid movement of varied ions during conduction of signals. Such rapid variation in diffusion is common in nerve cell and muscles.

For example, in a nerve cell the signal is conducted within the nerve fibre due to the movement of potassium and sodium ions outside and inside the cell. Hence for discussion, let us consider a nerve cell as shown in Fig. 1.16. The concentration of potassium ions is more within the cell than its exterior during normal resting condition. There are other substances like proteins and ions which keep the internal environment negatively charged compared to the other side of the membrane. The rate of diffusion of the potassium ions is controlled by the charge at both the sides. When adequate potassium ions migrate, the positive potential across the membrane

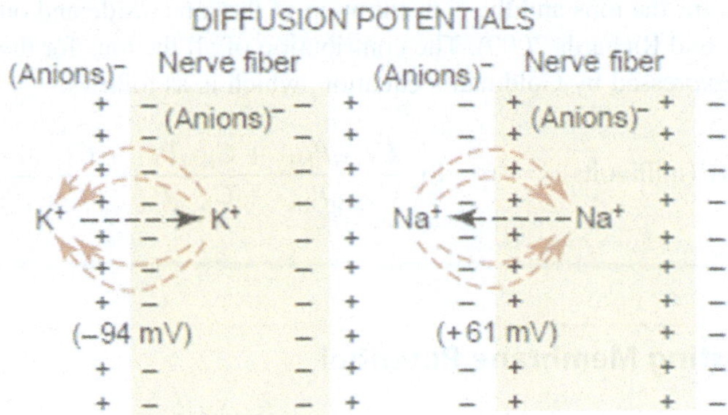

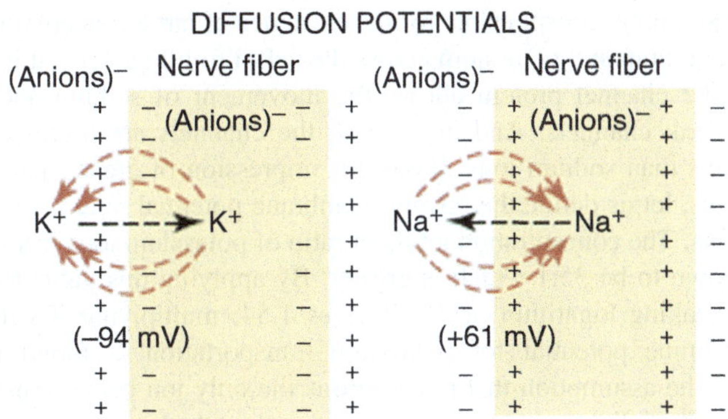

Fig. 1.16 Diffusion potential across nerve fibre

becomes near equal which resists further movement of potassium ions out of the cell. This rapid variation of the charge across the membrane due to diffusion increases the membrane potential which will block further movement of ions. In a mammalian nerve fibre, this potential difference required is about 94 millivolts with positive inside the cell. The same phenomenon can be applied for the mobility of sodium ions across the membrane which leads to a vice versa situation, where the potential difference required is about 61 millivolts with negative inside the cell.

We can now understand that the membrane potential of the cell is contributed by many ion movements. The important ions involved in the generation of membrane potential are sodium, potassium and chlorine. Hence, we need to consider the following three factors of the ions involved when we calculate the membrane potential. The factors are the polarity of the charge of the ions, membrane's permeability for the ions and the concentration of the ions inside and outside of the cell (Pocock and Richards 2009). The contribution of all the ions for the membrane potential is expressed by Goldman's equation, which is as follows:

$$\mathrm{EMF}\,(\text{millivolts}) = -61 \cdot \log \frac{C_{\mathrm{Na}^+{}_l} P_{\mathrm{Na}^+} + C_{\mathrm{K}^+{}_l} P_{\mathrm{K}^+} + C_{\mathrm{Cl}^-{}_o} P_{\mathrm{Cl}^-}}{C_{\mathrm{Na}^+{}_o} P_{\mathrm{Na}^+} + C_{\mathrm{K}^-{}_o} P_{\mathrm{K}^+} + C_{\mathrm{Cl}^-{}_l} P_{\mathrm{Cl}^-}} \quad (3)$$

1.11 Resting Membrane Potential

When the neuron or nerve cell is not transmitting any signal to its immediate environment, it is known as resting potential. The active participants of generation of membrane potential are sodium and potassium ions. At resting potential, when the nerve cell is not conducting any signal, there exists negative potential within the cell membrane due to the fact that more positive ions are expelled outside the cell than that enter the cell. That is, for three sodium ions pumped outside the cell, only two potassium ions enter, causing negative charge of the internal environment of the cell.

The Na^+-K- pump is responsible for the existence of large concentration gradient difference of sodium and potassium across the cell. From Fig. 1.17, it is clear that a stimulus to the channel protein allows the movement of sodium and potassium through the leak channels. And in general, the channels are more permeable to potassium ions than sodium ions. From this impression of greater permeability of potassium ions, let us define the resting membrane potential in terms of leakage of potassium ions. The concentration gradient ratio of potassium ions inside to outside of cell is known to be 35:1 which is greater. By applying this value to the Nernst equation, by taking logarithm of 35, it gives 1.54; multiplying it with −61, the resting membrane potential for potassium transportation is found to be −94 millivolts for the assumption that potassium is the only ion being transported. The Nernst potential of slightly permeable sodium ions is calculated to be +61 millivolts.

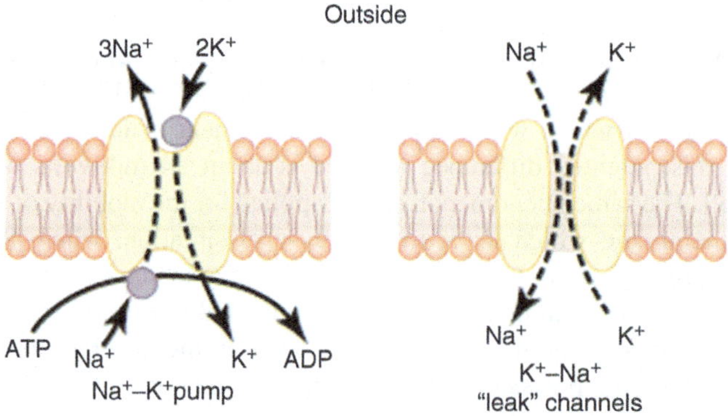

Fig. 1.17 Characteristics of Na^+-K^- pumps

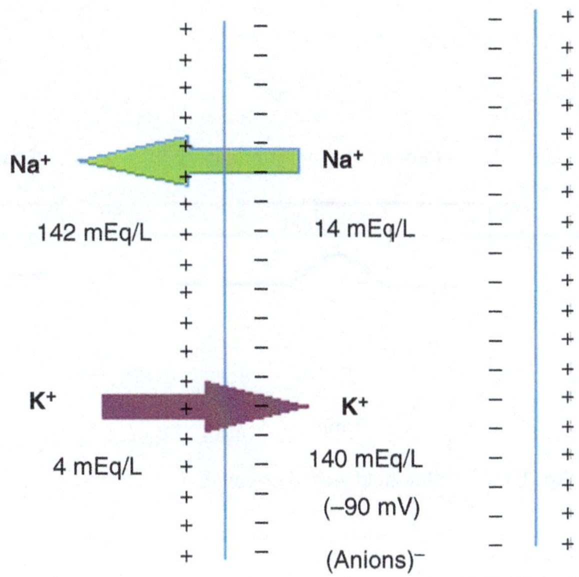

Fig. 1.18 Contribution of sodium and potassium ions for the membrane potential

The permeability of ions across the membrane is shown in Fig. 1.18. When the considerations mentioned in the Goldman equation are counted for calculation, we arrive at the value of –86 millivolts as the resting potential due to the potassium ions. The additional -4 millivolts is contributed by the leakage of ions at rest. Hence the total resting potential is –90 millivolts (Rossier et al. 2002).

1.12 Action Potential

Nerve cells communicate with other cells through the conduction of action potential. The signals are conducted across cell by the conduction of action potential which causes a series of events occurring in the membrane.

Graded potentials are those subthreshold potential developed as a local response of the cell membrane that does not cause a true action potential. They produce almost two types of physiochemical disturbances.

- Local, graded, non-propagated potentials
- Action potentials or nerve impulses which are propagated down the axon to cause the release of neurotransmitters

The graded potentials are local and varying magnitude signals that occur only in a very small segment of the plasma membrane. As shown in Figs. 1.19 and 1.20, the potential that occurs below the threshold can be reinforced. It can be hypopolarized or hyperpolarized based on the stimulus.

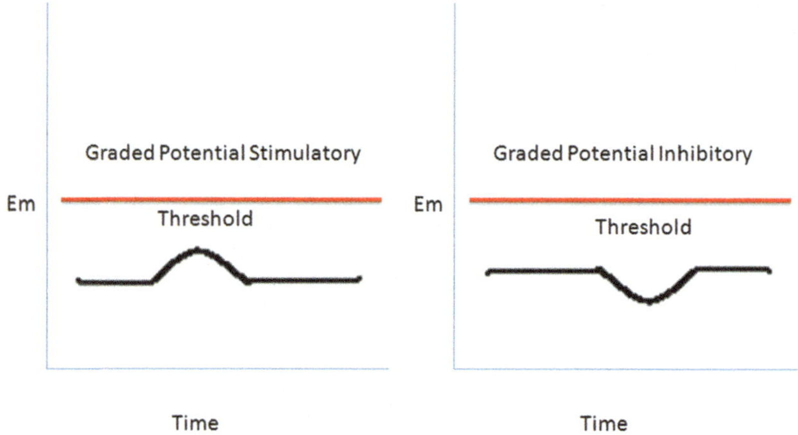

Fig. 1.19 Subthreshold graded potential

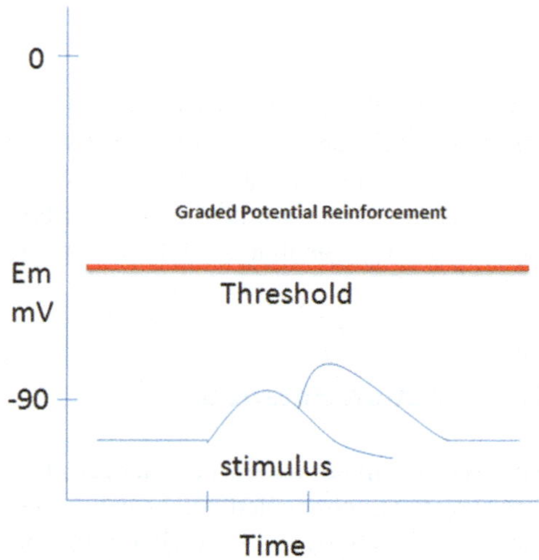

Fig. 1.20 Graded potential reinforcement

Such graded potentials live only a short distance from the origin of the stimulus (less than 1mm distance) due to leakage potential which decreases the membrane charge. We should keep in mind that if the graded potential crosses the threshold potential then it is no more a local potential but it turns into action potential which is conducted across the membrane. Table 1.6 shows various types of graded potential which is the only mode of signalling in certain neurons and cell.

Table 1.6 Types of graded potentials

Type	Description
Receptor potential	Sensory receptors respond to stimuli from the following:
	Mechanoreceptors
	Thermoreceptors
	Nociceptors (pain)
	Chemoreceptors
	Electromagnetic receptors (vision)
Pacemaker potential	Specialized coronary muscle cells in SA (Sinu atrial) node have leaky ion channels producing true cardiac action potential
Postsynaptic membrane potentials	Potential developed at the postsynaptic neuron. If threshold crosses it will conduct
EPP (end-plate potential)	Postsynaptic graded potential developed at the neuromuscular junction

1.12.1 Conduction of Action Potential

Those local graded potentials that reach the threshold are conducted as action potential which can travel long distance without any propagational loss of the signal as it was with graded potential. That is, the size and shape of the action potential do not change as it propagates through the nerve fibres.

So the nerve signals are transmitted by action potential. Action potential conduction in a nerve cell causes rapid changes in the cell membrane potential leading to rapid conduction of charged ions in and out of the cell along with rapid variation in the polarity of the cell. It is known that the resting membrane potential of nerve cell is −90 millivolts which is negative. On the arrival of action potential, the negative potential is changed to positive; as the signal propagates through the cell, it becomes more positive, and as the signal is conducted to the next cell, the potential reverses back to near negative as that of original. There are three stages of events that occur in a nerve cell during the conduction of the action potential. They are:

- Resting stage
- Depolarization stage
- Repolarization stage

Sodium and potassium ions mobility is the driving key for the conduction of the action potential. The permeability of the sodium and potassium pump is highly influenced by the stimulus which leads to the permeability of the ions resulting in action potential conduction as shown in Fig. 1.21.

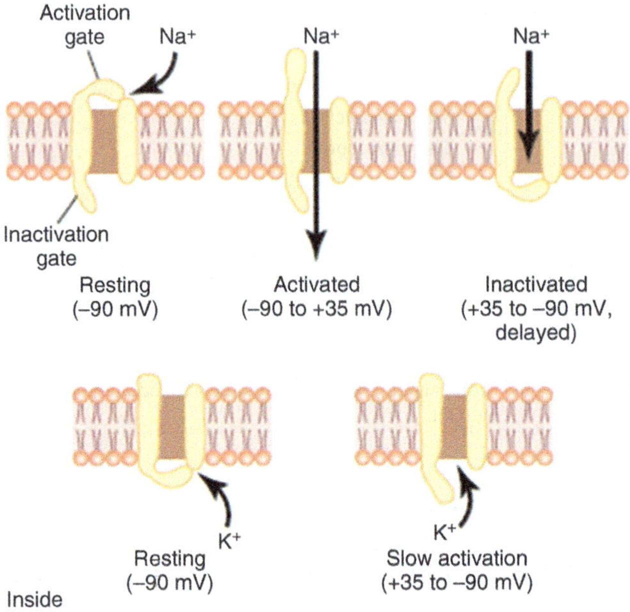

Fig. 1.21 Characteristics of voltage-gated sodium and potassium channels of the cell

1.12.2 Resting Stage

As the name itself implies, at this stage the cell membrane is at its resting membrane potential before conducting the action potential which is negative. The resting membrane potential for various types of cell is as follows: for nerve cell it is −90 mV, for heart pacemaker it is −60 mV, and for skeletal muscle it is −83 mV.

1.12.3 Depolarization Stage

The action potential is conducted into the cell which is marked by the opening of the channel proteins of sodium pump, thus allowing sodium ions to diffuse into the cell resulting in increase in the internal positive charge of the cell. This is called depolarization of the cell. More sodium ions influx results in more positivity inside the cell which neutralizes the internal environment and causes overshoot of the cell. In most of the nerve cell, the charge reaches near zero, and in some large nerve cell, the charge crosses zero to have a small positive peak.

1.12.4 Repolarization Stage

When the polarity of the cell is changed due to the influx of the sodium ions within 10,000th of a second, the permeability of sodium ions had increased which causes the sodium channel to close. The high positive potential of the cell will automatically allow the potassium channel to open so that the potassium ions efflux the cell into the external environment in order to re-establish the original resting negative potential of the membrane.

As shown in Fig. 1.22, the voltage-gated sodium and potassium pumps have a two-way gate. The outer gate is a fast moving gate and the inner gate is a slow gate.

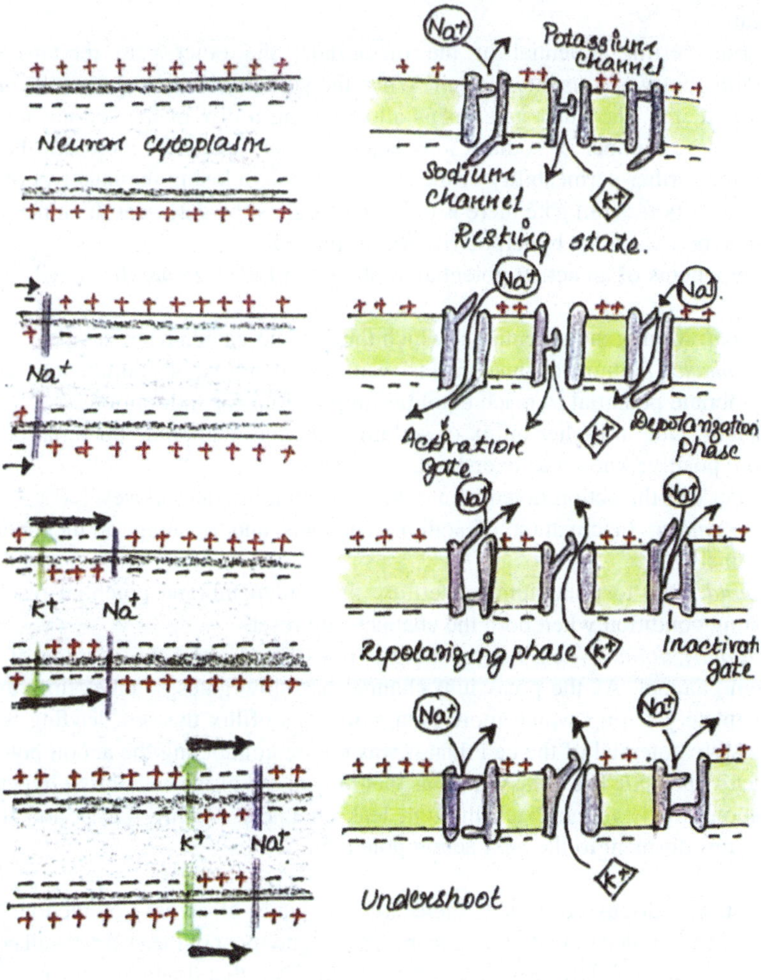

Fig. 1.22 Various events in cell membrane during action potential conduction

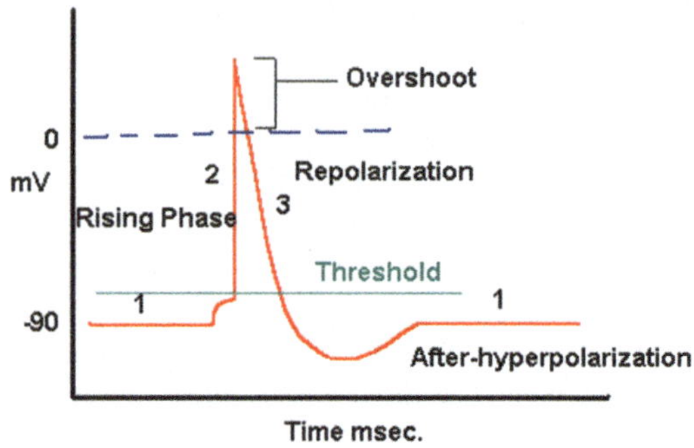

Fig. 1.23 Components of action potential

In other words, the outer gate is an activation gate and the inner gate is an inactivation gate.

During resting potential of the membrane, the outer gate remains closed preventing entry of ions into the cell. When the voltage changes as the cell conducts action potential, the outer gate opens allowing the influx of the sodium ions, and when a large number of sodium ions enter, the cell positivity increases; hence, it ceases the further permeability of the sodium ions by closing the inactivation gate.

But with potassium gate there is only one single slow gate near the inner surface which is the reason for hyperpolarization of the cell.

Components of an action potential as shown in Fig. 1.23 are discussed.

Threshold: Membrane potential at which the voltage-gated channels open.

Rising phase: Sodium channel opens resulting in influx of sodium ions causing membrane potential to reach equilibrium potential for sodium ions.

Overshoot: Due to higher influx of sodium ions, the membrane potential becomes more positive known as overshoot.

Peak: Peak of the action potential at which the sodium gates closes.

Repolarization: Inactivation of sodium channels and opening of the potassium channels.

Threshold: Due to potassium ion efflux, the cell membrane potential reaches its resting condition when both the channels are reset.

Hyperpolarization: When the threshold is re-established by the opening of potassium channel. As the potassium channel has slow gates, it takes time to close completely during which more potassium ions efflux the cell, leading to more negative potential of the cell than it was before conducting the action potential.

Refractory period: The potassium ions that moved out of the cell during hyperpolarization slowly retract back through leak channels. During this period the cell remains dormant to the next action potential.

As it was discussed earlier, there are two ways of signal conduction: action potential and graded potential. All the nerve cell and most of the skeletal cell conduct by action potential. The most fascinating mechanism that draws our interest is how

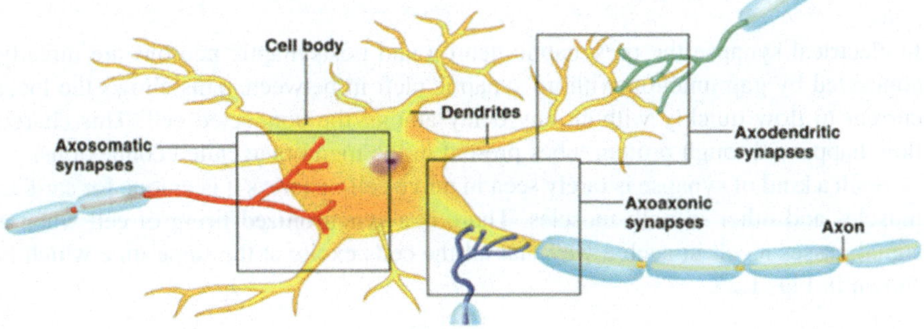

Fig. 1.24 Various modes of contact in synapse

the heart is pumping continuously? What is the underlying principle for the involuntary activity of the cardiac muscle? How cardiac muscles differ from that of other skeletal muscles? The answer for all these questions can be obtained if we know the cardiac action potential.

Before we discuss the cardiac action potential, we should be aware of the mode of conduction of the action potential in the cell (Muller and Nikonenko 2003). The nerve cell to cell communication can happen at a junction between them. This junction where the communication is established is known as synapse. The following are the components of a synapse.

Presynaptic neuron – conducts impulses towards the synapse
Postsynaptic neuron – transmits impulses away from the synapse
Synaptic cleft – junction between presynaptic neuron and postsynaptic neuron

Synapse as mentioned earlier is the gap in which a nerve cell connects to another nerve cell or a skeletal cell. This contact can be established in any of the combination of regions which is shown in Fig. 1.24. They are:

- Axodendritic synapse: where the presynaptic nerve cell axon contacts the dendrite of the postsynaptic nerve cell
- Axosomatic synapse: where the presynaptic nerve cell axon contacts the cell body of the postsynaptic nerve cell
- Axoaxonic synapse: where the presynaptic nerve cell axon contacts the axon of the postsynaptic nerve cell

There are two types of synapse.

- Electrical synapse
- Chemical synapse

1.13 Electrical Synapse

In electrical synapse the presynaptic neuron and postsynaptic neurons are directly connected by gap junction with no synaptic cleft in between. This allows the local current to flow quickly without any delay among the connected cell. This charge flow happens through protein tubes present in the membrane called connexions.

Such a kind of synapse is rarely seen in nerve cell, whereas it is unique for cardiac muscles and other smooth muscles. There is a synchronized firing of cell when a signal passes a cell in such a way that all the cells excite at the same time which is shown in Fig. 1.25.

1.13.1 Cardiac Action Potential

To understand this stance of electrical synapse, we need to know the mechanism of pumping of the heart. The heart is a conical muscular organ made up of special type of cell known as cardiac cell. The heart consists of four chambers, two right and two left chambers, separated by septum. The upper chambers are known as auricles or atrium and the lower chambers are known as ventricles.

The right side of the heart handles deoxygenated blood and feeds the pulmonary circulation and the left side of the heart handles oxygenated blood from lungs and circulates it to the body tissues through systemic circulation. The function of the heart is to pump blood into the pulmonary, systemic and coronary circulation network of blood vessels. The contraction and relaxation of the heart muscle should be in high coordination in order to effectively pump the blood. The cell does not receive the control signals from the nerve cell. They are controlled by auto-generating cell located at the right atrium of the heart which is a self-excitatory cell known as sinu atrial node (SA node) which generates the cardiac potential. They

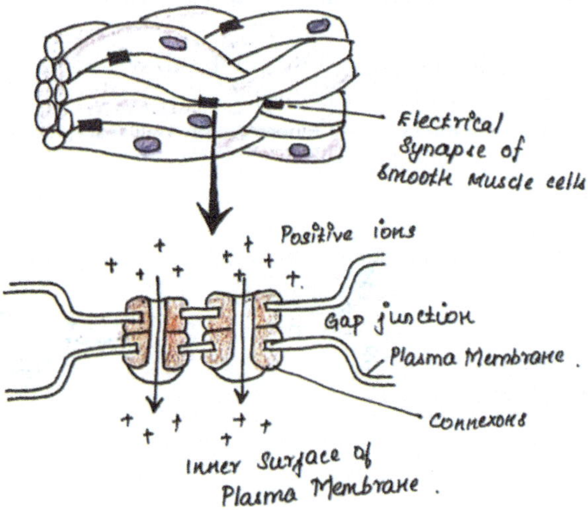

Fig. 1.25 Electrical synapse of the heart

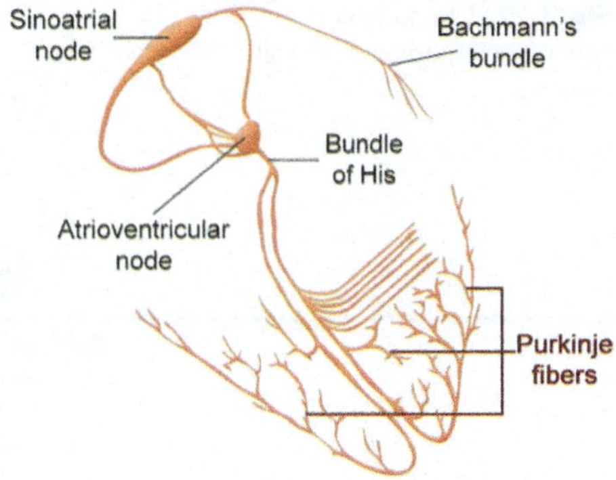

Fig. 1.26 Electrical activity of the heart

produce about 60–100 action potentials per minute and hence the resting heart rate is 60–100 beats per minute. The cardiac muscle has electrical synapse which spreads the excitation potential evenly across all the cells of the atrium at the same time which allows the atrium muscles to contract in synchronize leading the blood to be pumped mechanically from the right and left atrium to enter into their corresponding right and left ventricles. There is sequence of events that happens in the heart due to the propagation of this action potential which can be recorded as electrocardiogram (Schram et al. 2002).

To continue with this, the heart's electrical activity (Fig. 1.26) is carried out by the propagation of action potential across the cell of the heart. The signal from SA node reaches the AV node from which a single bundle of special cell known as bundle of his carries the signal towards the periphery through two branch fibres (purkinje system) which extensively branch to cover the ventricles on both anterior and posterior sides. Thus, as the signal spreads from SA node to AV node, the atrium contracts. The signal conducted down the purkinje system will start spreading from apex to base resulting in contraction of ventricles, and mechanically the blood is pushed into the gain vessels connected to them.

1.14 Chemical Synapse

It is the most common type of synapse which consists of all the components of a typical synapse. Electrical signal is conducted through chemical neurotransmitters (NTs) released by the presynaptic neuron. The NTs bind to the receptors of the postsynaptic neurons which activates the channel protein to open, thus allowing the ions to enter the postsynaptic neuron and hence the action potential.

The following events occur during conduction of action potential in a chemical synapse.

Events at a chemical synapse as shown in Fig. 1.27 are explained below.

1. Action potential at the presynaptic neuron transmits down the axon tail towards the terminal, and the event is set by the opening of voltage-gated Ca^{++} channels.
2. Ca^{++} influx into presynaptic terminal which increases the Ca^{++} ion concentration within the axon terminal.

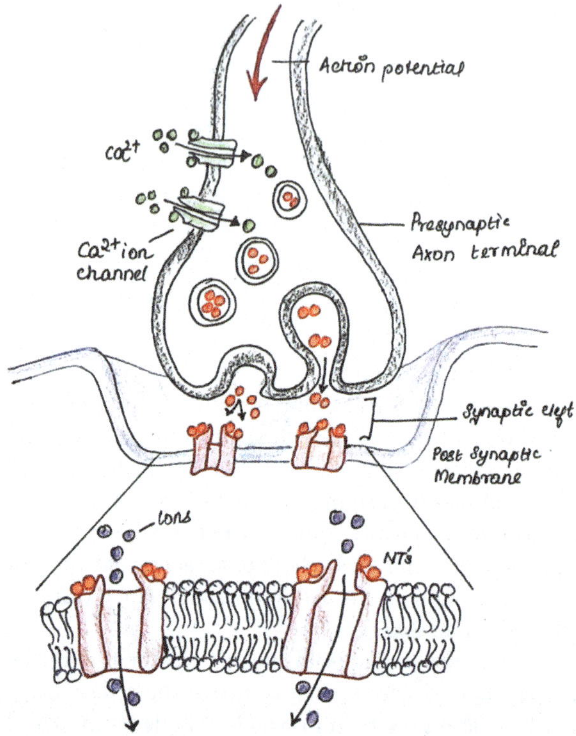

Fig. 1.27 Conduction of action potential in a chemical synapse

3. Ca^{++} acts as a stimulator initiating the vesicle formation, wherein the neurotransmitters are bound into the vesicles which will be transported out of the membrane by exocytosis.
4. Ca^{++} is removed from synaptic knob by mitochondria or calcium pumps.
5. The neurotransmitters diffuse across synaptic cleft and bind to receptor centres in the postsynaptic membrane.
6. Receptor modifies the shape of the ion channel allowing the transportation of charged ions through the membrane.
7. NT is quickly destroyed by enzymes or taken back up by *astrocytes* or presynaptic membrane.

For every nerve impulse reaching the presynaptic terminal, about 300 vesicles are emptied into the cleft. Each vesicle contains about 3000 molecules.

Such a kind of action potential conduction happens in nerve cell and neurons. Next, let us discuss about nerve cell and central nervous system.

1.15 Nervous System

It is the central control system of all the voluntary and involuntary activities carried out in the body. The nervous system is classified as shown in Fig. 1.30.

Neuron: It is the basic unit of the nervous system which receives, integrates and transmits information. They operate through electrical impulses and communicate with other neurons through chemical signals.

Neuron as shown in Fig. 1.28 consists of dendrite branched structure which conducts "signal" towards the cell body. They are characterized as short, numerous and highly branched structures (Ganong 2010). The signal from the adjacent neuron can enter the cell by dendrites in an axon dendrite synaptic neuron. Axon is a single long fibre of the neuron which is involved mainly in the conduction of the nerve impulses. All the discussion we made in the action potential happens here in this axon which is also called neuron tail. Axon ending forms the presynaptic neuron end of the synaptic cleft which is a clustered structure with hundreds to thousands of branches, each ending with a bulb-like synaptic knob. It relays signal to the next neuron. There are about 100 billion neurons and 10× more glial cells. Glial cells support neurons by providing physical support and nutrients, cover the neuron with myelin and clean the debris. Myelin sheaths cover the axon tail which is made up of fatty material of glial cell. It allows rapid movement of electrical impulses along the axon. Nodes of Ranvier are the gap junctions of myelin sheaths where the action potential is transmitted. The speed of the impulse ranges from 2 to 200 +mph which is shown in Fig. 1.29.

There are three types of neurons

Sensory neurons: Neurons that send signals from the senses, skin, muscles and internal organs to the CNS

Motor neurons: Neurons that transmit commands from the CNS to the muscles, glands and organs

Interneurons: Interpretation of sensory signals within processing centres

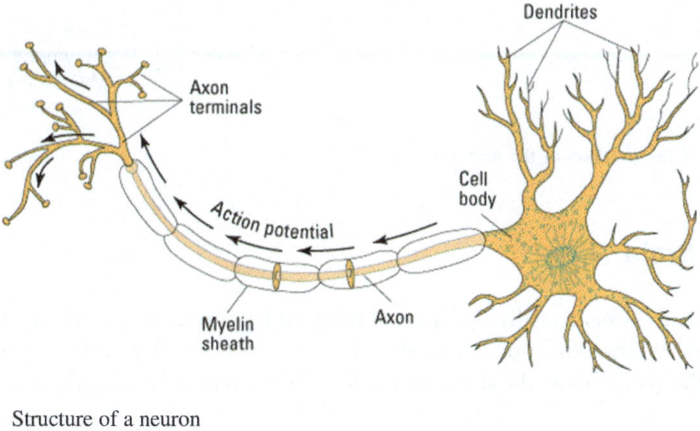

Fig. 1.28 Structure of a neuron

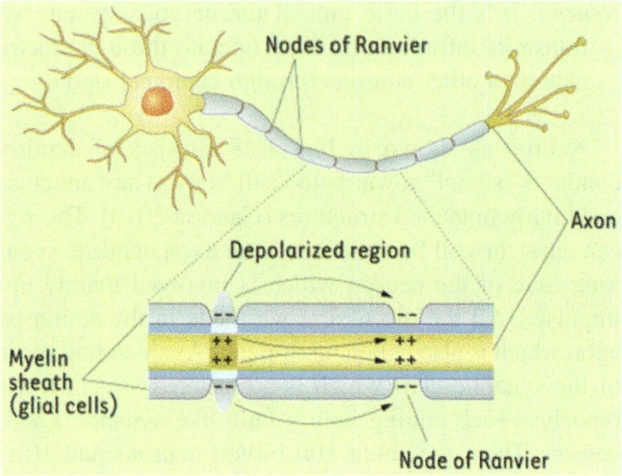

Fig. 1.29 Neuron showing conduction of impulse through myelin sheath

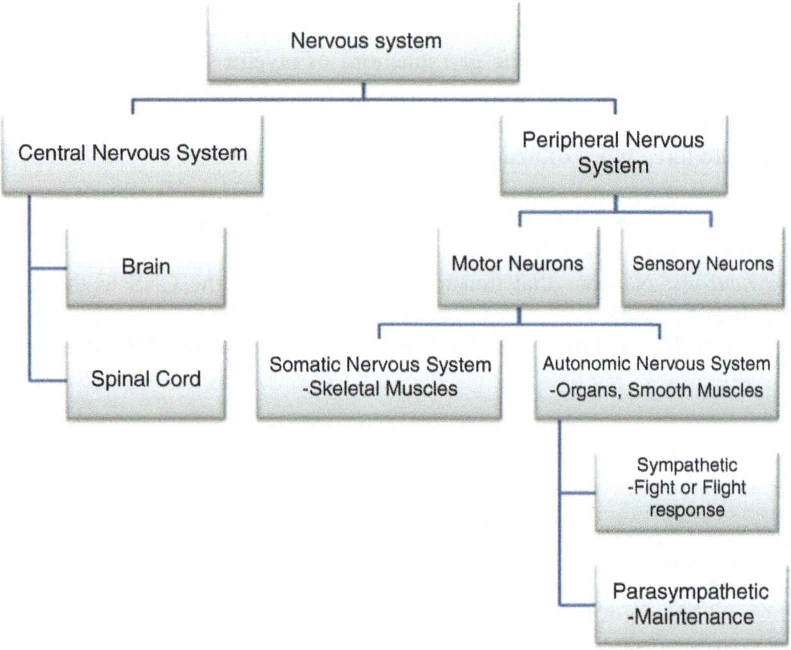

Fig. 1.30 Classification of the nervous system

1.16 Brain

The nervous system is divided into central and peripheral nervous system (Fig. 1.30). Embryonic development of the brain as shown in Fig. 1.31 begins from a neural tube which then classified into subdivisions such as forebrain, midbrain and

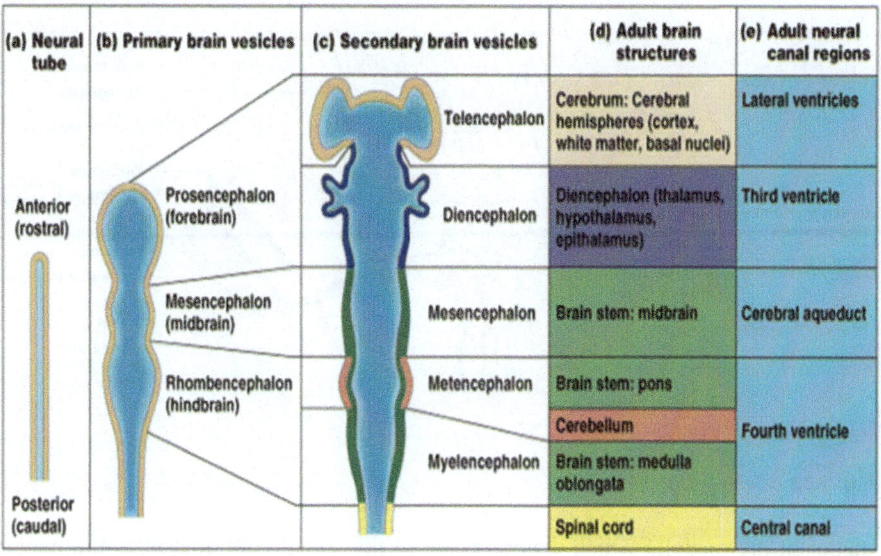

Fig. 1.31 Embryonic development of the brain

Fig. 1.32 Structure of the brain

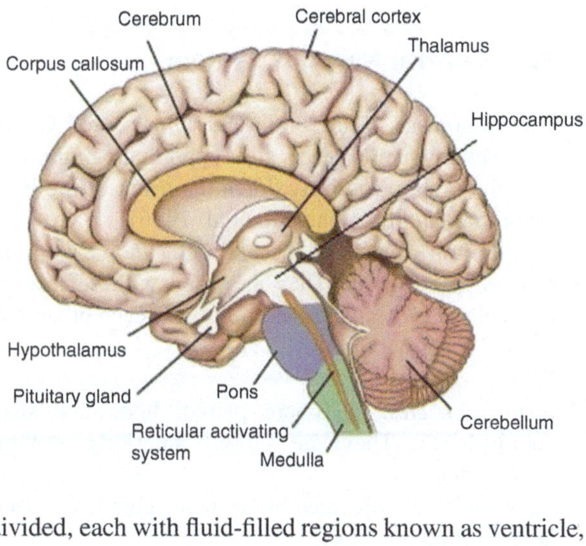

hindbrain, which is further divided, each with fluid-filled regions known as ventricle, aqueduct or canal.

1.17 Anatomical Classification of the Brain

Structurally the brain (shown in Fig. 1.32) consists of the cerebrum, cerebellum, brain stem and diencephalon. The brain is covered by meninges which act as cushion between the brain and the skull. The layers are dura mater, arachnoid mater and pia mater as shown in Fig. 1.33. The brain has the grey matter in the cortex and white

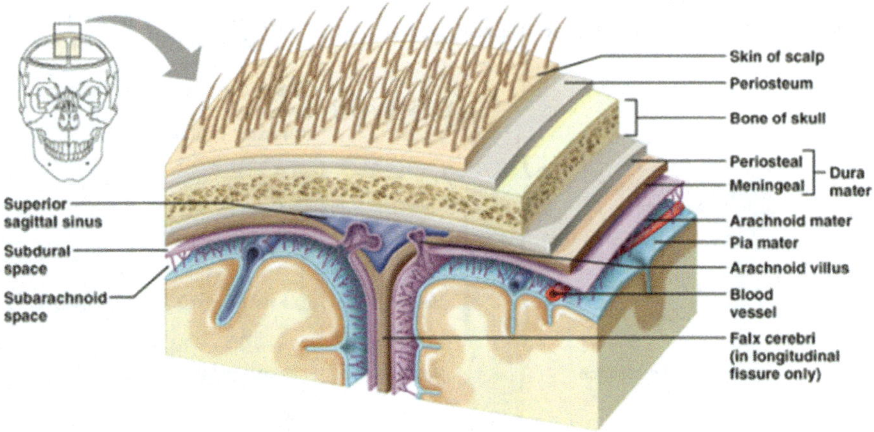

Fig. 1.33 Meninges of the brain

matter in the medulla which gets reversed in the spinal cord, that is, white matter outside and grey matter inside.

1.18 Central Nervous System

The CNS consists of the brain and spinal cord. The brain is contained within the cranial cavity which protects and supports the brain, and the spinal cord is continuation of the brain posteriorly down the trunk of the body protected within vertebral column (Sarada Subramanyam et al. 1996).

The central cavities of the brain are known as ventricles. There are about four ventricles in the brain. They are filled with cerebrospinal fluid secreted from the choroid plexus. They are continuous with each other and with the central canal of the spinal cord (shown in Fig. 1.34).

Lateral ventricles: They are paired, horseshoe shape located within cerebral hemispheres. They are closed anteriorly separated only by thin *septum pellucidum.*

Third ventricle: It is located in the diencephalon connecting interventricular foramen and cerebral aqueduct.

Fourth ventricle: This is located in the brainstem dorsal to pons and top of medulla. Here there are holes that connect it with subarachnoid space of the meninges, hence allowing the circulation of cerebrospinal fluid within the brain and the surfaces.

When we look at the surface of the brain, it consists of several elevated ridges known as gyri which are separated by deep and shallow grooves known as sulci or sulcus. The deep grooves are known as fissures.

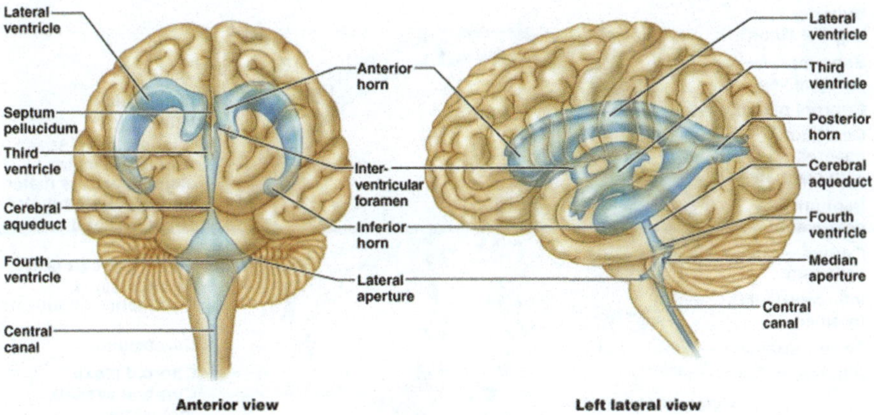

Fig. 1.34 Ventricles of the brain

1.19 Cerebrospinal Fluid

They are similar in composition to that of blood plasma but do not contain formed cell in it. CSF is secreted in the choroid plexus and circulated through ventricles, median and lateral apertures, subarachnoid space and arachnoid villi and into the blood of the superior sagittal sinus and finally drained into arachnoid villus. The circulation of CSF is shown in Fig 1.35.

The cerebral hemisphere is divied into 4 lobes frontal, parietal, occipital and temporal. Functionals brain is classified into three areas sensory, motor and associative. The lobes of the brain are shown in Fig. 1.36.

1.19.1 Cerebral Cortex

It is responsible for executive functioning capability of the body. It is made up of grey matter of neuron cell bodies, dendrites and short unmyelinated axons. There are about 100 billion neurons with average of 10,000 contacts each 2–4 mm thick (about 1/8 inch). All the neurons are interneurons which have to synapse with other neurons before the information reaches the peripheral nerves. There are three kinds of functional areas in cerebral cortex: motor, which controls movements; sensory, which controls perception; and association areas, which integrate diverse information and convert to an action.

Motor areas: movement
Sensory areas: perception
Association areas: integrate diverse information to enable purposeful action

Homunculus as shown in Fig. 1.37 also known as little man is how human body is spatially represented on cortex upside down.

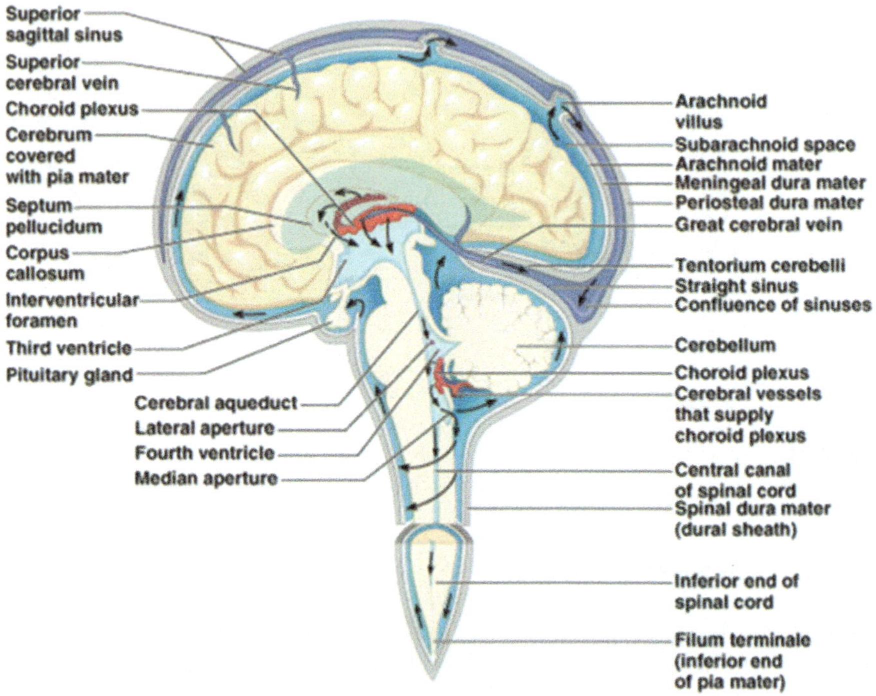

Fig. 1.35 Circulation of cerebrospinal fluid

Cerebral white matter is involved in extensive communication between areas of cortex and cortex with other part of CNS. White matter is composed of myelinated axon fibres bundled into tracts.

Commissures – interconnect right and left hemispheres of cerebrum.

Association fibres – connect different parts within the *same* hemisphere.

Projection fibres – they run vertically down to the spinal cord. It carries motor impulses down the tract and sensory impulses up to the receptor regions in the cerebral cortex.

Cerebellum – separated structure from the cerebrum by fourth ventricle. It consists of two hemispheres with three lobes each, anterior, posterior and flocculonodular, and vermis is the midline that interconnects the hemispheres. The structure of cerebellum is shown in Fig. 1.38.

Functions of cerebellum: They help in smooth, coordination in bodily movement. They help to maintain body posture and helps in maintaining equilibrium.

Midbrain: It consists of pons and medulla oblongata, and they are responsible for regulation of involuntary activities such as respiration, heart rate and regulation of blood pressure as they are programmed for automatic behaviour for survival. It is the passage way for fibre tracts running between cerebrum and other parts. They provide innervations for face and head.

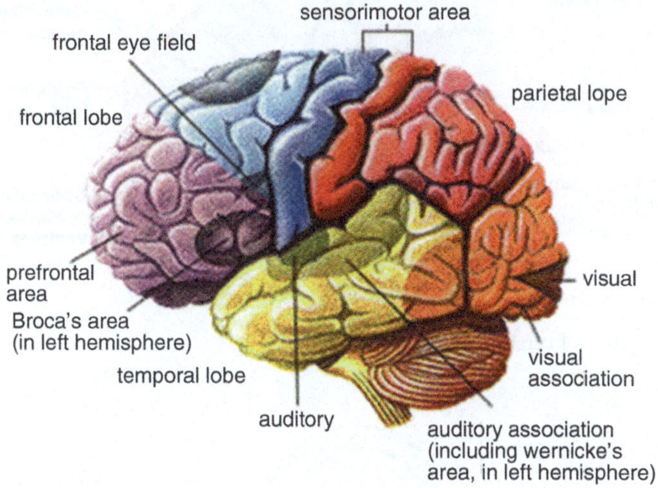

Fig. 1.36 Lobes of brain

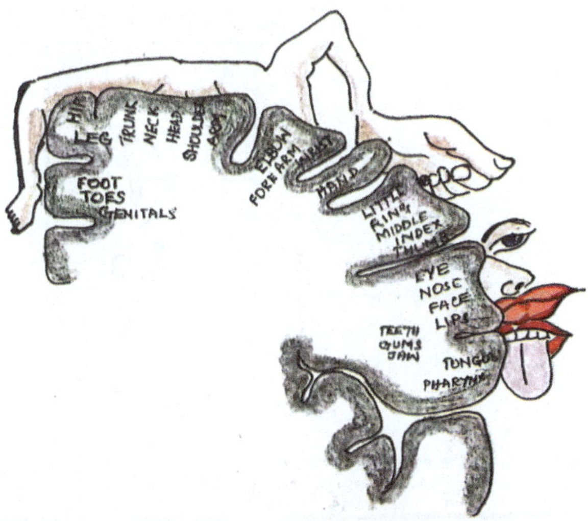

Fig. 1.37 Homunculus body mapping of cerebral cortex

Let us discuss in brief the regulation of respiration. Respiration is the process of providing the oxygen to the body through the mechanism of breathing. Respiration is an involuntary action and the signalling for regulation respiration is provided from pontomedullary centres located in the midbrain (West 1996). If the blood oxygen level falls below normal, the information of conditions like hypoxia, hypercapnia and pH were provided by the baroreceptors and chemoreceptors located in aortic and carotid bodies. This information will stimulate the inspiratory centres located in medulla oblongata. They provide an inspiratory signal to the respiratory organs of lungs. The mechanism is shown in Fig. 1.39.

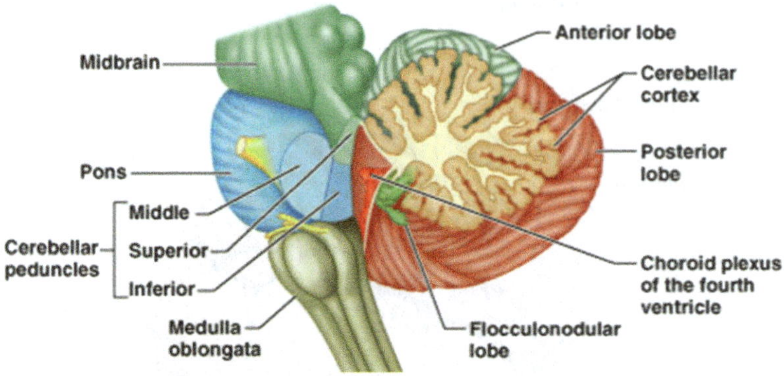

Fig. 1.38 Structure of cerebellum

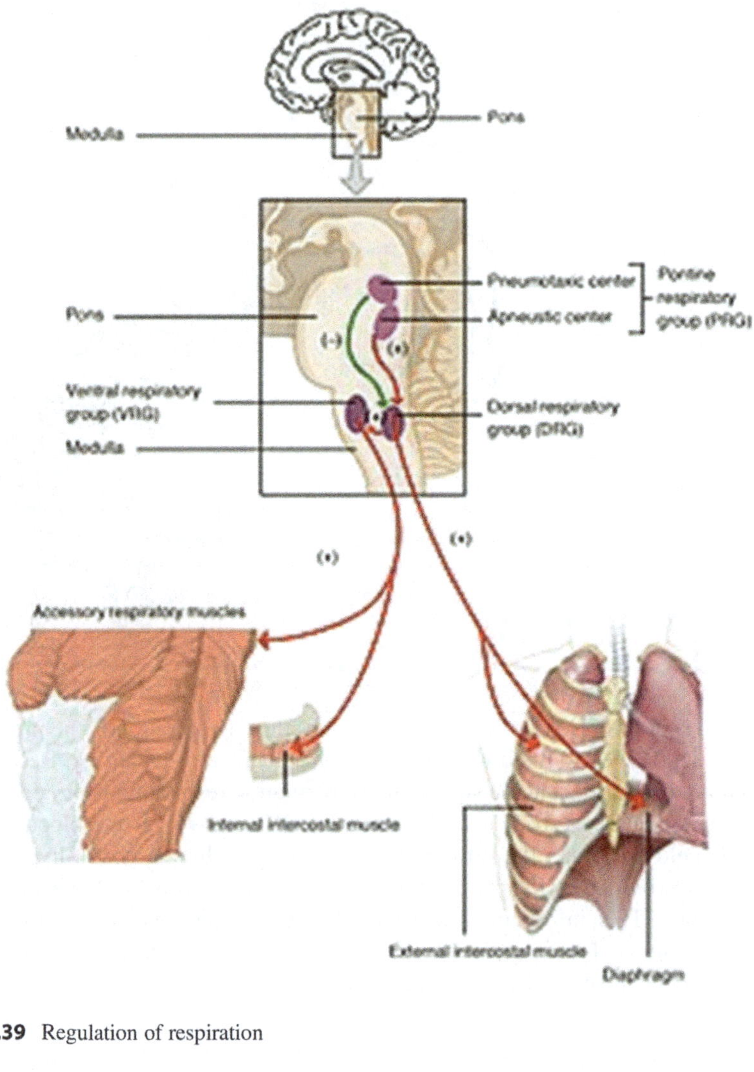

Fig. 1.39 Regulation of respiration

1.20 Autonomous Nervous System

It is a part of the nervous system which coordinates the functions of various organs like blood vessels, stomach, intestine, lungs, pupils, heart, etc.

They have two pathways:

Sympathetic nervous system – stimulatory function
Parasympathetic nervous system – inhibitory in function

To explain it clearly, let us discuss the regulation of blood pressure by autonomous nervous system. Blood pressure is the pressure exerted by the blood upon the walls of the blood vessels as they flow through. When it increases beyond the normal range it has to be controlled. Constriction of the blood vessels increases blood pressure which can be reduced by the parasympathetic nervous system which will lower the heart rate and hence the cardiac output and promotes vasodilation, thus reducing the blood pressure. If the blood pressure is below normal, then the sympathetic nervous system will cause vasoconstriction and increases the heart rate and thus increases the cardiac output. The process of regulation of blood pressure is shown in Fig. 1.40.

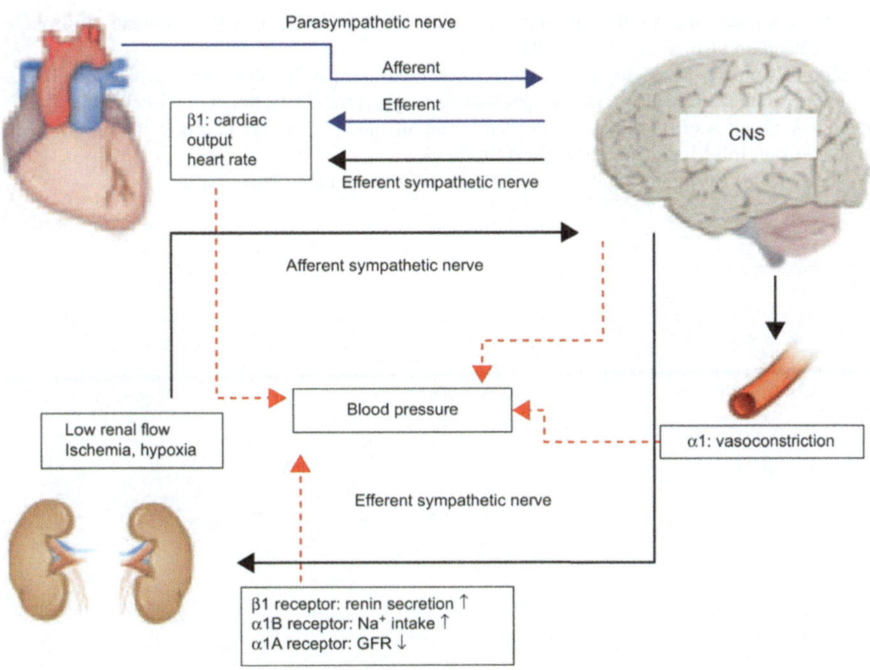

Fig. 1.40 Regulation of blood pressure

1.21 Conclusion

For a biomedical engineer, it is essential to have a thorough understanding of the underlying science of anatomy and physiology. This chapter has provided an overview of certain mechanisms like membrane potential, action potential generation and conduction which are vital to understand biosignal acquisition and processing systems that further helps in the development of healthcare devices.

References

Decoursey TE (2003) Voltage-gated proton channels and other proton transfer pathways. Physiol Rev 83:475

Ganong WF (2010) Review of medical physiology.", 22nd. McGraw Hill, New York

Hall JE (2013) Guyton and Hall, "Text book of medical physiology", 13th edn. Saunders, Philadelphia

Marieb EN (2007) Essentials of human anatomy and physiology, 8th edn. Pearson Education, London

Muller D, Nikonenko I (2003) Dynamic presynaptic varicosities: a role in activity-dependent synaptogenesis. Trends Neurosci 26:573

Pocock G, Richards CD (2009) The human body – an introduction for biomedical and health sciences. Oxford University Press, New York

Rossier BC, Pradervand S, Schild L, Hummler E (2002) Epithelial sodium channel and the control of sodium balance: interaction between genetic and environmental factors. Annu Rev Physiol 64:877

Sarada Subramanyam, Madhavan Kutty K, Singh HD (1996) Text book of human physiology. S. Chand and Company, New Delhi

Schram G, Pourrier M, Melnyk P, Nattel S (2002) Differential distribution of cardiac ion channel expression as a basis for regional specialization in electrical function. Circ Res 90:939

Waugh A, Grant A (2014) Ross and Wilson "Anatomy and physiology in health and illness", 12th edn. Churchill Livingstone Elsevier, Edinburgh

West JB (1996) Respiratory physiology. Oxford University Press, New York

Anatomically Real Microwave Tissue Phantoms

2

Gyanendra Sheoran and Vineeta Kumari

Abstract

Artificial phantoms or bio-models are the replicas of human body tissues, which are utilized for modeling of microwave propagation in tissues. These phantoms are highly needed while developing, optimizing, and evaluating microwave imaging systems. Such tissue imitations should ideally reflect the 3D structure of the human tissues. Overall, the phantoms play a key factor for initial design and development of systems. The phantoms/models act as a bridge for transferring the microwave-based lab setups to the commercial systems. This chapter includes all the aspects of the phantoms: dielectric properties, types of tissues, materials, structural properties, prospect of durability and reproducibility of phantoms with complementary properties, and applications.

Electromagnetic (EM) spectrum varies from X-rays to radiowave radiations. The existing and developing equipment and imaging systems operate at different regions of EM spectrum. Each equipment/imaging system needs the tissue-simulating object to mimic the properties of human tissue referred to as phantom; for the validation of their performance and calibration before clinical trials with the approval of regulatory bodies. The developed system can be tested for its repeatability, accuracy, limitations, safety issues and reproducibility. Overall the phantoms play a key factor for initial design and development of systems, improving the ratio of signal and noise in the existing systems and quality control of the device and to compare the performance of the systems with the existing one. The parameters of the instrument

G. Sheoran (✉)
Department of Applied Sciences, NIT Delhi, New Delhi, India
e-mail: gsheoran@nitdelhi.ac.in

V. Kumari
Department of Electronics and Communication Engineering, NIT Delhi, New Delhi, India

S. Paul (ed.), *Biomedical Engineering and its Applications in Healthcare*,
https://doi.org/10.1007/978-981-13-3705-5_2

can only be verified and suitable for clinical studies if they have been experimented on the phantoms that possess the similar properties as the original tissues does.

2.1 Introduction

Tissue phantoms are test objects used for diagnosis, medical imaging and demonstrating properties of living tissue samples. Tissue phantoms are the keynote of research and development for microwave-based systems. The obvious and quantitative validation, evaluation of safety and implementation of the microwave-based systems/devices are the critical aspects in the development of technological processes. Although a microwave system can be modelled numerically and simulated, it cannot contemplate the realistic parameters which may be unshielded to various perturbations like mechanical, electrical and environmental etc. Hence, the availability of the living tissues can provide the optimal measuring parameters for the above systems. However, the utilization of live human volunteers for testing and validating the devices unveils the whole procedure to various inherent uncertainties, like movement of the respiratory system, cardiovascular vibration and variation in skin moisture additionally with the safety concerns associated with new instruments/devices. Thus, the utilization of artificial tissue-mimicking (ATM) phantoms is much advantageous for evaluating the microwave-based instrument/system (Mobashsher and Amin 2015).

2.2 Classification of Phantoms

Several numerical phantoms have been investigated for theoretical analysis and computational simulations of the microwave system. The phantoms used for theoretical analysis are generally simple in shapes, whereas the phantoms used in computational simulations are composed of the small voxel which are designated as realistic numerical phantoms (Koichi 2007).

2.2.1 Theoretical Phantoms

Theoretical phantoms are homogeneous and flat-layered phantoms. These phantoms are spherical and cylindrical structure in nature and are used to evaluate the EM systems and dosimetry (Tell 1972). Spherical phantoms are used for the head and eyes (Weil 1975), whereas cylindroid phantoms are used as whole-body models (Massoudi et al. 1979; Nishizawa and Hashimoto 1999).

2.2.2 Voxel Phantoms

These are composed of a three-dimensional array of voxels. The advancement in computational domain of the medical imaging modalities, such as X-ray computed tomography (CT) and magnetic resonance (MR) imaging, empowers the researchers to evolve the high-resolution voxel models for the head, breast and whole body. In order to develop an anatomically real numerical phantom for human organs/tissues, it is anticipated to have several features such as precise dielectric properties for the desired frequency range and longer shelf life. Such models can be utilized for the assessment of the interactions between EM spectrum and human body tissues computationally by using the finite difference time-domain simulations. The electrical properties, i.e. dielectric permittivity 'ε_r' and conductivity 'σ', of human body tissues vary in wide range according to their types and operating microwave frequencies. The properties for various tissues of the body are measured by various researchers experimentally (Johnson and Guy 1972; Watanabe et al. 1996; Gabriel et al. 1996a, b; Lazebnik et al. 2007). The range of electrical properties, i.e. relative permittivity, and conductivity of different living tissues of human body (Gabriel et al. 1996a, b; Lazebnik et al. 2007) are indexed in Table 2.1.

2.2.3 Experimental Phantoms

These are self-developed phantoms that could be of homogeneous/inhomogeneous structure in nature. The exposure of microwaves on the human body tissues is having major dependency on the shape of phantoms that's why the shape is a main concern while designing and developing the phantoms (Kumari et al. 2018; Scapaticci et al. 2014; Davis and Balzano 2009). These phantoms are composed of materials/polymers having properties similar to the dielectric/electrical properties of human body tissues. It is reported in the literature that during an experimental study, different ATMs are employed for particular applications and purposes. Such typical experimental phantoms are:

- Liquid phantoms – saline and sugar solution, ethanol, etc.
- Gel phantoms – agar/gelatin, TX-150, TX-151, polyethylene powder, etc.
- Semi-solid phantoms – silicon rubber, carbon fibre, etc.
- Solid phantoms – ceramics, resins, acrylonitrile butadiene styrene (ABS), etc.

2.3 Development of Tissue Phantoms

2.3.1 Head Phantoms

The human head is a highly convoluted structure in the human body. Emulating it is a strenuous job because of the complex distribution of the various head tissues. Human head tissues consist of grey matter, white matter, cerebrospinal fluid (CSF)

Table 2.1 Ranges of dielectric properties for different human body tissues

Tissues	Muscle	Blood	Adipose	Fibroglandular	Grey matter	White matter	CSF	Skin	Tumour
ε_r	42.8–56.5	45.1–63.3	2–6	35–47	38.1–56	28–41	52.4–70.1	31.3–45	53–57
σ	0.8–10.6	1.38–13.1	0.2–0.8	4–6	0.8–10.3	0.47–7.3	2.3–15.4	0.73–8	3.1–6.4

and the skull. The head phantoms could be of simple rectangular shape, plastic skull models with layered tissues and models with tissue-mimicking materials, which uniformly represents the overall dielectric properties of the head tissues. The dielectric properties of head tissues are calculated by fitting the single- and two-pole Cole-Cole and Debye dispersion model parameters shown in Eqs. 2.1 and 2.2, respectively (Lazebnik et al. 2007; Kumari et al. 2018).

$$\varepsilon^*(\omega) = \varepsilon_\infty + \sum_{i=1}^{n} \frac{\Delta\varepsilon_i}{1 + (j\omega\tau_i)^{1-\alpha_i}} + j\frac{\sigma}{\omega\varepsilon_0} \tag{2.1}$$

$$\varepsilon^*(\omega) = \varepsilon_\infty + \sum_{i=1}^{n} \frac{\Delta\varepsilon_i}{1 + (j\omega\tau_i)} + j\frac{\sigma}{\omega\varepsilon_0} \tag{2.2}$$

where different dielectric parameters, i.e. permittivity at high frequency (ε_∞), difference of static permittivity and high-frequency permittivity ($\Delta\varepsilon$), static conductivity (σ_s) and exponential parameter of Cole-Cole (α), are used to calculate the frequency-dispersive properties.

The phantoms developed for head tissues can be categorized into liquid, gel and solid.

(i) **Liquid phantoms:** Homogeneous liquid phantoms are commonly synthesized using sugar-saline solutions with various preservatives (Scapaticci et al. 2014; Davis and Balzano 2009; Mochizuki et al. 2007; Hombach et al. 1996) or by utilizing the readily available commercial liquids as tissue-mimicking materials (Filho et al. 2009; Picher et al. 2012; Monebhurrun 2010; Saraereh et al. 2004). Along with the sugar-saline solutions (Davis and Balzano 2009), a mixture of deionized water and propylene glycol can also be used to develop the homogeneous liquid phantoms (Rodrigues et al. 2014). The receptacle of all the liquid phantoms are moulded using low permittivity polyvinyl chloride (PVC) materials or polyester. The coalescing of the polymers while fabricating liquid materials to mimic tissue can be resolved utilizing the water-soluble materials like hydroxyethyl cellulose (HEC). This is a viscosity enhancement agent which also improves both the bacterial resistance and shelf life of the phantom (Davis and Balzano 2009).

(ii) **Gel/semi-solid phantoms:** These phantoms are mostly made up of a gelling agent, e.g. agar, for both narrowband (Zhang et al. 2009; Jofre et al. 1990) and wideband frequency spectrum (Ito et al. 2001). Other polymers are also utilized to obtain expected properties for both narrowband and short-wave range (Bini et al. 1984). Semi-solid phantoms are frequently utilized in the broadband applications such as miniaturization of antenna, microwave imaging system for stroke detection and specific absorption rate (SAR) evaluation. These phantoms are fabricated from various homogeneous tissue emulating materials that are put to use to construct layered/heterogeneous semi-solid tissue phantoms (Schwerdt et al. 2012a; Mohammed et al. 2014; Mobashsher et al. 2014; Mobashsher and Abbosh 2014; Okano et al. 2000; Ito et al. 1998;

Mohammed and Abbosh 2014; Mustafa et al. 2013; Schwerdt et al. 2012b; Ishido et al. 2004). These phantoms primarily used agar for maintaining the phantom's shape. Also, some other gelling agents, such as TX-151 is frequently utilized in phantoms for increasing the viscidity of the mimicking materials (Hombach et al. 1996; Schwerdt et al. 2012a; Ito et al. 1998; Schwerdt et al. 2012b; Ishido et al. 2004; Levick et al. 2011).

(iii) **Solid phantoms:** There are very few solid phantoms for head tissues that have been investigated in state of the art due to their sophisticated fabrication procedure and expensive materials as compared to the other tissue-mimicking material counterparts. These solid phantoms are primarily developed for the applications in narrowband frequency range such as wearable communication systems (Ito et al. 1998; Loh et al. 2014; Looi and Chen 2005; Looi et al. 2005). However, a solid ATM phantom is also reported for ultra-high-frequency (UHF) band applications (Moon et al. 2000). According to the tissue distribution accuracy, the maximum reported phantoms are found to be anthropomorphic in nature. Some of the multi-layer human head tissue phantoms with mimicking materials are reported in state of the art. Apart from this, various studies are there, where the three-dimensional (3D) printed phantoms (skull) filled with homogeneous and multi-layer realistic semi-solid tissue compositions are described (Mobashsher et al. 2014; Mobashsher and Abbosh 2014). Here Table 2.2 shows a summary of the head phantoms along with their respective applications and materials.

2.3.2 Breast Phantoms

Human breast tissues are composed of adipose tissues (determine the exact structure and shape of the breast and also act as a protective layer for them) and fibroglandular tissues (composed of connective tissues, lobules and milk ducts). In the literature, the breast tissue phantoms are majorly propounded and developed for the testing, validation and evaluation of microwave-based instruments/systems for early-stage detection of cancerous tissues in the breast (Kumari et al. 2018).

Several materials, viz. ABS and resins, are used to form solid and TX-150 and TX-151 are used to form semi-solid phantoms. The procedure for fabricating multi-layer phantoms is simple and rapid using aforesaid materials. Among these materials, gelatin is more appealing because of the ease in fabrication and its steady mechanical properties. They are of two types: (i) oil-in-gelatin and (ii) water-in-gelatin.

Development of a breast phantom has been commonly done by oil-in-gelatin mixtures (Koichi 2007). Along with the basic shape (hemispherical) breast models, various anatomically realistic breast phantoms based on oil-in-gelatin mixture are also described. These breast phantoms are heterogeneously constructed numerically by assigning the dielectric properties calculated by Debye and Cole-Cole model

Table 2.2 Details of the tissue phantoms of human head

Reference	Year	P or T	Freq.	Phantom type	Involved tissues	Phantom structure	Applications
Bini et al. (1984)	1984	T	13.6, 27, 40.7 or 2450 MHz	HO, G	Fat, muscle, skin, brain,	–	Hyperpyrexia
Jofre et al. (1990)	1990	P	2.45 GHz	S, G, HE	Skin, cerebellum, grey matter, cortical bone, blood and muscle	CT scan-based realistic shape, box-shaped, stylized	Microwave tomography, wearable systems
Looi and Chen (2005)	2005						
Looi et al. (2005)	2005						
Hombach et al. (1996)	1996	P	0.9 GHz	L, HO	Brain, bone, muscle, skin	Realistic/ANM	SAR evaluation
Guy and Chou (1986)	1986	P	0.9 GHz, 1.5 GHz	HO, S	Head tissue equivalent	ANM	SAR measurement
Kobayashi et al. (1993)	1993						
Watanabe et al. (1996)	1996						
Tamura et al. (1997)	1997						
Okano et al. (2000)	2000						
Kawai et al. (2007)	2007						
Kawamura et al. (2009)	2009						
Picher et al. (2012)	2012						

(continued)

Table 2.2 (continued)

Reference	Year	P or T	Freq.	Phantom type	Involved tissues	Phantom structure	Applications
Ito et al. (1998)	1998	P	0.2 GHz–3 GHz	ML, SS	Skull, brain tissue	Cubic, spherical	Specific absorption analysis
Okano et al. (2000)	2000						
Ogawa and Matsuyoshi (2001)	2001	P	0.9 GHz	HO, L	Brain tissue	Stylized	Analysis of antenna performance
Ito et al. (2001)	2001	P	0.3GHz–2.5 GHz	HO, G	Brain, muscle	Cubic	Hyperpyrexia
Youngs et al. (2002)	2002	P	1 MHz-10GHz	S, ML	Fat, muscle, grey matter	Spherical and layered	Measurement of absorption rate
Saraereh et al. (2004)	2004	P	0.9, 1.75, 1.95 GHz	HO, L	Brain tissue	ANM	SAR analysis
Monebhurrun (2010)	2010						
Mochizuki et al. (2007)	2007	P	0.402 GHz	HO, L	Skin tissue	Cubic, ANM	Body implantable devices
Zhang et al. (2009)	2009	P	7 MHz	G, HO	Brain	ANM	Implantable devices and medical sensors
Chen et al. (2009)	2009	P	0.4 MHz	L	Grey and white matter	ANM	Body implants
Levick et al. (2011)	2011	T	3.4 GHz	HE, SS	Neonatal brain tissue	Simulated and modelled	Microwave passive imaging, i.e. radiometry
Mohammed et al. (2012)	2012	P	0.2–20 GHz	SS, HE	Grey matter, white matter, CSF and blood	Realistic	Microwave imaging modalities
Karathanasis et al. (2012)	2012	P	2.4 GHz	L, HO	Brain stimulant	ANM	Hyperpyrexia

Mobashsher et al. (2014)	2014	P	0.5–4 GHz	HE, SS	Spinal cord, skull, grey matter, white matter, dura mater, CSF, blood	MRI developed phantom, anatomically real	Brain imaging, SAR analysis
Mobashsher and Abbosh (2014)	2014						
Mohd et al. (2015)	2015	P	1–6 GHz	G, SS, HE	CSF, grey matter, white matter, blood, skin	Realistic	Microwave imaging application and system
Lee et al. (2016)	2016	P	0.5–2 GHz	S, HE	Blood, cerebrospinal fluid	3D printed	Brain stroke localization
Mcdermott et al. (2017)	2017	P	1–8.5 GHz	S, HE	Blood, brain	Realistic	Microwave imaging
Nadine et al. (2018)	2018	P	0.5–6 GHz	L, S	Brain, CSF	ANM, 3D printed	Microwave imaging

HO homogeneous, *HE* heterogeneous, *T* tissue, *P* phantom, *G* gel, *ML* multi-layered, *SS* semi-solid, *L* liquid, *ANM* anthropomorphic, *S* solid parameters in the MR and CT images of actual human breasts. These numerical phantoms can be printed using 3D printers (O'Halloran et al. 2014; Mashal et al. 2011; Bakar et al. 2011a; Croteau et al. 2009). The solid materials used for such phantoms are dispersive in nature w.r.t. frequency. Although the verified CT images presented in (Mashal et al. 2011) exhibit better fabrication accuracy of the breast phantom as compared to MR images, they have limited shelf life, i.e. up to 8 weeks (Lazebnik et al. 2005). It is to be mentioned here that the polyacrylamide material is also not suitable for fabricating the phantoms due to its shorter shelf life. That's why, the liquid phantoms based on water and Triton X-100 (Joachimowicz et al. 2014; Henriksson et al. 2010; Romeo et al. 2011) have become popular. This type of mixture has a stable shelf life, i.e. around 1 year (Joachimowicz et al. 2014). Moreover, the dielectric permittivity and conductivity of such mixture are more diverse and can be easily moulded according to the water content corresponding to the real tissues. Thus, such mixture is suitable as the filling material for 3D printed breast phantoms consists of solid materials that have frequency-dispersive properties (Burfeindt et al. 2012). The tissue-mimicking phantoms for breast tissues are primarily developed for wideband operations (O'Halloran et al. 2014; Mashal et al. 2011; Croteau et al. 2009; Lazebnik et al. 2005; Joachimowicz et al. 2014; Romeo et al. 2011; Burfeindt et al. 2012; Klemm et al. 2009; Porter et al. 2010, 2011; Hahn and Noghanian 2012; Modiri and Kiasaleh 2013; Ostadrahimi et al. 2009; Li et al. 2004; Klemm et al. 2010; Craddock et al. 2005). Solid breast phantoms are also well documented (Bourqui et al. 2010, 2012); however, due to their non-dispersive dielectric properties, i.e. dielectric properties are not dependent on frequency, they are not widely adopted.

The various phantoms for breast tissues are indexed in Table 2.3 in accordance to their reported frequency bands.

2.3.3 Limb Phantoms

The various reported limb phantoms are developed for the broadband applications, viz. hyperthermia, electrical impedance tomography and body wearable communication systems (Loh et al. 2014; Yuan et al. 2012; Chou et al. 1991; Cuyckens 2010; Gabriel 2007). The frequency range of such mimicking phantoms can vary from kHz to GHz and is application-specific. According to the type of application, various tissue/skin equivalent materials like ceramics, silicon rubber, etc. (Chou et al. 1991; Cuyckens 2010; Gabriel 2007) are utilized to construct the limb phantoms homogeneously. However, the heterogeneous structure of some limb tissue phantoms can also be built (Loh et al. 2014; Yuan et al. 2012; Chou et al. 1991). The shelf life of such phantoms is defined as 2–4 weeks. The first inhomogeneous limb phantoms were developed (Chou et al. 1991) to study the heating rate patterns, where the cylindrical structures of a mixture of polyethylene powder, water, TX-151 and NaCl were developed as phantoms for arms and thigh tissues.

In Table 2.4, the limb tissue-mimicking phantoms are listed according to the application-specific frequencies.

Table 2.3 Details of the outlined breast tissue phantoms

Ref.	Year	P or T	Frequency band	Phantom type	Defined tissues	Structure/shape of the breast phantom	Applications
Gajda et al. (1979)	1979	T	2.45 GHz	HO, L	Fat, muscle	--	Pattern description of electric field
Klemm et al. (2010)	2004	P	1–11 GHz	L, ML	Adipose tissue, skin and malignant tissue	Shaped and stylized	Imaging systems
Craddock et al. (2005)	2005	P	3GHz–10 GHz	HE, SS	Fibroglandular, tumour, healthy tissue, skin tissue	ANM	Breast cancer detection
Lazebnik et al. (2005)	2005	T	0.5GHz–20 GHz	SS, HE	Muscle, dry skin, wet skin, malignant lesions	–	Microwave devices and applications
Zastrow et al. (2008)	2008	P	0.5–20 GHz	Simulation	Breast tissues	Anatomically real	Breast cancer detection
Ostadrahimi et al. (2009)	2009	P	0.5GHz–8 GHz	SS, HE	Skin, fat, fibroglandular and malignant tumour tissues	Cylindrical, stylized	Microwave breast imaging
Croteau et al. (2009)	2009	P	1GHz–13 GHz	HE, G	Skin, fibroglandular, adipose, tumour	Realistic	Radar-based microwave imaging
Zhou (2009)	2009	P		Simulation, HE	Breast	Hemispherical	Breast cancer detection
Henriksson et al. (2010)	2010	P	2.45 GHz	HO, L	Tumour, breast tissue	Stylized, cylindrical	Microwave breast system
Porter et al. (2010)	2010	P	0.2GHz–6 GHz	HE, SS	Skin tissue adipose, fibroglandular, tumour	ANM, hemispherical	Microwave imaging
Porter et al. (2011)	2011						
Bourqui et al. (2010)	2010	P	2GHz–12 GHz	HO, S	Breast tissues	ANM	Ultra-wideband imaging system
Bourqui et al. (2012)	2012						

(continued)

Table 2.3 (continued)

Ref.	Year	P or T	Frequency band	Phantom type	Defined tissues	Structure/shape of the breast phantom	Applications
Romeo et al. (2011)	2011	T	0.5GHz–12 GHz	HO, L	Healthy and malignant tissues	--	Cancer detection in breast
Bakar et al. (2011a)	2011	P	3GHz–11 GHz	HE, SS	Adipose, lesions, fibroglandular	Hemispherical	Ultra-wideband imaging
Bakar et al. (2011b)	2011						
Mashal et al. (2011)	2011	P	1GHz–6GHz	SS, HE	Adipose, fibroglandular, skin	Realistic, ANM, CT images	Microwave imaging experiments
Burfeindt et al. (2012)	2012	P	0.5GHz–3.5 GHz	HE, S	Adipose, fibroglandular	MRI derived realistic	Microwave breast imaging
Hahn and Noghanian (2012)	2012	P	0.5GHz–6 GHz	HE, SS	Fibroglandular tissue, adipose and skin	Stylized, cylindrical	Microwave imaging
Modiri and Kiasaleh (2013)	2013						
Joachimowicz et al. (2014)	2014	T	0.5GHz–6 GHz	HO, L	Healthy and malignant tissues	–	Microwave imaging
O'Halloran et al. (2014)	2014	P	0.5GHz–4 GHz	HE, SS	Fibroglandular tissue, adipose skin and tumour	MRI derived realistic	Microwave imaging
Garrett et al. (2015)	2015	P	0.5GHz–10 GHz	S	Breast tissue with outer skin layer	3D printed	Microwave breast imaging
Tunçay and Akduman (2015)	2015	P	0.5GHz–20 GHz	Simulation	Breast adipose, glandular, skin	Anatomically real	Microwave imaging system
Kiarashi et al. (2015)	2015	P		S	Breast tissues	3D printed	Cancer detection
Wang and Niu (2017)	2017	P	10GHz–20 GHz	SS	Breast with lesions	–	Holographic microwave imaging system

Omer and Fear (2017)	2017	P	0.5GHz–15 GHz	Simulation	Breast (adipose, fibroglandular)	Anatomically real using MRI	Microwave breast imaging
Joachimowicz et al. (2017)	2017	P	1GHz–6 GHz	S, L	Breast tissues	3D printed with liquid solutions	Microwave imaging
Faenger et al. (2017)	2017	P	0.5GHz–7 GHz	S	Breast tissue with skin	Anatomically real, 3D printed	Microwave imaging system
Barbara et al. (2018)	2018	P	1GHz–8 GHz	S	Breast tumours	Realistic	Microwave breast imaging
Islam et al. (2018)	2018	P	3.01GHz–11 GHz	S, HE	Breast tissue	ANM	Microwave imaging antenna sensor
Kumari et al. (2018)	2018	P	0.5GHz–20 GHz	Simulation	Adipose, fibroglandular, skin	Anatomically real	Breast cancer detection
Fiaschetti et al. (2018)	2018	P	0.5GHz–4 GHz	SS	Breast	ANM	Microwave and US imaging modality
Nadine et al. (2018)	2018	P	0.5GHz–6 GHz	L, S	Fibroglandular, tumour, adipose	ANM, 3D printed	Microwave imaging

Table 2.4 Details of the outlined limb tissue phantoms

Reference	Year	P or T	Freq.	Type of phantom	Included tissues	Limb phantom structure	Applications
Gabriel (2007)	2007	P	0.6GHz–6GHz	HO, S	Blood, fat, muscle, sinew, bone, skin, hand tissues	Realistic	SAR analysis
Yuan et al. (2012)	2012	P	0.08GHz–0.5GHz	HE, SS	Fat, muscles, tumours, bone marrow	ANM	MRI thermal monitoring and RF heating
Chahat et al. (2012, 2013)	2012 2013	P	55GHz–65GHz	HO, SS	Human skin	ANM	Body networks

2.3.4 Other Body Phantoms

Apart from the above-explained tissue phantoms, some whole-body phantoms are also reported in the literature, e.g. the tailor-made phantoms developed in 2016 for different human tissues up to 18 GHz and particularized for the main current body area network (BAN) operating bands (Castelló-palacios et al. 2016). A full-body phantom of a woman with embryo/foetus model is developed for SAR evaluations (Nagaoka et al. 2007) for a range of frequencies (10 Hz–2 GHz). Apart from this, some other body phantoms (muscle tissue, bone and skin) are reported for ultra-wideband communication applications for a frequency ranging from 900 MHz to 10 GHz (Takimoto et al. 2007; Garrett and Fear 2014). Also, a phantom for monitoring of blood glucose was reported (Vrba et al. 2015), where two distinct phantoms, viz. pig blood-glucose solutions and physiological saline-glucose, were compared for making the phantom corresponding to blood glucose. It was found that physiological saline-glucose doesn't resemble to blood glucose at 1–3 GHz frequency range and cannot be used as phantom.

2.4 Materials for Phantom

The list of materials used in the development of various phantoms is mentioned in Table 2.5.

Table 2.5 List of phantom materials

Materials	Specifications	Advantages/key roles
Water (H_2O)	Main ingredient for tissue-mimicking materials	Principally accords highly dispersive properties
Sugar	White crystalline, odourless powder	Used for tuning the 'ε_r', while also increases the 'σ'
Sodium chloride (NaCl)	Crystalline salt	To adjust the conductivity tissue phantoms
Triton X-100	Non-ionic detergent which is viscous in nature	Utilized as surfactant in ATM liquid solutions and to minimize the 'ε_r'
Triton X-150	Powder that is soluble in water	To raise viscidity and uniformity of phantoms
Sodium azide (NaN4)	Toxic white powder	Usually acts as a preservative and to escalate the 'σ' in solution of water
Glycerine	Lossy coupling liquid	Reproduce both high- and low-water content organs replicate the organs having high and low water density, provides moisturizing to phantoms, acts as preservative
Polyethylene powder	Powder which is additive to unsaturated polyesters	Used to adjust the relative permittivity 'ε_r'
Ceramic powder (MgO, $BaTiO_3$)	Lossy material	To increase 'ε_r' of the tissue emulating materials
Agar, gelatin	Gelling agents; agar is powder that is yellowish white in colour and gelatin is colourless	Retention of self-shaping by gelling agent, acts as coagulant and tune the dielectric properties
Graphite (carbon) powder	A semiconductor material that is enveloped with adhesive like polyvinylidene difluoride (PVDF)	To increase 'σ' of the tissue-mimicking materials, used to make dry phantom
Hydroxyethyl cellulose	A non-ionic water-soluble polymer (gelling agent)	Increases viscosity of water-based amalgams
Oil (paraffin, grape seed oil, kerosene, safflower oil, etc.)	A hydrophobic viscous liquid	To fabricate low water content tissue emulating materials
Carbon fibre and aluminium powder	Thin carbon fibres and powdered aluminium	Enables refinement of the process in different semi-solid tissue-mimicking materials
Silicone rubber (not to be confused with silicon (Si))	A heat-resistant, elastomer, synthetic polymer	For holding the active ingredient; with required amount of carbon to give right consistency without water
Polyacrylamide (C_3H_5NO)	Synthetic materials	Used in lowering down the 'ε_r' and slightly increases 'σ', enhances the robustness and slightly impairs transparency

(continued)

Table 2.5 (continued)

Materials	Specifications	Advantages/key roles
Bactericide	Bacterial resistance substances	Prevention of decomposition of tissue-mimicking materials from microbial growth; enhances shelf lifetime
TX-151	Polysaccharide and propyl-paraben preservative	Gelling agent; can simulate high-water content organs, mimics tissue texture and increases stickiness
Polyvinyl chloride (PVC)	White polymer powder	For exponentially decaying 'σ' and linearly decreasing 'ε_r'

2.5 Conclusion

This chapter extends a very detailed overview on the development of anatomically real microwave phantoms that are capable of emulating the electrical, mechanical and structural properties of the tissues. The overview included all the aspects of the phantoms: dielectric properties, types of tissues, materials, structural properties, prospect of durability and reproducibility of phantoms with complementary properties and applications. It is also discussed that different materials are used to mimic the tissue texture and each material has its own strength and weakness, and some are used for enhancement of shelf lifetime, while others are used to increase the mechanical strength and viscosity and hold the shape of the tissues. Since phantoms are the replica of living human tissue, therefore, it is envisaged that the phantoms are critical factors and play an influential role in the development and validation of any microwave-based instruments/systems. It acts as a bridge for transferring the microwave-based lab setups to the commercial systems.

Acknowledgement The authors would like to acknowledge the assistance of Ms. Neelam Barak and Mr. Aijaz Ahmed of NIT Delhi in writing the book chapter.

References

Bakar AA, Abbosh A, Sharpe P, Bialkowski ME, Wang Y (2011a) Heterogeneous breast phantom for ultra-wideband microwave imaging. Microw Opt Technol Lett 53(7):1595–1598

Bakar AA, Abbosh A, Bialkowski M (2011b) Fabrication and characterization of a heterogeneous breast phantom for testing an ultrawideband microwave imaging system. In: Proceedings of Asia-Pacific Microw Conference, Dec. 2011, pp 1414–1417

Barbara OL, Declan LO, Martin HO, Porter E (2018) Microwave breast imaging: experimental tumour phantoms for the evaluation of new breast cancer diagnosis systems. Biomed Phys Eng Express 4(2):1–13

Bini MG, Ignesti A, Millanta L, Olmi R, Rubino N, Vanni R (1984) The polyacrylamide as a phantom material for electromagnetic hyperthermia studies. IEEE Trans Biomed Eng 3:317–322
Bourqui J, Campbell MA, Williams T, Fear EC (2010) Antenna evaluation for ultra-wideband microwave imaging. Int J Antennas Propag 2010:1–8
Bourqui J, Sill JM, Fear EC (2012) A prototype system for measuring microwave frequency reflections from the breast. J Biomed Imaging 2012:1–12
Burfeindt MJ, Colgan TJ, Mays RO, Shea JD, Behdad N, Veen BDV, Hagness SC (2012) MRI-derived 3D printed breast phantom for microwave breast imaging validation. IEEE Antennas Wireless Propag Lett 11:1610–1613
Castelló-palacios S, Garcia-pardo C, Fornes-leal A, Cardona N, Vallés-lluch A (2016) Tailor-made tissue phantoms based on acetonitrile solutions for microwave applications up to 18 ghz. IEEE Trans Microw Theory Technol 64(11):3987–3994
Chahat N, Zhadobov M, Sauleau R (2012) Broadband tissue-equivalent phantom for BAN applications at millimeter waves. IEEE Trans Microw Theory Technnol 60(7):2259–2266
Chahat N, Leduc C, Zhadobov M, Sauleau R (2013) Antennas and interaction with the body for body-centric wireless communications at millimeter-waves. In: Proceedings of European Conference on Antennas and Propagation, Apr. 2013, pp 772–775
Chen ZN, Liu GC, See TS (2009) Transmission of RF signals between MICS loop antennas in free space and implanted in the human head. IEEE Trans Antennas Propag 57(6):1850–1854
Chou CK, Mcdougall JA, Chan KW, Luk KH (1991) Heating patterns of microwave applicators in inhomogeneous arm and thigh phantoms. Med Phys 18(6):1164–1170
Craddock J, Nilavalan R, Leendertz J, Preece A, Benjamin R (2005) Experimental investigation of real aperture synthetically organised radar for breast cancer detection. In: Proceedings of IEEE Antennas Propagation Society International Symposium, July 2005, pp 179–182
Croteau J, Sill J, Williams T, Fear E (2009) Phantoms for testing radar-based microwave breast imaging. In: Proceedings of ANTEM/URSI International Symposium on Antenna Technology Application Electromagnetic Canadian Rad. Sci. Meeting, Feb. 2009, pp 1–4
Cuyckens T (2010) Influence of the human body on the behaviour of monopole and patch antennas, Ph.D. dissertation, University Gent. [online]. Available: http://lib.ugent.be/fulltxt/RUG01/001/418/705/RUG01-001418705_2010_0001_AC.pdf
Davis CC, Balzano Q (2009) The international intercomparison of SAR measurements on cellular telephones. IEEE Trans Electromag Compat 51(2):210–216
Faenger B, Ley S, Helbig M, Sachs J, Hilger I, Jena F (2017) Breast phantom with a conductive skin layer and conductive 3D-printed anatomical structures for microwave imaging. In: 11th European Conference on Antennas and Propagation March 2017, pp 1065–1068
Fiaschetti G, Browne JE, Cavagnaro M, Farina L, Ruvio G (2018) Tissue mimicking materials for multi-modality breast phantoms. In: 2018 2nd URSI Atl. Radio Sci. Meet.(International Union of Radio Science URSI, 2018), pp 1–6
Filho RHC, Oliveira RMD, Sobrinho CLSSD, Almeida AMD (2009) Parallel-FDTD and experimental results of SAR for flat and head phantoms@ 900 MHz. In: Proceedings of SBMO/IEEE MTT-S International Microwave and. Optoelectronics Conference, pp 373–378
Gabriel C (2007) Tissue-equivalent material for hand phantoms. Phys Med Biol 52(14):4205–4210
Gabriel S, Lau RW, Gabriel C (1996a) The dielectric properties of biological tissues: II. Measurements in the frequency range 10 Hz to 20 GHz. Phys Med Biol 41(11):2251–2269
Gabriel S, Lau RW, Gabriel C (1996b) The dielectric properties of biological tissues: III. Parametric models for the dielectric spectrum of tissues. Phys Med Biol 41(11):2271–2293
Gajda G, Stuchly MA, Stuchly SS (1979) Mapping of the near-field pattern in simulated biological tissues. Electron Lett 15(4):120–121
Garrett J, Fear E (2014) Stable and flexible materials to mimic the dielectric properties of human soft tissues. IEEE Antennas Wireless Propag Lett 13:599–602
Garrett J, Member S, Fear E, Member S (2015) A new breast phantom with a durable skin layer for microwave breast imaging. IEEE Trans Antennas Propag 63(4):1693–1700

Guy AW, Chou C-K (1986) Specific absorption rates of energy in man models exposed to cellular UHF mobile-antenna fields. IEEE Trans Microw Theory Technol 34(6):671–680

Hahn C, Noghanian S (2012) Heterogeneous breast phantom development for microwave imaging using regression models. J Biomed Imaging 2012:1–12

Henriksson T, Joachimowicz N, Conessa C, Bolomey J-C (2010) Quantitative microwave imaging for breast cancer detection using a planar 2.45 GHz system. IEEE Trans Instrument Meas 59(10):2691–2699

Hombach V, Meier K, Burkhardt M, Kuhn E, Kuster N (1996) The dependence of EM energy absorption upon human head modeling at 900 MHz. IEEE Trans Microw Theory Technol 44(10):1865–1873

Ishido R, Onishi T, Saito K, Uebayashi S, Ito K (2004) A study on the solid phantoms for 3–6 GHz and evaluation of SAR distributions based on the thermographic method. In: Proceedings of IEEE International Symposium on Electromagnetic Compatibility, pp 577–580

Islam MT, Samsuzzaman M, Kibria S, Singh MJ (2018) A homogeneous breast phantom measurement system with an improved modified microwave imaging antenna sensor. Sensors 18(9):1–23

Ito K, Okano Y, Hase A, Ida I (1998) A tissue-equivalent solid phantom for estimation of interaction between human head and handset antenna. In: Proceedings of IEEE-APS Conference Antennas and Propagation for Wireless Communications, Nov. 1998, pp 89–92

Ito K, Furuya K, Okano Y, Hamada L (2001) Development and characteristics of a biological tissue-equivalent phantom for microwaves. Electron Commun Japan (Part I: Commun) 84:67–77

Joachimowicz N, Conessa C, Henriksson T, Duchene B (2014) Breast phantoms for microwave imaging. IEEE Antennas Wireless Propag Lett 13:1333–1336

Joachimowicz N, Duchêne B, Conessa C, Meyer O (2017) Reference phantoms for microwave imaging. In: 2017 11th European Conference on Antennas and Propagation Ref, pp 2719–2722

Jofre L, Hawley MS, Broquetas A, Reyes EDL, Ferrando M, Elias-Fuste AR (1990) Medical imaging with a microwave tomographic scanner. IEEE Trans Biomed Eng 37(3):303–312

Johnson CC, Guy AW (1972) Nonionizing electromagnetic wave effects in biological materials and systems. Proc IEEE 60(6):692–718

Karathanasis KT, Gouzouasis IA, Karanasiou IS, Uzunoglu NK (2012) Experimental study of a hybrid microwave radiometry hyperthermia apparatus with the use of an anatomical head phantom. IEEE Trans Inf Technol Biomed 16(2):241–247

Kawai H, Tanaka S, Wake K, Watanabe S, Taki M, Uno T (2007) Localized exposure using an 8-shaped loop antenna system with a director for animal study in 1.5 GHz band. In: Proceedings of IEEE Asia-Pacific Microwave Conference, Dec. 2007, pp 1–4

Kawamura T, Saito K, Kikuchi S, Takahashi M, Ito K (2009) Specific absorption rate measurement of birdcage coil for 3.0-T magnetic resonance imaging system employing thermographic method. IEEE Trans Microw Theory Technol 57(10):2508–2514

Kiarashi N, Nolte AC, Sturgeon GM, Segars WP, Ghate SV, Nolte LW, Samei E, Lo JY (2015) Development of realistic physical breast phantoms matched to virtual breast phantoms based on human subject data. Med Phys 42(7):4116–4126

Klemm M, Craddock IJ, Leendertz JA, Preece A, Benjamin R (2009) Radar-based breast cancer detection using a hemispherical antenna array experimental result. IEEE Trans Antennas Propag 57(6):1692–1704

Klemm M, Leendertz JA, Gibbins D, Craddock IJ, Preece A, Benjamin R (2010) Microwave radar-based breast cancer detection: imaging in inhomogeneous breast phantoms. IEEE Antennas Wireless Propag Lett 8:1349–1352

Kobayashi T, Nojima T, Yamada K, Uebayashi S (1993) Dry phantom composed of ceramics and its application to SAR estimation. IEEE Trans Microw Theory Technol 41(1):136–140

Koichi I (2007) Human body phantoms for evaluation of wearable and implantable antennas. In Second European Conference on Antennas and Propagation, EuCAP 2007, pp 1–6

Kumari V, Sheoran G, Tirupathiraju K, Vyas R, Rao SA (2018) Development and analysis of anatomically real breast phantoms using different dispersion models. J Electron Imaging 27(5):051208(1–12)
Lazebnik M, Madsen EL, Frank GR, Hagness SC (2005) Tissue-mimicking phantom materials for narrowband and ultrawideband microwave applications. Phys Med Biol 50(18):4245–4258
Lazebnik M, McCartney L, Popovic D, Watkins CB, Lindstrom MJ, Harter J, Hagness SC (2007) A large-scale study of the ultrawideband microwave dielectric properties of normal breast tissue obtained from reduction surgeries. Phys Med Biol 52(10):2637–2656
Lee J, Bang J, Choi J (2016) Realistic head phantom for evaluation of brain stroke localization methods using 3D printer. J Electromagn Eng Sci 16(4):254–258
Levick A, Land D, Hand J (2011) Validation of microwave radiometry for measuring the internal temperature profile of human tissue. Meas Sci Technol 22(6):1–8
Li X, Davis SK, Hagness SC, Weide DWVD, Veen BDV (2004) Microwave imaging via space-time beamforming: experimental investigation of tumor detection in multilayer breast phantoms. IEEE Trans Microw Theory Technol 52(8):1856–1865
Loh TH, Cheadle D, Rosenfeld L (2014) Radiation pattern measurement of a low-profile wearable antenna using an optical fibre and a solid anthropomorphic phantom. Electronics 3(3):462–473
Looi CK, Chen ZN (2005) Design of a human head equivalent phantom for ISM 2.4-GHz applications. Microw Opt Technol Lett 47(2):163–166
Looi CK, See TSP, Chen ZN (2005) Study of human head effects on the planar inverted-F antenna. In: Proceedings of IEEE International Workshop on Antenna Technology, Mar. 2005, pp 223–226
Mashal A, Gao F, Hagness SC (2011) Heterogeneous anthropomorphic phantoms with realistic dielectric properties for microwave breast imaging experiments. Microw Opt Technol Lett 53:1896–1902
Massoudi H, Durney CH, Barbar PW, Iskander MF (1979) Electromagnetic absorption in multi-layered cylindrical models of man. IEEE Trans Microwave Theory Tech 27(10):825–830
Mcdermott B, Porter E, Santorelli A, Divilly B, Morris L, Jones M, McGinley B, O'Halloran M (2017) Anatomically and dielectrically realistic microwave head phantom with circulation and reconfigurable lesions. Prog Electromagn Res B 78:47–60
Mobashsher AT, Abbosh A (2014) Three-dimensional human head phantom with realistic electrical properties and anatomy. IEEE Antennas Wireless Propag Lett 13:1401–1404
Mobashsher AT, Amin AM (2015) Artificial human phantoms. IEEE Microw Mag 16(6):42–62
Mobashsher AT, Abbosh AM, Wang Y (2014) Microwave system to detect traumatic brain injuries using compact unidirectional antenna and wideband transceiver with verification on realistic head phantom. IEEE Trans Microw Theory Technol 62(9):1826–1836
Mochizuki S, Wakayanagi H, Hamada T, Watanabe S, Taki M, Yamanaka Y, Shirai H (2007) Effects of ear shape and head size on simulated head exposure to a cellular phone. IEEE Trans Electromag Compat 49(3):512–518
Modiri A, Kiasaleh K (2013) Experimental results for a novel microwave radiator structure targeting non-invasive breast cancer detection. In: Proceedings of IEEE Global Hum Technology Conference., Oct. 2013, pp 203–208
Mohammed BJ, Abbosh AM (2014) Realistic head phantom to test microwave systems for brain imaging. Microw Opt Technol Lett 56(4):979–982
Mohammed B, Abbosh A, Henin B, Sharpe P (2012) Head phantom for testing microwave systems for head imaging. In: Cairo International Biomedical Engineering Conference, pp 191–193
Mohammed BJ, Abbosh AM, Mustafa S, Ireland D (2014) Microwave system for head imaging. IEEE Trans Instrument Meas 63(1):117–123
Mohd SMS, Norhudah S, Noor SR, Tharek RA (2015) Modeling of gelatin-based head phantom based on its electrical properties for wideband microwave imaging application. Appl Mech Mater 781:608–611
Monebhurrun V (2010) Conservativeness of the SAM phantom for the SAR evaluation in the child's head. IEEE Trans Magn 46(8):3477–3480

Moon KS, Choi HD, Lee AK, Cho KY, Yoon HG, Suh KS (2000) Dielectric properties of epoxy dielectrics carbon black composite for phantom materials at radio frequencies. J App Poly Sci 7(6):1294–1302

Mustafa S, Mohammed B, Abbosh A (2013) Novel preprocessing techniques for accurate microwave imaging of human brain. IEEE Antennas Wireless Propag Lett 12:460–463

Nadine J, Bernard D, Conessa C, Olivier M (2018) Anthropomorphic breast and head phantoms for microwave imaging. Diagnostics 8(4):1–12

Nagaoka T, Togashi T, Saito K, Takahashi M, Ito K, Watanabe S (2007) An anatomically realistic whole-body pregnant-woman model and specific absorption rates for pregnant woman exposure to electromagnetic plane wave from 10 MHz to 2 GHz. Phys Med Biol 52:6731–6745

Nishizawa S, Hashimoto O (1999) Effective shielding analysis for three-layered human models. IEEE Trans. Microwave Theory Technol 47(3):277–283

O'Halloran M, Lohfeld S, Ruvio G, Browne J, Krewer F, Ribeiro CO, Pita VCI, Conceicao RC, Jones E, Glavin M (2014) Development of anatomically and dielectrically accurate breast phantoms for microwave breast imaging applications. In: Proceedings of SPIE 9077 Radar Sensor Technology May 2014, pp 1–7

Ogawa K, Matsuyoshi T (2001) An analysis of the performance of a handset diversity antenna influenced by head, hand, and shoulder effects at 900 MHz: part I – effective gain characteristics. IEEE Trans Vehic Technol 50(3):830–844

Okano Y, Ito K, Ida I, Takahashi M (2000) The SAR evaluation method by a combination of thermographic experiments and biological tissue-equivalent phantoms. IEEE Trans Microw Theory Technol 48(11):2094–2103

Omer M, Fear EC (2017) Automated 3D method for the construction of flexible and reconfigurable numerical breast models from MRI scans. Med Biol Eng Comput 56(6):1027–1040

Ostadrahimi M, Reopelle R, Noghanian S, Pistorius S, Vahedi A, Safari F (2009) A heterogeneous breast phantom for microwave breast imaging. In: Proceedings of IEEE Annual International Conference of the IEEE Engineering in Medicine and Biology Society, Sept. 2009, pp 2727–2730

Picher C, Anguera J, Andujar A, Puente C, Kahng S (2012) Analysis of the human head interaction in handset antennas with slotted ground planes. IEEE Antennas Propag Mag 54(2):36–56

Porter E, Fakhoury J, Oprisor R, Coates M, Popovic M (2010) Improved tissue phantoms for experimental validation of microwave breast cancer detection. In: Proceedings of Fourth European Conf. Antennas Propag., Apr. 2010, pp 1–5

Porter E, Santorelli A, Bourdon A, Coulibaly D, Coates M, Popovi M (2011) Time-domain microwave breast cancer detection: experiments with comprehensive glandular phantoms. In: Proceedings of IEEE Asia-Pacific Microw Conference, Dec. 2011, pp 203–206

Rodrigues DB, Maccarini PF, Salahi S, Oliveira TR, Pereira PJS, Limao-Vieira P, Snow BW, Reudink D, Stauffer PR (2014) Design and optimization of an ultra-wideband and compact microwave antenna for radiometric monitoring of brain temperature. IEEE Trans Biomed Eng 61(7):2154–2160

Romeo S, Donato LD, Bucci OM, Catapano I, Crocco L, Scarfi MR, Massa R (2011) Dielectric characterization study of liquid-based materials for mimicking breast tissues. Microw Opt Technol Lett 53(6):1276–1280

Saraereh OA, Jayawardene M, McEvoy P, Vardaxoglou JC (2004) Simulation and experimental SAR and efficiency study for a dual-band PIFA handset antenna (GSM 900/DCS 1800) at varied distances from a phantom head. In: Proceedings of IEE Antenna Meas. SAR, May 2004, pp 5–8

Scapaticci R, Bellizzi G, Catapano I, Crocco L, Bucci OM (2014) An effective procedure for MNP-enhanced breast Cancer microwave imaging. IEEE Trans Biomed Eng 61(4):1071–1079

Schwerdt HN, Miranda FA, Chae J (2012a) A fully passive wireless backscattering neurorecording microsystem embedded in dispersive human-head phantom medium. IEEE Electron Device Lett 33(6):908–910

Schwerdt HN, Chae J, Miranda FA (2012b) Wireless performance of a fully passive neurorecording microsystem embedded in dispersive human head phantom. In: Proceedings of IEEE Antennas Propagation Society International Symposium., July 2012, pp 1–2
Takimoto T, Onishi T, Saito K, Takahashi M, Uebayashi S, Ito K (2007) Characteristics of biological tissue equivalent phantoms applied to UWB communications. Electron Commun. Japan (Part I: Commun.) 90(5):48–55
Tamura H, Ishikawa Y, Kobayashi T (1997) A dry phantom material composed of ceramic and graphite powder. IEEE Trans Electromag Compat 39(2):132–137
Tell RA (1972) Microwave energy absorption in tissue, Twinbrook Research Laboratory technical report. EPA, Washington, DC
Tunçay AH, Akduman I (2015) Realistic microwave breast models through T1-weighted 3-D MRI data. IEEE Trans Biomed Eng 62(2):688–698
Vrba J, Karch J, Vrba D (2015) Phantoms for development of microwave sensors for noninvasive blood glucose monitoring. Int J Antennas Propag 2015:1
Wang L, Niu J (2017) Development of a biological tissue-like phantom for microwave imaging systems. Int J Sci Eng Appl Sci 3(4):62–69
Watanabe S, Taki M, Nojima T, Fujiwara O (1996) Characteristics of the SAR distributions in an head exposed to electromagnetic fields radiated by a hand-held portable radio. IEEE Trans Microw Theory Technol 44(10):1874–1883
Weil CM (1975) Absorption characteristics of multi-layered sphere models exposed to UHF/microwave radiation. IEEE Trans Biomed Eng 22(6):468–476
Youngs IJ, Treen AS, Fixter G, Holden S (2002) Design of solid broadband human tissue simulant materials. IEEE Proc Sci Meas Technol 149(6):323–328
Yuan Y, Wyatt C, Maccarini P, Stauffer P, Craciunescu O, MacFall J, Dewhirst M, Das SK (2012) A heterogeneous human tissue mimicking phantom for RF heating and MRI thermal monitoring verification. Phys Med Biol 57(7):2021–2037
Zastrow E, Davis SK, Lazebnik M, Kelcz F, Van Veen BD, Hagness SC (2008) Development of anatomically realistic numerical breast phantoms with accurate dielectric properties for modeling microwave interactions with the human breast. IEEE Trans Biomed Eng 55(12):2792–2800
Zhang F, Liu X, Hackworth SA, Sclabassi RJ, Sun M (2009) In vitro and in vivo studies on wireless powering of medical sensors and implantable devices. In: Proceeding of IEEE/NIH Life Sci. Syst. Appl. Work, Apr. 2009, pp 84–87
Zhou H (2009) A breast imaging model using microwaves and a time domain three dimensional reconstruction method. Prog Electromagn Res 93:57–70

Biophotonics in Disease Diagnosis and Therapy

3

Shrutidhara Biswas, Vlad Bogdan Gavra, Anand Kant Das, and Umakanta Tripathy

Abstract

Biophotonics is the multidisciplinary domain of science that uses light, in the visible and near-visible range, to study biological materials. Most biological tissues are sensitive to light, and these interactions can be harnessed for their imaging, detection, and manipulation. With the advent of advanced lasers, optics, spectroscopy, and microscopy tools, biophotonics find widespread application in biological and clinical research. Here, we provide an overview of how the field has expanded in the area of disease diagnosis and therapy with particular emphasis on label-free harmonic generation imaging microscopy, label-free multiphoton fluorescence imaging microscopy, spectroscopy, and tomography tools for clinical use. We have discussed, in brief, the fundamentals and principles behind each of the biophotonics method, the specific advantages and disadvantages of the tools, and the latest development in its use for improving diagnosis and therapy of various disease conditions. We intend to motivate the readers to draw inspiration for research, development, and translation of biophotonics to address the unmet clinical needs of humanity.

S. Biswas
Department of Bioscience and Bioengineering, IIT Guwahati, Guwahati, Assam, India

V. B. Gavra · A. K. Das (✉)
Institute of Applied Physics, Vienna University of Technology, Vienna, Austria
e-mail: das@iap.tuwien.ac.at

U. Tripathy (✉)
Department of Applied Physics, IIT (ISM), Dhanbad, Jharkhand, India
e-mail: utripathy@iitism.ac.in

S. Paul (ed.), *Biomedical Engineering and its Applications in Healthcare*,
https://doi.org/10.1007/978-981-13-3705-5_3

3.1 Introduction

Biophotonics is the field of photon-based technology involving interaction of light with biological molecules, cells, tissues, and organisms (Krafft 2016). In brief, it involves the fusion of photonics and biology. Biophotonics has emerged as a highly active, broad, and interdisciplinary area of research involving bioassays, biosensors, bioimaging, and photonic devices for in vitro disease diagnosis, medical imaging, and therapy (Marcu et al. 2017). Apart from its widespread use in basic research, biophotonics has emerged as a powerful tool in the area of clinical research encompassing both diagnostics and therapeutics. Diagnostic biophotonics involves technologies for the detection of disease states and monitoring efficacy of proposed treatments. On the other hand, therapeutic biophotonics involves altering biological processes and treating disease conditions. There are concerted efforts in the field to develop highly sensitive, safe, and low-cost solutions to clinical problems by exploitation of the field of photonics.

The chapter is divided into three main sections. Section 3.2, "Harmonic generation imaging microscopy," presents a summary of the concepts of second and third-harmonic generation imaging microscopy for advanced disease diagnosis. This is followed by "Multiphoton fluorescence microscopy" in Sect. 3.3 where we discuss in detail the concept of multiphoton imaging and the widespread use of label-free imaging in biomedical research. Section 3.4 deals with "Tomography and spectroscopy" where we discuss concepts of various tomography and spectroscopy methods that find widespread application in the diagnosis of various diseases and for its intervention.

3.2 Harmonic Generation Imaging Microscopy

Harmonic generation processes are parametric processes where the initial and final quantum mechanical states of the atoms or molecules are the same, and no real state is present between the initial and final states. These processes are best described under the banner of "nonlinear optics"—a magical branch of physics. For greater clarity, we need to start our journey from "linear optics". Therefore, under the regime of linear optics, when an electromagnetic light wave with electric field E(t) (for simplicity, scalar notations are used in this chapter) interacts with any matter, the polarization P(t) depends linearly upon the electric field E(t), so the following expression can describe it:

$$\mathrm{P(t)} = \epsilon_0\, \chi^{(1)} \mathrm{E(t)} \tag{3.1}$$

where $\chi^{(1)}$ is the linear susceptibility and ϵ_0 is the permittivity in the free space. The above approximation expression is valid for lower strength of the applied electric field, i.e., lower than the interatomic field strength (~10^{11} V/m). Here, the dipoles oscillate harmonically and dipolar displacement is directly proportional to the restoring force. When the applied electric field strength is higher than the interatomic field strength, then the dipolar oscillation becomes anharmonic and the displacement of dipoles is not directly proportional to the restoring force. Therefore, the approximation expression (Eq. 3.1) that describes the relation between P(t) and E(t) no longer holds true and hence warrants modification. Under this condition, nonlinear optics dominates. Therefore, in nonlinear optics, the polarization is expressed as a power series in terms of electric field and is given by the following expression:

$$\begin{aligned} P(t) &= \epsilon_0\chi^{(1)}E(t) + \epsilon_0\chi^{(2)}E^2(t) + \epsilon_0\chi^{(3)}E^3(t) + \cdots \\ &= P^{(1)}(t) + P^{(2)}(t) + P^{(3)}(t) + \cdots \end{aligned} \tag{3.2}$$

where $\chi^{(2)}$ is the second-order nonlinear susceptibility, $\chi^{(3)}$ is the third-order nonlinear susceptibility, and so on and so forth. $P^{(2)}$ and $P^{(3)}$ are the second- and third-order nonlinear polarizations, respectively. $\chi^{(2)}$ is responsible for processes like second harmonic generation (SHG), sum frequency generation (SFG), difference frequency generation (DFG), optical parametric oscillation (OPO), optical parametric amplification (OPA), etc. Similarly, $\chi^{(3)}$ is responsible for the third harmonic generation (THG), self-phase modulation (SPM), optical phase conjugation, four-wave mixing, Kerr effect, self-focusing/self-defocusing, stimulated Raman scattering, etc.

As described above, under nonlinear regime when the electromagnetic wave interacts with any matter that possesses nonlinearity (mostly anisotropic medium), the electromagnetic wave and the generated nonlinear polarization wave (P^{NL}) travel in the medium that governs by the wave equation:

$$\nabla^2 E - \frac{n^2}{c^2}\frac{\partial^2 E}{\partial t^2} = \frac{1}{\epsilon_0 c^2}\frac{\partial^2 P^{NL}}{\partial t^2} \tag{3.3}$$

where c is the velocity of light in vacuum and n is the refractive index of the medium. As we learned, the generated processes are nonlinear in origin, i.e., high peak power from the excitation laser sources is required to evoke these processes. Therefore, tightly focused femtosecond (fs) pulsed laser sources are used to gain high peak power. Also, for an efficient generation of nonlinear signal, the phase-matching condition (the speed of nonlinear polarization wave and harmonic generated electromagnetic wave should be the same) has to be satisfied. In general, people use either birefringence (angle or temperature tuning) or quasi-phase-matching techniques to get an efficient nonlinear harmonic signal. Also, along with phase-matching, we need to establish a coupling between the electromagnetic wave and nonlinear susceptibility matrix (specific to the nonlinear material or medium). We can satisfy this by choosing the proper direction of the excitation electromagnetic wave so that it will establish coupling with the nonzero value of the nonlinear susceptibility matrix of the material under study. However, in cases where the size of the investigating material is smaller than the excitation wavelength, we need to satisfy neither the phase-matching condition nor the coupling to generate an efficient nonlinear signal. In this section, we will only focus on processes like SHG and THG. These nonlinear tools form the basis of harmonic generation imaging microscopy.

Harmonic generation imaging microscopy, i.e., both SHG and THG imaging microscopy, are techniques which combine the advanced optical tools of nonlinear laser scanning microscopy with multiphoton excitation of longer wavelength to capture high-resolution and three-dimensional (3D) images of samples under study. The signal contrast is obtained from variations in the ability of a specimen, and exogenous probes are not required. It features striking advantages over traditional confocal microscopy for ex-vivo, in-vitro and in-vivo 3D imaging of cells and tissues. Multiphoton excitation, which occurs only at the focal point of the microscope, minimizes the effects of photodamage/photobleaching, which are the vital limiting factors in imaging live and fixed cells. These above-outlined advantages allow for investigations on thick, living, and fixed tissue specimens that would otherwise not have been possible with conventional imaging tools, thereby promoting its utility in disease diagnosis and therapy.

3.2.1 Second Harmonic Generation (SHG) Imaging Microscopy

Principle Let us consider an electromagnetic plane wave with the following electric field,

$$\mathrm{E(t)} = \mathrm{E}e^{-i\omega t} + \mathrm{E}^{\star} e^{i\omega t} \tag{3.4}$$

and that wave is incident on any nonlinear material with a nonzero $\chi^{(2)}$ (non-centrosymmetric structure). By using Eq. 3.4 in Eq. 3.2, the second-order polarization ($\mathrm{P}^{(2)}(\mathrm{t})$) will be

$$\mathrm{P}^{(2)}(\mathrm{t}) = 2\epsilon_0\chi^{(2)}\mathrm{EE}^{\star} + \epsilon_0\chi^{(2)}\left(\mathrm{E}^2 e^{-2i\omega t} + (\mathrm{E}^{\star})^2 e^{2i\omega t}\right) \tag{3.5}$$

In Eq. 3.5, it should be noted that the first term does not lead to the generation of an electromagnetic wave since its derivative with respect to time is zero (independent of frequency). It only leads to a process known as optical rectification, in which a static electric field is created across the nonlinear material/medium. However, the second term, which has 2ω frequency, generates second harmonic electromagnetic wave under specific circumstances. Figure 3.1 shows a Jablonski energy level diagram illustrating the process of SHG.

Here, each arrow represents a single photon with assigned energy $\hbar\omega$, which is proportional to its length. The upward arrows signify the two incident photons that are simultaneously annihilated to convert into one photon (energy $= \hbar\omega'$) of emitted SHG light with twice the incident photon energy ($2\hbar\omega$) denoted by the downward arrow. The time scale for this type of event is ~10^{-15} s. In Fig. 3.1, the dashed lines symbolize virtual energy states. According to Heisenberg's uncertainty principle, the time that an atom/molecule stays in the virtual state is $\sim\frac{\hbar}{\delta E}$, where δE is the energy gap between the virtual state and the nearest real energy level. These time scales are in

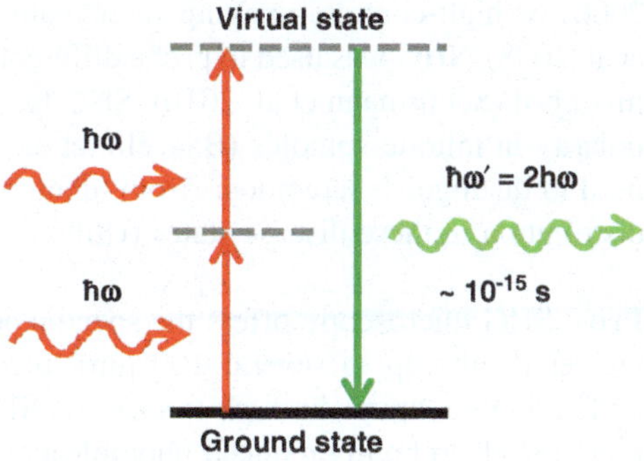

Fig. 3.1 Jablonski energy level diagram for the second harmonic generation (SHG) process. The upward arrows represent the incident photons, while the downward arrow represents the SHG generated photon

the order of fs. These nonlinear tools led to the development of SHG imaging microscopy.

Over the last three decades, SHG has emerged as an imaging technique of choice for a wide range of materials, including medicinal and biological samples. Progress in SHG imaging microscopy made it possible to carry out noninvasive 3D visualization of samples with high spatial resolution and greater penetration depth. Due to its virtual energy conservation characteristic, SHG does not deposit energy into the sample under study, thereby providing the necessary optical noninvasive property essential for microscopy applications. Also, in-focus damage and photobleaching are expected to be minimal. Here, the generated SHG intensity is proportional to the square of the intensity of excitation light. The localized excitation due to the nonlinear dependency is ideal for inherent optical sectioning. SHG is highly useful for structural studies because of the coherent generation that is sensitive to biological structures, i.e., materials with non-centrosymmetric molecular structures give rise to the SHG signal. SHG photons travel both in the forward and backward direction from the direction of excitation photon.

Background In 1971, SHG microscopy was first proposed by Fine and Hansen, where a Q-spoiled ruby laser was used to monitor the SHG signal in collagenous tissues (corneas, scleras, tendon, and skin of rabbit and dog) (Fine and Hansen 1971). Subsequently, the first SHG image of a biological sample was obtained by Freund and Deutsch from rat tail tendon using a Q-switched Nd:Yag laser in 1986 (Freund and Deutsch 1986). Also, the need for distinguishing the coherent and incoherent SHG imaging was raised. Type I collagen is a crucial protein whose structural change is associated with aging, tumor, and cancer. Traditionally, histological staining of collagen in tumor tissue biopsies was used for disease diagnosis. However, SHG imaging being a noninvasive technique made it possible to image collagen in tumors without the need for staining (Brown et al. 2003). SHG imaging also revealed alterations in collagen microstructure organization with subsequent matrix reconstruction, a mechanism dominant in wound healing processes (Yeh et al. 2004). Over the years, other applications of SHG microscopy emerged including imaging of deep skeletal muscles (actin-myosin) (Plotnikov et al. 2006) or probing action potential in neurons (Dombeck et al. 2004; Sacconi et al. 2006; Nuriya et al.

2006) or high-contrast imaging of myosin filaments in cardiomyocytes (Wallace et al. 2008). SHG was used to probe different states of motor protein interaction with myofibril (Schürmann et al. 2010). SHG has also been used to evaluate microtubule polarity in mitotic spindles (Bancelin et al. 2017). Of late, SHG imaging has been used to distinguish sarcoidosis from tuberculoid leprosy by evaluating the collagen distribution in these disease states (Utino et al. 2018).

Pros SHG microscopy offers the specific advantage of high-resolution imaging at greater depth (up to several 100 μm), thereby promoting direct visualization of underlying structures in thick tissues. As SHG signals arise from induced polarization, it leads to highly reduced photobleaching and phototoxic effects (Campagnola 2011).

Cons There is no doubt that SHG imaging microscopy is a powerful nonlinear tool for biological investigations. However, as with any technology, it also has certain limitations. As SHG is a nonlinear technique, it requires expensive femtosecond pulsed lasers to generate the SHG signals. Availability of other compact systems, like the catheter-sized probe, affects the broader application of SHG imaging microscopy. Penetration depths in the range of millimeters within the tissue or organs are still not achievable. Lastly, the biggest limitation of SHG microscopy is its capability to assess only a few structural proteins.

3.2.1.1 Application of SHG in Disease Diagnosis and Therapy

Tumor-associated collagen signatures (TACS) for tumor progression	Provenzano et al. (2008), Provenzano et al. (2006)
Arterial wall structure and composition	Le et al. (2007)
Pathologies of the cornea	Matteini et al. (2009)
Diagnosis of human ovarian cancer	Nadiarnykh et al. (2010)
Diagnosis of breast and colon cancer	Provenzano et al. (2006), Roy et al. (2008)
Microstructural characterization of vocal cords	Miri et al. (2013), Miri et al. (2012)
Osteogenesis imperfecta (OI)	Nadiarnykh and Campagnola (2009), Nadiarnykh et al. (2007)
Skin cancer imaging	Cicchi et al. (2007), Cicchi et al. (2008)
Spinal damage	Reiser et al. (2007)
Muscle disorders	Nadiarnykh and Campagnola (2009)

3.2.2 Third Harmonic Generation (THG) Imaging Microscopy

Principle When an electromagnetic plane wave is propagating with the following electric field,

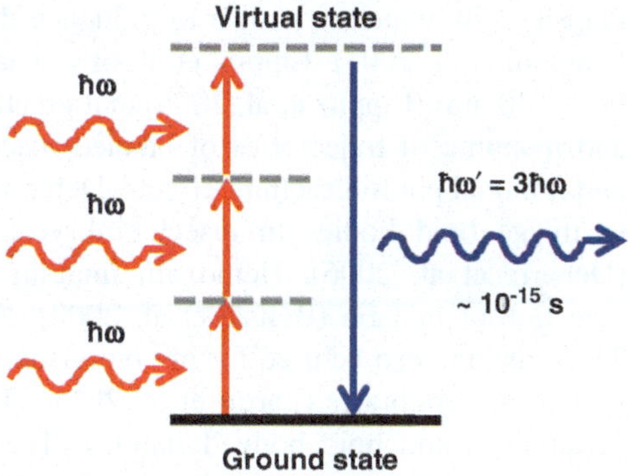

Fig. 3.2 Jablonski energy level diagram for the third harmonic generation (THG) process. The upward arrows represent the incident photons while the downward arrow represents the THG generated photon

$$E(t) = Ee^{-i\omega t} + E^{\star} e^{i\omega t} \tag{3.6}$$

and is incident on any material with a nonzero $\chi^{(3)}$, then by using Eq. 3.6 in Eq. 3.2, the third-order polarization ($P^{(3)}(t)$) is obtained as.

$$P^{(3)}(t) = 3\chi^{(3)}\left(E^2E^{\star}e^{-i\omega t} + E(E^{\star})^2 e^{i\omega t}\right) + \chi^{(3)}\left(E^3e^{-3i\omega t} + (E^{\star})^3 e^{3i\omega t}\right) \tag{3.7}$$

As we can see, the first term in Eq. 3.7 does not contribute to the THG as its frequency is ω, but it surely adds a nonlinear contribution to the polarization at the frequency of the incident electric field. This also leads to a nonlinear contribution to the refractive index and explains the self-focusing/self-defocusing of light. The second term that has frequency 3ω is responsible for the generation of THG light.

Figure 3.2 shows the Jablonski energy level diagram that illustrates the THG process wherein three incident photons (represented by upward arrows), each having the energy $\hbar\omega$, are simultaneously annihilated to create a single photon (energy $= \hbar\omega'$) having three times the incident photon energy ($3\hbar\omega$) and are represented by the downward arrow. This effect also occurs at a similar time scale as SHG, i.e., $\sim 10^{-15}$ s. THG photons mostly travel in the forward direction from the direction of excitation photon.

Background The THG technique led to the development of THG imaging microscopy. Over the last two decades, this imaging modality has been rigorously used to investigate biological and nonbiological samples. In 1997, it was first introduced by Barad and coworker for transparent imaging of objects through variations in their third-order susceptibilities (Barad et al. 1997). One year later, the first biological sample, *Chara* plant rhizoid, was imaged. Unlike fluorescent dyes, the THG signal does not bleach even after hours of continuous exposure (Fittinghoff et al. 1998). Live neurons were imaged using THG microscopy by Yelin and Silberberg (1999). Millard et al. imaged the flowing red blood cells by utilizing the THG signal from heme (Millard et al. 1999). Canioni and coworkers imaged the calcium dynamics in glial cells and studied THG intensity as a function of time after injecting thapsigargin, which is known to influence the concentration of Ca^{2+} in the cytosol (Canioni et al. 2001). Lippitz et al. observed the THG signal in gold nanoparticles (size ~40 nm) (Lippitz et al. 2005) that greatly improved the single-particle tracking and mapping of trajectories of labeled macromolecules with nanometer precision and minimal photobleaching effects. Debarre and coworkers used THG microscopy to image lipid bodies in insect embryos, plant seeds, and mammalian tissues (Debarre et al. 2006). Hemozoin imaging using THG has led to the sensitive detection of malaria (Bélisle et al. 2008). Schaffer's group has demonstrated that THG imaging can be used for in vivo visualization of myelin in the central nervous system of vertebrates (Farrar et al. 2011). THG has also been used to image lipid metabolism and lipid body dynamics (Tserevelakis et al. 2014). Recently, THG imaging modality has been utilized to probe the alterations in the collagen network in fibrotic disease conditions (Ricard-Blum et al. 2018).

Pros THG helps in noninvasive 3D visualization of biological and nonbiological specimens at high penetration depth. The virtual energy conservation characteristic leads to minimal photobleaching effects even with long exposures. Here, the incident light intensity has a cubic dependence on the generated THG intensity, which is ideal for intrinsic optical sectioning and localized excitation. The coherent nature of the THG process is very insensitive to the local biological structures and hence can be utilized as a general-purpose microscopy technique for morphological studies.

Cons Like SHG, THG is also a nonlinear technique that requires expensive femtosecond pulsed lasers to generate the THG signal. The samples need to be thin as the THG imaging microscope generally operates in a transmission mode. Also, the scattering coming from the short-wavelength THG light interferes with the imaging process. Like SHG, the penetration depths in the range of several millimeters within the material are still not achievable with THG. The major limitation of THG imaging microscopy is its capability to provide only the structural information, unlike the coherent anti-stokes Raman scattering and autofluorescence that can provide molecular information from the sample.

3.2.2.1 Application of THG in Disease Diagnosis and Therapy

Detection of malaria	Bélisle et al. (2008), Tripathy et al. (2013)
Imaging of human articular cartilage	Tsai et al. (2009)
Visualization of mineralized bone matrix	Oron et al. (2004)
Calcified teeth	Chen et al. (2008), (2015)
3D organization and interface in fat tissues	Tsai et al. (2012), Weigelin et al. (2012)
Blood flow dynamics	Chen and Liu (2012)
Tumor cell migration	Weigelin et al. (2012)
Liver tissue analysis	Tsai et al. (2011)
Myelin organization in neurodegenerative diseases	Rehberg et al. (2011), Lim et al. (2014)
Melanoma lesions	Weigelin et al. (2012)

3.3 Multiphoton Fluorescence (MPF) Microscopy

Multiphoton (two-photon, three-photon, etc.) fluorescence (MPF) microscopy has emerged as a powerful tool to carry out stain-free imaging.

Principle MPF works on the principle of multiphoton excitation, which is a nonparametric process where atoms or molecules jump from one real energy level to another. In two-photon or three-photon absorption, the excitation of an atom or molecule takes place from the ground state to an excited state by the absorption of two photons or three photons, respectively. Figure 3.3 (A) and (B) shows the Jablonski energy level diagram for two-photon and three-photon absorption (~10^{-15} s), respectively followed by vibrational relaxation (~10^{-15} s) and fluorescence emission (~10^{-9} s). Here the dashed lines represent the virtual energy states. The two- and three-incident photons of energy $\hbar\omega$ each are represented by upward arrows, and downward arrows represent the emitted fluorescence photons. The fluorescence photon has energy lower than $2\hbar\omega$ and $3\hbar\omega$ in case of two-photon and three-photon fluorescence, respectively, due to energy losses in the non-radiative processes (diagonally oriented curly arrows). The longer time span of two-photon and three-photon effect makes it possible to temporally separate it from the parametric effects such as SHG and THG in the eventuality of overlap in the wavelength (e.g., time-gated detection).

Background Multiphoton absorption was first predicted by Goppert-Meyer in 1931 and eventually observed by Singh and Bradley in 1964 (Singh and Bradley 1964). Subsequently, Watt Webb's group at Cornell University developed the multiphoton fluorescence microscopy technique in the 1990s (Denk et al. 1990). In MPF microscopy, fluorophores are excited by absorption of either two/three photons of twice/thrice the wavelength instead of single-photon excitation. Using the MPF microscopy, it was shown that action potentials invade the dendritic spines of pyramidal neurons. Furthermore, the detected pre- and postsynaptic activity indicated that spines are the functional units of neuronal integration (Yuste and

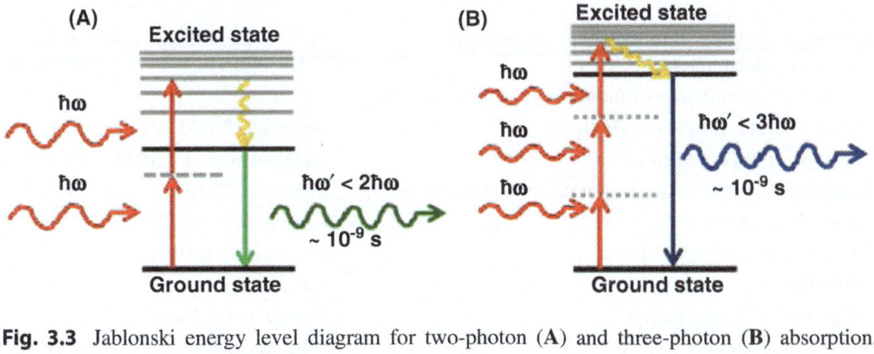

Fig. 3.3 Jablonski energy level diagram for two-photon (**A**) and three-photon (**B**) absorption, vibrational relaxation, and fluorescence process. The upward arrows represent the absorbed photons, while the down arrow represents the emitted photons

Denk 1995). The in vivo imaging of the primary cortex of rats was carried out using multiphoton excitation (Svoboda et al. 1997). The MPF microscopy was also used to image thioflavine S to monitor the evolution of senile plaque in Alzheimer's disease mice model (Christie et al. 2001). Brown et al. studied the alterations in tumors using multiphoton imaging (Brown et al. 2001). MPF imaging has been used in preclinical studies for live imaging of skin cancers (Sun et al. 2017) and nonmelanoma skin abnormalities (Paoli et al. 2008). Of late, intravital MPF microscopy has been used to study renal pathophysiology (Sandoval and Molitoris 2017) and metabolic changes in the kidney proximal tubules (Bugarski et al. 2018).

Pros For commonly used fluorescent dyes, single-photon absorption usually occurs in the visible region of the electromagnetic spectrum. Therefore, infrared (IR) ultrafast short pulse (~fs pulse width) lasers are typically used as excitation sources in MPF microscopy. IR light undergoes less scattering in biological samples compared to the visible light and hence can be used for deep tissue imaging. Additionally, multiphoton excitation involves the simultaneous absorption of multiple photons and is dependent on the intensity of the incident beam. As the excitation is limited to a small focus, MPF microscopy does the require pinholes, unlike the confocal microscopy.

Cons MPF microscopy requires expensive femtosecond pulsed laser sources to generate the signal. Long duration imaging of samples using multiphoton excitation can lead to extensive photobleaching and photodamage due to absorption at the excitation laser wavelength. Additionally, under multiphoton excitation, specific excitation of molecules of interest can be tricky as the excitation spectrum of the multiphoton laser beam is usually much broader, thereby leading to nonspecific excitation of unwanted molecules.

3.3.1 Application of Multiphoton Imaging in Disease Diagnosis and Therapy

Label-free imaging of monoamine neurotransmitters (serotonin, dopamine)	Maiti et al. (1997a), Sarkar et al. (2014), (2012)
Vesicular concentration of monoamine neurotransmitter	Das et al. (2017)
Atherosclerotic plaque development	Wu et al. (2017)
Metabolic imaging of tumors	Zagaynova et al. (2018)
Intraoperative tool for urothelial carcinoma in situ (CIS)	Jain et al. (2015)
Ophthalmic imaging	Gibson et al. (2011)
Diagnosis of colorectal cancer	Matsui et al. (2017)
Liver fibrosis	Stanciu et al. (2014)
Imaging of skin physiology	Yew et al. (2014)
Inflammatory bowel disease	Schürmann et al. (2013)

3.4 Spectroscopy and Tomography

Spectroscopy deals with the science of measurement and investigation of different types of the spectrum obtained when materials interact with electromagnetic radiations, thereby elucidating the structure and property of matter. Tomography, on the other hand, involves sectional imaging of a specimen by use of penetrating waves. Both spectroscopy and tomography are biophotonics tools that find widespread applications in biomedical engineering, disease diagnosis, and therapy.

3.4.1 Fluorescence Correlation Spectroscopy

Fluorescence correlation spectroscopy (FCS) is a fluctuation spectroscopy technique that has emerged as a promising tool for disease diagnosis with extremely high sensitivity and efficacy.

Principle FCS was developed by Watt Webb and coworkers (Magde et al. 1972) in the Cornell University that involves correlative statistical analysis of spontaneous fluorescence intensity fluctuations resulting from Brownian motions of molecules in a small observation volume. FCS does not depend on absolute intensity values but fluctuation in fluorescence intensity. The autocorrelation of the observed variation can then be used to determine the temporal progression of a system. This function can be fitted to an appropriate diffusion model involving discrete components or model-free approach such as the maximum entropy method (MEM). A typical confocal FCS setup is shown in Fig. 3.4. FCS has been widely used over the last few decades to determine diffusion coefficients, oligomerization state, concentrations, triplet state lifetimes, chromatin organization, protein-protein interactions, receptor-ligand interactions, and scores of molecular parameters with high precision (Maiti et al. 1997b; Shahzad and Köhler 2011; Singh and Wohland 2014; Machan and Wohland 2014; Abhyankar et al. 2012; Das et al. 2015).

Pros FCS has several advantages. A typical FCS measurement utilizes very small volumes (<10 μl) and concentrations (picomolar) of the sample and operates at single-molecule sensitivity. Additionally, one can directly measure particle number without the need for any calibration.

Cons FCS can detect only mobile fluorescent molecules and is silent on immobile ones (Diekmann and Hoischen 2014). This technique has some limitations as one can obtain misleading calculations under different environments arising mostly due to the lack of appropriate interpreting models (Enderlein et al. 2004; Marrocco 2004; Nishimura and Kinjo 2004; Hac et al. 2005).

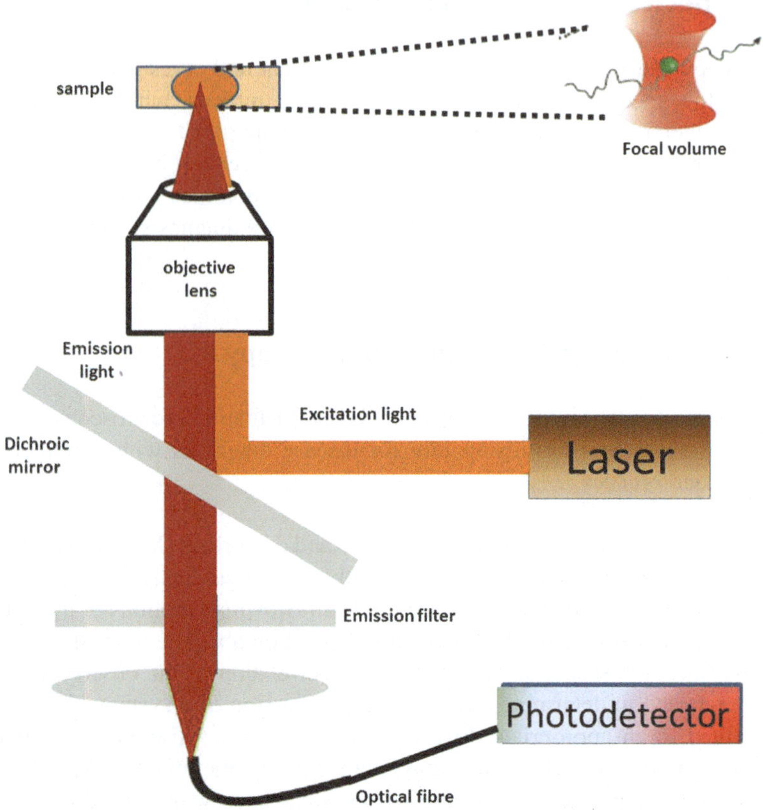

Fig. 3.4 A typical confocal FCS system. A high numerical aperture objective focuses the laser light to a diffraction-limited spot. The same objective collects the fluorescence. A pinhole is placed in the conjugate plane to reduce out of focus light

3.4.1.1 Application of FCS in Disease Diagnostics and Therapy

Glucocorticoid receptors and imbalance of hypothalamus-pituitary-adrenal axis	Maier et al. (2005)
Alzheimer's amyloid-β aggregation in cerebrospinal fluid	Pitschke et al. (1998)
Diagnostic tool in medical microbiology	Shahzad et al. (2009)
Diagnosis of prion diseases	Fujii et al. (2007)
Molecular pharmacology of G protein-coupled receptors	Kilpatrick and Hill (2016)
Biophysical characterization of viruses	Wruss et al. (2007)
Cell-free DNA analysis in schizophrenia	Jiang et al. (2018)
Contactin-2 levels in Alzheimer's patients	Chatterjee et al. (2017)
Monitoring drug nanocarriers in human blood	Negwer et al. (2018)
Diagnosis of type 1 von Willebrand disease	Torres et al. (2012)

3.4.2 Endoscopy

Endoscopy is the minimal invasive biophotonics tool that is introduced in the body to visualize internal parts to facilitate disease diagnosis and directed therapies (Martin et al. 2014). The field of endoscopy has revolutionized the diagnosis of gastrointestinal diseases and their therapeutic interventions.

Principle Endoscopy uses light to illuminate tissues and collects reflected light from inner surfaces of the gastrointestinal tracts aiding in the visualization of the inner-organ surfaces, as schematically represented in Fig. 3.5. Endoscopy is often used with other diagnostic modalities for an accurate understanding of the disease states.

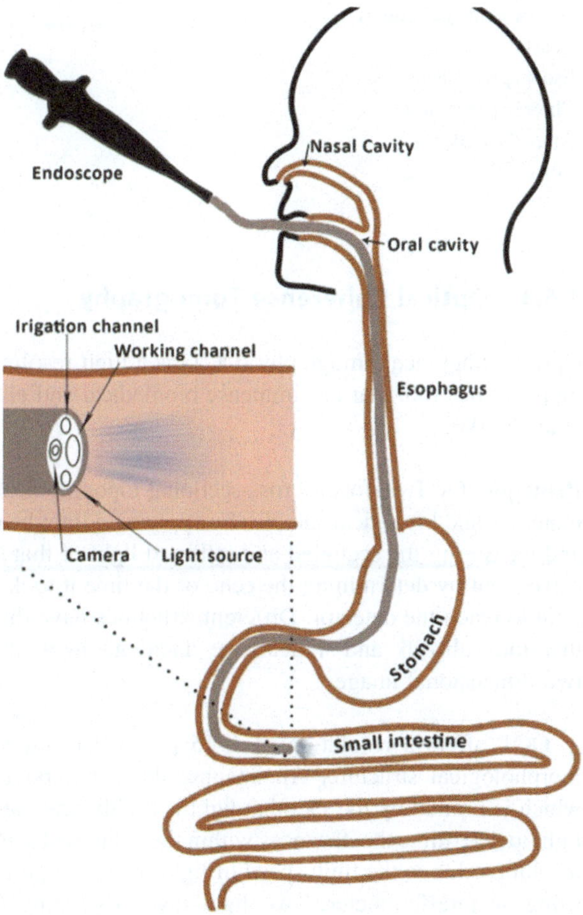

Fig. 3.5 A schematic showing the use of endoscopy in diagnosis and intervention in a large number of gastrointestinal disease conditions. The light source helps to illuminate the GI tract, and the camera records the reflected light from the inner walls of the tract

Pros Endoscopy procedure is minimally invasive and has extremely low rates of morbidity and mortality in animals (Moore 2003). Endoscopy is extremely sensitive to the diagnosis of mucosal diseases of the stomach, esophagus, duodenum, ileum, and colon (Moore 2003). It possesses great therapeutic potential, for example, in its utility for removal of a foreign body from the GI tract.

Cons Endoscopy cannot detect diseases in the deep submucosa. The procedure is not recommended if bowel perforation is suspected. It has limited use in diagnosing diseases of the small intestine.

3.4.2.1 Application of Endoscopy in Disease Diagnostics and Therapy

Diagnosis of gastrointestinal neoplasia	Yao et al. (2008)
Gastrointestinal cancer	Coda et al. (2015)
Inflammatory bowel disease	Leighton et al. (2006)
Peptic ulcer disease	Dooley et al. (1984)
Gastric epithelial polyps	Hirota et al. (2006)
Caustic injury	Hirota et al. (2006)
Esophageal cancer	Jacobson et al. (2003)
Chronic pancreatitis	Adler et al. (2005)
Ampullary adenoma	Adler et al. (2006)
Lower GI bleeding	Davila et al. (2005)

3.4.3 Optical Coherence Tomography

Optical coherence tomography (OCT) is a high-resolution, cross-sectional imaging biophotonics tool that has immense biomedical and clinical applications (Fujimoto et al. 2000a).

Principle OCT performs cross-sectional topographical imaging of the structure of materials and biological samples by applying a light beam from a source to a sample and measuring the scattered and reflected light of that sample. The measurement is carried out by determining the echo or the time it took for the reflected or scattered light to reach the detector. Different structures have different optical properties and therefore absorb and reflect the incident light differently. The result is a two-dimensional image.

OCT offers a resolution of 1–15 μm which allows for good visualization of morphological structure. An imaging depth of about 2–3 mm can be achieved, which is limited by the fact that light is highly scattered in biological tissue. Using light to determine the distances within the examined sample requires a high temporal resolution due to the high speed of light. High temporal resolution can be achieved using an interferometer. Two light rays originating from a common source are

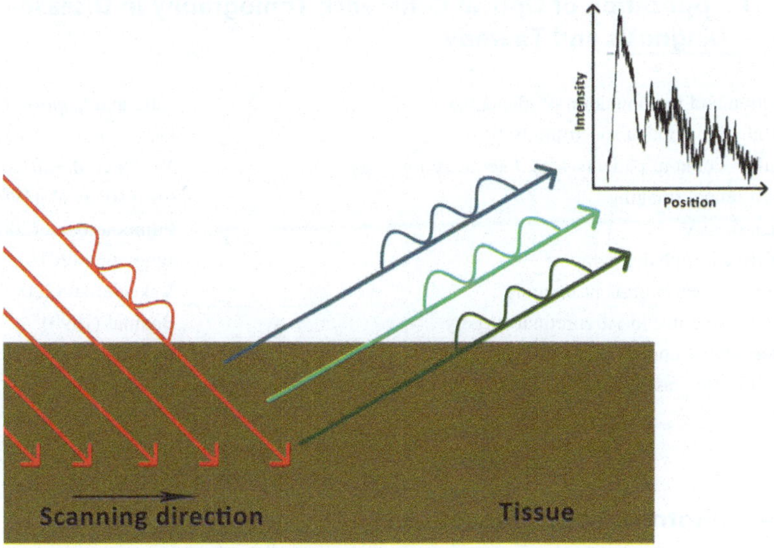

Fig. 3.6 A schematic illustrating the working principle of optical coherence tomography

interfered to measure the intensity modification. The light ray coming from the sample will have some delay, and its intensity decreased. Delay and intensity modification will vary depending on the different structures on the sample under measurement. The probe measures the backscattered intensity in one dimension. The incident beam from the light source is then scanned across the surface of the sample to obtain multiple one-dimensional images which can then be combined to form the final cross-section as shown in Fig. 3.6. OCT can also offer real-time imaging, making it a powerful tool for surgery. A delivering image system in the form of a catheter can be developed which delivers images from underneath the tissue in real time and allows for surgery to be guided.

Pros OCT is a faster and easier method for disease diagnosis. One of the advantages of OCT is that one can measure without the need to excise the specimen. For example, in ophthalmology, biopsy on the retina is impossible, but OCT can be used to determine the morphology of the retina for diagnosing or monitoring certain diseases. OCT can also be used to detect neoplastic changes before metastasis occurs.

Cons Since OCT utilizes light waves, the opacities of the sample can interfere with the optical imaging. OCT machines are expensive and have limited penetration power.

3.4.3.1 Application of Optical Coherence Tomography in Disease Diagnosis and Therapy

Detection and manipulation of glaucoma	Jaffe and Caprioli (2004)
Neuromyelitis optica spectrum disorders	Oertel et al. (2018)
Macular edema in patients with diabetic retinopathy	Virgili et al. (2015)
In vivo retinal imaging	Swanson et al. (1993)
Optical biopsy	Fujimoto et al. (2000b)
Label-free lymphangiography	Jung et al. (2010)
Tumor microenvironment imaging	Vakoc et al. (2009)
Intraoperative use in vitreoretinal surgery	Baumal (1999)
Central serous chorioretinopathy	Baumal (1999)
Subretinal neovascularization	Baumal (1999)

3.4.4 Photoacoustic Tomography

Principle Photoacoustic tomography (PAT) is based on the photoacoustic effect (Fig. 3.7), where an incident electromagnetic wave is absorbed in the tissue. The absorbed energy induces a slight local temperature increase, causing a thermoelastic expansion which produces transient ultrasonic waves.

The transient ultrasonic waves act as an acoustic source and can produce acoustic waves that propagate in tissue in all directions. These propagating acoustic waves reach the tissue surface and can be recorded using an ultrasound transducer. When the electromagnetic wave reaches the tissue surface, it will penetrate to a certain depth because light scatters in the tissue. Frequencies used for the electromagnetic (EM) pulse are in the range of visible light to near IR or radiofrequency. These frequencies have been chosen as such due to their good penetration depths in tissue and because they are nonionizing. Higher frequencies would mainly be absorbed by

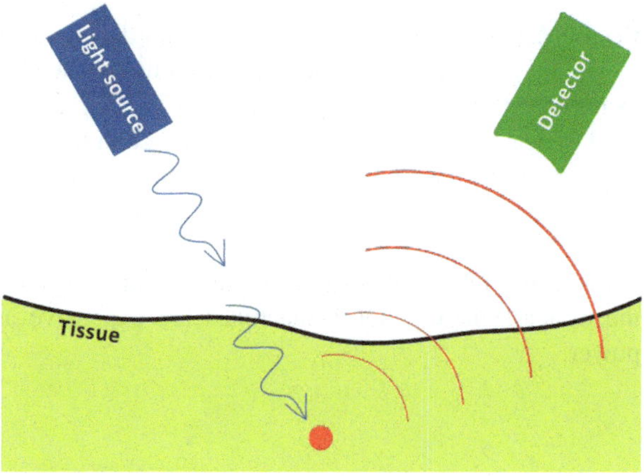

Fig. 3.7 Generation of acoustic waves through tissue by thermoelastic expansion (red dot). Light penetrates the tissue to a certain depth where it causes a small temperature increase, leading to thermoelastic expansion

tissue and therefore not achieve a deep penetration depth. Even higher frequencies, in the range of X rays and gamma rays, would be harmful to the tissue, due to the high energy it possesses.

In PAT, an ultrasonic transducer scans the surface of the tissue for the acoustic waves produced by the EM pulse. By scanning the surface, cross-sectional images are obtained that can be used to model a 3D image. The optical properties (such as absorption or scattering) of tissues are dependent on the tissue structure. If an anomaly is present, it would affect the optical properties of that tissue and therefore absorb light differently.

Pros The strength of PAT lies in its ability to image molecular and functional alterations in live animals at significant tissue depth.

Cons The sensitivity of PAT is affected by the scattering of light within tissues. It takes a longer time to scan to achieve high lateral resolution. PAT resolution is limited by the inverse relationship between the imaging depth and the frequency of the ultrasound.

3.4.4.1 Application of Photoacoustic Tomography in Disease Diagnosis and Therapy

Detection of glioblastoma	de la Zerda et al. (2010)
Colon carcinoma	Li et al. (2009)
Prostate cancer	Agarwal et al. (2007)
Squamous cell carcinoma	Li et al. (2008)
Brain lesions	Wang et al. (2003)
Hemoglobin concentration and oxygenation	Wang et al. 2006)
Human inflammatory arthritis	Jo et al. (2017)
Bladder microvasculature	Xie et al. (2011)
Melanoma imaging	Yao et al. (2011)
Ovarian cancer	Kumavor et al. (2013)

3.5 Conclusions and Perspectives

The field of medical optics and biophotonics is ever expanding and looks immensely promising. It will be too intriguing to see how design and development of novel biophotonics devices and combining some of the modalities will facilitate newer clinical applications, provide novel insights into disease mechanisms, and help the field grow towards low-cost personalized medicine and therapy.

References

Abhyankar, R., et al. Amyloid diagnostics: probing protein aggregation and conformation with ultrasensitive fluorescence detection. SPIE BiOS. 2012. SPIE

Adler DG et al (2005) ASGE guideline: the role of ERCP in diseases of the biliary tract and the pancreas. Gastrointest Endosc 62(1):1–8

Adler DG et al (2006) The role of endoscopy in ampullary and duodenal adenomas. Gastrointest Endosc 64(6):849–854

Agarwal A et al (2007) Targeted gold nanorod contrast agent for prostate cancer detection by photoacoustic imaging. J Appl Phys 102(6):064701

Bancelin S et al (2017) Probing microtubules polarity in mitotic spindles in situ using interferometric second harmonic generation microscopy. Sci Rep 7(1):6758–6758

Barad Y et al (1997) Nonlinear scanning laser microscopy by third harmonic generation. Appl Phys Lett 70(8):922–924

Baumal CR (1999) Clinical applications of optical coherence tomography. Curr Opin Ophthalmol 10(3):182–188

Bélisle JM et al (2008) Sensitive detection of malaria infection by third harmonic generation imaging. Biophys J 94(4):L26–L28

Brown EB et al (2001) In vivo measurement of gene expression, angiogenesis and physiological function in tumors using multiphoton laser scanning microscopy. Nat Med 7(7):864–868

Brown E et al (2003) Dynamic imaging of collagen and its modulation in tumors in vivo using second-harmonic generation. Nat Med 9(6):796–800

Bugarski M et al (2018) Multiphoton imaging reveals axial differences in metabolic autofluorescence signals along the kidney proximal tubule. Am J Physiol Ren Physiol 315(6): F1613–F1625

Campagnola P (2011) Second harmonic generation imaging microscopy: applications to diseases diagnostics. Anal Chem 83(9):3224–3231

Canioni L et al (2001) Imaging of ca(2)+ intracellular dynamics with a third-harmonic generation microscope. Opt Lett 26(8):515–517

Chatterjee M et al (2017) Detection of contactin-2 in cerebrospinal fluid (CSF) of patients with Alzheimer's disease using fluorescence correlation spectroscopy (FCS). Clin Biochem 50(18):1061–1066

Chen C-K, Liu T-M (2012) Imaging morphodynamics of human blood cells in vivo with video-rate third harmonic generation microscopy. Biomed Opt Express 3(11):2860–2865

Chen S-Y, Hsu C-YS, Sun C-K (2008) Epi-third and second harmonic generation microscopic imaging of abnormal enamel. Opt Express 16(15):11670–11679

Chen Y-C et al (2015) Third-harmonic generation microscopy reveals dental anatomy in ancient fossils. Opt Lett 40(7):1354–1357

Christie RH et al (2001) Growth arrest of individual senile plaques in a model of Alzheimer's disease observed by in vivo multiphoton microscopy. J Neurosci 21(3):858–864

Cicchi R et al (2007) Multidimensional non-linear laser imaging of Basal Cell Carcinoma. Opt Express 15(16):10135–10148

Cicchi R et al (2008) Nonlinear laser imaging of skin lesions. J Biophotonics 1(1):62–73

Coda S et al (2015) Biophotonic endoscopy: a review of clinical research techniques for optical imaging and sensing of early gastrointestinal cancer. Endos Int Open 3(5):E380–E392

Das AK, Pandit R, Maiti S (2015) Effect of amyloids on the vesicular machinery: implications for somatic neurotransmission. Philos Trans R Soc Lond Ser B Biol Sci 370(1672):20140187

Das AK et al (2017) Label-free Ratiometric imaging of serotonin in live cells. ACS Chem Neurosci 8(11):2369–2373

Davila RE et al (2005) ASGE guideline: the role of endoscopy in the patient with lower-GI bleeding. Gastrointest Endosc 62(5):656–660

de la Zerda A et al (2010) Ultrahigh sensitivity carbon nanotube agents for photoacoustic molecular imaging in living mice. Nano Lett 10(6):2168–2172

Debarre D et al (2006) Imaging lipid bodies in cells and tissues using third-harmonic generation microscopy. Nat Methods 3(1):47–53
Denk W, Strickler JH, Webb WW (1990) Two-photon laser scanning fluorescence microscopy. Science 248(4951):73–76
Diekmann S, Hoischen C (2014) Biomolecular dynamics and binding studies in the living cell. Phys Life Rev 11(1):1–30
Dombeck DA, Blanchard-Desce M, Webb WW (2004) Optical recording of action potentials with second-harmonic generation microscopy. J Neurosci 24(4):999–1003
Dooley CP et al (1984) Double-contrast barium meal and upper gastrointestinal endoscopy: a comparative study. Ann Intern Med 101(4):538–545
Enderlein J et al (2004) Art and artefacts of fluorescence correlation spectroscopy. Curr Pharm Biotechnol 5(2):155–161
Farrar MJ et al (2011) In vivo imaging of myelin in the vertebrate central nervous system using third harmonic generation microscopy. Biophys J 100(5):1362–1371
Fine S, Hansen WP (1971) Optical second harmonic generation in biological systems. Appl Opt 10(10):2350–2353
Fittinghoff DN et al (1998) Collinear type II second-harmonic-generation frequency-resolved optical gating for use with high-numerical-aperture objectives. Opt Lett 23(13):1046–1048
Freund I, Deutsch M (1986) Second-harmonic microscopy of biological tissue. Opt Lett 11(2):94
Fujii F et al (2007) Detection of prion protein immune complex for bovine spongiform encephalopathy diagnosis using fluorescence correlation spectroscopy and fluorescence cross-correlation spectroscopy. Anal Biochem 370(2):131–141
Fujimoto JG et al (2000a) Optical coherence tomography: an emerging technology for biomedical imaging and optical biopsy. Neoplasia 2(1–2):9–25
Fujimoto JG et al (2000b) Optical coherence tomography: an emerging technology for biomedical imaging and optical biopsy. Neoplasia 2(1):9–25
Gibson EA et al (2011) Multiphoton microscopy for ophthalmic imaging. J Ophthalmol 2011:11
Hac AE et al (2005) Diffusion in two-component lipid membranes--a fluorescence correlation spectroscopy and Monte Carlo simulation study. Biophys J 88(1):317–333
Hirota WK et al (2006) ASGE guideline: the role of endoscopy in the surveillance of premalignant conditions of the upper GI tract. Gastrointest Endosc 63(4):570–580
Jacobson BC et al (2003) The role of endoscopy in the assessment and treatment of esophageal cancer. Gastrointest Endosc 57(7):817–822
Jaffe GJ, Caprioli J (2004) Optical coherence tomography to detect and manage retinal disease and glaucoma. Am J Ophthalmol 137(1):156–169
Jain M et al (2015) Multiphoton microscopy: a potential intraoperative tool for the detection of carcinoma in situ in human bladder. Arch Pathol Lab Med 139(6):796–804
Jiang J et al (2018) Analysis of the concentrations and size distributions of cell-free DNA in schizophrenia using fluorescence correlation spectroscopy. Transl Psychiatry 8(1):104
Jo J et al (2017) A functional study of human inflammatory arthritis using photoacoustic imaging. Sci Rep 7(1):15026
Jung Y, Zhi Z, Wang RK (2010) Three-dimensional optical imaging of microvascular networks within intact lymph node in vivo. J Biomed Opt 15(5):050501–050501
Kilpatrick LE, Hill SJ (2016) The use of fluorescence correlation spectroscopy to characterize the molecular mobility of fluorescently labelled G protein-coupled receptors. Biochem Soc Trans 44(2):624–629
Krafft C (2016) Modern trends in biophotonics for clinical diagnosis and therapy to solve unmet clinical needs. J Biophotonics 9(11–12):1362–1375
Kumavor PD et al (2013) Co-registered pulse-echo/photoacoustic transvaginal probe for real time imaging of ovarian tissue. J Biophotonics 6(6–7):475–484
Le TT et al (2007) Label-free molecular imaging of atherosclerotic lesions using multimodal nonlinear optical microscopy. J Biomed Opt 12(5):054007

Leighton JA et al (2006) ASGE guideline: endoscopy in the diagnosis and treatment of inflammatory bowel disease. Gastrointest Endosc 63(4):558–565
Li P-C et al (2008) In vivo photoacoustic molecular imaging with simultaneous multiple selective targeting using antibody-conjugated gold nanorods. Opt Express 16(23):18605–18615
Li M-L et al (2009) In-vivo photoacoustic microscopy of nanoshell extravasation from solid tumor vasculature. SPIE
Lim H et al (2014) Label-free imaging of Schwann cell myelination by third harmonic generation microscopy. Proc Natl Acad Sci 111(50):18025–18030
Lippitz M, van Dijk MA, Orrit M (2005) Third-harmonic generation from single gold nanoparticles. Nano Lett 5(4):799–802
Machan R, Wohland T (2014) Recent applications of fluorescence correlation spectroscopy in live systems. FEBS Lett 588(19):3571–3584
Magde D, Elson E, Webb WW (1972) Thermodynamic fluctuations in a reacting system---measurement by fluorescence correlation spectroscopy. Phys Rev Lett 29(11):705–708
Maier C et al (2005) G-protein-coupled glucocorticoid receptors on the pituitary cell membrane. J Cell Sci 118(Pt 15):3353–3361
Maiti S et al (1997a) Measuring serotonin distribution in live cells with three-photon excitation. Science 275(5299):530–532
Maiti S, Haupts U, Webb WW (1997b) Fluorescence correlation spectroscopy: diagnostics for sparse molecules. Proc Natl Acad Sci U S A 94(22):11753–11757
Marcu L et al (2017) Biophotonics: the big picture. SPIE
Marrocco M (2004) Fluorescence correlation spectroscopy: incorporation of probe volume effects into the three-dimensional Gaussian approximation. Appl Opt 43(27):5251–5262
Martin T, Schwab K, Singh S (2014) Principles of gastrointestinal endoscopy. Surgery (Oxford) 32(3):139–144
Matsui T et al (2017) Non-labeling multiphoton excitation microscopy as a novel diagnostic tool for discriminating normal tissue and colorectal cancer lesions. Sci Rep 7(1):6959
Matteini P et al (2009) Photothermally-induced disordered patterns of corneal collagen revealed by SHG imaging. Opt Express 17(6):4868–4878
Millard AC et al (1999) Third-harmonic generation microscopy by use of a compact, femtosecond fiber laser source. Appl Opt 38(36):7393–7397
Miri AK et al (2012) Nonlinear laser scanning microscopy of human vocal folds. Laryngoscope 122(2):356–363
Miri AK et al (2013) Microstructural characterization of vocal folds toward a strain-energy model of collagen remodeling. Acta Biomater 9(8):7957–7967
Moore LE (2003) The advantages and disadvantages of endoscopy. Clin Tech Small Anim Pract 18(4):250–253
Nadiarnykh O, Campagnola PJ (2009) Retention of polarization signatures in SHG microscopy of scattering tissues through optical clearing. Opt Express 17(7):5794–5806
Nadiarnykh O et al (2007) Second harmonic generation imaging microscopy studies of osteogenesis imperfecta, vol 12. SPIE, p 051805
Nadiarnykh O et al (2010) Alterations of the extracellular matrix in ovarian cancer studied by second harmonic generation imaging microscopy. BMC Cancer 10:94–94
Negwer I et al (2018) Monitoring drug nanocarriers in human blood by near-infrared fluorescence correlation spectroscopy. Nat Commun 9(1):5306
Nishimura G, Kinjo M (2004) Systematic error in fluorescence correlation measurements identified by a simple saturation model of fluorescence. Anal Chem 76(7):1963–1970
Nuriya M et al (2006) Imaging membrane potential in dendritic spines. Proc Natl Acad Sci U S A 103(3):786–790
Oertel FC et al (2018) Optical coherence tomography in neuromyelitis optica spectrum disorders: potential advantages for individualized monitoring of progression and therapy. EPMA J 9(1):21–33

Oron D et al (2004) Depth-resolved structural imaging by third-harmonic generation microscopy. J Struct Biol 147(1):3–11
Paoli J et al (2008) Multiphoton laser scanning microscopy on non-melanoma skin cancer: morphologic features for future non-invasive diagnostics. J Investig Dermatol 128(5):1248–1255
Pitschke M et al (1998) Detection of single amyloid beta-protein aggregates in the cerebrospinal fluid of Alzheimer's patients by fluorescence correlation spectroscopy. Nat Med 4(7):832–834
Plotnikov S et al (2006) Optical clearing for improved contrast in second harmonic generation imaging of skeletal muscle. Biophys J 90(1):328–339
Provenzano PP et al (2006) Collagen reorganization at the tumor-stromal interface facilitates local invasion. BMC Med 4(1):38
Provenzano PP et al (2008) Collagen density promotes mammary tumor initiation and progression. BMC Med 6:11
Rehberg M et al (2011) Label-free 3D visualization of cellular and tissue structures in intact muscle with second and third harmonic generation microscopy. PLoS One 6(11):e28237
Reiser KM et al (2007) Quantitative analysis of structural disorder in intervertebral disks using second harmonic generation imaging: comparison with morphometric analysis. SPIE
Ricard-Blum S, Baffet G, Théret N (2018) Molecular and tissue alterations of collagens in fibrosis. Matrix Biol 68-69:122–149
Roy HK et al (2008) Spectral slope from the endoscopically-normal mucosa predicts concurrent colonic neoplasia: a pilot ex-vivo clinical study. Dis Colon Rectum 51(9):1381–1386
Sacconi L, Dombeck DA, Webb WW (2006) Overcoming photodamage in second-harmonic generation microscopy: real-time optical recording of neuronal action potentials. Proc Natl Acad Sci U S A 103(9):3124–3129
Sandoval RM, Molitoris BA (2017) Intravital multiphoton microscopy as a tool for studying renal physiology and pathophysiology. Methods (San Diego, Calif) 128:20–32
Sarkar B et al (2012) The dynamics of somatic exocytosis in monoaminergic neurons. Front Physiol 3:414
Sarkar B et al (2014) Label-free dopamine imaging in live rat brain slices. ACS Chem Neurosci 5(5):329–334
Schürmann S et al (2010) Second harmonic generation microscopy probes different states of motor protein interaction in myofibrils. Biophys J 99(6):1842–1851
Schürmann S et al (2013) Label-free imaging of inflammatory bowel disease using multiphoton microscopy. Gastroenterology 145(3):514–516
Shahzad A, Köhler G (2011) Fluorescence correlation spectroscopy (FCS): a promising tool for biological research. Appl Spectrosc Rev 46(2):166–173
Shahzad A et al (2009) Emerging applications of fluorescence spectroscopy in medical microbiology field. J Transl Med 7(7):99
Singh S, Bradley LT (1964) Three-photon absorption in Napthalene crystals by laser excitation. Phys Rev Lett 12(22):612–614
Singh AP, Wohland T (2014) Applications of imaging fluorescence correlation spectroscopy. Curr Opin Chem Biol 20:29–35
Stanciu SG et al (2014) Experimenting liver fibrosis diagnostic by two photon excitation microscopy and bag-of-features image classification. Sci Rep (4):4636
Sun TY, Haberman AM, Greco V (2017) Preclinical advances with multiphoton microscopy in live imaging of skin cancers. J Invest Dermatol 137(2):282–287
Svoboda K et al (1997) In vivo dendritic calcium dynamics in neocortical pyramidal neurons. Nature 385(6612):161–165
Swanson EA et al (1993) In vivo retinal imaging by optical coherence tomography. Opt Lett 18(21):1864–1866
Torres R, Genzen JR, Levene MJ (2012) Clinical measurement of von Willebrand factor by fluorescence correlation spectroscopy. Clin Chem 58(6):1010–1018
Tripathy U et al (2013) Optimization of malaria detection based on third harmonic generation imaging of hemozoin. Anal Bioanal Chem 405(16):5431–5440

Tsai M-R, Chen C-H, Sun C-K (2009) Third and second harmonic generation imaging of human articular cartilage. 40
Tsai M-R et al (2011) In vivo optical virtual biopsy of human oral mucosa with harmonic generation microscopy. Biomed Opt Express 2(8):2317–2328
Tsai C-K et al (2012) Virtual optical biopsy of human adipocytes with third harmonic generation microscopy. Biomed Opt Express 4(1):178–186
Tserevelakis GJ et al (2014) Label-free imaging of lipid depositions in C. elegans using third-harmonic generation microscopy. PLoS One 9(1):e84431
Utino FL et al (2018) Second-harmonic generation imaging analysis can help distinguish sarcoidosis from tuberculoid leprosy. SPIE 23:1
Vakoc BJ et al (2009) Three-dimensional microscopy of the tumor microenvironment in vivo using optical frequency domain imaging. Nat Med 15:1219
Virgili, G., et al., Optical coherence tomography (OCT) for detection of macular oedema in patients with diabetic retinopathy. The Cochrane database of systematic reviews, 2015. 1: p. CD008081-CD008081
Wallace SJ et al (2008) Second-harmonic generation and two-photon-excited autofluorescence microscopy of cardiomyocytes: quantification of cell volume and myosin filaments. J Biomed Opt 13(6):064018
Wang X et al (2003) Noninvasive laser-induced photoacoustic tomography for structural and functional in vivo imaging of the brain. Nat Biotechnol 21:803
Wang X et al (2006) Noninvasive imaging of hemoglobin concentration and oxygenation in the rat brain using high-resolution photoacoustic tomography, vol 11. SPIE, p 024015
Weigelin B, Bakker G-J, Friedl P (2012) Intravital third harmonic generation microscopy of collective melanoma cell invasion: principles of interface guidance and microvesicle dynamics. Intravital 1(1):32–43
Wruss J et al (2007) Attachment of VLDL receptors to an icosahedral virus along the 5-fold symmetry axis: multiple binding modes evidenced by fluorescence correlation spectroscopy. Biochemistry 46(21):6331–6339
Wu Z et al (2017) Multi-photon microscopy in cardiovascular research. Methods 130:79–89
Xie Z et al (2011) Evaluation of bladder microvasculature with high-resolution photoacoustic imaging. Opt Lett 36(24):4815–4817
Yao K et al (2008) Clinical application of magnification endoscopy and narrow-band imaging in the upper gastrointestinal tract: new imaging techniques for detecting and characterizing gastrointestinal neoplasia. Gastrointest Endosc Clin N Am 18(3):415–433
Yao J et al (2011) Label-free oxygen-metabolic photoacoustic microscopy in vivo. J Biomed Opt 16(7):076003–076003
Yeh AT et al (2004) Imaging wound healing using optical coherence tomography and multiphoton microscopy in an in vitro skin-equivalent tissue model. J Biomed Opt 9(2):248–253
Yelin D, Silberberg Y (1999) Laser scanning third-harmonic-generation microscopy in biology. Opt Express 5(8):169–175
Yew E, Rowlands C, So PTC (2014) Application of multiphoton microscopy in dermatological studies: a mini-review. J Innov Opt Health Sci 7(5):1330010–1330010
Yuste R, Denk W (1995) Dendritic spines as basic functional units of neuronal integration. Nature 375(6533):682–684
Zagaynova E, et al Metabolic imaging of tumor for diagnosis and response for therapy. *SPIE BiOS*. 2018. SPIE

Part II

Basics of Bioelectronics

Overview of Medical Electronics for Physically Disabled

4

Vinay Kumar Pandey and Sudip Paul

Abstract

About one-tenth of the total world population is living with active or passive disability which accounts a lot of money in training and creation of skilled manpower to rehabilitate them. The disability is of the various forms that can be sensory, motor, mixed, or cognitive that makes them inefficient in performing daily task for normal living. This may be acquired or developmental or got before birth or during pregnancy. Older people with chronic conditions are prone to suffer from cognition and memory disorders. In such scenario, it is very difficult to prepare workforce for every individual, but it can be done through the technology-enabled diagnostic, intervention and treatment in combination with the medical professionals. Medical electronics is doing the same thing to provide therapy, assistance, and guidance to the patient and clinicians.

Medical electronics is changing from analog domain to the digital domain with more accuracy and easy handling. Digital advancement is having the potential to change the quality of life of the disabled by providing assistive devices, adaption, and accessible support. Assistive technology is designed to enhance the functional ability of physically disabled community. It is having a wide range of services like IT-enabled support including speech, communication, prosthetics, and rehabilitation. It can be a primary care system for children with developmental delay and autistic population. It opened a way for new technologies like virtual reality and augmented virtual reality that are a new and growing field in today's world.

Keywords

Medical electronics · Rehabilitation · Therapy · Technology · Device · Virtual reality · Game

V. K. Pandey · S. Paul (✉)
Department of Biomedical Engineering, North-Eastern Hill University, Shillong, Meghalaya, India

S. Paul (ed.), *Biomedical Engineering and its Applications in Healthcare*,
https://doi.org/10.1007/978-981-13-3705-5_4

4.1 Introduction

Medical electronics is a branch of electronics where we apply electronics principles and design concepts in the field of human well-being. The field of rehabilitation is one of the fields where we make a bridge between medical and engineering to serve people having some form of disability. Creation of such bridges is the combined efforts of health professionals, doctors, physician, researcher, research centers, hospitals, engineers and industry sector. The ultimate goal of medical electronics is to reduce the disability and disease to improve the quality of everyday living of the individuals and to provide assistive devices and infrastructure to be used in cardiology, emergency medicine, orthopedics, communication disorders, mental disorder, neuroscience, pathology, pediatrics, pathology, radiology, psychiatry, imaging and surgery (Iniewski 2009). The use of electronics is not limited only to the biomedical instrumentation but it is expanding in the field of biophysics, biomaterials, biostatistics, biosystem, biosignals, biomechanics, biotransportation, clinics, pathology, rehabilitation, tissue, plasma, membrane and cellular engineering (Carr and Brown 2001).

Today's world of electronics is very fast and rapidly changing. It has shown its application in every field in our day-to-day life. Early in the morning when we got up, we find electronic items near us like mobile phones, TV, refrigerator, LED lights, and a lot more. Life without electronic things brings us back to the era of the eighteenth and nineteenth century when people were living without mobile phones and internet.

In recent years, advances have been seen in automated and semi-automated therapeutic and rehabilitation systems to support in treatments and management of disability (Holden 2005). Most of these systems are focused on mental health, cognition, oral motor disorder, speech language pathology and neurodegenerative disorders. This system aims to recover neurological, motor, and sensory functions of a patient. A disabled person requires a well-controlled therapeutic, assistive, or recovery system that is reliable, compact, easy to operate and handle, and cost-effective that can provide quick response to the stimulus (Scherer 2002). The recovery may take several days to several months of rehabilitation depending upon the level of severity of the disease like mild, moderate or severe.

People having disability face many problems in their day-to-day life in the form of exclusion from society which may lead to social isolation causing to feel lonely and limit their participation in social meets, community, and family. Disabilities isolate them from public health services, education and commonly available services for the citizens in their respective country. Such type of isolation in early childhood and later in adulthood impact their social behavior in communication and participation in their societal life. Electronics is playing and will play a crucial role in fulfilling the gap created between a healthy person and a disabled person by providing supportive services in enabling the disabled to participate in social activities to contribute for their family and for society also (Schultheis and Rizzo 2001).

Assistive technology and prosthetics are a subcategory of rehabilitation engineering, dealing with the instruments, devices, and products related to support disabled people to live normal life. Such type of instruments may include mental health, communication, hearing, mobility, coordination, self-care, day-to-day life tasks, cognition, memory, and learning engagement. These devices are used to enhance the functional and neuronal changes to improve the quality of living.

Utilizing the potential of electronics in disability and rehabilitation can give way to fulfill the needs and necessary requirements of a child. The purpose of medical electronics is to provide diagnostic results to the physician as to access functional and performance analysis of the organs of the body (McCord and Grant 2000). Stethoscope, X-ray, and electrocardiography are the key inventions that led the future of electronics in medical industry. These instruments are used to hear the heart sound, to see internal body organs and the signal pattern followed by the heart in defining malfunctioning of the body parts. Today's medical electronics need to address the problems of diagnostics and therapeutics along with important societal issues like worldwide availability of health care at an affordable cost (Lloyd et al. 2008).

4.2 Medical Devices in the Twentieth and Twenty-First Century

The application of physics theory and principles in service of mankind played a vital role from several decades (Drucker 2012). One of the most important breakthroughs is the invention of X-ray technique that laid down the foundation of radiology in diagnostic and treatment. The knowledge of electromagnetic radiation and its nature along with the fundamental principle of X-ray production makes the basis of imaging technology (Keevil 2012). Its potential to reveal the internal body structure visualization makes it a suitable candidate in medical application. With advanced technologies, it can produce the internal structures of bones, blood vessels, and teeth and intensity of the matter in different body parts. This is used by the medical professionals in diagnosis and prediction of injury and can also be used in imaging of soft tissues. The interaction of these rays with body matter and its effect sparked the imagination and concept of ray therapy or radiation therapy. The spectra of radio waves in which it exists are shown in Fig. 4.1.

Some hundred years ago, electronics in medical application was very less, and only few medical devices were available. Devices like stethoscope, electrocardiogram and X-ray were available which greatly impacted the practice of medicine from that day till today and made the basis of electronics industry in medicine (Nokes et al. 1995). Earlier stethoscope was made from hollow wooden tube along with a funnel-like structure placed on the patient's skin, and another end was kept near the physician's ear. Today it became a symbol of medical professionals like doctors and clinicians.

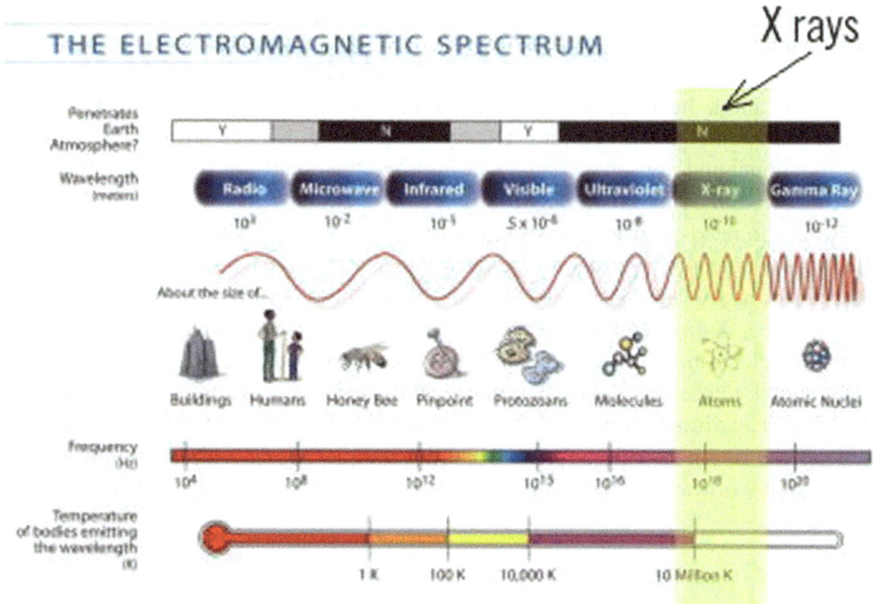

Fig. 4.1 Spectra of electromagnetic radiation. (Image source: https://www.explainthatstuff.com/xrays.html)

Another device known as X-ray was invented in 1895 by Wilhelm Conrad Rontgen. It is used to see the internal skeletal structure of human body. This basic discovery of medical imagining technology made the basis of modern magnetic resonance imagining and other techniques. It aims to diagnose without harmful side effects and is able to go inside the body without any surgical procedure. Imaging technique can be used in treatment and diagnostic purposes.

4.2.1 Determination of Disease or Disorder Through Various Medical Tests (Fig. 4.2)

In treatment it is used in therapy and stimulation by using the various spectrums of radio frequencies. The various forms of medical imaging that are used in diagnosis are as given below:

A. X-ray
B. Ultrasound
C. MRI
D. CT
E. PET

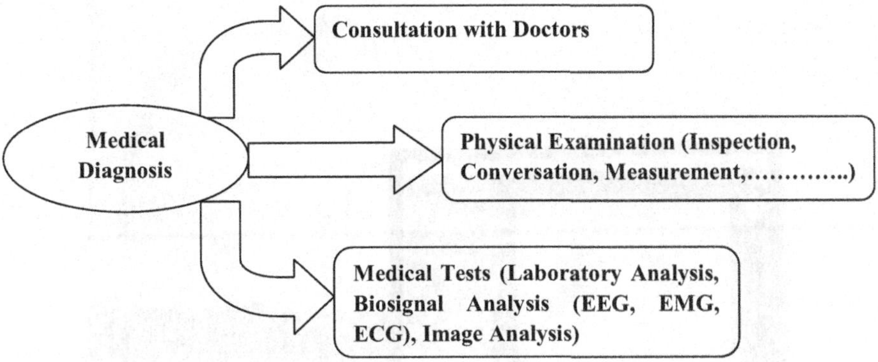

Fig. 4.2 Various medical tests for intervention and treatment

4.2.2 X-Ray

It is the first and most frequently used in diagnosis till date. The development of X-ray images was first done by Roentgen on his wife in his laboratory when he showed her the images of her bones and ring fingers. After that this discovery spread all over the globe with potential use in medical application. In X-ray imaging, a linearly propagating wave forming a divergent beam is partly transmitted and partly absorbed by the object under diagnosis showing the structure by varying thickness and density. The physical setup of an X-ray machine involves a transmitter and a receiver unit which is known as X-ray source and detector or image receptor or radiation detector, and the object is placed in between them. X-ray (X-ray tube) source consists of two electrodes named as anode (positive) and cathode (negative). Cathode is heated up to emit electrons, and these are accelerated toward anode by applying 20–200 kV potential in a vaccumized tube. On arrival of electrons to anode, they give their energy to atoms of anode by ionization and excitations. The two basic principles that take place are photoelectric absorption and scattering (incoherent and coherent). To increase the contrast of the images, X-ray absorbing element is introduced which can absorb the radiation very effectively (Dolhem 2008) (Fig. 4.3).

4.2.3 Ultrasound

It is also known as ultrasonography which uses high-frequency sound waves to scan and capture live action of different internal body organs (O'Desky et al. 1990). It allows doctors and clinician to diagnose the problems in vessels, tissues, and other body parts without surgical procedures (noninvasive). Generally this test is performed when you are having pain in the gallbladder, bladder, kidney, liver, and blood vessels. It is also used to detect problems of the uterus, ovaries, testicles, thyroid, spleen, and pancreas. Before doing ultrasonography, the clinician instructs

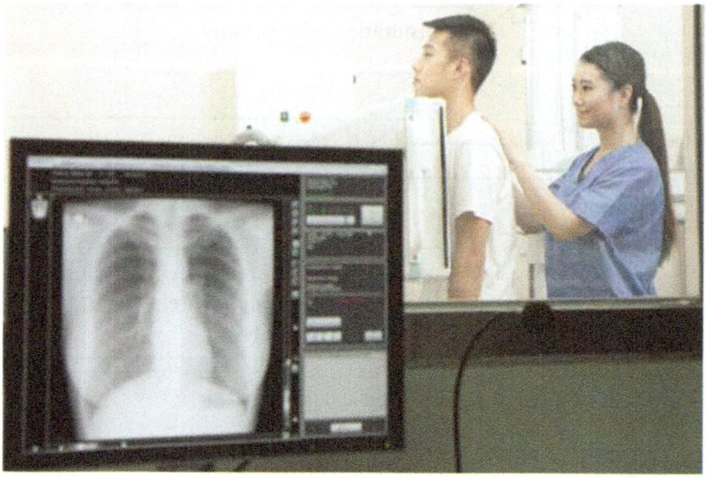

Fig. 4.3 Radiographer taking chest X-ray. (https://www.radiologyinfo.org/en/info.cfm?pg=chestrad)

the patient like to fast in the evening or after breakfast or another type of instruction depending on the body part which is to be diagnosed. A jelly-like paste is applied before ultrasonography by the technician to reduce friction and to pass sound wave effectively.

4.2.4 Magnetic Resonance Imaging

Magnetic resonance imaging techniques come late in the twentieth century to examine the live activities of human body parts. It applies the physics phenomena in order to see the atomic activities under magnetic field of internal body structure for diagnostic purposes (Beltran 1995). It becomes a lifesaving technique within a few years of its evolution. It is a kind of scanning internal body organ without surgical procedure using radio waves and magnetism to produce clear image of human anatomy (Goldstein et al. 2017). The medical application of MRI is diverging from one dimension to multidimensional field. This applies the principle of radio frequency energy absorption by the nuclei of some material like ^{1}H as hydrogen is principle constituent of human organs. After getting energy from radio frequency, it starts showing resonance; as a result, the RF pulses interact differently as different body parts are having different chemical compositions (Basser et al. 2000). This provides a possible cause to detect the deep organ image of the selected body parts. Preclinical MRI has the advantage of exploring the disease progression and regression and the effect of various therapeutic modalities applied to the disabled to assess the efficacy of therapeutic intervention and bio-pathological mechanisms. The greatest advantage is that it can detect small signal changes produced by brain neuronal activities over other available techniques.

Nuclear magnetic resonance (NMR) is one of the fields of magnetic resonance application where it is used in determination of molecular structure and quality control along with biophysical features. NMR and MRI techniques are very useful in various diseases, including motor neuron disorder and neurological diseases like epilepsy, stroke, Parkinson disease (PD), etc. NMR has great strides in understanding the structure of molecules and proteins. This utilizes the nuclear spin and splitting of energy levels in a magnetic field to get anatomical images. But both are based on different principles. As in NMR, the radiation is given by a RF emitter in the form of very short pulses of linearly polarized waves, whereas in fluorescent spectroscopy the object is energized from electromagnetic radiation. The combined application of both NMR and MRI is having great advantage in biomedical, biomedicine and clinical practice. MRI is a powerful tool as compared to NMR spectroscopy as it can produce the images of the water and fat molecules also. In general, 1.5 T and 3 T MRI setup is available in clinical practice having normal range of application from 0.2 T to 3 T. Usually, we prefer high-intensity magnetic field as low-intensity magnetic field produces a weak signal, so for clear projection of voxels, strong magnetic field is needed as it is less affected by noise and the signal-to-noise ratio will be high which is desirable in every case. For safety purpose, we cannot use higher strength magnetic field above the approved level, but research is going on for safer use of 6 T machine and higher order. To achieve uniform magnetic field, it must be shielded against all kind of electromagnetic radiation and external magnetic field. Advanced MRI helped in various domains including cognitive neuroscience, behavioral neuroscience, and neurosurgery with improvement in day-to-day living of the patient (Shenton et al. 2001). It is used in the treatment and management of functional recovery and reorganization of brain function like morphological, structural and functional study of PD and stroke patient before and after the treatment. Functional MRI is also used in studying the human brain behavior in order to diagnose and predict developmental delays in children and bipolar disorders. The below figure shows the setup and the scanning process of patient under diagnosis as given, and it is an active technique where the patient under diagnosis is himself active differing from other technique where the object is passive. In comparison to other techniques, it provides a three-dimensional image of object responding in a different way to the radiation which can be seen on a computer screen installed with machine setup. These images are available for later use for the medical professionals to view, refer, and predict the liaison present in the specific part of our body and are saved in the computer memory, and two-dimensional views of these images can be taken on paper or in the soft forms in pen drive or in hard disk. It is very easy to handle these images as it can be transported to any part of the globe without physical movement that makes easy to take expert advice from a different country or from a distant geographical region.

4.2.5 Computer Tomography (CT)

It is also known as slice imaging technique that can project layer-by-layer images by using X-rays to generate images of internal organs. Earlier it was not possible to visualize the structural deformity and lesion, but nowadays CT makes it simple to visualize it with remarkable clarity. Unlike X-ray imaging, it rotates around the body to take multiple projection of the body at different angles, whereas in X-ray the whole image overlaps and it is very difficult to distinguish different parts of an organ. In this technique, the source and detector move around the body simultaneously to take measurement, and all the projected slices are vertical to the long body axis. Due to this it is sometimes called computer axial tomography or computer-assisted tomography (Woods et al. 1993). Computer tomography utilizes a virtual pile of reconstructed images by using normal X-ray images. The width of X-ray beam and the detector determine the resolution of the scanned images. It allows computer system to draw complete image of the body organ. It provides a better picture to the doctors and clinicians to diagnose and manage the disease. It is used in the diagnosis of cause of pain, blood clotting in lungs, tissue swallowing, and appendicitis.

4.2.6 Positron Emission Tomography (PET)

It is a noninvasive scanning technique for diagnostic examination to find information and activity of different body parts. It uses a special substance that is considered by the body as sugar and carried into cells (Vardi et al. 1985). This substance is known as a tracer and has almost the same properties as sugar, but a small radioactive substance is added to it. The image formed on scanning is based on the radiation and detection of positron from the radioactive material. Image obtained in this is used in detection of various diseases like tumor, cancer, etc.

It works on the principle of cellular changes occurred in the body in disease condition. In PET scan, the tracer is given to the patient which breaks down quickly in order to release the positron. The most commonly known tracer is FDG (fluorodeoxyglucose) and is similar to glucose (sugar), and it travels to the location having higher concentration of sugar and breaks down to release positron that combines with an electron and releases radioactive waves. These waves are detected by the PET scanner and it is converted into electrical signal to be analyzed by the computer. The installed computer system creates images of the specific tissue activity either in color code or in black and white images (Petersen et al. 1988). This color code or varying brightness of image tells us the degree and level of tissue functionality. Darker in the case of color coding and brightness in black and white area shows more concentration of radioactive substance that indicates tissue is growing much faster than the normal one.

PET scans are commonly used for cancerous cell detection and demonstration as it can detect biochemical (metabolic) changes associated with body due to cancer and provides effective diagnostic way for treatment (Lowe et al. 1998). Another application of PET scan is in heart disease to determine blood flow and may be used

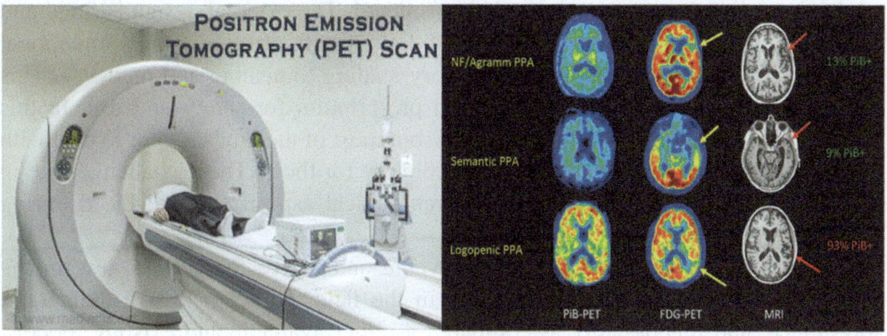

Fig. 4.4 (**a**) PET scanner and (**b**) PET images along with MRI images. (Image source: https://www.medindia.net/patients/patientinfo/positron-emission-tomography-scan.htm and https://www.researchgate.net/figure/Changes-in-Brain-MRI-FDG-PET-and-PiB-PET-in-Three-Subtypes-of-PPA_fig 6_268802122)

to determine seizures, memory loss, brain tumors, etc. It is different from CT and MRI as it determines biochemical changes, whereas CT and MRI provide anatomical changes (Fig. 4.4).

4.3 Prosthetics

It is a branch of medicine which deals with the replacement of unavailable body parts that is used in restoring normal body functions in order to live comfortably. Broadly, it is involved in the design and development of artificial body parts and is a new field of biomedical engineering that applies engineering principle in the service of mankind to provide a better way of living for the disabled. Prosthetic limbs are artificially engineered dynamic entities operated by body sensation and movement (Biddiss and Chau 2007). They can be bioelectrically operated or can be static or manually active mechanical instruments in service of the disabled. As our body is a master tool box that is able to repair small issues occurred in our body, sometimes and in some cases it is not able to repair and reconstruct the system in that place, and we need external support in order to maintain daily living. Sometimes this external support is in the form of prosthetics or in the form of assistive technology which is the result of technology advancement. In prosthetics point of view, every technological state-of-art design extends their hands to serve humans by continuing their abilities beyond the bodily absent and inappropriate functions. Technical design is used in replacing, extending, and functionality rebuilding tool in the service of natural toolbox provided by the nature (Antfolk et al. 2013).

Modern technology aims to provide normal life to the individuals without much support from others. It looks simple in a straightforward way, but restoration of function activity with artificial arrangement is a complex task which needs lots of research and critical analysis (Bogue 2009). If a person is missing a limb or loses due to accidents, then fixing the same body part with artificially designed and customized fabrication requires many factors to be considered. The external design is fitted and the person is trained for operation to use in daily life. In order to achieve functional recovery, it may include amputation surgery leading to healing and recovery. This procedure is followed by assessment and prescription for safety and use. In this process, the patient faces lots of challenges, but most of the time they try to deal with these life changes and these things are secondary for them. These challenges are in the form of physical, emotional, and psychological aspects that are faced during prosthetic restoration and rehabilitation.

Bionic hands having individually movable and active fingers are new in the market. These hands make a revolution in the field of prosthetics in their look, appearance, and functionality as compared to other available and existing models. Touch Bionics, the first company, has given its name as "iLimb" in which an individual finger is known as digits. Every component is having their own motor and gearbox including fingers (Clement et al. 2011). The degree of freedom is five which is achieved by the use of cable, spring, and myoelectric sensors.

4.4 Electronics, Rehabilitation, and Virtual Reality in Treatment and Management of Disorder

Rehabilitation is a broad term consisting of different mechanisms for different cases. Home-based rehabilitation is the need of the present days, and the new trend of virtual reality is a powerful technique in fulfilling the present demand by providing the virtual real environment. It is a new and consistently growing field. It gives freedom to the user to interact with the virtually created natural objects to get natural sense of touch, vision, audible hearing and other activities (Schultheis and Rizzo 2001). It is having inherent property of stimulation bringing natural world in the form of controlled environment. Due to this, virtual reality platforms are implemented in therapy and postsurgery rehabilitation by using biofeedback in real-time operation. It is also used in performance analysis through the stores data of the used from the first day to the last day of the session including all the parameters like session duration, force applied, active and passive participation, etc. that are used by the experts, clinician, and doctors. They can adjust different parameters of the system depending upon the need of the patient for self-practicing. Another form that can be utilized for home-based rehabilitation is tele-rehabilitation that can provide services to remote locations (Wilson et al. 1997) (Fig. 4.5).

4.4.1 Virtual Reality (VR) in Rehabilitation

Developing virtual reality system is a complex design task, but they are very helpful for disabled people and health professionals. It is having a great scope in health-care and therapeutic applications due to its ability to stimulate day-to-day life tasks. The ability to create an environment like real life makes users feel immersed by interacting artificial surroundings. This field is suffering from high cost, but at the

Fig. 4.5 Human interaction to virtual reality. (Image source: https://virtual.reality.news/news/vtime-brings-human-interaction-virtual-reality-0171730/)

same time it can benefit lots of the field like health care, education, military, manufacturing, and entertainment. Its potential to change parameters and update allows individuals to explore and interact according to their level and choice. This is not possible in the real life in therapeutic point of view. These systems are more useful where visualization is required for training and treatment. Virtual reality is an advanced and emerging research field.

As compared to the traditional way of therapy, this technique is having many advantages over manual and physical therapy given by therapeutic clinics where children feel uncomfortable and react differently. When the same type of therapy is given in the home premises either by the parents or by self-monitoring systems, they feel relaxed and comfortable, and also they cooperate in therapeutic sessions. Other advantages of the virtual systems include ease of accessibility, individual and personalized setup, easy adjustment of various parameters, measurement and analysis ability, feedback features, etc. (Ojha 1994). These setups are application and software-based, so it is easy to install and free from space allocation (Fig. 4.6).

Specifically designed software for measurement, evaluation and analysis are integrated to work as a single unit to give best possible result. Its stimulation property by using virtual reality brings physical world in controlled setups (Lee et al. 2003). Initial data is stored of an individual for the reference, evaluation, and analysis to be used by the experts in determining the level of severity in order to advise appropriate self-performing activity to be done at home as per patient need and requirement. The expert can provide needful advice to the patient through tele-rehabilitation facility that allows services to remote locations. A basic characteristic of rehabilitation is repetition which can be fulfilled by virtual reality systems by providing stimulus to the individuals; otherwise, they feel bored during the sessions. It is very useful for the child as it can attract for repetitive therapeutic sessions which is much needed for effective training to increase efficiency and to get better recovery

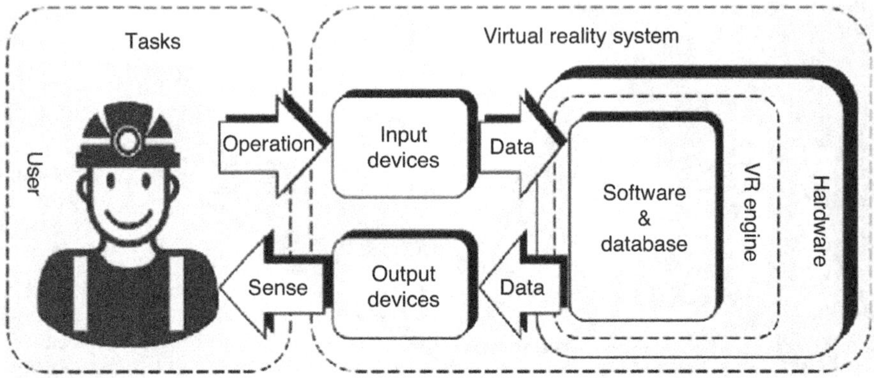

Fig. 4.6 Components of virtual reality systems. (Image source: Head-mounted display-based intuitive virtual reality training system for the mining industry by Hui Zhang)

result. It may include real life task, challenges or games in order to take their attention and involvement for completing the training sessions.

4.4.2 Stroke Rehabilitation Through Virtual Reality

Stroke is the third leading cause of mental disability in the world. A stroke survivor suffers a lot from various problems leading to paralysis, limb stiffness, brain damage, balance and coordination issues (Lucca 2009). It affects control and motor movements of the limbs, cognition, memory, speech and intelligence. Proper rehabilitation may save them from severe disability by rebuilding functional ability to live normal life. Depending on the stroke impact and severity, the rehabilitation practices can be short or extensive, continuous, and repetitive to maximize activity and to gain long-term outcomes. This aims to increase their social interaction ability, mobility, coordination, and communication (Goude et al. 2007). This process takes a long time to regain the functionality in months to years. Repetitions of activities are used to refine the learned skills in order to maintain healthy living. Time and type of activities vary from person to person depending upon the affected lime, brain portion and guidance provided by specialists and doctors. In this way progress and recovery of a patient vary from one to another and may be unique. Even specialized center exist in all over the world but still a large population is not able to get these services mostly in developing and undeveloped countries. The biggest barrier for them is money; transportation and lack of awareness are also a factor. In this scenario, home-based and self-directed rehabilitation plays an important role for stroke survivors through rehabilitation practices within home premises (Jack et al. 2000).

It can utilize tele-conferencing, motion capture camera, sensors, display device, computer, mobile, etc. to make connection with virtually created environment or from remotely located user or therapist in order to get instructions, advice, and support (Kowalczewski et al. 2011). Motion capture cameras are used to take user's arm, leg, or other body parts' movement, which is used by the expert in the analysis and treatment and also by the system in creating task to be performed by the system. Sensors may be placed either at head location or at arms, shoulder, or elbow to get muscle signals which are used as biofeedback to the system. These procedures are used to train patients to regain and relearn the functional activity that can be motor, sensory, or mixed. The motor skills can be recovered through various activities like phone dialing for finger activity, virtual writing, painting, and gaming (Zampolini et al. 2008). Telephone dialing can be brought in regular routine to activate fingers which are very important for daily livings and day-to-day activities. The application developed for finger training is focused to train hand movement in coordination with fingers with force, precision, and speed. Writing is a skilled motor activity involving great coordination of eye-hand movement. Writing skill training systems are little bit different language like Hindi, English, Japanese, Chinese, etc. because all are having different characters and different trajectories. These systems guide the users in real-time writing and provide the results in terms of completeness, force applied, and movement. In the same way, virtual painting is used to train controlled movement of hand. It involves a virtual image composed of triangles, square, circle, curved lines, etc. The user has to complete the painting task with virtual pen and paint. Time spent during the task is given on the screen with other parameters like coordinate position, movement, speed, time, etc. Interactive gaming in virtual environment is also a good technique in developing motor skills in the disabled. It includes virtual cricketing or hockey environment with stimulus to attract users. In this view virtual systems are useful in gesture analysis for Parkinson's patients, rebuilding hand and arm movement for stroke patients, and cognitive and psychological rehabilitation.

4.4.3 Augmented Reality (AR) Systems

Augmented reality is basically a combination of virtual reality with the real world creating a mixture of both the real environment and virtual environment (Burke et al. 2010). This technique superimposes virtual objects upon the real-world objects. The fundamental characteristics of AR systems comprise of the combination of VR and natural with interaction in real time (Aung and Al-Jumaily 2012). Therapeutic rehabilitation is a multidisciplinary field. It demands a diversified technique to fulfill the request of various activities. Such demand can be fulfilled by computers by providing several therapeutic support by providing countless services to the disabled in terms of communication, autonomy, personal growth, and social and interpersonal skills. Computers and software programs make it easy for the therapist and clinician in the treatment and management of the disabled. Personalized treatment approach can be adopted for individual patient for safety and effectiveness. Several parameters can be changed at a time seeking to decrease or increase the difficulty level for better impact. Moreover, the therapist can guide the intervention process and make a treatment plan. All these things are possible due to technological advancement. Virtual and augmented reality systems are examples that are making conventional hardware in creation of differentiated environment. As the research and development progresses, AR systems become more accessible in intervention and clinical settings.

AR system's core components are sensors, processor, and display device. Sensor gathers data from realistic natural world. This can be a camera to take images from real world that can be processed as per the need and requirement, and these are determined by computer programs with respect to environment. Visual cues, used in pointing things, are also known as fiducial markers that can show relative location and orientation of an object with reference to other objects. This information is analyzed by the processor gathered by sensors, cameras, and other input devices. These processors are so strong that they are able to process real-time signals (Correa et al. 2007). The processed data is displayed by the display device.

4.4.4 Assistive Technology

IT-enabled advanced technologies enhanced the assistive need of disabled and old-age population. As its name suggests, it is a technique to provide assistance to the disabled by performing such tasks that they are not able to do. By using this technology, they are independent from intensive family care and support for their daily need and care (Parette and Scherer 2004). It became important as we know that the whole world is facing aging problems, and aged population is growing day by day in global scenario. In this way, the need suggests to think for assistive technology in order to serve them. Hence, lots of research and development are in progress to get better result with precision and accuracy.

4.5 Conclusion

Rehabilitation with safety is the current need of the world. In this, mechatronics and virtual reality are key fields that can accomplish this task with reliability and comfortability. Virtual reality along with stimulation can recover the functional ability of a limb in less time duration and with less cost. So, we can save both time and money by this way. Interactive gaming attracts the child to perform actions that lead to other passive benefits like cognitive development and effective use of time.

References

Antfolk C, D'Alonzo M, Rosén B, Lundborg G, Sebelius F, Cipriani C (2013) Sensory feedback in upper limb prosthetics. Expert Rev Med Devices 10(1):45–54

Aung YM, Al-Jumaily A (2012) AR based upper limb rehabilitation system. In: Biomedical Robotics and Biomechatronics (BioRob), 2012 4th IEEE RAS & EMBS International Conference on IEEE, pp 213–218

Basser PJ, Pajevic S, Pierpaoli C, Duda J, Aldroubi A (2000) In vivo fiber tractography using DT-MRI data. Magn Reson Med 44(4):625–632

Beltran J (1995) MR imaging of soft-tissue infection. Magn Reson Imaging Clin N Am 3 (4):743–751
Biddiss E, Chau T (2007) Upper-limb prosthetics: critical factors in device abandonment. Am J Phys Med Rehabil 86(12):977–987
Bogue R (2009) Exoskeletons and robotic prosthetics: a review of recent developments. Ind Rob Int J 36(5):421–427
Burke JW, McNeill MDJ, Charles DK, Morrow PJ, Crosbie JH, McDonough SM (2010) Augmented reality games for upper-limb stroke rehabilitation. In: 2010 second international conference on games and virtual worlds for serious applications. IEEE, pp 75–78
Carr JJ, Brown JM (2001) Introduction to biomedical equipment technology, vol 4. Prentice Hall, Upper Saddle River
Clement RGE, Bugler KE, Oliver CW (2011) Bionic prosthetic hands: a review of present technology and future aspirations. Surgeon 9(6):336–340
Correa AGD, De Assis GA, do Nascimento M, Ficheman I, de Deus Lopes R (2007) Genvirtual: an augmented reality musical game for cognitive and motor rehabilitation. In: Virtual Rehabilitation. IEEE, pp 1, 2007–6
Dolhem R (2008) The history of electrostimulation in rehabilitation medicine. Annales de readaptation et de medecine physique: revue scientifique de la Societe francaise de reeducation fonctionnelle de readaptation et de medecine physique 51(6):427–431
Drucker P (2012) Management challenges for the 21st century. Routledge, New York
Goldstein JI, Newbury DE, Michael JR, Ritchie NWM, Scott JHJ, Joy DC (2017) Scanning electron microscopy and X-ray microanalysis. Springer
Goude D, Björk S, Rydmark M (2007) Game design in virtual reality systems for stroke rehabilitation. Stud Health Technol Inform 125(2007):146–148
Holden MK (2005) Virtual environments for motor rehabilitation. Cyberpsychol Behav 8(3):187–211
Iniewski K (ed) (2009) Medical imaging: principles, detectors, and electronics. Wiley
Jack D, Boian R, Merians A, Adamovich SV, Tremaine M, Recce M, Burdea GC, Poizner H (2000) A virtual reality-based exercise program for stroke rehabilitation. In: Proceedings of the fourth international ACM conference on assistive technologies. ACM, pp 56–63
Keevil SF (2012) Physics and medicine: a historical perspective. Lancet 379(9825):1517–1524
Kowalczewski J, Chong SL, Galea M, Prochazka A (2011) In-home tele-rehabilitation improves tetraplegic hand function. Neurorehabil Neural Repair 25(5):412–422
Lee JH, Jeonghun K, Cho W, Hahn WY, In YK, Lee S-M, Kang Y et al (2003) A virtual reality system for the assessment and rehabilitation of the activities of daily living. Cyber Psychol Behav 6(4):383–388
Lloyd C, Waghorn G, Williams PL (2008) Conceptualising recovery in mental health rehabilitation. Br J Occup Ther 71(8):321–328
Lowe VJ, Fletcher JW, Gobar L, Lawson M, Kirchner P, Valk P, Karis J et al (1998) Prospective investigation of positron emission tomography in lung nodules. J Clin Oncol 16(3):1075–1084
Lucca LF (2009) Virtual reality and motor rehabilitation of the upper limb after stroke: a generation of progress? J Rehabil Med 41(12):1003–1006
McCord JF, Grant AA (2000) Prosthetics: pre-definitive treatment: rehabilitation prostheses. Br Dent J 188(8):419
Nokes L, Jennings DFTT, Flint T, Turton B (1995) Introduction to medical electronics applications. Butterworth-Heinemann
O'Desky RI, Ball MJ, Ball EE (1990) Computers in health care for the 21st century. Methods Inf Med 29(02):158–161
Ojha AK (1994) An application of virtual reality in rehabilitation. In: Southeastcon'94. Creative technology transfer-a global affair., Proceedings of the 1994 IEEE, pp 4–6. IEEE, 1994
Parette P, Scherer M (2004) Assistive technology use and stigma. Educ Train Dev Disabil:217–226
Petersen SE, Fox PT, Posner MI, Mintun M, Raichle ME (1988) Positron emission tomographic studies of the cortical anatomy of single-word processing. Nature 331(6157):585

Scherer MJ (2002) Assistive technology: matching device and consumer for successful rehabilitation. American Psychological Association, Washington, DC

Schultheis MT, Rizzo AA (2001) The application of virtual reality technology in rehabilitation. Rehabil Psychol 46(3):296

Shenton ME, Dickey CC, Frumin M, McCarley RW (2001) A review of MRI findings in schizophrenia. Schizophr Res 49(1–2):1–52

Vardi Y, Shepp LA, Kaufman L (1985) A statistical model for positron emission tomography. J Am Stat Assoc 80(389):8–20

Wilson PN, Foreman N, Stanton D (1997) Virtual reality, disability and rehabilitation. Disabil Rehabil 19(6):213–220

Woods RP, Mazziotta JC, R. Cherry a S (1993) MRI-PET registration with automated algorithm. J Comput Assist Tomogr 17:536–536

Zampolini M, Todeschini E, Guitart Bernabeu M, Hermens H, Ilsbroukx S, Macellari V, Magni R et al (2008) Tele-rehabilitation: present and future. Ann Ist Super Sanita 44(2):125–134

Advances in Diagnostic Techniques for Therapeutic Intervention

5

Geetesh Verma, Radhika Kesharwani, Pabbala Veeresh, Harpreet Kaur, Deepaneeta Sarmah, Vignesh Kotian, Leela Mounica, Anupom Borah, Kiran Kalia, and Pallab Bhattacharya

Abstract

The ability to get amazing real images from inside the human body has been made possible through the advent of medical imaging techniques. A lifesaving diagnostic tool began with the accidental discovery of X-rays. The last quarter of the nineteenth century saw an explosion of technological advancement in medical imaging from ultrasound to the magnetic resonance imaging (MRI) which can peek inside the human body to a greater extent with improved image quality and shorter examination time at a reasonable cost. The aim of this chapter is to provide a concise information regarding the major imaging modalities available, their applications, benefits, and risks. Towards the end a comparison of all modalities are summarized in a table which will help to provide a better understanding of these medical imaging techniques to encourage further interest in the use and development of these techniques to benefit all human beings.

Keywords

X-ray · Piezoelectric effect · Positron emission tomography (PET) · Single photon emission computed tomography (SPECT) · Ultrasound · Magnetic resonance imaging (MRI)

G. Verma · R. Kesharwani · P. Veeresh · H. Kaur · D. Sarmah · V. Kotian · L. Mounica · K. Kalia · P. Bhattacharya (✉)
Department of Pharmacology and Toxicology, National Institute of Pharmaceutical Education and Research-Ahmedabad (NIPER-A), Gandhinagar, Gujarat, India
e-mail: pallab.bhattacharya@niperahm.ac.in

A. Borah
Department of Life Science and Bioinformatics, Assam University, Silchar, Assam, India

S. Paul (ed.), *Biomedical Engineering and its Applications in Healthcare*,
https://doi.org/10.1007/978-981-13-3705-5_5

5.1 Introduction

Till date, pathology remains the gold standard tool for disease diagnosis, substantially because it gives exact morpho-structural information about a disease. However, pathology alone cannot provide a correct diagnosis, and sometimes results produced can be a false positive (where test result indicates the presence of disease or condition but in actual it does not) or a false negative (where result indicates the absence of disease but in actual it does hold). A false-positive result leads to unnecessary treatment, and a false-negative result leads to a false diagnostic (Sacerdoti et al. 2016). In this scenario, the accuracy of diagnosis can be improved using medical imaging techniques which may provide more accurate early detection, diagnosis, and treatment of disease (Dougherty 2009).

5.2 Conventional Radiography (X-Ray, Plain Films)

Professor Rector Wilhelm Conrad Roentgen in the late 1890s started working with a cathode tube placed inside a glass container. When the tube was exposed to a beam of electrons, the tube started to show fluorescence, which Prof. Rector covered with a cardboard to prevent its decay. Also, he noticed a strange phenomenon where another screen of barium platinocyanide, placed outside the tube, started fluorescing. In other words, invisible rays had passed through the cardboard and interacted with the barium screen. The rays were named as X-rays, as Prof. Rector had no idea what those rays were. He eventually won the Nobel Prize in Physics for the same in 1901. However, the news was received by the public with a mixture of disbelief and apprehension. But doctors saw the possibilities of this new ray in the diagnosis of disease in humans and began to use them (Assmus 1995). Before the discovery of X-rays, medical diagnosis made by physicians was based on verbal description to elicit out patient's symptoms and history (Berger 1999). In the early years of radiology, patients subjected to X-ray suffered burns due to poor handling and its unexplored role in a biological system. As scientists came across the detrimental effects of ionizing radiation, actions were taken to protect operators from radiation and to reduce patient exposure time from minutes to milliseconds. However, it was not until 1930 that their use became certain and reliable (Sansare et al. 2011). Today, X-rays have dramatically changed the diagnostic procedure across the world and have contributed towards a better understanding of illnesses and injuries in billions of patients each year (Chen et al. 2018).

5.2.1 Working Principle

The X-ray tube is a vacuum tube that produces X-rays. It consists of a glass envelope which is made up of Pyrex glass that protects and provides a vacuumed environment. The glass has a target window, the cathode is made up of tungsten filament and has a high melting point, and the anode is also made up of tungsten filament both of which are connected to the high voltage output. The cathode is connected to the negative end and anode to the positive end. When the cathode is heated by electric current, it produces electron through thermoionic emission. The produced electron travels from the cathode towards the anode and hits the anode. Once the anode is hit, two effects come into place: (1) photoelectric effect and (2) Bremsstrahlung effect. Thus, X-rays are produced when an electron strikes the target (anode), 99% of which is converted into heat, while only 1% is converted into X-rays. The area of the anode from where X-rays are emitted is called the focal spot. Therefore, in modern X-ray tubes rotating anode is present which will have hits at various positions, therefore will have less heat generation or less heat capacity. However, there are some exceptions like portable radiographic and dental X-ray. X-rays have a very short wavelength between 0.01 and 10 nm (suitable for medical imaging) unlike visible light and are able to pass right through most tissues in the human body. Since bone contains calcium, denser material than most other tissues, they are able to stop some of the X-rays and thus cast a shadow behind it that gets registered on film or on a digital sensor. The process of getting an image on film is called radiography, and the finished film is known as a radiograph. Bones are clearly visible, while soft tissue such as muscle is shadowy; as a result the two-dimensional (2-D) images produced by the X-rays are quite useful in revealing the structures within the body and can easily display fractures of bone (Assmus 1995).

5.2.2 X-Ray Radiography Benefits

- Non-invasive, quick, and painless.
- Guides medical professionals in inserting stents, catheter inside the patient, as well as helps in planning surgical and medical treatment.
- Helps in identifying foreign objects inside or around the bones.

5.2.3 X-Ray Radiography Risks

- Radiation exposure increases the chances of developing cancer in the long run.
- High levels of radiation exposure cause skin burns and hair loss, which occur at relatively high doses.
- Interaction with lighter element is not so strong.
- It does not provide three-dimensional (3-D) information.

5.2.4 X-Ray Radiography Medical Applications

- X-ray radiography is used in dentistry for finding cavities and identifying broken bones.
- Chest X-rays are used for diagnosing pneumonia, tuberculosis, and lung cancer.
- Arthrography is used to diagnose arthritis and gout.
- Fluoroscopy radiographs are used for displaying the movement of organs and diagnosing large airway diseases.
- Mammography is a screening tool used to diagnose and screen the breast tissue.
- Bone densitometry is used to diagnose osteoporosis and mineral content in bone.

5.3 Computed Tomography

Computed tomography (CT) (Claesson 2001) or computerized axial tomography (CAT) is a revolutionized diagnostic imaging technique widely used in hospitals to produce cross-sectional images of the human body. The CT scanner was discovered by Dr. Alan Cormack in 1963 and first clinical CT scanner was presented by Hounsfield in the year 1972 (Claesson 2001).

5.3.1 Working Principle

A CT scanner uses a series of narrow X-ray beams to make a detailed study of the body. These X-ray beams move like an arc through the object and emit succession of narrow beams. It measures the projection of an object from different angles and generates cross-sectional images of the body using the attenuation coefficient of X-ray beams which results in a 3-D description of an object (Hsieh 2009). Recently four-dimensional (4-D) CT has been introduced to generate both temporal and spatial information on organ mobility (Keall et al. 2005).

CT uses a series of X-ray beams and gives a more detailed final image than X-ray picture. CT provides digital images and measurements made by a detector that are proportional to the sum of the attenuation coefficients. Dense objects like bones are easy to see in the CT scan, but in the case of soft tissues, CT does not provide good images. Hence some contrast materials (dyes) are required to highlight the soft tissues. Contrast materials are usually made of barium sulfate or iodine and may be administered in one or more ways including oral, injection, or through enema.

5.3.2 Spiral CT

Spiral CT, also called helical CT, was introduced in the 1990s. This technique allows quick detection, and it can even detect small nodules which cannot be detected by normal X-ray. Compared to conventional CT, it produces a lower dose of radiation

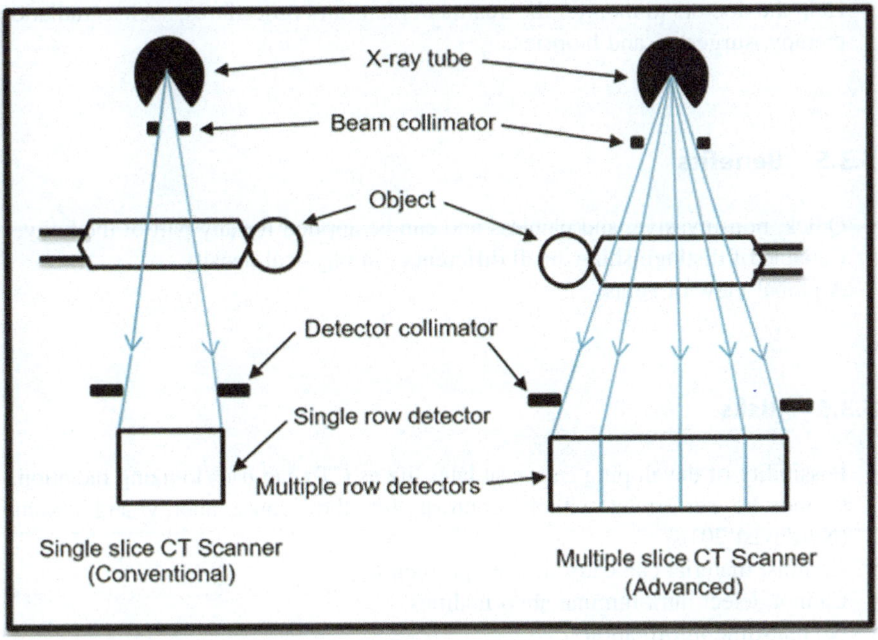

Fig. 5.1 Difference between conventional and advanced CT scanner

(Kalender 1994). Another advantage of spiral CT is that it gives clear 3-D images by acquiring data from patient anatomy, all in one position (Jiang et al. 2001).

5.3.3 Multislice CT

Spiral CT was found to be very hard for X-ray tubes and produced a lot of heat. Multislice CT (MSCT) is one of the approaches to overcome such problems. As MSCT has more than one detector, it reduces the total number of rotations and thereby preventing heat generation (Fig. 5.1; Kalender 1994).

5.3.4 Applications

- CT scan can examine various parts of the human body such as the soft tissues, lungs, abdomen, brain, teeth, etc. (What's the difference between an X-ray, CT scan and MRI? 2015).
- The preferred way to diagnose bone fractures and joint problems. Small bones are also clearly visible in the CT scan.
- Detect the presence and location of tumors (Nordqvist 2018).
- Provide valuable data on blood flow and help to identify internal injuries, blood clot, and bleeding.

- Help the doctors to monitor the treatment plans and procedures, such as radiation therapy, surgeries, and biopsies.

5.3.5 Benefits

- Quick, non-invasive, and painless and can be applied for any part of the body.
- Capable of distinguishing small differences in physical density.
- A global view of veins.

5.3.6 Risks

- Possibility of developing cancer at later life as CT scan uses ionizing radiation.
- Cannot be performed without contrast, but they cause allergy and toxicity (Nordqvist 2018).
- Contrast material can cause kidney problems.
- Cannot detect intra-luminal abnormalities.
- No real-time information.

5.4 Magnetic Resonance Imaging

Magnetic resonance imaging (MRI) is the most informative and versatile imaging tool used to create detailed images of the body. MRI is also called nuclear magnetic resonance (NMR) as it relies on the physical principles of NMR (McRobbie et al. 2017). Research on MRI started way back in the 1970s. The first clinical magnetic resonance images were produced in Nottingham and Aberdeen in 1980 (Hawkes et al. 1980). In 1986, Magnetic resonance angiography was introduced which was capable of imaging the blood vessels throughout the body (Thedens 2009). MRI scan uses a strong magnetic field, magnetic field gradients, and radio waves to generate images of the organs of the body.

5.4.1 Working Principle

MRI utilizes the magnetic properties of the atomic nucleus. The human body mostly contains water. Water molecules have hydrogen nuclei (proton) which oscillate in the presence of an applied magnetic field. This magnetic field is applied by the use of radio waves (Fig. 5.2). An MRI scanner forms a temporarily strong magnetic field (about 0.2–3 Tesla), which makes the atoms to spin and subsequently oscillate in the magnetic field while returning to equilibrium. Simultaneously, the excited hydrogen atoms emit a radio signal, which is detected by receiving coils, which can further be

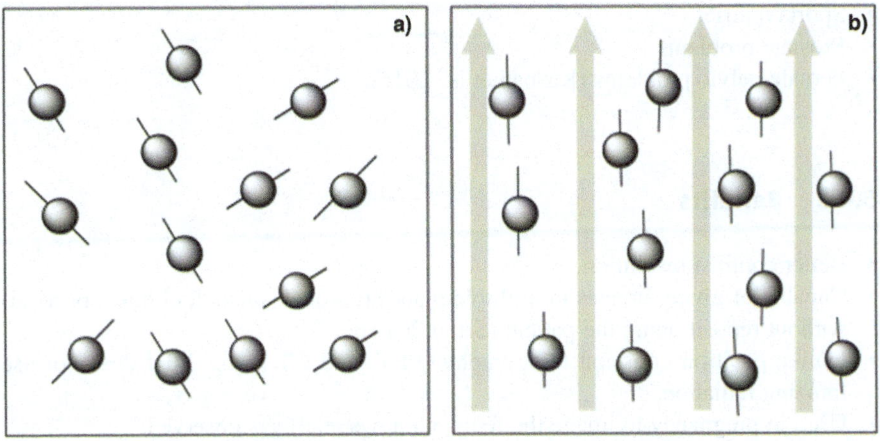

Fig. 5.2 (**a**) Random arrangement of hydrogen atoms and (**b**) under a strong magnetic field

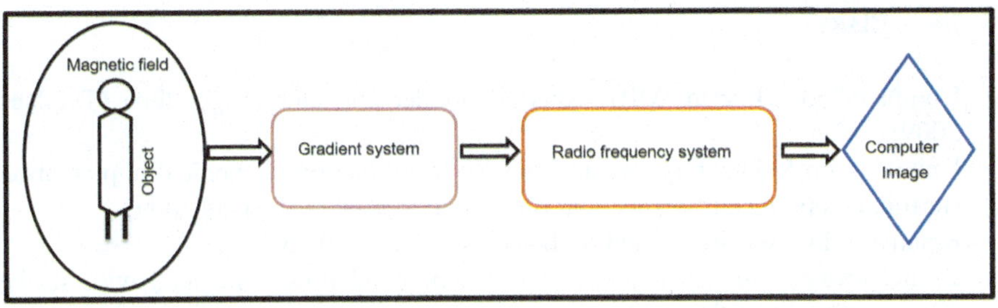

Fig. 5.3 Illustration of basic components of MRI

processed to form images of the body tissues (Magnetic resonance – basic principles 2018; Tanya Lewis 2017). The contrast between different tissues can be determined by the rate at which excited atoms return to the equilibrium state. Some exogenous contrast agents are used in order to get clear images (Koivula 2016).

MRI hardware consists of a magnet, gradient system, radio frequency system, and a computer to process the data (Fig. 5.3).

5.4.2 Applications

MRI has a wide range of applications, some of which include:

- To diagnose ailments related to brain and spinal cord (Kasban et al. 2015).
- Problems related to the musculoskeletal system.
- Certain gastrointestinal tract conditions.
- Nose, ear, and throat conditions.
- Abnormalities in the vasculature.

- Sports injuries.
- Prostate problems.
- Female pelvic problems (Kasban et al. 2015).

5.4.3 Benefits

- Better contrast resolution.
- Capable of giving images in multiple planes (coronal, sagittal, oblique, or axial) without repositioning the patient (Srinu 2015).
- Unlike positron emission tomography (PET) and CT scans, MRI does not use ionizing radiation.
- Used in pregnancy to image the fetus; no adverse effects observed.

5.4.4 Risks

- Compared to CT scan, MRI scans are louder and take longer time (Thedens 2009).
- Unable to do MRI safely on patients who have undergone medical implantation (ferromagnetic metals, stents, and orthopedic screws) (Thedens 2009).
- Relatively low sensitivity and high cost (Kasban et al. 2015).
- Brain perfusion to detect areas of reduced blood flow associated with stroke (Constantinesco et al. 2005).
- Hypoxia imaging (Krohn et al. 2008).
- Imaging of stem cells (Chin et al. 2003).
- To study the basic mechanism of neurodegeneration in Parkinson's disease and Alzheimer's disease (Lorberboym et al. 2004).
- Useful in preclinical neuroimaging (Sharma and Ebadi 2008).
- Useful in the drug discovery process (Vanderheyden 2009).
- To determine the accurate method of image reconstruction (Khalil et al. 2011).
- Estimation of renal function and thyroid function (Khalil et al. 2011).
- Useful in bone imaging (Khalil et al. 2011).

5.5 Positron Emission Tomography (PET)

Positron emission tomography is a non-invasive imaging technique. It is considered as a molecular or functional imaging system due to its application in the study of the biologic function of healthy and disease condition, in contrast to MRI and CT. It involves three-dimensional imaging of the positrons emitted from the radiopharmaceutical (radiotracer) administered. Radiotracer used in PET imaging is labeled with positron emitting isotope such as ^{11}C, ^{18}F, ^{15}O, ^{68}G, ^{82}Rb, etc. (Basu et al. 2014).

The radiotracer used in PET does not have any harmful effect as it is not associated with any significant change in the physiologic or pharmacologic activity of the body. Great advancement has been achieved in this field, which started with its use in measuring the cerebral blood flow, followed by the introduction of ^{18}F-fluorodeoxyglucose (FDG) in the body for understanding glucose metabolism, approved by USFDA. This was followed by the development of ^{18}F-amyvid, a USFDA approved PET-Tracer which allows detection of β-amyloid plaques in Alzheimer's disease (Lameka et al. 2016).

5.5.1 Working Principle

PET is a molecular imaging technique which detects the distribution of radiotracer. When a small amount of radiotracer is administered to the patient, it rapidly distributes within the organs of the body (Lameka et al. 2016). The proton of the radionuclide nucleus is converted to neutron by the emission of a positron and a neutrino. The neutrino is electrically neutral having negligible mass and it escapes without interacting in the surrounding material while the positron is highly interactive. The emitted positron combines with an electron to form the unstable positronium. The positronium then produces two gamma-ray photons which are emitted in the opposite direction approximately 180°. The PET detector detects these emitted photons. Positron emission is also known as β^+-decay. It is an isobaric decay as it involves a change in atomic mass but not in atomic number (Fig. 5.4; Basu et al. 2014).

5.5.2 Benefits

- Capable of diagnosis of diseases by detecting changes in biochemical functions before anatomy gets affected.
- Capable of studying metabolic functions. Can provide an alternative to biopsy.
- Can distinguish between malignant and benign tumors.

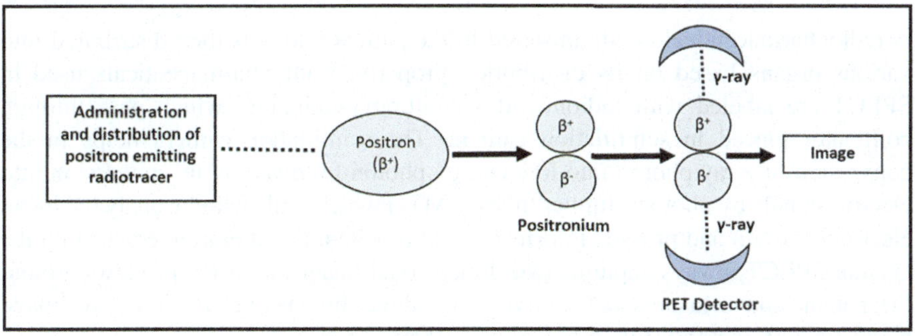

Fig. 5.4 Schematic working diagram of positron emission tomography (PET)

5.5.3 Risks

- Movement of the patient in PET imaging creates confusion in determining the actual position of the origin of the photon.
- Restriction in the movement of patients during the entire study.
- Attenuation correction is required.
- Localization of tracers in the small region is not possible.
- Not suitable for pregnant or breastfeeding women.
- Requires cyclotrons to generate radiotracers.
- Is an expensive technique.

5.5.4 Applications

- Diagnosis, staging, and follow-ups of non-small cell lung cancer, lymphoma, colorectal, oesophageal cancer (Krohn et al. 2008).
- Characterization of pulmonary nodules (Lameka et al. 2016).
- ^{18}F-fluorodeoxyglucose measures glucose metabolism in the brain (Kapoor et al. 2004).
- ^{18}F-amyvid detects β-amyloid plaques (Khalil et al. 2011).

5.6 Single Photon Emission Computed Tomography (SPECT)

Single photon emission computed tomography is a molecular imaging technique based on tomographic reconstruction method and nuclear imaging method. It detects γ-rays at different angles with the use of two or three gamma cameras rotating around the patient. It has now become a routine procedure in the Department of Nuclear Medicine. SPECT provides functional information about a patient. Result of SPECT is based on the concentration of radiotracers (Council 1996).

5.6.1 Working Principle

A radiopharmaceutical is administered to the patient which is then distributed into various organs based on its distribution property. Radiopharmaceuticals used in SPECT are labeled with radionuclides emitting γ-radiation, which pass through collimator placed in scintillation camera. The scintillation camera helps in the conversion of γ-ray photon into low-energy photon followed by its conversion into electric signals by photomultiplier tubes (PMTs) which will again be analyzed by an electronic circuit and provide information about where the photon interacts with the crystal. SPECT always captures two-dimensional images at different views around the patient and measures radioactivity in a three-dimensional form via an image reconstruction method (Fig. 5.5; Council 1996).

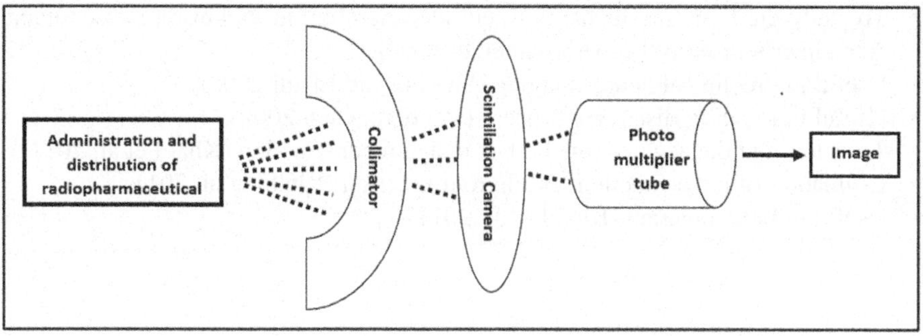

Fig. 5.5 Concept of single-photon emission computed tomography (SPECT)

5.6.2 Benefits

- Low amount of photons are required.
- Increased detection efficiency.
- Provides physiological information (metabolic activity, blood flow) through functional imaging.
- Intrinsic lesion localization.
- 3-D imaging.

5.6.3 Risks

- γ-Radiations are very harmful due to high-ionization potential.
- Long scan time (30–40 mins).
- Radiopharmaceuticals can induce allergic reactions.
- Severe shortage in supply of radiopharmaceuticals.

5.6.4 Applications

- Useful in anatomical localization of the disease.
- Myocardial perfusion imaging to detect coronary artery disease or myocardial infarction (Tsui and Kraitchman 2009).
- Detection and differentiation of tumor cells (Khalil et al. 2011).
- Understanding glucose metabolism (Khalil et al. 2011).
- Brain perfusion to detect areas of reduced blood flow associated with stroke (Constantinesco et al. 2005).
- Hypoxia imaging (Krohn et al. 2008).
- In imaging of stem cells (Chin et al. 2003).

- To study the basic mechanism of neurodegeneration in Parkinson's disease and Alzheimer's disease (Lorberboym et al. 2004).
- Useful in preclinical neuro-imaging (Sharma and Ebadi 2008).
- Useful in the drug discovery process (Vanderheyden 2009).
- To determine the accurate method of image reconstruction (Khalil et al. 2011).
- Estimation of renal function and thyroid function (Khalil et al. 2011).
- Useful in bone imaging (Khalil et al. 2011).

5.7 Ultrasound Imaging

Ultrasound is an important imaging modality taking tomographic images of the body. The sound frequency above the audible range is called the ultrasound, ranging from 20 Hz to 20 kHz. Sound waves in contrast to electromagnetic waves cannot travel in a vacuum and need a medium to propagate. For ultrasound imaging, a transducer generates a sound that acts like a loudspeaker which along a narrow beam sends out an acoustic pulse in a particular direction. Following which the transducer begins to act like a microphone for recording those acoustic echoes generated by the tissue. The emission and recording of the acoustic echoes take place in the same transducer. This is different from other techniques like CT, where the X-ray tube and the detector are placed on the opposite side of the patient (Wilhjelm et al. 2013).

5.7.1 Working Principle

An ultrasound machine is a computer-integrated diagnostic tool that uses ultra-high-frequency sound waves in the megahertz range (millions of cycles per second) to produce cross-sectional images of the internal body organs. These sound waves are generated, transmitted, and received by an ultrasound transducer (a hand-held probe) that is directly placed on and moved over the surface of the body to sense sound waves. The transducer contains several piezoelectric crystals that are excited by an electric current. An electric current applied to a piezoelectric crystal causes it to expand and contract to create a pressure wave. These pressure waves propagate through the tissue in the body. The echo reflected from the tissue returns to the transducer crystals and is processed to create an ultrasound image. The piezoelectric crystal is most frequently a combination of $PbZrO_3$ and $PbTiO_3$ molecules. The transformation of mechanical energy into electric energy is referred to as a piezo-electric effect. Doppler imaging is used to assess blood flow or tissue vibration in the body, such as in carotid artery (Kasban et al. 2015).

5.7.2 Modes of Ultrasound

Four different modes of ultrasound are used in medical imaging:

1. **A-mode (amplitude mode):** A-mode is the original display mode of ultrasound based on the pulse-echo principal used to measure the depth of organ. It only gives one-dimensional information.
2. **B-mode (2-D):** B stands for brightness current display mode of choice where you scan a plane through the body that can be viewed as a 2-D image on a screen in a gray scale of brightness.
3. **M-mode:** M stands for motion. M-mode measures the motion of a particular structure over time. A gait is placed on the ultrasound image, and then M mode demonstrates any motion along with the structure of that gait. M mode is often used for cardiac measurements and to determine fetal heart rate and valve closure rates and assess function of valves opening and closing and wall thickness.
4. **Doppler mode:** This mode uses Doppler effect which gives information about the direction and velocity of flow through a structure, like measuring and visualizing blood flow in blood vessels (Jauhiainen 2009).

Ultrasound pulse velocity, V: Determined by the material.

$$V = \frac{\sqrt{\text{Stiffness}}}{\text{Density}}$$

1. Soft tissue = average 1540 m/s.
2. Air = 330 m/s.
3. Fat = 1460 m/s.
4. Bone = 4080 m/s.

5.7.3 Benefits

- Non-invasive and painless; it does not involve the use of ionizing radiation.
- Process high-resolution real-time information.
- The real-time moving image can be used to support drainage and biopsy procedures.
- Derive images of soft tissue that even does not come in the X-ray.

5.7.4 Risks

- The modality highly depends on the operator's skills, unlike CT and MRI, which produce cross-sectional images in a reasonably programmed fashion.
- It cannot produce an image of the intestinal tract and normal lung tissue (apart from demarcating pleural effusions) due to the interference of air or gas and neither of bone because they are too dense to penetrate.
- A global view of vein and arteries is not possible.

5.7.5 Medical Applications

- **Cardiology:** To diagnose deep vein thrombosis (thrombosonography). Echocardiography is a fundamental tool to diagnose the function of heart valves and ventricles.
- **Endocrinology:** To visualize thyroid nodules or cysts, abdominal ultrasound helps in identifying peritoneal effusion and in visualizing the solid organs of the gastrointestinal tract such as the pancreas, liver, gall bladder, bile ducts, kidneys, spleen, etc.
- **Neurology:** Assessing blood flow through arteries using transcranial Doppler (TCD), detecting vasospasm after ischemic stroke, and in screening of peripheral nerve tumor.
- **Obstetrics:** Obstetrical ultrasound is commonly used during pregnancy for monitoring fetus development, diagnosing birth defect, localization of placenta, and also to diagnose conditions like breech birth or placenta previa.
- **Urology:** For determining the amount of residual volume in bladder, size, and shape of the kidney. Transrectal ultrasound is performed to examine prostate pathologies like prostatic carcinoma, prostatitis, and Mullerian duct cyst (Carovac et al. 2011).

5.8 Conclusion

A more comprehensive and targeted view of internal organs can be achieved by medical imaging modalities. This chapter discusses some of the common medical imaging modalities which help in the precise diagnosis of a condition within few minutes. The need of the hour is to identify and discover new techniques for imaging and diagnosis which can overcome the limitations of the available imaging modalities (Table 5.1).

Table 5.1 Comparison of currently available medical imaging modalities

Modalities	X-ray film	CT	MRI	PET	SPECT	Ultrasound	Reference
Types of radiation	Ionizing radiation	Ionizing radiation	Non-ionizing radiation	Ionizing radiation	Ionizing radiation	Non-ionizing radiation	Iniewski (2009)
Probe	High energy photons (X-ray)	High energy photons (X-ray)	Low energy photons (r.f.)	High energy photons (gamma ray)	High energy photons (gamma ray)	High-frequency sound waves	lalitesh (2012)
Source	X-ray tube and generator	X-ray tube and generator	Magnet, r.f. transmitter, and antenna coil	Radioisotope introduced into the body	Radioisotope introduced into the body	Transmitter, transducer	Wolbarst (1999)
Spatial resolution (mm)	1 mm	0.5 mm	0.5 mm	1–2 mm	0.5–1 mm	1 mm	Kasban et al. (2015), Patching (2015)
Detected/ imaged by	Screen and film	Array of X-ray detector	Antenna coil, r.f. receiver	Gamma camera	Gamma camera	Transducer, receiver	Wolbarst (1999)
Detection	Soft tissue and fluid	Soft tissue and bone	Soft tissue and bone	Soft tissue and bone	Soft tissue and bone	Soft tissue and some bone	Wolbarst (1999)
Dimension	2-D	2-D/3-D	2-D/3-D	2-D	2-D	2-D/3-D	Iniewski (2009)
Soft tissue contrast	Low	High (compared to MRI low)	High	Good to high	Good to high	High	lalitesh (2012)
Shows anatomy/ physiology	Anatomy	Anatomy	Anatomy	Physiology	Physiology	Both	lalitesh (2012)
Real-time information	Yes	No	No	No	No	Yes	Iniewski (2009)
Cost	Medium	High	High	High/medium	High/medium	Low	Iniewski (2009)

References

Assmus A (1995) Early history of X rays. Beam Line 25(2):10–24
Basu S et al (2014) The basic principles of FDG-PET/CT imaging. PET Clinics 9(4):355–370
Berger D (1999) A brief history of medical diagnosis and the birth of the clinical laboratory. Part 1—Ancient times through the 19th century. MLO Med Lab Obs 31(7):28–30
Carovac A, Smajlovic F, Junuzovic D (2011) Application of ultrasound in medicine. Acta Informatica Medica 19(3):168
Chen Q et al (2018) All-inorganic perovskite nanocrystal scintillators. Nature 561(7721):88
Chin B et al (2003) 111In oxine labelled mesenchymal stem cell SPECT after intravenous administration in myocardial infarction. Nucl Med Commun 24(11):1149–1154
Claesson T (2001) A medical imaging demonstrator of computed tomography and bone mineral densitometry. Universitetsservice US AB, Stockholm
Constantinesco A et al (2005) Assessment of left ventricular perfusion, volumes, and motion in mice using pinhole gated SPECT. J Nucl Med 46(6):1005
Council NR (1996) Mathematics and physics of emerging biomedical imaging. National Academies Press, Washington, DC
Dougherty G (2009) Digital image processing for medical applications. Cambridge University Press, Cambridge
Hawkes R et al (1980) Nuclear magnetic resonance (NMR) tomography of the brain: a preliminary clinical assessment with demonstration of pathology. J Comput Assist Tomogr 4(5):577–586
Hsieh J (2009) Computed tomography: principles, design, artifacts, and recent advances. SPIE, Bellingham
Iniewski K (2009) Medical imaging: principles, detectors, and electronics. Wiley, Hoboken
Jauhiainen, J., The physical principles of medical imaging. 2009
Jiang M et al (2001) Blind deblurring of spiral CT images. In: Signals, systems and computers, 2001. Conference record of the thirty-fifth Asilomar conference on. IEEE
Kalender WA (1994) Technical foundations of spiral CT. In: Seminars in ultrasound, CT and MRI. Elsevier, New York
Kapoor V, McCook BM, Torok FS (2004) An introduction to PET-CT imaging. Radiographics 24(2):523–543
Kasban H, El-Bendary MAM, Salama DH (2015) A comparative study of medical imaging techniques. Int J Inform Sci Intell Syst 4(2):37–58
Keall PJ et al (2005) Four-dimensional radiotherapy planning for DMLC-based respiratory motion tracking. Med Phys 32(4):942–951
Khalil MM et al (2011) Molecular SPECT imaging: an overview. Int J Mol Imaging 2011:1
Koivula, L., Magnetic resonance imaging-based radiation therapy treatment planning. 2016
Krohn KA, Link JM, Mason RP (2008) Molecular imaging of hypoxia. J Nucl Med 49(Suppl 2):129S–148S
lalitesh (2012) Comparison between different imaging modalities-physics-report, projects for physics. Allahabad University
Lameka K, Farwell MD, Ichise M (2016) Positron emission tomography. Handb Clin Neurol, Elsevier:209–227
Lorberboym M et al (2004) 123I-FP-CIT SPECT imaging of dopamine transporters in patients with cerebrovascular disease and clinical diagnosis of vascular parkinsonism. J Nucl Med 45(10):1688–1693
Magnetic resonance – basic principles. 2018
McRobbie DW et al (2017) MRI from picture to proton. Cambridge University Press, Cambridge
Nordqvist, C., *CT scan or CAT scan: How does it work?* 2018
Patching SG (2015) Roles of facilitative glucose transporter GLUT1 in [18F] FDG positron emission tomography (PET) imaging of human diseases. J Diagn Imaging 2(1):30–102
Sacerdoti FM, Giordano A, Cavaliere C (2016) Advanced imaging techniques in clinical pathology. Springer, New York

Sansare K, Khanna V, Karjodkar F (2011) Early victims of X-rays: a tribute and current perception. Dentomaxillofac Radiol 40(2):123–125
Sharma S, Ebadi M (2008) SPECT neuroimaging in translational research of CNS disorders. Neurochem Int 52(3):352–362
Srinu, D. (2015 March 5). MRI introduction,components.
Tanya Lewis, S.W., What is an MRI (Magnetic Resonance Imaging)? 2017
Thedens, D.R., Basic principles of MRI. 2009
Tsui BM, Kraitchman DL (2009) *Recent advances in small-animal cardiovascular imaging.* Journal of nuclear medicine: official publication. Soc Nucl Med 50(5):667
Vanderheyden J (2009) The use of imaging in preclinical drug development. QJ Nucl Med Mol Imaging 53(4):374
What's the difference between an X-ray, CT scan and MRI? 2015
Wilhjelm JE, et al (2013). Medical diagnostic ultrasound physical principles and imaging. By Andersen Biomedical Engineering, DTU Elektro Technical University of Denmark, Ver, 3(2)
Wolbarst AB (1999) Looking within: how X-ray, CT, MRI, ultrasound, and other medical images are created, and how they help physicians save lives. University of California Press

Part III

Biosensors and Transducers

Biosensors and its Transducers

6

Janani Viswananthan and Gopu Govindasamy

Abstract

The practice of medicine has changed dramatically over the past decades with the drastic development in the field of biosensors. It had a significant impact on all aspects of the healthcare sector. The concern about the detection of chemical or biological important compounds and their need is expanded from clinical diagnostics to personal care/point care devices. Assessment of health and functioning is one of the growing apprehensions in the common population. With the expanding of technology every day, it is possible to take healthcare technology to the doorstep of the patient. Consider a few examples that are important in today's lifestyle: a diabetic monitor/glucometer, pregnancy strips, insulin drug administrator, biochips, etc. Other significant developments include the use of protein/other biomolecules as a device component (biorecognition element) rather than constituents of a living cell and the application of production methods such as screen printing that can manufacture a very large number of sensors with a high degree of reproducibility. These studies had further become a prerequisite for design and development in biosensors field to emerge as a commercial product from the laboratory. The application of biosensors now has also broadened with time from clinical diagnostics to environmental monitoring, food industry, agriculture, textile industry, water management and many more.

Keywords

Biochips · Clinical diagnostics · Glucometer · Insulin drug administrator · Point care devices · Pregnancy strips

Janani Viswananthan · G. Govindasamy (✉)
Department of Biomedical Engineering, Sri Ramakrishna Engineering College, Coimbatore, Tamil Nadu, India
e-mail: janani.bme@srec.ac.in; gopu.govindasamy@srec.ac.in

S. Paul (ed.), *Biomedical Engineering and its Applications in Healthcare*,
https://doi.org/10.1007/978-981-13-3705-5_6

6.1 Biosensors

According to IUPAC, "biosensor is a self-contained integrated device, which is capable of providing specific quantitative or semi-quantitative analytical information using a biological recognition element which is in direct spatial contact with a transducer element" (Thévenot et al. 2001).

Biosensors are the analytical device used for the quantitative and qualitative analysis of an analyte compound (e.g. urea, creatinine, glucose) which combines with a biorecognition element/biologically derived material (e.g. protein, enzymes, nucleic acid, antibodies, etc.) with a transducer (e.g. optical, electrochemical, piezo-electric, etc.).

The construction of biosensor is a significantly important and promising aspect for the accurate, reliable and precise detection in areas of the food industry, clinical diagnostics, drug development, etc. The pictorial representation of biosensors is represented in Fig. 6.1. With the definition, a chemical sensor can sound the same as biosensor, but a chemical sensor is a device that transforms chemical information into an analytically useful signal. Biosensors are chemical sensors which adopt a biochemical mechanism for the recognition of an analyte in the sample (Ju et al. 2011). Bioanalytical methods require added processing steps like reagent addition, etc., whereas biosensors are one step method with no processing steps. With these developments of sensor platforms, biosensors find a variety of applications in different fields like clinical diagnostics, biological assays, process control, food analysis, environmental parameter monitoring, space applications and as biomarkers for determination of certain molecules (Eggins 2008).

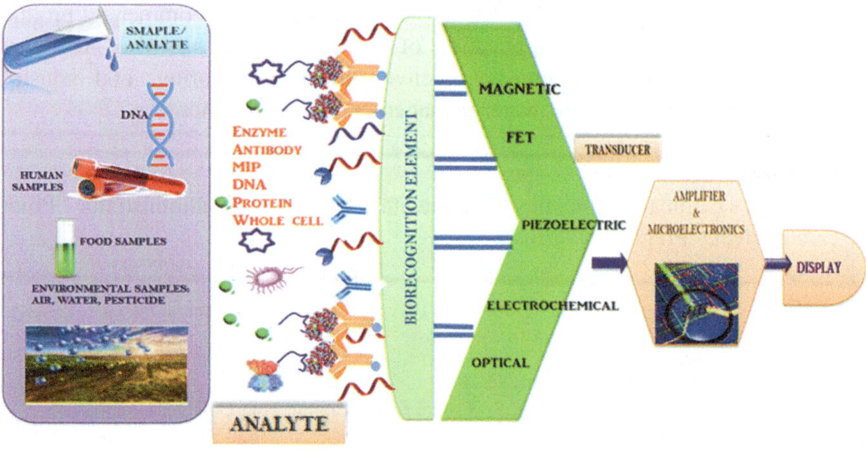

Fig. 6.1 Schematic sketch depicting the construction of a biosensor

6.2 Major Components of Biosensors

As biosensors are defined it is a self-contained device with different components as shown in Fig. 6.1. It consists of three parts: the sensitive biological element, the transducer and the associated electronics or signal processors that are primarily responsible for the display of the results in a user-friendly mode.

The basic structural element of a biosensor is the bioreceptor or biorecognition platform (Sect. 6.2.1), which helps in the selective specific recognition of analyte or analytes of our interest among the other interfering molecules. The interaction between the bioreceptors and the analytes produces some changes which are converted into corresponding electrical or optical signal by using a transducer which is in direct spatial contact with the receptor material. The output of the transducer is further amplified and fed into a microelectronic unit for data processing, and the results are obtained on the display unit. Even the biosensor obeys the pattern of generalised measurement units which consists of a data acquisition unit followed by data processing and data display unit.

6.2.1 Biorecognition Elements

The biorecognition elements are considered to be the heart of the biosensor device as they are used to find the target. Target is the analyte compound, which binds with the receptor likes enzyme, protein, deoxyribonucleic acid (DNA) that are immobilised on the surface of the sensor for specific binding. The performance and efficiency of any biosensor depend mainly on its capability to differentiate the target among other analyte molecules present in the sample. As such, various biorecognition elements including proteins, nucleic acids, cells, bacteriophages and antibodies have been employed by researchers over the years. This section seeks to review these biorecognition elements, target binding and signal generation with various immobilisation techniques.

From Fig. 6.2, it is well described how a biomolecule is immobilised on the substrate for the selective and specific binding of the target. As schematically the

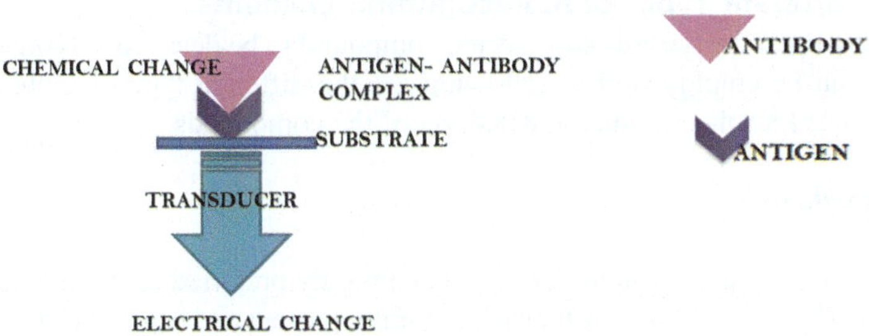

Fig. 6.2 Generalised measurement of antigen-antibody interaction by means of a transducer

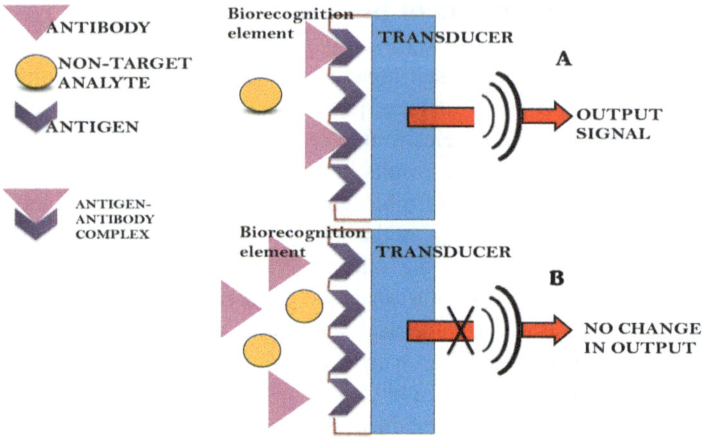

Fig. 6.3 Measurement of analytes with respect to the binding of biomolecules

receptors are made to bind on the substrate and made to interact with the sample. When the target of the analyte binds, there is a chemical reaction induced which in turn is converted into an electrical signal by integrated circuits and processors which are in direct contact with the biorecognition element. Thus, this converted electrical signal is then measured as a quantitative parameter. The intensity of the measured output parameter directly says about the concentration of analyte that is present in the sample.

There is also another probability that nontarget molecule can interfere in the analytical analysis of the sample as well.

In order to differentiate between target and nontarget, few sensors employ a labelled binding approach that only produces the signal on target-specific interaction. To implement such sensitive sensors, a biorecognition element has to be chosen in such a way that it should be highly specific in order to retain sensor performances. Other non-specific and nontarget molecules are not able to bind on the bioreceptor as shown in Fig. 6.3b. There is only output generated due to binding (Fig. 6.3a) and vice versa (Fig. 6.3b).

6.2.1.1 Different Types of Biorecognition Elements

Depending upon the type of biomolecule compound to be detected, a biorecognition element can be employed. The following are the different type of biorecognition elements used for the volumetric detection of the compounds.

(a) ***Antibodies***

Since the past, these antibodies have been highly preferred as the biorecognition elements. These are chosen, as they are naturally present and can be produced by the immune system as a response. They are made of glycoprotein and sensors which employ antibodies as sensing elements and are termed as immuno-biosensors. Antibody-based biosensor platforms with different transduction like electrochemical (Lai et al. 2018), fluorescent-based biosensors (Feng et al. 2012) and colourimetric assays (Patel et al. 2016) are used for a wide range of applications. The antibodies produce changing stringency of interactions depending on whether they are monoclonal or polyclonal. Figure 6.4 is an example of immune chromatographic test with visual detection.

As the figure represents, at the last stage, a colour change is visually noted. This is a lateral flow technique where DNA probes that target species are immobilised on the gold substrate. When a sample containing the target is introduced in the channel, the molecules hybridise and migrate which again hybridise causing a colour change. A classic example following the pattern is pregnancy test strips and HIV test strip.

(b) ***Nucleic Acid Probes***

Nucleic acid probe-based biosensors predominantly use deoxyribonucleic acid (DNA), peptide nucleic acid (PNA), ribonucleic acid (RNA) and aptamers (both DNA and RNA) as oligonucleotide probes. The fundamental principle behind these type of sensors is Chargaff's rules of base pairing (for DNA, A=T, G C) except aptamers sensors. The typical working of DNA sensors is shown in Fig. 6.5. Nucleic acid probe-based detection continues to obtain huge attention from researchers because of their association with various genetic-based diseases and disorders. In past research, various pathogens are also found to produce unique nucleic acid sequences that have been utilised in biosensors for their detection. Nucleic acids including single-strand DNA, peptide nucleic acids, locked nucleic acids, G-quadruplexes and DNA enzymes have been utilised as biorecognition elements to construct DNA biosensors (Sun et al. 2016).

6.2.2 Immobilisation Techniques of Biorecognition Elements

Immobilisation is imprisonment of biomolecules, bioreceptor on the substrate, or the sensor platform to increase the functional efficiency and reproducibility of the process. With the studies that have been undergone, researches have come out with various methods of immobilisation techniques as shown in Fig. 6.6. The primary immobilisation methods are listed below.

Parameters	Adsorption	Covalent bonding	Entrapment	Copolymerisation/ cross-linking	Encapsulation
Definition	In this method, the biorecognition elements like enzymes get adsorbed to the external substrate There is no permanent bond formation. The bond formed in this method is a weak bond	Formation of a chemical bond between the chemical group of molecules and chemical group in the substrate	Compounds are physically entrapped inside the porous material/matrix which a water-soluble polymer. Bonds involved in stabilising can be covalent or non-covalent	Immobilised molecules are directly linked with covalent bonds between each other via polyfunctional compounds like diazonium salts	Enclosing the biomolecules in a membrane capsule made of semipermeable membrane like cellulose
Types	Static process	Diazoation	Inclusion in gels, fibres and microcapsules	–	–
	Dynamic process	Peptide bond			
	Reactor loading	Poly functional reagents			
	Electrode positioning process				
Advantages	Simple and easy	Strong linkage to the substrate	Fastest method of immobilisation	This method is cheap and simple	Inexpensive and large quantity can be immobilised
	No reagents required	No leakage	Pore size can be altered.		
	Less disruption to molecules attached	A variety of support with a different functional group can be made	Fewer chances of conformational changes		
Disadvantages	Efficiency is less	Chemical modification of enzymes leading to loss of functional conformation of molecules	Leakage of molecules	The polyfunctional compounds used will denature and not pure for longer uses	Pore size limitation
	More leakage		Pore diffusion limitation		
	Desorption of molecules from the substrate				

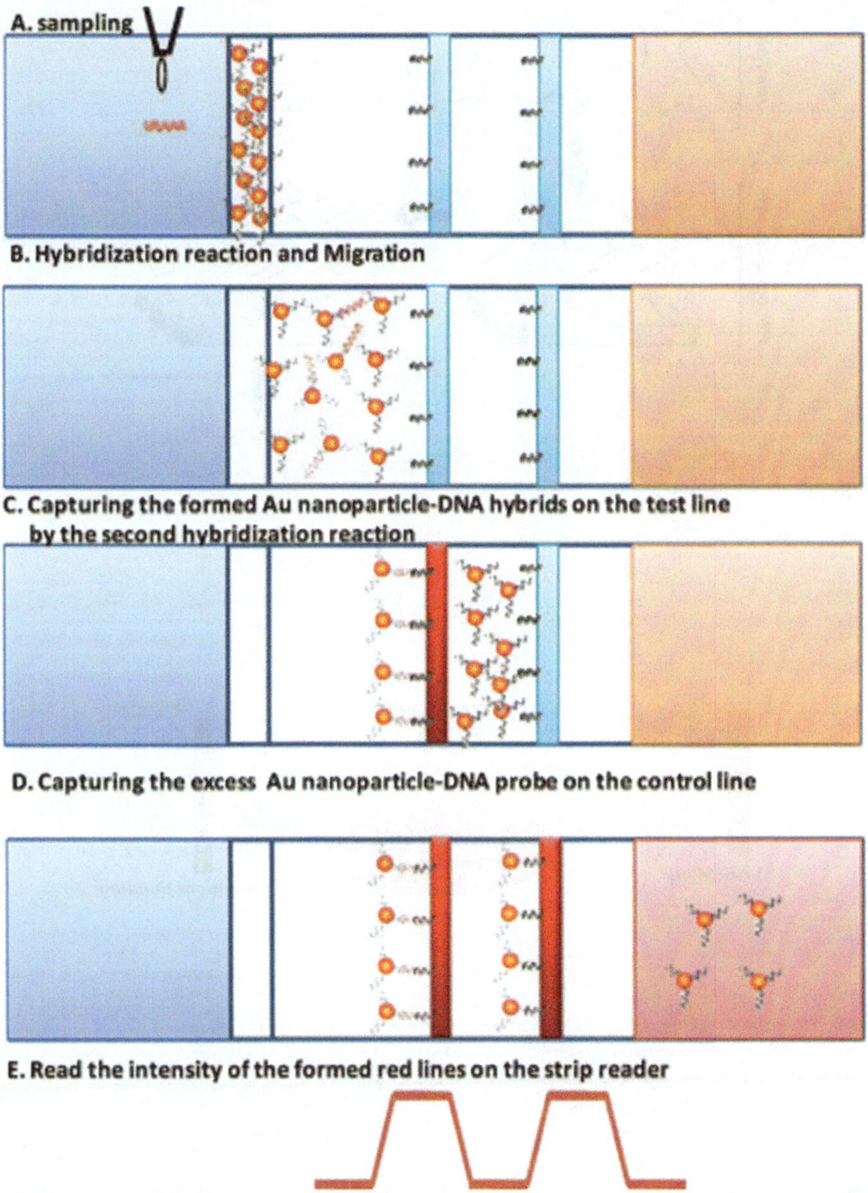

Fig. 6.4 DNA sensor working mechanism with visual detection in lateral flow test strips. (Mao et al. 2009)

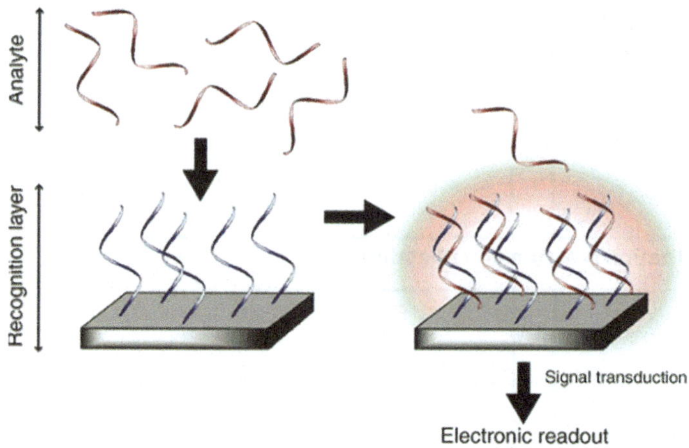

Fig. 6.5 DNA sensor working mechanism

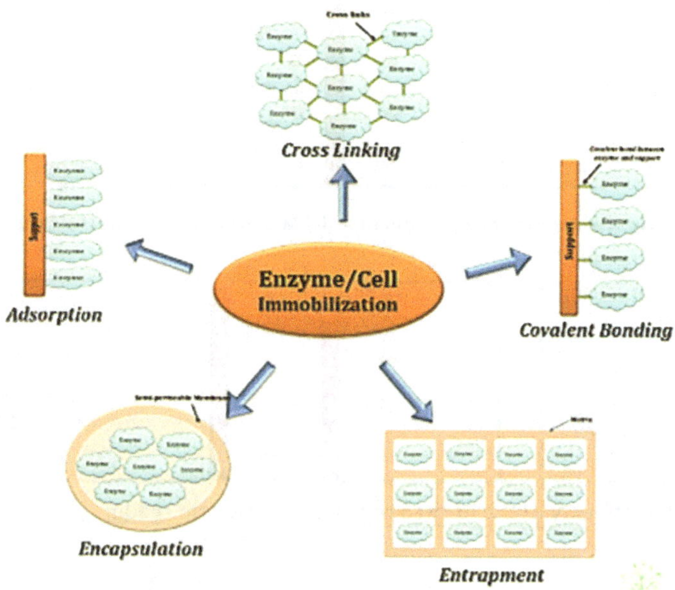

Fig. 6.6 Various immobilisation techniques for biorecognition element. (https://www.easybiologyclass.com/enzyme-cell-immobilization-techniques)

6.3 Classification of the Biosensor

Classification of the biosensor based on the bioreceptor, transducer and its size (Thévenot et al. 2001) is schematically represented in Fig. 6.7.

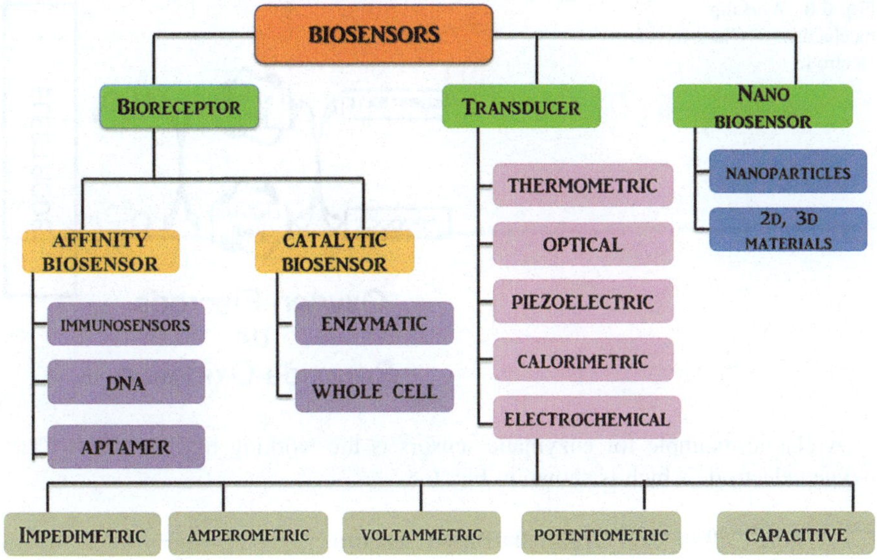

Fig. 6.7 Classification of biosensors

6.3.1 Based on Bioreceptors

Biosensors can also be classified into different types based on the bioreceptor (sensing element). Bioreceptor is basically a biological component, which may be an antibody, enzyme, protein, nucleic acid, or even whole cells that involve a biochemical conversion for the further recognition process. Thus, these bioreceptor forms the primary sensing element in a biosensor. Based on the type of biological entity employed for sensing, biosensors can be categorised as follows:

(a) **Catalytic Biosensors**

These sensors measure the concentrations of species or analyte formed or lost during a biocatalytic reaction. Under this category the most important biosensors are:

(i) **Enzymatic Sensors**

These sensors are based on the catalytic property of enzymes. They selectively and specifically react with the analyte of interest through the biorecognition process. These are the most studied and frequently applied biosensors since the pioneering invention by Clark and Lyon in 1962 for the detection of glucose by immobilising glucose oxidase (GO) on an amperometric oxygen electrode (Wang 2008). Since then this field has undergone drastic developments and has extended to the detection of various other analytes also.

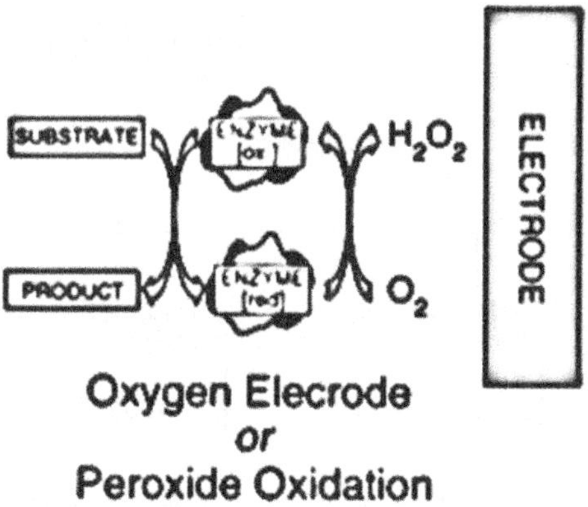

Fig. 6.8 Working mechanism of Clark oxygen electrode

A classic example for enzymatic sensors is the working mechanism of Clark oxygen electrode which is shown in Fig. 6.8.

(ii) **Whole Cell or Cell Organelle-Based Sensors**

In these types of sensors, microorganisms like bacteria, yeast, or eukaryotic cells or cell organelles are used for biorecognition mechanism by immobilising them on the sensor. The transduction of cellular signals can be done by different approaches, namely, measures of cell metabolism, intracellular potential, extracellular potential and impedance (Pancrazio et al. 1999).

(b) **Affinity Biosensors**

In this type of sensors, the analyte molecules bind irreversibly to the receptor molecules that result in a physicochemical change detectable by the transducer. The main categories under affinity sensors are discussed below:

(i) **Immunosensors**

Immunosensors are based on highly specific antigen-antibody interactions. Antibodies are generally immobilised on the sensing platform, though there are variants that incorporate antigen on the sensing platform as explained in Sect. 6.2.1.1. They help in determining the concentration of an analyte in real time without consuming supplementary reagents unlike the traditional immunoassays like enzyme-linked immunosorbent assay (ELISA) and colourimetric pregnancy test strips. Antibodies are currently employed in two different formats: for the detection of a single analyte as one prospective and for simultaneous detection of a large number of different proteins for proteomics applications on the other hand (Gorton 2005). In the application of immunosensors, the major challenges are inherent antibody instability, limited reversibility of binding and antibody manufacturability. One of the most commercially used examples is ELISA.

ELISA

ELISA (enzyme-linked immunosorbent assay) is a popularly used diagnostic technique that is based on antibody-antigen interactions. In an ELISA, primarily an antigen is immobilised on a surface, and then they are made to bind with an antibody that is linked to an enzyme in the sample. Detection is undergone by analytically analysing the conjugated enzyme activity.

This can be done by incubating the enzymes with the substrate to produce a measurable quantity. After binding of the secondary antibody, the substrate for the enzyme is added which gets converted to a coloured product whose intensity is quantified as a measure of the analyte concentration. Generally, the enzyme employed in ELISA is horseradish peroxidase which can convert hydrogen peroxide to water and oxygen radicals.

There are three types of techniques which are used. They are as follows:

(a) **Direct ELISA**

In direct ELISA method, the desired antibody is initially immobilised on the surface, and then it is incubated with an enzyme-labelled antibody to the target compound (or specific antigen to the target antibody) as shown in Fig. 6.9a. After which the analytical tools are used for measuring the concentration of target protein/antibody bound on the substrate.

(b) **Indirect ELISA**

A target protein is immobilised on the substrate and then incubated as similar to direct method with an antibody to the target protein which is a primary antibody. This is followed by the interaction of the secondary antibody as shown in Fig. 6.9b. After these multiple processes, the activity of enzymes is indirectly measured.

A layered method is followed in this technique eliminating the need of labelled antibody, unlike direct method.

(c) **Sandwich ELISA**

In sandwich ELISA method, an antibody to a target protein is immobilised on the surface. Then, it is incubated with the target protein and then sandwiched with the second target protein. This second target is protein-specific labelled antibody-enzyme. This sandwich architecture is shown in Fig. 6.9c.

After washing, the activity of the enzyme is measured. Various clinical diagnoses like HIV test, etc. are made using ELISA. Recent research studies on different enzymes and performance are tested for different diseases.

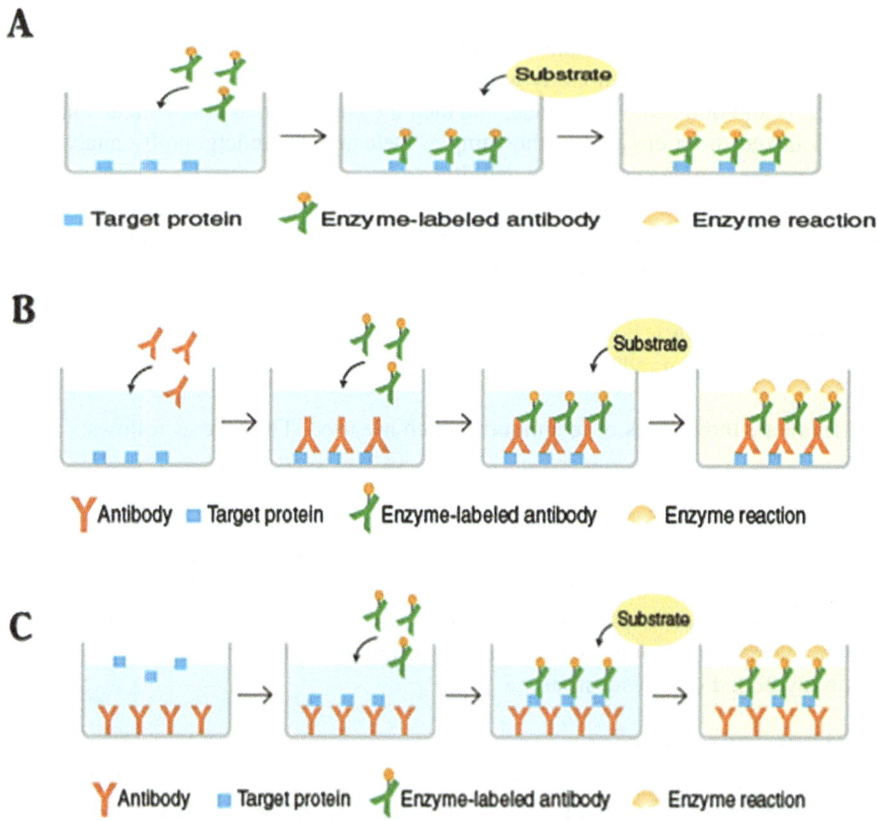

Fig. 6.9 Various ELISA techniques: (**a**) direct method, (**b**) indirect method and (**c**) sandwich method

(ii) **DNA Sensors**

The need and development of DNA sensors are one of the thirst areas of biosensing field, which steps towards the development of highly sensitive and selective DNA sensors for various applications. The important aspects of these types of sensors include DNA interactions, DNA specific binding, drug-related studies and discovery, oxidative DNA damage, analysis of specific target gene for genetic disease studies, etc. These sensors highly depend on DNA hybridisation, which is a natural phenomenon. In these types of sensors, the probes or sequences are generally immobilised on the electrode surface. This immobilised DNA on the substrate serves as a bioreceptor to bind with the target molecule. These on hybridisation with complementary base pair can be measured using optical, piezo-electric, or electrochemical transducers (Drummond et al. 2003).

Example

This section elucidates the working of the DNA hybridisation sensor. In these types of sensors, complementary DNA base pairing is the basis for the biorecognition. Commercially, the short base pairs are used as they are selective towards the target. The DNA fragments have to be immobilised in a way that retains better stability, reactivity, accessibility to the target analyte and optimal orientation as they can undergo conformational changes.

As a result of these binding, a measurable signal is produced. This itinerary is called hybridisation. The steps involved in this type of sensing are shown in Fig. 6.10. Initially, the DNA probes which are immobilised as probe DNA exhibit target-specific interaction with complementary DNA (sample) and produce interactions and bind to the sites as shown.

This is referred to as hybridisation and due to which conformal changes or colour changes are measured as output as shown in Fig. 6.10.

(iii) **Aptamer-Based Sensors**

Aptamers are described as artificial synthesised nucleic acid ligands. The aptamers are those which are chemically correlated to nucleic acid probes and mimic more like antibodies, these show high specificity and affinity towards their targets similar or equivalent antibodies, these show high specificity and affinity towards their targets similar or equivalent antibodies. Thus, aptamers have been selected as potential candidates as a replacement for the conventional antibodies in analytical methods and techniques. They possess several advantages like relatively simple techniques and apparatus required for their isolation, the number of alternative molecules that can be screened per experiment and their chemical simplicity (Baldrich 2010). The work of aptamer sensors is the same as antibody-based sensors.

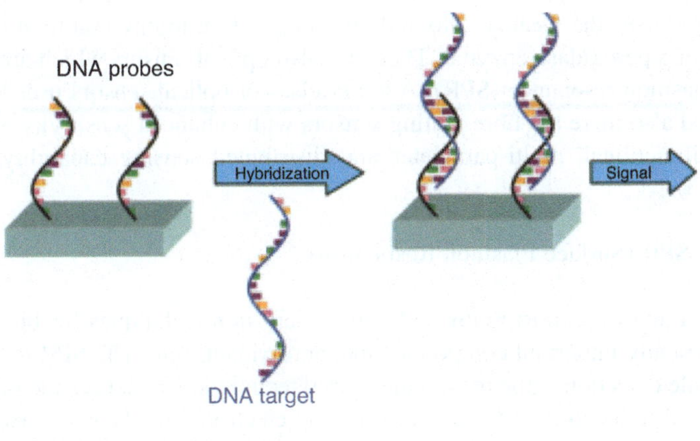

Fig. 6.10 Mechanism of DNA hybridisation

6.3.2 Classification of Biosensors Based on the Transducer

Transducer is another main component in a biosensor, which identifies biorecognition event and converts it into a measurable signal. Based upon transduction element, biosensor can classify into different classes, and the most common classes are electrochemical and optical biosensors. They are briefly explained below.

(a) **Optical Biosensors**

The biosensors which use light for sensing are called as optical biosensors. As they use optical energy conversion systems, they can be used for direct and real-time sensing of biological substances. The signal obtained after transduction is light. In most cases, the transduction process results in the change of parameters such as light polarisation, phase changes, frequency and amplitude of the incident input light. This change is induced due to physical or chemical change produced during the biorecognition process of the analyte. Optical methods are broadly classified as absorbance or reflectivity-based, fluorescence-based, chemiluminescence-based and electron-generated chemiluminescence-based methods. In optical biosensors, the chemical interactions are measured as photonic property like absorption, rarefaction, diffraction, fluorescence and luminescence or refractive index. The molecules are expected to flow along the surface of the sensor. The interaction of light with molecules modulates the light properties which are to be analysed. With these techniques, investigations on antigen-antibody, receptor-ligand, protein-DNA and DNA-related interactions have been experimented. The fluorescence detection in optical method uses labelled fluorescent dye. The output for these is measured based on the intensity of the light emitted by the labels. This colourimetric test is based on the colour produced by the end of reactions. Indicators are used in cases where the reaction products or intermediates are not coloured. These are mainly used in case of colourimetric test strips; they have a disposable cellulose pad impregnated with enzymes and reagents (chromogen). For example, in glucose detection, hydrogen peroxide, produced by the oxidation reaction of glucose in the presence of glucose oxidase, oxidises the weakly coloured chromogen to a highly coloured dye in the presence of a peroxidase enzyme. There are also optical sensors which are based on surface plasmon resonance (SPR). Another class of optical sensors uses fibre optic cables, and also there are fibre grating sensors with enhanced sensitivity, selectivity and multifunctional, multi-parameter and distributed sensing capability (Martins et al. 2013).

Example: SPR (Surface Plasmon Resonance)

SPR is one of the most renowned optical detection techniques for biomolecules and biologically important compounds like antioxidants, etc. The SPR technique of biomolecule detection is the most widely used technique which uses the principle of total internal reflection (TIR) which excites an electron, known as surface plasmon (Nguyen et al. 2015). SPR is an electromagnetic wave which oscillates

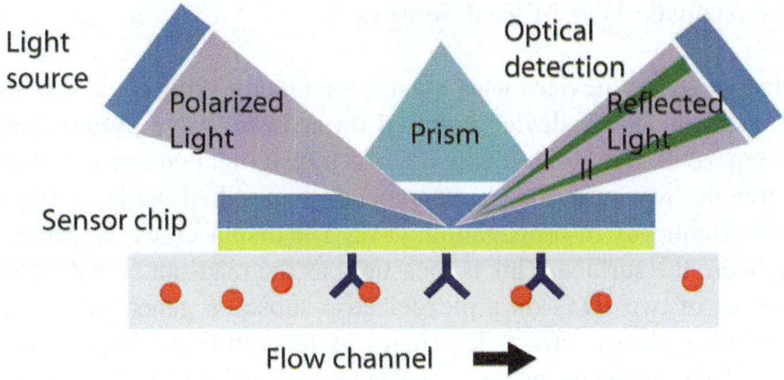

Fig. 6.11 Surface plasmon resonance technique. (Cliquet 2004)

longitudinally on the surface. The free electron on the interaction with dielectric medium and a metal follows this mode of oscillation. Gold and silver are the most used metals to produce these free electrons. The working of SPR follows the standard Kretschmann experimental setup, which consists of a metal film, light source and detector arrays. These surface plasmons are confined to the surface of the metal and result in polarities by interacting with the light source. The resonance energy of surface molecules (plasmons) is excited at the metal/air interface. In optics, the refractive index describes how light propagates through that medium. If there is any change in the refractive index of the surface of a conductive thin film due to the presence of a mass in the substrate, those signals are detected. This principle is exactly employed in a biosensing platform using SPR. This helps to detect the immobilisation of the biomolecules and ligands on any substrates. This change in mass with respect to the immobilisation causes a direct increase in the refractive index of the read-out signal as shown in Fig. 6.11.

(b) **Piezoelectric-Based Biosensor**

These sensors are based on the basic idea that crystal oscillation frequency of a piezoelectric crystal varies with respect to mass. Piezoelectric sensors are based on the coupling of a bio-element with the piezoelectric crystal modified electrode. Piezoelectric crystals can be made to oscillate at a particular resonant frequency by the application of electric signal of that frequency. Thus, the oscillation frequency is dependent on the applied electrical signal frequency. This property is being exploited in the fabrication of piezoelectric sensors. As molecules or analytes bind on the surface, the crystal oscillation frequency also changes and this can be measured electrically. Thus, a change in oscillation frequency can be directly correlated with the additional mass deposited on the surface. There are several piezoelectric materials available like quartz, tourmaline, lithium tantalate, oriented aluminium nitride, etc., but quartz is commonly used for analytical applications because of the properties of quartz (Kell 1988). One of the most common examples of this type of biosensors is surface acoustic wave (SAW)-based detection of biomolecules.

(i) Surface Acoustic Wave-Based Sensors

This kind of sensor devices uses wave propagation property as the measurand. The working of the SAW device is based on acoustic wave propagation near the surface of a piezoelectric solid. It consists of a transmitter and receiver that transmits and measures the wave that has been trapped or modified while propagating with respect to the frequency of propagating wave. The displacement is shifted exponentially away from the surface. This is measured as the read-out signal. A basic SAW device consists of two IDTs on a piezoelectric substrate generally like quartz that produces the piezoelectric effect. This helps in generating the acoustic waves.

These surface acoustic waves are generated electrically by means of an interdigital transducer (http://wirelesslifesciences.org/2015/03/oj-bio-at-wlsa-convergence-summit-2015/). The transmitter and receiver usually consist of sets of IDT. A radiofrequency applied to propagate from the transmitter produces mechanical vibrations in the crystal. The produced waves are Raleigh-type surface acoustic wave (SAW) which is received by the second set of electrodes. This collected wavelength is further used for measurement purposes as shown in Fig. 6.12. The surface wave penetrates into the crystal depending upon its energy, to a depth of about one wavelength (rather like evanescent optical waves), and reflects back to the receiver, and the received quantity is converted into electrical signal. Thus, a species immobilised on the surface will affect the transmission of the wave. This difference in wave intensity before and after immobilisation of biomolecules is measured. This measured quantity will indirectly say about the amount of biomolecules that are immobilised on the surface of the crystal. Thus, these methods are effectively used in biosensing platforms.

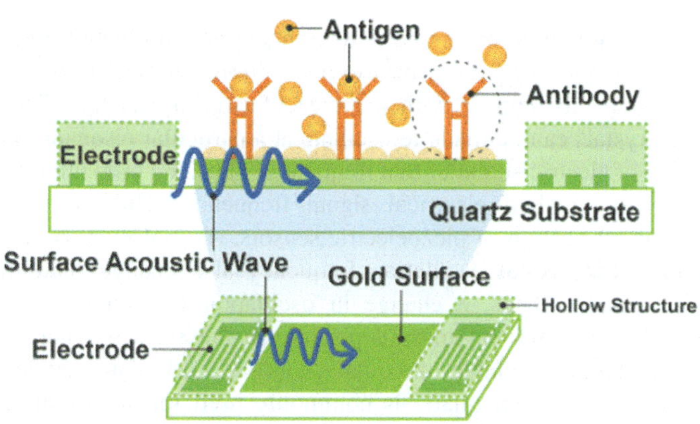

Fig. 6.12 Mechanism of SAW-based biosensing. (http://wirelesslifesciences.org/2015/03/oj-bio-at-wlsa-convergence-summit-2015/)

(c) **Calorimetric Sensors**

In these types of sensors, whenever an analyte or molecule comes in contact or binds with the biorecognition element, the reaction heat then generated is measured via thermistors. The generated or absorbed heat will be proportional to the concentration of analyte present in the sample. Frequent recalibrations are not required in the case of thermal biosensors. Thermal biosensors find applications in a variety of fields like food, cosmetics and pharmaceuticals.

(d) **Electrochemical Biosensors**

Similar to all biosensors, there will be a biological recognition element which classifies it under either affinity type or catalytic biosensors. But the transduction mechanism employed in these sensors would be electrochemical. The biorecognition element interacts with the specific biomolecule, and this specific chemical interaction is transduced by the transducer to an electrical quantity, generally current, voltage or impedance. Accordingly, there are different types of electrochemical biosensors based on the measured electrical quantity.

(i) **Amperometric/Voltammetric Sensors**

Amperometric/voltammetric techniques are based on the measurement of current/volume, respectively, produced by the oxidation or reduction of analyte or molecules by applying potential on the working electrode. This helps in quantitative/volumetric concentration of the analyte present in the sample. Amperometric sensors are the most commonly used class of biosensors. In this technique, a sweep of potential is applied across the working and reference electrode and the corresponding current response is measured. This current is then directly correlated with the concentration of the analyte. Amperometric sensors are more suitable for mass production than the other classes of sensors (Brett and Oliveira Brett 1993).

In voltammetric techniques, the potential is generally varied over a potential window and the current response is plotted as a function of applied potential. The main advantage of these techniques is that the oxidation/reduction response of more than one species can be studied at a time. Also, the current obtained is proportional to the concentration of the target analyte present in the solution. The main difference between amperometric and voltammetric techniques is that in amperometry constant potential is being applied and in voltammetry, the potential is scanned over the desired range.

(ii) **Impedimetric Sensors**

These sensors are based on the conductivity or resistivity change in the solution due to either consumption or evolution of ions due to certain biochemical activities or processes. Thus, the measured quantity using this transduction method is the conductance/resistance of the solution. Generally, these types of sensors support other electrochemical techniques.

(iii) **Potentiometric Sensors**

This transducer measures the potential of an electrode at virtually zero current flow. The potentiometric technique involves the measurement of potential difference between the working electrode and a reference electrode (Kissinger and Heineman 1983). Potentiometric sensors are generally based on ion-selective electrodes which can be easily mass-fabricated using advance silicon or thick film technologies in miniaturised formats. Due to the employment of species-selective working electrode used in the system, potentiometric biosensors have high sensitivity and selectivity. Examples of such sensors are pH sensor, gas sensors, urea sensor, etc. Electrochemical techniques are one of the most researched topics in the biosensing field. Most commercial biosensing mechanism involves either optical or electrochemical techniques.

Field-Effect Transistor (FET)-Based Sensor
Field-effect transistors (FETs) are biosensor devices based on electrochemical sensing mechanism, in which a semiconductor amplifier is used as a miniature transducer for the detection and measurement of potentiometric signals, produced by a potentiometric sensor process on the gate junction of the FET. The FET device is an integrated circuit system most importantly used for processing of the analytical data from three electrodes as shown in Fig. 6.13. The working electrode pairs a transistor with a biorecognition element that sensitively and specifically detects the target biomolecules such as nucleic acids and proteins. A FET system consists of a semiconductor field-effect transistor that acts as a transducer which is partitioned by an insulator layer (e.g. SiO_2) from the biological recognition element (e.g. receptors or probe molecules) which is selective to the desired analyte. Once when the analyte gets binds to the recognition element, the surface potential changes. These changes in the potential of the semiconductor layer change gate voltage in the transistor,

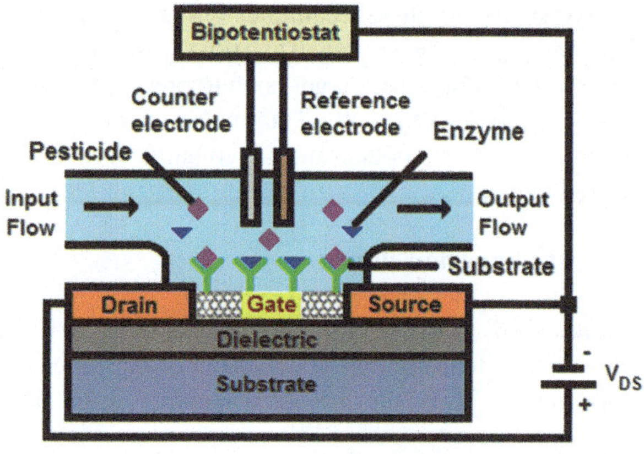

Fig. 6.13 Construction of FET-based biosensor. (Cliquet 2004)

i.e. the amount of current between the source and drain electrodes changes (Kaisti 2017). This change in current (or conductance) can be measured as equivalent to the concentration of analytes. The relationship between the current and analyte concentration depends upon the region of transistor operation. Similar to FET, ISFET (ion-selective field effect transistor), CHEMFET (chemical field effect transistors), etc. will work.

(iv) **Capacitive Sensors**

In these types of biosensors, the change in dielectric properties issued for analysis. The variation can be due to changes in dielectric constant or due to dielectric losses. By analysing these changes, the binding of an analyte to an immobilised element can be detected and directly measured. This binding causes changes in capacitance and it is measured. Capacitive biosensors can be used for the detection of proteins, DNA, antibodies, antigens and heavy metal ions (Hong et al. 2004).

6.3.3 Based on Nanomaterial

Nanoscience and technology are one of the growing fields of biosensing application, and it is also used in a variety of fields. The improvement of the nanotechnology field has given different dimensions for sensor technology. The past two decades come with different 2D and 3D materials like carbon-based carbon nanotubes, quantum dots and graphene and its composites.

6.4 Major Requisites of a Biosensor

Ideal biosensors must possess the following characteristics behaviours as follows:

- **Sensitivity**

The sensitivity of the biosensor is defined as a change in the magnitude of the sensor output signal with respect to change in concentration of an analyte compound of interest (i.e. this is the ability to measure only the target compound in the sample).

$$\text{sensitivity} = \frac{\Delta s\ (\text{magnitude ofoutput})}{\Delta c\ (\text{concentration of the analyte})}.$$

Usually the target analyte is not directly measured by an output signal; rather the changes in the concentration of the co-reactant in sample generating biological/chemical reaction are measured as shown in Fig. 6.14.

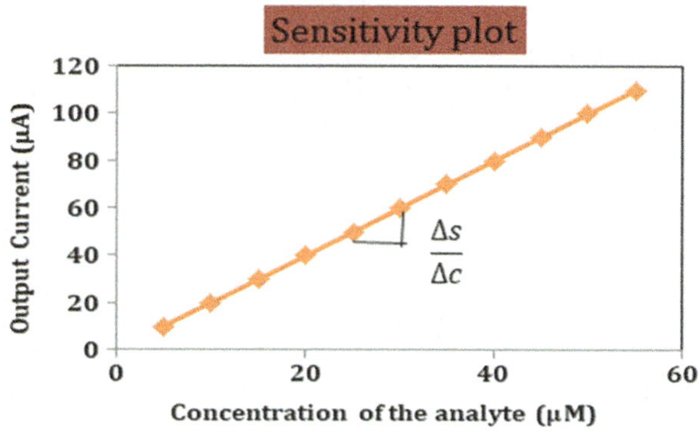

Fig. 6.14 Calibration plot

- **Linearity**

Linearity refers to the array of sample concentrations where the sensor response (output) changes linearly with concentration of the desired compound in the sample. Linearity of the sensor must be high and it should exhibit a dynamic linear range. This promotes wide-operating ranges of the sensors.

- **Selectivity**

This is the ability to selectively measure only the desired biological compound in the sample with minimum interference from many other species present in the sample. This selectivity is the most important parameter to be considered during designing of the biosensor. The higher degree of selectivity depends upon the biorecognition element that selectively binds with the affinity compound, e.g. antibody-antigen interaction.

- **Specificity**

The other most important parameter to be considered during the design of a good biosensor is its selectivity. It refers to the sensor which detects the desired analyte by only reacting/catalysing with specific molecules and remains non-reactive to other compounds present. Biosensors offer the inherent advantage of a high degree of specificity.

- **Response Time**

Another important feature of perturbing is the response time. It can be referred to as the time taken for the analyte to be detected, i.e. required for deriving the result from the detection assay. The transfer kinetics of the reaction is the important factor for the response time of a sensor.

- **Repeatability**

The other important characteristic of the sensor is repeatability. The sensor performance should be able to repeat for *n* number of times producing the output. The repeatability is highly based on reversible reaction. But antigen-antibody interactions are irreversible, and this fact restricts in the development of immuno-biosensors.

- **Reproducibility**

Reproducibility is the ability of the biosensor to produce distinguishable output responses for a repeated experimental procedure. It is characterised by the precision and accuracy of the biosensor system.

- **Limit of Detection (LoD)**

It can be defined as the lowest concentration of the analyte which can be detected by the developed sensor.

- **Limit of Quantification (LoQ)**

It is expressed as the lowest concentration difference which can be distinguished when the composition of the analyte varied continuously.

- **Interference**

The ability of the biosensor to clearly distinguish between the other interfering molecules which can possess the same behaviour that of the desired molecule is one of the key factors to be considered during sensing of biological samples. This is the important parameter to be considered during the development of a biosensor.

- **Stability**

It is the ability of the sensor to maintain its performance for a certain period of time.

- **Life Span**

This is an important parameter that determines the system's life cycle. Life span is the time of stay of the sensor to which it does not undergo any deteriorating results from original sensing performance. The life span of a biosensor is dependent on all internal systems.

6.5 Application of Biosensor

The classic example of biosensors traditionally the generation speaks about is glucose sensor evolution. The evolution of glucose electrodes began with the history of the evolution of Clark electrodes. Glucose sensor uses three electrode electrochemical systems. These are amperometric sensors. Glucose sensors are now a billion dollar market business that made more researchers seek attention towards glucose detection (Wang 2008).

6.5.1 Generations of Glucose Sensor

There are three generations of glucose-based biosensor and each generation has its own significance.

- **First generation**: It involves normal product formation from a chemical reaction that is converted into an equivalent electrical response.
- **Second generation**: It employs specific mediators for the oxidation of glucose with an improved response.
- **Third generation**: This involves direct and feasible detection of glucose.

(a) **First-Generation Glucose Sensors**

The first-generation glucose enzyme electrode depends primarily on the reaction layer/active layer that is involved in a chemical reaction. A thin layer of GO (glucose oxidase) is entrapped over an oxygen electrode through a semipermeable membrane. Detection of glucose is done by analysing the oxygen consumed by the enzyme-catalysed reaction. GO requires a redox cofactor – flavin adenine dinucleotide (FAD) – which serves as an electron acceptor initially and gets reduced into $FADH_2$. The cofactor is regenerated by reacting with oxygen leading to the formation of hydrogen peroxide.

The schematic depiction of first-generation glucose sensor working is shown in Fig. 6.15. Hydrogen peroxide is oxidised, and a number of electrons transferred are directly proportional to the number of glucose molecules present in the sample of

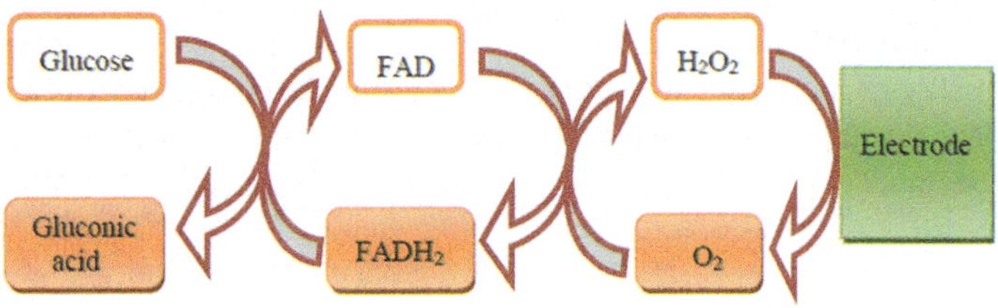

Fig. 6.15 Mechanism of the first-generation glucose sensor

blood. Usually, platinum electrodes are used. Technically, three strategies of electrochemical sensing of glucose can be done:

- By measuring hydrogen peroxide produced at the end of enzyme reaction
- By measuring oxygen consumption at the end of catalyst reaction
- By using a diffusible or immobilised mediator to transfer electrons from Gox to the electrode

Drawbacks

- High operating potential (0.7 V) to drive hydrogen peroxide is required.
- Deactivation of enzymes due to H_2O_2 generation.
- Oxygen dependence due to restricted solubility of oxygen occurred during the reaction.

(b) **Second-Generation Sensors**

The disadvantage of first-generation glucose sensors are high-applied potential that causes other interfering species to be oxidised, and it highly depends on oxygen consumption. To overcome these drawbacks, mediators were used in the second generation of glucose sensor. Figure 6.16 shows a schematic representation of a second-generation glucose sensor. Various redox mediators such as ferrocene, quinines, etc. were used to improve sensor performance. This usage of these mediators will eliminate the need of oxygen for electron transfer at the surface of the electrode, and lower redox potential (0.2 V) is enough for electrochemical sensing in this method.

Drawbacks

- The lower redox potential evoked the other electroactive species to interfere in the chemical reaction.
- High completion of electron transfer between mediator and oxygen had occurred.

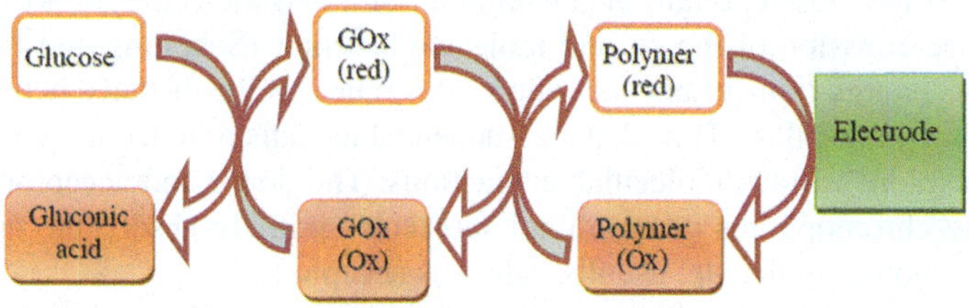

Fig. 6.16 Second-generation glucose sensor

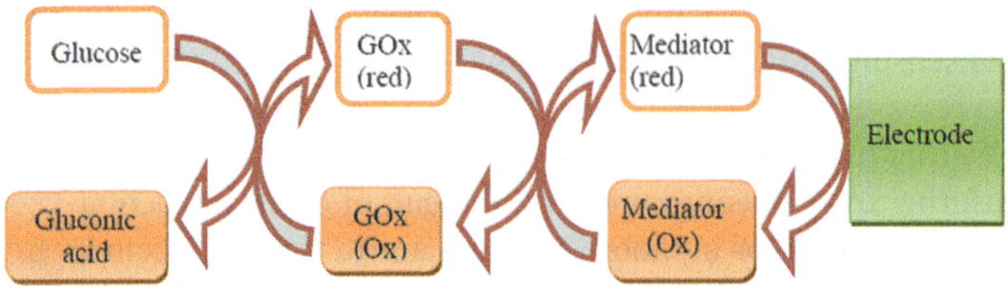

Fig. 6.17 Third-generation glucose sensors

- The small size and diffusion of mediators caused leaching of mediators from the substrate.

(c) **Third-Generation Glucose Sensors**

The third-generation glucose sensors are based on the direct electron transfer between the active centre of the enzyme and the electrodes. This generation of glucose sensor employs reagent-less glucose biosensor with low operating potential by eliminating the mediator. In this case, the electron is transferred directly from glucose to the electrode and the enzyme. The elimination of mediators is the greatest advantage of such third-generation biosensors, leading to operating on low potentials. The working mechanism of third-generation glucose sensors is shown in Fig. 6.17.

6.6 Other Biosensor Detection Techniques

Another interesting example of biosensors is FRET-based detection techniques. These kinds of biosensors are most fantasised and researched study of detection mechanism.

(a) **FRET**

FRET is a process of transferring energy from an excited molecule (donor) to another neighbouring molecule (acceptor). In this energy transfer process, the molecules that can emit fluorescence are called fluorophores. If two such fluorophores are close enough, then excitation of a first molecule (donor) results in fluorescence emission of the second molecule acceptor (Schöning and Poghossian 2002). This energy transfer between molecules relies on the distance between those participating molecules. This distance-dependable detection technique of FRET makes it explore various molecular interactions. The donor and acceptor molecule are usually chromophores where donor molecules initially absorb the energy and transfer to another molecule, usually called as acceptors.

In the mode of transfer of energy, acceptor's emission intensity increases with reduction in the donor's fluorescence intensity and its excited state lifetime. There are few optimised conditions that are required to undergo the resonance: (1) the distance dependency typically 10–100 Å range. (2) The absorption or excitation energy spectrum of the acceptor must overlap the fluorescence emission spectrum of the donor.

This transfer of energy to nearby acceptor is a non-radioactive reaction. This takes place through dipole-dipole interaction. FRET quantification is mostly based on measuring changes of fluorescence intensity upon time.

This technique of using fluorescent probes permits observation of the various properties of specific proteins in live cells. Figure 6.18 schematically represents the working of FRET.

The donor molecule is a chromophore, when sensory domain-substrate interaction takes place; the resonance takes place, and the acceptor molecule produces fluorescence due to conformational changes. This technique thus helps in the detection of various biomolecules. As shown in Fig. 6.19, the ON/OFF mechanism is implemented for target binding that perturbs the electron density along the CP backbone. In Fig. 6.19, various modes of detection are schematically shown. (A) It shows the changes in a conformational change in polymer chain resulting in CP fluorescence. (B) It shows the resonance mechanism due to intermolecular aggregation of polymer chains in the sensor platform. Fluorescence colour emission may occur due to change in electron density along the conjugated backbone of a CP in the

Fig. 6.18 Mechanism of working of FRET-based biosensors. (Lee et al. 2007)

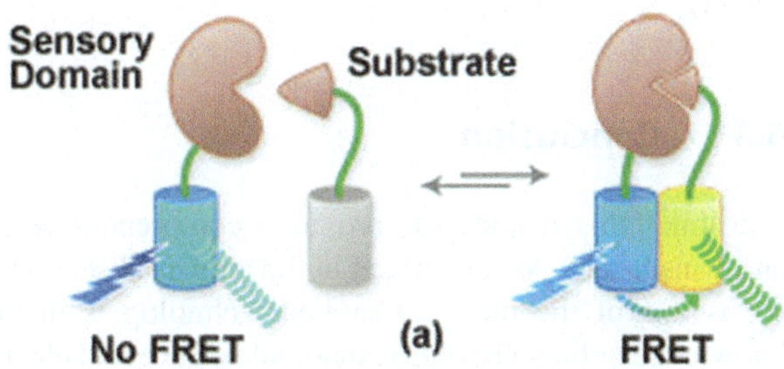

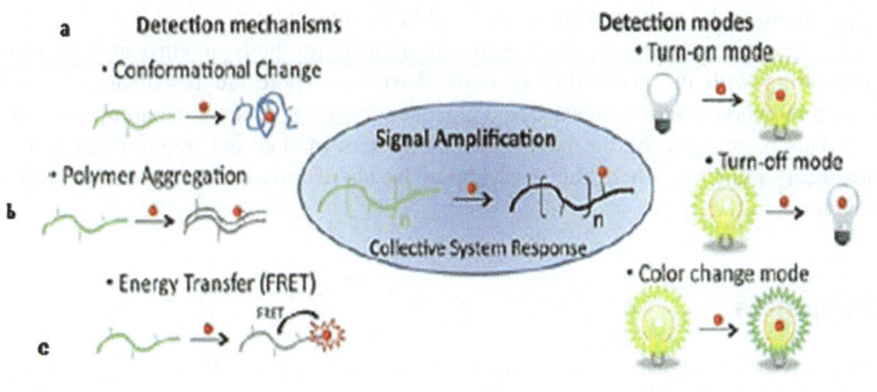

Fig. 6.19 Various detection modes using FRET. (Lee et al. 2007)

target site. (C) As a process, fluorescence is emitted by reporting fluorophore, by energy transfer mechanism between CP and a reporting fluorophore. Furthermore, for any of detection modes, the signals are detected and quantitatively sensed.

6.7 Point Care Devices

With the improvement in the areas of science and technology, medical biosensors are becoming an important expansion in the field of medicine. Therefore, recent research is concentrating towards a retrofit device with a miniaturised kit for use in household earliest screening. Individualised and personal healthcare is one of the biggest concerns in modern days. Healthcare to the doorstep is becoming the thrust among common public. One of the most important applications of biosensors is the point of care testing (POCT) at doorstep. POCT is the practice of performing a diagnostic procedure, meaning that the test itself has to be quick and easily performed without the use of expensive or complicated instrumentation, a skilled technician, a laborious process, hospital environment, etc. The significant point in POCT is that accurate diagnosis is mandatory which steps into early diagnosis of diseases. POCT is going to be the next-generation need of healthcare technology required. POCT is also the same as a prognostic test near the patient to provide rapid diagnosis involving various POCT techniques. One of those emerging technologies is lab-on-a-chip device.

6.8 Conclusion

The aim of this chapter is to provide a comprehensive view on the topic of biosensor and transducers. As mentioned in the above subsections, the development of biosensor is one of the most researched technologies in the healthcare industry. With knowledge in biotechnology, material science and electronics, the field of biosensors and bioelectronics is establishing various technologies in sensing for different applications. In recent years there has been growing apprehensions on other possible applications of biosensors in industry, healthcare, etc.

In the current scenario, there is a requirement for both in vitro and in vivo monitoring with the possibility of early detection. There are few examples like cancer biomarker sensor, sweat-based glucose sensor, urea sensor, creatinine sensor, etc. This is going to be the next high commercial market and requirement in the healthcare industry, where the overview of basics of biosensors is covered in this chapter.

References

Baldrich E (2010) Aptamers: versatile tools for reagentlessaptasensing. In: Recognition receptors in biosensors. Springer, New York, pp 675–722

Brett C, Oliveira Brett AM (1993) Electrochemistry: principles, methods, and applications (No. 544.6 BRE).
Cliquet P (2004) Development of immune detection assays for penicillins and sulfonamides in food of animal origin Doctor in Veter inary Sciences, UGent 2004. Promotors Cox E and Goddeeris B.
Drummond TG, Hill MG, Barton JK (2003) Electrochemical DNA sensors. Nat Biotechnol 21(10):1192
Eggins BR (2008) Chemical sensors and biosensors, vol 28. Wiley, Chichester
Feng L, Wu L, Wang J, Ren J, Miyoshi D, Sugimoto N, Qu X (2012) Detection of a prognostic indicator in early stage cancer using functionalized graphene based peptide sensors. Adv Mat 24 (1):125–131
Gorton L (ed) (2005) Biosensors and modern biospecific analytical techniques, vol 44. Elsevier, Amsterdam
Hong J, Yoon DS, Park MI, Choi J, Kim TS, Im G, Kim S, Pak YE, No K (2004) A dielectric biosensor using the capacitance change with AC frequency integrated on glass substrates. Jpn J Appl Phy 43(8R):5639
http://wirelesslifesciences.org/2015/03/oj-bio-at-wlsa-convergence-summit-2015/
https://www.easybiologyclass.com/enzyme-cell-immobilization-techniques
Ju H, Zhang X, Wang J (2011) NanoBiosensing: principles, development and application. Springer, New York
Kaisti M (2017) Detection principles of biological and chemical FET sensors. Biosen Bioelectron 98:437–448
Kell D (1988) Biosensors. Fundamentals and applications. Turner APF, Karube I, Wilson GS (eds), Oxford University Press, Oxford.
Kissinger PT, Heineman WR (1983) Cyclic voltammetry. J Chem Educ 60(9):702
Lai Y, Wang L, Liu Y, Yang G, Tang C, Deng Y, Li S (2018) Immuno sensors based on nanomaterials for detection of tumor markers. J Biomed Nanotechnol 14(1):44–65
Lee K, Povlich LK, Kim J (2007) Label-free and self-signal amplifying molecular DNA sensors based on bio conjugated polyelectrolyte. Adv Func Mat 17(14):2580–2258
Mao X, Ma Y, Zhang A, Zhang L, Zeng L, Liu G (2009) Disposable nucleic acid biosensors based on gold nanoparticle probes and lateral flow strip. Anal Chem 81(4):1660–1668
Martins TD, Ribeiro ACC, de Camargo HS, da Costa Filho PA, Cavalcante HPM, Dias DL (2013) New insights on optical biosensors: techniques, construction and application. In: State of the art in biosensors-general aspects. In Tech, Croatia
Nguyen H, Park J, Kang S, Kim M (2015) Surface plasmon resonance: a versatile technique for biosensor applications. Sensors 15(5):10481–10510
Pancrazio JJ, Whelan JP, Borkholder DA, Ma W, Stenger DA (1999) Development and application of cell-based biosensors. Ann Biomed Eng 27(6):697–711
Patel S, Nanda R, Sahoo S, Mohapatra E (2016) Biosensors in health care: the milestones achieved in their development towards lab-on-chip-analysis. Biochem Res Int 2016
Schöning MJ, Poghossian A (2002) Recent advances in biologically sensitive field-effect transistors (BioFETs). Analyst 127(9):1137–1151
Sun D, Lu J, Zhong Y, Yu Y, Wang Y, Zhang B, Chen Z (2016) Sensitive electrochemical aptamer cytosensor for highly specific detection of cancer cells based on the hybrid nanoelectrocatalysts and enzyme for signal amplification. Biosen Bioelectron 75:301–307
Thévenot DR, Toth K, Durst RA, Wilson GS (2001) Electrochemical biosensors: recommended definitions and classification. Anal Lett 34(5):635–659
Wang J (2008) Electrochemical glucose biosensors. Chem Rev 108(2):814–825

Advance Biomedical Sensors and Transducers

7

Harishchandra Digambar Jirimali

Abstract

Biosensors are the devices which detect the biological components or use of biological components for the detection of an analyte. Transducers are of different types, which process and amplify the signals produced by receptors and analyte interaction. The transducers are thermal, piezoelectric, optical and electrochemical. Redox polymers and the bio-elements like enzymes, DNA and proteins are used in biosensors. The biosensors are used for the detection of glucose, cholesterol, hydrogen peroxide, urea, bilirubin, biomarkers.

Keywords

Bilirubin · Biomarkers · Biosensor · Cholesterol · DNA · Electrochemical · Enzymes · Glucose · Imunosensors · Nanoparticles · Optical · Polymers · RNA · Transducer · Urea

7.1 Introduction

Sensors are devices that sense a physical or biological entity by using the physiochemical properties of that entity. The first thermostat came to the market in 1883 invented by Warren S. Johnson, and many consider this as the first modern man-made sensor. Infrared (IR) sensors have been available since the late 1940s. Motion detectors were invented by Samuel Bagno in 1950. The interest of the scientific community in detecting different entities like temperature and pressure, accordingly, leads to the development of different types of sensors. The commonly

H. D. Jirimali (✉)
Department of Chemistry, Uka Tarsadia University, Gopal-Vidya Nagar, Maliba Campus, Surat, Gujarat, India
e-mail: harishchandra.jirimali@utu.ac.in

S. Paul (ed.), *Biomedical Engineering and its Applications in Healthcare*,
https://doi.org/10.1007/978-981-13-3705-5_7

used sensors are speedometer in vehicles, thermometers, humidity sensors, remotes, and glucose meters. There are several components and principles of sensors, and accordingly, they are classified as chemical, biological, and physical sensors (Ponmozhi et al. 2012). Physical sensors are devices that convert physical entities in the form of electric current, light waves, displacement, and other physical forms of the detection. The commonly used physical parameters in sensing devices include resistance, capacitance, displacement, electromagnetic waves, and photo-thermoelectric parameters (Introduction, A visual temperature sensors with high thermal 2016). The physical sensors used in medical technology include the electrocardiogram, which measures the pulse rate and cardiac functioning. Electroencephalography is used to measure the electrical activity of the brain by applying an electrical pulse around the brain (Felix and Angnes 2017). Sonography or medical ultrasound is a technique based on ultrasound waves, a commonly used physical method for diagnostic imaging purpose. It is also used to detect the pathophysical conditions of the internal body organs, blood vessels, and tissues. Thermocouple converts temperature response in measurable signals and processes it electronically to give temperature as an output. Chemical sensors measure the physicochemical properties of a substance. The well-known examples of chemical sensors are carbon monoxide sensor and pH sensor. The principle of the pH sensor is based on the potential difference developed in glass electrode and reference electrode, which determines the pH of the solution. pH electrodes give information about the concentration of H^+ ions and OH^- ions present in the solution, as well as how alkalinity and acidity of solution is changing during the chemical reactions or dilution (Gayathri and Senthil Kumar 2014). Ion-selective electrodes bind to ions or measure the concentrations of ions. They are based on conductance or change in potential between the ion selective electrode and the reference electrode. The humidity sensor is also a classic example of the measurement of moisture in the environment or test commodities, which is based on the change in current, capacitance, or temperature. Chemical sensors adopted with the component for sampling of analytes to be detected and transported toward the detection parts or signal inducing parts. After the formation of signals, the processing of signals to data can be performed by electronic parts (Thevenot et al. 2013). The use of biological component for the detection of an analyte or biological elements in a chemical sensor is known as a biosensor; they cannot be described as a separate class because the principles on which they are working is similar to chemical sensors. They are also classified according to biological entities used for the detection. The bioelements used may be an enzyme, protein, antibodies, antigen, tissues, cells, organelles, membranes, and bacteria (Guide n.d.; Measurements 2008).

7.2 Biosensors

Biosensors are the devices which are used for the detection of biological components or use of biological components for the detection of an analyte of interest (Malhotra et al. 2017). The receptor element connected to transducers processes the signals produced by the receptor-analyte interaction. Transducers are of different types,

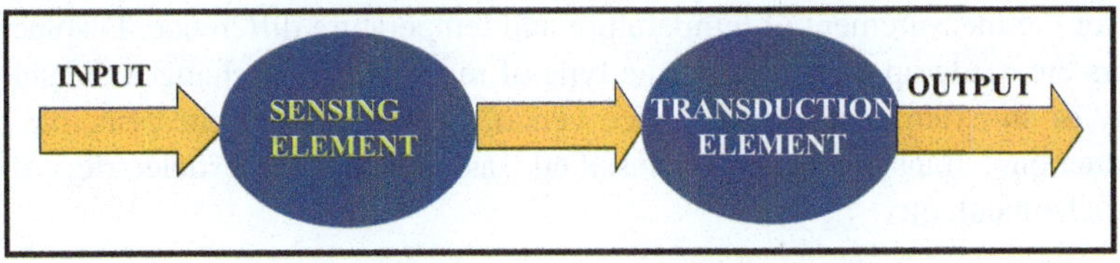

Fig. 7.1 Schematics of the transducers

which process and amplify the signals produced by receptors. Biosensors are used for the measurement of analyte information in medical, agricultural, and environmental fields. Their higher accuracy and easy handling make them an important tool for fast analytical measurements. The prototype biosensors were developed by L. Clarck for the measurement of oxygen concentration in blood serum, where enzyme was immobilized on the surface of the platinum electrode (Ponmozhi et al. 2012). Later on different methods have since been developed for electrode modifications and immobilizations of biocomponents. Biosensors have been developed from generations to generations; first-generation biosensors measure analyte response at transducer site, and second-generation biosensors use the mediators for analyte response measurement (Jirimali et al. 2013a). Third-generation biosensors use direct analytical response toward the transducer elements. Transducers used in biosensors are of different types depending on the principle of measurement, types of analyte, and measurement site (Fig. 7.1).

7.3 Transducers

Transducers are devices that convert energy from one form to measurable signals of another form. According to the different elements and principles of measurement, there are several types of transducers, e.g., thermal, piezoelectric, optical, and electrochemical. A transducer is made up of three elements input unit, sensor, and output unit. The input unit receives signals and transfers it to the sensor, where the sensor recognizes the analyte and generates a measurable response. The output unit displays or stores the generated information electronically. Some of the important properties of transducers are as follows: (i) transducers measure the lower concentration of the analyte in the environment; (ii) transducers can continuously measure the small change in the concentration of an analyte; and (iii) the response of the measured energy or analyte is persistent for wide ranges of concentrations and is constant at constant environmental conditions. The selection of the transducers depends upon the type of analyte to be measured, the concentration of analyte, and the principle of the measurement (Transducers for bio medical n.d.). Transducers can be classified on the basis of an energy source as passive and active used in the measurement of the analyte. Active transducers need an external energy source for the operation of the transducer, whereas passive transducers can operate without external energy supply. Transducers can be classified on the basis of the type of analyte measured, temperature, concentration, displacement, etc. Thermocouples are used for the measurement of temperature and temperature difference. Displacement sensors are made up of actuators (the type of material which changes the shape on supply of an external stimuli and vice versa). On the basis of the principle of the transduction, transducers are classified as thermal, piezoelectric, optical, electrochemical, etc.

7.3.1 Thermal Transducers

Temperature transducers measure the temperature and convert it into measurable signals in the form of electricity, resistance, or other entities. Major types of temperature transducers are resistance temperature detectors (RTD), thermistors, and thermocouples. RTD is basically a temperature-sensitive resistor having positive temperature coefficient and is used in electrical devices. RTD are used in medical laboratories, where low-temperature measurement is required. These devices provide electrical signals as an output, so they are used in the temperature-controlled systems working at low temperatures. According to the principle of resistors, temperature increases with a decrease in the resistivity, but thermistor is a device which shows a decrease in resistivity with an increase in temperature. This working principle of thermistors is used for the measurement of temperature of the surrounding. Based on the working principle, there are two types of thermistors, positive temperature coefficient (PTC) thermistors and negative temperature coefficient (NTC) thermistors. The resistance of the PCT thermistors increases linearly with a decrease in temperature and vice versa. The NTC thermistors are made up of semiconducting materials, which increases the electron flow with an increase in temperature where current is measured by an ammeter. Thermistors are used in 3D printers to control the melting temperature of the plastic to stop overheating of the materials. NTC thermistors are used to monitor the temperature of the incubators used in biomedical laboratories. Thermistors are incorporated in subcutaneous and esophageal medical probes used in the patients. Also they are used to guide catheters in different locations of the heart and measure the blood flow. The thermocouple is a thermoresponsive material made up of two different metals, where the joining end is called the hot junction and the other end is called the cold junction. The temperature difference in the hot junction and cold junction develops a potential, which results in the generation of EMF as a result of temperature. Due to their wide measurement range from 270–2000 °C, thermocouples are extensively used in industrial applications and medical devices.

There are many commercial applications of the thermal transducers in sophisticated analytical instruments like thermo-gravimetric analyzer, scanning calorimetry, and thermomechanical analyzers, apart from that they are used in the air conditioning and cooling devices. Infrared radiations can be used for the measurement of temperature of an environment; these radiations convert thermal energy into electrical measurable signals and vice versa; for this purpose, metal alloys of different compositions are used, e.g., nicrome alloy. Temperature transducers are used as a temperature controller in cryotechniques and thermoforming of dental products. The cryotechnology is known as very low-temperature technology, where working temperature is in the range of −150 °C. Reduction of the pain in the injured sportspersons can be achieved by using this cryotechnology, where the cold shock treatment is given to the injured part to stop or reduce the pain. Generally liquid nitrogen or solid CO_2 can be supplied at the working sides to reduce the temperature of the surrounding. Liquid CO_2 bags can be used to cool the inflammation part by reducing the temperature of the skin by −4 °C, and immediate relief can be provided to the patient. Thermoforming of dental and orthopedic products can be achieved by heating the thermoforming plastic material in controlled temperature zones. The desired shape of the material can be achieved by controlling the temperature; in this regard, continuous monitoring of the temperature is needed. Contact temperature sensors can be used in thermoforming; also advance IR sensors are getting importance due to noncontact and sensitive modes. Dry condensation devices in the ventilators can be used to provide pure and controlled air to the patients. Controlled anesthesia device is used to control the exact amount of anesthesia to the patient, and it can be controlled using a thermally controlled valve in automatic anesthetic chambers. Infrared (IR) transducers measure the electromagnetic radiations in the wavelength range of 0.7 μm to 14 μm. The radiations emitted by the objects are collected by the transducer and amplified by using photodetectors, and after amplification of the signals, they produce an output in the form of current. The current produced in a photodetector linearly increases with the temperature. The current is used to detect different temperature regions in the object, which cannot be differentiated by observing the visual rays. The infrared sensors are used to detect the temperature of different body parts. The perfect IR radiators are used to generate IR radiations. The emitted radiations by the body parts are measured as a function of temperature, by which different pathophysical conditions of the body can be detected (Li et al. 2007).

7.3.2 Piezoelectric Transducers

Piezoelectric transducers are a type of transducer, which converts pressure or force into the measurable electrical signal by using a piezoelectric material. The word piezoelectric means electricity produced by pressure. Quartz is a natural piezoelectric crystal, whereas lithium sulfate and di-potassium tartrate are the synthetic materials. Piezoelectric transducers work on the principle of piezoelectric effect. Application of voltage to piezoelectric materials results in the charge displacement, which causes a change in the dimension of a piezoelectric material. Application of force or pressure in piezoelectric materials can generate measurable signals in the form of charge. The piezoelectric transducers work on different principles as thickness shear, face shear, thickness expansion, and transverse expansion. Single-layer piezoelectric generators are made up of a single thin layer of the piezoelectric material, where energy generated is released very quickly, the voltage generated is very high, and the current is very low. Multilayer piezoelectric generators are made up of multiple thin layers of piezoelectric materials in a stack. Multilayer piezoelectric produces less voltage and higher current compared to the single layer generators. The piezoelectric transducers are utilized in the form of a patch for the measurement of pulse rate; this is a miniaturized sensor which can be used as a wearable biomedical device. Piezoelectric materials can be used for the penetration of deoxyribonucleic acid (DNA) in the mammalian cells by ultrasound method (Hung et al. 2014). Quartz crystal microbalance (QCM) is a device which is used to measure a small change in weight. Antigens immobilized piezoelectric substrates are used as recognition elements for specific antibodies. The imprinted polymer masked on the piezoelectric substances and the specific vibrations are used to detect an analyte; this is a quite promising method used for selective biomolecules detection. Concentration of serum albumin is performed by immobilizing antibodies on the quartz surface, where QCM oscillations are measured after and before the analyte interactions. Primary aliphatic alcohols like ethanol, propanol, and butanol are also quantified by using polyaniline (PANI)-modified QCM surface (Pohanka 2018).

7.3.3 Optical Transducers

Optical transducers are devices that convert optical energy, i.e., photo energy into the measurable electrical signals, using SPR, LSPR, interferometers, resonators, gratings, and refractometer. Surface plasmon resonance (SPR) occurs on the conducting surfaces, where the incident polarized light is interacting with the conducting surface, which is modified with ligands or biorecognition centers. The analyte of interest flows in the vicinity of the modified surface, which changes the refractive index or reflectivity and is measured through optoelectronic devices. Sometimes the binding can occur with a receptor and analyte element, which can be detected by SPR that includes 1:1 binding stoichiometry, nonspecific absorption of ligand- analyte binding, antigen-antibody interaction, and avidity of the molecules. The SPR is a major tool for the study of interaction in biological and pharmaceutical molecules (Silva et al. 2004). SPR imaging can be carried out on a spatially designed pattern and interaction of multiple molecules with the receptors; this method can be used to study the kinetics of drugs. Gold nanoparticles are used in localized surface plasmon resonance (LSPR), where it interacts with analyte and changes absorption of incident light. Fluorescent biosensors use fluorescence intensity and fluorescence quenching for the detection of an analyte (Saha and Sen 2013). Bioluminescence is used for the detection and quantification of the analyte by using labeled fluorescent antigens in ELISA (Kalyani and Sharma 2016). Polarization of the light is used to detect the influenza A virus in an ellipsometric biosensor.

7.3.4 Electrochemical Transducers

Electrochemical transducers convert analyte information into measurable electrical signals using electrochemical techniques; they include potentiometric, amperometric, conductometric, ion sensitive field effect transistors (ISFET), and enzyme FET. Potentiometric transducers measure the potential difference developed between the working electrode and the reference electrode. Accumulation of the ionic potential is a result of the concentration of an analyte species, and it is measured on the basis of the Nernst equation. Potentiometric sensors are used to monitor the concentration of

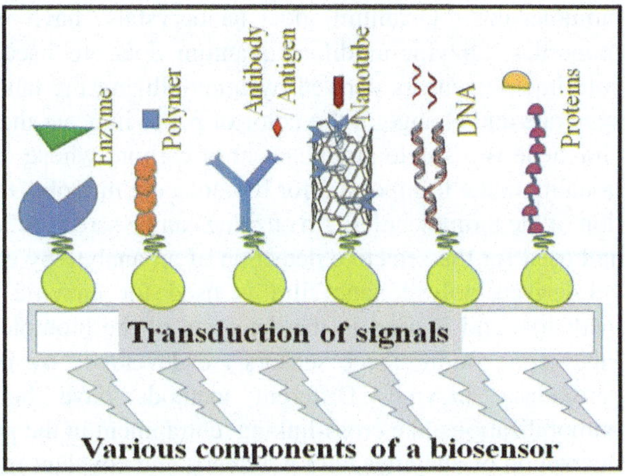

Fig. 7.2 Various components of biosensors

ions or ionic species and the pH of the biological fluids. Ion-selective electrodes can be used in the detection of the specific ions with high sensitivity. Amperometric devices measure the concentration of the analyte by quantifying the generated current in biochemical redox reactions. The current produced in a biochemical reaction is proportional to the concentration of an analyte. Current measured in controlled variation of the voltage is called voltammetry, and current measured at constant potential is called an amperometry (Thevenot et al. 2013). The redox active enzymes and redox polymers are used as mediators in amperometric detection of glucose, urea, cholesterol, and various biomolecules (Jirimali et al. 2011). Conductometric transducers are measuring the ability of the analyte or a medium to carry a current; this can be measured by using an impedometric and capacitive determination. This method is used to study enzyme substrate kinetics and determination of an analyte. Semiconductor technology is developed to detect the concentration of an analyte by using conductometry. An integrated array of electrodes is used to detect multiple analytes, like environmental pollutants, biological important species, and drugs (Fig. 7.2).

7.4 Bioelement-Based Biosensors

Biological elements like enzymes, DNA, antigen, antibody, etc. are used as a receptor for the constructions of biosensors, where receptors provide selectivity and sensitivity in the measurement of an analyte. Enzymes are biocatalysts, used for the selective activity of an enzyme and analyte. Enzymes like glucose oxidase, peroxidase, and dehydrogenases are used in the detection of glucose and other biomolecules (Beden et al. 2015). Enzyme sulfite oxidase is used for the detection of sulfite in the food products. Different enzymes are modified with redox active polymers, and modified electrodes are used for the selective detection of the biomolecules. Nanostructured materials like gold nanoparticles and carbon nanotubes (CNT) are used for enzyme immobilization and detection of biomolecules. Quantum dot nanocrystals have specific photoluminescence properties; enzyme-modified quantum dots are used in the FRET sensors. The colorimetric sensors derived by immobilizing the labeled molecules on the paper/ substrate and change in the color of paper indicate the concentration of an analyte. Graphene is a single atomic sheet of carbon, whose excellent properties and functionality make it important for biomolecule immobilizations for the selective detection of the biomolecules. An enzyme can be screen printed with polymeric materials and used for the selective detection of an analyte by electrochemical methods. Sol-gel-derived polysiloxane film is used for enzyme entrapment with the redox mediators and is used in the detection of the biomolecules (Jirimali et al. 2018a). Fiber optic fluorescence sensors are developed by immobilizing the horseradish peroxidase enzyme. Different methods have been developed for enzyme immobilizations like cross-linking, entrapment in the polymeric matrix, electrostatic interaction of enzymes with polymers, and covalent immobilization.

7.4.1 DNA-Based Sensors

DNA-based biosensors are getting importance due to their high specificity for the detection of gene hybridization. DNA, RNA, single-stranded DNA, aptamers, and other nucleic acids are directly used for the sensing purpose. Polymerase chain reaction (PCR) technique is used for DNA production as it produces a large amount of DNA in a short time. The synthetic DNA probes are modified by using a biorecognizing tag that can be used in a biosensor. DNA modified with enzymes, proteins, and fluorescent molecules is known as a labeled DNA. Single strands of DNA or aptamers are directly used for analyte detection; this method is known as a label-free biosensor. There are several methods of DNA immobilizations, among them monolayer formation on the gold surface, immobilization using functionalized silica, entrapment of DNA in a polymeric or ceramic scaffold, covalent immobilization of DNA are important. Colloidal gold nanoparticles are used for the immobilization of DNA in optical biosensors. Array-based sensors are prepared by using a lithographic technique, where DNA can be directly or covalently immobilized on the microarray structures. Molecular beacons are the fluorophore-modified DNA immobilized on the conductive surface, where the addition of target DNA strand unfolds the hairpin and gives signals in the form of fluorescence. DNA-modified microcantilevers can be used for human disease monitoring. Aptamer-modified gold electrodes are used for the electrochemical detection of the thrombin. DNA-modified electrodes are used for the development of several electrochemical sensors for the detection of DNA damage (Jagota 2015).

7.4.2 Proteins

Proteins are the basic building blocks in living organisms having specific functions. There are several kinds of proteins, such as fluorescent proteins, redox-active proteins, and membrane proteins, which are used for biosensing applications

(Ohba et al. 2013). Redox-active proteins are used in electrochemical biosensors for the purpose of electron transfer from analyte to the electrode surface for the detection of biomolecules. Some redox proteins directly immobilized on a gold electrode surface, which is electrocatalyzing an analyte. The metalloproteins contain a metal center, which shows reversible electron transfer reaction on the electrode surface, and also catalyze the reaction of analyte detection. In the FRET-based detection of an analyte, biomacromolecules like proteins can be engineered by transducer-active elements and used in the biorecognition. The natural fluorescent proteins are used for the detection biomolecules by FRET analysis. The fluorescent proteins are used in microphysiological sensing, where biosensing of stem cells, pathogenic cells, and normal cells is performed for the identification and diagnostic purpose. The proteins are immobilized on the conductive and nonconductive surfaces by various methods, like affinity, electrostatic interaction, entrapment in the polymer matrix, and covalent modification. Carbon-based nanostructures can be covalently modified with the proteins and used for the preparation of biosensing platform.

7.4.3 Antigen-Antibodies

Antigen-antibodies are proteins produced in the body to detect and destroy the foreign elements entered in the body. An antigen is a molecule capable of inducing an immune response. There is specific affinity in antigen and antibody, which is used for the detection of the diseases by using immunosensors. Immunosensors are biosensors that work on the affinity of the recognition molecules, which are binding to analyte resulting in the formation or dissociation of the complex. ELISA is a technique popularly used for the diagnostic purpose of detection of the diseases, by using antigen-antibody interaction. Piezoelectric immunosensors compete with ELISA due to its simplicity and ease of handling in lateral flow immunoassay. Direct detection of the microorganism is also possible by using piezoelectric immunoassay technology. Optical fluorescence sensor is developed for the veteran diagnostic antibodies immobilized porous glass substrate (Silva et al. 2004). Au (gold) nanoparticles show high affinity toward the antibodies; graphene-based magnetic nanoparticles modify gold nanoparticles and are used in the immobilization of the antibodies for the detection of alpha-fetoproteins (Wang et al. 2017). The immunosensor develops from immobilization of Antigen AFB1 along with BSA on CNT-modified electrode for high-sensitive detection of aflatoxin B1. The important feature of immunosensor is that it is acting on only bounded molecules, and unspecific molecules are not taking part in the reaction.

7.4.4 Redox Polymers

Redox polymers can shuttle electrons from analyte to electrode surface and catalyze several electrochemical reactions (Jirimali et al. 2013b). Redox polymers are composed of ligand-bonded organometallic complexes of the transition metals, which

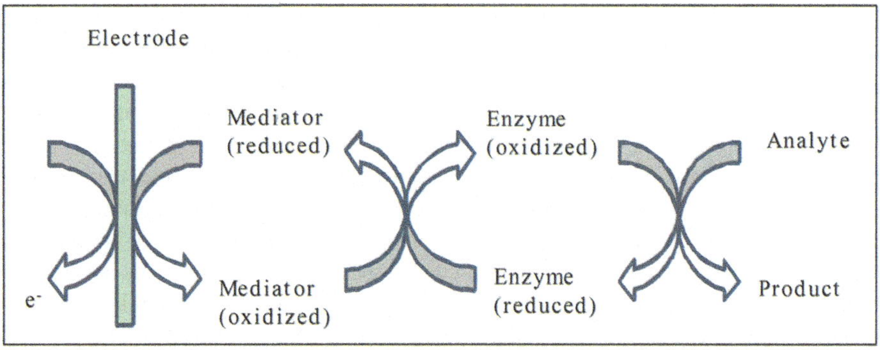

Fig. 7.3 Schematic representation of redox polymer mediated electron transfer

are showing reversible redox reactions (Jirimali et al. 2018b). The well-known example of redox polymer is osmium polymers, and they have been used in the sensing of various analytes like glucose, vitamin C, sulfite, and biofuel cell. Ferrocene polymers are extensively studied for the sensing activity as a mediator and also as an electro-catalyst. Redox-active polymers can contain an organic redox center, like quinone, hydroquinone, viologen, and conducting polymers used as mediators (Jirimali et al. 2015). Redox polymers can be cross-linked within their matrix or with other compounds for immobilization. The sensitivity of the electrodes can be enhanced by incorporating the nanoparticles like carbon nanotubes, graphene oxide, gold nanoparticles, and metal oxide nanoparticles in the polymer matrix (Fig. 7.3).

7.5 Examples of Biosensors

7.5.1 Glucose

Glucose measurement is one of the tedious tasks for diabetes patients, requiring considerable attention and care in clinical laboratories. Biomedical sensors used for glucose monitoring are now getting importance due to their ease of operation, time-saving, and cost. There are a variety of glucose sensors based on different principles and use in glucose-level monitoring. The biosensor development was carried out in three generations as first, second, and third. The first-generation glucose sensors were developed by Clark in 1962 based on the use of the natural oxygen substrate and production of hydrogen peroxide. The concentration of hydrogen peroxide produced is proportional to the glucose concentration. The amperometric first-generation glucose sensors worked at higher potential, and electro-oxidation of the interfering species like ascorbic acid, dopamine, and uric acid occurs. To minimize these interferences, the second-generation glucose biosensors have been developed, where the redox mediators were used for electron shuttling. Enzyme redox centers are buried deep inside the protein sheath, so electron transport is very sluggish at the electrode surface. Wiring of the redox polymer with an enzyme effectively shuttles the electrons from analyte to electrode surface. Third-generation biosensors are reagentless biosensors, where the toxic redox polymers are replaced by the organic charge transfer complexes, nanomaterials, and enzyme. Only a few enzymes had shown the ability to transfer an electron at normal conditions, and the addition of nanomaterials increases the sensitivity and selectivity.

7.5.2 Cholesterol

Cholesterol is an important constituent in the human body, where it is involved in the formation of biological barriers and precursor for the synthesis of several vital chemicals. Development in the socioeconomic status and modern life patterns created medical issues like hypertension, cardiovascular diseases, and malfunctioning of vital organs. Monitoring the concentration of cholesterol in the blood is becoming important for better management of these diseases. As earlier discussed, electrochemical, optical, piezoelectric, and immunosensors have been developed for monitoring of cholesterol. Cholesterol oxidase and cholesterol esterase are the two enzymes which are used in the detection of cholesterol. For the construction of cholesterol biosensors, it is important to immobilize enzyme on active surfaces by physical adsorption, entrapment, electrodeposition, and covalent binding. Layer-by-layer method has also been established for enzyme immobilization on the electrode surface. The optical biosensors are developed by using a scratched optical fiber and used for direct cholesterol monitoring in the blood.

7.5.3 Hydrogen Peroxide

Hydrogen peroxide is known for its cytotoxic effect, but it is also an important molecule for signal transduction in a biological system. Hydrogen peroxide is produced in many important enzymatic reactions and synthesis. It is important for antioxidant activity and modulating peroxide signal transduction. Hydrogen peroxide is produced in altered physiological condition and leads to immune response and apoptosis. Several methods of hydrogen peroxide sensing have been developed; among them, enzyme-based methods are commonly used. HRP is a popular enzyme used for the detection of hydrogen peroxide by electrochemical methods, where the enzyme is immobilized on the electrode surface. Metal complexes comprising transition metals and polymers are also used for the hydrogen peroxide detection without the enzyme. Nanomaterial-modified electrodes are designed for the highly sensitive detection of the hydrogen peroxide by enzymatic and nonenzymatic methods. The nanomaterials like iron oxide, silver oxide and silver nanoparticles, copper oxide, and cobalt oxides are used for the enzyme-free hydrogen peroxide detection. Fluorescent probes can be used for the detection of hydrogen peroxide.

7.5.4 Urea

Urea is the final detoxification product of ammonia in the liver, where proteins and nitrogenous compounds metabolize to produce urea. Urea concentration in a normal person ranges from 2.5 to 7.5 mM. Protein-rich food and catabolism increase urea concentration in the body. A drastic change in urea concentration indicates kidney dysfunction. Urea level can be lowered down by hemodialysis or peritoneal dialysis of blood. There are several methods for the determination of urea based on spectrophotometric detection, gas chromatography, and conductometry. Urea is detected by the enzyme-catalyzed reaction of urease, where the enzyme is coupled with the acrylate-based microspheres in the optical biosensor. Polymeric hydrogels are also used in the immobilization of the urease enzyme for urea monitoring. The field effect transistors are used for the enzyme-based urea detection using modified FTO glass (Velychko et al. 2016). The flexible urea sensor is demonstrated by using electrodes modified with molecularly imprinted polymer-modified CNTs. Nonenzymatic detection of urea is carried out by electrode modified with silver nanoparticle-decorated CNT. The simultaneous determination of the urea and creatinine can be carried out by the surface plasmon resonance method using gold nanoparticle-modified surface.

7.5.5 Bilirubin

Bilirubin (BLR) is a yellow-colored pigment found in the blood, which is the degradation product of hemoglobin present in red blood cells. Increased level of bilirubin results in jaundice and may also lead to liver failure and brain damage. Bilirubin is found in two forms; one is free BLR, which can bind to serum albumin, and other bounded to glucuronic acid. The small amount of free bilirubin in blood indicates bilirubin toxicity and results in jaundice. In clinical laboratories the bilirubin is detected by color change of sulfanilic acid in diazotization reaction with bilirubin. Bilirubin oxidase is an enzyme which catalyzes the free bilirubin. The bilirubin oxidase-modified nanoparticles co-immobilized on the electrode surface and are used for electrochemical sensor fabrication. The enzyme and non-enzyme-based biosensors are prepared by using metal oxides and carbon-based nanostructures. The protein labeled with the fluorescence probe is used for the preparation of the optical sensor for bilirubin detection. Electrochemical methods are also used for the rapid and sensitive detection of the bilirubin using the nanostructured-modified electrodes.

7.5.6 Biomarkers

Biomarkers are an intended measurable characteristic that indicates the physiological and pathological process or pharmacological response to a therapeutic intervention. Biomarkers are classified in different ways, as type zero, type one, and type two. Type zero biomarkers (natural history biomarker) help to measure the natural history of disease. Type one biomarkers (drug activity biomarkers) measure the effect of drug intervention and are also classified as efficacy biomarkers (indicating the therapeutic effect of the drug), mechanism biomarkers (giving information about the mechanism of action of the drug), and toxicity biomarkers (indicating the toxicological effect of the drug). Type two biomarkers serve as clinical markers and help to diagnose the effect of an intervention. There are various biomarkers used for clinical analysis like prognostic biomarker, predictive biomarker, and pharmacodynamic biomarkers. Different diseased conditions produce different biomarkers, indicating different physiological state. Various methods have been used for the detection of biomarkers, which depends on the physiochemical characteristics and disease condition. There are several biomolecules, like DNA, RNA, enzymes, proteins, antigens, and carbohydrates, which can be considered as biomarkers. ELISA is a useful technique for the detection of biomarkers by using antigens and antibodies. Immunosensors developed by using electrochemical methods are applied to biomarker detection using gold nanoparticle-modified electrodes. SPR is used to detect PSA (prostate-specific antigen) biomarkers, and an electrochemical method is used to detect FAM134B protein in colon cancer cell extracts by using an antibody-modified screen-printed electrode. Carbon nanotube-based substrate modified with the specific antibodies is used to detect the cancer blood cells present in blood samples, by optical method. Kidney cancer biomarkers are detected by using bioplasmonic gold nanoparticle-modified filter paper. Chronic kidney disease (CKD) can be detected by using the screen-printed CNT-modified electrode probe using differential pulse voltammetry. The kidney injury molecule (KIM-1) and high mobility group box-1 (HMGB-1) are organ injury biomarkers detected by using SPRI technique (Rizvi and Kashani 2017).

7.6 Commercial Biosensors and Their Importance

Biosensors are getting popular due to their ease of handling, pain-free operation, and quick results. The first biosensor floated in the market was a glucose biosensor used for diabetes and blood glucose-level monitoring. Worldwide key players in the biosensor market are Abbott Diagnostics; Pinnacle Technologies, Inc.; 77 Electronica Kft; Sanofi; and Life Scan (Johnson and Johnson Company) that engage in the glucose sensor market. FreeStyle is a glucose meter developed by Abbot Diagnostics. Counter flex is the glucose meter manufactured by the BAYAR Ltd., OneTouch UltraEasy glucometer by Johnson and Johnson. SensoLite Nova and SensoLite Nova Plus are manufactured by 77 Elektronica Kft. Blood glucose meter with blood ketone meter/cholesterol td-4140 is manufactured by Tai-doc Taiwan. Nova vet/Nova veterinary glucose/ketone meter is produced by Nova Biomedical. The Health Check Systems, Inc. has a wide range of products such as CardioChek blood testing device for blood cholesterol level, and CardioChek Professional test strips include glucose, total cholesterol, HDL cholesterol, blood ketone, triglyceride, LDL cholesterol, and more that are sold separately.

7.7 Future Biosensors and Their Importance

Detection of biomolecules for diagnostic purpose is an interesting field due to its easier handling and cheaper methods. The biosensors available at present are mostly suffering from the interferences of the other biological species present in the matrix. Biosensors either detect a single analyte or need to take samples outside the body for analysis. It is important to develop the devices having multi analyte sensing capacity on a single cheap miniaturized implantable biosensing device needed in order to get the real-time measurement. The real-time monitoring of the biological species along with the drug delivery systems will help for the development of artificial organs. The biomarker analysis can help in the diagnosis of the deadly diseases like cancer and help the society to be alert. The existing prices of the biosensing devices could be lowered down by promoting interdisciplinary research, using advance electronics and engineering technology. Definitely the last person of the society could get the benefit of this modern biosensor technology in the new era of science.

Acknowledgments H. D. Jirimali thanks to DST SERB young scientist scheme, file number SB/FT/CS-037/2014, and RPS Uka Tarsadia University UTU/RPS/1264/2018. Sincere thanks to Prof. Woonsup Shin who always inspire and guide me to work in the field of biomedical technology for the betterment of mankind.

Glossary

Antigen In immunology, an antigen is a molecule capable of inducing an immune response in the host organism. Sometimes antigens are part of the host itself in autoimmune disease. Antigens are "targeted" by antibodies

Antibody An antibody, also known as an immunoglobulin, is a large, Y-shaped protein produced mainly by plasma cells that are used by the immune system to neutralize pathogens such as pathogenic bacteria and viruses. The antibody recognizes a unique molecule of the pathogen, called an antigen

Catabolism It is the set of metabolic pathways that breaks down molecules into smaller units, which are either oxidized to release energy or used in other anabolic reactions

Electrocardiogram ECG is a record or display of a person's heartbeat produced by electrocardiography

Electro-catalyst An electro-catalyst is a catalyst that participates in electrochemical reactions

Electroencephalography EEG is an electrophysiological monitoring method to record the electrical activity of the brain

Field effect transistors FET is a transistor that uses an electric field to control the electrical behavior of the device

Fluorescence resonance energy transfer FRET is a distance-dependent radiationless transfer of energy, from an excited donor fluorophore to a suitable acceptor fluorophore, and is one of the few tools available for measuring nanometer-scale distances and the changes in distances, both in vitro and in vivo

Thermistors It is a solid-state temperature sensing device that acts a bit like an electrical resistor but is temperature sensitive

Thermoforming Thermoforming is a manufacturing process where a plastic sheet is heated to a pliable forming temperature, formed to a specific shape in a mold, and trimmed to create a usable product.

Temperature coefficient It is a coefficient expressing the relationship between a change in the physical property and temperature that causes it

References

Beden N, Shin W, Ludwig R, Peterbauer CK (2015) iMedPub journals bioelectrochemical behavior of the composite PVP-Os/chitosan as a mediator with different types of enzymes at graphite electrode,pp 1–7

Felix FS, Angnes L (2017) Immunosensors – a powerful tool for analytical. Biosens Bioelectron 102:470. https://doi.org/10.1016/j.bios.2017.11.029

Gayathri P, Senthil Kumar A (2014) Electrochemical behavior of the 1,10-phenanthroline ligand on a multiwalled carbon nanotube surface and its relevant electrochemistry for selective recognition of copper ion and hydrogen peroxide sensing. Langmuir 30:10513–10521

Guide P. Introduction to transducers, sensors and actuators

Hung W, Feng G, Cherng J (2014) Ultrasonics sonochemistry fabrication of piezoelectric components for a tunable and efficient device for DNA delivery into mammalian cells. Ultrason – Sonochemistry 21:819–825

Introduction, A visual temperature sensors with high thermal (2016) pp 6602–6613. https://doi.org/10.1039/C6TC01841F

Jagota (2015) A DNA biosensors. pp 1–27

Jirimali HD, Nagarale RK, Lee JM, Saravanakumar D, Shin W (2011) Preparation of PEG tethered ferrocene modified polyacrylic acid/silica composite as an electroactive polymeric platform for biosensors. Electroanalysis 23:2109

Jirimali HD, Nagarale RK, Lee JM, Saravanakumar D, Shin W (2013a) Chitosan-cross-linked osmium polymer composites as an efficient platform for electrochemical biosensors. Chem Phys Chem 14:2232–2236

Jirimali HD, Nagarale RK, Saravanakumar D, Lee JM, Shin W (2013b) Hydroquinone modified chitosan/carbon film electrode for the selective detection of ascorbic acid. Carbohydr Polym 92:641–644

Jirimali HD, Saravanakumar D, Shin W (2015) Preparation of catechol-linked chitosan/carbon nanocomposite-modified electrode and its applications. Bull Kor Chem Soc 36

Jirimali HD, Nagarale RK, Saravanakumar D, Shin W (2018a) Ferrocene tethered polyvinyl alcohol/silica film electrode for electrocatalytic sulfite sensing. Electroanalysis 30:828. https://doi.org/10.1002/elan.201700459

Jirimali, H. D., Saravanakumar, D. & Shin, W. Chitosan-Cu-salen/carbon nano-composite based electrode for the enzyme-less electrochemical sensing of hydrogen peroxide. 9, 1–7 (2018b)

Kalyani VL, Sharma V (2016) Optical sensors and their use. In: Medical field, vol 564, pp 3–6

Li F, Winnik MA, Mandelis A (2007) Polypyrrole nanoparticles as a thermal transducer of NIR radiation in hot-melt adhesives. pp 4309–4315. https://doi.org/10.1039/b708707a

Malhotra S, Verma A, Tyagi N, Kumar V (2017) Biosensors: principle, types and applications, pp 3639–3644

Measurements E (2008) Transducers and sensors, pp 1–18

Ohba Y, Fujioka Y, Nakada S, Tsuda M (2013) Fluorescent protein-based biosensors and their clinical applications, vol 113. Elsevier

Pohanka M (2018) The determination of human albumin by a quartz crystal microbalance immunosensor. Int J Electrochem Sci 13:8471–8480

Ponmozhi J, Frias C, Marques T, Frazão O (2012) Smart sensors/actuators for biomedical applications: review. Measurement 45:1675–1688

Rizvi MS, Kashani KB (2017) Biomarkers for early detection of acute kidney. 2:1–14. https://doi.org/10.1373/jalm.2017.023325

Saha P, Sen M (2013) Toward optical sensors : review and applications. https://doi.org/10.1088/1742-6596/423/1/012064

Silva M, Cruz H, Rossetti O, Arese A, Oliva A (2004) Development of an optical immunosensor based on the fluorescence of Cyanine-5 for veterinarian diagnostics. Biotechnol Lett 26:993–997

Thevenot, D. et al. Electrochemical biosensors : recommended definitions and classification to cite this version : HAL Id : hal-00862855 Electrochemical Biosensors : Recommended. (2013)
Transducers for bio medical
Velychko TP et al (2016) A novel Conductometric urea biosensor with improved analytical characteristic based on recombinant urease adsorbed on nanoparticle of silicalite. Nanoscale Res Lett 11:3–8. https://doi.org/10.1186/s11671-016-1310-3
Wang Y et al (2017) Ultrasensitive label-free electrochemical Immunosensor based on multifunctionalized graphene nanocomposites for the detection of alpha fetoprotein. Nat Publ Gr 7:1–10. https://doi.org/10.1038/srep42361

Dr. Harishchandra D. Jirimali is working as assistant professor at the Department of Chemistry, Uka Tarsadia University. He received an M.Sc. degree from the School of Chemical Sciences, North Maharashtra University, Jalgaon, Maharashtra (India). He received a PhD degree at Interdisciplinary Program of Integrated Biotechnology from the Department of Chemistry, Sogang University, Seoul, South Korea. Under the guidance of Prof. Woonsup Shin, he joined the same lab for his postdoctorate. After the completion of postdoctoral work, he has joined the School of Chemical Sciences, North Maharashtra University, as a young scientist on Fast Track research scheme funded by DST-SERB India. His expertise is in the field of biosensors, electro-osmotic pumps, redox polymers, capacitive deionization, biofuel cell, biomaterials, nanomaterials and anticorrosive coatings, electroluminescence, and pesticide residue analysis.

Part IV

Biomaterials and Its Medical Applications

Introduction to Ideal Characteristics and Advanced Biomedical Applications of Biomaterials

8

Govinda Kapusetti, Namdev More, and Mounika Choppadandi

Abstract

Biomaterial intervention in healthcare is inevitable, rather required for a better life. They are well practiced from ancient times, and the successive evolution made them more potent, versatile, and easy for clinical practice. However, the mimicking of the materials is not at all absolute, and in some cases it is very minimal as compared to the native tissues. The course of development of a biomaterial needs a strong understanding of the basic characteristics of the material behavior in bioenvironmental. The chapter discusses the various types of biomaterials starting from polymers to composites, and it presents the detailed information about the required ideal characteristics of a biomaterial like biocompatibility, bio-inertness, bioactivity, bioabsorbable pattern, bio-adaptability, sterilization, etc. In contemporary medical technology, biomaterials play a major role to answer many complications with high accuracy. The chapter primarily focuses on the latest advancements in biomaterials for major areas like orthopedics, cardiovascular, ophthalmology, neuronal, etc. Further, the chapter gives special emphasis on tissue engineering aspect of biomaterials to the regeneration of tissues and therapy.

8.1 Introduction

Numerous materials abound on earth with a unique identity and application. Among biomaterials are a class of entities can able to interact with bioenvironment and tissues for various requisites. Biomaterials are evolving exponentially from the last 50 years, through the combined aspects of medicine, biology, chemistry, and

G. Kapusetti (✉) · N. More · M. Choppadandi
Department of Medical Devices, National Institute of Pharmaceutical Education and Research (NIPER)-Ahmedabad, Gandhinagar, Gujarat, India

S. Paul (ed.), *Biomedical Engineering and its Applications in Healthcare*,
https://doi.org/10.1007/978-981-13-3705-5_8

materials science. Majorly, the engineered material participated in disease diagnosis or therapy or both, and sometimes it acts as therapeutic as well. In this journey, the material must exhibit essential physicochemical, mechanical, and biological properties in order to augment or replace or support an organ or a tissue or a part of the body, to make the diseased/unhealthy part functional. To fulfill the function, the candidate must be biostable, biocompatible, and bio-tolerable as the immune system may treat the substance as a foreign material. The definition of biomaterials was framed in different ways, but most prominently "Biomaterials are a class of materials- be it natural or synthetic, alive or lifeless, and usually made of multiple components that interact with the biological systems. They are often used in medical applications to augment or replace a natural function."

Materials' intervention in human health is not a modern aspect since in ancient times multiple materials are used in the reconstruction of the human body. An ancient Hindu sacred book *Rigveda* (3500–1800 BC) was compiled with all sorts of medical amputations and augmentations with biomaterials. Sutures made from animal tendons, cotton, horsehair, leather, vegetable fiber, etc. were used for surgeries in those days. In ancient days, the Chinese first introduced gold in the dentistry, and now it is a well-established material for dental fillings. Earlier in the seventeenth century, brass sutures were used in fracture repairs. Later, nickel-, aluminum-, and platinum-coated devices were most prominent in orthopedics as plates and screws, whereas in the twentieth century, the carbon steel was being used in this persistence. Dr. Otto Rohm, a German chemist, developed poly(methyl methacrylate) (PMMA) in 1901 and patented as Plexiglas in 1933 (Arora et al. 2013; Hosseinzadeh et al. 2013). PMMA is widely used as a bone cement that fills the bone defects and also fills the space between the bone and prosthesis which ultimately results in uniform stress distribution. Hydroxyapatite (HA) is the highly consumed biomaterial being used from the last 50 years for musculoskeletal tissue regeneration and therapy. Collagen is an animal-derived biomaterial with a triple helical structure. Most of the connective tissues, such as bone, tendon, ligament, and skin, contain collagen; it is also commercially available as a wound-healing product (Agrawal 1998). In 1940, the cellulose acetate was first used in dialysis tubing. Dacron and polyether urethanes are used in vascular grafts and artificial pacemaker, respectively (Langer and Tirrell 2004). However, the biomaterials have been classified into a wide range of classification from ancient days, and till date there are a large number of materials and composite materials raised as a biomaterial for multiple applications in multiple approaches. When the material is implanted into the biological system, it may regenerate or repair or augment the organ or tissue, but they also produce some foreign body giant reactions or host reactions between the biological tissue and implant.

Biomaterials are used for several applications, such as joint replacements, bone plates, bone cement, artificial ligaments and tendons, dental implants for tooth fixation, blood vessel prostheses, heart valves, artificial tissue, contact lenses, and breast implants. In the future, biomaterials are expected to enhance the regeneration of natural tissues, thereby promoting the restoration of structural, functional, metabolic, and biochemical behavior as well as biomechanical performance.

8.2 Biomaterial Host Reactions

Biological response of a biomaterial on the target site of implantation is the most important factor to be understood. The interaction between the implant and biological system leads to the host defense mechanism in which it primarily causes nonspecific inflammation, infection, and specific immunological reactions such as immunity, inflammation, and blood coagulation to protect the body from foreign organisms. Generally, after implantation, the material is treated as a foreign substance and leads to the formation of nonspecific inflammatory reactions. Primarily, the foreign body reactions are initiated by macrophages. These macrophages activated in the process of interacting with the implant may intricate cytokines, which stimulate fibrosis. Biocompatibility is the important characteristic of biomaterial to decide the fate of the material for its medical use. It is opposite to the inflammation, more the biocompatible lesser the inflammation. Biocompatibility of a material is determined by various in vitro and in vivo assays (Schoen 2013; Samavedi et al. 2014). Biocompatibility of material is regulated by the International Standards Organization (ISO), Geneva, Switzerland. It comprises of standards for "Biological Evaluation of Medical Devices" in which ISO 10993 part-3 describes the in vitro and in vivo tests for genotoxicity, carcinogenicity, and reproductive toxicity. Genotoxicity tests also follow "organization for economic co-operation and development" guidelines (OECD 474, 1997; 475, 1997). ISO 10993 part-4 explains the selection of tests for interaction with blood such as thrombosis, coagulation, platelets and platelet function, hematology, and immunology. ISO 10993 Part-5 explicates about the in vitro cytotoxicity tests, i.e., tests on extracts (L929 elution test, neutral red uptake test, colony formation test, MTT, and related tests), direct contact test, and indirect contact tests (agar diffusion test, filter diffusion test). Biomaterial/medical device testing also follows the American Society for Testing and Material (ASTM) international standards.

8.3 Classification of Biomaterials

For a better understanding of applications of biomaterials by key characteristics, they are broadly classified into four major categories: polymers, ceramics, metals, and composites (Fig. 8.1). These materials are used in various processes of synthesizing to form an implantable medical device that is biocompatible, bioactive, and also sterilizable. Based on the severity of risk, medical devices are classified into four classes: they are class I (low risk), class IIa (medium risk), class IIb (medium/high risk), and class III (high risk). Class I medical devices are noninvasive devices, class IIa and IIb are invasive devices, and class III are active devices (Cheng 2003). Biomaterials are also classified based on the duration of contact with the tissue and named as class A (limited period ≤24 h), class B (prolonged period 24 h–30 days), and class C (permanent >30 days). By modifying bulk raw material/material composites, it will be converted to an implantable medical device. Generally, in the synthesis of biomaterials from polymers or copolymers, ceramics, metals or

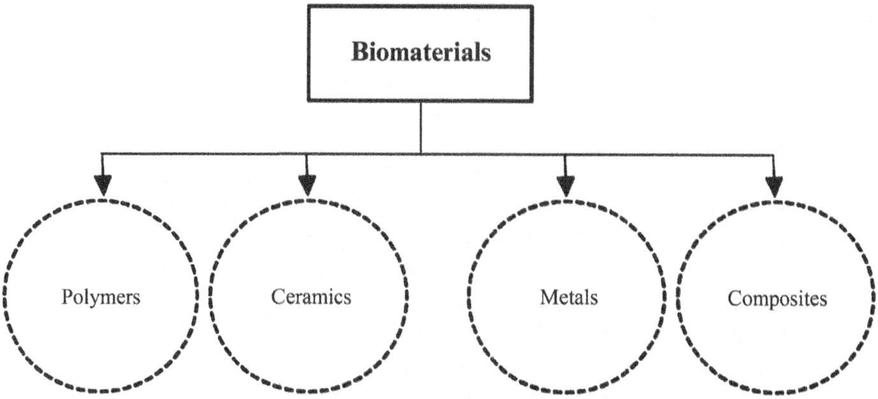

Fig. 8.1 The classification of biomaterials based on structural features (Teoh 2004; Kulinets 2015)

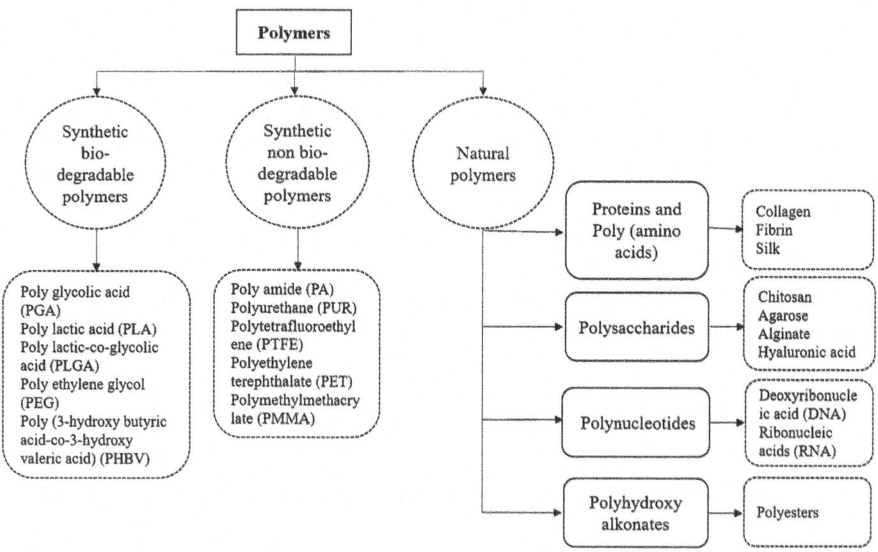

Fig. 8.2 The classification of polymers (Numata and Kaplan 2011; Pattanashetti et al. 2017; Teo et al. 2016)

alloys, and composites of polymers, ceramics, and metals are used individually or in combination to improve the physicochemical, mechanical, and biological properties.

By simplifying further, polymers are classified as natural, synthetic biodegradable, and synthetic nonbiodegradable polymers (Fig. 8.2). Polymers are the repetitive units of a single monomer. They are not only used as a scaffold material but they are also extensively used in the drug delivery systems. These natural polymers are rich in plant and animal source, and these materials have higher biocompatibility as they already have binding sites from cells and adhesion molecules. Synthetic biodegradable polymers mostly help in repair and regeneration, whereas synthetic nonbiodegradable polymers are used as fillers, bone-fixing agents, and bone-supporting materials. In the case of synthetic biomaterials, the biocompatibility is questionable, and the material acts as a foreign substance, which leads to immune reactions. However, natural, synthetic, and the combination of natural and synthetic composite materials are used in biomedical applications (Numata and Kaplan 2011; Pattanashetti et al. 2017). Most of the polymers used in the biomedical applications are approved by FDA, namely, polylactic acid (PLA), polyglycolic acid (PGA), polylactic-*co*-glycolic acid (PLGA), polyethylene glycol (PEG), polyvinyl alcohol (PVA), poly-*N*-vinyl pyrrolidone (PVP), poloxamer, starch, hyaluronate, gelatin, alginic acid, and collagen (Mansour et al. 2010).

Ceramics are natural or synthetic inorganic, nonmetallic, polycrystalline materials. Similarly, bioceramics are categorized into three classes: bioinert, bioactive, and bioresorbable ceramics (Fig. 8.3). Ceramics are mostly used as a dental-filling agent, vertebral bone cement (vertebroplasty and kyphoplasty), and cranial and maxillofacial fillings. Maximum ceramics are bioactive; they form a connection with the bone tissue processes known as osseointegration (Bohner 2008). Ceramics also play a major role in the biomedical applications such as alumina and zirconia yttria are used in top hip replacement, apatite wollastonite is used in vertebral reconstruction in tumor patient, and HA-based ceramics are used for teeth and bone (Bohner 2008; Kokubo 1991).

The metals are further classified into biodegradable and nonbiodegradable, and metals are known for high-strength applications (Fig. 8.4). Metals are mainly used in

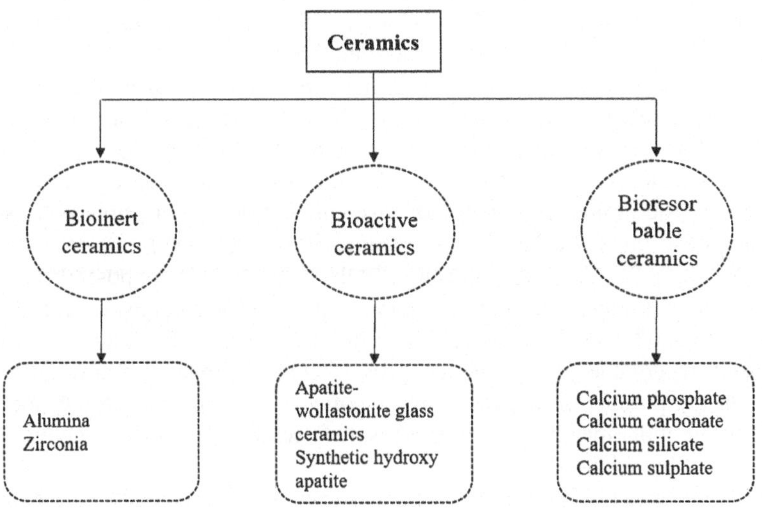

Fig. 8.3 The classification of ceramics with respect to interaction with bioenvironment (Bohner 2008)

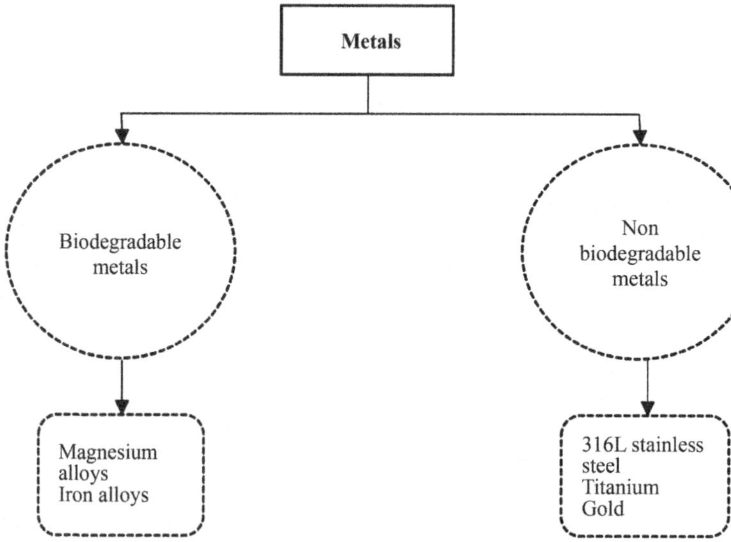

Fig. 8.4 The classification of metals based on biointegration (Niinomi 2002)

joint replacement, bone replacement, maxillofacial implants, dentistry, and cardiac stents. To avoid local inflammation or rejection, generally, metals are surface modified with a biocompatible material, which reduces the formation of biofilm. In the orthopedics fixation, plates, screws, and pins are used for bone healing, and they are made from 316 L SS, Ti (titanium), and Co-Cr-Mo (cobalt chromate and molybdenum) alloy. Amalgam, Au (gold), and Ti are used in dentistry. Cardiac stents are made from 316 L SS, Co-Cr-Mo, and Ti. The artificial eardrum is made up of 316 L SS (Hermawan et al. 2011; Niinomi 2002).

Composites (Fig. 8.5) are engineered materials composed of two or more different materials, often ones that have very different properties. The composites are intended for improvement of physicochemical, mechanical, and biological properties of the material in all aspects. Synthesis of composite materials provides with all specific properties in terms of osseointegration, degradation, porous structure formation, and adhesion property. Composites offer tuned properties based on the composition for required application; on the contrary, it is not possible in a single material. For instance, calcium phosphate and PLGA composite scaffold were prepared to modulate the degradability from poor degradable calcium phosphate and rapid degradable PLGA (Wang 2016). PLGA degrades hydrolytically and it results in the production of lactic and glycolic acid monomers. Due to the acidic nature of the monomers, the calcium phosphate cement degrades in the by-product environment by acidic dissolution.

All the Figures (1–5) described the classification of biomaterials with examples; the source of origin, structural features, ideal properties of biomaterial, and biomedical applications are explained in detail hereafter.

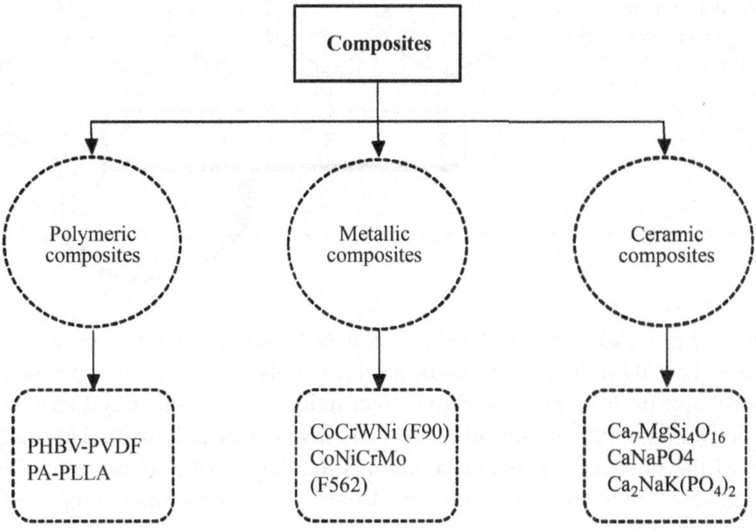

Fig. 8.5 Classification of composite materials from their matrix materials (Kulinets 2015; Ramakrishna et al. 2001)

8.4 Source of Origin

Natural biomaterials are obtained from the plant, marine, and animal source. For instance, the materials obtained from these sources are cellulose from a plant source, chitosan from a marine source, and collagen from an animal source. Metals are obtained from sources such as magmatic, sedimentary, and metamorphic rocks, soil, surface and groundwaters, and in the atmosphere (Bradl 2005), whereas ceramics are found from the geological origin, i.e., rocks and their interaction with elements (Velde and Druc 2012). Synthetic biomaterials are produced in the laboratories synthetically.

8.5 Structural Features

Structural features of polymers include general structure, polymeric structure, molecular weight, and tacticity. The general structure describes the building block of polymer like "mer" as the single unit of the polymeric chain and polymer is "many mers". The polymer is the repetitive unit of a single monomer. The polymeric structures may be linear, branched, cross-linked, and networked. The linear structure has the linkage of mers in a series fashion from one end to the other. During the synthesis procedure, the results in branches and this branched polymeric chain are known as a branched structure. The polymers that cross-link at a particular point

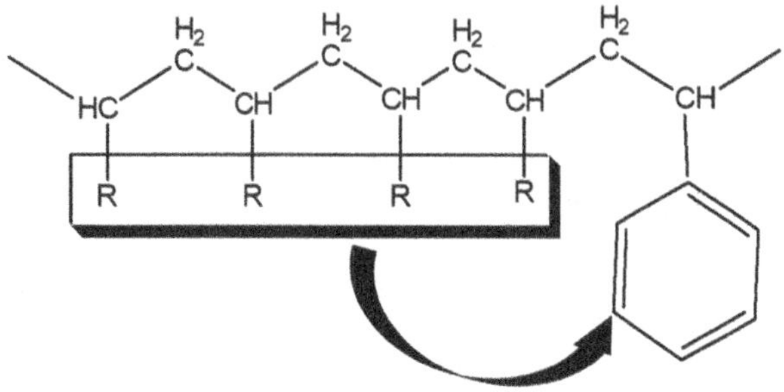

Fig. 8.6 Polystyrene structure explaining tacticity with R group

forming 3D polymeric structure or by covalent bonding is known as cross-linking polymers. The 3D polymeric network produces polymeric structures for the multi-functional approach in which multiple "mer units" bind together and form complicated network structures and are known as networked polymers. The molecular weight of the polymeric structure represents the degree of polymerization and the chain length of polymers. Generally, the degree of polymerization is represented by the letter "n" in the formula. The structural configuration of polymeric molecule is the primary part of the molecule and it can be modified only by reforming or breaking of primary bonds. In Scheme 6, "R" indicates the benzene ring in polystyrene. Consider "R" as the functional group; if the functional groups of "R" are arranged on the same side of the chain, then it is known as isotactic configuration. If the functional groups are in the alternative positions on either side of the chain, it is known as the syndiotactic configuration. If the "R" groups are situated randomly in the molecule, then it is said to be atactic configuration (Fig. 8.6).

Structural features of ceramics include the nature of the bond, crystal structure, high melting temperature, and low electrical conductivity. In the ceramics, the arrangement of bond linkage is with both ionic and covalent bonds. Due to this ionic and covalent bond, ceramics are responsible for unique properties like highest hardness, highest melting point, lowest thermal expansion, good chemical resistance, and brittleness (this leads to fractures unless the material is not reinforced with some reinforcing agent) (Society 2018).

Structural features of metals include crystalline nature and metallic bonding. Crystal structures are explained on the basis of unit cells. The unit cell is the configuration of atoms with a repetition of the crystal unit. Common crystal lattice structures in metals are face-centered cubic (FCC), body-centered cubic (BCC), and hexagonal close-packed (HCP) structures (Fig. 8.7). As the metallic bonding is nondirectional, there are some broad varieties of atomic configuration that creates crystal structure. Metals are widely used because of their structural properties such as strength, high melting point, ductility, toughness, and thermal and electrical conductivity. Therefore, these materials are extensively used for biomedical applications in orthopedics.

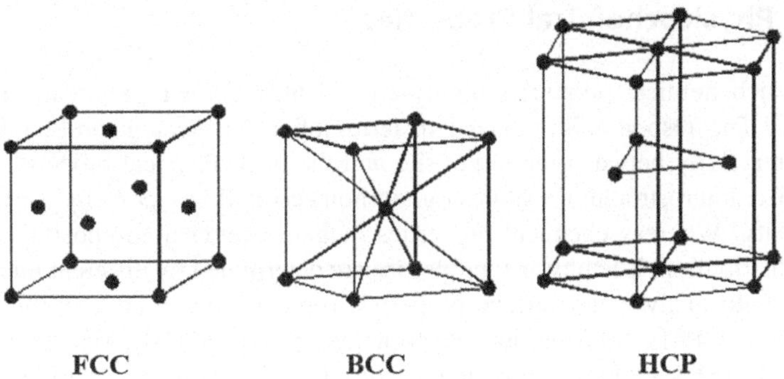

Fig. 8.7 Illustrations of different crystal lattice structures of metals

Fig. 8.8 The flow chart of the ideal characteristics of a biomaterial

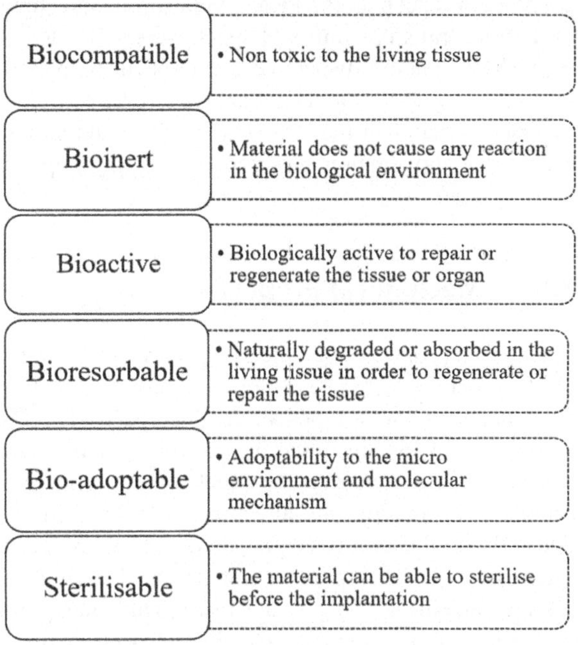

8.6 Ideal Characteristics of Biomaterials

Ideally, biomaterial should be biocompatible, bioinert, bioactive, bioresorbable (biodegradable), bio-adoptable, and sterilizable (Fig. 8.8). The degree of the characteristics signifies the ability of the material for the biomedical application. Importantly, the ideal characteristics also influence in mimicking the microenvironment for the therapy or care.

8.7 Physicochemical Properties

The physicochemical properties highly control the microenvironment for cellular activities. The response is reflected in terms of cell adhesion, proliferation, and surface reactivity between the target site and the host. Physical properties include shape, size, microstructural features (amorphous/crystalline), porosity, surface area, and density. Whereas chemical properties include chemical composition and elemental distribution. The physical properties are determined by different microscopic techniques to analyze the surface properties (optical microscopy, scanning probe microscopy (SPM), transmission electron microscopy (TEM), scanning electron microscopy (SEM), atomic force microscopy (AFM), particle size (dynamic light scattering, DLS), porosity (porosimetry), surface area (gas adsorption), surface energy (hydrophilic/lipophilic property—contact angle measurements), and chemical properties such as chemical composition, and elemental distribution is evaluated by Fourier-transform infrared spectroscopy (FTIR), X-ray photoelectron spectroscopy (XPS), energy-dispersive X-ray spectroscopy (EDAX), and carbon, hydrogen, nitrogen, and sulfur elemental analysis (C/H/N/S elemental analysis, respectively). In order to maintain the structural stability and structural activity of a biomaterial, each of the abovementioned properties is very important (Ramalingam et al. 2016).

8.8 Mechanical Properties

Mechanical properties of material play a vital role in the biomedical applications. An implant may experience different kinds of functional loads, i.e., compression, tension, shear, and at times all the three parallelly. The compression strength is the application of loading force on both ends of the sample from the same direction, whereas tensile strength is the equal amount of force applied on the two ends of the sample from the opposite directions, and the shear force causes angular distortion due to the application of tangential force from the alternative surface. When all three forces are experiencing by the implant the distortion, frequency increases. Universal testing machine (UTM) is the instrument mainly used to check the mechanical properties of the biomaterials with respect to the ASTM standards. Hardness, strength, toughness, and viscosity elasticity of the material can be determined. The biomaterials of four different classes, i.e., polymers, ceramics, metals, and composites react differently to each of the applied loading forces. Polymers, metals, and composites exhibit both brittle and ductile nature based on their composition but most of the ceramics are in ductile nature. The scaffolds (films/fibers/sutures) are checked for the tensile strength; definitely shaped implants (metallic/polymeric/bioceramic/composites) are checked for the compression strength and shear stress (Roeder 2013).

8.9 Biological Property

The material which is used for the medical purpose be it organ replacement/tissue growth/drug release must possess characteristic biological properties such as biocompatibility, inertness, and biofunctionality. In addition to this, biomaterial must be sterilizable to avoid implant rejection, inflammation, and irritation. Biological characterization of a biomaterial is carried by ISO 10993 guidelines. The list of biological tests includes cytotoxicity, sensitization, hemocompatibility, pyrogenicity, implantation, genotoxicity, carcinogenicity, reproductive and developmental toxicity, and degradation assessments (DOS Santos et al. 2017; Bandyopadhyay and Bose 2013).

8.10 Biocompatibility

Biocompatibility is a field of medical implants that explains the beneficial and risky interactions with the host tissue. Compatibility of the material with the biological system is known as biocompatibility and it is described as "the ability of a material to perform with an appropriate host response in a specific application" (Naahidi et al. 2017; Williams 2009). It can be evaluated by both in vitro and in vivo tests. In vitro studies give a rough idea about the appropriate cell type that survives in the presence of biomaterial. Although the biomaterial has shown good cell viable results in in vitro studies, it may not show the same effect in vivo due to host defense mechanism and may lead to inflammation at the site. In some cases, the cell interaction with host can induce the release of inflammatory chemotactic mediators at the target site (Naahidi et al. 2017; Kohane and Langer 2010; Williams 2008; Ziats et al. 1988). If the material is toxic, the surface modification of the material helps in reduction of toxicity. The surface modification is achieved by either physicochemical modification or surface coating with a bioactive material. The physicochemical modification is attained by etching, mechanical polishing, and chemical reactions such as oxidation, reduction, acetylation, and utilization of organosilanes. Surface coatings of the biomaterial include thin film deposition, grafting, and covalent and non-covalent coatings. This surface modification of biomaterial improves implantable device and tissue interaction and also provides biocompatibility and shows its bioactivity (DOS Santos et al. 2017).

8.11 Bioactivity

Bioactive materials are the materials that form chemical bonds with bone tissue known as osseointegration. It is a process through which the implant is connected to the bone tissue. There are some materials that show bioactivity in terms of regeneration, i.e., calcium phosphate and hydroxyapatite (DOS Santos et al. 2017). For instance, calcium phosphate bone cement produces great bioactivity; it was shown by the formation of apatite on the calcium phosphate cement surfaces when soaked in SBF for 1 week (Xu et al. 2017; Sadiasa et al. 2014). When the calcium phosphate cement was implanted into rabbit femur, it shows the increased healing effect and bone ingrowth. These types of cement have the osteoconductive property that facilitates the attachment, ingrowth, proliferation, migration, and phenotypic expression of bone cells, which leads to the formation of new bone. Ideally, the scaffolds must have 60–80% interconnected porosity with 150–500 μm range pore size by which nutrient exchange, waste elimination, and cell penetration take place. Biodegradability of calcium phosphate scaffolds occurs by active (via cell-mediated process) and passive resorption (by chemical dissolution). Physical factors in degradability include bulk property of scaffold, crystallinity, porosity, and surface area, whereas chemical factors include chemical composition and ionic substitutions. Biological factors include activation of osteoclasts and macrophages (Lu et al. 2002). Calcium phosphate and PLGA-based scaffold are implanted into rabbit femoral bone defect model. The study reveals enhanced bone regeneration of >13% with >55% degradation in 6 weeks. Further, it shows >40% bone formation along with 90% degradation in 26 weeks (Grosfeld et al. 2016). The osteoconductive, osteoinductive, and biodegradability properties altogether explain the bioactivity of a biomaterial.

8.12 Bio-tolerability

Most of the metals and synthetic polymers are bio-tolerable; they form a fibrous layer surrounding the metallic or polymeric implant which forms by the release of ions, chemicals, and corrosive products. Some of the bio-tolerable materials are implanted as dental fillers, bone-supporting materials, and bone-fixing materials (DOS Santos et al. 2017). The chemical treatment enhances the bio-tolerability (electropolishing, acid cleaning, and chemical passivation); the chemical composition of 316 L stainless steel is C = 0.028%, Ni = 10.15%, Mo = 2%, and Fe, whereas 304 L stainless steel composition consists of C = 0.017%, Cr = 18.32%, Ni = 8.03%, Mo = 0.29%, and Fe. The bio-tolerable materials are considered as corrosion resistant and bioinert (Ghanavati et al. 2016).

8.13 Applications

Material interference in human healthcare is not a prerequisite; it is necessary for modern society lifestyle. The applications start from simple syringes to complex devices like artificial pacemakers and artificial organs. In ancient times, natural material like the wood was used to augment or repair the tissue or part of the tissue. In 21st century, the revolutionary changes in life science and materials science research are the fundamental base for the development of novel biomaterials. The biomaterials not only mimic the structure of the tissues but also exhibit the suitable microenvironment for fitting tissue response. Based on the applications, the biomaterials are broadly classified into various categories like tissue engineering,

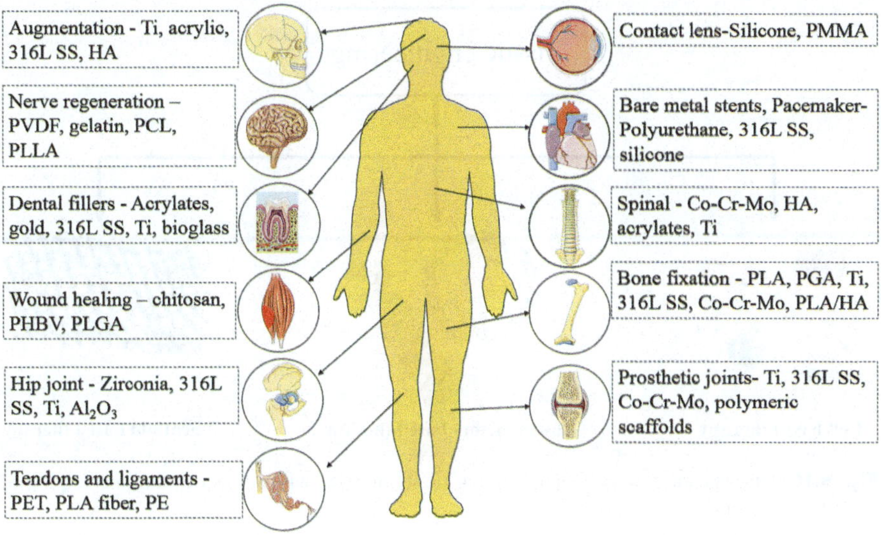

Fig. 8.9 The illustration of notable applications of biomaterials in human health

biosensors, and advanced drug delivery application. Figure 8.9 shows some of the applications of biomaterials in human health (O'brien 2011; Hubbell 1995).

8.14 Tissue Engineering

Tissue engineering is the groundbreaking field in healthcare which evolved from the biomaterials and acts as supporting structure for tissue regeneration and therapy. The practice is established by the application of the scaffold, cells, and bioactive molecule in the suitable microenvironment. Tissue engineering technology is broadly categorized into three classes based on the practice and regenerating tissues and the classification shown in Fig. 8.10. The major applications of tissue engineering serve major areas like orthopedics, cardiovascular, ophthalmology, etc. (Lutolf and Hubbell 2005).

8.15 Orthopedic Applications

Bone and cartilage are the key components of the skeletal system, of which 80% of the system is composed of bone and it contributes 18% of the total body weight (Ubelaker 1984). Bone serves a variety of functions like structural support, protection of internal organs, hematopoiesis, mineral homeostasis, and storage of triglyceride. Bone is a composite material composed of extracellular matrix and cellular system. The extracellular matrix contains inorganic component like calcium phosphate and calcium carbonate, and the organic components are collagen,

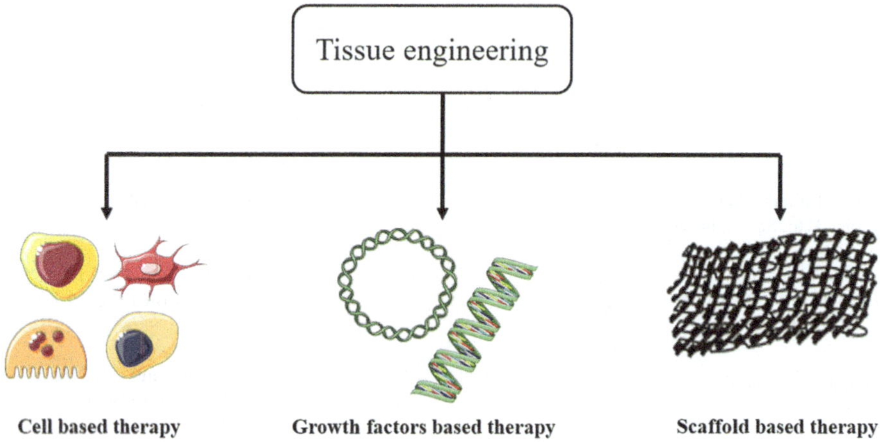

Fig. 8.10 Classification of tissue engineering by its practice with respect to the intervention of materials, molecules, and cells

proteoglycan, glycosaminoglycan (GAG), and other non-collagenous proteins. The cellular component comprises osteoblast, osteoclast, osteocytes, and osteogenic progenitor cells (Bilezikian et al. 2008). Osseous tissue continuously modifies in response to various stimuli such as electrical, mechanical, piezoelectric, etc.; every year 10% get replaced with a newly regenerated bone (Leppik et al. 2018; Robling et al. 2006). Though bone has self-healing capability, still the regeneration and repair are much more challenging and urge new therapeutic innervation. The major bone grievances occur due to mechanical trauma, accidental damage, and disease conditions like osteoarthritis, Paget's disease, osteomalacia, and cancer (Quarles 2008).

At the initial phase of tissue engineering, materials like metal and alloy such as iron and steel were employed for reconstruction of the osseous tissue. In the later stage, advanced metallic prosthetics were developed from cobalt-chromium alloy, gold-gold alloy, Mg-Mg alloy, and Ti-Ti-based alloy to minimize the adverse effects. However, the developed prosthetics indeed shows biological toxicity and poor mechanical performance. These issues are answered with ASTM-developed multicomponent stainless steel-based implantable material (Burg et al. 2000).

As the biomaterials are classified in accordance with tissue interaction, the first generation is intended to support the tissue, and the role of the second generation is to support the tissue and utilize as a carrier for the bioactive molecule. The third-generation material is biologically active material and has the intrinsic ability to stimulate the regeneration of the tissue. In order to enhance the bioactivity, the conventional metallic implants were surface modified and functionalized with osteoinductive material and functionalized the metal surface with a suitable chemical group. Most prominently, the osteoconductivity of an implant system was enhanced by simple coating of bioactive materials like bioactive glass (bioglass) (Hench 2006), HA (Swetha et al. 2010), carbon nanomaterials (CNT, nanodiamond, and graphene) (Bhong et al. 2019), perovskites, calcium phosphate (Meroni and Ardizzone 2018), etc. Furthermore, the polymers such as chitosan, PCL, PLGA, PHBV, PLLA, collagen, silk fibroin, PMMA, etc. (Zhang et al. 2018; Choi et al. 2018; Ogueri et al. 2018; Ahmad et al. 2019) were also extensively used in bioactive coatings (Sharma et al. 2017). Apart from the conventional prosthetics, the engineered scaffolds for tissue engineering show overwhelming results but the clinical acceptance will be gone long. Tissue engineering is practiced in various categories like cell implanted, growth factor implanted, and smart material-based scaffolds. The development of the stem cell-seeded scaffold is the decedent choice for bone tissue repair and regeneration. Mostly, osteoprogenitor cells such as mesenchymal stem cells (MSCs), embryonic stem cells (ESCs), pluripotent stem cells (iPSC), adult stem cells, and osteoblasts are directly incorporated or with carrier embedded at the site. The autologous chondrocyte implantation and the matrix-associated chondrocyte implantation are cell-based therapies which possess clinical importance in bone and cartilage regeneration (Meijer et al. 2007). In the second choice, the addition of growth factors in scaffolds is to stimulate cell growth and differentiation with certain biochemical cues. The growth factors such as bone morphogenic protein (BMP-2, BMP-7), transforming growth factor (TGF-α), vascular endothelial growth factor (VGEF), and insulin-like growth factor (IGF) are key candidates for bone and cartilage engineering (Lee et al. 2010; Cancedda et al. 2003). The growth factor-impregnated scaffolds demonstrated excellent preclinical results but limited for clinical application due to practical challenges. The challenges arise primarily dose optimization, unpredictable regeneration of the tissue, and an ethical concern associated with embryonic stem cells and the stability (More and Kapusetti 2017).

In order to astound the major glitches of traditional cell and growth factor-based strategies, in this contest, the polymeric scaffolds have huge potential to show a path for better therapy. The fabricated scaffolds exhibit tailored properties in porosity, mechanical strength, surface area, structure, surface architecture, etc. by its fabrication techniques (Petite et al. 2000; Hutmacher 2000; Hollister 2005). Various techniques are employed for fabrication which include freeze-drying, solvent casting, self-assembly, particulate leaching, injection molding, electrospinning, rapid prototype synthesis, template synthesis, and most advanced 3D printing (Roseti et al. 2017; Liu and Ma 2004). FDA-approved polymers involved in orthopedic applications include PLGA in the fabrication of the screw, nails, pins, and plates. Examples of the marketed products are FixSorb, NeoFix Resorption, Leadfix, MacroSorb System, etc. The PCL offers good mechanical strength, biocompatibility, and most interestingly long-lasting biodegradation. Apart from the synthetic polymers, the naturally derived polymers such as collagen, chitosan, silk fibroin, alginate, cellulose, and hyaluronic acid are the best choices of candidates for tissue engineering (Bose et al. 2012; Swetha et al. 2010). Among them, collagen is the key component of the extracellular matrix of the bone to provide structural integrity, and it plays a pivotal role in the maintenance of the mechanical strength. The collagen scaffolds are fabricated from a simple film casting technique to advanced 3D printing (Inzana et al. 2014). The scaffold promotes cell attachment and proliferation, thereby

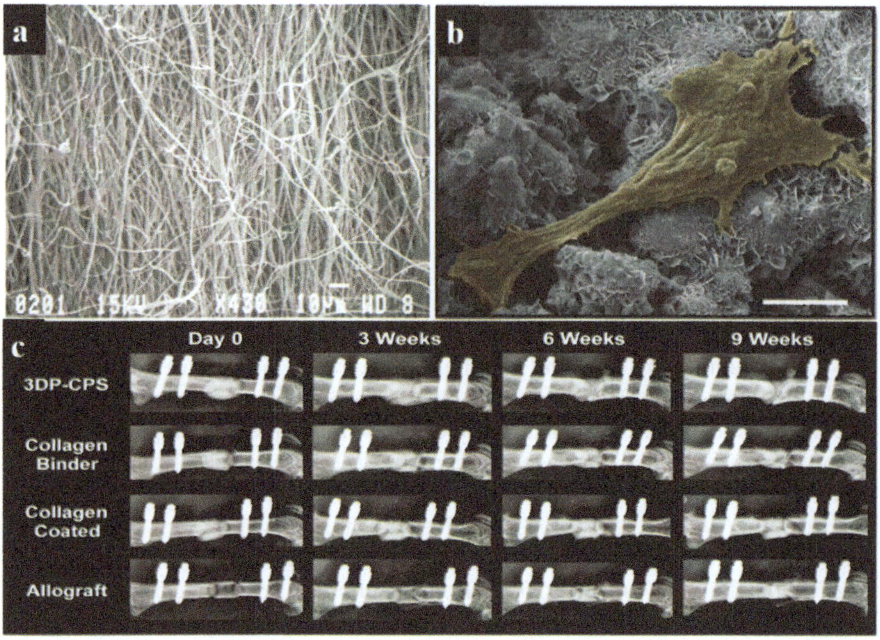

Fig. 8.11 (**a**) Fibrous morphology of the electrospun scaffold of collagen by SEM analysis (Matthews et al. 2002). (**b**) Well surface integrated osteoblast morphology on 3D-printed collagen-calcium phosphate scaffold (SEM image). (**c**) The X-ray images of murine femur implanted with various scaffolds including 3D-printed calcium phosphate-collagen scaffold (Inzana et al. 2014)

promoting tissue regeneration (Ferreira et al. 2012). Fig. 8.11a shows the electrospun scaffold of collagen with structural similarities with natural tissues (Matthews et al. 2002), and Fig. 8.11b and c shows the favorable osteoblast cell attachment on the 3D-printed collagen scaffold and X-ray digital images of scaffold implanted bone. The study reveals that the 3D-printed scaffold shows a similar kind of regeneration capacity as allograft (Inzana et al. 2014).

Unlike collagen, the silk scaffolds are fabricated by a combination of salt leaching and gas-foaming techniques for guided bone regeneration with an optimum pore size (Kim et al. 2005). Even though pristine polymeric scaffolds exhibit good passive regeneration capacity, still it lacks the bioactivity. The best possible method to enhance the bioactivity is to make a composite by the addition of bioactive fillers. The hybrid materials indeed enhance the bioactivity and also provide enhanced mechanical properties (Mieszawska et al. 2010; Gong et al. 2015). The inorganic fillers like calcium phosphate, bioglass, HA, CNT, graphene, nacre, barium titanate ($BaTiO_4$), zinc oxide (ZnO), wollastonite, and tricalcium phosphate are highly explored for bioactive composite preparation (Hajiali et al. 2018; Bhong et al. 2019; Rezwan et al. 2006). The amine functionalized graphene was reinforced in PMMA bone cement to enhance the osteoconductivity by Sharma et al. for better

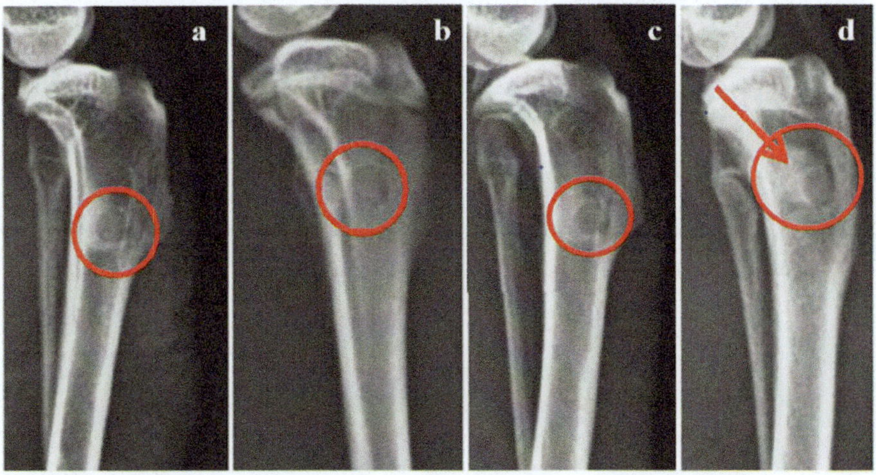

Fig. 8.12 The X-ray images of rabbit tibia at day 0. (**a**) Pristine PMMA bone cement, (**b**) Amine functionalized graphene reinforced PMMA bone cement, (**c**) Pristine PMMA bone cement, and (**d**) Amine functionalized graphene-reinforced PMMA bone cement at day 20 (Sharma et al. 2017)

integration in total joint arthroplasty (Sharma et al. 2017). Fig. 8.12 shows the X-ray digital images of rabbit tibia implanted with pristine bone cement and aminated graphene reinforced bone cement at day 0 (a and b) and day 20 postsurgery, respectively. The group reported that the reinforced bone cement-implanted tibia was healed completely within 20 days by clear observation calcium deposition, whereas in pristine bone cement the impression remains same as day 0.

Apart from the conventional biomaterials, the most advanced smart material approach is used for guided bone regeneration. The smart material or intelligent material has an intrinsic ability to respond to the external stimuli and help to tailor the biological environment. The piezoelectric, ferroelectric, magnetic, and the shape memory materials are considered as smart materials. As bone is a piezoelectric material which one of the biomechanical cues for the bone regeneration. The piezoelectricity is defined as the generation of electricity against mechanical pressure and vice versa. Piezoelectric materials such as PVDF, PLLA, collagen, PHBV, cellulose, and alginate along with inorganic materials like $BaTiO_4$, lead zirconate, and ZnO are used for the orthopedic application (Wei et al. 2011; Jacob et al. 2018; Tandon et al. 2018; Vasquez-Sancho et al. 2018).

8.16 Cardiovascular Application

The management of cardiovascular diseases like atherosclerosis, coronary heart myocardial infraction, and dysfunctioning of the heart valves is a major challenge in modern medicine. Conventionally, the cardiac diseases are treated with the drug intervention, but results are limited (Mohan 2005). The biomaterial used in the

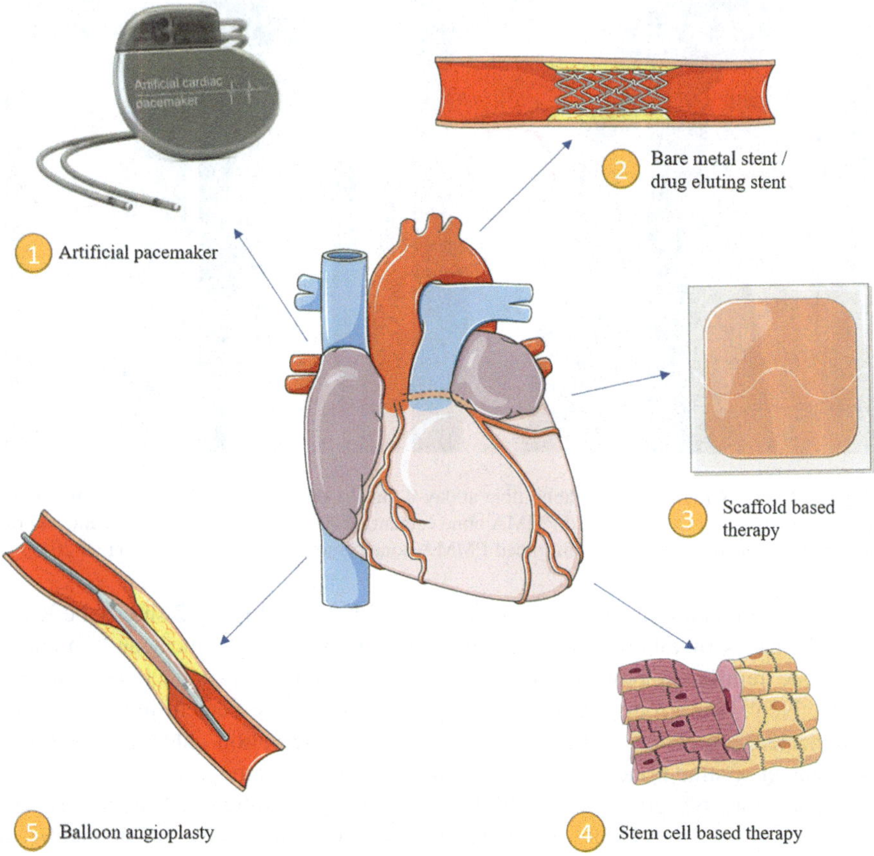

Fig. 8.13 The illustration shows the material intervention for various cardiovascular applications

management of the cardiac disease ranges from the cardiac patch to a coronary stent. Fig. 8.13 summarizes the various cardiovascular applications of the biomaterials.

Heart failure associated with myocardial infarction (MI) is a leading cause of death. Cardiomyocyte death associated with MI is a critical condition which is unable to be treated with the conventional drug therapy, but the cardiac patch has given meaningful results. Chitosan, cellulose, alginate, hyaluronic acid, gelatin, polyester urethane urea (PEUU), and PLGA are extensively used for fabrication of cardiac patches (D'Amore et al. 2016; Chi et al. 2013). Despite advancements in the polymer technology, the patches are compromised in mechanical strength. The new paradigm shifted to the development of composites to counter the limitations in cardiac patches, which not only improves the mechanical strength but also induces bioactivity in the pristine polymer. In this context, composite of PLGA-reinforced carbon fiber (Stout et al. 2011; Dvir et al. 2011), gold nanoparticle-added electrospun PCL, and gelatin nanofibrous scaffolds (Shevach et al. 2013) and

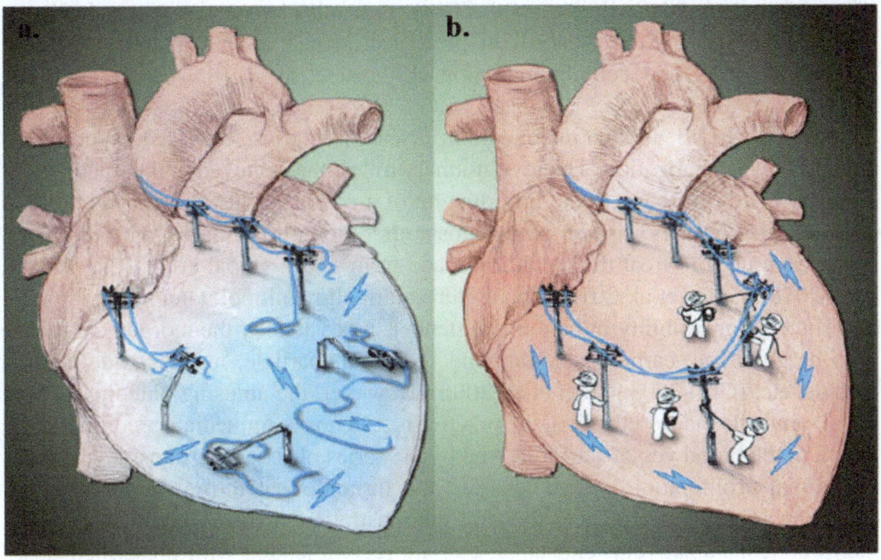

Fig. 8.14 The graphical representation of (**a**) Infracted heart and (**b**) Reconstruction of infarcted heart by using conducting polymer (Baino et al. 2014)

bioglass-reinforced PGS are developed (Souza et al. 2017). Similarly, bioactive Si patch (chitosan-reinforced calcium silicate) with enhanced conductivity was developed for the cardiac application (Wang et al. 2018). Figure 8.14 shows the graphic of infarcted heart and reconstruction by using conducting patch. However, the patches show significant enhancement in cardiac functions, but the complex surgical suturing limits the clinical applications. The development of 3D gold nanowire reinforced in albumin scaffold was demonstrated direct attachment on damaged without suturing by the eradication of NIR light. The NIR light absorption on material converts into thermal energy results in the strong attachment of the patch (Malki et al. 2018). Without thermal induction for the patch attachment, the paintable conductive patch was developed by using conductive polymer like polypyrol and dopamine through in situ polymerization (Liang et al. 2018).

Though bioinspired scaffolds have an ability to recover the damaged cardiac tissue due to lack of proliferation capacity, cardiomyocyte urges for the development of cell-loaded technologies. The skeletal myocytes, human embryonic stem cell (hESC), human-induced pluripotent stem cell (hiPSC), hematopoietic stem cells (HSCs), and human coronary artery endothelial cells (HCAECs) with suitable scaffolding material are used for the treatment of MI (Chaudhuri et al. 2017). In the following perspective, CNT-incorporated alginate framework with HCAECs seeded methacrylate collagen 3D scaffold was developed by the Izadifar and research group (Izadifar et al. 2018; Cui et al. 2016). Recently reported cell-based therapy includes MyoCELL® which comprises skeletal muscle myoblast cells and technology developed by BIOHEART, and it is in phase II/III trial in the USA

(Chaudhuri et al. 2017; Bejleri et al. 2018). Recently, two major research groups developed smart scaffolds based on piezoelectric and auxetic patches to avoid complex suturing and cell impregnation (Kapnisi et al. 2018; Arumugam et al. 2019).

In contrast to MI, another major cardiac disease is atherosclerosis which needs the coronary revascularization. The revolutionary milestone achieved in the treatment of coronary heart disease with the introduction of a cardiac stent. The major hurdle in treatment of coronary disease is restenosis after stenting. Initially, the bare metal stents were used to treat the condition made from 316 L (Iqbal et al. 2013; Bukka et al. 2018; Sabate et al. 2012), cobalt-chromium alloy, nitinol, tantalum, platinum, etc. The biocompatibility of bare metal stents is motivated by the inorganic coatings such as graphene, carbon nanodiamond, gold nanoparticle, nitride oxide, silicon carbide, etc. (Grill 2003; Chen et al. 2006). However, the amusing outcome of bare metal stents the major setback of the practice is restenosis. Immediate recoiling of the lumen of the blood vessel start followed by the series of sequence transpires leads to neointimal proliferation, which leads to the formation of thrombotic plaque. To avoid the initial consequence, the oral administration of pharmacological agents like probucol, rapamycin, cilostazol, etc. are recommended with prescribed doses (Hara et al. 2006). Due to lack of site specificity, conventional drug therapy is not the desired strategy. In the wake of drug-loaded medical devices, the drug-eluting cardiac stent came into existence. The antiproliferative drugs like paclitaxel, sirolimus, everolimus, ABT-578, tacrolimus, angiopeptine etc. were coated on the surface of the stent with suitable carriers for the desired realizing profile (Burt and Hunter 2006). The polymers such as PLLA, PLGA, chitosan, PCL etc. were well reported as carriers for drug-eluting stents. Table 8.1 summarizes major product of drug-eluting cardiac stents (Wache et al. 2003; Cui et al. 2016).

Another major cause of cardiac failure is dysfunctioning of the cardiac valve (mitral valve or tricuspid valve). The metallic and the plastic valves were used to treat the condition initially, but the poor clinical outcome has endorsed the development of the tissue-engineered valve. The idea involves the isolation of decellularized scaffolds from the allograft and xenograft procedures and conjugated with a bioactive ligand such as fibronectin, fibrin, etc., which helps in enhancement of cell adhesion and proliferation. Generally, decellularized scaffolds are obtained from

Table 8.1 The marketed product of drug-eluting stents with the brand and drug name

Sr. No.	Drug	Commercial name of the stent
1	Sirolimus	RAVEL, SIRIUS, SELECT
2	Paclitaxel	TAXUS 2, TAXUS4, ASPECT, ELUTES, EXPRESS, DELIVER, ACHIEVE
3	Everolimus	CHALLEGE
4	ABT-578	BiodivYsio
5	Dexamethasone	DEXAMET
6	Biolumis	BioMatrix flex™, BioMatrix™

human cadaver, pocrine, goat, and pig. The allograft and xenograft-derived cardiac valves exhibit significant drawbacks such as finite supply, infection, and transmission of the disease like HIV, HBV, immunogenic rejections, etc. Therefore, the most favorable strategy is a fabrication of polymeric scaffolds from highly biocompatible polymers like collagen, alginate, hyaluronic acid, polyalkenoate, etc. Similarly, PU- and PTEF-based valves also show good clinical performance. Some of the marketed polymeric cardiac valves are Sapien®, CoreValve®, Melody®, MAGNA MITRAL EASE VALVE, CENTERA, etc. (Claiborne et al. 2012).

8.17 Ophthalmology Applications

The eye is the most sensitive and complex organ and eases to surgical access by its anatomical position. The impairment vision arises basically from the malfunctioning of the eye portions like cornea, sclera, lenses, vitreous body, optical nerve damage, etc. Apart from that, impaired vision can be associated with the disease like impaired vision, myopia, hyperopia, astigmatism, presbyopia, cataracts, primary open-angle glaucoma, dry-eye syndrome, age-related macular degeneration, and diabetic retinopathy (Willoughby et al. 2010). Biomaterials are well recognized in the treatment of eye diseases in the form of contact lenses, intraocular lenses (IOL), keratoprosthesis, inlay-onlay, scleral buckling material, ophthalmic viscosurgical devices, vitreous replacement, and glaucoma shunt. PMMA is the best choice of biomaterial for the fabrication of contact lenses and other applications due to its better optical properties and biocompatibility. First, contact lenses and vitreous fluid are formulated from PMMA and PVP, respectively (Lloyd et al. 2001). The material requisite for the fabrication of contact lens is adequate oxygen permeability with the sufficient hydrophilicity, optically clear, and mechanical stability. PMMA, HEMA, silicone hydrogel, chitosan/gelatin, poly(4-methyl-1-pentene), and cellulose acetate butyrate are the best choice materials for the fabrication of contact lenses in contemporary times (Singh and Agrawal 1992; Mittal and Miranda 2018). The microbial keratitis is the potential risk associated with contact lens, and it has been countered by the development of antibiotic-loaded contact lenses. Silicon hydrogel loaded with drugs like levofloxacin, timolol, chlorohexidine, diclofenac, ketotifen fumarate, norfloxacin, vitamin A, etc. has been extensively reported for ophthalmic applications (Galante et al. 2018). Most advanced smart contact lenses are developed by using thermosensitive polymers, microfluidics techniques, and glucose sensing materials for controlled drug release (Park et al. 2018; Alvarez-Lorenzo et al. 2018). Similar to contact lenses, intraocular lenses (IOL) are used to treat myopia, cataract, and hyperopia. The choice of materials for the fabrication of IOL is PMMA, silicone hydrogel, polypropylene, polyimide, polyvinylidene fluoride (PVDF), and polysiloxane urea (Hayashi et al. 1997; Riehle et al. 2018; Stapleton et al. 2006). The practice keratoplasty is known as replacement of damaged or diseased cornea with artificial scaffold, and it is referred to as keratoprosthesis, PMMA, and PHEMA which are the choices for materials of keratoprosthesis. Biomaterials such as collagen immobilized with PVA hydrogel, HA-incorporated PVA hydrogel, graphite-

PVA hydrogel, and polyether ether ketone were used for keratoprosthesis (Myung et al. 2008; Pino et al. 2008).

Viscosurgical devices are the viscous non-active liquid materials used for the maintenance of fluidity in the anterior chamber of the eye during ophthalmic surgery, also known as phacoemulsification. Biomaterials such as hydroxypropyl methylcellulose, sodium hyaluronate, sodium chondroitin sulfate, and sodium hyaluronate are the best candidates for the viscosurgical devices (Bissen-Miyajima 2008). Glaucoma is the chronic eye disease reduces the intraocular pressure and results in permanent blindness. Glaucoma shunt is the best possible method when pharmacological and surgical innervation fails. The silicone and polypropylene are used to fabricate the shunts, and Ex-PRESS™ is the best know marketed shunt (Hendrick and Kahook 2008).

8.18 Wound Management

The injury or the physical alteration in the continuity to the biological tissue as the impact of mechanical trauma or certain pathological etiology leads to the wounds. The wounds are classified based on the duration of the healing as acute and chronic wounds. The mechanism of the wound healing is a dynamic and complex process, which involves a series of the sequential process, starts with hemostasis, inflammation, and proliferation and ends with tissue remodeling (Velnar et al. 2009). Although the physiologic mechanism contributes to wound healing, still it is a clinical challenge and needs for the development of new clinical approaches. The wound management is classified into two categories: passive and active intervention. The passive intervention is unable to regenerate the new skin cell, but it supports a local wound environment such as moisture, protection of prewound tissue, cleaning, and removal of dead tissue and minimizes the pain. Cotton, collagen patch, cellulose, and alginate are used for passive intervention. The active innervation is impregnated with active ingredients such as antibiotics, growth factor, bioactive molecule, etc. Tissue engineered scaffolds are fabricated by various methods such as freeze drying, particulate leaching, film casting, and 3D printing (Zhong et al. 2010). The chitin and chitosan are well known and gold standard polymer for wound management because of its strong antifungal and antibacterial activity (Dai et al. 2011; Ong et al. 2008). Apart from this, collagen, gelatin, fibrin, keratin, silk fibroin, eggshell membrane, PGA, PLLA, and polyacrylic acid are also extensively used for the active wound management (Zhong et al. 2010; Dainiak et al. 2010; Liu et al. 2017). In order to enhance active innervation, materials are reinforced with various additives like inorganic filler, drugs, and growth factors (Zhong et al. 2010). Table 8.2 presents the various polymers and additive for active wound management.

Table 8.2 The biomaterial with the bioactive agent used in active wound management

Sr. No.	Biomaterial	Bioactive agent	Method/type of scaffold	References
1	Chitosan	Montmorillonite	Intercalated film	Aguzzi et al. (2014)
2	Chitosan–PVP	Titanium dioxide nanoparticle		Archana et al. (2013)
3	PVA	Na-montmorillonite	Freeze–thawing scaffold	Kokabi et al. (2007)
4	Chitosan	Halloysite	Simple solid-dispersed scaffold	Sandri et al. (2017)
5	Eggshell membrane	Copper (Cu)-containing bioactive glass	Pulsed laser deposited scaffold	Li et al. (2016)
6	PCL-gelatin	*	Electrospun scaffold	Chong et al. (2007)
7	Chitosan	ZnO and castor oil	Casted film scaffold	Díez-Pascual and Díez-Vicente (2015)
8	PVA-chitosan	Minocycline	Freeze-thaw hydrogel	Sung et al. (2010)
9	Chitosan	Copper nanoparticle	Solid dispersion scaffold	Gopal et al. (2014)
10	Acrylic acid and *N,N′*-methylene bisacrylamide	Ag/graphene	Hydrogel	Fan et al. (2014)
11	Collagen	Gold nanoparticle	Cross-linked sponges	Akturk et al. (2016)
12	Collagen chitosan	Grafted *N*-acrylamide	Immobilized fabric	Wang et al. (2008)
13	Chitosan	AgZnO	Lyophilized sponges	Lu et al. (2017)
14	Gelatin	Polyurethane	Electrospun scaffold	Vedakumari et al. (2015)
15	Chitosan	Fibrin, quercetin	Electrospun scaffold	Vedakumari et al. (2017)
16	Chitosan–PEO	VEGF, platelet-derived growth factor-BB	Electrospun scaffold	Xie et al. (2013)
17	Cellulose	Kaolin	Film-casted scaffold	Wanna et al. (2013)
18	PVA	Glucose oxidase MCNT	Electrospun scaffold	Santos et al. (2014)
19	PCL	ZnO	Electrospun scaffold	Augustine et al. (2014)

(continued)

Table 8.2 (continued)

Sr. No.	Biomaterial	Bioactive agent	Method/type of scaffold	References
20	Chitosan-PVA	Curcumin-silver nanoparticle	Casted film	Vimala et al. (2011)
21	PVA	Calcium alginate	Electrospun scaffold	Tarun and Gobi (2012)
22	PVA	Cellulose nanowhiskers	Freeze–thawing	Gonzalez et al. (2014)
23	Polyurethane	PVDF	Electrospun scaffold	Guo et al. (2012)
24	Poly (3-hydroxybutyrate-*co*-3-hydroxyvalerate)	Collagen, graphene oxide	Electrospun scaffold	Zine and Sinha (2017)
25	Silk	Epidermal growth factor	Electrospun scaffold	Schneider et al. (2009)
26	Peg	Plasmid bFGF polyplex	Electrospun scaffold	Yang et al. (2011)
27	PLGA	Collagen	Electrospun scaffold	Liu et al. (2010)

* No information available
VEGF Vascular endothelial growth factor

8.19 Nerve Regeneration

The nerve injury or damage is still a major clinical challenge due to lack of self-repairing capability. Similar to bone nerve, tissue engineering is categorized as cell, scaffold, and the growth factor-based approaches. Earlier, allografts (epineural sheath, tendon, vein-decellularized muscle, skeletal muscle-filled Schwann cell) were frequently used to treat the neurological problem associated with nerve damage (Johnson et al. 2005; Ide et al. 1983). Undesired reactions and immunological rejections are the bottlenecks for the practice. The researchers have developed several biodegradable biomaterials to support nerve cell repair and regeneration since the last 50 years. PLGA and collagen are choice of material for fabrication of nerve conduction because of their good biocompatibility and biodegradability. The polymers like PCL, PCL-gelatin, PLLA, chitosan, etc., were also reported for the nerve regeneration (Ghasemi-Mobarakeh et al. 2011; Yang et al. 2005). In addition, the bioactive scaffolds such as Schwan cell-loaded PCL and PLGA foam encapsulated with the Schwan cell are also well established for nerve repair (Hadlock et al. 2000). Growth factors such as human glial cell-derived neurotropic factor (GDNF), brain-derived neurotropic factor (BDNF), insulin-like growth factor (IGF-1, IGF-2), platelet-derived growth factors, fibroblast growth factor, and ciliary neurotropic factor encapsulated into the engineered scaffolds for enhanced nerve regeneration (Dodla and Bellamkonda 2008; Wood et al. 2009).

In contrast to conventional polymer, the conducting polymers like polypyrrole (PPy), PDLA, poly(3,4-ethylenedioxythiophene), and polyaniline participated in nerve signal conduction and regeneration (Ghasemi-Mobarakeh et al. 2011). Smart polymer PVDF is utilized for its piezoelectricity for nerve regeneration (Rajabi et al. 2015).

8.20 Cosmetic Applications

The cosmetic and aesthetic rejuvenation of the organs like lips, a facial fold, breast, and wrinkles are most frequently performed. The biomaterials like silicon, e-PTEF, collagen sponges, fats, and hyaluronic acid are extensively used in the aforementioned process (Niamtu 2006). Among them, the hyaluronic acid is mostly used as dermal filler in facial fine lines such as lip rhytides, extended radial cheek lines, and horizontal forehead furrows etc. (Fagien 2010). Surgisis is an acellular collagen (surgisis is derived from the ECM of the small intestine of the pig) reported for breast reconstruction, mastopexy, lip augmentation, nasolabial fold, labiomandibular folds, and glabellar folds etc. (Centeno 2009). Similarly, adipose-derived cells were used for the aesthetic surgery of lips and hand rejuvenation (Mehrabani et al. 2013). Beautification of an eye was being improved by wearing cosmetic contact lenses. Generally, the lenses are fabricted with PMMA in addtion of FDA approved pigment. An ocular prosthesis is a class of device used to replace the damaged eye with enhanced appearance. Initially, ocular prosthesis or damaged eye is directly replaced with pig eye; later on, it was fabricated from the glass, known as the glassy eye. Today, most of the orbital implants or ocular prostheses are made up of PMMA (Baino et al. 2014). Fig. 8.15 shows the PMMA ocular prosthesis, and biomaterials such as silicon, polyacrylamide gel, PLGA scaffold, and collagen were extensively used in breast augmentation (Siggelkow et al. 2003).

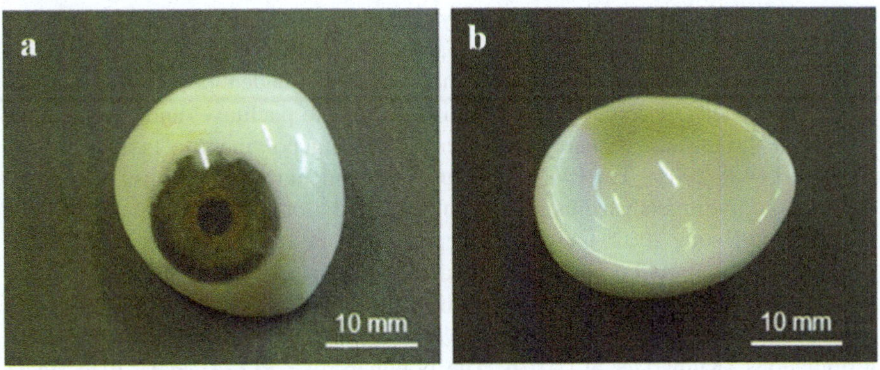

Fig. 8.15 The digital image of the hand-painted aesthetic PMMA orbital implant looks similar to normal eye: (**a**) Frontal view and (**b**) Lateral view of the prosthesis (Baino et al. 2014)

8.21 Biosensor and Bioelectronics

The biosensor is defined as the devices used to detect the biological substance into a measurable physical quantity. The key components of the biosensor are bioanalyte, bioreceptor, transducer, and the detector. There are different types of biosensors such as electrochemical, immunosensor, enzymatic, piezoelectric, microfluidic biosensor, etc. More emphasis of the biosensor is out of the scope of the present chapter. In the fabrication of electrochemical biosensor, conductive polymers like polypyrrole, polyaniline, and chitosan are used along with additive like CNT, ZnO, graphene, nanodiamond, etc. Some of the best-reported biosensors are chitosan combined with carbon nanotube and glucose peroxidase, cellulose zinc oxide composite, and chitosan-MCNT for glucose detection. Similarly, carboxymethylcellulose/ZnCdS, cellulose-graft-poly(*p*-dioxanone), and chitosan-folic acid-combined sensor were deployed for the detection of cancer. Graphene-modified cellulose-based biosensor was used to detect the HIV-I (Pérez et al. 2018).

Biomaterials in the wearable electronics have key attention due to flexibility, good mechanical properties, renewability, and biodegradability. The polymers like PLLA, which are obtained from the natural source like soybean, potato, and beets, have unique piezoelectric property and are used in the piezoelectric biosensor. Flexible polymers like polyurethane (PU) and polyhydroxyalkanoate (PHB, PHBV) polymers exhibit good flexibility, so they are used in the pressure sensor. Natural polymers like silk fibroin, cellulose, chitin, and chitosan are reported for the fabrication of flexible electronic devices (Sun et al. 2018).

8.22 Drug Delivery

A large group of biomaterials is vastly employed as drug carriers, and enormous advancements are reported on every day for drug delivery. Primarily, the polymers have a major stack in this category and demonstrated controlled and targeted drug delivery. The trending approach nowadays is drug-loaded implants for onsite delivery with controlled release. The approach is successfully employed in various applications, primarily drug-coated stents, antibiotic-loaded bone cement, drug-loaded hydrogels, etc. The antiproliferative drugs like paclitaxel and sirolimus are coated on the surface of bare metal stents by using polymers to treat restenosis (Oberhoff et al. 2002). The PVA, PEG, and poloxamer are the polymers used in the synthesis of hydrogels used as implantable scaffolds for drug delivery applications (Peppas and Huang 2002; Langer and Peppas 2003). From the last 10 years, plenty of literature is reported on nanostructured materials for drug delivery, but very few are accepted for clinical applications. The nanoformulations of polymers are well established for controlling drug delivery commercially against conventional formulations (Kumari et al. 2010). The polymers like PLGA, PCL, PLA, PCL, chitosan, gelatin, and poly(alkyl-cyanoacrylates) (PAC) are extensively explored for drug delivery in the form of nanoformulations.

References

Agrawal CM (1998) Reconstructing the human body using biomaterials. JOM 50:31–35

Aguzzi C, Sandri G, Bonferoni C, Cerezo P, Rossi S, Ferrari F, Caramella C, Viseras C (2014) Solid state characterisation of silver sulfadiazine loaded on montmorillonite/chitosan nanocomposite for wound healing. Colloids Surf B: Biointerfaces 113:152–157

Ahmad M, Manzoor K, Ikram S (2019) Chitosan nanocomposites for bone and cartilage regeneration. In: Applications of nanocomposite materials in dentistry. Elsevier

Akturk O, Kismet K, Yasti AC, Kuru S, Duymus ME, Kaya F, Caydere M, Hucumenoglu S, Keskin D (2016) Collagen/gold nanoparticle nanocomposites: a potential skin wound healing biomaterial. J Biomater Appl 31:283–301

Alvarez-Lorenzo C, Anguiano-Igea S, Varela-García A, Vivero-Lopez M, Concheiro A (2018) Bioinspired hydrogels for drug-eluting contact lenses. Acta Biomater

Archana D, Singh BK, Dutta J, Dutta P (2013) In vivo evaluation of chitosan–PVP–titanium dioxide nanocomposite as wound dressing material. Carbohydr Polym 95:530–539

Arora M, Chan EK, Gupta S, Diwan AD (2013) Polymethylmethacrylate bone cements and additives: a review of the literature. World J Orthop 4:67

Arumugam R, Srinadhu ES, Subramanian B, Nallani S (2019) β-PVDF based electrospun nanofibers–a promising material for developing cardiac patches. Med Hypotheses 122:31–34

Augustine R, Dominic EA, Reju I, Kaimal B, Kalarikkal N, Thomas S (2014) Electrospun polycaprolactone membranes incorporated with ZnO nanoparticles as skin substitutes with enhanced fibroblast proliferation and wound healing. RSC Adv 4:24777–24785

Baino F, Perero S, Ferraris S, Miola M, Balagna C, Verné E, Vitale-Brovarone C, Coggiola A, Dolcino D, Ferraris M (2014) Biomaterials for orbital implants and ocular prostheses: overview and future prospects. Acta Biomater 10:1064–1087

Bandyopadhyay A, Bose S (2013) Characterization of biomaterials. Newnes

Bejleri D, Streeter BW, Nachlas AL, Brown ME, Gaetani R, Christman KL, Davis ME (2018) A bioprinted cardiac patch composed of cardiac-specific extracellular matrix and progenitor cells for heart repair. Adv Healthc Mater 7:1800672

Bhong SY, More N, Choppadandi M, Kapusetti G (2019) Review on carbon nanomaterials as typical candidates for orthopaedic coatings. SN Appl Sci 1:76

Bilezikian JP, Raisz LG, Martin TJ (2008) Principles of bone biology. Academic Press

Bissen-Miyajima H (2008) Ophthalmic viscosurgical devices. Curr Opin Ophthalmol 19:50–54

Bohner M (2008) Bioresorbable ceramics. Elsevier, Degradation rate of bioresorbable materials

Bose S, Roy M, Bandyopadhyay A (2012) Recent advances in bone tissue engineering scaffolds. Trends Biotechnol 30:546–554

Bradl H (2005) Sources and origins of heavy metals. Elsevier, Interface Science and Technology

Bukka M, Rednam PJ, Sinha M (2018) Drug-eluting balloon: design, technology and clinical aspects. Biomed Mater 13:032001

Burg KJ, Porter S, Kellam JF (2000) Biomaterial developments for bone tissue engineering. Biomaterials 21:2347–2359

Burt HM, Hunter WL (2006) Drug-eluting stents: a multidisciplinary success story. Adv Drug Deliv Rev 58:350–357

Cancedda R, Dozin B, Giannoni P, Quarto R (2003) Tissue engineering and cell therapy of cartilage and bone. Matrix Biol 22:81–91

Centeno RF (2009) Surgisis acellular collagen matrix in aesthetic and reconstructive plastic surgery soft tissue applications. Clin Plast Surg 36:229–240

Chaudhuri R, Ramachandran M, Moharil P, Harumalani M, Jaiswal AK (2017) Biomaterials and cells for cardiac tissue engineering: current choices. Mater Sci Eng C 79:950–957

Chen MS, John JM, Chew DP, Lee DS, Ellis SG, Bhatt DL (2006) Bare metal stent restenosis is not a benign clinical entity. Am Heart J 151:1260–1264

Cheng M (2003) Medical device regulations: global overview and guiding principles. World Health Organization

Chi N-H, Yang M-C, Chung T-W, Chou N-K, Wang S-S (2013) Cardiac repair using chitosan-hyaluronan/silk fibroin patches in a rat heart model with myocardial infarction. Carbohydr Polym 92:591–597

Choi JH, Kim DK, Song JE, Oliveira JM, Reis RL, Khang G (2018) Silk fibroin-based scaffold for bone tissue engineering. Springer, Novel Biomaterials for Regenerative Medicine

Chong E, Phan T, Lim I, Zhang Y, Bay B, Ramakrishna S, Lim C (2007) Evaluation of electrospun PCL/gelatin nanofibrous scaffold for wound healing and layered dermal reconstitution. Acta Biomater 3:321–330

Claiborne TE, Slepian MJ, Hossainy S, Bluestein D (2012) Polymeric trileaflet prosthetic heart valves: evolution and path to clinical reality. Expert Rev Med Devices 9:577–594

Cui Z, Yang B, Li R-K (2016) Application of biomaterials in cardiac repair and regeneration. Engineering 2:141–148

D'Amore A, Yoshizumi T, Luketich SK, Wolf MT, Gu X, Cammarata M, Hoff R, Badylak SF, Wagner WR (2016) Bi-layered polyurethane–extracellular matrix cardiac patch improves ischemic ventricular wall remodeling in a rat model. Biomaterials 107:1–14

Dai T, Tanaka M, Huang Y-Y, Hamblin MR (2011) Chitosan preparations for wounds and burns: antimicrobial and wound-healing effects. Expert Rev Anti-Infect Ther 9:857–879

Dainiak MB, Allan IU, Savina IN, Cornelio L, James ES, James SL, Mikhalovsky SV, Jungvid H, Galaev IY (2010) Gelatin–fibrinogen cryogel dermal matrices for wound repair: preparation, optimisation and in vitro study. Biomaterials 31:67–76

Díez-Pascual AM, Díez-Vicente AL (2015) Wound healing bionanocomposites based on castor oil polymeric films reinforced with chitosan-modified ZnO nanoparticles. Biomacromolecules 16:2631–2644

Dodla MC, Bellamkonda RV (2008) Differences between the effect of anisotropic and isotropic laminin and nerve growth factor presenting scaffolds on nerve regeneration across long peripheral nerve gaps. Biomaterials 29:33–46

DOS Santos V, Brandalise RN, Savaris M (2017) Biomaterials: characteristics and properties. Springer, Engineering of Biomaterials

Dvir T, Timko BP, Brigham MD, Naik SR, Karajanagi SS, Levy O, Jin H, Parker KK, Langer R, Kohane DS (2011) Nanowired three-dimensional cardiac patches. Nat Nanotechnol 6:720

Fagien S (2010) Variable reconstitution of injectable hyaluronic acid with local anesthetic for expanded applications in facial aesthetic enhancement. Dermatol Surg 36:815–821

Fan Z, Liu B, Wang J, Zhang S, Lin Q, Gong P, Ma L, Yang S (2014) A novel wound dressing based on ag/graphene polymer hydrogel: effectively kill bacteria and accelerate wound healing. Adv Funct Mater 24:3933–3943

Ferreira AM, Gentile P, Chiono V, Ciardelli G (2012) Collagen for bone tissue regeneration. Acta Biomater 8:3191–3200

Galante R, Oliveira AS, Topete A, Ghisleni D, Braga M, Pinto TJ, Colaço R, Serro AP (2018) Drug-eluting silicone hydrogel for therapeutic contact lenses: impact of sterilization methods on the system performance. Colloids Surf B: Biointerfaces 161:537–546

Ghanavati S, Shishesaz MR, Farzam M, Danaee I (2016) Effects of surface treatment on corrosion resistance of 304L and 316L stainless steel implants in Hank's solution. Iran J Oil Gas Sci Technol 5:65–72

Ghasemi-Mobarakeh L, Prabhakaran MP, Morshed M, Nasr-Esfahani MH, Baharvand H, Kiani S, Al-Deyab SS, Ramakrishna S (2011) Application of conductive polymers, scaffolds and electrical stimulation for nerve tissue engineering. J Tissue Eng Regen Med 5:e17–e35

Gong T, Xie J, Liao J, Zhang T, Lin S, Lin Y (2015) Nanomaterials and bone regeneration. Bone Res 3:15029

Gonzalez JS, Ludueña LN, Ponce A, Alvarez VA (2014) Poly (vinyl alcohol)/cellulose nanowhiskers nanocomposite hydrogels for potential wound dressings. Mater Sci Eng C 34:54–61

Gopal A, Kant V, Gopalakrishnan A, Tandan SK, Kumar D (2014) Chitosan-based copper nanocomposite accelerates healing in excision wound model in rats. Eur J Pharmacol 731:8–19

Grill A (2003) Diamond-like carbon coatings as biocompatible materials—an overview. Diam Relat Mater 12:166–170

Grosfeld E-C, Hoekstra JWM, Herber R-P, Ulrich DJ, Jansen JA, VAN DEN Beucken JJ (2016) Long-term biological performance of injectable and degradable calcium phosphate cement. Biomed Mater 12:015009

Guo H-F, Li Z-S, Dong S-W, Chen W-J, Deng L, Wang Y-F, Ying D-J (2012) Piezoelectric PU/PVDF electrospun scaffolds for wound healing applications. Colloids Surf B: Biointerfaces 96:29–36

Hadlock T, Sundback C, Hunter D, Cheney M, Vacanti JP (2000) A polymer foam conduit seeded with Schwann cells promotes guided peripheral nerve regeneration. Tissue Eng 6:119–127

Hajiali F, Tajbakhsh S, Shojaei A (2018) Fabrication and properties of polycaprolactone composites containing calcium phosphate-based ceramics and bioactive glasses in bone tissue engineering: a review. Polym Rev 58:164–207

Hara H, Nakamura M, Palmaz JC, Schwartz RS (2006) Role of stent design and coatings on restenosis and thrombosis. Adv Drug Deliv Rev 58:377–386

Hayashi K, Hayashi H, Nakao F, Hayashi F (1997) Reduction in the area of the anterior capsule opening after polymethylmethacrylate, silicone, and soft acrylic intraocular lens implantation. Am J Ophthalmol 123:441–447

Hench LL (2006) The story of bioglass®. J Mater Sci Mater Med 17:967–978

Hendrick AM, Kahook MY (2008) Ex-PRESS™ mini glaucoma shunt: surgical technique and review of clinical experience. Expert Rev Med Devices 5:673–677

Hermawan H, Ramdan D, Djuansjah JR (2011) Metals for biomedical applications. Biomedical engineering-from theory to applications. InTech

Hollister SJ (2005) Porous scaffold design for tissue engineering. Nat Mater 4:518

Hosseinzadeh HRS, Emami M, Lahiji F, Shahi AS, Masoudi A, Emami S (2013) The acrylic bone cement in arthroplasty. Arthroplasty-Update, InTech

Hubbell JA (1995) Biomaterials in tissue engineering. Biotechnology 13:565

Hutmacher DW (2000) Scaffolds in tissue engineering bone and cartilage. The Biomaterials: Silver Jubilee Compendium. Elsevier

Ide C, Tohyama K, Yokota R, Nitatori T, Onodera S (1983) Schwann cell basal lamina and nerve regeneration. Brain Res 288:61–75

Inzana JA, Olvera D, Fuller SM, Kelly JP, Graeve OA, Schwarz EM, Kates SL, Awad HA (2014) 3D printing of composite calcium phosphate and collagen scaffolds for bone regeneration. Biomaterials 35:4026–4034

Iqbal J, Gunn J, Serruys PW (2013) Coronary stents: historical development, current status and future directions. Br Med Bull 106:193–211

Izadifar M, Chapman D, Babyn P, Chen X, Kelly ME (2018) UV-assisted 3D bioprinting of nanoreinforced hybrid cardiac patch for myocardial tissue engineering. Tissue Eng Part C Methods 24:74–88

Jacob J, More N, Kalia K, Kapusetti G (2018) Piezoelectric smart biomaterials for bone and cartilage tissue engineering. Inflammation and Regeneration 38:2

Johnson EO, Zoubos AB, Soucacos PN (2005) Regeneration and repair of peripheral nerves. Injury 36:S24–S29

Kapnisi M, Mansfield C, Marijon C, Guex AG, Perbellini F, Bardi I, Humphrey EJ, Puetzer JL, Mawad D, Koutsogeorgis DC (2018) Auxetic cardiac patches with tunable mechanical and conductive properties toward treating myocardial infarction. Adv Funct Mater 28:1800618

Kim K-H, Jeong L, Park H-N, Shin S-Y, Park W-H, Lee S-C, Kim T-I, Park Y-J, Seol Y-J, Lee Y-M (2005) Biological efficacy of silk fibroin nanofiber membranes for guided bone regeneration. J Biotechnol 120:327–339

Kohane DS, Langer R (2010) Biocompatibility and drug delivery systems. Chem Sci 1:441–446

Kokabi M, Sirousazar M, Hassan ZM (2007) PVA–clay nanocomposite hydrogels for wound dressing. Eur Polym J 43:773–781

Kokubo T (1991) Bioactive glass ceramics: properties and applications. Biomaterials 12:155–163

Kulinets I (2015) Biomaterials and their applications in medicine. Regulatory affairs for biomaterials and medical devices. Elsevier

Kumari A, Yadav SK, Yadav SC (2010) Biodegradable polymeric nanoparticles based drug delivery systems. Colloids Surf B: Biointerfaces 75:1–18

Langer R, Peppas NA (2003) Advances in biomaterials, drug delivery, and bionanotechnology. AICHE J 49:2990–3006

Langer R, Tirrell DA (2004) Designing materials for biology and medicine. Nature 428:487

Lee K, Silva EA, Mooney DJ (2010) Growth factor delivery-based tissue engineering: general approaches and a review of recent developments. J R Soc Interface 8:153–170

Leppik L, Zhihua H, Mobini S, Parameswaran VT, Eischen-Loges M, Slavici A, Helbing J, Pindur L, Oliveira KM, Bhavsar MB (2018) Combining electrical stimulation and tissue engineering to treat large bone defects in a rat model. Sci Rep 8:6307

Li J, Zhai D, Lv F, Yu Q, Ma H, Yin J, Yi Z, Liu M, Chang J, Wu C (2016) Preparation of copper-containing bioactive glass/eggshell membrane nanocomposites for improving angiogenesis, antibacterial activity and wound healing. Acta Biomater 36:254–266

Liang S, Zhang Y, Wang H, Xu Z, Chen J, Bao R, Tan B, Cui Y, Fan G, Wang W (2018) Paintable and rapidly bondable conductive hydrogels as therapeutic cardiac patches. Adv Mater 30:1704235

Liu X, Ma PX (2004) Polymeric scaffolds for bone tissue engineering. Ann Biomed Eng 32:477–486

Liu S-J, Kau Y-C, Chou C-Y, Chen J-K, Wu R-C, Yeh W-L (2010) Electrospun PLGA/collagen nanofibrous membrane as early-stage wound dressing. J Membr Sci 355:53–59

Liu M, Duan X-P, Li Y-M, Yang D-P, Long Y-Z (2017) Electrospun nanofibers for wound healing. Mater Sci Eng C 76:1413–1423

Lloyd AW, Faragher RG, Denyer SP (2001) Ocular biomaterials and implants. Biomaterials 22:769–785

Lu J, Descamps M, Dejou J, Koubi G, Hardouin P, Lemaitre J, Proust JP (2002) The biodegradation mechanism of calcium phosphate biomaterials in bone. J Biomed Mater Res 63:408–412

Lu Z, Gao J, He Q, Wu J, Liang D, Yang H, Chen R (2017) Enhanced antibacterial and wound healing activities of microporous chitosan-Ag/ZnO composite dressing. Carbohydr Polym 156:460–469

Lutolf M, Hubbell J (2005) Synthetic biomaterials as instructive extracellular microenvironments for morphogenesis in tissue engineering. Nat Biotechnol 23:47

Malki M, Fleischer S, Shapira A, Dvir T (2018) Gold Nanorod-based engineered cardiac patch for suture-free engraftment by near IR. Nano Lett 18:4069

Mansour HM, Sohn M, Al-Ghananeem A, Deluca PP (2010) Materials for pharmaceutical dosage forms: molecular pharmaceutics and controlled release drug delivery aspects. Int J Mol Sci 11:3298–3322

Matthews JA, Wnek GE, Simpson DG, Bowlin GL (2002) Electrospinning of collagen nanofibers. Biomacromolecules 3:232–238

Mehrabani D, Mehrabani G, Zare S, Manafi A (2013) Adipose-derived stem cells (ADSC) and aesthetic surgery: a mini review. World J Plast Surg 2:65

Meijer GJ, DE Bruijn JD, Koole R, VAN Blitterswijk CA (2007) Cell-based bone tissue engineering. PLoS Med 4:e9

Meroni D, Ardizzone S (2018) Preparation and application of hybrid nanomaterials. Multidisciplinary Digital Publishing Institute

Mieszawska AJ, Fourligas N, Georgakoudi I, Ouhib NM, Belton DJ, Perry CC, Kaplan DL (2010) Osteoinductive silk–silica composite biomaterials for bone regeneration. Biomaterials 31:8902–8910

Mittal S, Miranda O (2018) Recent advancements in biodegradable ocular implants. Curr Drug Deliv 15:144–154

Mohan H (2005) Textbook of pathology. Jaypee Brothers Medical Publishers, New Delhi

More N, Kapusetti G (2017) Piezoelectric material–a promising approach for bone and cartilage regeneration. Med Hypotheses 108:10–16

Myung D, Duhamel PE, Cochran JR, Noolandi J, Ta CN, Frank CW (2008) Development of hydrogel-based keratoprostheses: A materials perspective. Biotechnol Prog 24:735–741

Naahidi S, Jafari M, Logan M, Wang Y, Yuan Y, Bae H, Dixon B, Chen P (2017) Biocompatibility of hydrogel-based scaffolds for tissue engineering applications. Biotechnol Adv 35:530–544

Niamtu J (2006) Advanta ePTFE facial implants in cosmetic facial surgery. J Oral Maxillofac Surg 64:543–549

Niinomi M (2002) Recent metallic materials for biomedical applications. Metall Mater Trans A 33:477

Numata K, Kaplan D (2011) Biologically derived scaffolds. Advanced wound repair therapies. Elsevier

Oberhoff M, Herdeg C, Baumbach A, Karsch KR (2002) Stent-based antirestenotic coatings (sirolimus/paclitaxel). Catheter Cardiovasc Interv 55:404–408

O'brien FJ (2011) Biomaterials & scaffolds for tissue engineering. Mater Today 14:88–95

Ogueri KS, Jafari T, Ivirico JLE, Laurencin CT (2018) Polymeric biomaterials for scaffold-based bone regenerative engineering. In: Regenerative engineering and translational medicine. Springer, New York, pp 1–27

Ong S-Y, Wu J, Moochhala SM, Tan M-H, Lu J (2008) Development of a chitosan-based wound dressing with improved hemostatic and antimicrobial properties. Biomaterials 29:4323–4332

Park J, Kim J, Kim S-Y, Cheong WH, Jang J, Park Y-G, Na K, Kim Y-T, Heo JH, Lee CY (2018) Soft, smart contact lenses with integrations of wireless circuits, glucose sensors, and displays. Science Adv 4:eaap9841

Pattanashetti NA, Heggannavar GB, Kariduraganavar MY (2017) Smart biopolymers and their biomedical applications. Procedia Manufacturing 12:263–279

Peppas NA, Huang Y (2002) Polymers and gels as molecular recognition agents. Pharm Res 19:578–587

Pérez JAC, Sosa-Hernández JE, Hussain SM, Bilal M, Parra-Saldivar R, Iqbal HM (2018) Bioinspired biomaterials and enzyme-based biosensors for point-of-care applications with reference to cancer and bio-imaging. Biocatal Agric Biotechnol 17:168–176

Petite H, Viateau V, Bensaid W, Meunier A, DE Pollak C, Bourguignon M, Oudina K, Sedel L, Guillemin G (2000) Tissue-engineered bone regeneration. Nat Biotechnol 18:959

Pino M, Stingelin N, Tanner K (2008) Nucleation and growth of apatite on NaOH-treated PEEK, HDPE and UHMWPE for artificial cornea materials. Acta Biomater 4:1827–1836

Quarles LD (2008) Endocrine functions of bone in mineral metabolism regulation. J Clin Invest 118:3820–3828

Rajabi AH, Jaffe M, Arinzeh TL (2015) Piezoelectric materials for tissue regeneration: a review. Acta Biomater 24:12–23

Ramakrishna S, Mayer J, Wintermantel E, Leong KW (2001) Biomedical applications of polymer-composite materials: a review. Compos Sci Technol 61:1189–1224

Ramalingam M, Kumar TS, Ramakrishna S, Soboyejo WO (2016) Biomaterials: a nano approach. CRC Press

Rezwan K, Chen Q, Blaker J, Boccaccini AR (2006) Biodegradable and bioactive porous polymer/inorganic composite scaffolds for bone tissue engineering. Biomaterials 27:3413–3431

Riehle N, Thude S, Götz T, Kandelbauer A, Thanos S, Tovar GE, Lorenz G (2018) Influence of PDMS molecular weight on transparency and mechanical properties of soft polysiloxane-urea-elastomers for intraocular lens application. Eur Polym J 101:190–201

Robling AG, Castillo AB, Turner CH (2006) Biomechanical and molecular regulation of bone remodeling. Annu Rev Biomed Eng 8:455–498

Roeder RK (2013) Mechanical characterization of biomaterials. In: Characterization of biomaterials. Elsevier, Amsterdam

Roseti L, Parisi V, Petretta M, Cavallo C, Desando G, Bartolotti I, Grigolo B (2017) Scaffolds for bone tissue engineering: state of the art and new perspectives. Mater Sci Eng C 78:1246–1262

Sabate M, Cequier A, Iñiguez A, Serra A, Hernandez-Antolin R, Mainar V, Valgimigli M, Tespili M, Den Heijer P, Bethencourt A (2012) Everolimus-eluting stent versus bare-metal stent in ST-segment elevation myocardial infarction (EXAMINATION): 1 year results of a randomised controlled trial. Lancet 380:1482–1490

Sadiasa A, Sarkar SK, Franco RA, Min YK, Lee BT (2014) Bioactive glass incorporation in calcium phosphate cement-based injectable bone substitute for improved in vitro biocompatibility and in vivo bone regeneration. J Biomater Appl 28:739–756

Samavedi S, Poindexter LK, van Dyke M, Goldstein AS (2014) Synthetic biomaterials for regenerative medicine applications. In: Regenerative medicine applications in organ transplantation. Elsevier, London

Sandri G, Aguzzi C, Rossi S, Bonferoni MC, Bruni G, Boselli C, Cornaglia AI, Riva F, Viseras C, Caramella C (2017) Halloysite and chitosan oligosaccharide nanocomposite for wound healing. Acta Biomater 57:216–224

Santos JCC, Mansur AA, Ciminelli VS, Mansur HS (2014) Nanocomposites of poly (vinyl alcohol)/functionalized-multiwall carbon nanotubes conjugated with glucose oxidase for potential application as scaffolds in skin wound healing. Int J Polym Mater Polym Biomater 63:185–196

Schneider A, Wang X, Kaplan D, Garlick J, Egles C (2009) Biofunctionalized electrospun silk mats as a topical bioactive dressing for accelerated wound healing. Acta Biomater 5:2570–2578

Schoen FJ (2013) Introduction: "biological responses to biomaterials". Biomaterials Science, 3rd edn. Elsevier

Sharma R, Kapusetti G, Bhong SY, Roy P, Singh SK, Singh S, Balavigneswaran CK, Mahato KK, Ray B, Maiti P (2017) Osteoconductive amine-functionalized graphene–poly (methyl methacrylate) bone cement composite with controlled exothermic polymerization. Bioconjug Chem 28:2254–2265

Shevach M, Maoz BM, Feiner R, Shapira A, Dvir T (2013) Nanoengineering gold particle composite fibers for cardiac tissue engineering. J Mater Chem B 1:5210–5217

Siggelkow W, Faridi A, Spiritus K, Klinge U, Rath W, Klosterhalfen B (2003) Histological analysis of silicone breast implant capsules and correlation with capsular contracture. Biomaterials 24:1101–1109

Singh J, Agrawal K (1992) Polymeric materials for contact lenses. J Macromol Sci Polym Rev 32:521–534

Society, T. A. C. 2018. The American Ceramic Society [online]. Available: https://ceramics.org/about/what-are-engineered-ceramics-and-glass/structure-and-properties-of-ceramics

Souza MT, Tansaz S, Zanotto ED, Boccaccini AR (2017) Bioactive glass Fiber-reinforced PGS matrix composites for cartilage regeneration. Materials (Basel, Switzerland) 10:83

Stapleton F, Stretton S, Papas E, Skotnitsky C, Sweeney DF (2006) Silicone hydrogel contact lenses and the ocular surface. Ocul Surf 4:24–43

Stout DA, Basu B, Webster TJ (2011) Poly (lactic–co-glycolic acid): carbon nanofiber composites for myocardial tissue engineering applications. Acta Biomater 7:3101–3112

Sun Q, Qian B, Uto K, Chen J, Liu X, Minari T (2018) Functional biomaterials towards flexible electronics and sensors. Biosens Bioelectron 119:237

Sung JH, Hwang M-R, Kim JO, Lee JH, Kim YI, Kim JH, Chang SW, Jin SG, Kim JA, Lyoo WS (2010) Gel characterisation and in vivo evaluation of minocycline-loaded wound dressing with enhanced wound healing using polyvinyl alcohol and chitosan. Int J Pharm 392:232–240

Swetha M, Sahithi K, Moorthi A, Srinivasan N, Ramasamy K, Selvamurugan N (2010) Biocomposites containing natural polymers and hydroxyapatite for bone tissue engineering. Int J Biol Macromol 47:1–4

Tandon B, Blaker JJ, Cartmell SH (2018) Piezoelectric materials as stimulatory biomedical materials and scaffolds for bone repair. Acta Biomater 73:1

Tarun K, Gobi N (2012) Calcium alginate/PVA blended nano fibre matrix for wound dressing. Indian J Fibre Text Res 37:127–132

Teo AJ, Mishra A, Park I, Kim Y-J, Park W-T, Yoon Y-J (2016) Polymeric biomaterials for medical implants and devices. ACS Biomater Sci Eng 2:454–472
Teoh SH (2004) Introduction to biomaterials engineering and processing—an overview. In: Engineering materials for biomedical applications. World Scientific, Singapore
Ubelaker, D. 1984. Human skeletal remains. Excavation, analysis, interpretation. Taraxacum. Federation Dentaire Internnationale (1982). Nouveau Système de Dèsignation des dents. Bulletin Et Memoires de la Sociedad d'Antropologie de Paris. Serie, 12, 83–85
Vasquez-Sancho F, Abdollahi A, Damjanovic D, Catalan G (2018) Flexoelectricity in bones. Adv Mater 30:1705316
Vedakumari WS, Prabu P, Sastry TP (2015) Chitosan-fibrin nanocomposites as drug delivering and wound healing materials. J Biomed Nanotechnol 11:657–667
Vedakumari WS, Ayaz N, Karthick AS, Senthil R, Sastry TP (2017) Quercetin impregnated chitosan–fibrin composite scaffolds as potential wound dressing materials—fabrication, characterization and in vivo analysis. Eur J Pharm Sci 97:106–112
Velde B, Druc IC (2012) Archaeological ceramic materials: origin and utilization. Springer, London
Velnar T, Bailey T, Smrkolj V (2009) The wound healing process: an overview of the cellular and molecular mechanisms. J Int Med Res 37:1528–1542
Vimala K, Yallapu MM, Varaprasad K, Reddy NN, Ravindra S, Naidu NS, Raju KM (2011) Fabrication of curcumin encapsulated chitosan-PVA silver nanocomposite films for improved antimicrobial activity. J Biomater Nanobiotechnol 2:55
Wache H, Tartakowska D, Hentrich A, Wagner M (2003) Development of a polymer stent with shape memory effect as a drug delivery system. J Mater Sci Mater Med 14:109–112
Wang Y (2016) Bioadaptability: an innovative concept for biomaterials. J Mater Sci Technol 32:801–809
Wang CC, Su CH, Chen CC (2008) Water absorbing and antibacterial properties of N-isopropyl acrylamide grafted and collagen/chitosan immobilized polypropylene nonwoven fabric and its application on wound healing enhancement. J Biomed Mater Res Part A 84:1006–1017
Wang X, Wang L, Wu Q, Bao F, Yang H, Qiu X, Chang J (2018) Chitosan/calcium silicate cardiac patch stimulates Cardiomyocyte activity and myocardial performance after infarction by synergistic effect of bioactive ions and aligned nanostructure. ACS Appl Mater Interfaces 11:1449–1468
Wanna D, Alam C, Toivola DM, Alam P (2013) Bacterial cellulose–kaolin nanocomposites for application as biomedical wound healing materials. Adv Nat Sci Nanosci Nanotechnol 4:045002
Wei Y, Zhang X, Song Y, Han B, Hu X, Wang X, Lin Y, Deng X (2011) Magnetic biodegradable Fe3O4/CS/PVA nanofibrous membranes for bone regeneration. Biomed Mater 6:055008
Williams DF (2008) On the mechanisms of biocompatibility. Biomaterials 29:2941–2953
Williams DF (2009) On the nature of biomaterials. Biomaterials 30:5897–5909
Willoughby CE, Ponzin D, Ferrari S, Lobo A, Landau K, Omidi Y (2010) Anatomy and physiology of the human eye: effects of mucopolysaccharidoses disease on structure and function–a review. Clin Exp Ophthalmol 38:2–11
Wood MD, Moore AM, Hunter DA, Tuffaha S, Borschel GH, Mackinnon SE, Sakiyama-Elbert SE (2009) Affinity-based release of glial-derived neurotrophic factor from fibrin matrices enhances sciatic nerve regeneration. Acta Biomater 5:959–968
Xie Z, Paras CB, Weng H, Punnakitikashem P, Su L-C, Vu K, Tang L, Yang J, Nguyen KT (2013) Dual growth factor releasing multi-functional nanofibers for wound healing. Acta Biomater 9:9351–9359
Xu HH, Wang P, Wang L, Bao C, Chen Q, Weir MD, Chow LC, Zhao L, Zhou X, Reynolds MA (2017) Calcium phosphate cements for bone engineering and their biological properties. Bone Res 5:17056
Yang F, Murugan R, Wang S, Ramakrishna S (2005) Electrospinning of nano/micro scale poly (L-lactic acid) aligned fibers and their potential in neural tissue engineering. Biomaterials 26:2603–2610

Yang Y, Xia T, Chen F, Wei W, Liu C, He S, Li X (2011) Electrospun fibers with plasmid bFGF polyplex loadings promote skin wound healing in diabetic rats. Mol Pharm 9:48–58

Zhang D, Wu X, Chen J, Lin K (2018) The development of collagen based composite scaffolds for bone regeneration. Bioactive Mater 3:129–138

Zhong S, Zhang Y, Lim C (2010) Tissue scaffolds for skin wound healing and dermal reconstruction. Wiley Interdiscip Rev Nanomed Nanobiotechnol 2:510–525

Ziats NP, Miller KM, Anderson JM (1988) In vitro and in vivo interactions of cells with biomaterials. Biomaterials 9:5–13

Zine R, Sinha M (2017) Nanofibrous poly (3-hydroxybutyrate-co-3-hydroxyvalerate)/collagen/graphene oxide scaffolds for wound coverage. Mater Sci Eng C 80:129–134

Nanostructured Materials and Their Biomedical Application

9

Sudip Mondal and Junghwan Oh

Abstract

Nanoscience and nanotechnology are at the forefront of current research. The advances of nanostructure-based therapeutics, sensors, diagnosis, and imaging facility have an incredible interest and tremendous potential for nanomedicine and biomedical application. Nanotechnology has shown non-exhaustive applications including diagnosis, imaging, therapy, drug delivery, implants, enzyme mimetic, tissue engineering, and many more. This book chapter will provide a broad and in-depth coverage of the nanostructured materials and their significant applications. The proposed book chapter contains six sections but not exhaustive on varying features of nano-structured materials. This chapter's significant major perception helps to understand the readers who might get a concept regarding nanostructured materials. This book chapter will definitely be useful to the beginners and also for the experienced personnel in the field of nanomaterials.

S. Mondal (✉) · J. Oh (✉)
Marine-Integrated Bionics Research Center, Pukyong National University, Busan, Republic of Korea
e-mail: jungoh@pknu.ac.kr

S. Paul (ed.), *Biomedical Engineering and its Applications in Healthcare*,
https://doi.org/10.1007/978-981-13-3705-5_9

Graphical Abstract

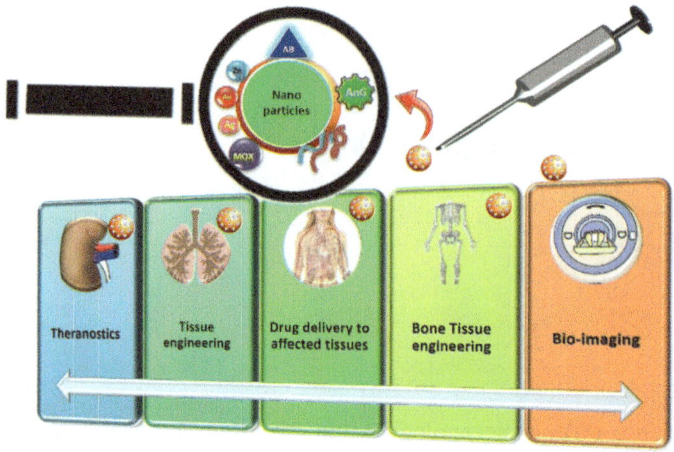

Nanostructured materials and their multifunctional biomedical application

Keywords
Nanomaterials · Nanostructured materials · Metallic nanoparticles · Ceramics nanoparticles · Nanocomposites · Biomedical applications · Drug delivery · Photothermal therapy · Bio-imaging · Contrast agents

9.1 Introduction

Recent progress in nanotechnology opens a new window for an expected solution to many complex unsolved problems in healthcare and biomedical field. The comprehensive application of nanotechnology in diverse fields like energy harvesting, storage, electronics, medicine, and healthcare industry already puts many success stories. The great impact of nanoscience in nanomedicine revolutionized the biomedical field through its multifunctional application such as advanced imaging, diagnosis, drug delivery, tissue engineering, etc. The last few decades witnessed the progress of multidisciplinary research in combination of materials science, chemistry, physics, biology, and medicine. Different synthetic routes paved the fabrication of nanostructured materials with diverse characteristics. The post-synthesis surface modifications have changed the entire property and functionality of synthesized materials. For biomedical application, many toxic materials become biocompatible after surface modifications; similarly for therapeutic or imaging purposes, nanomaterials acquired selective targeting, enhanced stability, solubility for drug delivery, or enhanced imaging properties. For biomedical application, nanomaterials could be classified into four major groups: (i) polymers,

(ii) ceramics, (iii) metals, and (iv) composites (in a combination of either three). All these classified biomaterials are mostly suitable for different kinds and site-specific application. Polymers are mostly suitable for drug delivery in soft tissues or soft tissue engineering, whereas metals and ceramics are useful for hard tissues such as bone and teeth and other purposes. The current development is more focused on multifaceted nano objects, with desired morphologies and enhanced controlled chemical and physical structures. The research in nanotechnology must exhibit facile techniques with reproducibility and tunable functionalities. In the starting age of pharmaceutical production, polymers always play a dominant character in drug delivery. The succeeding nanotechnology revitalized the enactment of metal, ceramics, and composites for biomedical application purposes. The application of nanomaterials in healthcare and biomedical purpose is not very successful until their toxicity level is acceptable. Many pharmaceutical industries used nanoparticles as a drug carrier to reduce toxicity, but unfortunately the drug carrier molecule itself imposes toxic effects over patients (De Jong and Borm 2008). The morphology, shape, and size have a great impact over different types of application (Fig. 9.1). The nanostructured materials from gold or silver with different shapes and morphologies

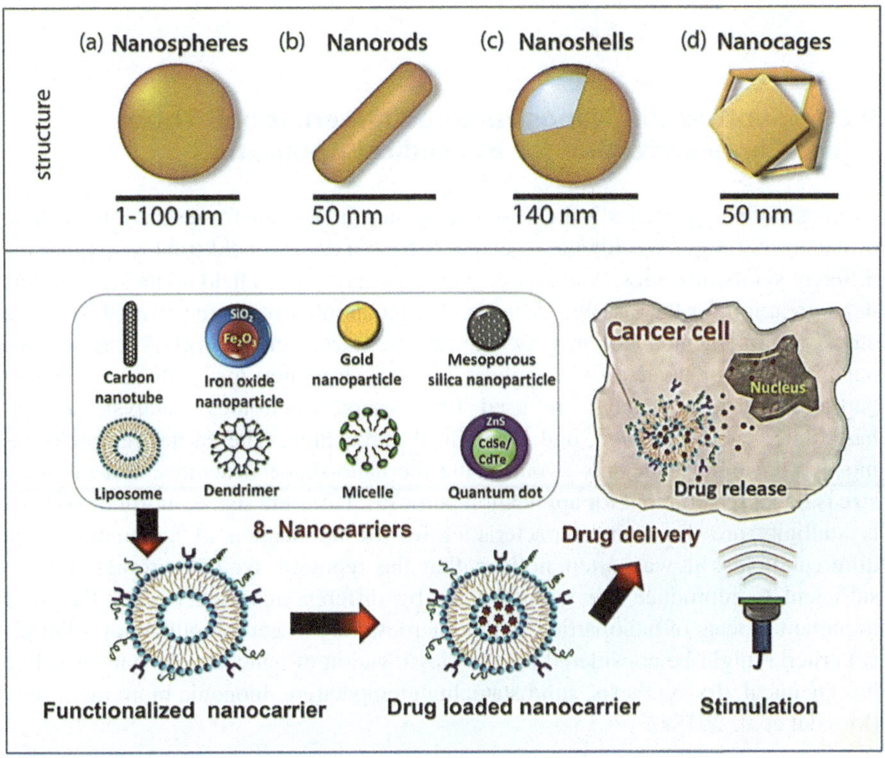

Fig. 9.1 Schematic representation of the nano-carriers and their potential use for biomedical application. (Reproduced with permission from Hossen et al. (2018)

have different applications. Gold nanorod, shell, or cages have NIR range light absorbing capacity, thus useful for photothermal therapy (PTT) for cancer treatments. On the other hand, the same gold nanoparticles with spherical structure do not have that property for photothermal treatment. In nanomaterials for specific application, shape, size, and morphology are very important. For therapeutic purposes, nanomaterials could be useful to treat cancer by a multifunctional approach such as drug delivery, hyperthermia, photodynamic, photothermal therapy, etc. Similarly, different types of disease even congenital disorders or age-associated anomalies such as Parkinson's, Alzheimer's, etc., could be treated. The future direction of nanomedicine goes to nano-surgery in a molecular level. With a concept in the near future, it would be possible to edit genes inside a chromosome or altered the base pairs in DNA by nano-surgery to improve human health.

In this book chapter, we have comprehensively summarized recent advancements in nanomedicine including different nanomaterial synthetic routes. Similarly, different applications of nanoparticles and their importance in the field of sensing, targeting, imaging, and therapeutic aspects are mentioned. The present scenario of nanoscience and nanotechnology with its drawback and future prospects are also discussed. The present status of nano-based healthcare products are summarized in Table 9.1.

9.2 Synthesis of Nanostructured Materials and Their Characterization for Biomedical Application

Synthesis of nanoparticles is the most important primary step for research in the field of nanotechnology. Worldwide researchers from a diverse field tried to explore the different synthetic routes. Among all, chemistry and physics field related researchers did a great job for the last two decades. A huge number of synthesis methods were already implemented to modify the shape, size, and morphologies of the nanoparticles for diverse field application. Nanostructured materials are the smart particles which nowadays are used for sensors, electronics, catalysis, energy harvesting, energy storage, and biomedical application. The earlier research was mostly concerned with only synthesizing the nano-shape structures. But the nano size is not only sufficient for application purposes. Different shape, morphology, and crystallinity are the main characteristics for the application of nanomaterials in different fields. It was often noticed that the reported synthetic routes are not sufficient to reproduce the nanoparticles by different researchers. So, the most important aspects of nanoparticle synthesis routes are to get reproducibility. Different criteria might be considered for the classification of nanoparticle syntheses like wet chemical, dry synthesis, solid state, high temperature, biogenic biomimetic, etc. (Mondal et al. 2018a).

Table 9.1 Lists of nanomedicine for biomedical application (FDA approved or under clinical trials)

Application(s)	Nanomaterials/ nanostructure	Manufacturing company	Product name	FDA approval year or status	Reference
Bone substitute	Hydroxyapatite nanocrystals	Angstrom Medica, Inc.	NanOss	2005	Gil et al. (2010))
Bone tissue engineering/ substitute	100-nm calcium-phosphate nanocrystals	Orthovita	Vitoss	2003	Wagner et al. (2006))
Bone tissue engineering/ substitute	Hydroxyapatite nanocrystals	IsotisOrthobiologicals US	OsSatura	2003	Roszek et al. (2005))
Dental application	Nanoparticles	Sybron Dental Specialties	Premise	2003	(Wagner et al. 2006)
Dental fillers	Ceramic nanoparticles	Dentspley	Ceram X Duo	2005	Etheridge et al. (2013))
Dental fillers composites	Silica and zirconium nanoparticles	3 M Company	Filtek	2008	Schmid et al. (1999))
Tissue engineering scaffold	30-nm titanium coating	GfEMedizintechnik GmbH	TiMESH	2004	Etheridge et al. (2013))
Coating materials	Antimicrobial nanosilver	I-Flow Corporation/ AcryMed, Inc.	ON-Q SilverSoaker/ SilvaGard™	2005	Hayman (2009))
Coating materials	Calcium phosphate nanocrystal coating	Biomet	NanoTite Implant	2008	Etheridge et al. (2013))
Bio-medical dressing	Antimicrobial nanosilver	Smith & Nephew, Inc.	Acticoat®	2005	Lu et al. (2008))
Filter for dialysis	Nanoporous membrane	NephroCare	Fresenius Polysulfone® Helixone®	1998	Etheridge et al. (2013))
Ovarian/breast cancer drug	Pegylated doxorubicin	Orthobiotech, Schering-Plough	Doxil (Caelyx)	1995	Pillai (2014))

(continued)

Table 9.1 (continued)

Application(s)	Nanomaterials/ nanostructure	Manufacturing company	Product name	FDA approval year or status	Reference
Liver cancer	Polymeric nanoparticles	Pharmacia & Upjohn Inc.	PK2	Phase I	Seymour et al. (2002)
Metastatic pancreatic cancer	Paclitaxel nano-spheres conjugated with albumin	Abraxis Bioscience, Astrazeneca	Abraxane	2005	Pillai (2014)
Acute lymphocytic leukemia	PEG-asparaginase	Enzon	Oncaspar	1994	(Pillai 2014)
HIV-related Kaposi sarcoma	Liposome encapsulated daunorubicin	Gilead Science	DaunoXome	1996	(Pillai 2014)
Aurolace therapy of carcinoma	Gold nanoshells	Nanospectra Biosciences	Auroshell	Clinical trial 1	(Pillai 2014)
Breast cancer/pancreatic cancer	Paclitaxel	Medigene/SynCore Biotechnology	EndoTAG-I	Clinical trial II	(Pillai 2014)
Head/neck/breastcancer	Liposomal cisplatin	Regulon	Lipoplatin	Clinical trial III	(Pillai 2014)
Breast cancer/small cell lung cancer	Paclitaxel-conjugated polymeric micelle	Samyang	Genexol-PM	Marketed in Europe, Korea	(Pillai 2014)
Cutaneous T-cell	Diphtheria toxin	Seragen, Inc	Ontak	1999	(Pillai 2014)
Philadelphia chromosome negative lymphoblastic leukemia	Vincristine	Talon Therapeutics	Marqibo	2012	(Pillai 2014)
Non-Hodgkin lymphoma	Liposomal vincristine	Inex	Onco-TCS	Clinical trial I/II	(Pillai 2014)
In vitro assay	Colloidal gold	Nymox	NicAlert	2002	(Wagner et al. 2006)
In vitro assay	Magnetic nanoparticles	Immunicon Corporation	CellTracks®	2003	(Wagner et al. 2006)
In vitro assay	Dendrimers	Dade Behring	Stratus CS	2003	(Roszek et al. 2005)

Fluorescent emission in vitro imaging	Quantum dot	Invitrogen Corporation	Qdot Nanocrystals	Research use only	(Etheridge et al. 2013)
MRI contrast for in vivo imaging	Iron oxide nanoparticles	Advanced Magnetics	Feridex IV, Ferumoxtran-10	1996 and Phase II	(Etheridge et al. 2013; Wang et al. 2001)
In vitro magnetic cell separation	Iron oxide nanoparticles	Veridex, LLC (Johnson & Johnson)	CellSearch® EpithelialCell Kit	2004	(Seymour et al. 2002)

9.2.1 Synthesis by Chemical Process

Synthesis routes for nanomaterials are broadly categorized in two primary phenomena such as (i) "top-down" and (ii) "bottom-up" approaches. The "top-down" method mainly comprised of physical methods which produce a bulk number of nanoparticles, whereas "bottom-up" approach is chemical/wet chemical approach with a very few production yields and superior homogeneous nanostructured property. The well-adopted method for nanoparticle synthesis is wet chemical precipitation. This methodology is very simple where ionic solutions of different precursor chemicals are mixed in different environments (different pH, temperature, pressure, etc.) to precipitate synthesized nanoparticles (Mondal et al. 2017a). The wet chemical precipitation stage may involve several steps such as (i) pre-nucleation, (ii) nucleation, (iii) growth, and (iv) crystalline (Gulati et al. 2018). The wet chemical precipitation includes a broad range of chemical reaction including sol–gel, polyol, micro-emulsion, hydrothermal, etc. Each methodology is adopted for different purposes to regulate morphology, shape, size, etc. In the polyol method, different reducing agents such as $NaBH_4$, citric acid, ascorbic acid, etc. are used for reducing purposes. This polyol methodology is a perfect bottom-up approach for nanoparticle synthesis (Xiong et al. 2011). In the sol–gel synthesis procedure, nanoparticles were synthesized with similar morphology and with well crystallinity (Bigi et al. 2004). In a micro-emulsion technique, nanoparticles are protected and covered by non-polar phases. In this process, basically three phases were used such as polar, non-polar, and suitable surfactant. For drug delivery application, this technology is widely used by the researchers (Lu and Yin 2012). Hydrothermal or solvothermal technique is employed to synthesize crystalline stable materials for catalyst, magnetic, or capacitor application. Generally, the solvents used in these purposes are heated to their critical point in a sealed reactor to obtained homogeneous nanoparticles (Yang et al. 2013). The sonochemical method was considered as non-conventional method for nanoparticle synthesis and categorized as a physical method. This application introduced ultrasound with different power density to process acoustic cavitation which initiates bubble formation growth and collapses in liquid media. The solvent with different pressure used in this reaction plays a key role in synthesizing nanoparticles (Compton et al. 1997). Dry chemical techniques are another choice of interest for researchers to synthesize nanoparticles. This dry chemical synthesis includes mechanochemical reaction, solid-state synthesis, etc. In dry chemical synthesis, generally no calcination is required, and the morphology of the synthesized nanoparticles is mostly of irregular shape (Silva et al. 2007). Materials scientists often synthesized good crystallinity nanoparticles by employing high-temperature synthesis such as thermal decomposition, combustion, pyrolysis, etc. The benefit of this type of synthetic route is high crystallinity with single phase materials being synthesized (Mondal et al. 2010, 2012; Sahni et al. 2013; Cho and Kang 2008; Sasikumar and Vijayaraghavan 2008). Nowadays, the most popular synthetic route for nanoparticles is biogenic/biomimetic synthesis. Mostly the natural sources such as plant, microbes, bio fluids, etc. are used as a precursor to synthesize nanoparticles. These approaches may produce ecofriendly, less toxic,

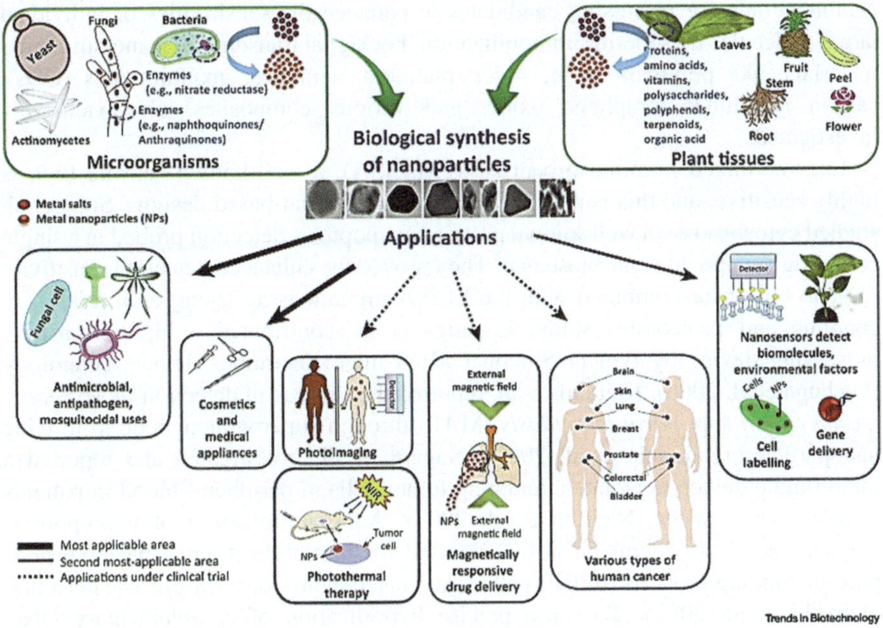

Fig. 9.2 Synthesis and multifunctional (imaging, sensing, therapeutic, drug carrier, etc.) biomedical application of nanoparticles. (Reproduced with permission from Singh et al. (2016)

and cost-effective products. But for practical biomedical application, biomimetic synthesis requires more control parameters to regulate the size shape of the final product (Klaus-Joerger et al. 2001; Mondal et al. 2014, 2016a; Manivasagan et al. 2018a). Some non-conventional methods such as microwave assisted (Mondal et al. 2018b), pulsed plasma assisted (Karpov et al. 2014), and laser ablation based (Zeng et al. 2012) are also widely popular for specific applications. Overall the synthetic technologies are investigated according to the application of the final product (Fig. 9.2).

9.3 Biomedical Application of Nanoparticles for Sensing and Targeting

Present-day biosensors are the most important field of study for detection, quantification, and many qualitative assays. Nanoparticles play a key role to enhance the sensing property of the desired molecule. This characteristic is owing to the large volume ratio compared to the available surface area of nanostructured materials which could address the sensitivity issues related to biosensors. The prime challenge for biosensor is the efficient signal capture which is termed as transduction. Transducers decipher the interaction of the analyte and biological element into electrochemical, optical, electro-chemiluminescent, or magnetic signals.

Nanomaterials are promising candidates to enhance the sensitivities of individual target molecules by superior immobilization. For signal transduction nano-structured materials, like polymers, gold, silver, platinum, ceramics, quantum dots (QDs), carbon nanotubes, graphene oxide, and various composites, are extensively investigated.

Enzyme-linked immunosorbent assay (ELISA), a well-known sensing tool, is highly sensitive, and this is possible due to nanomaterial-based designs. Song et al. studied cytochrome c a well-known protein for apoptosis detection probed in a single cell using the optical nano-biosensor. They proved the enhanced detection sensitivity of nano-biosensor combined with the ELISA immunoassay (Song et al. 2004). In genomic and proteomics study, Lechuga et al. contributed a highly sensitive biosensing device by typical Silicon CMOS microelectronics device technology (Lechuga et al. 2006). Kaittanis et al. reported the successful detection of *Mycobacterium avium* spp. *paratuberculosis* (MAP) through superparamagnetic iron oxide nanoparticles (Kaittanis et al. 2007). Nagarth et al. investigated and reported a microfluidic device to isolate circulating tumor cells in peripheral blood in patients affected with cancer (Nagrath et al. 2007). Another invention of nano-pore is reagent-free electrochemical biosensor, used for immobilization of bacterial periplasmic binding proteins (bPBP) within Au-coated nano-pores for glucose detection (Tripathi et al. 2007). To sense precise hybridization of complementary DNA structures Bio-SAMs of PNA (Peptide Nucleic Acid)-based technology was stated by Mateo-Marti et al. (Mateo-Martí et al. 2007). Riboh et al. suggested the localized surface plasmon resonance nano-biosensor involving biotin and anti-biotin which works on the different refractive index and monitored by optical biosensors with the LSPR extinction maximum through UV-Vis spectrophotometry (Riboh et al. 2003). Wang et al. described a gold nanoparticle-associated ATP assay detection method (Wang et al. 2007). Li et al. proposed a DNA sensing strategy based on photoelectrochemical"signal-off" by using photoactive CdS quantum dots sensitized titanium oxide nanorod array. The enhanced photo-electrochemical enactment of the designed DNA assay might propose a super low DNA detection limit of 5.2 aM with a broad range from 10 am to 100 pm (Fig. 9.3).

9.3.1 Targeting

The efficiency to treat disease is directly associated with the capability to target affected cells. Novel nano-structured molecules/therapeutics should assist precise directing toward affected tissues, hence diminishing the undesirable effects of the drug/molecules to non-targeted normal tissues. This specific targeting could be attained by fastening therapeutics to independently tailored carriers. Some of the very successful drug carriers are liposomes (Biswas et al. 2012; Kim et al. 2003), dendrimer nano-carriers (Shi et al. 2011; Yu et al. 2011), carbon nanomaterials (nanotubes, nanohorns, etc.) (Chen et al. 2011; Tripisciano et al. 2010), silica (Moorthy et al. 2017, 2018a), hydroxyapatite (Mondal et al. 2016b, 2017b, 2018c; Kim et al. 2018a, 2018b), magnetic nanoparticles (Arias et al. 2010; Bajpai and

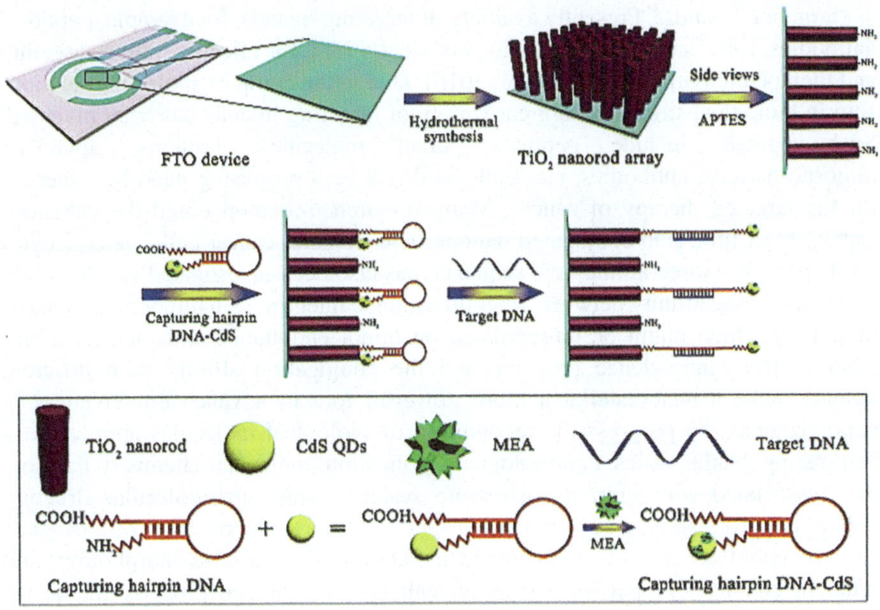

Fig. 9.3 Schematic representation of photoelectrochemical DNA sensing by "signal-off"strategy. (Reproduced with permission from Li et al. (2015)

Gupta 2011), etc. For effective molecular therapy, the targeted drugs remain sedentary until it achieves its target site (Chari 1998). Monoclonal antibodies which were described as its unique property to be able to bind specifically with tumor antigens could be a promising material for this purpose (Köhler and Milstein 1975). In cancer therapy, drugs could be targeted through targeting angiogenesis (Brannon-Peppas and Blanchette 2012), targeting tumor vasculature (Chen et al. 2001), and delivery of therapeutic agents such as paclitaxel, doxorubicin, 5-fluorouracil, antineoplastic agents, gene delivery, etc. Formulation schemes such as liposome formation, micellization, co-solvent systems, and emulsification have been well described by several researchers. Wang et al. investigated the biodegradable nanoparticle preparations using poly lactic-co-glycolic acid (PLGA) (Wang et al. 1996). Doxorubicin, the most effective and extensively studied anti-cancer drug, works on the nucleic acids synthesis inhibition within cancer cells (Yoo et al. 2000). FDA have approved the camptothecin-related drugs, such as astopotecan and irinotecan, which are used most frequently either in conjunction with 5-fluorouracil. Important ligands like transferrin (Tf) and epidermal growth factor (EGF) have selective targeting properties to cancer cells. Different DNA delivery complexes composed of polyethylenimine linked to polyethylene glycol further coated with either Tf or EGF were reported with a widely distributed size range of 49–1200 nm (Kircheis et al. 1997; Blessing et al. 2001). For cell-specific drug delivery application, communications between nano-structured carriers and target molecules must be ideal. Molecules attached on nanoparticles for these types of application are coined as "targeting ligands." Presently a variety of targeting ligands, for example, peptides, antibodies, folic acid, transferrin, etc., covalently bind for effective tumor targeting and therapeutic purpose (Davis et al. 2010). Due to the complex molecular relationship in biological structures, a great variety of targeting ligands has been invented. Such ligands include peptides, small molecules, proteins, aptamers, oligosaccharides, antibodies, etc. Folic acid is a very promising nano-biomaterials for the targeted therapy of cancer. Many researchers demonstrated the enhanced capability of folic acid conjugated nanoparticles to enter cancer cells via endocytosis. Dipeptide a three-amino-acid sequence has also been investigated as a targeting ligand. A strong affinity between dipeptide and the integrin v3, a tumor angiogenesis biomarker which might be up-regulated on tumor endothelial cells acts as a biomarker. The multivalence property and the conjugation affinity with different ligands make a nanomaterial a more powerful tool to advance effectiveness in nanomedicine. To progress the recognition of biological molecules applicable in cell tracing, health issues, and biological application, molecular chemistry has now been associated with supra-nanostructure systems. This supramolecular structure could be a promising system in nano-design for different recognition of biological actions, could act as ions and organic molecules with variable morphology and inherent chemical properties and as an enhancer by surface property alteration, bio-conjugation, etc. (Fig. 9.4).

9.3.2 Imaging

Biomedical application of nanomaterials already proved its potential for different therapeutic applications including theranostics, personalized medicine, preventive medicine, imaging, etc. Prior to treatment of any disease, proper diagnosis is most important. Nanoparticles perform a vital role in this purpose. The location of affected tissues, or tumors, or the drug molecules distribution inside the body system could be possible to monitor and quantify by the help of recently developed technologies in nanoscience. Among different therapeutic approaches, image-guided techniques are more reliable and effective. The modern imaging techniques compatible with X-ray, magnetic resonance imaging (MRI) (Li et al. 2012a), photoacoustic imaging (PAI) (Phan et al. 2018; Bharathiraja et al. 2018; Moorthy et al. 2018b), computed tomography (CT), up-conversion luminescence (UCL), single-photon emission computed tomography (SPECT), positron emission tomography (PET), etc. making diagnosis easy, less time consuming, less or noninvasive and more accurate. The most important characteristics of nanoparticles as an imaging molecule or contrast agent are its efficiency and toxicity. Nanoparticles are designed for very specific targeted application in the body system. The final fate of the nanoparticles might be elimination from the body system without hampering normal metabolism or biodegradation. The passage of nanoparticles through different biological barriers are also very important. Nanoparticles with smaller size (<100 nm) are mostly considered as a biocompatible material with easy circulation throughout the body system, whereas a very small size (<10 nm) may cause cell damage due to their easy circulation

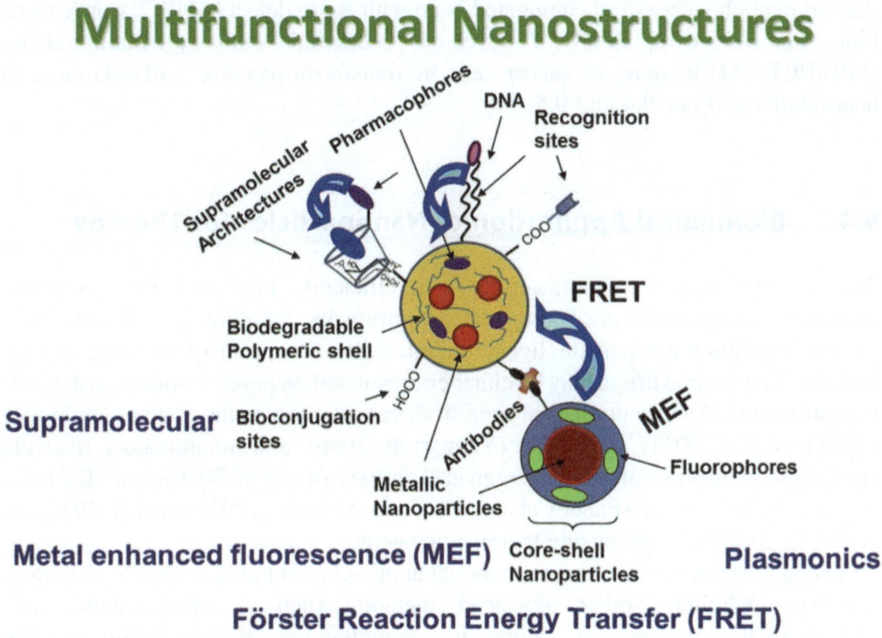

Fig. 9.4 Multifunctional nanostructure for guided drug delivery and bio-imaging purpose. (Reproduced with permission from Gontero et al. (2017)

throughout the different pores of cells which may cause damage of cellular organelles. Also different nanostructured shapes affect their hydrodynamic flow with variable in vivo retention time. Wen et al. reported ultra-small superparamagnetic iron oxide (USPIO) nanoparticles layered with polyethylene glycol and anti-mouse OxLDL polyclonal antibody loaded for magnetic resonance imaging at 7.0 Tesla. The study was performed with an atherosclerotic diseased carotid perivascular collar mice model. The interesting results suggest the 8–24 h MRI signal susceptibility of the developed nanoparticles (Wen et al. 2012). Qiao et al. studied lactoferrin (Lf) with PEG-coated Fe_3O_4 nanoparticles to access the blood–brain barrier by receptor-mediated transcytosis targeted to cerebral endothelial cells. MRI study for in vivo experiments confirms the enhanced penetration of Fe_3O_4-Lf nanoparticles through blood–brain barrier (Qiao et al. 2012). PET imaging is a non-invasive diagnostic approach. Generally, the antibodies are labeled with isotopes (such as ^{11}C half-life 20 min, ^{15}O half-live 2 min, ^{18}F half-life 2 h, ^{68}Ga half-life 68 min, ^{64}Cu half-life 13 h, and ^{76}Br with half-life of 16 h), and their emission was monitored. SPECT is another type of nuclear medicine tomography imaging method which provides 3D inspection by using gamma rays. Hong et al. employed PET imaging using ^{66}Ga as the radiolabel to quantitatively evaluate the pharmacokinetics and tumor targeting efficacy of nano-graphene (Hong et al. 2011). Lee et al. reported a metastatic biomarker, Au nanoparticle-quenched probe, by conjugating a Cy5.5-labeled matrix metalloproteinase-2 (MMP-2) substrate, on the

Au nanoparticles via a thiol conjugated bond with Au to detect MMP-2, a prognostic biomarker (Lee et al. 2008). Yang et al. successfully explained multimodality (MRI/PET/PAI) imaging of cancer cells by transferrin receptor 1 (TfR1) targeted nano-platform (Figs. 9.4 and 9.5).

9.4 Biomedical Application of Nanoparticles for Therapy

The last few decades, nanomaterials were efficiently fabricated for therapeutic purposes. Nanoparticles are conjugated with drugs by adopting various techniques such as adsorption mechanism, ligand bound, with a porous structure, ionic interaction, etc. The desired drugs may be either encapsulated on porous, hollow, polymeric nanostructures. A large number of drug or therapeutic molecules such as antibiotics (Yukihara et al. 2011), anticancer (Yang et al. 2006), anti-inflammatory (Paavola et al. 2000), neurotransmitters (Afergan et al. 2008; Losic et al. 2010), anti-HIV (Dev et al. 2010), anti-fungal (Pandey et al. 2005), anti-Alzheimer (Wilson et al. 2010) are already reported as a promising therapeutic agent.

Therapy or treatment with nanomaterial approaches is not restricted to only drug delivery. Definitely, other advanced methods such as photodynamic and photothermal therapies are similar to chemotherapy. For example, metallic nanoparticles are able to produce heat from light, and this characteristic has been comprehensively studied for photo-based cancer therapy. Up-conversion nanoparticles (UCNPs) which are able to convert NIR light into UV–vis for deep tissue therapy are also being studied. Therapeutic agents are directed to specific target sites for more successful treatment with less ailments. Different types of carrier molecules are already prepared to specific target sites. A few of them are, for example, ocular (Attama et al. 2007), dermal (Arayachukeat et al. 2011), pulmonary (Liu et al. 2008), osseous (Gu et al. 2013), rectal (Sznitowska et al. 2001), etc. Hatakeyama et al. reported MMP-cleavable peptide assembled to a PEGylated liposome for gene delivery mediated cancer therapy (Hatakeyama et al. 2007). This developed nanostructure exhibits threefold higher gene transfection efficiency. In contrast to the synthetic micelles, the polyelectrolyte complex micelles are broadly applicable for intravenous and intracellular delivery of numerous bio-molecules comprising proteins, nucleic acids, and peptides (Mao et al. 2006). The discovery and potential application of RNA interference (RNAi) in mammalian cells has advanced a novel healing implement for human disease therapies (Elbashir et al. 2001). Small-interfering RNA (siRNA) is a promising biomaterial which can modulate gene expression sequence specifically at a post-transcriptional level. This characteristic attracts the promising therapeutic approach for different kinds of cancers in human. Holle et al. studied Bcl-2 targeting siRNA expression by a T7 vector molecule which efficiently prevents in vitro tumor cell growth (Holle et al. 2004). Immunotherapy is an interesting phenomenon to treat cancers by activating the body's immune system which serves as a "security/police." Different approaches

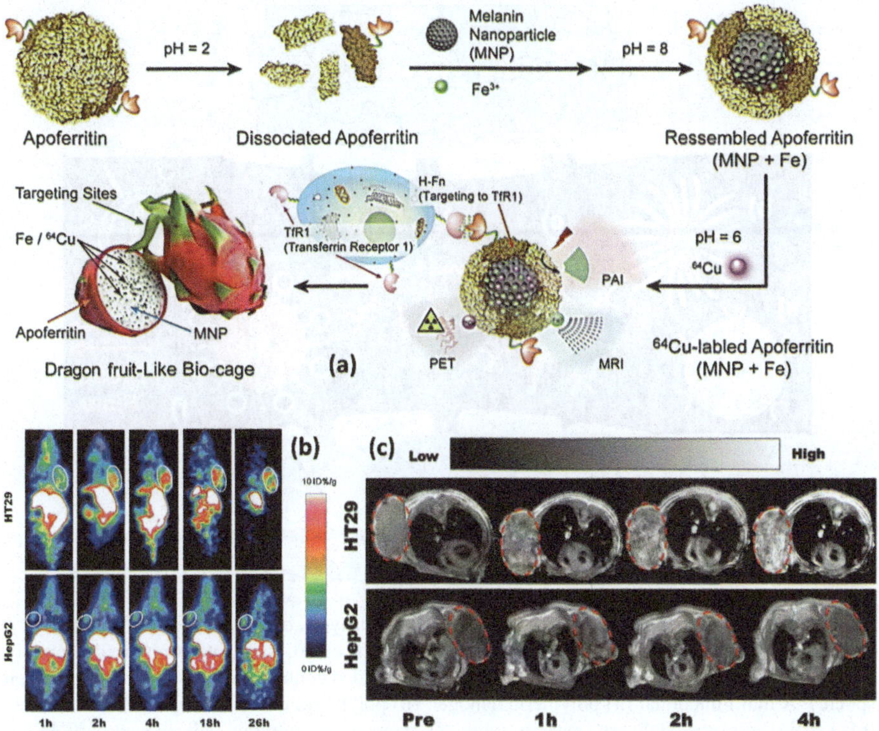

Fig. 9.5 (**a**) Graphical representation of nano-cage synthesis. (**b**) PET images of HT29 (top) and HepG2 (bottom) tumor-bearing mice at different time intervals after administration of ^{64}Cu-AMF. (**c**) Magnetic resonance images (T1) of HT29 and HepG2 tumor model after nanomaterials administration; tumor area represented by a red circle. (Reprinted with permission from Yang et al. (2015)

are taken such as (i) designed monoclonal antibody, (ii) cancer vaccination, and (iii) immune checkpoint inhibitors to treat cancers through immune therapy (Sandin et al. 2014; Zhu et al. 2017; Topalian et al. 2015). Photodynamic therapy (PDT) is now approved for a clinical trial to treat cancers like esophageal, non-small cell lung cancers, skin cancer, etc. (Fan et al. 2016; Yano et al. 2011). The three major elements (i) light, (ii) photosensitizer (PS), and (iii) oxygen are making a synchronous effect to produce selective reactive oxygen species (ROS) to treat cancer cells. Like PDT, photothermal therapy is another approach which generates heat to kill cancer cells. Different nanoparticles are used for PTT treatment such as Au nanorods, carbon nanotubes, silver triangles, CuS, metal oxides, FeS2, polymeric nanoparticles such as polypyrrole, NIR absorbing dyes such as ICG, IR825, Prussian blue, etc. (Manivasagan et al. 2018b; Panchanathan et al. 2018; Phan et al. 2019; Huang et al. 2015; Cheng et al. 2016; Chakravarty et al. 2008). A nano-therapeutic approach is demonstrated and discussed with different types of theranostics approaches (Fig. 9.6).

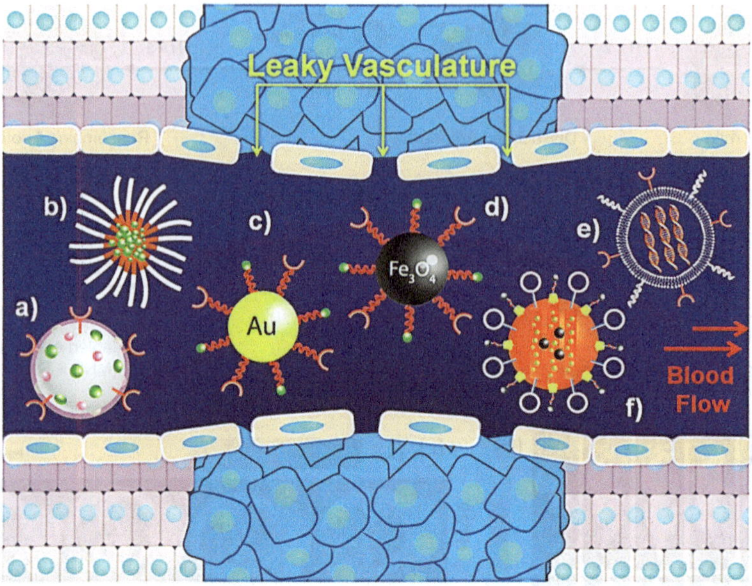

Fig. 9.6 Schematic representation illustrates the different types of theranostic nano-medicines in "attack mode" on a targeted tumor. Green spheres depict the cargo, conjugated or housed internally. Circles/semi-circles represent the targeting ligands. Imbedded contrast agents represented by purple spheres. A multifunctional (**a**) polymeric nanogel, (**b**) polymeric micelle, (**c**) gold, (**d**) iron oxide, (**e**) siRNA-liposome delivery system, and (**f**) capped mesoporous silica nanoparticle are represented. (Reproduced with permission from Li et al. (2012b)

9.5 Limitation and Prospects of Nanostructured Materials for Biomedical Applications

The terms "nanotechnology" and "nanomedicine" are very ambiguous and broad spectrum to be classified and accepted universally. Not all nanostructured biomaterials are suitable for biomedical application. Effective optimization of nanoparticles toward future possible application must be studied carefully using interdisciplinary approaches prior to biomedical application. The foremost important criterion for biomedical application is to select the material based on its toxicity and biocompatibility. In recent days by employing different synthetic routes each day researchers through worldwide synthesized a variety of nanoparticles. But a few of them are really acceptable for biomedical application after long-term exhaustive study. To reduce such precious time consumption, materials selection is very important. The second most important characteristic is their bio-absorbability and physical properties which help to determine their interaction capabilities with other materials such as peptides, proteins, drugs, or biomolecules. To overcome these issues, advanced synthesis methodologies with reproducible nanostructure should be standardized with controllable chemical, physical, and topographical properties. The surface properties are also very important for selective targeting of nanoparticles inside the body system. Plasmonic nanoparticles show much potentiality for constructing a multifaceted system with high regulation of drug delivery at precise target sites. Another most challenging aspect of nanomedicine is low signal from fluorescence materials used for bio-imaging due to fluorescence quenching. The innovative technologies associated with molecular detection through nanostructured advanced materials may overcome such drawback with powerful "search–target–treat" mechanism. All this development is possible nowadays because of the great association of researchers from diverse disciplines to advance such complex inter-disciplinary difficulties. Immense development in biomedical research surely brings beneficiary effects to mankind. Some of the issues are discussed below to be primarily focused before beginning of new smart era of future medicine:

- Future medicine requires theranostics nanoplatforms for synchronized diagnosis and treatment.
- For secure therapy with nanomaterials, molecular profiling is also necessary for individual patients. All nanoparticles might not be suitable for all types of patients.
- Diagnosis is important prior to treat any disease. Noninvasive diagnostic techniques are the foremost priority for this purpose.
- Another most important approach for nanomedicine is on-demand therapy at the onset of any ailments. This type of treatment mainly focused on the pre-introduced nanoparticles immediately activated when necessary by external stimuli.
- Stimulating agent for activation and therapy purposes either be chemicals (which may generate reactive oxygen species, ROS) or may be any physical agents such as infrared stimuli, heat, ultrasound, magnetic field, etc.

9.6 Conclusions

Application of nano-structured systems or nanomaterials in the field of biomedicine has immense potential for future therapeutic application. The interdisciplinary approach of nanomaterials in the nanomedicine landscape is defined in this article. Nanomedicine-related applications in healthcare which are undergoing clinical trials or therapeutics already approved by the US Food and Drug Administration (FDA) or equivalent agencies worldwide are discussed. The bio-sensing and imaging technologies exceling concerning superior sensitivity, ranging from cellular to organelles imaging and molecular level detection. Before considering clinical level applications of developed nano-structured systems or nanomaterials, the primary duty is to evaluate the biocompatibility of these materials. The prospect of tailored medicine focuses on further improvement of theranostics systems, superior nano-platform, enhanced detection capabilities, and effective biomarkers. Future nanostructured therapeutic systems could be devised to remotely regulate the therapeutic release kinetics with minimal toxic effects. Application of nanomaterials for precise therapeutic purposes encompasses definite functionalities. Sometimes these are challenging to acquaint all properties with its large-scale quantity. Progress of nanotechnology unlocked this opportunity for varied morphologies' nanostructures. Future investigations to improve design, properties, and synthetic routes are necessary for fabricating patient-specific optimum structures. In this chapter, we tried to high point most of the new ideas reported in the literature, emphasizing their assumptions with the advancement of our understanding, related to nano-structured materials, associated with advanced biomedical application.

References

Afergan E, Epstein H, Dahan R, Koroukhov N, Rohekar K, Danenberg HD, Golomb G (2008) Delivery of serotonin to the brain by monocytes following phagocytosis of liposomes. J Control Release 132:84–90

Arayachukeat S, Wanichwecharungruang SP, Tree-Udom T (2011) Retinyl acetate-loaded nanoparticles: dermal penetration and release of the retinyl acetate. Int J Pharm 404:281–288

Arias JL, López-Viota M, Delgado ÁV, Ruiz MA (2010) Iron/ethylcellulose (core/shell) nanoplatform loaded with 5-fluorouracil for cancer targeting. Colloids Surf B: Biointerfaces 77:111–116

Attama A, Schicke B, Paepenmüller T, Müller-Goymann C (2007) Solid lipid nanodispersions containing mixed lipid core and a polar heterolipid: characterization. Eur J Pharm Biopharm 67:48–57

Bajpai A, Gupta R (2011) Magnetically mediated release of ciprofloxacin from polyvinyl alcohol based superparamagnetic nanocomposites. J Mater Sci Mater Med 22:357–369

Bharathiraja S, Bui NQ, Manivasagan P, Moorthy MS, Mondal S, Seo H, Phuoc NT, Phan TTV, Kim H, Lee KD (2018) Multimodal tumor-homing chitosan oligosaccharide-coated biocompatible palladium nanoparticles for photo-based imaging and therapy. Sci Rep 8:500

Bigi A, Boanini E, Rubini K (2004) Hydroxyapatite gels and nanocrystals prepared through a sol–gel process. J Solid State Chem 177:3092–3098

Biswas S, Dodwadkar NS, Deshpande PP, Torchilin VP (2012) Liposomes loaded with paclitaxel and modified with novel triphenylphosphonium-PEG-PE conjugate possess low toxicity, target mitochondria and demonstrate enhanced antitumor effects in vitro and in vivo. J Control Release 159:393–402

Blessing T, Kursa M, Holzhauser R, Kircheis R, Wagner E (2001) Different strategies for formation of pegylated EGF-conjugated PEI/DNA complexes for targeted gene delivery. Bioconjug Chem 12:529–537

Brannon-Peppas L, Blanchette JO (2012) Nanoparticle and targeted systems for cancer therapy. Adv Drug Deliv Rev 64:206–212

Chakravarty P, Marches R, Zimmerman NS, Swafford AD-E, Bajaj P, Musselman IH, Pantano P, Draper RK, Vitetta ES (2008) Thermal ablation of tumor cells with antibody-functionalized single-walled carbon nanotubes. Proc Natl Acad Sci 105:8697

Chari RV (1998) Targeted delivery of chemotherapeutics: tumor-activated prodrug therapy. Adv Drug Deliv Rev 31:89–104

Chen Q-R, Zhang L, Gasper W, Mixson AJ (2001) Targeting tumor angiogenesis with gene therapy. Mol Genet Metab 74:120–127

Chen Z, Pierre D, He H, Tan S, Pham-Huy C, Hong H, Huang J (2011) Adsorption behavior of epirubicin hydrochloride on carboxylated carbon nanotubes. Int J Pharm 405:153–161
Cheng L, Shen S, Shi S, Yi Y, Wang X, Song G, Yang K, Liu G, Barnhart TE, Cai W (2016) FeSe2-decorated Bi2Se3 nanosheets fabricated via cation exchange for chelator-free 64Cu-labeling and multimodal image-guided photothermal-radiation therapy. Adv Funct Mater 26:2185–2197
Cho JS, Kang YC (2008) Nano-sized hydroxyapatite powders prepared by flame spray pyrolysis. J Alloys Compd 464:282–287
Compton RG, Eklund JC, Marken F (1997) Sonoelectrochemical processes: a review. Electroanalysis 9:509–522
Davis ME, Chen Z, Shin DM (2010) Nanoparticle therapeutics: an emerging treatment modality for cancer. Nanosci Technol A Collect Rev Nat J World Sci:239–250
De Jong WH, Borm PJ (2008) Drug delivery and nanoparticles: applications and hazards. Int J Nanomedicine 3:133
Dev A, Binulal N, Anitha A, Nair S, Furuike T, Tamura H, Jayakumar R (2010) Preparation of poly (lactic acid)/chitosan nanoparticles for anti-HIV drug delivery applications. Carbohydr Polym 80:833–838
Elbashir SM, Harborth J, Lendeckel W, Yalcin A, Weber K, Tuschl T (2001) Duplexes of 21-nucleotide RNAs mediate RNA interference in cultured mammalian cells. Nature 411:494
Etheridge ML, Campbell SA, Erdman AG, Haynes CL, Wolf SM, McCullough J (2013) The big picture on nanomedicine: the state of investigational and approved nanomedicine products. Nanomedicine 9:1–14
Fan W, Huang P, Chen X (2016) Overcoming the Achilles' heel of photodynamic therapy. Chem Soc Rev 45:6488–6519
Gil PR, Hühn D, Loretta L, Sasse D, Parak WJ (2010) Nanopharmacy: Inorganic nanoscale devices as vectors and active compounds. Pharmacol Res 62:115–125
Gontero D, Lessard-Viger M, Brouard D, Bracamonte AG, Boudreau D, Veglia AV (2017) Smart multifunctional nanoparticles design as sensors and drug delivery systems based on supramolecular chemistry. Microchem J 130:316–328
Gu W, Wu C, Chen J, Xiao Y (2013) Nanotechnology in the targeted drug delivery for bone diseases and bone regeneration. Int J Nanomedicine 8:2305
Gulati S, Sachdeva M, Bhasin K (2018) Various synthetic routes for the preparation of nanoparticles. AIP Conf Proc, AIP Publishing:030215
Hatakeyama H, Akita H, Kogure K, Oishi M, Nagasaki Y, Kihira Y, Ueno M, Kobayashi H, Kikuchi H, Harashima H (2007) Development of a novel systemic gene delivery system for cancer therapy with a tumor-specific cleavable PEG-lipid. Gene Ther 14:68
Hayman ML (2009) The emerging product and patent landscape for nanosilver-containing medical devices. Nanotech L & Bus 6:148
Holle L, Hicks L, Song W, Holle E, Wagner T, Yu X (2004) Bcl-2 targeting siRNA expressed by a T7 vector system inhibits human tumor cell growth in vitro. Int J Oncol 24:615–621
Hong H, Goel S, Zhang Y, Cai W (2011) Molecular imaging with nucleic acid aptamers. Curr Med Chem 18:4195–4205
Hossen S, Hossain MK, Basher M, Mia M, Rahman M, Uddin MJ (2018) Smart nanocarrier-based drug delivery systems for cancer therapy and toxicity studies: a review. J Adv Res
Huang P, Gao Y, Lin J, Hu H, Liao H-S, Yan X, Tang Y, Jin A, Song J, Niu G (2015) Tumor-specific formation of enzyme-instructed supramolecular self-assemblies as cancer theranostics. ACS Nano 9:9517–9527
Kaittanis C, Naser SA, Perez JM (2007) One-step, nanoparticle-mediated bacterial detection with magnetic relaxation. Nano Lett 7:380–383
Karpov I, Ushakov A, Fedorov LY, Lepeshev A (2014) Method for producing nanomaterials in the plasma of a low-pressure pulsed arc discharge. Tech Phys 59:559–563
Kim JK, Choi SH, Kim CO, Park JS, Ahn WS, Kim CK (2003) Enhancement of polyethylene glycol (PEG)-modified cationic liposome-mediated gene deliveries: effects on serum stability and transfection efficiency. J Pharm Pharmacol 55:453–460

Kim H, Mondal S, Jang B, Manivasagan P, Moorthy MS, Oh J (2018a) Biomimetic synthesis of metal–hydroxyapatite (Au-HAp, Ag-HAp, Au-Ag-HAp): Structural analysis, spectroscopic characterization and biomedical application. Ceram Int 44:20490–20500

Kim H, Mondal S, Bharathiraja S, Manivasagan P, Moorthy MS, Oh J (2018b) Optimized Zn-doped hydroxyapatite/doxorubicin bioceramics system for efficient drug delivery and tissue engineering application. Ceram Int 44:6062–6071

Kircheis R, Kichler A, Wallner G, Kursa M, Ogris M, Felzmann T, Buchberger M, Wagner E (1997) Coupling of cell-binding ligands to polyethylenimine for targeted gene delivery. Gene Ther 4:409

Klaus-Joerger T, Joerger R, Olsson E, Granqvist C-G (2001) Bacteria as workers in the living factory: metal-accumulating bacteria and their potential for materials science. Trends Biotechnol 19:15–20

Köhler G, Milstein C (1975) Continuous cultures of fused cells secreting antibody of predefined specificity. Nature 256:495

Lechuga LM, Tamayo J, Calle A, Domínguez C (2006) Nanobiosensors based on optoelectronic and nanomechanical transducers for genomic and proteomic applications. Revista mexicana de física 52:94–98

Lee S, Cha EJ, Park K, Lee SY, Hong JK, Sun IC, Kim SY, Choi K, Kwon IC, Kim K (2008) A near-infrared-fluorescence-quenched gold-nanoparticle imaging probe for in vivo drug screening and protease activity determination. Angew Chem 120:2846–2849

Li M, Kim HS, Tian L, Yu MK, Jon S, Moon WK (2012a) Comparison of two ultrasmall superparamagnetic iron oxides on cytotoxicity and MR imaging of tumors. Theranostics 2:76

Li Z, Barnes JC, Bosoy A, Stoddart JF, Zink JI (2012b) Mesoporous silica nanoparticles in biomedical applications. Chem Soc Rev 41:2590–2605

Li S, Wang Y, Gao C, Ge S, Yu J, Yan M (2015) "Signal-off" photoelectrochemical DNA sensing strategy based on target dependent DNA probe conformational conversion using CdS quantum dots sensitized TiO2 nanorods array as photoactive material. J Electroanal Chem 759:38–45

Liu J, Gong T, Fu H, Wang C, Wang X, Chen Q, Zhang Q, He Q, Zhang Z (2008) Solid lipid nanoparticles for pulmonary delivery of insulin. Int J Pharm 356:333–344

Losic D, Yu Y, Aw MS, Simovic S, Thierry B, Addai-Mensah J (2010) Surface functionalisation of diatoms with dopamine modified iron-oxide nanoparticles: toward magnetically guided drug microcarriers with biologically derived morphologies. Chem Commun 46:6323–6325

Lu Z, Yin Y (2012) Colloidal nanoparticle clusters: functional materials by design. Chem Soc Rev 41:6874–6887

Lu S, Gao W, Gu HY (2008) Construction, application and biosafety of silver nanocrystalline chitosan wound dressing. Burns 34:623–628

Manivasagan P, Bharathiraja S, Santha Moorthy M, Mondal S, Seo H, Dae Lee K, Oh J (2018a) Marine natural pigments as potential sources for therapeutic applications. Crit Rev Biotechnol 38:745–761

Manivasagan P, Bharathiraja S, Santha Moorthy M, Mondal S, Nguyen TP, Kim H, Phan TTV, Lee KD, Oh J (2018b) Biocompatible chitosan oligosaccharide modified gold nanorods as highly effective photothermal agents for ablation of breast cancer cells. Polymers 10:232

Mao S, Neu M, Germershaus O, Merkel O, Sitterberg J, Bakowsky U, Kissel T (2006) Influence of polyethylene glycol chain length on the physicochemical and biological properties of poly (ethylene imine)-graft-poly (ethylene glycol) block copolymer/SiRNA polyplexes. Bioconjug Chem 17:1209–1218

Mateo-Martí E, Briones C, Pradier C-M, Martin-Gago JA (2007) A DNA biosensor based on peptide nucleic acids on gold surfaces. Biosens Bioelectron 22:1926–1932

Mondal S, Mahata S, Kundu S, Mondal B (2010) Processing of natural resourced hydroxyapatite ceramics from fish scale. Adv Appl Ceram 109:234–239

Mondal S, Bardhan R, Mondal B, Dey A, Mukhopadhyay SS, Roy S, Guha R, Roy K (2012) Synthesis, characterization and in vitro cytotoxicity assessment of hydroxyapatite from different bioresources for tissue engineering application. Bull Mater Sci 35:683–691

Mondal S, Mondal A, Mandal N, Mondal B, Mukhopadhyay S, Dey A, Singh S (2014) Physicochemical characterization and biological response of Labeo rohita-derived hydroxyapatite scaffold. Bioprocess Biosyst Eng 37:1233–1240

Mondal S, Pal U, Dey A (2016a) Natural origin hydroxyapatite scaffold as potential bone tissue engineering substitute. Ceram Int 42:18338–18346

Mondal S, Dey A, Pal U (2016b) Low temperature wet-chemical synthesis of spherical hydroxyapatite nanoparticles and their in situ cytotoxicity study. Adv Nano Res 4:295–307

Mondal S, Reyes MEDA, Pal U (2017a) Plasmon induced enhanced photocatalytic activity of gold loaded hydroxyapatite nanoparticles for methylene blue degradation under visible light. RSC Adv 7:8633–8645

Mondal S, Manivasagan P, Bharathiraja S, Santha Moorthy M, Nguyen VT, Kim HH, Nam SY, Lee KD, Oh J (2017b) Hydroxyapatite coated iron oxide nanoparticles: a promising nanomaterial for magnetic hyperthermia cancer treatment. Nano 7:426

Mondal S, Dorozhkin SV, Pal U (2018a) Recent progress on fabrication and drug delivery applications of nanostructured hydroxyapatite. Wiley Interdiscip Rev Nanomed Nanobiotechnol 10:e1504

Mondal S, Hoang G, Manivasagan P, Moorthy MS, Phan TTV, Kim HH, Nguyen TP, Oh J (2018b) Rapid microwave-assisted synthesis of gold loaded hydroxyapatite collagen nano-bio materials for drug delivery and tissue engineering application. Ceram Int

Mondal S, Hoang G, Manivasagan P, Moorthy MS, Nguyen TP, Phan TTV, Kim HH, Kim MH, Nam SY, Oh J (2018c) Nano-hydroxyapatite bioactive glass composite scaffold with enhanced mechanical and biological performance for tissue engineering application. Ceram Int 44:15735

Moorthy MS, Subramanian B, Panchanathan M, Mondal S, Kim H, Lee KD, Oh J (2017) Fucoidan-coated core–shell magnetic mesoporous silica nanoparticles for chemotherapy and magnetic hyperthermia-based thermal therapy applications. New J Chem 41:15334–15346

Moorthy MS, Hoang G, Manivasagan P, Mondal S, Phan TTV, Kim H, Oh J (2018a) Chitosan oligosaccharide coated mesoporous silica nanoparticles for pH-stimuli responsive drug delivery applications. J Porous Mater:1–10

Moorthy MS, Hoang G, Subramanian B, Bui NQ, Panchanathan M, Mondal S, Tuong VPT, Kim H, Oh J (2018b) Correction: Prussian blue decorated mesoporous silica hybrid nanocarriers for photoacoustic imaging-guided synergistic chemo-photothermal combination therapy. J Mater Chem B 6:5476–5477

Nagrath S, Sequist LV, Maheswaran S, Bell DW, Irimia D, Ulkus L, Smith MR, Kwak EL, Digumarthy S, Muzikansky A (2007) Isolation of rare circulating tumour cells in cancer patients by microchip technology. Nature 450:1235

Paavola A, Kilpeläinen I, Yliruusi J, Rosenberg P (2000) Controlled release injectable liposomal gel of ibuprofen for epidural analgesia. Int J Pharm 199:85–93

Panchanathan M, Jun SW, Thanh PTN, Hoang G, Mondal S, Moorthy MS, Kim H, Tuong VPT, Doan VHM, Kim C-S (2018) Multifunctional near-infrared laser-triggered drug delivery system using folic acid conjugated chitosan oligosaccharide encapsulated gold nanorods for targeted chemo-photothermal therapy. J Mater Chem B

Pandey R, Ahmad Z, Sharma S, Khuller G (2005) Nano-encapsulation of azole antifungals: potential applications to improve oral drug delivery. Int J Pharm 301:268–276

Phan TTV, Bui NQ, Cho S-W, Bharathiraja S, Manivasagan P, Moorthy MS, Mondal S, Kim C-S, Oh J (2018) Photoacoustic imaging-guided photothermal therapy with tumor-targeting HA-FeOOH@ PPy nanorods. Sci Rep 8:8809

Phan TTV, Hoang G, Nguyen TP, Kim HH, Mondal S, Manivasagan P, Moorthy MS, Lee KD, Junghwan O (2019) Chitosan as a stabilizer and size-control agent for synthesis of porous flower-shaped palladium nanoparticles and their applications on photo-based therapies. Carbohydr Polym 205:340–352

Pillai G (2014) Nanomedicines for cancer therapy: an update of FDA approved and those under various stages of development. SOJ Pharm Pharm Sci 1(2):13

Qiao R, Jia Q, Hüwel S, Xia R, Liu T, Gao F, Galla H-J, Gao M (2012) Receptor-mediated delivery of magnetic nanoparticles across the blood–brain barrier. ACS Nano 6:3304–3310

Riboh JC, Haes AJ, McFarland AD, Ranjit Yonzon C, Van Duyne RP (2003) A nanoscale optical biosensor: real-time immunoassay in physiological buffer enabled by improved nanoparticle adhesion. J Phys Chem B 107:1772–1780

Roszek B, De Jong W, Geertsma RE (2005) Nanotechnology in medical applications: state-of-the-art in materials and devices

Sahni G, Gopinath P, Jeevanandam P (2013) A novel thermal decomposition approach to synthesize hydroxyapatite–silver nanocomposites and their antibacterial action against GFP-expressing antibiotic resistant E. coli. Colloids Surf B: Biointerfaces 103:441–447

Sandin LC, Tötterman TH, Mangsbo SM (2014) Local immunotherapy based on agonistic CD40 antibodies effectively inhibits experimental bladder cancer. Oncoimmunology 3:e27400

Sasikumar S, Vijayaraghavan R (2008) Effect of metal-ion-to-fuel ratio on the phase formation of bioceramic phosphates synthesized by self-propagating combustion. Sci Technol Adv Mater 9:035003

Schmid G, Bäumle M, Geerkens M, Heim I, Osemann C, Sawitowski T (1999) Current and future applications of nanoclusters. Chem Soc Rev 28:179–185

Seymour LW, Ferry DR, Anderson D, Hesslewood S, Julyan PJ, Poyner R, Doran J, Young AM, Burtles S, Kerr DJ (2002) Hepatic drug targeting: phase I evaluation of polymer-bound doxorubicin. J Clin Oncol 20:1668–1676

Shi L, Fleming CJ, Riechers SL, Yin N-N, Luo J, Lam KS, Liu G-y (2011) High-resolution imaging of dendrimers used in drug delivery via scanning probe microscopy. J Drug Deliv

Silva C, Graça M, Valente M, Sombra A (2007) Crystallite size study of nanocrystalline hydroxyapatite and ceramic system with titanium oxide obtained by dry ball milling. J Mater Sci 42:3851–3855

Singh P, Kim Y-J, Zhang D, Yang D-C (2016) Biological synthesis of nanoparticles from plants and microorganisms. Trends Biotechnol 34:588–599

Song JM, Kasili PM, Griffin GD, Vo-Dinh T (2004) Detection of cytochrome C in a single cell using an optical nanobiosensor. Anal Chem 76:2591–2594

Sznitowska M, Gajewska M, Janicki S, Radwanska A, Lukowski G (2001) Bioavailability of diazepam from aqueous-organic solution, submicron emulsion and solid lipid nanoparticles after rectal administration in rabbits. Eur J Pharm Biopharm 52:159–163

Topalian SL, Drake CG, Pardoll DM (2015) Immune checkpoint blockade: a common denominator approach to cancer therapy. Cancer Cell 27:450–461

Tripathi A, Wang J, Luck LA, Suni II (2007) Nanobiosensor design utilizing a periplasmic E. coli receptor protein immobilized within Au/polycarbonate nanopores. Anal Chem 79:1266–1270

Tripisciano C, Costa S, Kalenczuk R, Borowiak-Palen E (2010) Cisplatin filled multiwalled carbon nanotubes–a novel molecular hybrid of anticancer drug container. Eur Phys J B 75:141–146

Wagner V, Dullaart A, Bock A-K, Zweck A (2006) The emerging nanomedicine landscape. Nat Biotechnol 24:1211

Wang YM, Sato H, Adachi I, Horikoshi I (1996) Preparation and characterization of poly (lactic-co-glycolic acid) microspheres for targeted delivery of a novel anticancer agent, taxol. Chem Pharm Bull 44:1935–1940

Wang Y-XJ, Hussain SM, Krestin GP (2001) Superparamagnetic iron oxide contrast agents: physicochemical characteristics and applications in MR imaging. Eur Radiol 11:2319–2331

Wang J, Wang L, Liu X, Liang Z, Song S, Li W, Li G, Fan C (2007) A gold nanoparticle-based aptamer target binding readout for ATP assay. Adv Mater 19:3943–3946

Wen S, Liu D-F, Liu Z, Harris S, Yao Y-Y, Ding Q, Nie F, Lu T, Chen H-J, An Y-L (2012) OxLDL-targeted iron oxide nanoparticles for in vivo MRI detection of perivascular carotid collar induced atherosclerotic lesions in ApoE-deficient mice. J Lipid Res:M018895

Wilson B, Samanta MK, Santhi K, Kumar KS, Ramasamy M, Suresh B (2010) Chitosan nanoparticles as a new delivery system for the anti-Alzheimer drug tacrine. Nanomedicine 6:144–152

Xiong J, Wang Y, Xue Q, Wu X (2011) Synthesis of highly stable dispersions of nanosized copper particles using L-ascorbic acid. Green Chem 13:900–904

Yang J, Park S-B, Yoon H-G, Huh Y-M, Haam S (2006) Preparation of poly ε-caprolactone nanoparticles containing magnetite for magnetic drug carrier. Int J Pharm 324:185–190

Yang Q, Lu Z, Liu J, Lei X, Chang Z, Luo L, Sun X (2013) Metal oxide and hydroxide nanoarrays: hydrothermal synthesis and applications as supercapacitors and nanocatalysts. Progr Nat Sci Mat Int 23:351–366

Yang M, Fan Q, Zhang R, Cheng K, Yan J, Pan D, Ma X, Lu A, Cheng Z (2015) Dragon fruit-like biocage as an iron trapping nanoplatform for high efficiency targeted cancer multimodality imaging. Biomaterials 69:30–37

Yano S, Hirohara S, Obata M, Hagiya Y, Ogura S-i, Ikeda A, Kataoka H, Tanaka M, Joh T (2011) Current states and future views in photodynamic therapy. J Photochem Photobiol C: Photochem Rev 12:46–67

Yoo HS, Lee KH, Oh JE, Park TG (2000) In vitro and in vivo anti-tumor activities of nanoparticles based on doxorubicin–PLGA conjugates. J Control Release 68:419–431

Yu H, Nie Y, Dohmen C, Li Y, Wagner E (2011) Epidermal growth factor–PEG functionalized PAMAM-pentaethylenehexamine dendron for targeted gene delivery produced by click chemistry. Biomacromolecules 12:2039–2047

Yukihara M, Ito K, Tanoue O, Goto K, Matsushita T, Matsumoto Y, Masuda M, Kimura S, Ueoka R (2011) Effective drug delivery system for duchenne muscular dystrophy using hybrid liposomes including gentamicin along with reduced toxicity. Biol Pharm Bull 34:712–716

Zeng H, Du XW, Singh SC, Kulinich SA, Yang S, He J, Cai W (2012) Nanomaterials via laser ablation/irradiation in liquid: a review. Adv Funct Mater 22:1333–1353

Zhu G, Zhang F, Ni Q, Niu G, Chen X (2017) Efficient nanovaccine delivery in cancer immunotherapy. ACS Nano 11:2387–2392

Ceramics for Biomedical Applications 10

Shyamal Mandal

Abstract

This study is focused on development of ceramics-based different types of body implants like the spinal cord, femoral head, and elbow joints and dental implants. Due to low fracture toughness and high hardness, sintered stage machining is not preferable for ceramics. Advancement of gelcasting and protein coagulation casting of ceramics is promoting green stage machining in recent time due to their high green strength. Interestingly, custom design and fabrication are essential due to variation in size and shape of body implants of different peoples of crown like many other biomedical objects. The surface quality of the machined sample was evaluated through surface profilometry and optical imaging. The Computerized Numerical Control (CNC) machining parameters of diamond-coated tools for the manufacturing of dental crown were optimized for superior surface quality in a minimum time period. Finally, different ceramic implants of varied size and shape were fabricated and characterized by using different tools.

Keywords

Gelcasting · Protein coagulation · Sintered · CNC machine

10.1 Introduction

Nowadays, different types of ceramics materials are used in biomedical medical applications. We can divide them into three classes according to their chemical reactivity with the sounding tissue or body fluids: (i) completely resorbable, (ii) surface reactive, and (iii) nearly inert.

S. Mandal (✉)
Biomedical Engineering Department, North-Eastern Hill University, Shillong, Meghalaya, India

S. Paul (ed.), *Biomedical Engineering and its Applications in Healthcare*,
https://doi.org/10.1007/978-981-13-3705-5_10

The completely resorbable ceramics means it's more reactive in our body like calcium phosphate; it releases ions from the surface for a long time as well as provides protein bond site. The ion-released mechanism promotes hydroxyapatite nucleation for mineralized bone, and it helps to fix the implant properly with the host tissue. Surface reactive bio-glass ceramics have some different behaviors as an implant. In these ceramics, surface provide major role for the proteinaceous constituent of soft tissue and cell membrane and it also producing tissue adherence. Nearly inert ceramics like alumina, zirconia even after thousands of hours deep into physiological fluid and there for show minimal interfacial bond with host tissue.

10.1.1 Replacement of the Spinal Cord

10.1.1.1 Introduction

The spinal cord, the downward continuation of the medulla oblongata, is contained in the vertebral canal. It has more or less uniform structure throughout. In cross section it is seen to contain the central canal in the middle, around which is a mass of gray matter (spinal gray matter) and in the periphery is the white matter. The spinal cord is divisible into 31 functional segments, each of which connected with a pair of spinal nerve. The whole of the central nervous system (CNS) is enclosed by a special connective tissue. The innermost one is the pia matter which intimately follows the elevations and depression on the surface of the brain and spinal cord. The middle layer is the arachnoid matter which just goes around, hence leaving some space under it called subarachnoid spaces which are filled up with cerebrospinal fluid (CSF).

10.1.1.2 Importance of Ceramics in the Spinal Cord

1. Tricalcium phosphate ($Ca_3(PO_4)_2$) including hydroxyapatite (HA) are being popular biomaterials for orthopedic and spinal cord replacement.
2. They are osteoconductive; tensile and bending strength are almost similar to achieve the optimal strength.
3. Tricalcium phosphate ($Ca_3(PO_4)_2$) have been used to increase the bone density and is also compatible with the local host tissue.
4. The compositions of tricalcium phosphate ($Ca_3(PO_4)_2$) and hydroxyapatite (HA) are easy to make appropriate shapes and size and also low cost.
5. The composition has been used for giving structural geometry of any bone which is similar to the cancellous bone that makes it highly osteoconductive and also helps to grow cells inside the implants (Figs. 10.1, 10.2 and 10.3).

10.1.1.3 Types of Spinal Implants

Spinal implants are divided into several parts which are discussed below:

1) Cages: It is also known to provide support to the interspace of two spinal disks. These kinds of implants are fixed by the surgeon in between two vertebrae. The implants basically have highly porous walls, so that the cells can easily grow

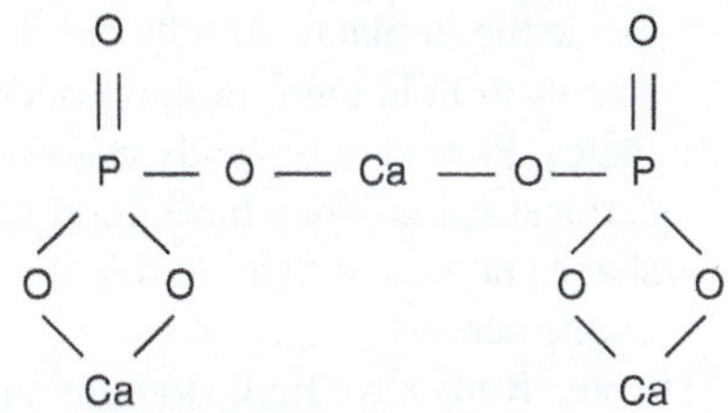

Fig. 10.1 Structure of tricalcium phosphate ($Ca_3(PO_4)_2$)

Fig. 10.2 SEM image of tricalcium phosphate

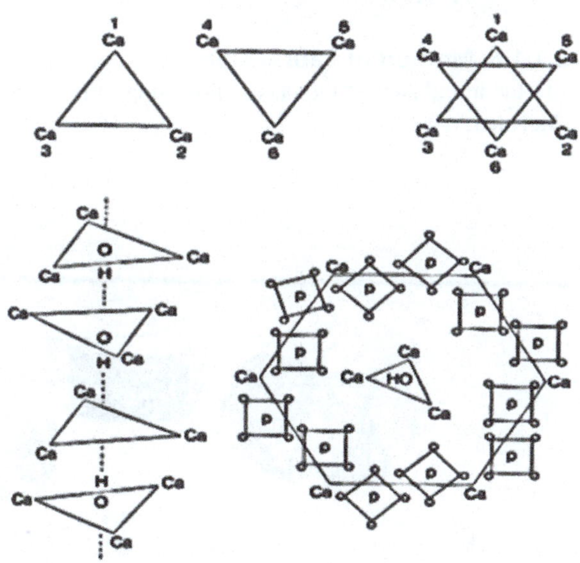

Fig. 10.3 Molecular structure of hydroxyapatite (HA)

inside the implants with its provided support and stability. So there is no need of screws to hold them in the proper place.

2) Plates: Plates are typically ceramic implant mainly used to provide support to the cervical spine. They have good flexibility to bend with the spine and can modify shape and size to fit an individual patient. Usually, plates are fitted with vertebrae by the screws.
3) Rods: Rods have high strength and flexibility. They help to stabilize and support the broken spinal cord so that it is properly aligned. Typically, rods are fixed to the vertebrae with either hooks or pedicle screws.

10.1.1.4 Anterior Interbody Cages

Anterior interbody cages look like cylinder used to support the spinal disk. The cages are made of porous materials to allow the bone graft to grow the cells from the vertebral column through the cage and connect into the next vertebral column. The cages offer excellent support to the vertebral column, so most patients do not need additional instrumentation for postoperative back braces to support the spinal cord (Fig. 10.4).

10.1.1.5 Artificial Lumbar Disks

Lumbar disks are very important organ in our body; maximum spinal load is carried by the lumbar disk. Usually, it is made of metal or plastic-like flexible (biopolymer) materials, or composite materials, but ceramics due to high hardness and high strength may be the promising biomaterials for replacement of lumbar disk. Ceramics can also function inside the body for long years without changing its size and shape (Fig. 10.5).

10.1.1.6 Method of Fabrication

1. In the initial step of the design procedure, the restrictions and design basis were decided.

Fig. 10.4 Anterior interbody cages

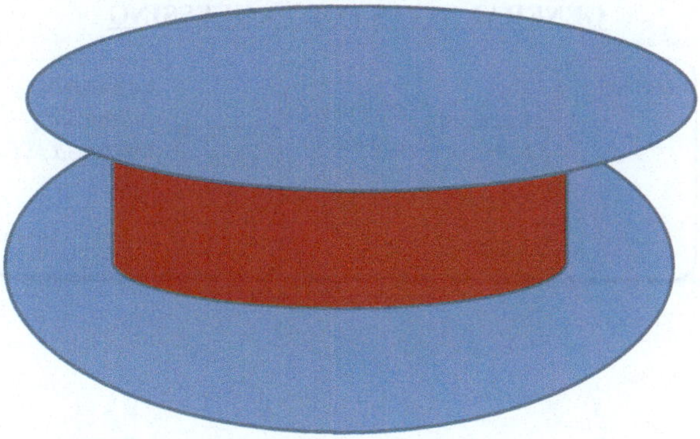

Fig. 10.5 Artificial lumbar disks

2. Images of the vertebral segment with a reference ruler were taken perpendicular to the disk surface and used as major design inputs.
3. By the help of the ruler, calculations of real distance-pixel on the image ratios were performed.
4. By using image processor software, place and rough drafts of spinal cage designs were done on the vertebral disk sections.
5. The disk region of the vertebrae was colored with blue.
6. Dimensional conversions of the blue areas were calculated using the real distance-pixel ratio of the corresponding image, and they were converted into defined drawing contours.
7. According to contours, solid models of each segment were created using 3D design software.
8. By using computer drawing software, 2D technical drawings of each spinal cage were composed. Solid models and technical drawings of each spinal cage were shown in (Fig. 10.6).

10.2 General Steps for Processing

10.2.1 Femoral Head of the Hip Bone

10.2.1.1 Introduction

The femoral head is an important part of the hip bone. It is located below the pelvis girdle, and it is also connected to the femoral neck. According to the scientist, corrosion is the real issue for damaging the femoral head. Metals like titanium and Co-Cr alloy are probable to highly damage due to the continuous changing of pH of the body fluids. Corrosion can be reduced by the highly surface-finished femoral head. But research has shown that the ceramic head has more lasting power than the cobalt chrome head. Some of the metal implants are also known to be more

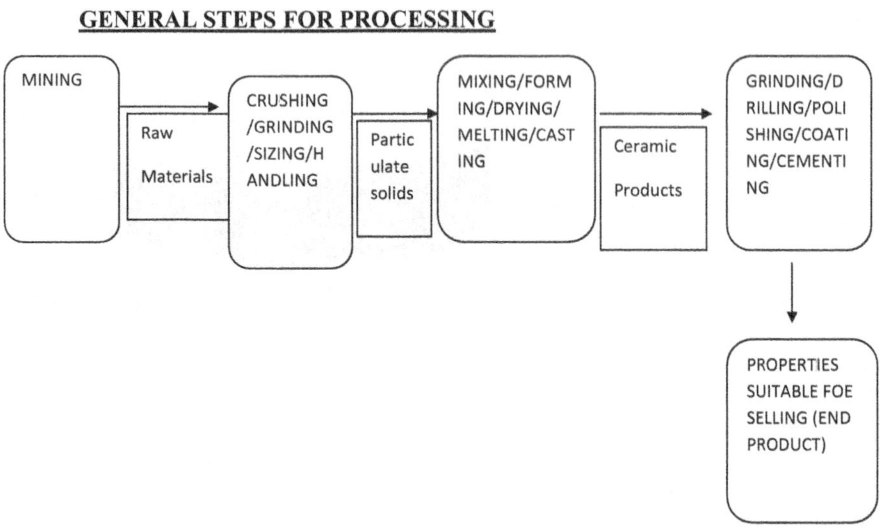

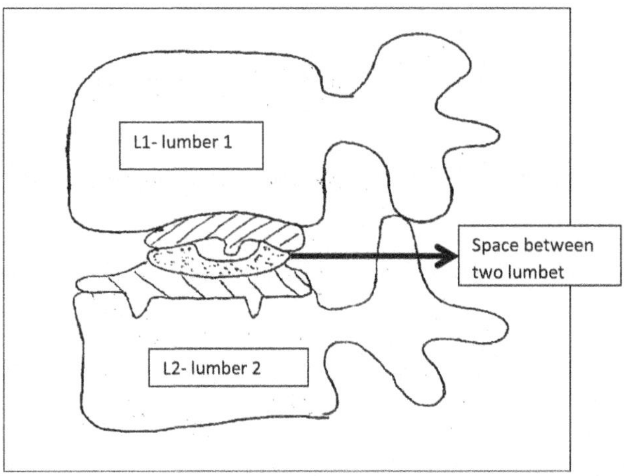

Fig. 10.6 Picture of artificial lumbar

cytotoxic. The research comes out with the statement that the ceramic head also has a slight advantages because of cost-effectiveness.

10.2.1.2 Ceramics Are Used in the Femoral Head

1. Ceramics on metal (COM)
2. Ceramics on ceramics (COC)
3. Ceramics on polyethylene (COP)

Ceramic heads are made of almost same ceramic powders used for manufacturing of ceramic plates and bowls.

Ceramic on metal: Before the twenty-first-century, maximum implants are made of metal on metal or ultra-molecular weight polyethylene (UHMPE) due to high hardness and high strength. But ceramics on metal may be a promising device for biomedical applications. Ceramics also have good biocompatible property and also have less corrosion rate inside the body fluid.

Ceramic on ceramic: Artificial hip joint is nowadays made of high-purity alumina (Al_2O_3) ceramic. The shape and size are given in the green stage of ceramics because after sintering it's difficult to give the exact size and shape of the ceramics. Green machining is also a promising step to make the proper size and shape of the ceramic implant. The hip joint is a ball and socket joint, and ceramics also have good bio-inert property, so it also helps to lessen corrosion.

Ceramic on polyethylene: Another good invention of the scientist was ceramics on polymers. Everybody know polymers are very soft materials and ceramics are very hard materials. But after the invention of ultra-high molecular weight polyethylene (UHMWP), its strength and other properties are almost similar to ceramics. The total hip replacement implants have some parts like head made of ceramics with neck and tail parts made of polymers. Sometimes polymer films are coated on the ceramics to deduce surface roughness.

10.3 Parts of the Implants

The hip replacement consists of three parts:

- Acetabulum cup
- The femoral head
- The labrum

Acetabulum: It is an organism which can be thought of as a "hip-socket." The perforate acetabulum is a cup-shaped opening on each side of the pelvis girdle formed where the ischium, ilium, and pubis all meet and into which the head of the femur inserts.

Femoral head: The head of the femur, which articulates with the acetabulum of the pelvic bone, comprises two-thirds of the sphere. It has a small groove, or fovea, connected through the round ligament to the side of the acetabulum notch. The head of the femur is connected to the shift through the neck or column.

Labrum: The acetabula labrum is a ring of cartilage that surrounds the acetabulum of the hip. The anterior portion is most vulnerable when the labrum tears.

10.3.1 Method of Fabrication

A femoral head made from a ceramic-metal composite having a multilayer construction and related method of manufacture is disclosed. The femoral head includes an inner metal core bonded to a relatively thin ceramic outer layer that is used as the articulating surface with an acetabular cup during total hip arthroplasty. In another

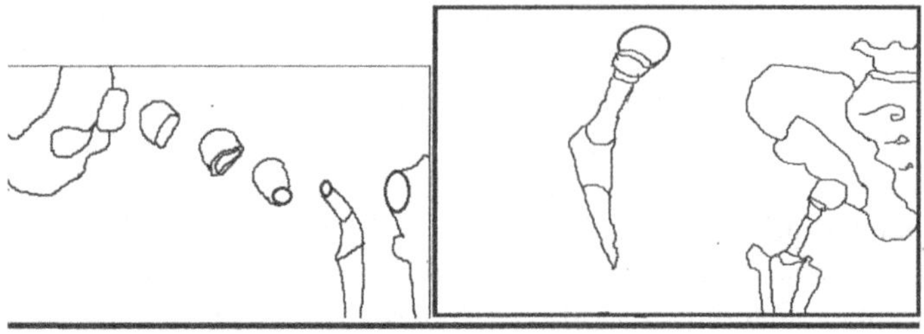

Fig. 10.7 Artificial hip joints

embodiment, an interface layer having a ceramic and metal mixture may be laminated between the inner metal core and the exterior ceramic layer (Fig. 10.7).

10.3.2 Clinical Significance

If there is a fracture of the neck of the femur, the blood supply through the ligament becomes crucial. In orthopedic surgery, the head of the femur is important because it can undergo avascular necrosis and it's also take load of the other part of body. The femoral head is removed in total hip replacement surgery.

10.3.3 Elbow Joints

Elbow joint is a complex joint that allows movement in several directions. It joints between arm and forearm, consisting of humeroulnar, humeroradial, and proximal radioulnar articulation. Total elbow replacement is a surgical procedure during which the elbow joint is replaced with an artificial joint. This joint is made up of a humeral component and an ulna component that are connected to each other.

10.3.4 Materials for Elbow Joints

For a material to be considered a biomaterial it must be able to safely and reliably replace with an appropriate physiological response. In other words, materials must be biocompatible.

1). Alumina
2). Zirconium
3). Calcium phosphate
4). Stainless steel

10.3.5 Fabrication Process

1. Zirconia-toughened alumina
2. Osteoceramic
3. Cold isostatic pressing

1. In the initial step of the design procedure, the restrictions and design basis were decided.
2. Images of elbow segment with a reference ruler were taken to perpendicular on the disk surface and used as major design inputs.
3. By the help of the ruler, calculations of real distance-pixel on the image ratios were performed.
4. By using image processor software, place and rough drafts of spinal cage designs were done on the elbow joints.
5. 70–90% alumina and 10–20% zirconia were mixed through a roller.
6. By using computer drawing software 2D technical drawings of elbow joints were composed. Solid models and technical drawings of elbow joints are fabricated through computerized numerical control machine (CNC).
7. When stress is induced by the transformation mechanism, zirconia plays a major role to lead the load.

10.3.6 Risks of Elbow Joint Replacement

1. Bone rupture and nerve or blood vessel injury can occur during the procedure.
2. Blood loss during and after the operation usually does not exceed 500 ml. Greater blood loss requires extra blood.
3. Microbial infection can be occurring after implantation that is why antibiotic is prescribed by the physician.

10.3.7 Dental Implants

Dental crown is an outer projection/cap of tooth usually used for restoration of dentistry which may encircle the existing impaired tooth or dental root implant and is typically bonded to the tooth/root implant using dental cement. Crowns are often used to improve function or aesthetic appearance of teeth. Crowns can be made from diverse materials, which are usually fabricated using indirect methods. Due to low fracture toughness and high hardness, sintered stage machining is not preferable. Advancement of gelcasting and protein coagulation casting of ceramics is promoting green stage machining in recent time due to their high green strength. Interestingly, custom design and fabrication are essential due to variation in size and shape of dental crown like many other biomedical objects. Thus mold-free rapid tooling via green stage machining is a viable option for the fabrication of unsymmetrical objects like crown [1–10].

This study is focused on development of alumina-based dental crown using a process involving direct machining of green ceramics (DCM) by computer numerical control (CNC) machine which allows manufacturing of ceramic crown including dental bridges. Green cylindrical alumina ceramics are prepared using protein coagulation casting followed by drying. The fully dried samples are characterized for green density, green strength and hardness. The sintered samples were characterized for binder wt% in green body, sintered density, mechanical strength, hardness and anisotropic shrinkage. Further, 3D image of collected tooth was generated through 3D scanning which was used as model image. Using this model image, crown was fabricated through optimization of machining parameters. Finally, the finished crown was sintered and evaluated for surface quality, microstructure and anisotropic shrinkage study. The surface quality of machined sample was evaluated through surface profilometry and optical imaging. The machining parameters of diamond-coated tools were optimized for superior surface quality in a minimum time period. Finally, different alumina crowns of varied size and shape were fabricated using a diamond-coated tool [11–25].

10.3.8 Structure Analysis of Dental Crown

A tooth is mainly made of four major tissues like enamel, dentine, pulp cavity, and cementum (Fig. 10.8). Enamel is the outer part of the teeth; its main content is calcium. Dentine is the main segment of the tooth alive tissue like bone not as hard or resistant to decay as enamel. Pulp cavity is where the nerve endings are placed which

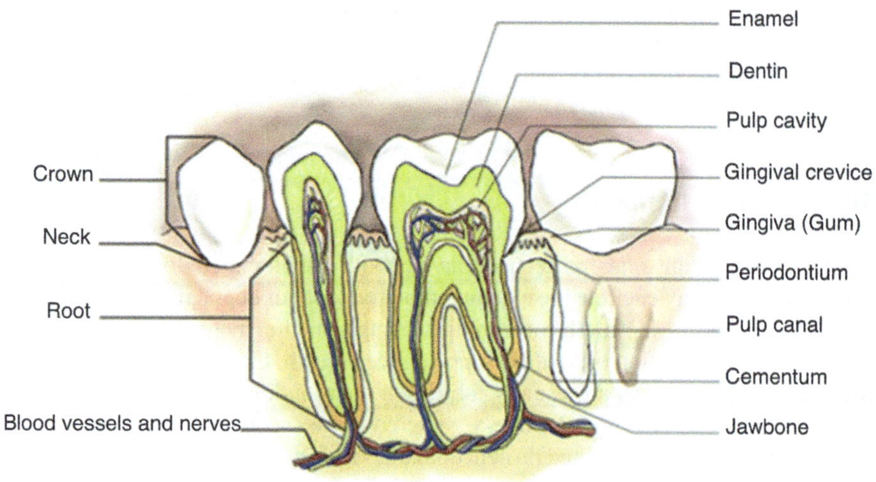

Fig. 10.8 Different parts of dental crown

is connected to the brain. These blood vessels carry oxygen and nutrients for cells and remove waste products. Cement is the rough layer, to which periodontal fibers are devoted, holding the tooth in place and jaw bone with plugs for the tooth.

10.3.8.1 Necessity of Dental Crown

A dental crown may be needed in the following situations:

- Protection of weak tooth (for instance, from decay) from breaking or to hold together parts of a crack tooth
- Restoration of a broken tooth or a tooth that has been severely worn out
- Covering and support of tooth with a large filling when enough crown part of the tooth is not left
- Holding a dental bridge in place of missing tooth in between
- Covering severely discolored teeth
- Covering metallic dental implants

10.3.9 Challenges

The major challenge in fabrication of dental implant is their variation in size and shape of one to another from person to person. The conventional fabrication route is to prepare different molds for different teeth. Accordingly, they create mask using alginate/dental wax as impression materials following conventional molding and demolding for one off custom products. However the technique of molding and demolding is costly and time-consuming specially for one off custom products like dental crown where size and shape of the samples vary from one to another. Newly developed bottom-up approach, rapid prototyping technique, though uses mold-free fabrication using 3D model image is not suitable for ceramic fabrication owing to its high sensitivity to notch and its low fracture toughness. Thus a top-down approach may be preferable especially where mold fabrication step is possible to avoid. In this context, different CAD-CAM technologies followed by rapid tooling may be explored for crown fabrication. But sintered stage machining of crown is extremely difficult and costly affair. Fine polishing is essential to avoid notch like defects due to its low fracture toughness. Green stage machining was impossible due to their low green body strength. Atwood studied on investment casting of titanium crown. Tsitrou evaluated the marginal fit of three margin designs of resin composite crowns using a CAD/CAM technique. Huang et al. emphasized on bio-inspired design of multilayer dental crown. Zhang et al. introduced a novel method for titanium crown casting using a combination of wax patterns fabricated by a CAD/CAM system and a non-expanded investment. Brien et al. developed lucite composite-based dental crown by using investment casting. Phira et al. developed lithia disilicate-based dental implant by a hot pressing technique, but it does not always work because the wear volume of ceramics molar crown is lower than the enamel. C.A. Michal et al. used Zink phosphate bonding alloy for developing dental implant by impressing record and customizing casting but the procedure failed due to the resin-modified glass incomer cement, and resin composite cement failed at a significant higher load than the zinc phosphate. J. Zhou developed dental crown by dental cement composite bonded to glass substrates, but this technology failed due to incising interracial toughness when loading angle increases. Sohelia Eteadi et al. developed resin-bonded porcelain veneer which failed over 5 years with the occurrence of 14.5%. Massimiliano Guazzato et al. used bi-layered porcelain zirconia (Y-TZP) by iso-statically hot-pressing for dental implants. Toroglu et al. used duel cured resin composite (Panavia F) and self-tapping screw for reappearing of fracture teeth by reattachment technique which provides immediate esthetic and functional rehabilitation of fracture teeth.

10.3.10 Current Technologies

A number of research groups are exploring the development of dental crown using a rapid tooling technique based on a 3D model image. For example, Filser et al. have developed fabrication of dental crown using CNC machining of bisque fired glass ceramics. This technique is not commercially well accepted due to a high rejection rate of fragile bisque fired glass ceramics. Very recently, Zel et al. have developed computer integrated ceramic reconstruction (CICERO) system for CAD/CAM fabrication of full-ceramic crown. However, due to a high rejection rate, green stage machining of ceramics may be explored due to improved green strength with predictable shrinkage. Further, different created notch during machining may be reduced or healed significantly due to shrinkage associated with sintering. Thus, green machining of ceramics can be explored using CNC machine owing to high green strength of ceramics compacts.

10.3.11 Search of Different Materials for Application of Dental Crown

Different materials like metal (titanium, gold), glass-ceramics (alumina, zirconia-toughened alumina, porcelain), and polymers (acrylics, resin) are explored for suitability in the application of dental crown. However, none of these materials are perfect for these intended applications due to the mismatch in different mechanical properties. Now, several new materials like leucite, lithium disulicate, and macor and especially different composites are exploring for suitability due to their closeness with mechanical properties like hardness, modulus, and strength. In this context, mechanical properties of different materials including different parts of crown are given in Table 10.1.

Table 10.1 Mechanical properties of tooth and progressive materials

	Density gm/cc	Elastic modulus (GPa)	Compressive strength (MPa)	Tensile strength (MPa)	Flexural strength (MPa)	Vickers hardness (GPa)
Cementum	2.03	—	—	—	—	—
Dentine	2.14	15	297	105	—	—
Enamel	2.97	75–90	384	10–20	—	3.3–3.9
HA	3.1	110	509	—	113	6
Ti/alloys	4.5	117	—	550–930	—	25
Leucite	2.47	63	—	—	173	5.3
Lithium Disilicate	2.4	—	—	—	—	5.7

10.3.12 Different Raw Materials Which Were Used for Sample Preparation Are as Follows

- Ceramic powders alumina (CT 3000Sg Alcoa)
- Dispersants Darvan 821A (R. T. Vandetbuilt, USA)
- Binders (egg white, sugar)
- Zirconia milling media (Jyoti Ceramics, Nasik, India) for slurry preparation

10.3.13 Machine and Software Which Were Used for Fabrication

- Vibration table for deairing
- 3D laser scanner for creation of a 3D model image (Roland, PICZA, LPX-600, Japan)
- CNC machine for rapid tooling (MDX-540, Japan)
- PVC pipe for casting cylindrical samples
- Diamond tools for machining of ceramics
- Surface profilometer for surface roughness characterization
- High-temperature furnace for sintering ceramics (OKAY, Electronics furnace, Kolkata)

10.3.14 Fabrication of Green Ceramics

The dense compacts alumina is prepared via protein coagulation casting (PCC) involving the preparation of aqueous alumina slurries with a mixed binder of egg white protein-sucrose. Darvon 821A (40 wt% solution) was used as a dispersant. In the present study, 55% alumina (Alcoa CT 3000 SG, USA) loading is used with

8 vol.% of the egg white protein along with 3 wt% sucrose on dry alumina powder weight basis. 0.2 wt% n-octanol is used as an antifoaming agent. The slurries are prepared by 24 h ball milling the above mix in the presence of spherical Zirconia media in 250 ml polypropylene bottles with 100 ml of slurry. Then the slurry was filtered through a sieve to separate the zirconia milling media. Then it is put into vibration table for 1 h until the bubbles come out on the top surface of the slurry; after that again it is put into the ball milling for 1 h. The de-aired slurries were casted into 20 mm inner diameter of cylindrical molds. Then the molds were covered and put on a flat table and remain for 24 h with some water. Then the mold is placed into the oven for 12 h at 50 °C. Then the green parts are removed from the mold and dried at 80 °C for 12 h.

10.3.15 Mold Selection

For the selection of mold, it is necessary to decide the shape and size of samples to be prepared which will be used for machining directly for dental crown fabrication. As machining will be carried out in two modes, (a) rotary axis and (b) vertical, accordingly the cylindrical samples will be suitable for machining in both the modes.

Initially, 24-well plates were chosen for casting cylindrical samples (Fig. 10.9). However, the samples length and diameter were shorter and larger respectively which resulted in longer machining time. Finally, PVC pipe was cut to the intended length as per the requirement depending on diameter and length of the crown which was fixed on to Perspex sheet using plastic glue (Fig. 10.10). The inner diameter of PVC pipe cylindrical mold with 17 mm and length 25 mm by four round sample casting. It was possible to prepare good cylindrical samples for machining in a relative shorter period of time. These molds were coated with petroleum jelly prior to casting which acts as mold releasing agent for easy removal of components. Also, rectangular bar samples were cast in silicon rubber mold for mechanical testing of ceramics.

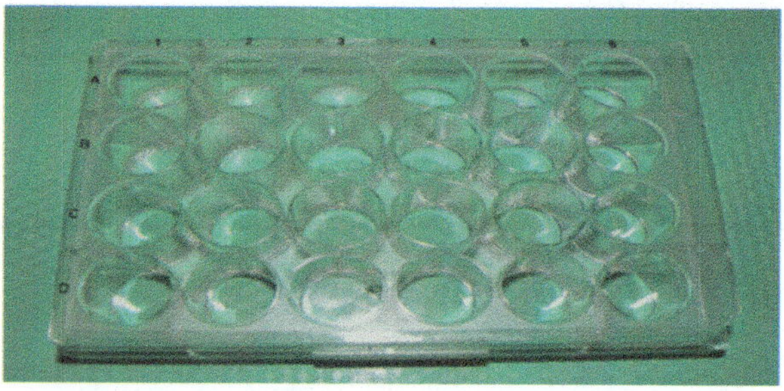

Fig. 10.9 24 well plates was used for casting of samples

Fig. 10.10 Mould fabricated from PVC pipe

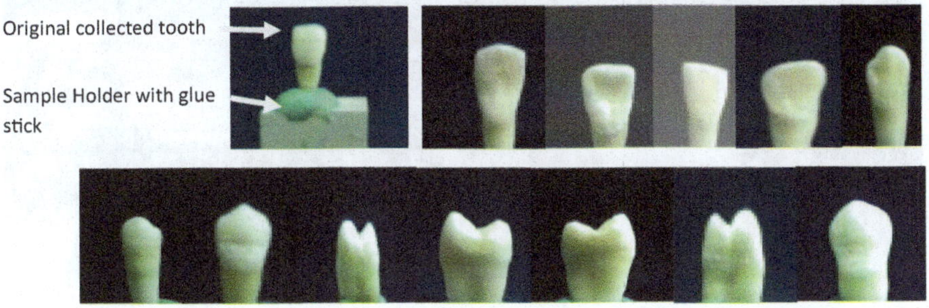

Fig. 10.11 Original crowns collected as standard samples for development of 3D image

10.3.16 Selection of the Objects for 3D Scanning

Different teeth were collected from a dental hospital for 3D scanning as shown in Fig. 10.7. Sample holder arrangement is made for fixing of samples for scanning in a 3D laser scanner. Here the sample is fixed in a sample holder by glue stick. This is shown in Fig. 10.11.

10.3.17 Creation of 3D Model Image

Different collected teeth were placed in 3D laser scanner for creation of a model image. The objects were scanned at different angles and merged using XOR/XOV following filtration and healing to create a final model image. This image was further saved in ∗.stl format for CNC machining (Fig. 10.12).

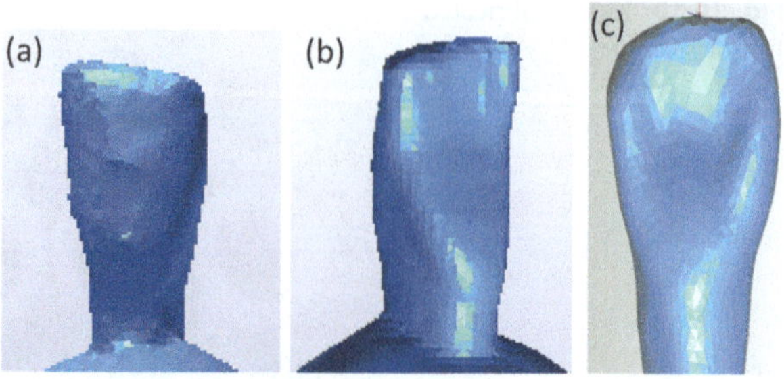

Fig. 10.12 (**a**) Vertical scanned (**b**) Horizontal scanned (**c**) Final filtered modified image

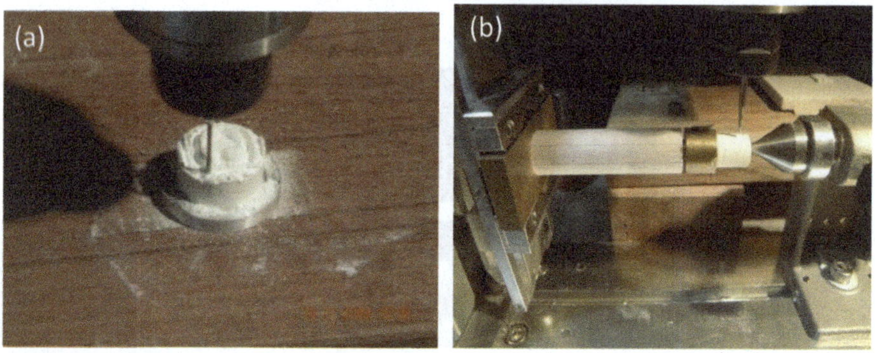

Fig.10.13 (**a**) Vertical axis machining (**b**) Rotary axis machining

10.3.18 CNC Green Machining of Ceramics

A Computerized Numerical Control milling machine (MDX 540, Roland DG Ltd., Japan) was used for manufacturing the dental crown. It was a fully programmed computer-controlled milling machine, which could be functioned at a variable speed of (4500–15,000) rpm with maximum *x–y* speed 50 mm/sec and *z* speed of 30 mm/s. In the present research, the *x–y* and *z* direction speeds were adjusted to obtain smooth samples at a minimum machining time. Green machining was carried out using diamond-coated flat end milling and conical tools with different tool size from 0.2 to 3 mm. Green ceramic samples were mounted on a designed brass sample holder and fitted into the rotary axis. The green sample was attached to the sample holder by using wax and Fevi kwik.

After the rotation was completed the sample was placed on the base plate for the vertical machining. To manufacture complex-shaped ceramic items, either CAD or scanned imaging files were shifted to standard STL files, which were then inputted to the CNC machine and a machining path simulation was accomplished before the machining operation. The ceramics were green machined through a surfacing roughening and finishing operation, respectively, to obtain good surface finishing (Fig. 10.13).

10.3.19 Green Body Characterization

The fully dried green body was fractured and gold coated for SEM analysis. The green rectangular bar samples were tested for flexural strength using a universal testing machine, and the green samples were machined, while samples are mounted using a double-sided tape by CNC machine. The surface quality of machined samples before and after sintering was investigated and hardness of green and sintered samples is also studied.

The flexural strength of samples was calculated as follows:

$$\sigma = \frac{3FL}{2bd^2}$$

where F, L, b, and d are load (force) at the fracture point, length of the support span, width, and thickness, respectively.

The sintering shrinkage was calculated in different directions using the following formula:

$$\text{Sintering shrinkage }(\%) = \frac{H_{\text{green}} - H_{\text{sintered}}}{H_{\text{sintered}}} \times 100$$

where H_{green} and H_{sintered} are dimensions of green and sintered ceramics in each direction respectively.

10.3.20 Sintering

Sintering is a method for making objects from powder, by heating the material in a sintering furnace below its melting point. The machine sample was put into high-temperature furnace at 900°C for bisque firing. On this temperature organic binder was burned out that is why some weight loss was found in the sample. After that the sample was put at 1550°C furnace for sintering. The sintered sample has good compactness and better surface finish. A final sintered crown is shown in Fig. 10.14.

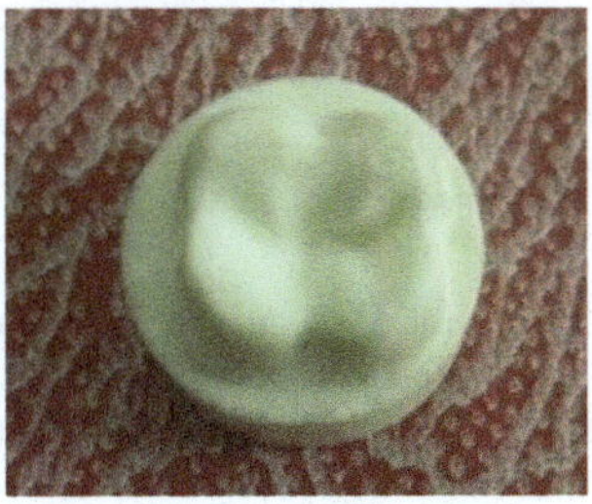

Fig. 10.14 Sintered molar crown

10.3.21 Picture of Fabricated Crowns (Fig. 10.15)

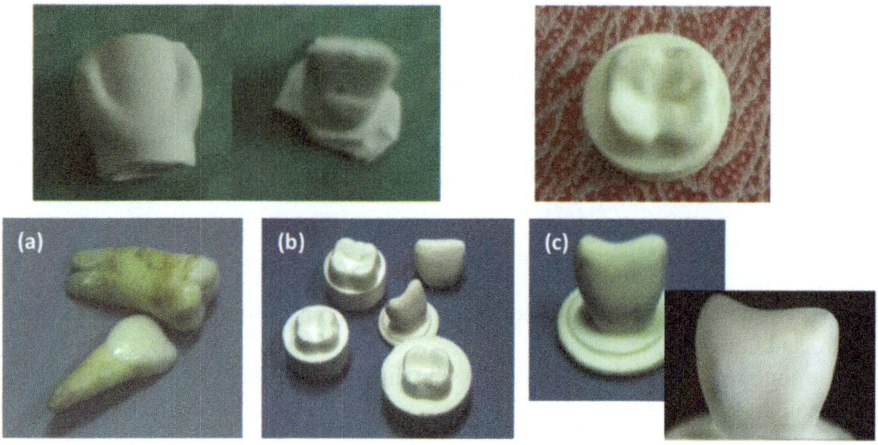

Fig. 10.15 (**a**) Original dental crown (**b**) Fabricated alumina crown and (**c**) Closed view of a fabricated crown

10.4 Conclusion

- The anatomical diameter of the femoral head offers distinct theoretical advantages in total hip arthroplasty. These short-term results are promising, and further study of this new skill on a larger scale with a longer follow-up time period is required.
- Using collected teeth, the 3D model image was created using 3D laser scanning following filtration and modification.
- Green ceramics with cylindrical samples were prepared using colloidal casting, and samples were defect-free, machinable using a specialized tool.
- The sintered density and microstructures are suitable for the intended application in restoration of dentistry.
- Finally using a 3D model image, molar and incisor crowns were prepared with a good surface finish.

References

Allahverdi M, Danforth SC, Jafari M, Safari A (2001) Processing of advanced electroceramic components by fused deposition technique. J Eur Ceram Soc 21:1485–1490

Binger T, Rücker M, Spitzer WJ (2006) Dentofacial rehabilitation by osteodistraction, augmentation and implantation despite osteogenesis imperfect

Brady GA, Halloran JW (1997) Stereolithography of ceramic suspensions. Rapid Prototyping J 3:61–65
Cawley JD (1999) Solid freeform fabrication of ceramics. Curr Opin Solid State Mater Sci 4:483–489
Dhara S, Kamboj RK, Pradhan M, Bhargava P (November 2002) Shape forming of ceramics via gelcasting of aqueous particulate slurries. Bull Mater Sci 25(6)
Giordano R (2006) Materials for chairside CAD/CAM-produced restorations. JADA 137:14S–21S
Halloran JW (1999) Freeform fabrication of ceramics. Br Ceram Proc 59:17–28
Holand W, Schweiger M, Frank M, Rheinberger V (2000) Comparison of the microstructure and properties of the IPS Empress 2 and IPS Empress glass-ceramics. J Biomed Mater Res 53:297–303
Kamboj RK, Dhara S, Bhargava P (2003) Machining behaviour of green gelcast ceramics. J Eur Ceram Soc 23:1005–1011
Lewis JA, Smay JE, Stuecker J, Cesarano J III (2006) Direct ink writing of three-dimensional ceramic structures. J Am Ceram Soc 89:3599–3609
Lohbauer U, Petschelt A, Greil P (2002) Lifetime prediction of CAD/CAM dental ceramics. Appl Biomater 63:780–785
Luthardt RG, Holzhuter MS, Rudolph H, Herod V, Walter MH (2004) CAD/CAM-machining effects on Y-TZP zirconia. Dent Mater
Moon J, Grau JE, Knezevic V, Cima MJ, Sachs EM (2002) Ink-jet printing of binders for ceramic components. J Am Ceram Soc 85:755–762
Rekow ED, Thompson VP (2005) Near-surface damage—a persistent problem in crowns obtained by computer-aided design and manufacturing. Proc Inst Mech Eng J; J Eng Med 219:233–243
Schleier P, Hyckel P, Fried W, Beinemann J, Wurdinger J, Hinz M, Steen M, Schumann D (2006) Vertical distraction of fibula transplant in a case of mandibular defect caused by shotgun injury. Int J Oral Maxillofac Surg
Schulera M, Gethin Rh (2006) Biomimetic modification of titanium dental implant model surfaces using the RGDSP-peptide sequence: a cell morphology study
Sindel J, Petschelt A, GRellner F, Dierken C, Greil P (1998) Evaluation of subsurface damage in CAD/CAM machined dental ceramics. J Mater Sci Mater Med 9:291–295
Smay JE, GRatson G, Shepard RF, Cesarano J III, Lewis JA (2002) Directed colloidal assembly of 3D periodic structures. Adv Mater 14:1279–1283
Strub JR, Rekow ED, Witkowski S (2006) Computer-aided design and fabrication of dental restorations. JADA 137:1289–1296
Su B, Dhara S (2005a) Green ceramic machining: a top-down approach for the rapid fabrication of complex-shaped ceramics. J Appl Ceramic Tech
Su B, Dhara S (2005b) Green machining to net shape alumina ceramics prepared using different processing routes. J Appl Ceram Technol 2([3]):262–270
Thompson JY, Stoner BR, Piascik JR (2007) Ceramics for restorative dentistry: critical aspects for fracture and fatigue resistance. Mater Sci Eng C 27
Tinschert T, Natt G, Hassenpflug S, Spiekermann H (2004) Status of current CAD/CAM technology in dental medicine. Int J Comput Dent 7:25–45
Tözüm TF Şençimen M (2006) Diagnosis and treatment of a large periapical implant lesion associated with adjacent natural tooth: a case report
Zhao X, Evans JRG, Edirishinghe MJ (2002) Direct ink-jet printing of vertical walls. J Am Ceram Soc 85:2113–2115

Biopolymeric Scaffolds for Tissue Engineering Application

11

Nalini Ranganathan, A. Mugeshwaran, R. Joseph Bensingh, M. Abdul Kader, and Sanjay K. Nayak

Abstract

The scaffold is a three-dimensional (3D) substrate that works as a pattern for revival of the tissue. The tissue-engineered products comprise heart valves, cartilages, bones, muscle, nerves, liver, bladder, etc. The scaffold provides the necessary provision for the cells to attach, multiply, and sustain their distinguished functions. In recent years, polymeric scaffolds play a significant role in the tissue engineering application. A wide range of natural and synthetic biopolymers are being used for the fabrication of the scaffolds, which includes proteins, chitosan, polysaccharides, poly(lactic acid) (PLA), poly(glycolic acid) (PGA), poly (ε-caprolactone) (PCL), poly(lactic-co-glycolic) acid (PLGA), etc. In addition bioactive glass and ceramic particles are reinforced with biopolymers and are used in the form of biocomposites for fabricating scaffolds structures. The reinforced biopolymers are extensively used in the development of scaffolds, due to incredible properties such as mechanical strength, regulated pore size, biodegradability, biocompatibility, and renewability. The fabricating materials are opted based on the functionality of the scaffolds, i.e., either synthetic or biologic, in other way degradable or nondegradable. The scaffolds are also distinguished based on the specific period of performance; commonly they are classified as short-term and long-term scaffolds. They are either injectable or implantable; usually the short-term scaffolds are injected and are biodegradable, whereas the long-term scaffolds are implanted and may be degradable or nondegradable. Various fabrication techniques were being adapted by the researchers to develop the scaffolds, which includes solvent casting, particulate leaching, phase

Nalini Ranganathan (✉) · A. Mugeshwaran · R. Joseph Bensingh · M. Abdul Kader
Advanced Research School for Technology and Product Simulation (SARP), TVK Industrial Estate, Chennai, Tamil Nadu, India

S. K. Nayak
Central Institute of Plastics Engineering & Technology (CIPET), TVK Industrial Estate, Chennai, Tamil Nadu, India

S. Paul (ed.), *Biomedical Engineering and its Applications in Healthcare*,
https://doi.org/10.1007/978-981-13-3705-5_11

separation, electrospinning, gas foaming, freeze-drying, and rapid prototyping. The selection of the materials and manufacturing methodologies are critically important for tissue engineering in designing the artificially made extracellular matrices (scaffolds), which can provide the support for three-dimensional tissue formation. Furthermore, various post-process surface modification methods are performed on the scaffolds to improve their functional performance.

Keywords
Biopolymer · Biocomposite · Tissue engineering · Biodegradation · Biocompatibility

11.1 Introduction

The biopolymeric scaffold in tissue engineering has been developed significantly in the last 10 years and offers immense potential in regenerative medical approaches (Yoshida et al. 2011). Biopolymers and their composites emerge to be an attractive candidate for scaffolding due to their biocompatibility and pliable biodegradation kinetics. In addition, they also provide the essential mechanical properties and acceptable physicochemical behavior for direct cell–material or cell–cell interactions (Meng et al. 2010). In other words, the scaffolds with ideal features such as strength, rate of degradation, permeability, as well as their size and shape are more readily and reproducibly altered in polymeric scaffolds (Fuchs et al. 2001). Furthermore, they can replicate the actual morphology and structure of the damaged tissue and help upholding the integration with the neighboring biological environment (Chaikof et al. 2002; Guarino et al. 2007).

Typically, three individual collections of biomaterials – synthetic polymers, ceramics, and natural polymers – are used in the building of scaffolds for human tissue engineering. Each of these distinct biomaterial groups has their own specific advantages and disadvantages with respect to their use in the fabrication of scaffolds (O'brien 2011). The material originating from natural source possesses exceptional biocompatibility and shows potential viability but limited physico-mechanical stability; therefore, they are not chosen for load-bearing application such as bone and femur. Synthetic biomaterials are categorized as organic and inorganic; they represent a large group of biodegradable polymers PLA, PGA, PLGA, PCL, etc. The synthetic biopolymers could be tailored to produce better physical and mechanical properties; however, the major drawback of those materials is biocompatibility, i.e., the synthetic biomaterials have difficulty in attaching with the cells and tissue regeneration. Henceforth, to overcome the shortcomings of natural and synthetic biopolymers, the biocomposites are being developed with the incorporation of bioactive ceramics such as bioglass and hydroxyapatite (HA) into the biopolymer matrix (Nigam and Mahanta 2014). The biocomposite scaffolds comprising polymers and ceramic-coated materials may have excellent properties such as biocompatibility, biodegradation, and mechanical strength (Swetha et al. 2010).

In addition, several substantial surface alteration techniques such as physical adsorption, plasma irradiation, and grafting are being employed by the researchers to alter the surface properties of polymeric scaffolds. The porous, biodegradable, and bioresorbable scaffolds developed using biopolymers and their composites are utilized in numerous tissue engineering applications, and are formed typically using modern techniques like 3D printing, particulate leaching, gas foaming, solvent casting, freeze-drying, electrospinning, phase separation, and solid form fabrication (Nigam and Mahanta 2014; Subia et al. 2010).

This chapter focuses to review the biomaterials and composites used for the fabrication of scaffolds. The various fabrication techniques used to develop the polymeric scaffolds were also discussed in detail. Furthermore, the different surface modifications methods that are performed onto the scaffolds to improve functional performance have also been detailed.

11.2 Functions of Scaffolds in Biomedical Application

- Promote the biological (cell–material or cell–cell) interaction.
- Transportation of necessary nutrients and gases.
- Facilitating and regulating the proliferation and differentiation for cell growth.
- Biodegradation with respect to rate of tissue regeneration.
- Induce a nominal degree of swelling or toxicity in vivo.

11.3 Ideal Characteristics of Scaffold

- Sufficient porous three-dimensional interconnected system that should facilitate cell growth and nutrient transportation and carry out the metabolic waste.
- The scaffold should have proper surface chemistry for cell attachment, differentiation, and proliferation.
- Required to have adequate mechanical and biological properties to match the tissues at the site of injection or implantation.
- The tissues are processed to form a range of shapes and sizes.

11.4 Biopolymers in Development of Scaffold

The biopolymers are a certain type of materials that are generated from renewable sources, such as plants, or can be chemically blended from sugars or starch. These polymers are bio-safe and nontoxic (Smiya Mushtaq and Muhammad Irfan Majeed 2017); their modification yields interesting functional properties, such as improved biodegradability, mechanical properties, energy absorption, compatibility, and heat and moisture resistance. The biopolymers are classified based on two major factors, one is the source of origin and the other is the degradability, i.e., the polymer is biodegradable or non-biodegradable; in other words, they are called as natural biopolymers and synthetic biopolymers.

11.4.1 Natural-Based Biopolymers for Scaffold

Natural polymers are generally found in living organisms, and they are generally more biocompatible (e.g., cellulose, starch, collagen, sugar). These polymers are categorized based on the monomers and structure of polymers. The monomer-based natural polymers are nothing but the polymers that are derived from animal proteins (such as silk, collagen, and gelatin), polynucleotides, and polypeptides (Yadav et al. 2015). On the other hand, the structure-based natural polymers are extracted from microbes such as bacteria and fungi which includes polyesters (polyhydroxyalkanoates, PHA) and polysaccharides (starch, cellulose, and chitin) (Mano et al. 2007).

11.4.1.1 Collagen

Collagen is a protein that is found in large number in the body. More than 20 genetically distinct forms have been identified. It is the most investigated biomaterial for biomedical applications (Mano et al. 2007; Hayashi 1994). Collagen is a fibrous structural protein and plays a significant role in sustaining the biological and structural integrity of the extracellular matrix (ECM) and provides physical support to tissues (Dong and Lv 2016). It has widespread sources such as bone cartilage, tendon, blood vessel, ligament, nerve, and skin, as it is considered as the core structural protein of most hard and soft tissues (Burgeson and Nimni 1992). Collagen scaffolds have lower mechanical strength and structural stability; however, they provide low immunogenicity, a porous structure, permeability, biocompatibility, and biodegradability. Further, it functions to adjust the morphology, migration, adhesion, and differentiation of cells (Chevallay and Herbage 2000; Wolf et al. 2003).

Mizuno et al. developed demineralized bone powder (DBP)/collagen sponge scaffolds with the pore size of 120–200 μm which is used for in vitro analysis of endochondral bone of human dermal fibroblasts (Mizuno and Glowacki 1996). These scaffolds were fabricated by emulsification freeze-drying/lyophilization and cross-linked by using ultraviolet light irradiation to produce the highly porous lattice structure. Shamugasundaram et al. have also developed chitosan-based scaffold in the form of interpenetrating polymeric network (IPN) for in vitro culture of human epidermoid carcinoma cells (HEp-2; Cincinnati) (Shanmugasundaram et al. 2001). Similarly, several chitosan-based bioactive scaffolds such as carboxymethyl chitosan (CMC) and magnesium gluconate (MgG) were fabricated by the researches and used for various applications such as bone treatments, drug delivery, anti-inflammatory, and wound healing (Levengood and Zhang 2014; Ahmed and Ikram 2016; Adhikari et al. 2016).

11.4.1.2 Polyhydroxyalkanoates (PHA)

PHA is biodegradable polyester derived from microorganisms (Lim et al. 2017). PHA has ester linear linkages structures which are produced by different types of bacteria/monomers as an intracellular storage material containing substantial volume of carbon and energy. PHA is long-chain hydroxyl fatty acid molecule and is hydrophobic in nature (Raza et al. 2018). PHA has a significant role in tissue engineering because of its unique properties like high biocompatibility and biodegradation is being investigated for regenerative medicine (repair devices, articular cartilage repair devices, guided tissue repair/regeneration devices, and bone marrow scaffolds).

Zhao et al. blended the microbial polyester PHA containing poly(hydroxybutyrate) (PHB) and poly(hydroxybutyrate-co-hydroxyhexanoate) (PHBHHx) and used as scaffolds for the generation of cartilage (Zhao et al. 2003). Several researchers utilized PHA particularly poly(hydroxybutyrate) (PHB), copolymers, and their composites in development of devices and scaffolds, which in turn are used for sutures, repair patches, slings, cardiovascular patches, adhesion barriers, orthopedic pins, stents, guided tissue repair/regeneration devices, articular cartilage repair devices, nerve guides, tendon repair devices, bone marrow scaffolds, wound dressings, etc. (Chen and Wu 2005; Grande et al. 2017).

11.4.1.3 Polysaccharides

Polysaccharides are made of long chains of monosaccharide units bound together by glycoside linkages and on hydrolysis give the constituent monosaccharide. Polysaccharides are linear to highly branched in structure. Most viable polysaccharides (e.g., cellulose, dextran, pectin, alginic acid, agar-agar, agarose, starch, carrageenan, and heparin) are either neutral or acidic, but chitosan is a basic polysaccharide (Ahmed and Ikram 2016). Starch, chitosan, and cellulose are widely used for development of scaffold in tissue engineering application (Swetha et al. 2010).

Several authors have developed chitosan scaffolds which are used in a wide range of tissue repair or regenerative applications. Ahmed et al. fabricated chitosan-based drug-loaded scaffolds for wound healing application (Ahmed and Ikram 2016). Kumar et al. fabricated titanium dioxide nanoparticle-doped chitosan scaffold for the application in bone tissue engineering (Kumar 2018). Similarly another reseracher prepared polysaccharide scaffolds by cross-linking of polysaccharide with chitosan or proteins and reported for cell carriers and cell growth (Ehrenfreund-Kleinman et al. 2003). DeSouza et al. have combined the polysaccharide chitosan with xanthan gum and investigated the scaffolds for soft tissue engineering (De Souza et al. 2019).

11.4.2 Synthetic-Based Biopolymers for Scaffold

The synthetic polymers for scaffolds are produced from chemical substances in the form of porous structure. The synthetic biopolymer is categorized based on the structures and chain group of copolymer with natural resources. Synthetic polymers are extremely used in tissue engineering because of their porosity, degradation time, and mechanical properties. Synthetic polymers are inexpensive compared to the natural polymers. The synthetic polymers such as saturated poly(α-hydroxy esters), poly(lactic acid) (PLA), poly(glycolic acid) (PGA), poly(lactic-co-glycolic) acid (PLGA), polyvinyl alcohol (PVA), poly (ε-caprolactone) (PCL), poly(hydroxybutyrate) (PHB), and poly(butylene succinate) (PBS) are used as biocompatible polymers in medical and tissue engineering applications (Okamoto and John 2013).

11.4.2.1 Poly(lactic Acid) (PLA)

PLA is most versatile biodegradable polyester extracted from renewable resources, such as corn starch or sugar. PLA has a wide range of properties such as it is biocompatible, biodegradable, and less toxic and comprises wide range of mechanical properties and the capacity to be molded into numerous shapes (Ghalia and Dahman n.d.). PLA has shown to be a very capable material in the field of bone fixation, drug delivery vehicle, tissue engineering, scaffold, and various other biomedical applications (Xiao et al. 2012). However, it possesses some shortcomings such as hydrophobicity, poor toughness, and lack of reactive side chains. In order to overcome these demerits, researchers adapt distinct surface modification methods (described in Sect. 11.5) to accomplish PLA's advantage in tissue engineering. PLA has been permitted by the Food and Drug Administration (FDA, USA) for use as a suture material because of features that offer crucial advantages. Since then they are utilized as biological material as well as for surgical implant material, drug delivery systems, and porous scaffolds for the growth of neo-tissue (Davis et al. 1996; Benicewicz and Hopper 1991).

Gregor et al. (2017) designed and fabricated PLA-based scaffolds using 3D printer for bone tissue replacement (Gregor et al. 2017). The printed scaffolds were characterized by Gremare et al. to evaluate their physicochemical and biological properties (Grémare et al. 2018). Further, the structural, chemical, mechanical, and biological performance were also characterized. Similarly, high-resolution PLA-based composite scaffolds were fabricated via 3D printing technology (Serra et al. 2013; Albannaa et al. 2016). These scaffolds were found to be not toxic to the cells and allowed the attachment of cells to the surface. NaOH treatment improved cell attachment (Wang et al. 2016). In addition to the 3D printing technique, the PLA scaffolds were also used for fabrication using electrospinning and solution casting methods. Gomez-Pachonet al. have fabricated PLA nanofiber scaffolds by electrospinning process. The impact of the processing constraints on structure including fiber alignment, take-up velocity, and post-thermal treatment was analyzed (Gómez-Pachón et al. 2014). Choudhury et al. have used different solvents for casting the porous PLA scaffolds (Choudhury et al. 2015).

11.4.2.2 Poly (ε-Caprolactone) (PCL)

PCL is a bioresorbable polymer, which is prepared by ring opening polymerization of the cyclic monomer ε-caprolactone. PCL degrades by the hydrolysis of its ester linkages in physical conditions (such as in the human body) and therefore received a deal of attention in biomedical application. The PCL potential application includes bone and cartilage repair, absorbable surgical sutures, drug delivery systems, nerve guides, and three-dimensional (3D) scaffolds for use in tissue engineering. PCL is an extensively studied biodegradable scaffold material for tissue engineering applications. It was permitted by the Food and Drug Administration (FDA) for use in humans. Degradation time of PCL is quite long, and thus it is used mainly in the replacement of rigid tissues in the body where curing also takes a prolonged period of time. It is also used in load-bearing tissues of the body by enhancing its stiffness. A number of manufacturing technologies have been applied to develop the PCL into 3D polymeric scaffolds of high porosity and surface area (Azimi et al. 2014). The PCL fabrication technique includes 3D printing or additive manufacturing (Choong et al. 2004; Williams et al. 2005; Yu et al. 2018), solvent casting (Guarino et al. 2002), bio-extrusion (Domingos et al. 2009), and electrospinning (Janmohammadi and Nourbakhsh 2018).

Choong et al. fabricated biodegradable PCL scaffolds by fused deposition modeling (FDM); the scaffolds were coated with calcium phosphate using a biomimetic method and seeded with human bone marrow osteogenic cells (Choong et al. 2004). Cai et al. and Chung et al. have developed 3D Poly (ε-caprolactone) scaffold using electrodynamic printing (jetting) for tissue regeneration (Cai et al. 2002; Chung et al. 2017). Williams et al. and Partee et al. have also computationally designed and printed the PCL scaffold using selective laser sintering (SLS) technique (Williams et al. 2005; Partee et al. 2006). Hutmacher has evaluated the mechanical properties and in vitro biocompatibility, i.e., cell cultural response of 3D-printed poly (ε-caprolactone) scaffolds (Hutmacher and Cool 2007). Domingos et al. have characterized chemical, morphological, and in vitro biological performance of the bio-extruded scaffolds (Domingos et al. 2009). The characterization of PCL scaffold revealed its high potential in terms of the chemical, physical, and biological properties.

11.4.2.3 Poly(glycolic Acid) (PGA)

PGA is the most widely used scaffolding polymer. Due to its hydrophilic nature, PGA degrades quickly in the aqueous medium or in vivo and loses the mechanical reliability between 2 and 4 weeks (Ma et al. 1995; Reed and Gilding 1981). Poly (glycolic acid) (PGA) and their copolymers are attractive materials for scaffolds as they degrade by random hydrolysis when implanted, and their degradation products are eventually ejected from the body as carbon dioxide and water (Whang et al. 1995). Vozzi et al. have fabricated poly(lactic-co-glycolic) acid (PLGA) scaffolds using two different technologies, namely, soft lithography and micro-syringe deposition method (Vozzi et al. 2002). Similarly, Oh et al. have used a blend of poly(lactic-co-glycolic) acid PLGA/polyvinyl alcohol (PVA) and fabricated the scaffold structure using melt-molding particulate-leaching method (non-solvent method) (Oh et al. 2003). The fabricated scaffold was examined by the author in vivo for tissue compatibility; the PLA copolymer scaffold has shown better cell bonding and growth than the control PLGA scaffold.

11.4.3 Bio-reinforcements for Composite Scaffolds

Generally, bio-reinforcements such as bioglass and hydroxyapatite are incorporated to the bio-matrix to develop biocomposites. The matrix opted to develop the biocomposites is based on the tissue engineering application. Various authors focused on developing the polymer/ceramic composites for the revival of soft and hard tissue in tissue engineering. Turnbull et al. discussed about the bioactive composite scaffold, which was fabricated by utilizing biodegradable polyesters and bioceramics (HA and bioglass) (Turnbull et al. 2018). Likewise, Zhitomirsky et al. developed the biopolymer/ceramic composites in tissue regeneration (Zhitomirsky et al. 2009).

11.4.3.1 Bioglass (BG)

The biodegradable and bioresorbable bioactive silicate glasses are synthesized with different ion release rates and are used in the application of tissue engineering. The silicates offer benefits as the inorganic constituents due to their high biocompatibility. It also has capability to blend with the biopolymers and creates suitability in soft and hard connective tissues (Boccaccini et al. 2003). Blaker et al. and Boccaccini et al. fabricated the PDLLA and bioglass composite scaffolds by using the thermally induced solid–liquid phase separation (TIPS); the phase separation process is an extensively used method for the development of polymer/BG composite scaffolds (Blaker et al. 2008; Boccaccini et al. 2003). Yang et al. have performed biological study on porous bioglass/PLA composite scaffold for human bone marrow stromal cell growth (Yang et al. 2006a). Prior to the in vitro test, the authors pre-treated the scaffolds with alkaline phosphatase to improve the functional performance, and in case of in vivo test, the formation of bone throughout the scaffold was inspected and reported in their studies. Similarly, Maquet et al. fabricated poly (D,L-lactic-co-glycolic acid) (PDLLA)/poly(lactic-co-glycolic) acid (PLGA)/BG composites using the TIPS process (Maquet et al. 2004). Further, the mechanical properties and biodegradation rate of the scaffold were studied and reported.

11.4.3.2 Hydroxyapatite (HA)

Hydroxyapatite (HA) is a ceramic filler developed from living organisms like human and animal bone. To develop the high-performance tissue engineering scaffold composites, the HA particulate (reinforcement) is used in micro-nano sizes (Okamoto and John 2013). Several researchers have developed composites for scaffold with better biocompatibility, good osteoconductivity, biodegradability, and mechanical strength (Rezwan et al. 2006; Wei and Ma 2004; Venkatesan and Kim 2010; Cao et al. 2010). Different types of techniques are tailored to develop biopolymers/HA composites (Kothapalli et al. 2005; Kim et al. 2006; Zhang and Ma 1999; Wiria et al. 2007; Nejati et al. 2008). Zhang et al. (Zhang and Ma 1999) developed PLLA/HA composites scaffold through solid–liquid phase separation; Lee et al. have also developed a bioceramic scaffold using SLS technique and coated with biopolymer to aid the restoration of bone tissue. The authors reported that the developed scaffold had better mechanical strength, regular micro-architecture, ultrafine structure, biocompatibility, etc., which favored its usage in tissue engineering (Lee and Barlow 1993).

11.5 Fabrication of Biopolymers for Scaffolds

Globally, researchers are adapting various technologies to fabricate scaffold for biomedical application. Enormous methods have been defined in the literature for preparing the porous structures to employ in tissue engineering; however, the successful scaffold fabrication equipment will be those that can produce good-quality scaffolds with consistency (Lee and Barlow 1993). Further, the capability of the technology to produce scaffolds in relatively large quantities at a nominal cost is also an important criterion. Researchers have adapted many techniques to fabricate scaffolds which includes solvent casting/particulate leaching method, phase separation, gas foaming process, freeze-drying, electrospinning, and additive manufacturing/3D printing method, out of which few techniques have received much attention due to their versatility.

11.5.1 Solvent Casting/Particle Leaching Method

Solvent casting is a simple and cost-effective technique to develop scaffold. Two solvent casting methods are adapted; in the first method, polymeric scaffold layers are formed by dipping the mold into the polymeric solution and second method is prepared by the addition of polymeric solution onto the mold followed by solvent evaporation to form the scaffold structure. Generally, the second method is widely adapted by the researchers, i.e., initially the biopolymer (PLLA or PLGA) is made to dissolve in a solvent such as in chloroform or methylene chloride, and then the solution is then poured onto a petri dish filled with the porogen. Thereafter, the solvent is allowed to disperse, leaving behind the polymer matrix with salty particles. Then, the composite is submerged in water where the salt filters out to produce a porous structure (Mikos and Temenoff 2000). The permeability of the scaffold is controlled by the amount of salt added, and the size of the pores will depend on the size of salt crystals. The porous foam scaffolds are prepared using this technique and are widely used in the tissue engineering application.

Liao et al. fabricated scaffold by mixing poly(lactic-co-glycolic) acid (PLGA) and sodium chloride particles; the mixture was casted using a special mold, were the organic solvent was passed to dissolve the PLGA particles under negative pressure (Liao et al. 2002). Yang et al. have fabricated PCL scaffolds by solvent casting/particulate leaching method; during the scaffold manufacturing, a series of operations such as pre-shaping, centrifuging, casting, and desolvating were employed (Yang et al. 2006b). Further, the scaffold's properties were also categorized, which includes microstructures, pore dimensions, porosity, and hydrophilicity. Similarly, the blends of biodegradable polymers with PCL and poly (D,L-lactic-co-glycolic acid) were prepared using this technique and evaluated its mechanical stability and degradation rates (Prasad et al. 2017). Lee et al.'s report has revealed that poly(lactic-co-glycolic) acid (PGCL) scaffolds produced using solvent casting/particle leaching method had open pore structures with a size of 250 μm and found its suitability for tissue engineering application. However, these scaffolds have some limitations such as limited mechanical property, residual solvents, and porogen material (Lee et al. 2003).

11.5.2 Phase Separation

Phase separation process is a widely adapted process for the production of porous scaffold structure. In this process, the phase separation of polymer solution occurs due to the induced temperature; in other words, changes in temperature onto the homogenous polymer solution creates de-mixing or multi-state system. The de-mixing of the solution occurs as low polymer concentration (polymer poor phase) and high polymer concentration (polymer rich phase). This de-mixing can be of solid–liquid phase and liquid–liquid phase (Conoscenti 2017). The used solvents are removed then by extraction, evaporation, and sublimation process; biopolyester-based polymers and composite scaffolds are prepared by this technique (Lu et al. 2013; Sachlos and Czernuszka 2003). Highly interconnected microporous scaffold structures are being developed using thermally induced phase separation (TIPS) technique.

Hua et al. fabricated biodegradable PLA macro-porous scaffold with steady and extremely interconnected structure with a size ranging from 50 to 150 μm (Hua et al. 2002). Likewise, several researchers have developed PLA-based scaffolds for tissue engineering application. Huang et al. have fabricated PLGA/NHA scaffolds, and the effects of solvent composition, polymer concentration, coarsening time, and coarsening temperature on the micro-morphology and mechanical properties were also evaluated and reported by the authors (Huang et al. 2008). Nam et al. developed porous foams scaffold using poly(L-lactic acid) and its copolymers with D-lactic acid and/or glycolic acid (Nam and Park 1999a; Nam and Park 1999b). Schugens et al. reported the preparation of biodegradable regularly connected microcellular foams of 1–10 μm in size via liquid–liquid phase separation (Schugens et al. 1996). Pavia et al. utilized ternary solution PLLA/dioxane/water to discover a reliable route to prepare TIPS porous, interconnected, and biodegradable scaffolds with morphology and properties suitable for tissue engineering applications (Pavia et al. 2008).

One of the advantages of this technique is that the process is fast and easily adaptable, and an extensive range of surface morphologies with a choice of characteristics such as open or close pores, fibrous structure, or membrane-like architecture are achieved (He et al. 2009; Yang et al. 2001).

11.5.3 Gas Foaming Process

In gas foaming method, a foaming agent such as sodium bicarbonate is added into the polymer phase to produce an inert gas such as N_2 or CO_2 at modest acidic solutions. The porous polymer structure is formed when the disseminated gas phase (discontinuous phase) is removed from the continuous phase of polymer (Dehghani and Annabi 2011). To be precise, the scaffold in this technique is developed by initially dissolving the polymer matrix in a identified solvent, i.e., in the state of gel and then it is allowed to precipitate by gentle mixing. The sodium bicarbonate particulates are then added onto the polymeric gel and blended to form homogenous mixture. The gel is then placed in a mold and allowed to partially evaporate under room temperature to obtain gelatinous or solidified mass. Then, the solidified mass is immersed in different concentration of citric acid solution to induce the gas foaming. Then, porous polymeric scaffolds formed are removed out of mold and washed with distilled water (Yoon and Park 2001). Yoon et al. reported that the porosities of these scaffolds could be controlled by the volume of ammonium bicarbonate incorporated to the polymer (Yoon and Park 2001). Barbetta et al. prepared a permeable structure of alginate foam by generating CO_2 through the reaction between tartaric acid and sodium bicarbonate in the presence of a suitable surfactant (i.e., Pluronic F-108) (Barbetta et al. 2009).

In other way, supercritical gas foaming technique is utilized to fabricate the scaffold. This technique utilizes depressurization CO_2 from high pressure vessel and is regulated using a back pressure regulator. Before inducing the pressure, the polymers or composite granules are kept in an open mold and loaded in the pressure vessel. Then, the thermodynamic instability tends to form by reducing the CO_2 gas pressure in the vessel to an ambient level, causing nucleation and expansion of dissolved CO_2, thus generating a porous structure within the polymer matrix (Montjovent et al. 2005; Zhu et al. 2008). The scaffold developed under this technique uses high pressure gases and avoids the use of organic solvents and high temperatures. Several gases (CO_2, N_2, He) were used in the building process, but only CO_2 resulted in the development of highly porous, mechanically intact matrices. However, the other parameters such as gas type, polymer composition, and molecular weight were found to affect the structural behavior of the scaffold (Sheridan et al. 2000). Tsivintzelis et al. used supercritical CO_2 as an anti-solvent for the production of porous poly(L-lactic acid) (PLLA) structure (Tsivintzelis et al. 2007). Since PLLA is moderately crystalline, the authors were not able to obtain constant porous structures with the gas foaming technique at temperatures lower than the melting point, whereas Davies et al.'s report revealed that amorphous poly (D,L-lactic-co-glycolic acid) (PDLLA) and PLGAs have larger free volume, and allow CO_2 more easily to diffuse into them for the formation of scaffold structure (Davies et al. 2008).

11.5.4 Electrospinning Process

In this process, the polymer fibers with normal radius are brought down to the nanometer range, or nanofibers are formed by exposing a fluid jet to a high electric field (Yoshimoto et al. 2003). Basically, the electrospinning setup consists of three major components: (1) high voltage power, (2) electrically conducting spinneret, and (3) a collector at a definite distance. The syringe, which clamps the polymer solution with a blunted tip needle as the spinneret, the solution will be pumped at a persistent and well-regulated fed rate. One of the electrodes from the power supply is coupled to the needle holding the spinning solution to charge the polymer solution and the other is attached to an opposite polarity collector (usually an earthed conductor). When the high voltage (typically in the range of 0–30 kV) is applied to the spinneret, the surface of the liquid droplet held by its own surface tension gets electro-statically stimulated at the spinneret tip. Thus, the drop comes out under the action of electrostatic forces before reaching the collector; the liquid elongates and evaporates the solvents, leading to the formation of nano/ultrathin fiber (Garg and Bowlin 2011; Huang et al. 2003). Various biopolymers have been successfully electrospun into ultrafine fibers scaffolds and used in tissue engineering applications which includes silk, fibrinogen, collagen, poly(lactic acid) (PLA), poly(glycolic acid) (PGA), poly(lactic-co-glycolic) acid (PLGA), and poly (ε-caprolactone) (PCL) (Zhang et al. 2004). These fibers possess distinctive physical features, such as high surface area to volume ratio and enhanced mechanical strength, associated to their micro-scaled counterparts (Li and Tuan 2009). Further, chemical, physical, and biological properties of electro-spun scaffolds are adjusted based on the application using the grouping of multi-component compositions (Liang et al. 2007). Electrospinning offers some unique benefits such as high surface to volume ratio, adaptable porosity of electro-spun structures, and the flexibility to spin into a variety of shapes and sizes (Karakaş 2015).

Several authors have developed biopolymeric scaffolds via electrospinning process and used them in a wide range of biomedical application such as wound dressing, tissue engineering, enzyme immobilization, and drug (gene) delivery. Li et al. (2006) have developed silk fibroin fiber scaffolds having bone morphogenetic protein 2 (BMP-2) and/or nanoparticles of hydroxyapatite (nHAP) via electrospinning process (Li and Tuan 2009). Bhattarai et al. have fabricated alginate-based nanofibers with an average diameter of ca. 75 nm via electrospinning process (Bhattarai and Zhang 2007); likewise gelatin (Gt) fibers have also been successfully produced and used in tissue engineering and regenerative medicine (Liu et al. 2013).

11.5.5 Freeze-Drying

Polymeric porous scaffolds are also prepared by freeze-drying method; this process is also called as drying process as this process converts solutions of liable materials into solids (Lu et al. 2010). In a first phase of this process, the polymer solution is chilled down to a definite temperature at which all materials are in a cold state and the solvent forms ice crystals, driving the polymer molecules to combine into the interstitial spaces. In the second phase, the solvent is removed by application of pressure, lower than that of the equilibrium vapor pressure of the frozen solvent. When the solvent is completely sublimated, a dry polymer scaffolds with controlled interconnected porous microstructure remains (Pikal 1990; Liapis and Bruttini 1994).

Several authors utilized directional freezing and freeze-drying to form aligned scaffolds from natural polymers such as silk and collagen. Lu et al. developed aligned porous gelatin scaffolds by unidirectional freeze-drying method (Lu et al. 2010). Ghalehet al. have fabricated gelatin-chitosan scaffolds for skin regeneration (Ghaleh et al. 2015). Similarly, Haugh et al. demonstrated the fabrication of collagen–glycosaminoglycan scaffolds via freeze-drying method (Haugh et al. 2011). The cellular structure of collagen–glycosaminoglycan scaffolds was designed based on the functional properties such as biocompatibility, pore size, degradability, pore structure, and exact surface area (O'Brien et al. 2005). These collagen-based scaffolds are used broadly in the biomedical industry for the repair and restoration of skeletal tissues and organs (Pawelec et al. 2014). The porosity of these scaffolds depends on the concentration of the polymer solution; the pore size dispersion is also affected by the cold temperatures.

11.5.6 Additive Manufacturing Process

Additive manufacturing or 3D printing or rapid prototyping is one of the versatile processes for the fabrication of biocompatible three-dimensional porous polymeric scaffold structures. Additive manufacturing can help build such a template in a layer-by-layer fashion; this process has the ability to produce the complex shape with help of CAD model. A number of 3D printing techniques are being used to fabricate the scaffolds, which includes fusion deposition method (FDM), selective laser sintering (SLS), binder jet (3DP), stereolithography (SLA), selective laser melting (SLM). Out all these techniques, FDM, SLS, and 3DP are broadly used for the fabrication of polymeric scaffold.

FDM is a nozzle-based technique wherein material in the form of strand is wound as a coil and mounted in the spool. It is then fed through the printer nozzle and filaments are extruded from a nozzle by the application of controlled heat. The extrudate is positioned relative to one another according to a pattern of the scaffold. Complex geometries and porous structures are then accomplished with a fully unified network of pores.

SLS and 3DP are classified as powder-based additive manufacturing techniques; they operate under the principal of laser sintering, i.e., the laser source is used to fuse the particles to form the porous scaffolds. These processes have the capability to produce scaffolds with accurate pore size. Further, the process parameters such as power, scanning speed, and layer thickness are optimized based on the printing material. The SLS technique fuses the particulars directly, whereas the binder jet process uses the liquid binder as a source which is selectively applied on the powder particle and then sintered with the help of heat source to form the desired porous structure. In 3DP technique, the optimization of process parameter is based on the binder and heat applied onto the surface. In addition, the binder should be carefully chosen in order to have better biocompatibility with the physiological environment (Do et al. 2015).

Using rapid prototyping methods, researchers have been able to alter scaffolds to mimic the biomechanical properties (in terms of structural integrity, strength, and microenvironment) of the organ or tissue to be fixed/changed quite closely (Hoque et al. 2012; Zein et al. 2002). Zein et al. developed a bioresorbable polymer poly(ε-caprolactone) (PCL)-based porous scaffold using FDM process. The material used was filament-modeling material that produces layers and gets directionally aligned using this computer-controlled extrusion and deposition process. Similarly, Wang et al. (2004) demonstrated the development of cellular poly (ε-caprolactone) tissue scaffold via extrusion deposition technique, and the effect of process parameters on poly (ε-caprolactone) (PCL) scaffold porosity was studied and reported by (Domingos et al. 2009). Likewise, researchers have also used the power-based technology for the fabrication of scaffold. Lam et al. have developed starch-based polymer powders (cornstarch, dextran, and gelatin) for the 3DP process; three-dimensional scaffolds were prepared using the blend and their characteristic was evaluated (Lam et al. 2002). Equally, porous PCL scaffolds were also computationally designed and then fabricated via selective laser sintering (SLS), technique. Williams et al.'s study have also verified the porous architecture of PCL scaffolds, which were found to have sufficient mechanical properties for bone tissue engineering applications (Williams et al. 2005).

11.6 Surface Modification of Scaffolds

An ideal scaffold developed through any of the fabrication techniques should have enough permeability for diffusion of nutrients and clearance of wastes and have acceptable mechanical stability to support and transfer loads. Another important parameter is successful regeneration of tissues and organs, for which the surface chemistry plays a vital role. The surface morphological characteristics include porosity size, shape, and texture of microstructure and macrostructure of the pores. The surface morphology of polymeric scaffolds has a key effect on protein and cell attachment. Multiple approaches have been adapted by the researches to alter the surface architecture of the scaffold in order to enable or enhance protein and cell interaction. The surface functionalization provides micrometer to nanometer alteration in the surfaces of the scaffolds which leads to improved biological performance. The surface modification or functionalization is generally classified as physical modification, chemical modification, radiation mediated modification, and grafting.

11.6.1 Physical Modification of Scaffold

Physical modification refers to modification of scaffolds by physical methods such as coating or adsorption, mechanical etching, polishing, patterning, and proper designing to improve the porosity and biomechanical property of materials/scaffolds and ultimately contribute toward biological performance.

The coating of biomaterial surfaces with cell adhesion proteins like fibronectin, vitronectin, collagen, laminin, or ECM-resembling molecules such as chitosan and gelatin is one of the most common methods for the development of improved surface of scaffolds (Lin et al. 2006; Sato et al. 2004). Surface behaviors bring preferred deviations in the surface chemistry; Mitra et al. studied the effects of surface micro/nano-patterning on the variation of bone cell response (Mitra et al. 2013). Similarly, Do cha et al. developed the micro-patterns using micro-electromechanical technique and evaluated the effect on cellular behaviors such as proliferation, adhesion, and osteogenic differentiation (Do Cha et al. 2012). Various surface physical patterns' relationship with respect to several cell types (fibroblast, macrophage, endothelial, and osteoblast) has been detailed in the report of (Lord et al. 2010).

11.6.2 Chemical Modification of Scaffold

Chemical modification of polymeric biomaterials is another major approach to modify or to bio-activate the surface. The process of chemical surface modification begins with surface activation, which involves the creation of functional modalities on the surface of the polymer (Katti et al. 2008; Jordan et al. 2016). Usually, a wide range of chemical reactions and reagents have been explored for this purpose (Goddard and Hotchkiss 2007). Jeewandara and Zamani et al. have performed hydrolytic reaction using sodium hydroxide (NaOH) on PCL 3D-printed scaffold; as a result the hydrophilicity of PCL was increased by creating surface carboxyl and hydroxyl groups, which in turn increased the cell attachment (Jeewandara et al. 2013; Zamani et al. 2018). Likewise, Mahjoubi et al. have adapted diazonium chemistry; in other words, the authors modified the surface of PLA-based scaffolds with phosphonate groups using diazonium, thereby allowing the biomaterial to covalently and homogeneously bind with a number of functional groups in order to avoid the degradation of the scaffold's polymeric matrix (Mahjoubi et al. 2014).

11.6.3 Radiation-Mediated Modification

The surface properties dominate the interaction between the material and biological environment. The surfaces of polymeric scaffolds are also modified via irradiation; mostly UV irradiation, ionized gas treatment, and plasma treatment are used. Surface modification of biomaterials via radiation technique is believed to be an efficient and potential technique, as it does not affect the bulk properties. Safinia et al. have modified the surface of poly (D,L-lactic-co-glycolic acid) (PDLLA) polymer scaffolds using air or water: ammonia plasma treatments. The authors reported that the surface treatment has increased the number of polar functional groups and enhanced the wetting behavior (Safinia et al. 2005). In the same way, Prabhakaran et al. have also improved the hydrophilicity of electrospun poly (ε-caprolactone) (PCL) nanofibrous scaffolds using simple plasma treatment process (Prabhakaran et al. 2008). Their study revealed that the plasma treatment had enhanced the cell adhesion, proliferation, and interaction with nanofibers for nerve tissue formation. Yang et al. have also adapted anhydrous ammonia plasma treatment to modify surface properties of macro porous poly(L-lactic acid) and poly(lactic-co-glycolic) acid (PLGA) cell scaffold (Yang et al. 2002). Slabko et al.'s study also revealed that the surface of PHA films after plasma treatment became more hydrophilic, owing to the attachment of greater number of endothelial cells (Slabko et al. 2010).

11.6.4 Modification via Grafting

Surface modification via grafting one of the versatile ways which provide new functionalities to the existing polymers such as hydrophilicity, adhesion, biocompatibility, conductivity, and lubrication (Uyama et al. 1998). Grafting of polymers could be achieved in several ways to produce polymers with detailed forms and chemistry. The grafting of polymer could be done in two different ways, namely, grafting-to and grafting-from techniques; the selection of either process depends on the fabrication technique. The grafting-to technique covalently fixes the polymer chain or polymer unit onto the responsive surface of the scaffold, while the grafting-from technique forms a polymer chain or polymer unit via polymerization from polymeric presenting moiety on the scaffold surface (Kim and Jung 2016). Surface alteration, explicitly by the grafting of PEG, has long been shown to be a promising strategy to improve the biocompatibility of materials. Zhang et al. grafted the surface of poly(l-lactide)(PLLA) with hydroxyapatite (HAP) which has shown many advantages for bone fixation materials due to its improved interface compatibility, mechanical property, and biocompatibility (Zhang et al. 2009).

11.7 Applications of Biopolymeric Scaffolds in Tissue Engineering

Biopolymer scaffolds play an vital role in tissue engineering due to their biocompatibility and biodegradable properties (O'brien 2011). Biopolymeric scaffolds are fabricated based on functionality in tissue engineering such as wound healing, cartilages repair, drug delivery, and bone replacement (Karageorgiou and Kaplan 2005). Figure 11.1 represents the biopolymeric scaffolds in different tissue engineering applications.

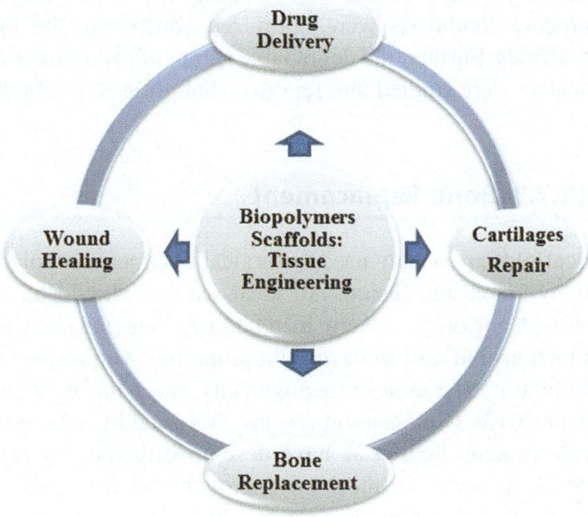

Fig. 11.1 Applications of biopolymeric scaffolds in tissue engineering

11.7.1 Wound Healing

Wound healing scaffolds are used to repair the injured or damaged skin tissue. Several authors fabricated biopolymeric scaffold for wound healing; mostly the electrospinning technique is widely adapted as it produces nanoscale scaffolds (Zhong et al. 2010; Thakur et al. 2008). Chong et al. (Chong et al. 2007) fabricated poly (ε-caprolactone) (PCL)/gelatin scaffold for antilogous layered dermal reconstitution (ALDR); Chandrasekaran et al. have also developed poly(L-lactic acid)-co-poly(ε-caprolactone) (PLACL) scaffold by electrospinning process. The biological performances of the scaffolds w.r.t. dermal skin regeneration were also evaluated by the authors (Chandrasekaran et al. 2011). Rho et al. have also investigated the different types of collagen nano-fibrous extracellular matrix for wound healing. The biological properties like cyto-compatibility, cell behavior, normal human oral keratinocytes (NHOK) of cell, and collagen interaction were also studied and reported (Rho et al. 2006).

11.7.2 Drug Delivery

Biopolymer scaffolds are used in controlled drug delivery systems; these scaffolds are either injected or seeded into the damaged tissue for tissue regeneration (Kohane and Langer 2010). Poly(lactic-co-glycolic) acid (PLGA) scaffolds are commonly used for drug delivery system due to their high biocompatibility and biodegradability (Makadia and Siegel 2011). These scaffolds are fabricated in either of the process as described in Sect. 11.5; however, 3D printing technique is mostly used. Salmoria et al. have developed the cellulose scaffold by SLS process and used it for drug delivery system. In their study, various sizes of power particles were utilized, and its process conditions were optimized concerning the mechanical properties of the scaffolds. Further, the degradable nature of cellulose of starch cellulose and cellulose acetate were studied and reported (Salmoria et al. 2009).

11.7.3 Bone Replacement

Scaffolds are widely used in surgical treatment to replace missing bone and to attain bony union and fusion. An ideal scaffold should have optimistic properties; they should not only concern in retaining, inducing, and bringing back the biological functions but also to have right properties with respect to degradation, cell binding, cellular uptake, non-immunogenicity, mechanical strength, and elasticity (Billström et al. 2013). Collagens, chitosans, PCL, and PLA-based composites are used in bone replacement. Seol et al. have developed the porous chitosan sponges using freeze-drying process for the bone regrowth and replacement (Seol et al. 2004). These scaffold had a pore size of 50–200 μm, which was used in the replacement of osteoblasts to regenerate bone tissues in vivo test. Similarly, several authors have also developed scaffolds and implanted them for growth of tissue (Williams et al. 2005; Petite et al. 2000; Meinel et al. 2004; Holy et al. 2000).

11.7.4 Cartilage Repair

Cartilage (soft tissue) is used to protect the joints at the ends of long bones. Due to the heavy load or injury, the cartilage may affect the tissue. A number of scaffolds fabricated using biodegradable and bioresorbable materials were experimentally studied for cartilage repair (Hutmacher 2000). Liu et al. have developed a biodegradable scaffold and used it in two major applications, namely, orthopedic surgery and neck reconstruction (Liu et al. 2013). Many researchers have fabricated similar types of scaffold and used them in in vivo cartilage reconstruction and studied its mechanical and biological performance (Nettles et al. 2004; Ghassemi et al. 2017; Svensson et al. 2005; Lee and Shin 2007; Dai et al. 2010).

11.8 Conclusion

In this chapter, various natural and synthetic biopolymers that are used in the fabrication of scaffolds such as proteins, chitosan, polysaccharides, poly(lactic acid) (PLA), poly(glycolic acid) (PGA), poly(lactic-co-glycolic) acid (PLGA), and poly (ε-caprolactone) (PCL) are discussed. In addition, bioactive glass and ceramic particles that are used as reinforcement in bio-matrix for fabricating scaffolds structures are also described. Further, the chapter discusses about different fabrication techniques of scaffolds using biopolymers and scaffolds, namely, solvent casting/particulate leaching, phase separation, gas foaming, electrospinning, freeze-drying, and rapid prototyping.

Moreover, this chapter has addressed various post-processing surface modification methods, namely, physical modification, chemical modification, radiation-medicated modification, and grafting, which improve the functional performance of the scaffold. The scaffold-fabricated biopolymer and composites are used in various tissue engineering applications. This chapter has also detailed about the use of biopolymer/composite scaffold in tissue engineering such as wound healing, drug delivery bone replacement, and cartilage repair.

References

Adhikari U, Rijal NP, Khanal S, Pai D, Sankar J, Bhattarai N (2016) Magnesium incorporated chitosan based scaffolds for tissue engineering applications. Bioactive Mater 1(2):132–139

Ahmed S, Ikram S (2016) Chitosan based scaffolds and their applications in wound healing. Achiev Life Sci 10(1):27–37

Albannaa RF, Kirkham J, Burke J, Liu C, Yang X (2016, October) 3D-printed PLA scaffolds for bone tissue regeneration: effect of scaffold structure on attachment and growth of human dental pulp stromal cells (HDPSCS). In Orthopaedic Proceedings (Vol. 98, No. SUPP_16, 10). The British Editorial Society of Bone & Joint Surgery

Azimi B, Nourpanah P, Rabiee M, Arbab S (2014) Poly (ε-caprolactone) fiber: an overview. J Eng Fibers Fabr 9(3):74

Barbetta A, Barigelli E, Dentini M (2009) Porous alginate hydrogels: synthetic methods for tailoring the porous texture. Biomacromolecules 10(8):2328–2337

Benicewicz BC, Hopper PK (1991) Polymers for absorbable surgical sutures—Part II. J Bioact Compat Polym 6(1):64–94

Bhattarai N, Zhang M (2007) Controlled synthesis and structural stability of alginate-based nanofibers. Nanotechnology 18(45):455601

Billström GH, Blom AW, Larsson S, Beswick AD (2013) Application of scaffolds for bone regeneration strategies: current trends and future directions. Injury 44:S28–S33

Blaker JJ, Knowles JC, Day RM (2008) Novel fabrication techniques to produce microspheres by thermally induced phase separation for tissue engineering and drug delivery. Actabiomaterialia 4(2):264–272

Boccaccini AR, Notingher I, Maquet V, Jérôme R (2003) Bioresorbable and bioactive composite materials based on polylactide foams filled with and coated by Bioglass® particles for tissue engineering applications. J Mater Sci Mater Med 14(5):443–450

Burgeson RE, Nimni ME (1992) Collagen types. Molecular structure and tissue distribution. Clin Orthopae Relat Res 282:250–272

Cai Q, Yang J, Bei J, Wang S (2002) A novel porous cells scaffold made of polylactide–dextran blend by combining phase-separation and particle-leaching techniques. Biomaterials 23(23):4483–4492

Cao H, Zhang L, Zheng H, Wang Z (2010) Hydroxyapatite nanocrystals for biomedical applications. J Phys Chem C 114(43):18352–18357

Chaikof EL, Matthew H, Kohn J, Mikos AG, Prestwich GD, Yip CM (2002) Biomaterials and scaffolds in reparative medicine. Ann N Y Acad Sci 961(1):96–105

Chandrasekaran AR, Venugopal J, Sundarrajan S, Ramakrishna S (2011) Fabrication of a nanofibrous scaffold with improved bioactivity for culture of human dermal fibroblasts for skin regeneration. Biomed Mater 6(1):015001

Chen GQ, Wu Q (2005) The application of polyhydroxyalkanoates as tissue engineering materials. Biomaterials 26(33):6565–6578

Chevallay B, Herbage D (2000) Collagen-based biomaterials as 3D scaffold for cell cultures: applications for tissue engineering and gene therapy. Med Biol Eng Comput 38(2):211–218

Chong EJ, Phan TT, Lim IJ, Zhang YZ, Bay BH, Ramakrishna S, Lim CT (2007) Evaluation of electrospun PCL/gelatin nanofibrous scaffold for wound healing and layered dermal reconstitution. Actabiomaterialia 3(3):321–330

Choong C, Triffitt JT, Cui ZF (2004) Polycaprolactone scaffolds for bone tissue engineering: effects of a calcium phosphate coating layer on osteogenic cells. Food Bioprod Process 82(2):117–125

Choudhury M, Mohanty S, Nayak S (2015) Effect of different solvents in solvent casting of porous PLA scaffolds—In biomedical and tissue engineering applications. J Biomater Tissue Eng 5(1):1–9

Chung TW, Lo HY, Chou TH, Chen JH, Wang SS (2017) Promoting cardiomyogenesis of hBMSC with a forming self-assembly hBMSC Microtissues/HA-GRGD/SF-PCL cardiac patch is mediated by the synergistic functions of HA-GRGD. Macromol Biosci 17(3):1600173

Conoscenti, G. (2017). PLLA-based scaffolds for osteochondral tissue regeneration via thermally induced phase separation technique.

Dai W, Kawazoe N, Lin X, Dong J, Chen G (2010) The influence of structural design of PLGA/collagen hybrid scaffolds in cartilage tissue engineering. Biomaterials 31(8):2141–2152

Davies OR, Lewis AL, Whitaker MJ, Tai H, Shakesheff KM, Howdle SM (2008) Applications of supercritical CO2 in the fabrication of polymer systems for drug delivery and tissue engineering. Adv Drug Deliv Rev 60(3):373–387

Davis SS, Illum L, Stolnik S (1996) Polymers in drug delivery. Curr Opin Colloid Interface Sci 1(5):660–666

De Souza RFB, de Souza FCB, Rodrigues C, Drouin B, Popat KC, Mantovani D, Moraes ÂM (2019) Mechanically-enhanced polysaccharide-based scaffolds for tissue engineering of soft tissues. Mater Sci Eng C 94:364–375

Dehghani F, Annabi N (2011) Engineering porous scaffolds using gas-based techniques. Curr Opin Biotechnol 22(5):661–666

Dhandayuthapani B, Yoshida Y, Maekawa T, Kumar DS (2011) Polymeric scaffolds in tissue engineering application: a review. Int J Polym Sci 2011

Do Cha H, Hong JM, Kang TY, Jung JW, Ha DH, Cho DW (2012) Effects of micro-patterns in three-dimensional scaffolds for tissue engineering applications. J Micromech Microeng 22(12):125002

Do AV, Khorsand B, Geary SM, Salem AK (2015) 3D printing of scaffolds for tissue regeneration applications. Adv Healthc Mater 4(12):1742–1762

Domingos M, Dinucci D, Cometa S, Alderighi M, Bártolo PJ, Chiellini F (2009) Polycaprolactone scaffolds fabricated via bioextrusion for tissue engineering applications. Int J Biomater 2009

Dong C, Lv Y (2016) Application of collagen scaffold in tissue engineering: recent advances and new perspectives. Polymers 8(2):42

Ehrenfreund-Kleinman T, Domb AJ, Golenser J (2003) Polysaccharide scaffolds prepared by crosslinking of polysaccharides with chitosan or proteins for cell growth. J Bioact Compat Polym 18(5):323–338

Fuchs JR, Nasseri BA, Vacanti JP (2001) Tissue engineering: a 21st century solution to surgical reconstruction. Ann Thorac Surg 72(2):577–591

Garg K, Bowlin GL (2011) Electrospinning jets and nanofibrous structures. Biomicrofluidics 5(1):013403

Ghaleh H, Abbasi F, Alizadeh M, Khoshfetrat AB (2015) Mimicking the quasi-random assembly of protein fibers in the dermis by freeze-drying method. Mater Sci Eng C 49:807–815

Ghalia MA, Dahman Y (n.d.) Synthesis and characterization of Green Poly (lactic Acid)-Based Biomaterial. CBS 2015, 32nd Annual Meeting of the Canadian Biomaterials Society

Ghassemi T, Saghatolslami N, Matin MM, Gheshlaghi R, Moradi A (2017) CNT-decellularized cartilage hybrids for tissue engineering applications. Biomed Mater 12(6):065008

Goddard JM, Hotchkiss JH (2007) Polymer surface modification for the attachment of bioactive compounds. Prog Polym Sci 32(7):698–725

Gómez-Pachón EY, Vera-Graziano R, Campos RM (2014) Structure of poly (lactic-acid) PLA nanofibers scaffolds prepared by electrospinning. IOP Conf Ser Mater Sci Eng 59:012003

Grande D, Ramier J, Versace DL, Renard E, Langlois V (2017) Design of functionalized biodegradable PHA-based electrospun scaffolds meant for tissue engineering applications. New Biotechnol 37:129–137

Gregor A, Filová E, Novák M, Kronek J, Chlup H, Buzgo M, Blahnová V, Lukášová V, Bartoš M, Nečas A, Hošek J (2017) Designing of PLA scaffolds for bone tissue replacement fabricated by ordinary commercial 3D printer. J Biol Eng 11(1):31

Grémare A, Guduric V, Bareille R, Heroguez V, Latour S, L'heureux N, Fricain JC, Catros S, Le Nihouannen D (2018) Characterization of printed PLA scaffolds for bone tissue engineering. J Biomed Mater Res A 106(4):887–894

Guarino V, Gentile G, Sorrentino L, Ambrosio L (2002) Polycaprolactone: synthesis, properties, and applications. Enc Polym Sci Technol:1–36

Guarino V, Causa F, Ambrosio L (2007) Bioactive scaffolds for bone and ligament tissue. Expert Rev Med Devices 4(3):405–418

Haugh MG, Murphy CM, McKiernan RC, Altenbuchner C, O'Brien FJ (2011) Crosslinking and mechanical properties significantly influence cell attachment, proliferation, and migration within collagen glycosaminoglycan scaffolds. Tissue Eng A 17(9-10):1201–1208

Hayashi T (1994) Biodegradable polymers for biomedical uses. Prog Polym Sci 19(4):663–702

He L, Zhang Y, Zeng X, Quan D, Liao S, Zeng Y, Ramakrishna S (2009) Fabrication and characterization of poly (l-lactic acid) 3D nanofibrous scaffolds with controlled architecture by liquid–liquid phase separation from a ternary polymer–solvent system. Polymer 50(16):4128–4138

Holy CE, Shoichet MS, Davies JE (2000) Engineering three-dimensional bone tissue in vitro using biodegradable scaffolds: Investigating initial cell-seeding density and culture period. J Biomed Mater Res A Off J Soc Biomater Jpn Soc Biomater Aust Soc Biomater Kor Soc Biomater 51(3):376–382

Hoque ME, Chuan YL, Pashby I (2012) Extrusion based rapid prototyping technique: an advanced platform for tissue engineering scaffold fabrication. Biopolymers 97(2):83–93

Hua FJ, Kim GE, Lee JD, Son YK, Lee DS (2002) Macroporous poly (L-lactide) scaffold 1. Preparation of a macroporous scaffold by liquid–liquid phase separation of a PLLA–dioxane–water system. J Biomed Mater Res A Off J Soc Biomater Jpn Soc Biomater Aust Soc Biomater Kor Soc Biomater 63(2):161–167

Huang ZM, Zhang YZ, Kotaki M, Ramakrishna S (2003) A review on polymer nanofibers by electrospinning and their applications in nanocomposites. Compos Sci Technol 63(15):2223–2253

Huang YX, Ren J, Chen C, Ren TB, Zhou XY (2008) Preparation and properties of poly (lactide-co-glycolide)(PLGA)/nano-hydroxyapatite (NHA) scaffolds by thermally induced phase separation and rabbit MSCs culture on scaffolds. J Biomater Appl 22(5):409–432

Hutmacher DW (2000) Scaffolds in tissue engineering bone and cartilage. In: The biomaterials: silver jubilee compendium. Elsevier Science, New York, pp 175–189

Hutmacher DW, Cool S (2007) Concepts of scaffold-based tissue engineering—the rationale to use solid free-form fabrication techniques. J Cell Mol Med 11(4):654–669

Janmohammadi M, Nourbakhsh MS (2018) Electrospun polycaprolactone scaffolds for tissue engineering: a review. Int J Polym Mater Polym Biomater:1–13

Jeewandara T, Waterhouse A, Wise S, Yin Y, Bilek M, Weiss A, Ng M (2013) Plasma based biofunctionalisation of cardiovascular stents. Heart Lung Circ 22:S46

Jordan AM, Viswanath V, Kim SE, Pokorski JK, Korley LT (2016) Processing and surface modification of polymer nanofibers for biological scaffolds: a review. J Mater Chem B 4(36):5958–5974

Karageorgiou V, Kaplan D (2005) Porosity of 3D biomaterial scaffolds and osteogenesis. Biomaterials 26(27):5474–5491

Karakaş H (2015) Electrospinning of Nanofibers and their applications. Istanbul Technical University, Textile Technologies and Design Faculty, Istanbul

Katti DS, Vasita R, Shanmugam K (2008) Improved biomaterials for tissue engineering applications: surface modification of polymers. Curr Top Med Chem 8(4):341–353

Kim W, Jung J (2016) Polymer brush: a promising grafting approach to scaffolds for tissue engineering. BMB Rep 49(12):655

Kim SS, Park MS, Jeon O, Choi CY, Kim BS (2006) Poly (lactide-co-glycolide)/hydroxyapatite composite scaffolds for bone tissue engineering. Biomaterials 27(8):1399–1409

Kohane DS, Langer R (2010) Biocompatibility and drug delivery systems. Chem Sci 1(4):441–446

Kothapalli CR, Shaw MT, Wei M (2005) Biodegradable HA-PLA 3-D porous scaffolds: effect of nano-sized filler content on scaffold properties. Actabiomaterialia 1(6):653–662

Kumar P (2018) Nano-TiO2 doped chitosan scaffold for the bone tissue engineering applications. Int J Biomater 2018

Lam CXF, Mo XM, Teoh SH, Hutmacher DW (2002) Scaffold development using 3D printing with a starch-based polymer. Mater Sci Eng C 20(1-2):49–56

Lee G, Barlow JW (1993, August) Selective laser sintering of bioceramic materials for implants. In Proceedings of the solid freeform fabrication symposium, Austin, TX, pp. 376–380

Lee SH, Shin H (2007) Matrices and scaffolds for delivery of bioactive molecules in bone and cartilage tissue engineering. Adv Drug Deliv Rev 59(4-5):339–359

Lee SH, Kim BS, Kim SH, Choi SW, Jeong SI, Kwon IK, Kang SW, Nikolovski J, Mooney DJ, Han YK, Kim YH (2003) Elastic biodegradable poly (glycolide-co-caprolactone) scaffold for tissue engineering. J Biomed Mater Res A Off J Soc Biomater Jpn Soc Biomater Aust Soc Biomater Kor Soc Biomater 66(1):29–37

Levengood SKL, Zhang M (2014) Chitosan-based scaffolds for bone tissue engineering. J Mater Chem B 2(21):3161–3184

Li WJ, Tuan RS (2009) Fabrication and application of nanofibrous scaffolds in tissue engineering. Curr Protoc Cell Biol 42(1):25–22

Li C, Vepari C, Jin HJ, Kim HJ, Kaplan DL (2006) Electrospun silk-BMP-2 scaffolds for bone tissue engineering. Biomaterials 27(16):3115–3124

Liang D, Hsiao BS, Chu B (2007) Functional electrospun nanofibrous scaffolds for biomedical applications. Adv Drug Deliv Rev 59(14):1392–1412

Liao CJ, Chen CF, Chen JH, Chiang SF, Lin YJ, Chang KY (2002) Fabrication of porous biodegradable polymer scaffolds using a solvent merging/particulate leaching method. J Biomed Mater Res A Off J Soc Biomater Jpn Soc Biomater Aust Soc Biomater Kor Soc Biomater 59(4):676–681

Liapis AI, Bruttini R (1994) A theory for the primary and secondary drying stages of the freeze-drying of pharmaceutical crystalline and amorphous solutes: comparison between experimental data and theory. Sep Technol 4(3):144–155

Lim J, You M, Li J, Li Z (2017) Emerging bone tissue engineering via Polyhydroxyalkanoate (PHA)-based scaffolds. Mater Sci Eng C 79:917–929

Lin Y, Wang L, Zhang P, Wang X, Chen X, Jing X, Su Z (2006) Surface modification of poly (L-lactic acid) to improve its cytocompatibility via assembly of polyelectrolytes and gelatin. Acta Biomater 2(2):155–164

Liu H, Ding X, Zhou G, Li P, Wei X, Fan Y (2013) Electrospinning of nanofibers for tissue engineering applications. J Nanomater 2013:3

Lord MS, Foss M, Besenbacher F (2010) Influence of nanoscale surface topography on protein adsorption and cellular response. Nano Today 5(1):66–78

Lu Q, Zhang X, Hu X, Kaplan DL (2010) Green process to prepare silk fibroin/gelatin biomaterial scaffolds. Macromol Biosci 10(3):289–298

Lu T, Li Y, Chen T (2013) Techniques for fabrication and construction of three-dimensional scaffolds for tissue engineering. Int J Nanomedicine 8:337

Ma PX, Schloo B, Mooney D, Langer R (1995) Development of biomechanical properties and morphogenesis of in vitro tissue engineered cartilage. J Biomed Mater Res 29(12):1587–1595

Mahjoubi H, Kinsella JM, Murshed M, Cerruti M (2014) Surface modification of poly (D, L-lactic acid) scaffolds for orthopedic applications: a biocompatible, nondestructive route via diazonium chemistry. ACS Appl Mater Interfaces 6(13):9975–9987

Makadia HK, Siegel SJ (2011) Poly lactic-co-glycolic acid (PLGA) as biodegradable controlled drug delivery carrier. Polymers 3(3):1377–1397

Mano JF, Silva GA, Azevedo HS, Malafaya PB, Sousa RA, Silva SS, Neves NM (2007) Natural origin biodegradable systems in tissue engineering and regenerative medicine: present status and some moving trends. J R Soc Interface 4(17):999–1030

Maquet V, Boccaccini AR, Pravata L, Notingher I, Jérôme R (2004) Porous poly (α-hydroxyacid)/Bioglass® composite scaffolds for bone tissue engineering. I: preparation and in vitro characterisation. Biomaterials 25(18):4185–4194

Meinel L, Karageorgiou V, Hofmann S, Fajardo R, Snyder B, Li C, Kaplan DL (2004) Engineering bone-like tissue in vitro using human bone marrow stem cells and silk scaffolds. J Biomed Mater Res A Off J Soc Biomater Jpn Soc Biomater Aust Soc Biomater Kor Soc Biomater 71(1):25–34

Meng D, Erol M, Boccaccini AR (2010) Processing technologies for 3D nanostructured tissue engineering scaffolds. Adv Eng Mater 12(9):B467–B487

Mikos AG, Temenoff JS (2000) Formation of highly porous biodegradable scaffolds for tissue engineering. Electron J Biotechnol 3(2):23–24

Mitra J, Tripathi G, Sharma A, Basu B (2013) Scaffolds for bone tissue engineering: role of surface patterning on osteoblast response. RSC Adv 3(28):11073–11094

Mizuno S, Glowacki J (1996) Three-dimensional composite of demineralized bone powder and collagen for in vitro analysis of chondroinduction of human dermal fibroblasts. Biomaterials 17(18):1819–1825

Montjovent MO, Mathieu L, Hinz B, Applegate LL, Bourban PE, Zambelli PY, Pioletti DP (2005) Biocompatibility of bioresorbable poly (L-lactic acid) composite scaffolds obtained by supercritical gas foaming with human fetal bone cells. Tissue Eng 11(11-12):1640–1649

Nam YS, Park TG (1999a) Porous biodegradable polymeric scaffolds prepared by thermally induced phase separation. J Biomed Mater Res A Off J Soc Biomater Jpn Soc Biomater Aust Soc Biomater Kor Soc Biomater 47(1):8–17

Nam YS, Park TG (1999b) Biodegradable polymeric microcellular foams by modified thermally induced phase separation method. Biomaterials 20(19):1783–1790

Nejati E, Mirzadeh H, Zandi M (2008) Synthesis and characterization of nano-hydroxyapatite rods/poly (l-lactide acid) composite scaffolds for bone tissue engineering. Compos A: Appl Sci Manuf 39(10):1589–1596

Nettles DL, Vail TP, Morgan MT, Grinstaff MW, Setton LA (2004) Photocrosslinkable hyaluronan as a scaffold for articular cartilage repair. Ann Biomed Eng 32(3):391–397

Nigam R, Mahanta B (2014) An overview of various biomimetic scaffolds: Challenges and applications in tissue engineering. J Tissue Sci Eng 5(2):1

O'Brien FJ, Harley BA, Yannas IV, Gibson LJ (2005) The effect of pore size on cell adhesion in collagen-GAG scaffolds. Biomaterials 26(4):433–441

O'brien FJ (2011) Biomaterials & scaffolds for tissue engineering. Mater Today 14(3):88–95

Oh SH, Kang SG, Kim ES, Cho SH, Lee JH (2003) Fabrication and characterization of hydrophilic poly (lactic-co-glycolic acid)/poly (vinyl alcohol) blend cell scaffolds by melt-molding particulate-leaching method. Biomaterials 24(22):4011–4021

Okamoto M, John B (2013) Synthetic biopolymer nanocomposites for tissue engineering scaffolds. Prog Polym Sci 38(10-11):1487–1503

Partee B, Hollister SJ, Das S (2006) Selective laser sintering process optimization for layered manufacturing of CAPA® 6501 polycaprolactone bone tissue engineering scaffolds. J Manuf Sci Eng 128(2):531–540

Pavia FC, La Carrubba V, Piccarolo S, Brucato V (2008) Polymeric scaffolds prepared via thermally induced phase separation: tuning of structure and morphology. J Biomed Mater Res A Off J Soc Biomater Jpn Soc Biomater Aust Soc Biomater Kor Soc Biomater 86(2):459–466

Pawelec KM, Husmann A, Best SM, Cameron RE (2014) Understanding anisotropy and architecture in ice-templated biopolymer scaffolds. Mater Sci Eng C 37:141–147

Petite H, Viateau V, Bensaid W, Meunier A, de Pollak C, Bourguignon M, Guillemin G (2000) Tissue-engineered bone regeneration. Nat Biotechnol 18(9):959

Pikal MJ (1990) Freeze-drying of proteins. Part II: Formulation selection. Bio Pharm 3(9):26–30

Prabhakaran MP, Venugopal J, Chan CK, Ramakrishna S (2008) Surface modified electrospun nanofibrous scaffolds for nerve tissue engineering. Nanotechnology 19(45):455102

Prasad A, Sankar MR, Katiyar V (2017) State of art on solvent casting particulate leaching method for orthopedic scaffolds fabrication. Mater Today Proc 4(2):898–907

Raza ZA, Riaz S, Banat IM (2018) Polyhydroxyalkanoates: Properties and chemical modification approaches for their functionalization. Biotechnol Prog 34(1):29–41

Reed AM, Gilding DK (1981) Biodegradable polymers for use in surgery—poly (glycolic)/poly (lactic acid) homo and copolymers: 2. In vitro degradation. Polymer 22(4):494–498

Rezwan K, Chen QZ, Blaker JJ, Boccaccini AR (2006) Biodegradable and bioactive porous polymer/inorganic composite scaffolds for bone tissue engineering. Biomaterials 27(18):3413–3431

Rho KS, Jeong L, Lee G, Seo BM, Park YJ, Hong SD, Min BM (2006) Electrospinning of collagen nanofibers: effects on the behavior of normal human keratinocytes and early-stage wound healing. Biomaterials 27(8):1452–1461

Sachlos E, Czernuszka JT (2003) Making tissue engineering scaffolds work. Review: the application of solid freeform fabrication technology to the production of tissue engineering scaffolds. Eur Cell Mater 5(29):39–40

Safinia L, Datan N, Höhse M, Mantalaris A, Bismarck A (2005) Towards a methodology for the effective surface modification of porous polymer scaffolds. Biomaterials 26(36):7537–7547

Salmoria GV, Klauss P, Paggi RA, Kanis LA, Lago A (2009) Structure and mechanical properties of cellulose based scaffolds fabricated by selective laser sintering. Polym Test 28(6):648–652

Sato T, Chen G, Ushida T, Ishii T, Ochiai N, Tateishi T, Tanaka J (2004) Evaluation of PLLA–collagen hybrid sponge as a scaffold for cartilage tissue engineering. Mater Sci Eng C 24(3):365–372

Schugens C, Maquet V, Grandfils C, Jérôme R, Teyssie P (1996) Polylactide macroporous biodegradable implants for cell transplantation. II Preparation of polylactide foams by liquid-liquid phase separation. J Biomed Mater Res A Off J Soc Biomater Jpn Soc Biomater Aust Soc Biomater Kor Soc Biomater 30(4):449–461

Seol YJ, Lee JY, Park YJ, Lee YM, Rhyu IC, Lee SJ, Han SB, Chung CP (2004) Chitosan sponges as tissue engineering scaffolds for bone formation. Biotechnol Lett 26(13):1037–1041

Serra T, Planell JA, Navarro M (2013) High-resolution PLA-based composite scaffolds via 3-D printing technology. Actabiomaterialia 9(3):5521–5530

Shanmugasundaram N, Ravichandran P, Reddy PN, Ramamurty N, Pal S, Rao KP (2001) Collagen–chitosan polymeric scaffolds for the in vitro culture of human epidermoid carcinoma cells. Biomaterials 22(14):1943–1951

Sheridan MH, Shea LD, Peters MC, Mooney DJ (2000) Bioabsorbable polymer scaffolds for tissue engineering capable of sustained growth factor delivery. J Control Release 64(1-3):91–102

Slabko VV, Volova TG, Krasnov PO, Kuzubov AA, Shishatskaya EI (2010) Surface modification of bioresorbable polymer scaffolds by laser treatment. Biophysics 55(2):234–238

Smiya Mushtaq, Muhammad Irfan Majeed (2017) Biopolymers for medical applications, Time Technology

Subia B, Kundu J, Kundu SC (2010) Biomaterial scaffold fabrication techniques for potential tissue engineering applications. In: Tissue engineering. InTech, London

Svensson A, Nicklasson E, Harrah T, Panilaitis B, Kaplan DL, Brittberg M, Gatenholm P (2005) Bacterial cellulose as a potential scaffold for tissue engineering of cartilage. Biomaterials 26(4):419–431

Swetha M, Sahithi K, Moorthi A, Srinivasan N, Ramasamy K, Selvamurugan N (2010) Biocomposites containing natural polymers and hydroxyapatite for bone tissue engineering. Int J Biol Macromol 47(1):1–4

Thakur RA, Florek CA, Kohn J, Michniak BB (2008) Electrospun nanofibrous polymeric scaffold with targeted drug release profiles for potential application as wound dressing. Int J Pharm 364(1):87–93

Tsivintzelis I, Pavlidou E, Panayiotou C (2007) Porous scaffolds prepared by phase inversion using supercritical CO2 as antisolvent: I. Poly (l-lactic acid). J Supercrit Fluids 40(2):317–322

Turnbull G, Clarke J, Picard F, Riches P, Jia L, Han F, Shu W (2018) 3D bioactive composite scaffolds for bone tissue engineering. Bioactive materials 3(3):278–314

Uyama Y, Kato K, Ikada Y (1998) Surface modification of polymers by grafting. In: Grafting/characterization techniques/kinetic modeling. Springer, Berlin/Heidelberg, pp 1–39

Venkatesan J, Kim SK (2010) Chitosan composites for bone tissue engineering—an overview. Mar Drugs 8(8):2252–2266

Vozzi G, Previti A, De Rossi D, Ahluwalia ARTI (2002) Microsyringe-based deposition of two-dimensional and three-dimensional polymer scaffolds with a well-defined geometry for application to tissue engineering. Tissue Eng 8(6):1089–1098

Wang F, Shor L, Darling A, Khalil S, Sun W, Güçeri S, Lau A (2004) Precision extruding deposition and characterization of cellular poly-ϵ-caprolactone tissue scaffolds. Rapid Prototyp J 10(1):42–49

Wang W, Caetano G, Ambler WS, Blaker JJ, Frade MA, Mandal P, Bártolo P (2016) Enhancing the hydrophilicity and cell attachment of 3D printed PCL/graphene scaffolds for bone tissue engineering. Materials 9(12):992

Wei G, Ma PX (2004) Structure and properties of nano-hydroxyapatite/polymer composite scaffolds for bone tissue engineering. Biomaterials 25(19):4749–4757

Whang K, Thomas CH, Healy KE, Nuber G (1995) A novel method to fabricate bioabsorbable scaffolds. Polymer 36(4):837–842

Williams JM, Adewunmi A, Schek RM, Flanagan CL, Krebsbach PH, Feinberg SE, Hollister SJ, Das S (2005) Bone tissue engineering using polycaprolactone scaffolds fabricated via selective laser sintering. Biomaterials 26(23):4817–4827

Wiria FE, Leong KF, Chua CK, Liu Y (2007) Poly-ε-caprolactone/hydroxyapatite for tissue engineering scaffold fabrication via selective laser sintering. Actabiomaterialia 3(1):1–12

Wolf K, Mazo I, Leung H, Engelke K, Von Andrian UH, Deryugina EI, Strongin AY, Bröcker EB, Friedl P (2003) Compensation mechanism in tumor cell migration: mesenchymal–amoeboid transition after blocking of pericellular proteolysis. J Cell Biol 160(2):267–277

Xiao L, Wang B, Yang G, Gauthier M (2012) Poly (lactic acid)-based biomaterials: synthesis, modification and applications. In: Biomedical science, engineering and technology. InTech, London

Yadav P, Yadav H, Shah VG, Shah G, Dhaka G (2015) Biomedical biopolymers, their origin and evolution in biomedical sciences: A systematic review. J Clin Diagn Res 9(9):ZE21

Yang S, Leong KF, Du Z, Chua CK (2001) The design of scaffolds for use in tissue engineering. Part I. Traditional factors. Tissue Eng 7(6):679–689

Yang J, Shi G, Bei J, Wang S, Cao Y, Shang Q, Wang W (2002) Fabrication and surface modification of macroporous poly (L-lactic acid) and poly (L-lactic-co-glycolic acid)(70/30) cell scaffolds for human skin fibroblast cell culture. J Biomed Mater Res A Off J Soc Biomater Jpn Soc Biomater Aust Soc Biomater Kor Soc Biomater 62(3):438–446

Yang XB, Webb D, Blaker J, Boccaccini AR, Maquet V, Cooper C, Oreffo RO (2006a) Evaluation of human bone marrow stromal cell growth on biodegradable polymer/Bioglass® composites. Biochem Biophys Res Commun 342(4):1098–1107

Yang Q, Chen L, Shen X, Tan Z (2006b) Preparation of polycaprolactone tissue engineering scaffolds by improved solvent casting/particulate leaching method. J Macromol Sci Part B Phys 45(6):1171–1181

Yoon JJ, Park TG (2001) Degradation behaviors of biodegradable macroporous scaffolds prepared by gas foaming of effervescent salts. J Biomed Mater Res A Off J Soc Biomater Jpn Soc Biomater Aust Soc Biomater Kor Soc Biomater 55(3):401–408

Yoshimoto H, Shin YM, Terai H, Vacanti JP (2003) A biodegradable nanofiber scaffold by electrospinning and its potential for bone tissue engineering. Biomaterials 24(12):2077–2082

Yu HS, Park J, Lee HS, Park SA, Lee DW, Park K (2018) Feasibility of polycaprolactone scaffolds fabricated by three-dimensional printing for tissue engineering of Tunica Albuginea. World J Men Heal 36(1):66–72

Zamani Y, Mohammadi J, Amoabediny G, Visscher DO, Helder MN, Zandieh-Doulabi B, Klein-Nulend J (2018) Enhanced osteogenic activity by MC3T3-E1 pre-osteoblasts on chemically surface-modified poly (ε-caprolactone) 3D-printed scaffolds compared to RGD immobilized scaffolds. Biomed Mater 14(1):015008

Zein I, Hutmacher DW, Tan KC, Teoh SH (2002) Fused deposition modeling of novel scaffold architectures for tissue engineering applications. Biomaterials 23(4):1169–1185

Zhang R, Ma PX (1999) Porous poly (l-lactic acid)/apatite composites created by biomimetic process. J Biomed Mater Res A Off J Soc Biomater Jpn Soc Biomater Aust Soc Biomater Kor Soc Biomater 45(4):285–293

Zhang Y, Huang ZM, Xu X, Lim CT, Ramakrishna S (2004) Preparation of core– shell structured PCL-r-gelatin bi-component nanofibers by coaxial electrospinning. Chem Mater 16(18):3406–3409

Zhang P, Hong Z, Yu T, Chen X, Jing X (2009) In vivo mineralization and osteogenesis of nanocomposite scaffold of poly (lactide-co-glycolide) and hydroxyapatite surface-grafted with poly (L-lactide). Biomaterials 30(1):58–70

Zhao K, Deng Y, Chen JC, Chen GQ (2003) Polyhydroxyalkanoate (PHA) scaffolds with good mechanical properties and biocompatibility. Biomaterials 24(6):1041–1045

Zhitomirsky D, Roether JA, Boccaccini AR, Zhitomirsky I (2009) Electrophoretic deposition of bioactive glass/polymer composite coatings with and without HA nanoparticle inclusions for biomedical applications. J Mater Process Technol 209(4):1853–1860

Zhong SP, Zhang YZ, Lim CT (2010) Tissue scaffolds for skin wound healing and dermal reconstruction. Wiley Interdiscip Rev Nanomed Nanobiotechnol 2(5):510–525

Zhu XH, Lee LY, Jackson JSH, Tong YW, Wang CH (2008) Characterization of porous poly (D, L-lactic-co-glycolic acid) sponges fabricated by supercritical CO2 gas-foaming method as a scaffold for three-dimensional growth of Hep3B cells. Biotechnol Bioeng 100(5):998–1009

Surface Modifications in Ti-Based Orthopaedic Implants

12

Sudip K. Sinha

Abstract

Research in the domain of various metallic and non-metallic biomaterials has been established to be of immense importance in the recent years since they can directly contribute to improve the quality and long life of human beings. Among a wide range of metallic biomaterials, Ti and its alloys have been extensively used to manufacture implantable components in numerous dental and orthopaedic applications. Their demand in this specific area of application arises owing to its superior biocompatibility resulting from negligible ion release when they come in close contact with body fluids, exceptional corrosion resistance, a specific combination of mechanical strength and toughness, lighter weight and many others.

In further developments, deposition of TiO_2 nanotubes (TNTs) by means of conventional electro-anodization technique on Ti-based metallic substrate has proved to be an excellent alternative for superior implant applications. As a modified surface, nanotubular surfaces promote cellular interaction compared with conventional flat or polished surfaces.

TiO_2-based modified nanotubular surfaces with distinct topography in the nanometric scale offer direct cellular interaction and show promises for tissue regeneration and bone substitute effects.

The present chapter offers a detailed discussion on the synthesis and surface preparation, growth mechanism, morphology and end application of TiO_2 nanotube arrays for orthopaedics.

Keywords

Biomaterials · TiO2 nanotubes · Electro-anodization · Tissue regeneration · Corrosion

S. K. Sinha (✉)
NIT Raipur, Raipur, Chhattisgarh, India
e-mail: sksinha.met@nitrr.ac.in

S. Paul (ed.), *Biomedical Engineering and its Applications in Healthcare*,
https://doi.org/10.1007/978-981-13-3705-5_12

12.1 Introduction

Use of natural or synthetic materials to replace/repair various human body parts dates back to ancient civilization, and there is profound evidence of the usage of natural materials like wood, etc. to structurally restore bones or tissues damaged to different diseases or traumas.

'Biomaterials', from materials science point of view, is typically described as follows: 'a substance that has been engineered to take a form which, alone or as part of a complex system, is used to direct, by control of interactions with components of living systems, the course of any therapeutic or diagnostic procedure' (Chen and Thouas 2015).

Therefore, biomaterials have significant contribution in many aspects of human life from the antiquity of civilization. Design and fabrication of new-generation biomaterials is a comprehensive process that requires knowledge of diverse field of biomedical sciences, engineering and basic science. The capacity to consider a biomaterial from an initial model to being a valid commercial product for clinical applications to human beings is a prolonged and expensive route, and this is indeed monitored by various administrative bodies. Apart from the compatible in vivo performance, the formulated biomaterial technology should satisfy a variety of criteria which includes but not limited to ease of fabrication and reproducibility, ability to scale up the technology used to produce the proposed biomaterial, stability in various physico-chemical conditions, the ease of implantation of the specific material that will be incorporated into the patient's body, cost-effectiveness and others. For the last several decades, considerable improvement has been done in the progress of sensitive, responsive and durable biomaterials, especially for implant and related applications.

In comparison with other class of materials, e.g. polymers and ceramics, metallic materials offer the best combination of properties essential from biomedical point of view which includes superior strength and toughness, fatigue limits and wear resistance. From historical perspective, the role of metallic biomaterials has found its proven record which might be discovered back to 200 A.D. when the Europeans developed the use of iron as dental implant materials into human bone (Ratner et al. 2004).

Taking consideration of the above advantages, it has been seen that at least 70% implants are based on metallic biomaterials. Among the various metallic biomaterials, orthopaedics (knee joint, total hip replacement, bone fixation plates, bone healing wires and other artificial joint components), cardiovascular and neurosurgical devices (heart valve elements, bare-metal stents, pacemakers, etc.) (Niinomi et al. 2012) are major fields of applications for this class of materials.

However, in spite of the numerous advantages, metallic biomaterials also experience various disadvantages like inferior corrosion resistance and poor biocompatibility. It is worth mentioning that lack of desired corrosion resistance property may end up in premature failure of the implant in its early stage and/or the unfavourable control on biological interactions. Moreover, owing to their limited biocompatibility, metallic biomaterials often exhibit degradation of joints to neighbouring tissues, which finally leads to loosening of bio-implants, causing serious suffering and/or mechanical injury to nearby tissues (Nasab et al. 2010; Navarro et al. 2008). The drawbacks arise from affected surfaces and therefore can rationally be solved by depositing a bio-metallic coating on the (Tomsia et al. 2005; Mirak et al. 2015; Salahinejad et al. 2013a, b) part/component to be implanted.

Among different schemes studied to influence these crucial factors, surface alterations of the implantable biomimetic species by means of biological, inert and composite layer deposits have confirmed to be an effective and stable alternative, therefore paying considerable interest in recent years. In order to build up a stable biocompatible surface factor like surface energy, chemical bonding, bioactivity, hydrophilicity, corrosion and surface deterioration in addition to their unique functionality, namely, drug delivery and cellular separation during cell-material interaction (Zhang et al. 2014), a versatile method should be extensively studied and must be adopted to modify the implant material surfaces.

Over the last few decades, Ti-based alloys have shown immense potential as orthopaedic implant materials owing to its low density, fitting elastic moduli and stability and compatibility in a suitable physiological environment (Geetha et al. 2009).

Metallic Ti and its α/β alloys are elastically stiff (Young's modulus: 115 GPa and T.S.: 210–1380 MPa), inherently strong, lightweight, corrosion resistant and found in abundance or could be fabricated without much difficulty.

In addition, Ti is nontoxic and biocompatible with human tissues, body fluids and bones. The superior strength-toughness properties in combination with exceptional corrosion protection and biocompatibility have established Ti and its alloys as the finest materials in the field of orthopaedic applications. Among the Ti-based alloys, Ti–13Nb–13Zr, Ti–6Al–4 V, Ti–6Al–7Nb, Ti–35.5Nb–7.3Zr–5.7Ta, etc. (Chen and Thouas 2015) offer the optimum properties simulating the natural bones in human. Figure 12.1 shows the materials used in a total hip joint replacement (Liu et al. 2004).

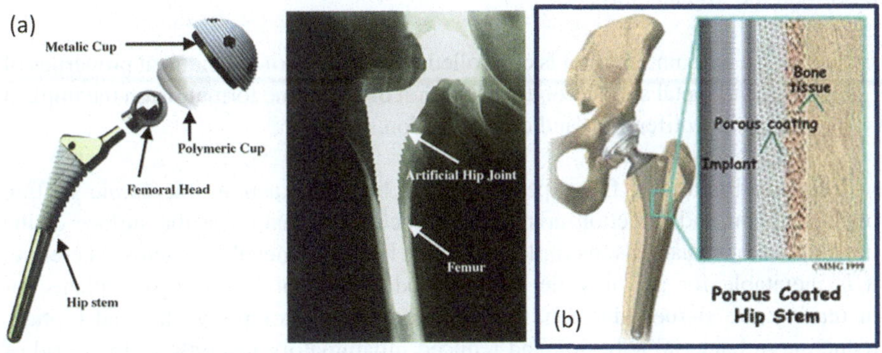

Fig. 12.1 Real and schematic diagram of artificial hip joint. (Reprinted with permission from Copyright 2004, MSE-R, Elsevier (Liu et al. 2004)). (**a**) Real image (X-ray of actual implant in patient shown as inset) (**b**) Representative image of total hip joint replacement

12.2 Surface Preparation of Ti and Its Alloys

In spite of a number of benefits of using Ti and its alloys for implant applications, numerous cases of implant failure have been taking place which eventually leads to the formation of wear particles, and this ultimately leads to loosening of the prosthesis. This is evident by suffering of the patient due to peri-prosthetic pain and a loss of agility in their movement (Langton et al. 2010; Urban et al. 2000). Usually, any implant surgery and its consequent care need severe healthcare expenses. In addition, the life risk factor also is a matter of concern which results from such surgeries. Keeping these in mind, it is obvious to say that strong and healthy implant integration is of paramount importance not only to reduce the requirement for revision surgery but also to ease related healthcare and social expenses. Commercially pure Ti and its alloys are the ultimate alternatives as implant (orthopaedic or dentistry) material since they are biocompatible with the adjacent tissues, blood vessels (haemocompatibility) and bone cells. In addition, they do not hinder curing and bone growth. The growth of a thermodynamically stable but steady oxide layer on the Ti-based implant surface transforms it into a naturally inert external layer in suitable bodily environments. Additionally, Ti-based implants suffer protein adsorption as well as interactions of neutrophils and macrophages under harsh atmosphere, and this may help in encapsulation by fibroblasts (Liu et al. 2004).

On the other hand, the most significant factor for a successful bone implant (orthopaedics and dental treatments) is to build up osseointegration, predominantly by creating a robust and enduring bond within the implant and peri-implant bone. The physico-chemical interaction eventually leads to a steady attachment in between bone and implants (Branemark et al. 1977).

The metal 'Ti' is closely connected with bone when the implantation takes place. However, an extremely fine layer of a soft tissue is developed within the Ti-based metallic implant and adjacent bone tissue which eventually arrest titanium from being in direct interaction with the bony substance (Xiao et al. 2001). The moment implantation took place, protein and other biologically active molecules start to absorb on the implant surface.

I. Successful osseointegration is controlled by a balance of mechanical properties of implants and metal surface-body interaction of new bone formation on the implant by bone cell proliferation and differentiation.

Inflammation arising from specific foreign body interaction, microscale shifting of the implant and infection arising from bacteria formation at the surface of the implant are the usual reasons for long implant failure in metallic systems. Therefore, it is inevitable for a nanoscale surface modification of biomedical implants for increasing the tissue adhesion, implant integration, control protein adsorption, reduction in bacterial adhesion and reduced inflammatory response or to get rid of

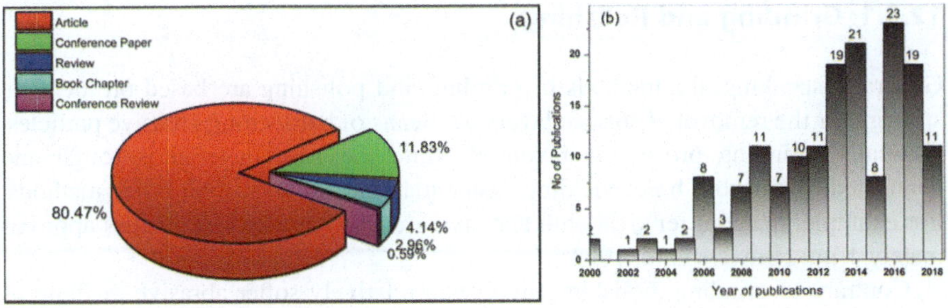

Fig. 12.2 (a) Pie chart showing distribution of published literature types. (b) Year-wise publication history in the time frame 2000–2018

reactions arising from foreign entities. In order to combat those adverse effects, orthopaedic coatings are considered to retain as long as the implant is not replaced within the body, and this might be within 20 and 30 years of time span. It can be construed from the above-mentioned facts that the coating material is constructed from the same material as of the implant, like Ti alloys or Co-Cr alloys. In the present form of scientific and technological development, the popular choice for attaining substantial advancement is by adjustment of the surface properties of the implants, either micro-structurally or by biochemical surface finishing.

During the last few decades, scientists and engineers have made extensive efforts to standardize the surface chemistry and surface topography of commercially available dental and orthopaedic implants by developing novel methodologies such as sandblasting, acid etching, electrochemical machining and anodizing which can directly trigger bone formation as well as diminish the hazard of infections. For example, the research findings on TiO_2-based surface modifications of Ti and its alloys between 2000 and 2018 can be obtained by looking at records indexed from the Scopus database (keyword: Titanium AND + AND TiO_2 AND implants AND orthopaedic). Figure 12.2a, b shows a total of 170 (including open access) research documents found from the Scopus database.

12.3 Mechanical Techniques

In order to obtain either a smooth or coarse outer surface, removal or surface abrasion processes are necessary. The specific intention of mechanical treatments is to acquire the precise topographical features of the implant surface with desirable roughness and eliminate surface impurity or dirt, which together results in superior adhesion in bonding, since the rough surface morphology initiates to help in bio-mineralization due to the enlarged surface area.

12.3.1 Grinding and Polishing

Generally speaking, the methods of grinding and polishing are based on identical strategy for the removal of surface layers by means of hard/strong abrasive particles. Typically, grinding process is favoured while the base material is rough and connected to a stable base, yielding faster material removal than other methods. For example, material removal with abrasive grade-60 results in R_a values approximately 1 μm (Bowers et al. 1992).

Contrary to grinding, polishing involves a relatively softer abrasive medium as backing plate. In this method, finer abrasive particles are successively applied in different orientations, sometimes with the help of lubrication, producing much smoother surfaces. SiC, alumina and diamond are regularly used as polishing media. Both of these methods help in descaling and finally direct in elimination of the native monolayer to obtain a flat and mirror-like surface finish.

12.3.2 Abrasive Blasting

In abrasive blasting (grit blasting), hard ceramic (e.g. alumina, silica, titania for Ti implants) particles are bombarded on the metal surface with high velocity. It is an extremely useful method for cleaning various stains/contaminations over the surface and, hence, is highly effective in superior surface preparation for the as-fabricated metallic biomaterials for implant applications.

Another upgraded version is known as shot peening, which is largely used to incorporate compressive residual stresses in the metal surface. The method of blasting is often applied as a pre-treatment procedure to remove surface impurities and dirts and roughen the surface of commonly found metallic prosthetic or dental implants and femoral stems, etc. Since its inception, this technique has been effectively demonstrated in various research efforts, and the properties of abrasive-blasted Ti metals/alloy surfaces and their influence on biomedical interactions (Buser et al. 1999; Schwartz et al. 1996) have been reported in peer-reviewed journals.

12.4 Chemical Methods

Ti and its alloys are often treated with chemical methods to provide enhanced biocompatibility, bioactivity and bio-corrosion resistance and to prepare a contamination-free surface. The list explains some of the popularly used techniques.

12.4.1 Acid Etching (Pickling)

Acid etching is one of the commonly used wet chemical etching techniques which is used to remove the native surface and is consist of oxide and a segment of the base metal, i.e. Ti. A rough surface can also be prepared this way. Moreover, etching processes increase the implants surface energy and area, allowing greater bone contact area, and this eventually leads to higher micromechanical absorption, in contrast to smooth surfaces. Among the frequently used acid pickling solutions for Ti and Ti-based alloys, a mixture of HNO_3 (10–30 vol. %) + hydrofluoric acid (HF, 1–3 vol. %) in distilled H_2O (ASTM Standard B600 1997) is universal. In this combination, HF shows a tendency to attack to TiO_2 and further affinity towards Ti for developing soluble fluorides of Ti (TiF_3 and TiF_4) and hydrogen.

In order to reduce the formation of nascent hydrogen, HNO3/HF ratio should be kept at 10:1 ASTM Standard B600 1997. Consequently, acid etching slowly develops a surface oxide layer, typically in thickness range of <10 nm.

12.4.2 Alkaline Etching

Another simple and useful wet chemical etching technique is the alkaline etching to modify Ti surfaces. This method is considered mainly as a prior method for apatite coating deposited by sol-gel route (Nishiguchi et al. 1999).

For example, treatment of Ti by 4–5 M of caustic soda (NaOH) at 60–80 °C for 24 h creates an external layer composed of sodium titanate (Na_2TiO_3) gel (Kim et al. 1999).

The effect of NaOH treatment on Ti substrate was explored by Balakrishnan et al. (Balakrishnan et al. 2007) for applying HAp coating on its bonding characteristics. Higher bonding strength with a value of 36.1 ± 5 MPa has been attained due to NaOH treatment on the Ti surface.

The typical thickness of these as-deposited layers is ~ 1 μm thick, conforming rough surface with open porosity of submicron scale all through the surface.

12.4.3 Nitric Acid Passivation

In order to obtain a homogeneously distributed oxide surface and also enhancing the corrosion resistance (by reducing ion release) for Ti and its alloys, nitric acid passivation treatments are often used in a systematic manner.

The process involves the immersion of Ti-based metals for a period of at least 30 min in a 20–40 vol. % HNO_3 solution at ambience. After neutralizing, the surface is carefully washed and dehydrated. Formation of a thin (2–6 nm) layer of oxide film is usually found predominantly of TiO_2 and/or other oxides from arising as alloying elements. Existence for the occurrence of minor quantity of suboxides (Ti_2O_3 and TiO) has also been observed in the interfaces (Oji et al. 1999).

Thermal treatments in air atmosphere or in close contact with boiling water are another way of passivating Ti-based metal surfaces.

12.4.4 Hydrogen Peroxide Treatment

Oxidation treatments by hydrogen peroxide (H_2O_2) are extensively used as an alternative method of surface modification due to its ease of application as well as its cost-effectiveness.

The in situ titanate gel layer, which is readily formed by NaOH treatment (previously discussed) and found to help in apatite layer formation as well as enhance its corrosion resistance, can also be deposited using H_2O_2 treatment on pure Ti metal substrate.

Recent results show (Karthega and Rajendran 2012) that the treatment of a Ti-15Mo alloy surface with H_2O_2 demonstrates high corrosion resistance in simulated body fluid in comparison with the unprocessed metal because of the creation of a porous TiO_2 nanostructured film.

It can be mentioned here that H_2O_2-treated Ti metals/alloys cause oxidation and hydroxylation of the base metal by eradicating the native deposit by means of removing the organic impurities and thereby purifies and disinfects the required product (Tengvall et al. 1989). Reports on H_2O_2-treated Ti alloys (TI6Al4V, TNTZ, etc.) are relatively scarce.

12.4.5 Electrochemical Treatment

12.4.5.1 Anodizing

It is found that metallic titanium generally shows a tendency for the formation of a TiO_2-based very thin layer (about 5–20 nm), and this layer provides a hard and shielding effect on the base metal.

This nanostructured TiO_2 film adhered onto the substrate surface establishes the propensity of the alloy (or c.p. Ti) to improve mechanical properties, corrosion resistance, bioactivity and bone conductivity in comparison with the virgin layer of metallic Ti and its alloys.

A 'bottom-up' approach is considered for the growth of these nanostructured surface-modified TiO_2 layers via this simple and cost-effective anodic oxidation process, and the reaction product can be formed with or without fluoride (F^-) ion-based electrolytes.

These porous TiO_2-based nanotubes display a wide range of dimensions with size range varying from ~ 20 to 100 nm (hole dia.), height ~ 50–6000 nm, while a tube wall in the depth/width range of ~ 6–34 nm (Minagar et al. 2012).

12.4.5.2 Anodizing Process

The anodic oxidation process takes place by applying suitable potential (potentiostatic) or current (galvanostatic) in a Ti implant (anode). A platinum counter electrode is used as a cathode in the electrolytic medium comprising an ionic solution. An oxidation reaction on the Ti surface starts taking place when a current is applied. The exact mechanism of formation of a thermodynamically stable TiO_2 nanotube at the anode fluoride-based electrolytic medium is explained as follows (Brammer et al. 2011).

First, electrolytic oxidation of metallic Ti forms TiO_2 layers:

$$Ti + 2H_2O \rightarrow TiO_2 + 4H^+ + 4e^- \tag{12.1}$$

TiO_2 nanotubes start forming in the subsequent reactions when the F^- ion starts reacting with as-deposited anodic TiO_2 and preferential etching starts to form on high-energy places as follows:

$$TiO_2 + 6F^- \rightarrow [TiF6]^- + 2O^{2-} \tag{12.2}$$

(i) Upon application of a constant potential, a nanosize pit is produced on the passive TiO_2 layer on the Ti surface.
(ii) The pit increases in size with the subsequent time and finally developed into a nanopore structure.
(iii) These nanopores and miniature pits start to form a permanent barrier layer.
(iv) Once the whole process completes after a predetermined time, thermodynamically stable TiO_2 nanotube layers are formed over the metal surface (Brammer et al. 2011).

Figure 12.3 (Shokuhfar et al. 2013) gives an overview of the morphology of TiO_2 nanotubes deposited on Ti foil substrate using FE-SEM. The TiO_2 nanotube shows amorphous and crystalline structure. Generally, the TiO_2 nanotubes span up to 10 mm and with inside diameters ranging from 50 to less than 200 nm. Table 12.1 shows the recent literature results of TiO_2-based nanotube formation for implant application.

12.5 Physical and Chemical Vapour Depositions

12.5.1 Chemical Vapour Deposition (CVD)

Chemical vapour deposition (CVD) is a widely used and versatile method to deposit thin corrosion protective layers and sometimes found to be advantageous for the uniform deposition of complex geometric coating on selected workpieces. At a primary level, the process exploits chemical reactions of a reactant gas in a sealed and heated compartment containing the implant (Nakayama et al. 1989). The reactant product thus grown on the exterior layer of the substrate is ultra-thin in nature, and the unwanted reactant products are flushed out as exhaust gas from the system. Deposition of TiO_2 on Ti-machined implants is often considered by the CVD (Giavaresi et al. 2004). The severity and affinity of osseointegration of the

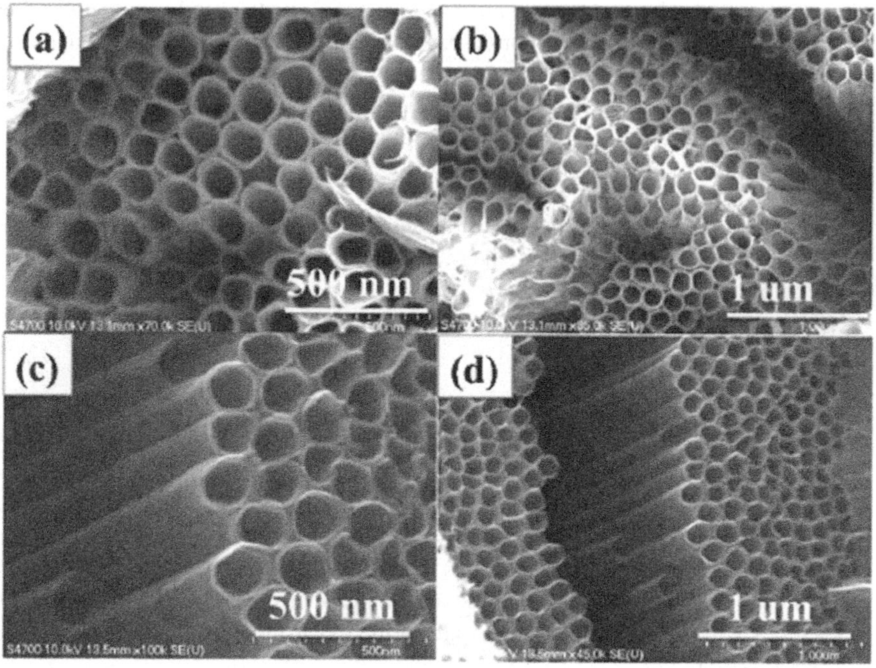

Fig. 12.3 (**a–d**): TiO_2 nanotubes synthesized by anodization technique using Cu electrode. Both amorphous (**a**) and crystalline (after annealing) TiO_2 nanotubes can be seen. [Reproduced by permission of The Royal Society of Chemistry Shokuhfar et al. (2013)].

TiO_2-coated Ti implants have been found to be significantly superior when compared with its bare counterparts in various bony matrixes tested for several weeks duration. These vapour-controlled deposition processes cannot be functional without sophisticated equipment, spares and vacuum chambers. As a result, the overall cost of the process goes up manifold. In addition to its enormous capital cost, a CVD process operating at high temperature requires reaction gases which further increase its operational cost.

12.5.2 Plasma Ion Coating

Here, plasma plumes arising due to extremely high temperature and kinetic energy of hydroxyapatite (HAp)-based ceramic particles are deposited onto the Ti-based implant surface (Duan and Wang 2006). The advantage of this method arises from its superior deposition rates, and its capability to form a large area deposit. High growth rate of apatites on the implant surface is evident in this method owing to its steady nucleation and increases the growth kinetics. The coated calcium hydroxyapatite coating on the implant surface results in outstanding bone-forming ability once the sample is immersed for 4 weeks in a simulated body fluid solution (SBF).

Table 12.1 An overview of anodization conditions on coating properties for surface modification of Ti and Ti alloys

Author	Coated material	Coating conditions	Coating characteristics	In vitro and in vivo test	Comments
1. D. Khudhair et al. (2017)	**Nitrogen-doped TiO_2 nanotubes**	Anode: Ti sheet	61 nm dia., 25 nm wall thicknesses, tube length of 2.25 µm. Superior biological interfacing, for instance, reduced impedance, high capacitance and excellent biocompatibility	Only cell culture test has been done	
		Cathode: Pt foil			
		Voltage, 10 V, 20 V, 30 V, 40 V, 50 V and 60 V. Tests were executed at 20 ∘C for 1 h			
		$(CH_2OH)_2$ with 0.5 wt.% NH_4F and 4 vol% D.I. water	Pore diameter varies between ~ 43 and 122 nm with > 0.65 in circularity while D.I. water mass enhanced		
		Anode: Ti sheet			
		Cathode: Pt foil			
		Voltage, 60V			
		Time, 1 h			
2. J. Kim et al. (2018), Scientific Reports	**TiO_2 nanotubes**	Electrolyte used:	Improves adhesion + better cytocompatibility + better resistance to corrosion + better antimicrobial property	No in vivo and in vitro has been reported	**Si** ***reduces toxic effect*** of Ag ions, whereas **Sr** addition also does the same with good cytocompatibility as well as good corrosion resistance
		Ethylene glycol containing 2, 4, 6, 8, 10 vol% D.I. water with 0.2 wt.% NH4F			
		Anode: Ti foil	**Adhesion strength : 15.7 MPa**	Cell morphology and cell proliferation test.	NO in vivo has been reported
		Cathode: Pt foil			
		Elect, 0.5 wt.% HF; V, 20; time, 40 min		a. MTT assay.	
		a) Ca(NO3)2·4H2O,AgNO3+ Sr(NO3)2+NH4H2PO4 +DI		b. MC3T3-E1cell test.	
		b). $Ca(NO3)_2$·4H2O (3.78×10^{-2} M)+		c. Antibacterial efficacy + gram-negative test+ *S. aureus*	
		Sr(NO3)2 (2.52×10^{-3} M) + $AgNO_3$			

(continued)

Table 12.1 (continued)

Author	Coated material	Coating conditions	Coating characteristics	In vitro and in vivo test	Comments
		$(1.68 \times 10^{-3}$ M) + $NH_4H_2PO_4$ $(2.5 \times 10^{-2}$ M)			
		c). Temperature : 65 °C			
		Time : 1500 sec			
		Current : 0.9 mA			
		Voltage : 0.67–1.13			
		pH, 4.5			
3. Y. Huang et al. (2017a)	**Sr-Ag TiO_2 nanotubes**	(a) Ethylene glycol with 5 vol. % D.I. water and 0.5 wt. % NH_4F	I. Ag released provides an antibacterial protection and prevents or restricts infections under in vivo conditions		
		(b) Anodizing parameter voltage/time conditions (10 V for 1 h) and (40 V for 40 min); **Electrolyte**	II. Sr's dual action of obstructing bone resorption and nurturing new bone formation has been widely explored both in vitro and in vivo		
		used for Sr and Ag:			
		0.02(M) $Sr(OH)_2$ solution (pH = 12.2) and the Sr-incorporated NTs samples absorbed in 1.5 M or 2.0 M AgNO3 solution (pH = 5.8) at room temperature			
		Elect, 0.5 wt.% HF;			
		V, 20; time, 40 min			

4. Cheng et al. /Acta Biomaterialia (2016)	**Sr-Ag-HAp/TN**	2.Coating Condition for AgHA/CS	Elect, 0.5 wt.% HF	Dawley rats with an average weight of 150 gm were used in all in vivo experiments	The primary obstacle of the study is the weak bond strength which links the film and substrate (AgHAp/CS) on Ti
		Electrolyte:	V, 20; time, 40 min		
		Calcium Nitrate + silver nitrate+ ammonium phosphate	2.Coating condition for AgHA/CS		
		3.For Si addition 1 wt.% Nano-SiO2 added with 4.2 pH	Elect:		
			Calcium Nitrate+silver nitrate + ammonium phosphate		
			3.For Si addition 1 wt.% nano-SiO_2 added with 4.2 pH		
5. Y. Huang et al. (2017b)	AgHA/ CS/TN			No in vivo test has been reported; MC3T3-E1 cells used for in vitro testing	
				a. In vitro tests give shows superior cell viability	
				b. Moreover increases cell differentiation ability	

However, the method possesses a number of disadvantages. It is not possible to coat intricate area with complicated geometries on the base metal uniformly. Moreover, the bonding strength arises from adhesion within the ceramic-metal interfaces is gradually decreases after long time exposure in SBFs. This arises as the intermolecular bonds in the dissimilar metal-ceramic interface become weaker which arises owing to the deterioration of the intermolecular bonds in the outer surface layer, signifying HAp coatings by this technique are not sustainable for a longer period of time.

12.6 Biochemical Surface Modifications

Apart from the so far established and widely practised physico-chemical surface modification techniques (as discussed earlier), other biochemical methods involving immobilization of bio-molecules on Ti-based surface to develop bio-reactive surface find promising alternative for implant applications.

The technique offers in situ physico-chemical signal generation that has the ability to control cell proliferation or differentiation, thus leading to generate more and more protein adsorption in the matrix and which finally promotes in creating a stable cell-extracellular matrix (ECM) interface on implant surfaces. In order to substitute themselves as promising implant materials, properties like adhesion, association, differentiation and mineralization in the matrix conforming osteoblasts and osteoprogenitor cells are essential.

Different studies show that biological approval and functionality of Ti implants could be enhanced by transforming their exterior with ECM species, and this shows promises in recent years as novel research directions. For example, it has been observed that ECM involves in the isolation of marrow stromal cells (MSCs) on osteoblast cells (Datta et al. 2005), once a Ti nanofiber-based scaffold is used in conjunction with progenitor cells. This study signifies that bone-mimicking ECM could augment the osteoblastic fragmentation of MSCs.

12.6.1 Collagen

Collagen is the most extensively distributed category of proteins that forms in the human body. During the last one decade, several attempts are anticipated for designing novel collagen-based biomaterials for application in areas like tissue scaffolding and regeneration owing to its hydrophilic tendency, negligible cytotoxicity, excellent haemostatic characteristics as well as abundance and biocompatibility (Meimandi-Parizi et al. 2013). This superior property makes it a candidate material for coating with Ti alloys for enhanced osteoblast bond and subsequent cell proliferation.

It is now established that collagen molecules actively participate to maintain the coherence from biological and physico-chemical structure aspect of the extracellular matrix (ECM) and supply material strength to the host tissue.

In a recently published work (de Assis et al. 2009), the osteoblastic phenotypic genes found in individual alveolar bone cell demonstrates the ability of cell growth for the coating comprises of collagen (type I) tissue on Ti surfaces. The study therefore indicates that collagen could play a significant role in bone healing and remodelling as an alternative surface modification technique.

12.6.2 Growth Factor

It is keenly observed that the bone formation ability in and around orthopaedic implants could be enhanced by using various growth factors (GFs) aiming at both cellular proliferation and differentiation.

GFs are considered to be soluble proteins that are discharged by cells with the capacity to control a vast range of cellular activities, for example, cell proliferation, migration and differentiation, by virtue of binding to specific transmembrane receptors on target cells. These elements offer a major contribution in tissue regeneration and therefore have been effective in new GF-based approach to improve the bone tissue-healing or replacement process.

Among the commonly known GFs, morphogenetic proteins (BMPs) in bones, transforming growth factor-β1 (TGF-β1), platelet-derived growth factor (PDGF), insulin-like growth factor 1 (IGF-1) and IGF-2, and fibroblast growth factor-fibronectin (FGF-FN) protein comprise the major and effective list of compositions for surface coatings of Ti-based implants which could selectively improve the bone healing process (Park et al. 2006).

12.6.3 Polypeptide

It is now established that both naturally occurred and laboratory-scale synthesized peptide-based biomaterials show promises to facilitate bone regeneration. This arises due to their ease of fabrication as 2D/3D scaffolds that closely resemble the extracellular matrix (ECM) of human bone as well as their systematic and efficient functionalization with chemical species that accelerates bone tissue growth.

In order to bio-functionalize the implant surface, proteins play a major role owing to their superior bioactivity. Nevertheless, the limited strength, rigidity and immunogenicity mark their restricted application in this area. In addition, proteins available from animals exhibit possible threat of pathogen transmission and variability (Liu et al. 2008).

The ECM proteins consist of a small functional domain which comprises amino acids, namely, arginine (R), glycine (G) and asparagine (D), frequently termed as RGD-motif and is considered as most bioactive substance used to functionalize Ti-based implant. This RGD-motif offers significant role in cell attachment and its subsequent growth. The as-coated RGD peptides may initiate cell reactions while in immediate contact.

In vitro studies show (Ardjomandi et al. 2012) that these RGD domains can be effectively used for cell adhesion, proliferation and differentiation in osteogenic lineage as an important bioactive substance used to functionalize Ti-based implant.

In another study (Germanier et al. 2006), comparison of polymer-modified RGD peptide-based implant surfaces is done by implanting in the maxillae of small pigs. This confirms its functionalization by promoting enhanced bone integration in the first phase of bone regeneration. Nevertheless, contradictory to these research findings, a new in vivo study shows that the existence of RGD chain does not help in Ti implant surface adhesion. The results reveal that still these coatings are active in experimental scale only, with showing some promises for applications in Ti surface implants.

12.6.4 DNA

Surface modification on Ti-based implants using DNA molecules is relatively rare and limited. However, the use of DNA molecules for this purpose offers some extra benefits, owing to their excellent biocompatibility and structure composition uniqueness. Therefore, use of these molecules in implant surfaces has been tried as an option for proteins or peptides. Van den Beucken et al. (2007) outlined the effect of application of layered DNA coatings for the development of mineralized surfaces from SBFs. Further, osteoblast-resembling cell responses in the presence of simulated body fluid have also been reported in their study.

12.7 Summary

A brief overview of the various surface preparation and modification procedures, namely, mechanical, chemical and electrochemical methods of Ti-based alloys, are provided for the purpose of improving the functional properties of orthopaedic implants.

A general description of each technique is discussed with current references where design and alteration of morphology and composition of the uppermost layers of Ti-based implants have been given priorities in view of its application. It is found that, generally, mechanical processes help in improving surface morphology and roughness. Usual chemical treatment techniques such as etching, chemical passivation, organic and inorganic coatings, etc. augment the metals' surface roughness and increase its corrosion and wear properties. All these factors lead to favourable osseointegration, cell adhesion and bone regeneration. The chemical techniques may lead to thin (< 10 nm) oxide layer formation based on TiO_2 (alloying elements from Ti-6Al-4 V type alloys).

Various surface oxide films comprising a wide range of nano- to micro-level microstructure, thickness, pore size distribution and novel composition can also be effectively formed with other alternative surface modification techniques involving chemical or electrochemical methods, namely, heat treatments, H_2O_2 reaction, apatite precipitation, electropolishing, anodizing, etc. In addition to previously mentioned Ti-based oxide layers, CaP, HA or recently developed HAp-CNT/TiO_2 double-layer coatings are also promising alternatives for implant surface mimicking bone biology, thereby eliminating the harmful effects of foreign species, and thereby enhancing cell adhesion and bioactivity. Finally, it may be concluded that opportunity of surface modifications of orthopaedic implants is promising and enormous. A suitable combination of the various physico-chemical surface modification techniques could show promises to develop stable implant materials mimicking orthopaedic structures.

Acknowledgements The author expresses his gratitude to Dr. A. Bit for fruitful discussions for writing up this chapter.

Conflict of Interest The authors confirm that this article content has no conflict of interest.

References

Ardjomandi N, Klein C, Kohler K, Maurer A, Kalbacher H, Niederlander J, Reinert S, Alexander D (2012) Indirect coating of RGD peptides using a poly-L-lysine spacer enhances jaw perios-teal cell adhesion, proliferation, and differentiation into osteogenic tissue. J Biomed Mater Res Part A 100(8):2034–2044

ASTM Standard B600 (1997) Standard guide for descaling and cleaning titanium and titanium alloy surfaces, Annual book of ASTM standards, Vol. 2.04. American Society for Testing and Materials, Philadelphia, pp 6–8

Balakrishnan A, Lee BC, Kim TN, Panigrahi BB (2007) Hydroxyapatite coatings on NaOH treated Ti-6Al-4V alloy using sol–gel precursor. Mater Sci Technol 23:1005–1007

Bowers KT, Keller JC, Randolph BA, Wick DG, Michaels CM (1992) Optimization of surface micromorphology for enhanced osteoblast responses in vitro. Int J Oral Maxillofac Implants 7:302–310

Brammer KS, Oh S, Frandsen CJ, Jin S (2011)Biomaterials and biotechnology schemes utilizing TiO2 nanotube arrays. In: Biomaterials science and engineering, Intechopen, London, UK, pp 193–210

Branemark P, Hansson B, Adell R, Breine U, Lindstrom J, Hallen O, Ohman A (1977) Osseointegrated implants in the treatment of edentulous jaw. Scand J Plast Reconstr Surg Suppl 16:1–132

Buser D, Nydegger T, Oxland T, Cochran DL, Schenk RK, Hirt HP, Snetivy D, Nolte LP (1999) Interface shear strength of titanium implants with a sandblasted and acid-etched surface: a biomechanical study in the maxilla of miniature pigs. J Biomed Mater Res 45(2):75–83

Chen Q, Thouas GA (2015) Metallic implant biomaterials. Mater Sci Eng R 87:1–57

Cheng H, Xiong W, Fang Z, Guan H, Wu W, Li Y, Zhang Y, Alvarez MM, Gao B, Huo K, Xu J, Xu N, Zhang C, Fu J, Khademhosseini A, Li F (2016) Strontium (Sr) and silver (Ag) loaded nanotubular structures with combined osteoinductive and antimicrobial activities. Acta Biomater 31:388–400

Datta N, Holtorf HL, Sikavitsas VI, Jansen JA, Mikos AG (2005) Effect of bone extracellular matrix synthesized in vitro on the osteoblastic differentiation of marrow stromal cells. Biomaterials 26:971–977 de Assis AF, Beloti MM, Crippa GE, de Oliveira PT (2009) M. Morra development of the osteoblastic phenotype in human alveolar bone-derived cells grown on a collagen type I-coated titanium surface. Clin Oral Implants Res 20:240–246

Duan K, Wang R (2006) Surface modifications of bone implants through wet chemistry. J Mater Chem 16(24):2309–2321

Geetha M, Singh AK, Asokamani R, Gogia AK (2009) Ti based biomaterials, the ultimate choice for orthopaedic implants – a review. Prog Mater Sci 54:397–425

Germanier Y, Tosatti S, Broggini N, Textor M, Buser D (2006) Enhanced bone apposition around biofunctionalized sandblasted and acid-etched titanium implant surfaces. A histomorphometric study in miniature pigs. Clin Oral Implants Res 17:251–257

Giavaresi G, Ambrosio L, Battiston GA et al (2004) Histomorphometric, ultrastructural and microhardness evaluation of the osseointegration of a nanostructured titanium oxide coating by metal-organic chemical vapour deposition: an in vivo study. Biomaterials 25(25):5583–5591

Huang Y, Zhang X, Zhang H, Qiao H, Zhang X, Jia T, Han S, Gao Y, Xiao H, Yang H (2017a) Fabrication of silver- and strontium-doped hydroxyapatite/TiO2 nanotube bilayer coatings for enhancing bactericidal effect and osteoinductivity. Ceram Int 43:992–1007

Huang Y, Xu Z, Zhang X, Chang X, Zhang X (2017b) Nanotube-formed Ti substrates coated with silicate/silver co-doped hydroxyapatite as prospective materials for bone implants. J Alloys Compd 697:182–199

Karthega M, Rajendran N (2012) Formation of nanoporous oxide layer over a binary β-phase titanium in simulated body fluid. J Porous Mater 19:573–577

Khudhair D, Amani Hamedani H, Gaburro J, Shafei S, Nahavandi S, Garmestani H, Bhatti A (2017) Enhancement of electro-chemical properties of TiO2nanotubes for biological interfacing. Mater Sci Eng C 77:111–120

Kim DG, Shin MJ, Kim KH, Hanawa T (1999) Surface treatments of titanium in aqueous solutions containing calcium and phosphate ions. Biomed Mater Eng 9(2):89–96

Kim J, Kim B, Oh C, Ryu J, Kim H, Park E, No K, Hong S (2018) Effects of NH4F and distilled water on structure of pores in TiO2 nanotube arrays. Sci Rep 8:12487

Langton DJ, Jameson SS, Joyce TJ, Hallab NJ, Natu S, Nargol AV (2010) Early failure of metal-on-metal bearings in hip resurfacing and large diameter total hip replacement: a consequence of excess wear. J Bone Joint Surg Br 92(1):38–46

Liu X, Chu PK, Ding C (2004) Surface modification of titanium, titanium alloys, and related materials for biomedical applications. Mater Sci Eng R Rep 47(3–4):49–121

Liu W, Merrett K, Griffith M, Fagerholm P, Dravida S, Heyne B, Scaiano JC, Watsky MA, Shinozaki N, Lagali N, Munger R, Li F (2008) Recombinant human collagen for tissue engineered corneal substitutes. Biomaterials 29:1147–1158

Meimandi-Parizi A, Oryan A, Moshiri A (2013) Role of tissue engineered collagen based tridimensional implant on the healing response of the experimentally induced large Achilles tendon defect model in rabbits: a long term study with high clinical relevance. J Biomed Sci 20:28

Minagar S, Berndt CC, Wang J, Ivanova E, Wen C (2012) A review of the application of anodization for the fabrication of nanotubes on metal implant surfaces. Acta Biomater 8(8):2875–2888

Mirak M, ALizadeh M, Salahinejad E, Amini R (2015) Zn–HA–TiO2 nanocomposite coatings electrodeposited on a NiTi shape memory alloy. Surf Interface Anal 47:176–183

Nakayama I, Suzuki A, Kusumoto Y, Takakuwa K, Ikuta T (1989) Method of forming a thin film by chemical vapor deposition, US 4800105 A, January 1989

Nasab MB, Hassan MR, Sahari BB (2010) Metallic biomaterials of knee and hip-a review. Trends Biomater Artif Organs 24:69–82

Navarro M, Michiardi A, Castano O, Planell J (2008) Biomaterials in orthopaedics. J R Soc Interface 5:1137–1158

Niinomi M, Nakai M, Hieda J (2012) Development of new metallic alloys for biomedical applications. Acta Biomater 8:3888

Nishiguchi S, Nakamura T, Kobayashi M, Kim HM, Miyaji F, Kokubo T (1999) The effect of heat treatment on bone-bonding ability of alkali-treated titanium. Biomaterials 20(5):49–500

Oji MO, Wood JV, Downes S (1999) Effects of surface-treated cp Ti and Ti6Al4V alloy on the initial attachment of human osteoblasts. J Mater Sci Mater Med 10:869–872

Park JM, Koak JY, Jang JH, Han CH, Kim SK (2006) Osseointegration of anodized titanium implants coated with fibroblast growth factor-fibronectin (FGF-FN) fusion protein. Int J Oral Maxillofac Implants 21:859–866

Ratner BD, Hoffman AS, Schoen FJ, Lemons JE (2004) Biomaterials science: an introduction to materials in medicine. Academic, New York

Salahinejad E, Hadianfard M, Macdonald D, Mozafari M, Walker K, Rad AT, Madihally S, Vashaee D, Tayebi L (2013a) Surface modification of stainless steel orthopaedic implants by sol–gel ZrTiO4and ZrTiO4–PMMA coatings. J Biomed Nanotechnol 9:1327–1335

Salahinejad E, Hadianfard M, Macdonald D, Mozafari M, Vashaee D, Tayebi L (2013b) A new double-layer sol–gel coating to improve the corrosion resistance of a medical-grade stainless steel in a simulated body fluid. Mater Lett 97:162–165

Schwartz Z, Martin JY, Dean DD, Simpson J, Cochran DL, Boyan BD (1996) Effect of titanium surface roughness on chondrocyte proliferation, matrix production, and differentiation depends on the state of cell maturation. J Biomed Mater Res 30(2):145–155

Shokuhfar T, Sinha-Ray S, Sukotjo C, Yarin AL (2013) Intercalation of anti-inflammatory drug molecules within TiO_2 nanotubes. RSC Adv 3:17380

Tengvall P, Elwing H, Sjöqvist L, Lundström I, Bjursten LM (1989) Interaction between hydrogen peroxide and titanium: a possible role in the biocompatibility of titanium. Biomaterials 10:118e20

Tomsia AP, Saiz E, Song J, Bertozzi CR (2005) Biomimetic bonelike composites and novel bioactive glass coatings. Adv Eng Mater 7:999–1004

Urban RM, Jacobs JJ, Tomlinson MJ, Gavrilovic J, Black J, Peoc'h ' M (2000) Dissemination of wear particles to the liver, spleen, and abdominal lymph nodes of patients with hip or knee replacement. J Bone Joint Surg Am 82(4):457–476

van den Beucken JJ, Walboomers XF, Leeuwenburgh SC, Vos MR, Sommerdijk NA (2007) Multilayered DNA coatings: in vitro bioactivity studies effects on osteoblast-like cell behavior. Acta Biomater 3:587–596

Xiao SJ, Kenausis G, Textor M (2001) In: Brunette DM, Tengvall P, Textor M, Thompson P (eds) Titanium in medicine: material science, surface science, engineering, biological responses and medical applications. Springer-Verlag, Berlin/Heidelberg, pp 417–455

Zhang BG, Myers DE, Wallace GG, Brandt M, Choong PF (2014) Bioactive coatings for orthopaedic implants—recent trends in development of implant coatings. Int J Mol Sci 15(7):11878–11921

Part V

Bioinstrumentation and Its Design Aspects

Biomedical Instrumentation: Focus Toward Point-of-Care Devices

13

Sandeep Choudhary, Gaurav Pandey, Rupsha Mukherjee, and Abhijeet Joshi

Abstract

Biomedical instrumentation (BMI) deals with the measuring, recording, and transmitting biological signals from and to the human body to enable development of new algorithms and instruments to solve health-related problems. Development of BMI involves interdisciplinary understanding in different fields such as engineering technologies, materials science, and medical science which finally aim to improve human health by timely diagnosis and suitable therapeutic measures. These have immense importance in medical procedures for data collection and their analysis to aid decision-making and planning of treatment regimen. Typical BMI assemblies consist of sensors to detect bioelectrical, biophysical, and biochemical parameters with a safe interface which couples the biological surface with arrangements to control different influencing parameters. Additionally, some of the BMI may have an actuator, which can deliver external agents via direct or indirect contact to help the therapeutic aspects of devices. Electronic interfaces are calibrated using computational units with different electrical characteristics like signal-to-noise ratio, efficiency, bandwidth, and safety. Other necessary components include the power supply, output unit which provides the display and storage of data. This review summarizes different technological interventions pertaining to the BMI, their classification, and applications in various diagnostic and therapeutic domains.

S. Choudhary · G. Pandey · A. Joshi (✉)
Discipline of Biosciences and Biomedical Engineering, Indian Institute of Technology Indore, Indore, Madhya Pradesh, India
e-mail: abhijeet.joshi@iiti.ac.in

R. Mukherjee
Discipline of Biological Engineering, Indian Institute of Technology, Gandhinagar, Gujarat, India

S. Paul (ed.), *Biomedical Engineering and its Applications in Healthcare*,
https://doi.org/10.1007/978-981-13-3705-5_13

Keywords
Biomedical instrumentation (BMI) · Point-of-care testing (POCT) · Monitoring and diagnosis instruments · Therapeutic instruments · Biosafety and electrical hazard

13.1 Introduction

Biomedical engineering is a multidisciplinary advanced field which uses a combination of knowledge and information from medical science, biological sciences, and other technological domains like electrical and mechanical engineering. The main aim is to enhance or develop new methods and devices for diagnosis and/or therapy for the betterment of human health. Biomedical instruments generally work in close conjunction with living tissues to measure the physiological signals, recording them, and transmitting the data obtained from them. Designing biomedical instrumentation should be considered as a priority for the performance of the instruments regarding speed, accuracy, resolution, range of detection or operation, energy or power consumption, etc. (Webster 2009). In spite of them using several electrical or mechanical systems, they should be able to fulfill the measurement needs regarding reliability, accuracy, safety, environmental conditions, etc. (Dym et al. 2005; Singh 2014). Many times, physical and chemical parameters measured in devices are transformed using mathematical/computational science/engineering principles to examine the behavior, biology, and clinical health (Hendee et al. 2002). The advancement in the existing technologies, algorithms, processes, and concepts brings the knowledge and understanding from molecular leads to organ system level for the improvement in the prevention, diagnosis, and treatment. Newer advances have led to developing instrumentation for artificial organs, biomaterials, and transplant materials which could be essential components of diagnostic and therapeutic strategies (Bronzino 1999).

A biomedical device could be as simple and as ancient as a "tourniquet." A tourniquet is believed to be used from ancient times since 1674, and it is utilized in medical treatment till date. Later than 1674, procedures of treatment have been upgraded in several aspects which are described in handbooks and manuals for providing medical aid like the *Handbook for the Medical Soldier* written by Tuttle after World War I. The history of biomedical devices and their usage has been updated and reformed several times in different reports concerning worldly events like World War II, Spanish Civil War, and so on. For example, there are vast adaptations in designing and usage of tourniquets along with reports of severe consequences of wrongly administered. Several medical practitioners who were allied with different armies of the same era had also criticized the practices and applications with tourniquets. This can be widely understood that modern biomedical tools and techniques have come a long way in the context of evolution and development from its failures in the past (Welling et al. 2012).

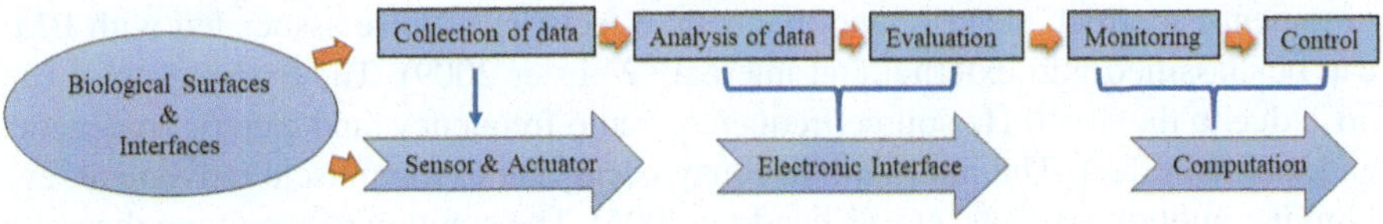

Fig. 13.1 Basic block diagram and components of biomedical instruments

Human physiological systems include the nervous system, musculoskeletal system, cardiovascular system, respiratory system, digestive system, etc. (Vashist et al. 2015). Typical biomedical instruments (BMI) may include highly sophisticated analytical methods for investigating health, physiology, or pathology of these. The modern sophisticated analytical methods may be employed for pathological, diagnostic, or therapeutic applications. BMI can be classified based on the sensing element or quantity, transducing principle, analyte or the organ system measurement, clinical characterization, etc.

Figure 13.1 describes the working principle of BMI. A sensor detects the bioelectrical, biochemical, or biophysical parameters and gives a secure interface with biological materials. It converts the information (physical quantity, condition, property, or energy) into another form, usually into electrical form. An actuator controls different parameters like bioelectrical and biochemical, delivers as an external agent using direct or indirect contact, and provides the safe interface with the external matrices. The electronics interface calibrates the output of the sensor with the input of the computational unit by maintaining the signal-to-noise ratio, sensor efficiency, and accuracy. The computational unit serves as a user-friendly interface by providing overall control and data storage system and displaying the signal. The output of the system can be in the form of visual, auditory, and text, or it can be temporary or permanent, discrete or continuous, numerical or graphical. Point-of-care measurement systems are those BMI which could be applied at both laboratory setup and in the hands of consumers, patients, or health practitioners like clinician or paramedical service providers. Point-of-care testing (POCT) devices are meant to develop testing mechanisms (bioelectrical, biochemical, physiological, etc.) on the site or nearby patient, whenever required and to evaluate the immediate testing/treatment requirements. They can provide preliminary information for a subsequent detail investigation (Renumadhavi et al. 2006).

13.1.1 BMI: Parameters and Properties

Considering the measurements of readings, the main aim of BMI is to detect the correct value accurately; however in practice the value is the sum of the actual value and the error associated in the measurement and computation. BMI are evaluated for specific parameters and features based on their capability and significance of the instrument or device. Analytical instruments are subject to several kinds of errors. The related error may occur due to the baseline shift; power line interface; movement of patients; thermal, electric, and magnetic effects; etc. Noise associated with BMI can be classified into external and internal (Webster 2009). The external noise can arise due to the 50/60 Hz source frequency, radio frequency, and electric, magnetic, and thermal effect. The internal noise may occur due to the muscle movement, eye blinking, motion artifacts, etc. (Khandpur 2005). The solution to removing the noise from the output signal is to use frequency filtering, as the desired signal will respond to the specific (Anandanatarajan 2011). Analytical parameters are described with parameters like linearity, accuracy, precision, resolution, selectivity, sensitivity, repeatability, reproducibility, etc., which are discussed below in detail. These parameters are essential for validating the performance of analytical methods (Anandanatarajan 2011) (Chandran and Singh 2007).

Linearity and Range It is defined as the increment or decrement in output with increment or decrement in input, respectively. The linearity of a system can be checked by dilution of the sample to different concentrations and plot the response against all the concentrations and validate through either visual inspection or regression coefficient (R^2). A correlation coefficient equals one is considered to be a linear system. The range of a system is defined as the interval (upper and lower limit) where concentrations of the analytes in a sample of a system are linear, accurate, and precise.

Accuracy It is a measure of total error, which means how close the measured value is with the actual value or reference value. The below formula can be used to calculate the percentage of error.

$$\%\text{Error} = \frac{\text{Reference value} - \text{Experimental value}}{\text{Reference value}} * 100.$$

Here, the reference value can also be termed as an actual value or real value. The accuracy of a system is checked by preparing three concentrations of short range, medium range, and higher range in triplicates and measures the response while comparing the theoretical value. Accuracy can also be expressed as percentage recovery which tells the closeness of calculated and actual values.

Precision It is defined as the number of distinct values or the alternatives from which the given result is selected. It is called as exactness in measurement. It is also defined as the closeness of the two or more measurements with each other. The precision of a system is validated by measuring the percentage of the relative standard deviation by doing the experiments either on different days, different analytes, or different types of equipment. Precision also constitutes parameters like repeatability and reproducibility. Repeatability is basically an intraassay precision, while reproducibility may be interlaboratory, interday, or interpersonal variations.

Limit of Detection (LOD) It is the measure of the smallest quantity being measured, which means how small a quantity will reliably be measured or detected by the instrument. The LOD regulates the signal of an apparatus to noise ratio (SNR). The lowest concentration of the analyte is detected in comparison to the blank or reference response.

Limit of Quantification (LOQ) The LOQ is determined by the concentration of the analyte above which the analyte can be reliably quantified with suitable precision and accuracy. The LOQ can be determined by reducing the concentration of the analyte at which the precision of the method is unacceptable.

Selectivity When measuring a solution containing the impurities, the result should not be altered with presence of the contaminant, e.g., the results of the two assays (one includes the analyte only and other contains the analyte and another test matrix with external stresses like heat, light, medium, etc.) should be same while comparing the selectivity.

Sensitivity and Response Time It defines the relation between the output and input, which means a change in recorded output with changes in unit input. Response time is defined as the time interval between saturation outputs from the addition of test sample.

13.2 Medical Laboratory Instruments (MLI)

Clinical laboratories maintain a set of tasks such as handling sample, test performance, sample discard, safety, and the management of information to store and analyze the data securely (Panel 2012). The primary objective of laboratory-based medical devices is to acquire the health information of a patient's specimens more accurately to provide for proper diagnosis. Medical laboratory instruments (MLI) can be classified as diagnostic instruments, laboratory instruments, medical imaging instruments, hospital equipment, and many other instruments based on the functionality and ability to perform tests, diagnosis, and their effectiveness for evaluating the treatment. Most commonly used MLI include the colorimeters, spectrophotometer, UV-visible, Raman, fluorescence and IR spectrophotometers, and various other optical sources and detection systems (Khandpur 2005). These characterization tools work on the principle of electromagnetic radiation interacting with the matter. When the interaction takes place, matter may absorb, reflect, scatter, or refract the electromagnetic radiation (Bronzino 1999). The type of communication is generally classified based on the kind of material, path length, and refractive index. These equipment are also useful in advanced research studies which are performed at various levels of development. The medical imaging instruments include magnetic resonance imaging (MRI), positron emission tomography (PET), single-positron emission-computed tomography (SPECT), computed tomography (CT) scanner, and X-ray machine, ultrasound, and, apart from the abovementioned MLI, various other pathological and clinical laboratory-based instruments such as particles' counters, ion-sensitive meters, centrifuges, microscopes, and auto-analyzers also classified under MLI. It also incorporates tools used in an operation theater, devices for maintenance, sterilization machines, and pathological devices (Wheeler 1998). The pathological tools used for the examination of blood, urine, and biofluid samples usually determine the composition of blood and the microbiological constituents to check the growth of the organism within the sample (Chaudhari et al. 2016; Kristensen et al. 2004). The specific principle and use of MLI are discussed in Table 13.1 (Plebani 2002; Powsner and Powsner 2008; Webster 2009).

13.3 Monitoring, Recording, and Diagnostic Instruments (MRDI)

13.3.1 Patient Monitoring Systems (PMS)

PMS are systems that continuously monitor the health of the patients and keep a record of the physical and chemical parameters throughout the treatment. Physical parameters mainly comprise the heart rate, pulse rate, temperature, and blood pressure, whereas blood gases, blood glucose, urea content in blood, or any biologically relevant markers are included in chemical parameters (Lemelson 1998). The classification of the PMS is done based on the mode of operation and settings required for the detection. The detecting parameters can further be classified as a single-parameter monitoring system and multiparameter monitoring system. Single-parameter measurement systems detect single parameter such as ECG monitoring system, electromyogram (EMG), electroencephalogram (EEG), oxygen saturation in the blood (Pulse oximeter), blood pressure monitor (Sphygmomanometer), etc.

Similarly, the multiple parameter-based system is the one which can detect more than one settings simultaneously like an ECG system with also provides information about the respiratory rate, blood pressure, and body temperature (Abdullah et al. 2015). Several electrodes have also been used clinically for capturing electric signals from different sets of the body. Different types of electrodes interface between the low-amplitude physiological signals detected from the body which are further amplified and displayed in the monitoring system. Some of the examples of the electrodes are skin electrodes for ECG, scalp electrodes for EEG, and needle electrodes for EMG (Ren et al. 2010). The characteristics of the physiological signals which are detected by the PMS systems are mentioned in Table 13.2 (Olson 1998).

13.3.2 Point-of-Care Testing (POCT) Devices for Measurement of Physical Parameters

Electrical Activity Measurements

When we are mentioning "electrical activity" in the context of human biology, it includes the study of all the electrical changes, properties, and processes within biological cells, tissues, or organs. All the attributes of electrical conduction like voltage, potential, current, etc. are taken into consideration, ranging from microscale

Table 13.1 Common medical laboratory instruments (MLI), their working principle, and applications

Instrument name	Type of MLI	Applications
Colorimeter	Laboratory instrument	Pathological, biological, and chemical laboratories for color solution assay, molecular weight determination, and various physiochemical studies. It is also used in pharmaceutical industries for aptamer and synthetic drug analysis
Spectrophotometer	Laboratory instrument	Pathological and clinical applications for biofluid analysis, used in pharmaceutical industries for studying the biochemistry of synthetic drugs
Particle counters	Diagnostic device	Radiology departments in hospitals and nuclear power plants for the detection of radiations
Centrifuges	Laboratory instrument	Biological laboratory, a pathological laboratory for separating microbes from the biofluids, blood bank, pharmaceutical industries for drug isolations
OT tools	Surgical instrument	Surgical and diagnostic application in medicine, biological labs for tissue engineering applications. It is used for cutting, dissecting, holding, grasping, providing oxygen supply in case of abnormalities
Autoanalyzer	Laboratory instrument	It has various chemical analytical applications and industrial applications for the analysis of water, soil, and fertilizer
X-ray	Medical imaging instrument	Clinical application in bone fracture detection, the presence of calcification due to kidney stones, and visualization of the lung for pneumonia infection
PET scanner	Medical imaging instrument	Clinical application for detecting cellular-level abnormalities leading to the brain and heart diseases and cancer
SPECT	Medical imaging instrument	Clinical application for monitoring brain and blood flow profiles at the site of stenosis
Ultrasound	Medical imaging instrument	Medical diagnostic application in cardiology, gynecology, obstetrics, and cancer detection
CT scanners	Medical imaging instrument	The medical diagnosis for blockage in a blood vessel, tumor in internal organs such as bones, ligaments, and soft tissues
MRI	Medical imaging instrument	Medical diagnosis of cancer or tumor in the brain, disk abnormalities in the spine
Sterilizer	Laboratory instrument	Pharmaceutical manufacturing unit for the disposal of vial products

by single-ion channels like small units to the magnitude of signals by a whole system propagating electrical signals, for understanding all the aspects of dynamics in biological systems. Some of the major systems and organs constituting the human biology have a large electrical activity like the brain is a powerhouse itself with several neurons generating electrochemical signals; apart from it, heart has a large

Table 13.2 Characteristics of the physiological signal detected using electrodes

Parameter	Measurement range	Frequency range	Measurement method/electrodes
Electrocardiogram (ECG)	0.5–4 mV	0.01–250 Hz	Skin electrodes
Electroencephalogram (EEG)	5–300 mV	DC –150 Hz	Scalp electrodes
Electromyogram (EMG)	0.1–5 mV	DC –10 KHz	Needle/skin electrodes
Electrooculogram (EOG)	50–3500 μV	DC –50 Hz	Contact electrodes
Body temperature	32–40 °C	DC –0.1 Hz	Thermistor/thermocouple
Respiratory rate	2–50 breaths/min	0.1–10 Hz	Strain gauge
Blood flow	1–300 mL/sec	DC –20 Hz	Ultrasonic flow meter
Blood pressure, arterial	10–400 mm Hg	DC –50 Hz	Strain-gage manometer cuff, auscultation
Direct	25–400 mm Hg		
Indirect			
Blood gases P_{O_2}	30–100 mm Hg	DC –2 Hz	Specific electrode, volumetric, or manometric
P_{Co_2}	40–100 mm Hg	DC –2 Hz	Specific electrode, volumetric or manometric
Blood pH	6.8–7.8 pH unit	DC –2 Hz	Specific electrode

electrical activity, every cell in the human body has ion channels generating a microlevel electrical activity. This makes the measurement and utilization of these electrical signals utmost essential to cater in the health domain for advanced healthcare facilitation.

ECG Monitoring Electrocardiogram (ECG) is a graphical description of the electrical activity of the heart. Electrical activity of the heart begins from sinoatrial (SA) node and through atrial muscle reaches atria ventricular (AV) node. From the AV node through a bundle of hiss (also called Purkinje fiber) arrives at the ventricular tissue, thereby completing the electric cardiac cycle. Every waveform of the ECG cycle represents some information, and change in the waveforms is a sign of abnormalities in the cardiac chambers (Lobodzinski and Laks 2012).

EEG Monitoring Electroencephalography (EEG) is a process to map the electrical signal generated by the brain's activity in different states of the patient. The procedure has a setup consisting of several electrodes placed within a band or cap with standard formations all over cranium with corresponding parts of the brain to be assessed. The EEG is a method to detect anomalies associated with brain functionality like epileptic seizures or sleep disorders. There are four waves, alpha, beta, theta, and delta, indicating brain activity in four different stages of brain activity.

The beta band gives the information of brain activity at full alert state, while alpha wave relates to the stage of relaxation, and theta band gives brain activity at partial arousal at just start of sleep state, whereas delta corresponds to the sleeping state brain activity. The recorded signals are plotted and compared to the normal-state signals for possible diagnosis of brain disorders (Monge-Pereira et al. 2017).

EMG Monitoring Electromyography is the method to detect changes in electrical potential in muscles. Muscle's electrical activity is measured by placing the electrodes over the region of passing or adjoining nerve and muscles for assessment of the state and functionality of a nerve or a muscle to diagnose the associated disorders. It is very helpful in the cases of paralysis or the muscular dystrophy-type disorders related to the muscular activity (Mills 2005).

EOG Monitoring Electrooculography (EOG) is performed by placing electrode to the extremities of frontal region and around eyes over the face of the patient. In this procedure, the signals are generated on the retinal layer due to movement of the eyeball in the eye socket. EOG is carried out to assess the sleep disorders and related functionalities associated with some parts of the brain for a possible diagnosis. The recorded signals are compared to the normal signal pattern for possible diagnosis of associated disorders. The EOG is also capable of tracking eyeball movement; thus, it is being further investigated to develop necessary rehabilitating aids for a disabled person (Aungsakul et al. 2012). Several point-of-care (PoC) devices are developed that can monitor electrical activity of the different organs based on voltage signal, in real time, and can record for later study display over the mobile screen and wireless transmission (through Bluetooth, Wi-Fi). There are several commercialized monitoring and recording devices available for electrical activity measurement, some of which are described in Table 13.3 (Bansal and Joshi 2018).

Pulse Oximeter For the proper functioning of the human body, an adequate amount of oxygen needed. The oxygen demand within the body is monitored by a device named pulse rate monitor which assesses the concentration of oxygen saturation in the blood. In normal conditions, saturated oxygen concentration ranges between 95 and 100% (Jian et al. 2013). The inability to meet the oxygen demand may lead to ischemia. So to detect the oxygen saturation level, pulse oximeter was introduced. A pulse oximeter uses an infrared light source, a photodetector, amplifier, and signaling unit to predict the oxygen level when it is below 90%; this condition is known as hypoxemia. A schematic representation is described below to show the working principle of the pulse oximeter in Fig. 13.2 (Webster 1997). The working principle of the pulse oximeters requires a source from where the beam light falls into the transparent region of the skin (e.g., finger or earlobe). The detector is positioned on the other side of the surface which detects the intensity of the transmitted light. The acquired light gives the concentration of the total light absorbed by the blood, hence providing the measure for the oxygen saturation (Pretto et al. 2014). The pulse oximeter has a vast number of applications and benefits in real time. For example, in case of the newborn baby, patient under anesthesia, the person with the chronic

Table 13.3 Commercially available electrical activity measurement devices

Measurement	Device	Location	Display/battery	Recording/storage	Application
ECG	Kardia Mobile	Chest, autonomic nervous system cardiovascular	Display on Android/iPhone/no built-in rechargeable battery	Recording and storage available	Cardiovascular health, fitness
ECG	REKA E100	Chest, autonomic nervous system cardiovascular	Does not display ECG recordings/ built-in rechargeable battery	Recording and storage device	Cardiovascular health, fitness
EEG	Neurostyle EEG	Forehead, mastoid, face	DC operated, fiber-optic coupled	Synchronous acquisition, multiple channel recording supported	Cognition, sleep, HMI
EMG	Delsys Trigno	Leg, arm, target muscle locations	Display on connected system/ rechargeable battery	Recording and storage	Rehabilitation, gait, HMI

HMI Human-Machine Interface

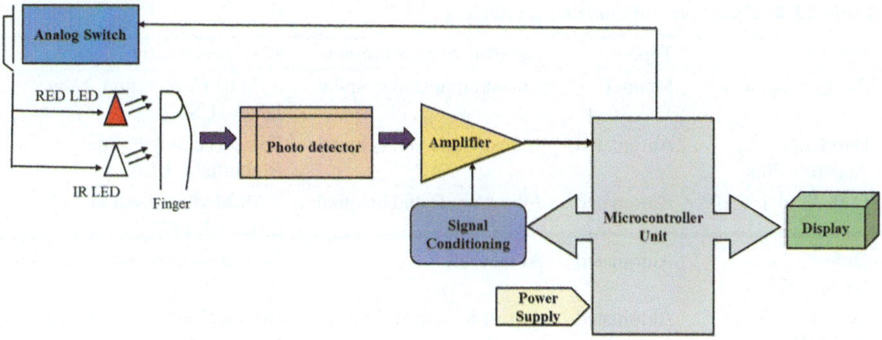

Fig. 13.2 Schematic of pulse oximeter

respiratory problem, sleep apnea, ventilators, etc. requires continuous monitoring of blood oxygen (King 2014; Schmitt et al. 1993). Some limitations of a pulse oximeter are that they may give incorrect results due to the reduction in the oxygen level in the surrounding tissue when exposed for a long duration. The change in position of the pulse oximeter may give inaccurate readings; it happens when the person is asleep and roll over at night (Murphy et al. 2016). A person may also experience uncomfortable, minor irritation, numbness, tingling, or change in skin color with continuous usage of this device.

Blood Pressure Monitoring The force applied by mobilizing blood on the walls of blood vessels is termed as blood pressure (BP) in the systemic circulation (Holmes 1915). The systolic blood pressure is the upper limit pressure during one heartbeat, while diastolic pressure is the lower limit pressure between two heartbeats and they are measured in millimeter of mercury (mmHg) in a sphygmomanometer (Parati 2005). The average resting value of blood pressure is 120 mmHg for systolic pressure and 80 mmHg for diastolic blood pressure (Master et al. 1950). The sphygmomanometer cuff is fastened to considerably above the known systolic pressure. The cuff pressure starts decreasing as the valve is open. Once the cuff's pressure matches the arterial systolic pressure, blood gets down to flow by the cuff, making blood flow turbulence and hearable sounds. By applying a stethoscope on the arm, these sounds can be heard, and the cuff's pressure is noted. The blood flow sounds will extend until the cuff's pressure falls under the arterial diastolic pressure (Perloff et al. 1993).

The noninvasive technique of blood pressure monitoring has their disadvantages as it tells the BP of a tiny volume of cardiac cycle; thus, this method is incapable of rendering the average BP of the complete day (Pickering et al. 2005). The blood pressure and blood oxygen saturation level can be detected simultaneously by the volume oscillometric method. This method is depending on the nonlinear behavior of the properties of the vascular walls (Jian et al. 2013). It uses a microphone and

Table 13.4 Blood pressure monitoring devices

Device	Type	Uses/site of measurement	Manufactured by
PyMaH mercury	Manual	At rest/clinic and hospital	PyMaH Corporation, New Jersey, USA
Datascope Accutorr Plus	Automated	At rest/clinic and hospital	Soma Technology Inc., Bloomfield, USA
CAS Model 9010	Automated	At rest/clinic and hospital	CAS Medical Systems, Branford, USA
Omron HEM-735C	Automated	At rest/self	Omron Corporation, Kyoto, Japan
Omron HEM-722C	Automated	At rest in elderly people	Omron Corporation, Kyoto, Japan
CH-DRUCK	Automated	At rest/ambulatory BP monitoring	Disetronic Medical Burgdorf, Switzerland

microprocessor for continuous monitoring of mean and arterial blood pressure (Imai et al. 1987). However, there are various blood pressure monitoring devices commercially available in the market, but few of them are recommended by the British Hypertension Society (BHS) and association for the advancement of medical instrumentation (AAMI) while keeping accuracy of the device as a prime factor. Table 13.4 (O'brien et al. 2001) provides some BP monitoring devices, recommended by BHS and AAMI.

The PyMaH is BP monitoring device manufactured by PyMaH Corporation, USA. It is a manual BP measuring technique which uses a cuff and mercury-based column to measure the arm BP (O'brien et al. 2001). The Datascope Accutorr Plus is a noninvasive, portable, and easy to use blood pressure monitor available with facultative infrared or prognostic temperature and recorder modules for cost-effective upgradability. It comes with lithium-ion battery technology for the entire runtime and battery life with minimal battery recharge (Anwar et al. 1997). The CAS Model 9010 is a well-established BHS and AAMI recommended electronic BP monitoring device made by Medical Device Assessment Ltd. (Medaval) used at clinics and hospitals (Alpert 1996). The Omron HEM-722C and Omron HEM-735C are the automated portable BP monitoring devices made by Omron Foundation, Inc., USA. The two models are operated on the rechargeable battery, where the former one is specially designed for elderly BP monitoring and the latter one is a fully automated with high capacity of data storage and integrated with the computer (Bortolotto et al. 1999). The CH-Druck BP monitoring device comprises a portable recorder, a random-access memory (RAM) card, left and right arm cuffs and bladders holding a microphone, two rechargeable batteries, and a carrying bag with strap and belt. The arrangement can be functioned manually by utilizing two buttons (O'Brien et al. 1993).

Temperature Monitoring Monitoring body temperature is critical in medicine. It is an important parameter determining the health of the person. The average human body temperature of an adult is considered as 37 °C (98.6 °F), and it depends upon the age, sex, reproductive stage of the subject, time of the day, etc. Thermometers are the most common device used to measure the temperature from various places of the body, like from mouth, under the arm, ear, nose, etc. The increase in body temperature can affect the number of white blood cell count, erythrocyte sedimentation rate (ESR), and C-reactive protein, and the chances of infection may grow in a person having a fever (Radder and Hermans 1990). The animal and human temperature can also be measured by near-infrared (NIR)-based sensing in which infrared thermography (IRT) is used. Infrared is the radiation emitted by the surface of a body whose temperature is higher than the absolute temperature (absolute zero). These emitted radiations are the function of temperature which can be detected by the thermal sensor, and the change in temperature results in a change in emitted thermal energy. Few degrees change in the body temperature is a sign of possible illness, and the IRT can accurately detect it. In case of a superficial tumor, the blood circulation is more around the tumor, e.g., breast cancer, and the increased blood flow results in an increase in tumor temperature that is easily detected by the IRT and thus serves as early detection of breast cancer (Usamentiaga et al. 2014). The in vivo and in vitro temperature sensing can also be done by the fluorescent nanothermometer which senses the temperature of the biological system, for instance, HeLa cancer cells or any solution. The working of nanothermometer is based on the temperature-dependent fluorescence of $NaYF_4 : Er^{3+}, Yb^{3+}$nanoparticles, where the change in temperature causes the change in the fluorescence intensity ratio of the green bands of the Er^{3+}dopant ions. The fluorescent nanothermometer can measure the temperature of the living cell from 25 °C to 45 °C, i.e., thermally induced death (Vetrone et al. 2010).

Heart Rate Monitoring Heart rate describes the rate of heartbeats in a minute. The standard heart rate of an adult is 72 beats per minute (60–100 bpm). A heart rate monitor (i.e., smartwatch, fit band) is a device that monitors and displays real-time heart rate or even records for later study. The heart rate below 60 bpm at rest is called as bradycardia, while the heart rate above 100 bpm at rest is termed as tachycardia (Malik et al. 1996). The variation in heart rate occurs due to physical activities like exercise, sleep, stress, anxiety, illness, etc. The pulse oximeter (with photoplethysmography (PPG)) can also be used to predict the heart rate along with the S_pO_2 and other parameters simultaneously. The working principle of PPG is described in Fig. 13.3 (Shelley 2007).

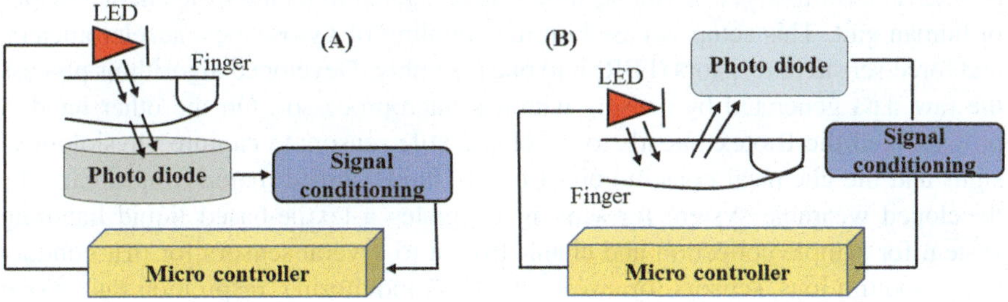

Fig. 13.3 (a) Transmission mode PPG; (b) Reflective mode PPG

The photoplethysmography (PPG) works by measuring the blood volume by finding light absorption by skin tissue. It uses a light-emitting diode (LED) as a source, signal processing unit, and a photodetector which detect the light after being absorbed by the skin tissue. PPG can work in refractive and reflective modes. In the former one, the transmitted light is being recognized by the detector, in which light source and detector are located opposite to each other. In the latter one, the reflected light is being detected by the detector, in which light source and detector remain on the same side (Saquib et al. 2015). ECG monitoring devices can also predict the heart rate by counting the number of squares between the two R waves.

Wearable PoC Devices Recent advancements with some of the ongoing projects in the field of medical devices like the Lifeguard Systems developed by NASA and Stanford, which includes various physiological sensors for heart rate, body temperature, ECG recording, SpO2 measurement, body temperature recording, blood pressure monitoring, and in a tiny box of easily wearable size, and its data can be communicated via Bluetooth to different storage modules. Similarly, INTREPID project has an objective of development of a multisensor wearable device for providing the therapy of anxiety-associated conditions crosschecking its efficiency in clinical settings in the collection of data along with reducing the anxiety-related symptoms in real time of occurrences. This device purposes biofeedback-driven virtual reality to bring the composure for a patient. The virtual reality relaxation process is supported by employing a mobile phone which can track and visualize the behavior of a user or a patient. Biosensors record the physiological data of a user or patient, and thus clinicians can control applied clinical procedures for the therapy of generalized anxiety disorder and sudden eruptions (Riva et al. 2009).

Another example is the AUBADE project, which is a wearable platform developed for analysis of emotional expressions in real-time scenario, using the collected signals from the facial region. This project utilizes a variety of healthcare-related programs designed with the information from the fields of psychology and neurology including speech disorders, depression, facial muscle disorders, stress assessment, and facial pain. The developed wearable platform has a mask that can transmit gathered signals (e.g., heart rate, ECG graph, respiration rate, and skin conductivity measures) wirelessly received from different sensors located on the face of the user connected back to a centralized system. It uses the more efficient procedure for multisensorial outputs by sensor management and data merger algorithms (Mahfouz et al. 2012). Similarly, a wearable shoe is investigated to identify the characteristics of human gait. This setup is based on the coupling of gyroscopes, accelerometers, and force-sensitive resistors (FSR) into one assembly. Developed algorithms process the raw data generated by sensors within a microprocessor. On the other hand, a project titled the Biotex intends to develop textile sensors to capture physiological signs and the chemical constitution of bodily fluids, aimed majorly on sweat. The developed wearable system for sensing integrates a textile-based liquid handling system for sample collection and channelizes it to several sensors for pH, conductivity, sodium ions, sensors for sweat rate, ECG monitoring, respiration rate, blood oxygenation levels, etc., and has been successfully deployed. With an adequate amount of sweat, the sensors are capable of far better results than conventional approaches (Coyle et al. 2010). Likewise, the ProeTEX project, participated by 23 research institutions, and related organizations in the field of emergency management along with some major universities in the European region, has developed a new generation of garments called as "smart" garments for rescue personals in the emergency-disaster domain. ProeTEX Garments has inbuilt portable sensors for continuous assessment of the risks surrounding the rescue personals in any operation. This smart setup enables detection of vital signs like heart rate, breathing rate, body temperature, SpO_2 levels, position, activity, and posture of the person wearing it, along with few environmental parameters like leakage of harmful gases, carbon monoxide (CO) and carbon dioxide (CO_2), external temperature, and the heat flux blowing over the garments (Curone et al. 2007; Curone et al. 2010). The collected data is processed and can be transmitted wirelessly to a remote station for better management of rescue operations (Cheng 2003) (Table 13.5).

13.3.3 Point-of-Care Testing (POCT) Devices for Measurement of Chemical Parameters

A biomarker is a characteristic that is aimed to assess and evaluate normal biological processes, pharmacologic responses, or any pathogenic progress, or applied therapeutic treatment if any. The measured response by the biomarker can be operational as well as physiological, biochemical at the level of the cell, or molecular fundamental interplay. Biomarkers are objective as well as quantitative characteristics of biological ongoings and detection of the chemical parameter like metabolism, electrolytes, blood gases, etc. Biomarkers act crucially in the improvement of the drug development projects as well as in the larger biomedical research endeavors. Figure 13.4 describes the working of the biomarker for chemical parameter measurement.

The analyte is being detected by the bio-receptor and converts the response into bioelectrical by means of the transducer and produces the calibrated output in the form of electrical signal, typically current (Chaudhari et al. 2012). There are several POCT devices available that can detect the chemical parameters. POCT devices are simple health checking or tests that can be done by the person themselves at home and provide result instantaneously. Point-of-care testing has been typically used for measurement of metabolites (e.g., lactate, ammonia, creatinine, urea, cholesterol) (Chaudhari et al. 2017; Prasad et al. 2011), features like hematocrit, hemoglobin, and basic electrolytes (e.g., sodium{Na}, potassium{K}), glucose, as well as blood gases (e.g., P_{O_2},P_{Co_2}) and acid-base balance (pH). Advanced portable point-of-care (POC) testing analyzers incorporate biochemical and silicon-chip techniques in self-calibrating, easily disposable units, such as used in the i-STAT System (Abbott Laboratories, Abbott Park, IL). The i-STAT setup evaluates sample assays using conductometric or amperometric and potentiometric circuits (Lee et al. 2011; Price

Table 13.5 Wearable POCT devices

Device brand name	Features	Mode of usage	Data communication	References
Fitbit Flex	Step count	Wrist-worn	Wireless connection to the mobile application only	Haghi et al. (2017)
Withings Pulse S	Step count, distance traveling, recording sleep time, measuring heart rate and percentage of the optimal sleep hours	Wrist-worn	Wi-Fi	Haghi et al. (2017)
Misfit Shine	Step count, distance measurement and daily calories burnt, sleep tracker monitoring and hours of light as well as deep sleep	Wrist-worn	Connection to a smartphone app	Haghi et al. (2017)
Jawbone	Sleep stages (REM, light and deep), HR, food and liquid intake, step count, distance traveled, running	Wrist-worn	Bluetooth (low energy)	Haghi et al. (2017)
(DuPont) IntexarTM	Reporting and sensing of heart rate, breathing rate, form awareness, and muscle tension	Sports Inner wears for the upper thorax	Real-time monitoring and data capture	Vagott and Parachuru (2018)
Hexoskin® Biometric® Shirt	Heart rate, heart rate variability, respiratory rate, step count, distance traveled, pace, maximal oxygen consumption, and calories burned.	Upper thorax wear	Bluetooth	Basholli et al. (2018)
Striiv® Fusion Bio Fitness Tracker	Heart rate, step count, distance traveled, calories burned, and quality of sleep	Wrist-worn	Bluetooth (low energy)	Basholli et al. (2018)
Microsoft® Band 2	Heart rate, calories burned, quality of sleep, calories and liquid intake, running, step count, elevation, climbing	Wrist-worn	Bluetooth (low energy)	Basholli et al. (2018)
Garmin vivosmart® HR Fitness Track	Heart rate, calories burned, quality of sleep, count of steps, swimming, climbing, and running	Wrist-worn	Bluetooth (low energy)	Basholli et al. (2018)

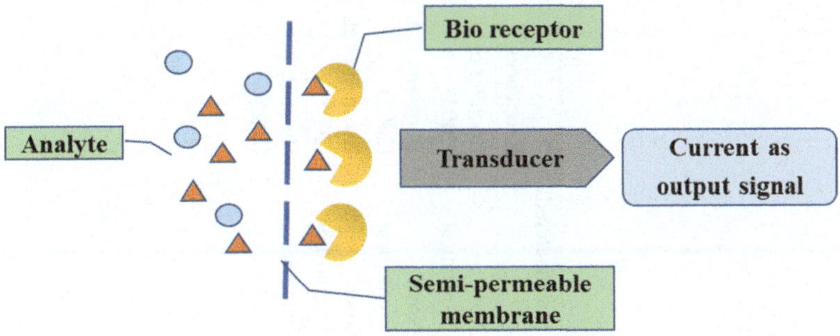

Fig. 13.4 Working principle of the biomarker for chemical parameter measurement

et al. 2004). Table 13.6 includes, but not limited to, POCT devices with their analyte measure, test time, type of sample, the method uses, and their manufacturer.

The StateSensor is a POCT device used for the rapid and accurate detection of the renal function (e.g., creatinine) in real time by fingerstick capillary blood sampling (1.2 μL). The working of the POCT device starts with finger sticking, blood drop is taken and then applied to the biosensor, and the creatinine reading is measured (Shephard 2011). The StateStrip is POCT device used for the detection of glucose/ ketone in a blood sample. Place the ketone strip in the POCT device; it will automatically turn the reader in the ketone measuring mode. A capillary blood sample is preferred over urine sample as it is simple to use and gives the rapid detection of ketone whenever the glucose level is higher than the 14 mmol/L (Cai et al. 2005). The OPTI CCA is POCT device that uses the innovative optical technique to analyze the blood gases and electrolytes with minimum maintenance. It contains disposable cassettes for whole blood, serum, and plasma samples. The results obtained are comparable to laboratory results with zero standby cost. The disposable sample collection unit locks safely after taking a measurement (Tusa and He 2005). A portable strip-based pregnancy test kit is used to assess the presence of human chorionic gonadotropin (hCG) in early pregnancy urine. Put three drops of urine collected in the morning after 2 weeks of missed menses and check the number of lines appear. If only C-line (control line) appears, then it means device works fine and the result of the pregnancy is negative. If both C-line and T-line appear, then it indicates the positive results of pregnancy. If no line appears, then it means the device is not working correctly (Butler et al. 2001).

Blood Glucose Monitoring Blood glucose monitoring devices are the most prominent diagnostic devices existing commercially. Point-of-care (POC) monitoring of blood glucose is required for a person with diabetes at a regular interval (Burrin and Alberti 1990). The average fasting blood glucose level in a healthy person has a range from 4.0 to 5.4 mmol/L (72–99 mg/dL) (Kiechle and Main 2000). Various glucose meters are commercially available in the market, e.g., Roche glucometer

Table 13.6 POCT devices for chemical parameter

S. N.	POCT device	Measured analyte	Test time	Test sample/test method	Manufactured by	References
01	StateSensor	Creatinine and hematocrit	30 s	Whole blood/Electrochemistry	Nova Biomedical	Boonlert et al. (2003)
02	StateStrip	Glucose, ketone, lactate	300 s	Whole blood, plasma, serum/potentiometry, amperometry, conductometry	Nova Biomedical	Boonlert et al. (2003)
03	OPTI CCA	Po_2, Pco_2, pH	Less than 120 s	Whole blood, plasma, serum/optical fluorescence	Roche Diagnostics	Boonlert et al. (2003)
04	Pregnancy test kit	Human chorionic gonadotropin (hCG)	Within 5 min	Urine	Prega news	Hu et al. (2014)

made up by Accu-Chek Aviva uses 0.6 μL of the blood sample to give glucose level in 5 s. Another glucose meter is Nova Biomedical Glucometer constructed by Nova Max which uses 0.3 μL of the blood sample and provides the result in 5 s. Glucose meter named OneTouch by LifeScan requires 1 μL of the blood sample and gives effect in 5 s (Yoo and Lee 2010). Most of the commercial glucose meters work on the principle of the amperometric biosensing. The glucose amperometric biosensors use two electrodes: one as the reference electrode (silver–silver chloride, Ag/AgCl) and one as the working electrode (platinum, Pt) for the blood glucose measurement. Glucose oxidase enzyme reacts with glucose present in blood in the presence of oxygen and produces hydrogen peroxide (H_2O_2), converting glucose ($C_6H_{12}O_6$) to gluconic acid ($C_6H_{12}O_7$). The amount of gluconic acid depends on the blood glucose level (Newman and Turner 2005). Considering the importance and market of glucose biosensor devices, researchers have developed several techniques of blood glucose measurement some of which are described in Fig. 13.5 (Oliver et al. 2009).

Different methods for glucose sensing can be classified into two categories, i.e., continuous glucose sensing and point sample glucose sensing. The continuous sensing is further divided into noninvasive, minimally invasive, and invasive type. The invasive technique of continuous glucose sensing can be classified by different optical imaging (e.g., fluorescence, Raman spectroscopy, NIR spectroscopy, etc.) and transdermal imaging (e.g., impedance spectroscopy, sonophoresis, and reverse iontophoresis). The point sample technique of glucose sensing is widely used and classified into two categories, i.e., finger prick glucometer and urine dipstick.

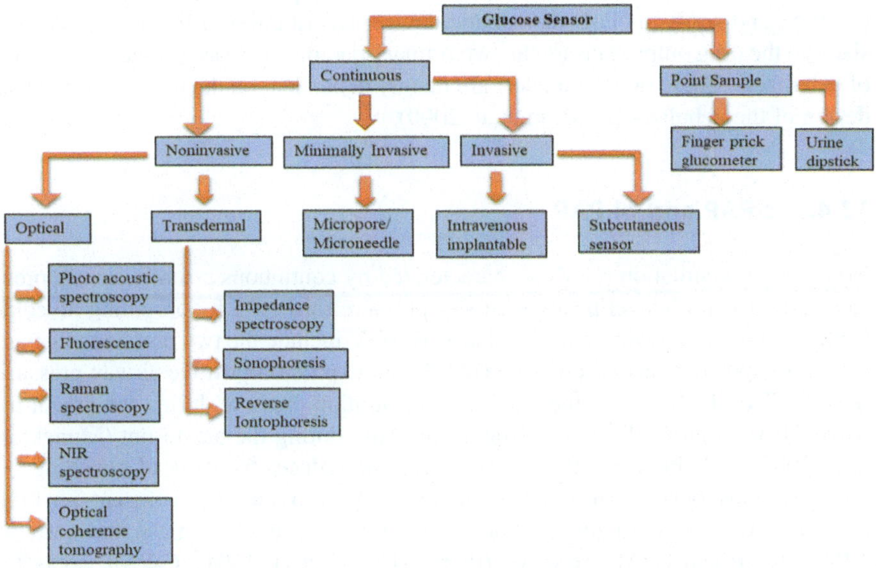

Fig. 13.5 Flowchart showing various glucose sensing techniques

13.4 Therapeutic Instruments (TI)

Therapeutic instruments are the medical devices that are used for the therapy and prevention of disease (e.g., continuous positive airway pressure (c-PAP), bilevel positive airway pressure (bi-PAP)), help in drug delivery (e.g., inhaler, nebulizer, auto-injector, microflux), and control the extent of disease (e.g., defibrillator, pacemaker). This section includes the details of the mentioned instruments with their working principle and applications.

13.4.1 Inhaler and Nebulizer

An inhaler (infusion pump) is a medical device used for the administration of medicine into the body through the respiratory tract. It is principally employed in the treatment of asthma and pulmonary diseases. There are four types of inhalers available in the market which include metered-dose inhaler (MDI), dry powder inhaler (DPI), nasal inhaler, and nebulizer. MDI are used for accurate contemporized drug delivery to minimize the deposition of medicine into the mouth and throat. Dry powder inhaler (DPI) is used for the administration of measured dry drug or medication into the lungs through the mouth and throat (Geller 2005). Nasal inhalers are used to remove the nasal congestions in the initial part of the respiratory tract by the administration of a liquid or spray of decongestant drugs. Nebulizer works on electric or battery power and converts the liquid droplets of anti-asthmatic drugs into the fine mist. The mist is delivered through the nasal or mouth cavity, into the lungs. All types of inhalers are easy to use since no complex working is involved in handling and serves as PoC therapeutic device. Distribution of the aerosol droplet size and the drug output rate are the two primary parameters to assess the functioning of nebulizers. These two parameters are mainly dependent on the user condition and design of the nebulizer (Le Brun et al. 2000).

13.4.2 c-PAP and bi-PAP

Noninvasive ventilation (NIV) is characterized by continuous positive airway pressure (c-PAP) and bilevel positive airway pressure (bi-PAP), by providing alveolar ventilation without invasive artificial airway. NIV include the two pressures that are inspiration positive airway pressure (IPAP) and expiration positive airway pressure (EPAP). The IPAP control the maximum inspiration pressure during the breath-in, while EPAP controls the end-expiratory pressure during the breath-out (Munckton et al. 2007). NIV improves the gas exchange and reduces the work of breathing.

Continuous positive airway pressure (c-PAP) provides the constant positive pressure (one pressure) during both inspiration and expiration. In the case of c-PAP, IPAP and EPAP are equal (Reeves-Hoche et al. 1994). C-PAP works by increasing functional residual capacity (FRC) of lungs and thereby improves the oxygenation (Ho and Wong 2006).

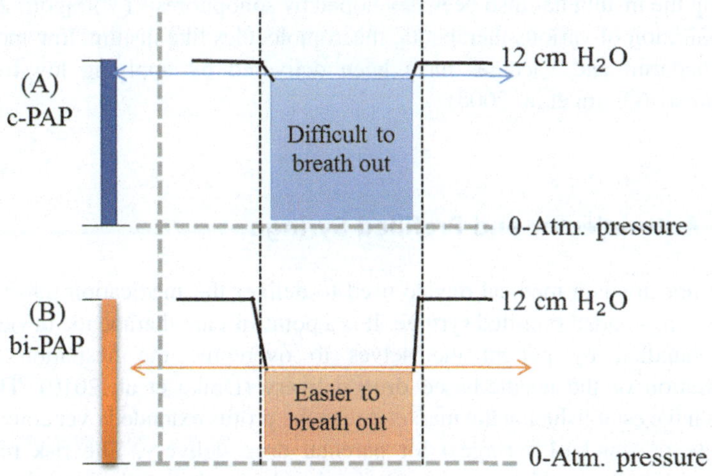

Fig. 13.6 Pressure level during (**a**) c-PAP (**b**) bi-PAP

Bilevel positive airway pressure (bi-PAP) works on two pressures, IPAP and EPAP in which IPAP is greater than EPAP. The primary use of bi-PAP is to reduce the expiratory pressure and used for the patients who face difficulties during exhalation while using c-PAP. The pressure support is the difference between the two pressures (IPAP and EPAP) and with an increase in pressure support the ease of ventilation increases (Hörmann et al. 1994).

Consider Fig. 13.6 (Antonescu-Turcu and Parthasarathy 2010) showing the pressure level during c-PAP and bi-PAP.

It is difficult to exhale during c-PAP as the pressure difference between IPAP and EPAP is zero, while it is easier to exhale during bi-PAP as pressure is maintained during inhalation and exhalation.

13.4.3 SonoPrep

SonoPrep is a portable low-frequency skin permeation device used to produce the micropores in the outermost layer (stratum corneum). The increase in permeability of the skin causes the molecules to cross the skin barrier with hundred times greater efficiency as compared to the intact skin (Gupta and Prausnitz 2009). The SonoPrep device is also used for the sampling of interstitial fluid (ISF) extraction from the generated micropores to examine the protein content at the time of sampling to predict the patient health status and also used for transdermal drug delivery (Lecomte et al. 2013). Enhanced skin permeability in the sonophoretic process provides extrication of diverse dissolved solutes like triglycerides, urea, dextran, lactate, albumin, calcium, etc. The application of ultrasound technology in the handling of diabetes by continuous monitoring of blood glucose level and simultaneously delivering the insulin has also been developed by sonophoresis (Mitragotri 2013). In vivo admiration of various therapeutic macromolecules like insulin, low-molecular-weight heparin, and vaccines have been delivered by applying low-frequency sonophoresis (Ogura et al. 2008).

13.4.4 Auto-injector and Prefilled Syringe

An auto-injector is a medical device used to deliver the medication into the body using a spring-loaded prefilled syringe. It is a point-of-care therapeutic device which can be handled by patient themselves to overcome the hesitation of self-administration of the needle-based drug delivery (Drake et al. 2010). The auto-injectors have established in the market due to the profits extended over conventional devices (e.g., standard syringes) for parental drug delivery. The risk related to established needle could be a chance of missed injection and discontinuation of the drug. These risks are overwhelming by auto-injectors and the cost of the healthcare is reduced, as there is a decrement in hospital visits. Hence, auto-injectors have assured for their improved compliance in patients for the handling of auto-injector in various diseases' treatment (Åmark and Bergens 2008). The diversity of auto-injector devices available in the market has increased and would develop considerably to encounter the necessities for urging invention in therapeutic areas. Auto-injection devices modify multiple sclerosis (MS) therapy and often provide extra characteristics to improve patient ease and adhesiveness. These characteristics incorporate reminder functions–as unknowingness is a significant obstacle for self-administered treatments, easiness, adjustability of injection speed, and depth. The electronic auto-injectors used for the MS are Betaconnect, RebiSmart, and ExtraviPro (mechanical) (Limmroth et al. 2017).

An auto-injector with prefilled syringe is shown in Fig. 13.7, having a head (includes ON/OFF switch) at one end and a prefilled syringe (PFS) at the other end. The auto-injector is integrating a needle assembly which does automatic penetration into the user's body, and complete drug delivery is assured by the head assembly which performs the medicament injection.

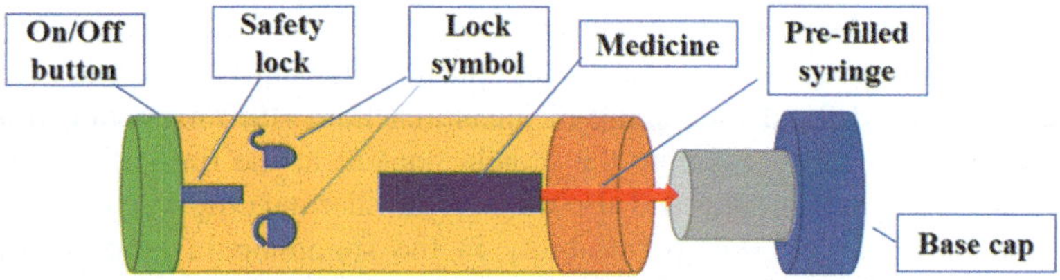

Fig. 13.7 Schematic of auto-injector (Stauffer et al. 2018)

13.5 Other Portable Devices

Some devices may be classified as portable monitoring or therapeutic tools, but they cannot be completely classified as point-of-care (POC) devices. For example, the pacemaker is a wearable therapeutic device and used for continuous monitoring and treatment. Likewise, the defibrillator can be used at the site of requirement, but for that, a trained person is required. Similarly, coagulation analyzer or micro-coagulation system can be used as a PoC device, if operated by a qualified person.

13.5.1 Defibrillators and Pacemakers

A defibrillator is a medical cum electrical device used to deliver electric shock or impulses to cure the fibrillation. Fibrillation is a term described to the abnormal electrical activity of the heart, whereas defibrillation describes the application of electric shock to the region of the heart which makes all heart muscle fibers to come in their refractory periods simultaneously, after which regular heart activity may resume. The defibrillator can be of three types, external, intravenous, and implantable cardioverter defibrillator (ICD), depending on the type of device required (Kramer et al. 2012). A pacemaker is a medical device used to deliver external electrical stimulation impulse to the heart muscle to modulate heart rate (generation of artificial pacing impulses). It uses low-current, low-duty-cycle pulse to deliver periodic electrical impulses conducted to electrodes placed within the heart cavity (Joshi et al. 2014). The cardiac pacemaker includes two units in which the former one is an electronic unit creating impulses of controlled rate and amplitude (pulse generator) for stimulus and the latter one is a lead assembly that carries the electrical pulses from pulse generator to heart. The pacemaker is broadly classified into two categories: one is external (used for the short duration of time) and the other is internal (used for an extended period). A pacemaker requires batteries and tiny wires to send the electric signals to the heart to help it pump in the right way (DiFrancesco 1993). The charge produced during stimulation is very small that one cannot feel while doing other activities. The batteries used in pacemakers had a limited lifetime and require a timely replacement which is very expensive and painful and risky for the patient. Therefore, various techniques are invented to harvest energy like chemical, electrical, thermal, and mechanical processes for the implanted pacemaker in the human body. These processes include the glucose oxidation, the electric potentials at the inner ear, mechanical movements of limbs, and the vibration of organs (Zheng et al. 2014). The pacemaker can also be used as a telemonitoring system for home ECG monitoring using an updated algorithm that can automatically detect the dysfunctional status of the pacemaker. This system uses the self-triggering of a pacemaker based on the ECG response and automatic ON/OFF of the device (Bai and Lin 1999).

13.5.2 Coagulation Analyzer/Micro-coagulation System

Coagulation analyzers use a rapid and straightforward process for the measurement of blood platelet levels. A coagulation test could provide diagnostic prevention from potential heart attack-inducing blood clots. A coagulation analyzer can be used to evaluate a coagulation pathway speed, apart from thrombin and thromboplastin levels in a few minutes. The coagulation system is used for the following test: (1) the prothrombin time (PT) test, which evaluates the clotting time of tissue factor-induced plasma; (2) the activated partial thromboplastin time (aPTT) test, called for testing monitoring unfreeze heparin (UFH) therapy; (3) the activated clotting time (ACT) test, currently used to monitor anticoagulation in patients receiving high doses of UFH and experiencing cardiopulmonary bypass (CPB); and (4) the thrombin time (TT) test, after the addition of thrombin to plasma, evaluating the conversion rate of fibrinogen to insoluble fibrin (Harris et al. 2013). The point-of-care coagulation analyzer uses the electrochemical sensors as they are simple to interface with the electronic instrumentation of the device and thus decrease the costs. Devices, such as the CoaguChek XS (Roche Diagnostics) and the i-STAT (Abbott Diagnostics), have used electrogenic substrates that result in the constitution of an electrochemically detectable production proportional to thrombin activity. The i-STAT is used for the PT/INR and ACT test, while the CoaguChek XS is aimed at the auto-test commercialization for PT/INR (Harris et al. 2013).

13.6 Safety of Medical Equipment and PoC Devices

The primary concern over the safety of biomedical devices is the electrical safety from electrical shock protection. Electrocution is the leading cause of injury or even death from devices. There are four primary causes to encounter an electrical shock: (1) poor technology interface with the patient or environment, (2) an inappropriate technology plan that does not hold to human factors and ergonomic principles, (3) unequal plan for carrying out a novel technology into practice, and (4) inadequate maintenance plan (Powell-Cope et al. 2008). Thus, approaches for prevention of such incidences like safe earthing (grounding), double-insulation low-voltage power supply, isolated power system, etc. need to be applied to all kinds of biomedical devices.

In recent times, medical devices are developed and operated within very stringent standards and benchmarks. To maintain the medical devices quality and standards, several agencies like the International Organization for Standardization (ISO) and International Electrotechnical Commission (IEC) are formed, and their prescribed regulations are widely accepted across many countries. These agencies keep on testing and validating the wide variety of innovations and industrial products including medical devices to publish the international standards for safety and reliability, periodically to keep updating them with recent findings. However, they are not the final authority and every country can develop its standards in a detailed manner for these medical devices, but they should grow in brief accordance of international rules laid by international agencies. For example, the US Food and Drug Administration (FDA) is a federal government agency in the country to put up law and compliance in the domain of food and drug administered to ordinary people. Similarly, ASTM International, formally known as the American Society for Testing and Materials (ASTM), is also one of the agencies for testing and validating the standards with technical specifications of products within the expertise of its constituting members. Similarly, for Canada, Canadian Standards Association (CSA) is the representative organization for ISO, and likewise, other countries also have similar other agencies to maintain the prescribed international quality and standards.

Every medical device is intended and purposed with some measurable goals. Even the operational conditions of the medical equipment are thought and planned with respect to its final purposes. The mode of application and course of action are different for every distinct medical device. Therefore, every medical equipment will be inheriting some of the degrees of risk along with them. Safety of a medical device is thus a relative term about ordinary usage, but they differ in other than ordinary circumstances. In general, the choices of material and process to incorporate in the medical device under the developmental phase are made with the idea of its possible working conditions, safety compliances, and to meet performance demands. Here, broadly two types of agreements can be justified as mechanical safety concerns and individual or environmental safety concerns. Agencies like ISO or ASTM International publish standards on a regular interval with the expertise of contributing members to suggest the best suitable materials, and other technological advancements keep the medical devices up to the latest benchmarks. With the availability of this data and benchmarks, the devices being developed already adapt to the latest and safest available materials and techniques. However, after the development, rigorous testing is done by several ends like the developers and later by various government and nongovernmental agencies to ensure the proposed safety and performance. This process also keeps evaluating the device, even after it gets initial acceptance to enter the consumer market. Data about performance and other vital aspects are taken into consideration to assess the medical instruments periodically to ensure compliance with international standards and proposed technical superiority concerning mechanical safety. On the other hand, the individual and environmental safety is also considered, while the device is under rigorous testing phase before reaching the consumer market. Every developed device must be presented before competent authorities with all the information, corresponding to the mode of action, usage requirements, associated risks, operational hazards, etc. by the developer's end. The experts of the device related area are then going to evaluate these provided technical specifications along with performance, reliability, and advancements. Any individual or operational hazard has to be declared in the technical specifications to authorities as well as to consumers in labeled and descriptive forms as a mandatory rule for developers to enter into the final consumer market (Cheng 2003).

13.7 Summary and Future Perspectives

As technology has advanced and occupied significantly our daily lives, health hazards like lifestyle diseases, mental stress, and related disorders have also grown rampantly. Disabilities associated with aging, the necessity of aid, and caretaking are also increasing with the same proportion, but the costs of living in old age and care homes are high as compared to individual establishments. Advancements and portability through miniaturization in biomedical engineering contribute enormously to healthcare. The high cost of establishment and investment of resources make healthcare impeding in reach to remotest locations. Biomedical instrumentation is nearing more and more toward portability and miniaturization by every day to fill the gaps in the availability of diagnosis and healthcare to individuals of developed countries as well as developing countries. The biomedical engineering allows electrical, electro-optical, electromechanical, electronic, and computer engineering to support clinical and biomedical applications. Biomedical engineering amends the field of healthcare diagnosis, monitoring, recording, and therapy. POCT evolve as a boon to biomedical instrumentation as it reduces the time and cost of testing and more importantly eases of diagnosis and treatment. We have reviewed a few point-of-care testing (POCT) tools for the monitoring, recording, and diagnosis as well as therapy. New opportunities and innovation are there for the betterment of existing devices with the use of nanoelectronics and micro-electromechanical systems (MEMS), to reduce the size and shape of the available tools with betterment in accuracy and low-cost development of POCT devices. POCT serves as a boon to the technology, and many more need to be developed for advance in healthcare. These advancements in the context of wearable sensors aid to the healthcare for remote diagnosis and real-time monitoring. They empower medical professionals and caretakers to tackle emergencies and healthcare-related anomalies in lesser time and a greater reach if a patient or athletes or any healthy subject is hooked up with an intelligent wearable care system much like a wearable monitoring system for vital features (Chan et al. 2012).

References

Abdullah A, Ismael A, Rashid A, Abou-ElNour A, Tarique M (2015) Real time wireless health monitoring application using mobile devices. Int J Comput Netw Commun IJCNC 7(3):13–30

Alpert B (1996) Validation of CAS model 9010 automated blood pressure monitor: children/adult and neonatal studies. Blood Press Monit 1(1):69–73

Åmark M, Bergens T (2008) Auto-injector. U.S. Patent 7,442,185, issued October 28, 2008

Anandanatarajan R (2011) Biomedical instrumentation and measurements. PHI Learning Pvt. Ltd, Delhi

Antonescu-Turcu A, Parthasarathy S (2010) CPAP and bi-level PAP therapy: new and established roles. Respir Care 55(9):1216–1229

Anwar YA, Tendler BE, McCabe EJ, Mansoor GA, White WB (1997) Evaluation of the datascope accutorr plus according to the recommendations of the association for the advancement of medical instrumentation. Blood Press Monit 2(2):105–110

Aungsakul S, Phinyomark A, Phukpattaranont P, Limsakul C (2012) Evaluating feature extraction methods of Electrooculography (EOG) signal for human-computer interface. Procedia Eng 32:246–252

Bai J, Lin J (1999) A pacemaker working status telemonitoring algorithm. EEE Trans Inf Technol Biomed 3(3):197–204

Bansal A, Joshi R (2018) Portable out-of-hospital electrocardiography: a review of current technologies. J Arrhythm 34(2):129–138

Basholli A, Lagkas T, Bath PA, Eleftherakis G (2018) Healthcare professionals' attitudes towards remote patient monitoring through sensor networks. Paper presented at the 2018 IEEE 20th international conference on e-health networking, applications and services (Healthcom)

Boonlert W, Lolekha PH, Kost GJ, Lolekha S (2003) Comparison of the performance of point-of-care and device analyzers to hospital laboratory instruments. Point Care 2(3):172–178

Bortolotto LA, Henry O, Hanon O, Sikias P, Mourad J-J, Girerd X (1999) Validation of two devices for self-measurement of blood pressure by elderly patients according to the revised British Hypertension Society protocol: the Omron HEM-722C and HEM-735C. Blood Press Monit 4(1):21–25

Bronzino JD (1999) Biomedical engineering handbook, vol 2. CRC Press, Boca Raton

Burrin J, Alberti (1990) What is blood glucose: can it be measured? Diabet Med 7(3):199–206

Butler SA, Khanlian SA, Cole LA (2001) Detection of early pregnancy forms of human chorionic gonadotropin by home pregnancy test devices. Clin Chem 47(12):2131–2136

Cai, X., Winarta, H., Vo, A., & Young, C. C. (2005). USPTO

Chan M, Estève D, Fourniols J-Y, Escriba C, Campo E (2012) Smart wearable systems: current status and future challenges. Artif Intell Med 56(3):137–156

Chandran S, Singh R (2007) Comparison of various international guidelines for analytical method validation. Die Pharmazie-An Int J Pharm Sci 62(1):4–14

Chaudhari RD, Joshi AB, Srivastava R (2012) Uric acid biosensor based on chemiluminescence detection using a nano-micro hybrid matrix. Sensors Actuators B Chem 173:882–889

Chaudhari RD, Joshi AB, Pandya K, Srivastava R (2016) pH Based Urea Biosensing Using Fluorescein Isothiocyanate (FITC)-Dextran Encapsulated Micro-Carriers of Calcium Alginate. Sens Lett 14(5):451–459

Chaudhari R, Joshi A, Srivastava R (2017) PH and urea estimation in urine samples using single fluorophore and ratiometric fluorescent biosensors. Sci Rep 7(1):5840

Cheng M (2003) Medical device regulations: global overview and guiding principles. World Health Organization, Geneva

Coyle S, Lau K-T, Moyna N, O'Gorman D, Diamond D, Di Francesco F et al (2010) BIOTEX—Biosensing textiles for personalised healthcare management. IEEE Trans Inf Technol Biomed 14(2):364–370

Curone D, Dudnik G, Loriga G, Luprano J, Magenes G, Paradiso R . . . Bonfiglio A (2007) Smart garments for safety improvement of emergency/disaster operators. Paper presented at the engineering in medicine and biology society, 2007. EMBS 2007. 29th annual international conference of the IEEE

Curone D, Secco EL, Tognetti A, Loriga G, Dudnik G, Risatti M et al (2010) Smart garments for emergency operators: the Proe TEX project. IEEE Trans Inf Technol Biomed 14(3):694–701

DiFrancesco D (1993) Pacemaker mechanisms in cardiac tissue. Annu Rev Physiol 55(1):455–472

Drake D, Jennings DI, Cooper SD, Barrow-Williams TD, Corrigan JP, Lee RT, Whitaker D (2010) Auto injector: Google Patents

Dym CL, Agogino AM, Eris O, Frey DD, Leifer LJ (2005) Engineering design thinking, teaching, and learning. J Eng Educ 94(1):103–120

Geller DE (2005) Comparing clinical features of the nebulizer, metered-dose inhaler, and dry powder inhaler. Respir Care 50(10):1313–1322

Gupta J, Prausnitz MR (2009) Recovery of skin barrier properties after sonication in human subjects. Ultrasound Med Biol 35(8):1405–1408

Haghi M, Thurow K, Stoll R (2017) Wearable devices in medical internet of things: scientific research and commercially available devices. Healthcare Inf Res 23(1):4–15

Harris LF, Castro-López V, Killard AJ (2013) Coagulation monitoring devices: past, present, and future at the point of care. TrAC Trends Anal Chem 50:85–95
Hendee WR, Chien S, Maynard CD, Dean DJ (2002) The National Institute of Biomedical Imaging and Bioengineering: history, status, and potential impact. Ann Biomed Eng 30(1):2–10
Ho KM, Wong K (2006) A comparison of continuous and bi-level positive airway pressure non-invasive ventilation in patients with acute cardiogenic pulmonary oedema: a meta-analysis. Crit Care 10(2):R49
Holmes W (1915) Blood pressure—its clinical significance. J Natl Med Assoc 7(1):30
Hörmann C, Baum M, Putensen C, Mutz N, Benzer H (1994) Biphasic positive airway pressure (BIPAP)--a new mode of ventilatory support. Eur J Anaesthesiol 11(1):37–42
Hu J, Wang S, Wang L, Li F, Pingguan-Murphy B, Lu TJ, Xu F (2014) Advances in paper-based point-of-care diagnostics. Biosens Bioelectron 54:585–597
Imai Y, Nihei M, Abe K, Sasaki S, Minami N, Munakata M et al (1987) A finger volume-oscillometric device for monitoring ambulatory blood pressure: laboratory and clinical evaluations. Clin Exp Hyperten Part A Theory Pract 9(12):2001–2025
Jian, Cui, Gao Shumei, Song Yilin (2013) Development of simultaneous monitoring system for non-invasive blood pressure and blood oxygen saturation. In 2013 IEEE 11th International Conference on Electronic Measurement & Instruments, Vol 2, pp 968–973. IEEE, 2013
Joshi AK, Tomar A, Tomar M (2014) A review paper on analysis of electrocardiograph (ECG) signal for the detection of arrhythmia abnormalities. Int J Adv Res Electr Electron Instrum Eng 3(10):12466–12475
Khandpur RS (2005) Biomedical instrumentation: technology and applications. Mcgraw-hill, New York
Kiechle FL, Main RI (2000) Blood glucose: measurement in the point-of-care setting. Lab Med 31(5):276–282
King D (2014) Marketing wearable home baby monitors: real peace of mind? BMJ 349:g6639
Kramer DB, Mitchell SL, Brock DW (2012) Deactivation of pacemakers and implantable cardioverter-defibrillators. Prog Cardiovasc Dis 55(3):290–299
Kristensen GB, Nerhus K, Thue G, Sandberg S (2004) Standardized evaluation of instruments for self-monitoring of blood glucose by patients and a technologist. Clin Chem 50(6):1068–1071
Le Brun P, De Boer A, Frijlink H, Heijerman H (2000) A review of the technical aspects of drug nebulization. Pharm World Sci 22(3):75–81
Lecomte MM, Atkinson KR, Kay DP, Simons JL, Ingram JR (2013) A modified method using the SonoPrep® ultrasonic skin permeation system for sampling human interstitial fluid is compatible with proteomic techniques. Skin Res Technol 19(1):27–34
Lee BE, Plitt S, Fenton J, Preiksaitis JK, Singh AE (2011) Rapid HIV tests in acute care settings in an area of low HIV prevalence in Canada. J Virol Methods 172(1–2):66–71
Lemelson JH (1998) Patient monitoring system: Google Patents
Limmroth V, Reischl J, Mann B, Morosov X, Kokoschka A, Weller I, Schreiner T (2017) Autoinjector preference among patients with multiple sclerosis: results from a national survey. Patient Prefer Adherence 11:1325
Lobodzinski SS, Laks MM (2012) New devices for very long-term ECG monitoring. Cardiol J 19(2):210–214
Mahfouz MR, Gary To, Kuhn MJ (2012) Smart instruments: wireless technology invades the operating room. In 2012 IEEE Topical Conference on Biomedical Wireless Technologies, Networks, and Sensing Systems (BioWireleSS), pp 33–36. IEEE, 2012
Malik M, Bigger JT, Camm AJ, Kleiger RE, Malliani A, Moss AJ, Schwartz PJ (1996) Heart rate variability: Standards of measurement, physiological interpretation, and clinical use. Eur Heart J 17(3):354–381
Master AM, Dublin LI, Marks HH (1950) The normal blood pressure range and its clinical implications. J Am Med Assoc 143(17):1464–1470
Mills K (2005) The basics of electromyography. J Neurol Neurosurg Psychiatry 76(suppl 2):ii32–ii35
Mitragotri S (2013) Devices for overcoming biological barriers: the use of physical forces to disrupt the barriers. Adv Drug Deliv Rev 65(1):100–103

Monge-Pereira E, Molina-Rueda F, Rivas-Montero F, Ibáñez J, Serrano J, Alguacil-Diego I, Miangolarra-Page J (2017) Electroencephalography as a post-stroke assessment method: an updated review. Neurologia 32(1):40–49

Munckton K, Ho K, Dobb G, Das-Gupta M, Webb S (2007) The pressure effects of facemasks during noninvasive ventilation: a volunteer study. Anaesthesia 62(11):1126–1131

Murphy D, Pak Y, Cleary JP (2016) Pulse oximetry overestimates oxyhemoglobin in neonates with critical congenital heart disease. Neonatology 109(3):213–218

Newman JD, Turner AP (2005) Home blood glucose biosensors: a commercial perspective. Biosens Bioelectron 20(12):2435–2453

O'Brien, E., Petrie, J., Littler, W., de Swiet, M., Padfield, P. L., Altman, D., . . . Atkins, N. (1993). The British Hypertension Society protocol for the evaluation of blood pressure measuring devices. J Hypertens, 11(Suppl 2), S43-S62

O'brien E, Waeber B, Parati G, Staessen J, Myers MG (2001) Blood pressure measuring devices: recommendations of the European society of hypertension. BMJ 322(7285):531

Ogura M, Paliwal S, Mitragotri S (2008) Low-frequency sonophoresis: current status and future prospects. Adv Drug Deliv Rev 60(10):1218–1223

Oliver N, Toumazou C, Cass A, Johnston D (2009) Glucose sensors: a review of current and emerging technology. Diabet Med 26(3):197–210

Olson WH (1998) Basic concepts of medical instrumentation. In: Medical Instrumentation Application and Design, pp 1–55

Panel BBR (2012) Guidelines for safe work practices in human and animal medical diagnostic laboratories. Morb Mortal Wkly Rep 61:1–102

Parati G (2005) Blood pressure variability: its measurement and significance in hypertension. J Hypertens 23:S19–S25

Perloff D, Grim C, Flack J, Frohlich ED, Hill M, McDonald M, Morgenstern BZ (1993) Human blood pressure determination by sphygmomanometry. Circulation 88(5):2460–2470

Pickering TG, Hall JE, Appel LJ, Falkner BE, Graves J, Hill MN et al (2005) Recommendations for blood pressure measurement in humans and experimental animals: part 1: blood pressure measurement in humans: a statement for professionals from the subcommittee of professional and public education of the American Heart Association Council on High Blood Pressure Research. Circulation 111(5):697–716

Plebani M (2002) Charting the course of medical laboratories in a changing environment. Clin Chim Acta 319(2):87–100

Powell-Cope G, Nelson AL, Patterson ES (2008) Patient care technology and safety. In: Patient safety and quality: an evidence-based handbook for nurses. Agency for Healthcare Research and Quality (US)

Powsner RA, Powsner ER (2008) Essential nuclear medicine physics. Wiley, New York

Prasad J, Joshi A, Jayant RD, Srivastava R (2011) Cholesterol biosensors based on oxygen sensing alginate-silica microspheres. Biotechnol Bioeng 108(9):2011–2021

Pretto JJ, Roebuck T, Beckert L, Hamilton G (2014) Clinical use of pulse oximetry: Official guidelines from the Thoracic Society of Australia and New Zealand. Respirology 19(1):38–46

Price CP, John AS, Hicks JM (2004) Point-of-care testing. AACC Press, Washington, DC

Radder A, Hermans J (1990) The significance of body temperature, sedimentation, C-reactive protein, leukocyte count and differential for the diagnosis of infections in an internal medicine emergency department. Ned Tijdschr Geneeskd 134(52):2536–2540

Reeves-Hoche MK, Meck R, Zwillich CW (1994) Nasal CPAP: an objective evaluation of patient compliance. Am J Respir Crit Care Med 149(1):149–154

Ren Y, Werner R, Pazzi N, Boukerche A (2010) Monitoring patients via a secure and mobile healthcare system. IEEE Wirel Commun 17(1):59

Renumadhavi C, Kumar SM, Ananth A, Srinivasan N (2006) A new approach for evaluating SNR of ECG signals and its implementation. Paper presented at the Proceedings of the 6th WSEAS international conference on simulation, modelling and optimization

Riva G, Gorini A, Gaggioli A (2009) The Intrepid project-biosensor-enhanced virtual therapy for the treatment of generalized anxiety disorders. Stud Health Technol Inform 142(155):155

Saquib N, Md Tarikul Islam Papon, Ahmad I, Rahman A (2015) Measurement of heart rate using photoplethysmography. In 2015 International Conference on Networking Systems and Security (NSysS), pp 1–6. IEEE, 2015

Schmitt HJ, Schuetz WH, Proeschel PA, Jaklin C (1993) Accuracy of pulse oximetry in children with cyanotic congenital heart disease. J Cardiothorac Vasc Anesth 7(1):61–65

Shelley KH (2007) Photoplethysmography: beyond the calculation of arterial oxygen saturation and heart rate. Anesth Analg 105(6):S31–S36

Shephard MD (2011) Point-of-care testing and creatinine measurement. Clin Biochem Rev 32(2):109

Singh M (2014) Introduction to biomedical instrumentation. PHI Learning Pvt. Ltd, Delhi

Stauffer VL, Sides R, Lanteri-Minet M, Kielbasa W, Jin Y, Selzler KJ, Tepper SJ (2018) Comparison between prefilled syringe and autoinjector devices on patient-reported experiences and pharmacokinetics in galcanezumab studies. Patient Prefer Adherence 12:1785

Tusa JK, He H (2005) Critical care analyzer with fluorescent optical chemosensors for blood analytes. J Mater Chem 15(27–28):2640–2647

Usamentiaga R, Venegas P, Guerediaga J, Vega L, Molleda J, Bulnes F (2014) Infrared thermography for temperature measurement and non-destructive testing. Sensors 14(7):12305–12348

Vagott J, Parachuru R (2018) An overview of recent developments in the field of wearable smart textiles. J Text Sci Eng 8(368):2

Vashist SK, Luppa PB, Yeo LY, Ozcan A, Luong JH (2015) Emerging technologies for next-generation point-of-care testing. Trends Biotechnol 33(11):692–705

Vetrone F, Naccache R, Zamarrón A, Juarranz de la Fuente A, Sanz-Rodríguez F, Martinez Maestro L et al (2010) Temperature sensing using fluorescent nanothermometers. ACS Nano 4(6):3254–3258

Webster JG (1997) Design of pulse oximeters. CRC Press, Boca Raton. ISBN 9780750304672

Webster JG (2009) Medical instrumentation application and design. Wiley, Hoboken

Welling DR, McKay PL, Rasmussen TE, Rich NM (2012) A brief history of the tourniquet. J Vasc Surg 55(1):286–290

Wheeler LA (1998) Clinical laboratory instrumentation. In: Medical instrumentation: application and design, p 486

Yoo E-H, Lee S-Y (2010) Glucose biosensors: an overview of use in clinical practice. Sensors 10(5):4558–4576

Zheng Q, Shi B, Fan F, Wang X, Yan L, Yuan W et al (2014) In vivo powering of pacemaker by breathing-driven implanted triboelectric nanogenerator. Adv Mater 26(33):5851–5856

Techniques/Tools to Study Epigenetic Biomarkers in Human Cancer Detection

14

Vivek Kumar, Alka Singh, Priyanka Gautam, and Manisha Sachan

Abstract

Epigenetic modifications such as DNA methylation and histone modification are essential for normal developmental process and regulation of tissue-specific gene expression in mammals. However, altered epigenetic processes contributed to change in gene expression and transform cell to malignant type. Global epigenetic changes such as DNA methylation (hypomethylation and hypermethylation), histone modification (acetylation and phosphorylation), and nucleosome repositioning are associated with cancer onset and progression. Expression of several tumour suppressor genes (CDH1, RB, and MLH1), oncogenes (BRCA1 and DAPK), and cell cycle regulatory genes (RB, RASSF1, and P27K1P1) was altered due to epigenetic modification. Several methods have been developed since the last two decades for studying epigenetic modification. Initially, bisulfite-based method was predominantly used for studying DNA modification; however, advancement in techniques led to the development of more robust methods (MeDIP assay, NGS-based method, mass spectrometry, and microarray analysis) for studying whole-genome methylome analysis, while ChIP (chromatin immunoprecipitation) is widely used for studying histone modification. Several DNA methylation-based biomarkers in context to cancer detection have been screened and purposed for its clinical utilization. Since purposed biomarkers are only proof of concept, however, their further validation in prospective studies on a larger sample subset still remains.

Keywords

DNA methylation · cfDNA · MSO-microarray · Histone modification · Cancer biomarker

V. Kumar · A. Singh · P. Gautam · M. Sachan (✉)
Department of Biotechnology, Motilal Nehru National Institute of Technology Allahabad, Prayagraj, Uttar Pradesh, India
e-mail: manishas@mnnit.ac.in

S. Paul (ed.), *Biomedical Engineering and its Applications in Healthcare*,
https://doi.org/10.1007/978-981-13-3705-5_14

14.1 Epigenetic Alteration

Epigenetics is the study of stable heritable changes in gene expression that occur without changing the DNA sequence and have enough potential to get reversed. Epigenetic events occur naturally, important in normal cellular functions but if get altered may result in altered gene expression and malignant cellular transformation. Epigenetic alternations play an important role in carcinogenesis. The initiation and progression of cancer are controlled by both genetic factors and epigenetic factors, and we need to understand how gene expression is controlled at both levels which consequently helps in expanding our knowledge and tools to manipulate human health. Within the cell, there are three key major epigenetic modifications: DNA methylation, histone modification, and nucleosome repositioning.

14.1.1 DNA Methylation

DNA methylation is one of the best known and extensively studied epigenetic modifications. DNA methylation is a highly specific process which involves covalent addition of methyl group to C5 of cytosine followed by guanine, called CpG dinucleotide, mediated by DNA methyltransferases. The distribution of CpG across the human genome is not uniform but instead, CpG found to be clustered in the promoter region (5′ end of the gene), called CpG islands which normally managed to escape methylation. The CpG island tracks three basic features described by Sati et al. (Satish et al. 2012): (i) length should be greater than 200 bp, (ii) GC content >50%, and (iii) observed/expected ratio of CpG > 0.6. About 60% of human gene promoter is covered with CpG islands. Normally, CpG islands remain unmethylated, whereas CpG sites present in repetitive regions and heterochromatin region are found to be methylated. DNA methylation is repressive in nature and leads to silencing of repetitive, transposable elements and other harmful genes which further enhances genomic stability. It is estimated that 70–80% of CpG dinucleotides are methylated throughout the genome. In mammalian cells, DNA methylation mostly occurs in symmetric CG context which is about 24%, to some extent 6.7% in CHG, and 1.7% for CHH context.

The methylation pattern is normally established at the time of implantation and selected CpG will, later on, get de novo methylated. This can be achieved through the polycomb proteins which recruit methyltransferase at CpG sites (Michaelson-Cohen 2011). For transcription to occur in the gene promoter region, DNA should be readily accessible to transcription factors, enhancers, and other regulatory units. But the same is not possible in methylated condition. The methyl group fits into the major groove of duplex DNA, changing the chromatin structure, thus making it inaccessible for transcription factors in its condensed form (Fig. 14.1).

For example, methylated DNA attracts methylated CpG binding domain which recruits various repressor proteins (MeCP2 or MeCP1) having transcription suppression capabilities, resulting in histone modification, thereby inhibiting transcription (Jones 2012). Generally, there is an inverse correlation between cytosine

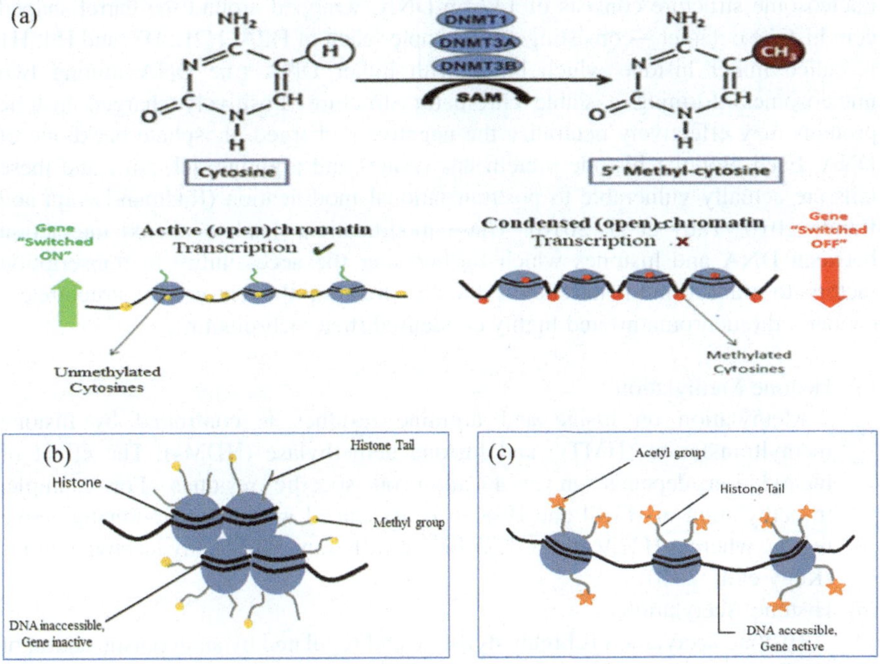

Fig. 14.1 Interplay between DNA methylation, gene transcription, and chromatin structure (**a**), methylation of DNA and histones causes nucleosomes condensation thus transcription repression (**b**), and histone acetylation results in loose packing of nucleosomes, thereby activating gene transcription (**c**)

methylation and CpG density. This can be depicted from the presence of dispersed CpG across the genome (98%) representing high levels of cytosine methylation, and CpG islands (2%) remain unmethylated (Ryan et al. 2009; Deaton and Bird 2011).

There are basically two types of methylation process: de novo methylation and maintenance methylation. DNA methylation is performed by a family of DNMTs that transfer methyl group from S-adenosyl methionine to 5C of cytosine to form 5mC. In de novo methylation, new methylation pattern is established to unmodified DNA by DNMT3a and DNMT3b independent of replication, whereas other enzyme DNMT1 first recognizes the methylation pattern of parental DNA and then copies this onto the newly daughter strand during replication (Sharma et al. 2010; Gun-Do et al. 2002; Okano et al. 1999; Yang et al. 2014; Esteller 2007).

14.1.2 Histone Modification

The next epigenetic alternation includes modification of the histone tail by methylation, acetylation, phosphorylation, ubiquitylation, and sumoylation. Histone proteins are basic proteins which form a nucleosome structure together with DNA. The nucleosome structure consists of 147 bp DNA, wrapped around the barrel-shaped core histone octamer—consisting of two copies each of H2A, H2B, H3, and H4. H1 is called linker histone which binds with linker DNA (the DNA joining two nucleosomes) forming a stable chromatin structure. Positively charged histone proteins may effectively neutralize the negatively charged phosphate backbone of DNA. Each globular histone protein has lysine- and arginine-rich tails, and these tails are actually vulnerable to posttranslational modification (Radman-Livaja and Rando 2010; Taby et al. 2010). These modifications lead to altered interaction between DNA and histones which further alter the accessibility of transcription factors to transcription start sites, thereby making the chromatin structureless condensed (euchromatin) and highly condensed (heterochromatin).

(a) Histone Methylation

Methylation on lysine and arginine residues is controlled by histone methyltransferase (HMTs) and histone demethylase (HDMs). The effect of methylation depends on its location on specific residues. For example, trimethylation of H3K4 and H3K36 is associated with transcriptionally active region, whereas H3K9 and H3K27 linked with transcriptionally inactive regions (Kelly et al. 2010).

(b) Histone Acetylation

Histone acetylation is highly dynamic and regulated by antagonistic action of HATs and HDACs (Yang et al. 2014). Acetyl groups are added onto the protruding lysine-rich tail out of histone. Histone acetyltransferases (HATs) catalyse the addition of acetyl group from acetyl CoA as a cofactor. This process neutralizes the lysine's positive charge and thereafter reduces the affinity of histone proteins to DNA. This leads to the formation of an open chromatin structure which is transcriptionally active. Histone deacetylase (HDACs) catalyses the removal of the acetyl group and forms a closed chromatin structure which is transcriptionally inactive and can be seen as heterochromatin.

14.1.3 Nucleosome Repositioning

The basic structural unit of chromatin is nucleosome, which compacts the DNA manifold. This compaction inhibits many processes like transcription, DNA repair, DNA replication, and chromosomal segregation and thereby enables little access to DNA. However, the above inhibition can be relieved by sliding or disassembling nucleosome. This disassembling or repositioning is achieved by chromatin remodelling complexes (remodellers) like ATP-dependent remodelling complexes (SWI/SNF and RSC chromatin remodelling complexes). These complexes use the energy of ATP hydrolysis to alter the packaging state of chromatin by ejecting, moving, and restructuring the composition of the nucleosome. Sometimes, this replacement of histone proteins by histone variants also plays an essential role in regulating chromatin structure by altering the accessibility of transcription factors to regulatory DNA sequences. These histone variants like H3.3 and H2A.Z are synthesized during S-phase of replication and preferentially bind to the promoter region of the gene (Kelly et al. 2010).

14.2 Epigenetic Events During Carcinogenesis

In human cancer, aberrant epigenetic alternation is known to accord various phases of tumour development—initiation, invasion, metastasis, and chemotherapy resistance. It has been shown that more than 300 genes and gene products are epigenetically altered in human cancer. During carcinogenesis, genome-wide hypomethylation and site-specific CpG island promoter hypermethylation occur simultaneously. The global hypomethylation is the first common epigenetic alternation of cancer cell and is found in all major human cancers like ovarian, liver, lung, and bladder. Hypermethylation is the most frequent epigenetic event occurring in the cancer cell. Hypermethylation normally affects promoter region containing CpG islands, thereby inactivating transcription. This transcriptional inactivation affects genes involved in DNA repair pathways (BRCA1, MGMT, MLH1,WRN), cell cycle regulators (Rb, $p16^{INK4a}$), carcinogen metabolism (GSTP1), metastasis (CDH1, CDH13,CDH10), and Ras signalling (RASSF1A) (Esteller et al. 2007). This inactivation imparts a growth advantage to tumour cells which further increases genome instability (Sellar et al. 2003; Shen et al. 2007; Kanwal and Gupta 2012; Pogribny 2010; Moss and Wallrath 2007). Hypermethylation also causes the disruption of many genes in key cellular pathways. Some of the hypermethylated genes and their affected pathways are given in Table 14.1.

Silencing of tumour suppressor genes by methylation is not only the mechanism for carcinogenesis. Besides this, silencing can also be caused by histone modifications especially acetylation, methylation, phosphorylation, biotinylation, and ubiquitylation. Similar to DNA methylation alternation, this also occurs on both genome-wide and gene-specific level. The histone modifications which occur genome-wide are H3K4me3, H3K4me2, and H3ac, while others show modification at gene-specific loci like H3K20me1 and H3K27me1. The most prominent histone modification seen in the cancer cell is the global reduction of monoacetylated H4K16 and HDACs (removes an acetyl group from histone) which have been found to be overexpressed in the cancer cell. The altered pattern of histone modification is mainly due to the overexpression of histone methyltransferase and histone demethylases (Dalgliesh et al. 2010) (Table 14.2).

14.3 Importance of Epigenetic Signature as a Cancer Biomarker

The biomarker is a measurable indicator of normal biological process or disease. The epigenetic biomarker has gained much attention due to their major role in cancer formation. We can exploit epigenetic profiles not only to differentiate between the different types of malignancies but also to know about different stages of cancer progression. These epigenetic modifications can be used as cancer biomarkers. In

Table 14.1 List of important hypermethylated genes and the affected pathways (Kanwal and Gupta 2012)

Pathway	Example of methylated gene	Example of cancer type
Growth signal autonomy	RASSF1A, SOCS1	Lung, bladder, ovarian, breast, lymphoma, gastric
Insensitivity to anti-growth signals	$p15^{INK4B}$ and $p16^{INK4A}$	Lymphoma, bladder, AML
Evading apoptosis	DAPK	Lymphoma
Tumour invasion and metastasis	CDH1, TIMP3	Gastrointestinal, oesophagus
Genomic instability	MGMT, LMNA, MLH1, CHFR	Lymphoma
Sustained angiogenesis	THBS1	T cell, lymphoma, neuroblastoma, endometrial cancer
Cell cycle	$P15^{INK4B}$,$p16^{INK4A}$, p14ARF, RB, RASSF1, p27KIP1	Gliomas, lung, colorectal, hepatic, head and neck, bladder, pancreatic, melanoma, prostate cancer, salivary gland, hepatocellular, gastrointestinal, haematological cancer
DNA repair	MLH1, MGMT, BRCA1, RASSF1, WRN	Colorectal, endometrial, gastric, squamous cell, melanoma, liver cancers
Cell adhesion	CDH1, CDH13	Breast, lung and stomach cancers, leukaemia
Cell growth and proliferation	MDG1	Breast and gastric cancer
Signal transduction	RARB2	Oesophageal and breast cancer
DNA damage	GADD45G	Liver, lung, skin, ovarian, cervical, breast, lymphoma cancer
Carcinogen detoxification	GSTP1	Prostate, breast and renal cancers

normal condition, all the above mechanisms control cell proliferation and differentiation, but their dysregulation contribute to cancer. DNA hypermethylation is one of the prevalent molecular changes in the cancer cell and can be easily detected. There are several advantages of DNA biomarker over others. DNA is a stable molecule, less prone to degradation than RNA, their reversible nature, earliest molecular changes in carcinogenesis, their restriction to limited regions, availability of highly sensitive and specific techniques for detecting methylation with a minimal amount of DNA. These techniques include pyrosequencing, methylation-specific PCR, and MethyLight assay (Eads et al. 2000; Fraga and Esteller 2002; Herman et al. 1996a; Egger et al. 2004). The presence of certain hypermethylated CDH13, MYOD1, MGMT, $p16^{INK4a}$, and RASSF1A genes varies significantly in different tumour types. These changes can be identified in plasma DNA and body fluids. For detecting bladder cancer, these genes can be detected in urine sediments (Egger et al.

Table 14.2 List of important hypomethylated genes in various human cancers (Pogribny 2010)

Tumour types	Genes
Breast	BCSG1, CDH1, NAT1, FEN1, SNCG, PLAU, CAV1, ZEB1, MUC3A, IL-10,
Uterus	S100A4, PAX2, DNMT3L, CAGE, ESR1
Lung	MUC3A
Ovarian	CLDN4, HNF1B, BORIS
Pancreatic	SERPINB5, TFF3, CLDN4, LCN2, MUC3A
Prostate	HPSE2, PLAU
Thyroid	SERPINB5
Kidney	CA9
Melanoma	TIMP1
Lymphoma	TNFRSF8, LGALS7
Leukaemia	BCL2, TCL1, PRAME, DDX43
Head-Neck	TKTL1
Stomach	MAGEA1, MAGEA3, XAGE1, CCND2, SERPINB5, MUC2
Colon	S100A4, CYP2W1, CDH3, BAGE, DCN, MAGEA1, MUC3A
Wilms tumours	GLIPR1/RTVP1

2004). Furthermore, combined hypermethylation assays for few numbers of genes such as RASSF1A, RARB2, APC, and GSTP1 aid in distinguishing between cancerous and benign changes in the prostate. Moreover, aberrant methylation is more common in the gene promoter and is easier to detect than the presence of mutations (Mulero-Navarro and Esteller 2008). These advantages enable investigators to use DNA methylation as a prognostic and diagnostic biomarker in many tumour types.

14.4 Overview of Methods for DNA Methylation Study in Human Cancer Detection

Several tools and techniques have been discovered for DNA methylation studies such as bisulfite PCR, pyrosequencing, MS-PCR, MethyLight assay, MBD capture assay, microarray, targeted deep-amplicon bisulfite sequencing, DNA methylation quantification by mass spectrometry, and method based on NGS. These methods can be further classified on the basis of genome-wide methylation profiling and gene-specific methylation study (Fig. 14.2).

14.4.1 Bisulfite PCR

Bisulfite PCR is among most popular technique used in DNA methylation study, and it has been considered as gold standard technique for studying DNA methylation because of its qualitative, quantitative, and coherent approach to discriminate

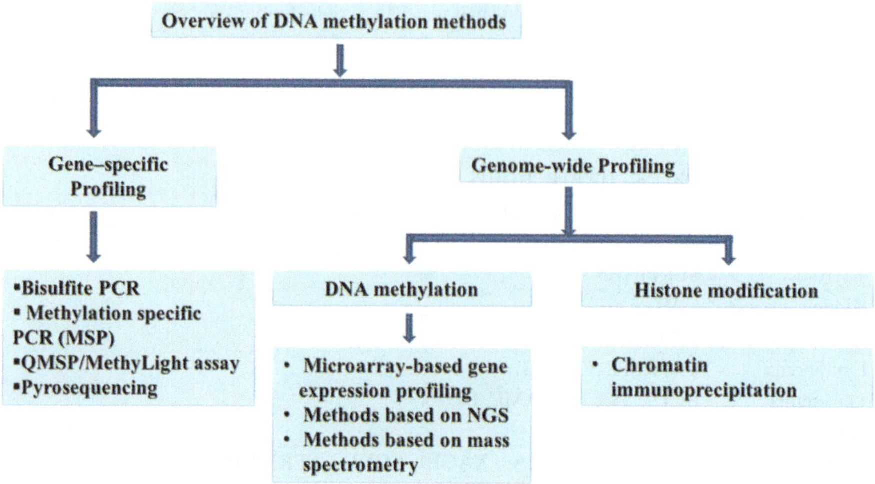

Fig. 14.2 Flowchart of techniques used in DNA methylation study

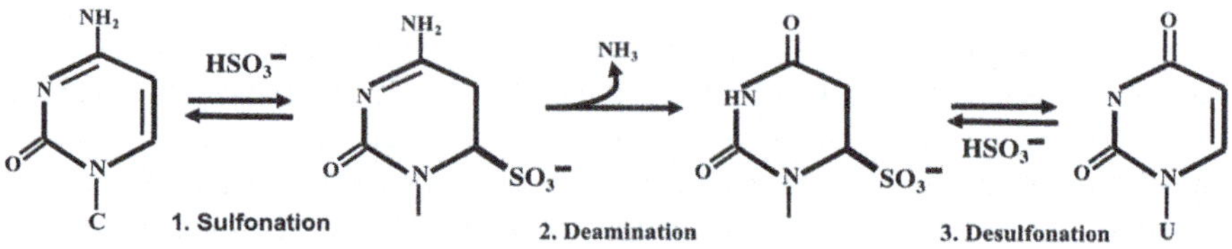

Fig. 14.3 Step towards deamination of cytosine to uracil using bisulfite method: (**a**) the first step is sulfonation of cytosine; (**b**) the second step leads to deamination; (**c**) in the third step, desulfonation of cytosine in alkaline pH leads to conversion of cytosine to uracil

between normal and 5-methylcytosine (5mC) at single base pair resolution. Method was initially introduced by Formmer et al.; they used sodium bisulfite to treat single-stranded DNA containing cytosine and 5-methylcytosine (5mC), and treatment with sodium bisulfite led to deamination reaction, which convert cytosine residues to uracil, while 5mC remains constant after this treatment (Fig. 14.3) (Mulero-Navarro and Esteller 2008; Shapiro et al. 1974; Hayatsu et al. 1976). Considering above mechanism, signature methylation pattern on DNA sequences is being identified. The method involves two steps; in the first step, total genomic DNA is denatured by alkali treatment and further treated with sodium bisulfite under such condition where cytosine is stoichiometrically converted to uracil, while 5mC remains nonreactive in reaction. The second step involves PCR amplification along with sequencing of PCR amplicons. PCR amplification requires two sets of methyl-sensitive primers followed by sequencing of obtained PCR products. Methylation status of sequences can be retrieved by two methods either direct sequencing of PCR product or sub-cloning of PCR product, followed by sequencing. The next step after bisulfite

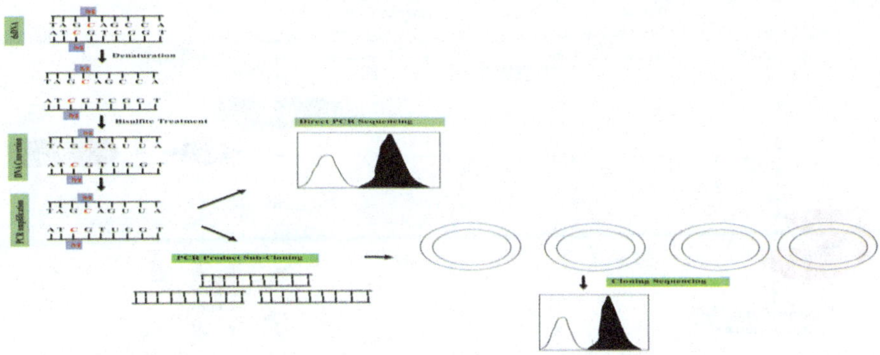

Fig. 14.4 Principle behind bisulfite-mediated methylation analysis involves conversion of cytosine residues into uracil, while methylated cytosine remains unaffected by conversion process. Further, PCR amplification coverts all uracils into thymines. DNA methylation is further predicted using direct PCR sequencing or cloning sequencing

PCR amplification or sub-cloning sequencing procedure is to identify methylation status of sequenced DNA. To identify DNA methylation status, sequenced DNA is compared with reference sequence on UCSC genome browser (Fig. 14.4).

All unmethylated cytosines (C) convert to thymine (T), while presence of methylated C (5mC) would give rise to a sharp peak. However, sometimes C- and T-peaks appear simultaneously, and it means incomplete bisulfite conversion or there is presence of partial methylation in DNA.

14.4.2 Pyrosequencing

Pyrosequencing technique is based on sequencing-by-synthesis method; it is designed to interrogate single-nucleotide polymorphism (SNP). However, bisulfite conversion of DNA allows us to measure DNA methylation locally and globally in real time. Methylation analysis using pyrosequencing is based on treatment of DNA sequence with sodium bisulfite which converts cytosine into uracil, while methylated cytosine (5mC) remains unaffected by treatment (Fig. 14.3). Further treated DNA is subjected to pyrosequencing. Pyrosequencing is based on sequencing-by-synthesis method, where single modified nucleotide is added at a time, incorporation of modified nucleotide in strands gives rise to bioluminometric signal which is detected by CCD camera, and remaining nucleotides are degraded prior to next nucleotide dispensation (Fig. 14.5).

Enzymatic cascade is required for the performance of pyrosequencing which includes template (single-stranded PCR amplicon), four different enzymes, and two substrates including DNA polymerase, ATP sulfurylase, luciferase and apyrase and adenosine 5′-phosphosulfate (APS), and luciferin (Ronaghi et al. 1998). Pyrosequencing-based methylome analysis interrogates degree of methylation at

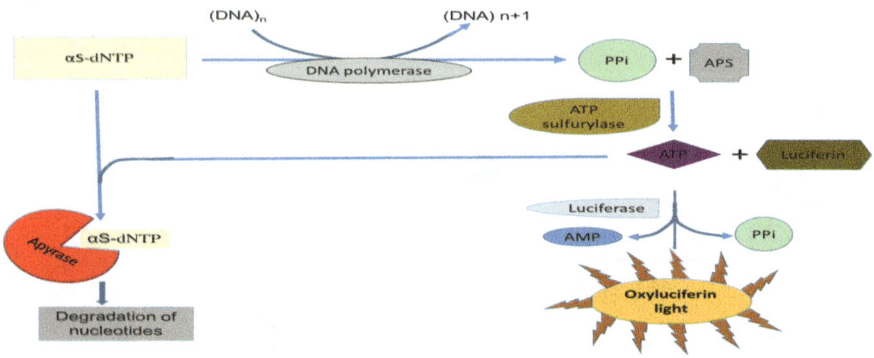

Fig. 14.5 Enzymatic events during pyrosequencing. Single nucleotide is added at a time by exonuclease-deficient DNA polymerase-I and a PPi is released during reaction which is converted into ATP by the enzyme ATP sulfurylase. Adenosine 5′ phosphosulfate is used as substrate for ATP sulfurylase. This reaction brings energy for luciferase to oxidize D-luciferin. Oxidation reaction produces oxyluciferin in excited state and its decay to ground level releases a photon (lambda = 560 nm) which is detected by an inbuilt charge-coupled device (CCD) camera. As dATP could act as substrate for luciferase, modified nucleotide αS-dATP is being used for extension. During reaction, unutilized dNTPs are degraded by apyrase before the next round of dNTPs addition

CpG sites after bisulfite treatment of DNA. Data is recorded and presented in the form of pyrogram (Bird et al. 2002; Dupont et al. 2004; Tost et al. 2003, 2006). Pyrogram plot is basically Relative Light Units (RLU) versus timescale (time taken by nucleotide dispensation). Percentage value (at the top of pyrogram) indicates average methylation level of respective CpG site in all sequenced PCR products under investigation. Pyrosequencing reaction would be termed as good if its peak is sharp, distinctive, and reaching at expected height in pyrogram. However, a bad quality pyrogram has decreased peak heights as shown in Fig. 14.6 (Tost et al. 2007).

14.4.3 Methylation-Specific PCR (MSP)

Methylation-specific PCR was first illustrated by Herman et al. in 1996. Methylation-specific PCR is an application of bisulfite sequencing method (Herman et al. 1996b). Its ability to differentiate between methylated from unmethylated CpG sites makes it very sensitive and specific for detection of any single block of CpG sites in stretch of CpG island. During MSP, DNA is initially denatured and treated with bisulfite (Fig. 14.3). Further, this modified DNA having signature methylated cytosine is amplified using sets of primer complementary only to the formerly methylated or unmethylated alleles (Licchesi et al. 2009). MSP can be interpreted based on PCR amplification of bisulfite-treated DNA and its expected size seen on agarose gel electrophoresis. Methylation status of sequence is dependent upon primer pair used, i.e. methylated or unmethylated. Limitation associated with MSP

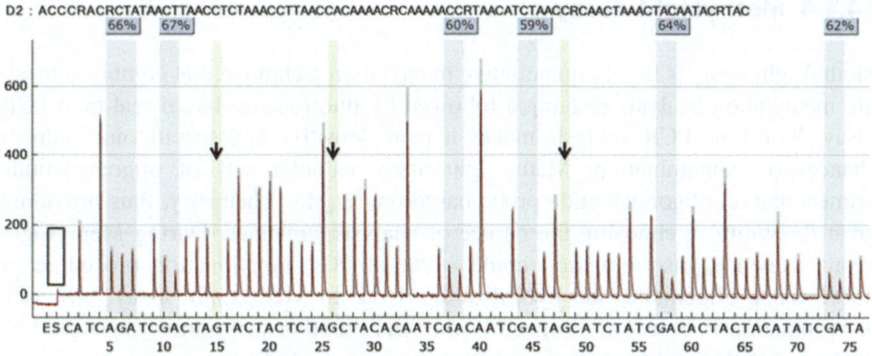

Fig. 14.6 Adopted from Ulrich Lehmann et al. (2015). This figure shows a high-quality assay, having distinguishable and sharp peak with expected height (exceeding 25 relative light units) (red line). Rectangular box at the beginning represents substrate peak, while each CpG site is represented with gray bar having numerical value

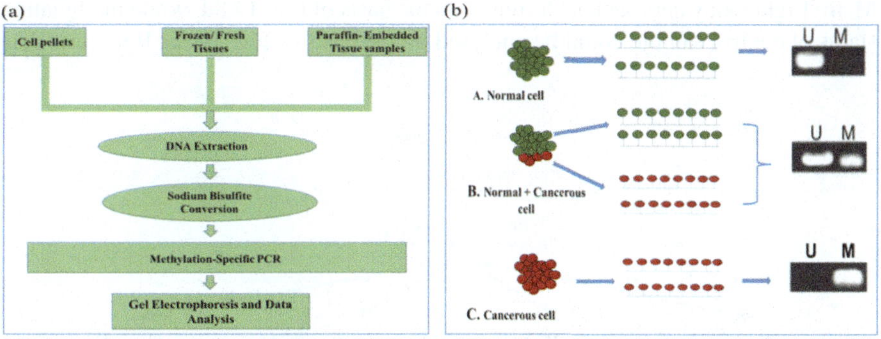

Fig. 14.7 MSP-based result and their interpretation. (**a**) Flow chart of methodology of MSP (**b**) Sample is composed of normal cell only. Normal cells usually have promoter region that is unmethylated in nature for tumor suppressor gene. So unmethylated primers have been raised for MSP analysis. When sample is composed of small amount of cancer and large proportion of normal cell, two primer sets one is for methylated and second is for unmethylated tumor suppressor gene are raised

is the low quantity and poor quality of starting DNA material (e.g. paraffin-embedded specimens). To overcome this problem, extension of MSP is developed and named as NM-MSP; in this technique, first round of amplification is done using primers unbiased towards the methylation status of gene followed by conventional MSP (House et al. 2003; van Engeland et al. 2003). Since the sensitivity of technique is largely dependent upon the designing of primer, MSP primers can be designed either manually or using available online tools such as MSP primer (http://www.mspprimer.org/cgi-mspprimer/design.cgi) (Fig. 14.7).

14.4.4 MethyLight Assay

MethyLight assay is highly quantitative methylation technique that combines bisulfite methylation analysis technique followed by fluorescence-based real-time PCR assay. Real-time PCR analysis makes it more sensitive and specific and reduces chances of contamination. MethyLight assay includes sets of oligonucleotide primers and an oligonucleotide probe based on TaqMan chemistry, thus providing great flexibility in choosing the choice of analysis strategies. During MethyLight assay sequence, discrimination could occur at PCR amplification procedure or fluorescence oligonucleotide hybridization level. In MethyLight assay, sequence discrimination is based on designed primer and probe or just primer that would be utilized in PCR amplification (Fig. 14.8.) (Eads et al. 2000).

After bisulfite treatment, fluorescence-based PCR is performed with primer and probe that either overlaps CpG sites or does not overlap CpG sites. MethyLight assay has four basic applications presented in Fig. 14.8. MethyLight assay is compatible with DNA of various qualities, and it can analyse genomic DNA isolated from fresh tissue, formalin-fixed, paraffin-embedded sections (Eads et al. 2001). Most of the MethyLight assay approach is designed on the basis of Fig. 14.8d, while methylation status of control sample could be analysed on the basis of Fig. 14.8a. Real-time data

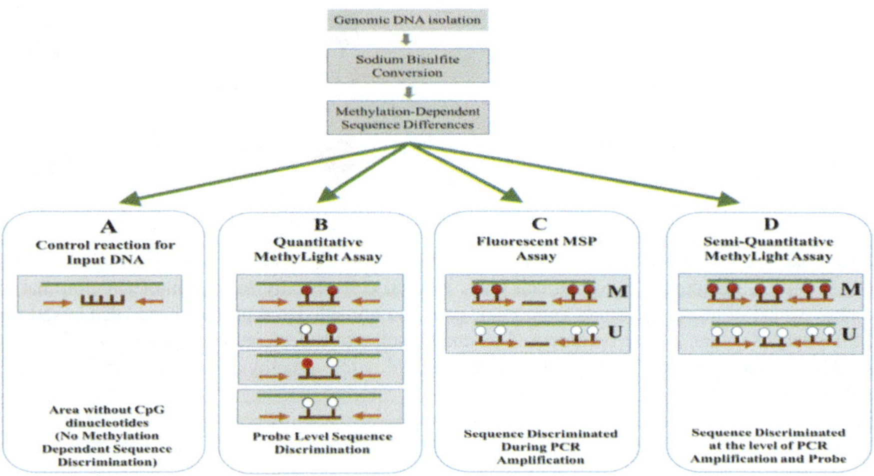

Fig. 14.8 Schematic representation of theoretical basis of MethyLight assay. In application (**a**) primer and probe do not have overlapping CpG sites, so PCR amplification is independent of methylation-dependent sequence discrimination; here sequence will be amplified without bias to methylation status and can serve as control for the input DNA. In application (**b**) probe is raised against methylated CpG sites that lead to sequence discrimination. Designing of probe is dependent upon methylation pattern of each sequence variant, in case of Fig. 14.8b two CpGs are illustrated so it requires $2^2 = 4$ probes that can be used in quantitative version of MethyLight technology. In case of (**c**) and (**d**), methylation-dependent sequence discrimination occurs at PCR amplification level. In case of (**c**) primer is designed to overlap or not overlap the methylation site, while in case of (**d**) primers and probe both are designed to overlap or not overlap the sequence

would be presented in the form of CT value, and it is entirely dependent upon initial amount of DNA template, capability of probe hybridization, primer amplification during the PCR, and release of the reporter dye. The CT is directly related to the methylation status and pattern of methylation present in original genomic DNA sample. Further absolute or relative quantification can be done to calculate PMR (Percentage of methylated reference) (Olek et al. 1996).

14.4.5 Methods Based on Next-Generation Sequencing

Next-generation sequencing (NGS), deep sequencing, or massively parallel sequencing is described as DNA sequencing platform used for whole-genome analysis. NGS platforms are used for whole-genome sequencing, and it requires short period of time to do so (Yong et al. 2016). NGS method of sequencing revolutionized translational biology and could be used for the study of genomic sequences and their alteration that cause mutations. Nowadays, extension of NGS platform is being used for the study of epigenetic landmark in genome (Laird et al. 2010). Recent understanding of NGS and its comprehensive data analysis make it more suitable for studying DNA methylome analysis (Anandhakumar et al. 2015). Several NGS-based techniques have been devised over decades for studying methylation pattern of genome. Techniques which are frequently used in genome-wide analysis of methylation are depicted in Table 14.3, while evolutionary overview of NGS-based methods for detection of epigenetic changes in DNA is depicted in Fig. 14.9.

Table 14.3 Comparative analysis of different genome-wide approaches for DNA methylation profiling

	Restriction enzyme-based methods	Affinity enrichment-based method	Bisulfite conversion-based method
Resolution	Single base	~ 150 bp	Single base
Reads/sample	~10 million reads	~30–50 million reads	>500 million reads
CpG covered	~2 million CpGs	~23 million CpGs	>28 million CpGs
Pros	Sensitivity is higher with low cost	No mutations introduced Low cost	Evaluation of every CpGs status
Cons	CpGs without restriction sites are not evaluated	Only hypermethylated regions are evaluated Can't predict absolute methylation	Cost is higher DNA input amount is higher DNA degradation after BS- treatment
Application	Site-specific nature and targeted study	Used for rapid, large-scale, and low-resolution study	Used only for large-scale study

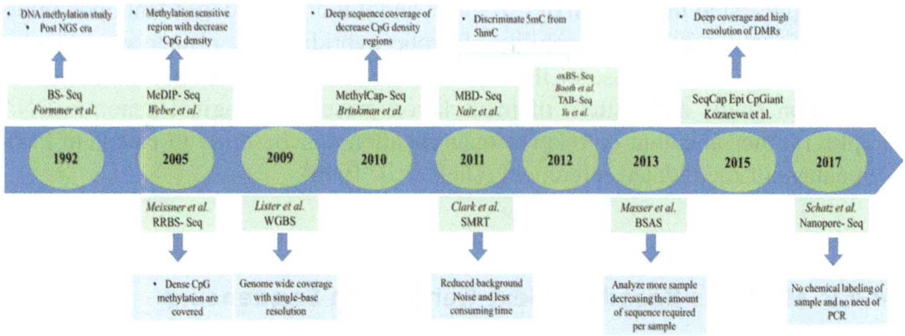

Fig. 14.9 Year-wise development of NGS-based techniques for DNA methylation profiling. BS sequencing, bisulfite sequencing; MeDIP-sequencing, methylated DNA immunoprecipitation sequencing; RRBS-seq, reduced representation bisulfite sequencing; WGBS, whole-genome bisulfite sequencing; MethylCap-seq, methylation capture sequencing; MBD-seq, methyl-CpG binding domain sequencing; oxBS-seq, oxidative bisulfite sequencing; TAB-seq, TET-associated bisulfite sequencing; BSAS, bisulfite amplicon sequencing

At different time spans, several sequencing techniques have been developed on the basis of above approaches (Table 14.3) for the analysis of DNA methylation. On the basis of affinity enrichment-based approach, MeDIP (Methylated DNA immunoprecipitation) and MBD (Methyl binding domain) sequencing have been developed, while WGBS (whole-genome bisulfite), and BSAS (bisulfite amplicon sequencing) are based on bisulfite conversion-based method; other methods like oxidative bisulfite sequencing and MethylCap-Seq are also being used for methylation analysis. Each method has its own advantage over others, and choice of method is entirely dependent upon experimental design and cost-effectiveness.

14.4.5.1 MeDIP Sequencing

MeDIP sequencing is a combination of methylated DNA immunoprecipitation and next-generation sequencing for the study of whole-genome methylation or studying given region of sequence. MeDIP is used to detect methylated cytosine (C) in mC and mCG; however, it can also detect hydroxyl methylcytosine (mCG), 5-formylcytosine, and 5-carboxylcytosine present in genome (Laird et al. 2010; Ficz et al. 2011; Pastor et al. 2011). In MeDIP sequencing, antibodies are raised against mC and mCG to precipitate DNA; further, these precipitated DNA fragments are sequenced using NGS platform. Antibodies against mC and mCG have equal specificity, so it offers unbiased and mutation-free analysis. MeDIP in combination with next-generation sequencing provides high-resolution methylome ranging from 100 to 300 bp. Several advantages are associated with MeDIP sequencing such as potential target (mC, mCG, or hmC) in whole genome or any region of interest, cost-efficiency and requirement of low-input DNA (Taiwo et al. 2012) (Fig. 14.10).

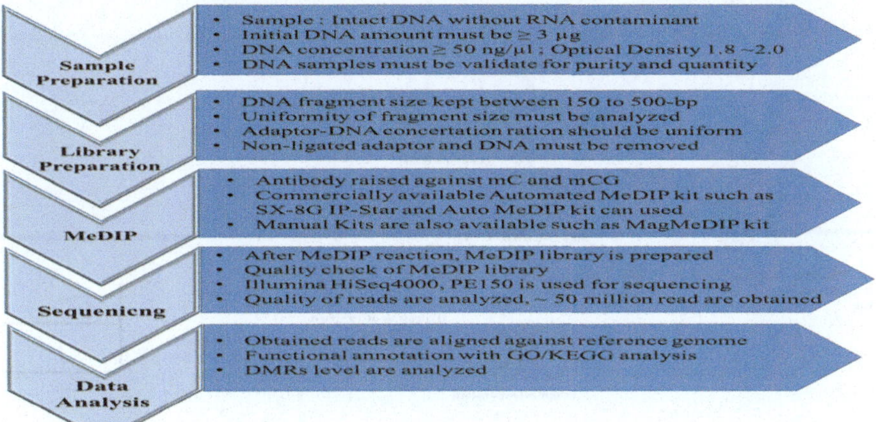

Fig. 14.10 Flowchart of methodology involved in MeDIP sequencing

14.4.6 DNA Methylation Quantification by Mass Spectrometry

DNA methylation study is currently based on three main techniques such as bisulfite modification, methylation-sensitive restricted enzymes, and affinity capture base, i.e. using methyl DNA-binding proteins (Gravina et al. 2015; Liu et al. 2013a; Cheow et al. 2015). However, these methods are labour-intensive and time-consuming. Recent development in mass spectrometry (MS) made it indispensable platform for study molecular and cellular biology with great specificity and sensitivity. MS offers accurate determination of molecular weight and structural information DNA/RNA (Angel et al. 2012; Alley et al. 2013). Mass spectrometry-based analysis of DNA/RNA oligomers has been explored recently, providing great sensitivity and specificity towards characterization of DNA/RNA damage, adducts, and epigenetic modification such as methylation of DNA (Wang et al. 2011; Liu et al. 2015; Chen et al. 2011; Fu et al. 2015). Several methods have been devised based on mass spectrometry for the study of whole-genome methylome analysis as well as targeted DNA methylation analysis. These include MS-TFA (mass spectrometry-based targeted fragmentation assay), MALDI-MS-based DNA methylation analysis, and liquid chromatography-electrospray ionization tandem mass spectrometry (LC-ESI-MS/MS) (Liu et al. 2013b, 2015; Tost et al. 2012). Out of these, some of the mass spectrometry-based quantifications of DNA methylation are illustrated below.

14.4.6.1 Mass Spectrometry-Targeted Fragmentation Assay

Mass spectrometry-targeted fragmentation assay (MS-TFA) is highly sensitive and accurate method for profiling methylation status of selected fragment. This method is based on selective cleavage of genomic DNA by restriction enzymes. Cleaved targeted fragment of DNA is further selectively captured using magnetic beads.

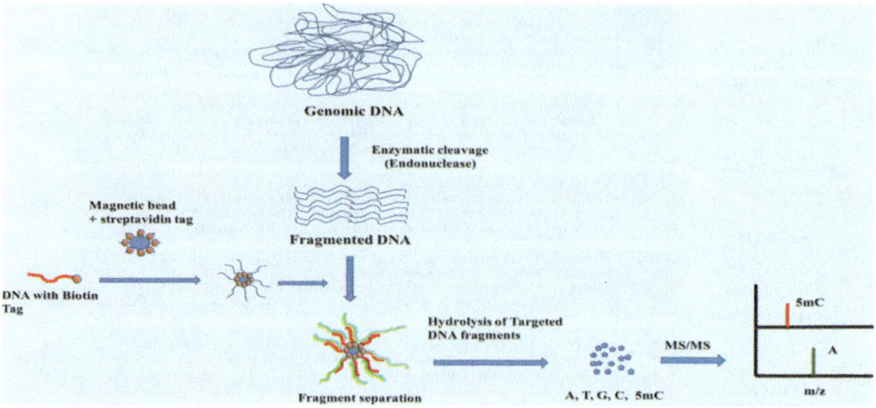

Fig. 14.11 Diagrammatic illustration of mass spectrometry-targeted fragmentation assay

Captured target fragments are further non-enzymatically degraded hydrolysed and are detected with highly sensitive technique, i.e. mass spectrometer (Fig. 14.11).

A unique feature of MS-TFA method is profiling of DNA methylation of specific DNA fragment. First step of MS-TFA is isolation of whole-genomic DNA and its cleavage into fragments using specific endonuclease enzymes. The first step ensures specificity of assay only by obtaining specific targeted fragment for downstream analysis. Use of specific endonuclease is the insurance of specific target ending with predefined composition of sequence. The next step is to capture DNA fragment using magnetic bead containing streptavidin-biotin probe. Capture DNA is collected and further processed for DNA methylation profiling. Collected DNA fragments are completely degraded into nucleobases using formic acid (Tang et al. 2012). In order to profile DNA methylation, degraded nucleobases are directly profiled using MS/MS analysis furnished with a nano-ESI as an ionization source. Result interpretation is based on peak intensity of methylated cytosine (5mC).

14.4.7 Methylation-Specific Oligonucleotide Microarray (MSO-Microarray)

Several devised variants of microarrays came into existence since its first use, and whole-genome methylation analysis can be also performed using microarray. Recent developments such as MSO-microarray (Methylation-Specific oligonucleotide) are best suited for high-throughput methylation analysis. It is based on bisulfite conversion of DNA, amplified using PCR, and hybridized on-chip containing short oligonucleotide of corresponding methylated and unmethylated alleles. Primers used for amplification are labelled with fluorescent dye. Here methylation analysis is based on sequence differences that would discriminate during annealing and hybridization. The general procedure for the methylation-specific oligonucleotide microarray (MSO) is given in Fig. 14.12. Microarray slide is prepared, which contains probes

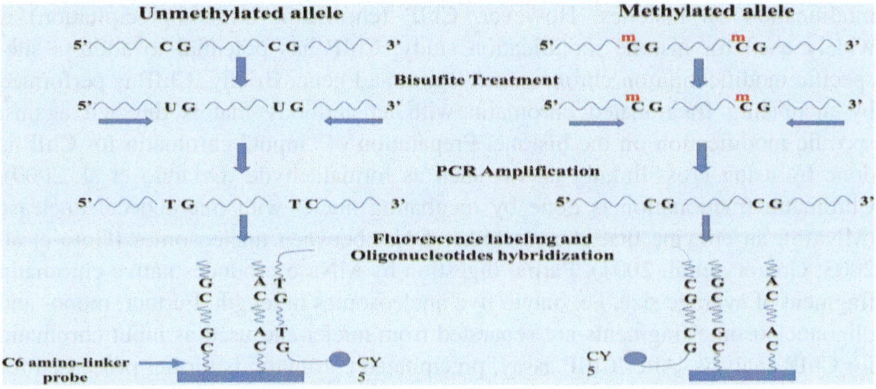

Fig. 14.12 Schematic illustration MSO-microarray-based DNA methylation analysis. Genomic DNA is treated with bisulfite and amplified by PCR using specific primer for CpG island. PCR-amplified product is labelled with fluorescence dye Cy5 and hybridized on array containing oligonucleotide probes. Two conditions are mentioned here: first one is for unmethylated allele, where oligonucleotide probes are designed to matched with unmethylated allele. However, in second condition, probes attached to array are designed to matched with methylated alleles

for methylated and unmethylated alleles of interest. Methylation analysis can be done by treating wild-type DNA with bisulfite. This modified DNA is taken as template for amplification of region of interest. Further amplified products of genes and CpG are purified and labelled with fluorescence dye and hybridized on microarray panel containing probes for them. After hybridization of target DNA on array, washing is carried out so that unbound DNA would flush out from the analysis (Gitan et al. 2018). Array is then exposed to laser beam, and fluorescence emission patter is recorded. Intensity of fluorescence would tell us about extent of methylation in particular gene or CpG fragment in target DNA. MSO-microarray has advantage of analysing large number of CpG sites that could be probed and interrogated for methylation with nanogram of DNA (Hacia et al. 1999).

14.5 Method for Histone Modification Study

Histone molecules have N-terminal (histone tails) globular domain protruding outside, and it has several functions (molecular memory, maintain integrity of nucleosome, act as docking site for proteins) (Strahl et al. 2000; Brower-Toland et al. 2005). Histone tails are prone to various modifications such as acetylation, methylation, phosphorylation, sumoylation, and ADP ribosylation, while ubiquitination takes place at C-terminus (Jason et al. 2002; Robzyk et al. 2000). However, covalent modifications such as methylation, acetylation, and phosphorylation have been studied widely. Histone modifications such as histone H3 lysine 9 acetylation and histone H3 lysine 9 trimethylation are believed to separate chromatin into active and inactive domains. Several approaches have been used to identify covalent modification on histones. However, ChIP (chromatin immunoprecipitation) is widely used for histone modification study. ChIP has potential to address site-specific modification on chromosome region and gene. Briefly, ChIP is performed by incubating fractionated chromatin with an antibody that is directed against specific modification on the histone. Preparation of "input" chromatin for ChIP is done by using cross-linking agents such as formaldehyde (Orlando et al. 2000). Chromatin fractionation is done by incubating nuclei with micrococcal nuclease (MNase), an enzyme that cleaves linker DNA between nucleosomes (Goto et al. 2003; Gregory et al. 2001). Partial digestion by MNase produces native chromatin fragment of average size, i.e. one to five nucleosomes in length. Further, mono- and oligonucleosome fragments are separated from nuclei and used as input chromatin for ChIP analysis. After ChIP assay, precipitated chromatin is further processed for DNA isolation, and qPCR or allele-specific PCR is used to find binding site of protein. qPCR-based determination of DNA sequences is difficult when target proteins are bound to it. To overcome this, primers are designed flanking to the potential regulatory region on DNA target. If candidate target gene or potential sites of interest are not available, then ChIP-on-chip and ChIP sequencing analysis can be the method of choice. Both are high-throughput techniques that provide huge data that could be used to evaluate genome-wide sequence (Tae Hoon et al. 2018) (Fig. 14.13).

14.6 Overview of Sample Consideration for DNA Methylation Detection

Association of aberrant DNA methylation with numerous diseases (Portela and Esteller 2010) and expanding research in its underlying mechanism along with technological advances has led to the identification of several disease-specific methylated loci (How Kit et al. 2012). Conversely, huge variation in the methylation frequencies has been reported among studies, for the same locus in same disease type as well as in related-disease subtype composition, emphasizing the need to limit technical inconsistency in DNA methylation analysis. For instance, promoter methylation frequencies varied from 19% to 97% for MGMT in gliomas, across several published reports (Havik et al. 2012). Additionally, this trend of divergence with respect to methylation frequencies has been also reported for various genes in several cancer types undertaking similar methylation analysis method, warranting the call for standardization (Xiong et al. 2014; Shi et al. 2015; Jiang et al. 2014). Here, we discuss the general considerations for samples undertaken in studies as well as the technical parameters which are important for consistent methylation results. Moreover, careful interpretation of result is also essential, as a certain level of variation in methylation data is much probable.

Tissue Samples Proper preservation of resected tissue sample is a crucial parameter. Deep freezing (either liquid nitrogen storage or − 80 °C deep freezers) of resected tissues is the best preservation method for DNA methylation analysis.

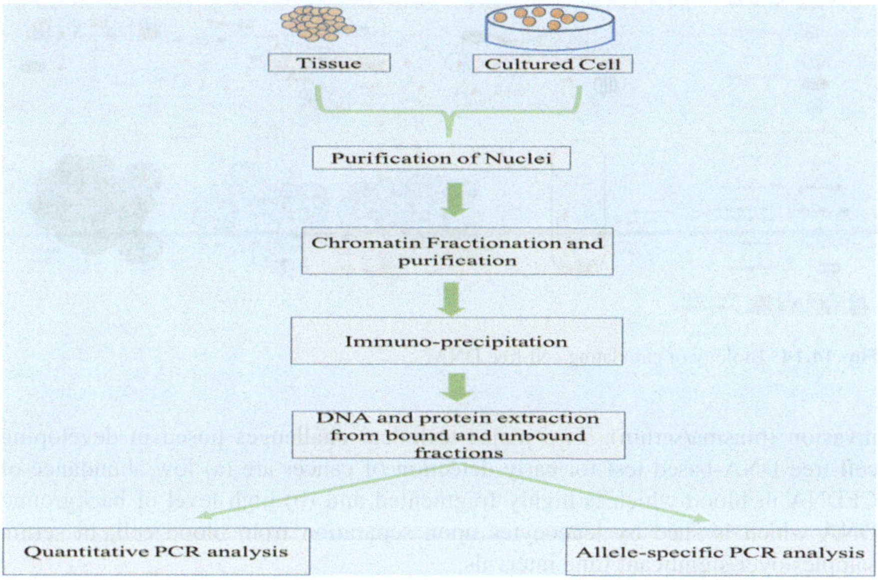

Fig. 14.13 Procedure used for studying site-specific covalent modification on histone. In brief, nuclei are extracted from fresh/frozen tissue sample or cell culture. Further, chromatin is fractionated using micrococcal nuclease and purified from the nuclei. Fractionated chromatin is comprised of fragments containing five nucleosomes in length; further, isolated fractions are incubated with antibody (against specific histone modification). Further, bound fraction is separated from unbound fraction and DNA is isolated from bound fragment and further subjected to quantitative and qualitative PCR for identification of gene or chromosomal region of interest

Formalin fixation maintains the morphological integrity of tissue but degrades the nucleic acid by fragmentation and results in DNA-protein cross-links (Campos and Gilbert 2012). Sample quality is severely influenced by the time of formalin fixation; longer formalin fixation of tissue results in higher degradation of DNA. However, these problems can be overruled by optimized and standardized fixation as well as isolation protocols. DNA fragmentation can be reduced by cold fixation at 4 degrees (Frankel 2012; Bussolati et al. 2011). Utility of reagents that partially reverse the DNA-protein cross-links in protocols for nucleic acid isolation from FFPE samples can substantially aid in extracting high-quality nucleic acids from FFPE samples.

Body Fluids Analysis of aberrant methylation pattern of genes/promoter in DNA fragments shed as a result of apoptosis, necrosis, etc., into the bloodstream, provides highly specific cancer signals (Fig. 14.14). The potential of circulating cell-free DNA in clinical diagnostics and management of several cancer types has been reported in numerous studies (Swarup and Rajeswari 2007; Widschwendter et al. 2017; O'Driscoll 2007; Salvi et al. 2016; Leygo et al. 2017). Body fluids can be obtained noninvasively (saliva, sputum, and urine) or through minimum

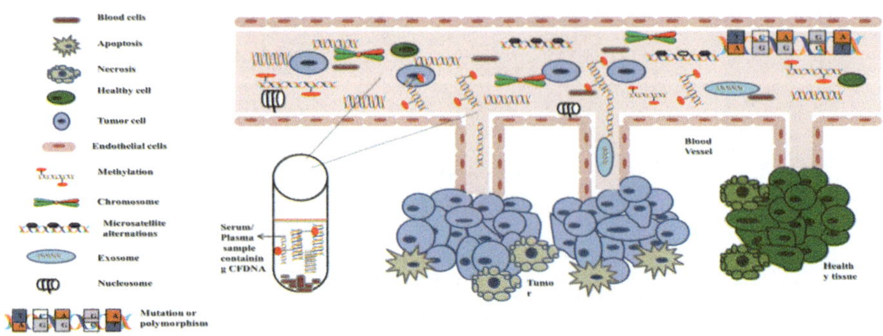

Fig. 14.14 Biology of circulating cell-free DNA

invasion (plasma/serum). Two major technical challenges posed in developing cell-free DNA-based test for early detection of cancer are (a) low abundance of CFDNA in blood which is highly fragmented and (b) high level of background DNA which is shed by leucocytes upon separation from blood cell, in serum samples over significant time intervals.

14.7 Introduction to Cell-Free DNA Biology for Epigenetic Biomarker-Based Cancer Detection

Advancement in the knowledge of molecular pathogenesis of cancer, along with the development of new molecular techniques, has facilitated the study of molecular alternations associated with cancer development at an early stage in body fluids. Circulating cell-free DNA which are believed to have derived from tumour cells reflect specific genetic and epigenetic alternations, and thus might serve as a potential blood-based biomarker for several cancers capable of providing valuable information regarding disease progression and response to therapy in real time.

To date, there is no standardized protocol for CFDNA extraction (Jahr et al. 2001; Holdenrieder et al. 2008; Su et al. 2004). Major challenge in extracting highly sufficient amount of CFDNA from liquid biopsies is faced because of high sample volume and low amount of short-sized fragmented DNA. Aberrant DNA methylation markers detected in serum/plasma/saliva originate from anywhere in the body, while those detected in urine/sputum are site-specific which is another important aspect to be considered. Therefore, markers which are identified in serum/plasma/saliva must be specific for a single or small group of disease enabling identification of site of malignancy for diagnostic point of view. Recently, based on literature and confirmatory experiments, various pre-analytical parameters for optimal blood sample handling prior to cell-free DNA extraction were defined by Messaoudi et al. (Table 14.4) (El Messaoudi et al. 2013).

Table 14.4 Guidelines for optimal pre-analytical blood sample handling prior to cfDNA isolation

1	Utility of plasma samples over serum to avoid contamination of gDNA from blood cells
2	Use of EDTA or cell-free DNA collection tubes for blood sample collection in order to prevent blood cell lysis
3	In order to retain DNA concentration and integrity, blood sample should be processed within 4 h after withdrawal
4	To remove any remaining cell, the second high-speed centrifugation should be included after the first blood sample centrifugation
5	To avoid freeze-thaw cycle, serum/plasma samples should be stored in aliquots. To preserve the DNA integrity, samples should be subjected to minimum freeze/thaw cycles
6	Serum/plasma samples should be stored at −80 °C and should be processed for DNA extraction within 9 months after sampling to preserve DNA integrity

14.8 Established Epigenetic Biomarker for Cancer Detection

The utility and clinical implementation of DNA methylation biomarkers for early screening and diagnosis, disease prognosis, and prediction is being currently tested and validated in various studies. Methylation of septin 9 (SEPT9) and vimentin (VIM) is used for early detection of colon cancer by analysing blood (SEPT9) or stool (VIM) samples of patients with improved sensitivity and specificity (deVos et al. 2009; Itzkowitz et al. 2008). Similarly, malignant and benign lung can be distinguished with the differential methylation of SHOX2 gene with a sensitivity of 78% and a specificity of 96% in bronchial aspirates (Kneip et al. 2011). The methylation of TWIST-1 and NID-2 along with other biomarkers is used to detect bladder cancer (Van Neste et al. 2012) and is associated with recurrence of bladder cancer (Nakayama et al. 2003). MGMT gene (encodes a DNA repair protein, O6-methylguanine DNA methyltransferases) methylation has been reported to be associated with survival benefit of glioblastoma patients after treatment with the temozolomide, thus highlighting its predictive potential in clinical settings (Hegi et al. 2005; Esteller et al. 2000).

14.9 Limitations

Several limitations in the methylation detection of cell-free serum DNA include extremely low amount of available cfDNA, missing bisulfite conversions as they are usually fragmented, low sensitivity demonstrated by a single marker, and time-consuming, complicated, and expensive conventional techniques for cfDNA isolation. MSP PCR (methylation-specific PCR) which is a bisulfite-conversion-based method is the most commonly used technique for methylation detection. The limitation of bisulfite conversion of cfDNA is the missing DNA. Because of the technical difficulties of DNA methylation analysis, only few DNA methylation-based markers have been identified to date, which apply only to a fraction of gynaecological cancers including breast, ovarian, and endometrial cancers (Lewis et al. 2015; Wittenberger et al. 2014).

Most of the studies undertaken provide proof of concept; however, their further validation in prospective studies on a larger sample subset still remains. Owing to the presence of both tumour and nonmalignant DNA in cell-free DNA, assays with high sensitivities are critically needed. On the contrary, it limits the utility of technologies such as whole-genome/whole-exome next-generation sequencing. Short, highly fragmented nature of cfDNA might also complicate further molecular analysis.

Rapid and point-of-care diagnostic applications of DNA methylation in liquid biopsy has been precluded as a result of cumbersome sample preparation with complicated conventional methods of isolation. New technologies which allow rapid identification of methylation signatures directly from blood will facilitate sample to answer solutions, thereby enabling next-generation point-of-care molecular diagnostics. Until addressed, clinical implementation is a far way to go.

References

Alley WR et al (2013) High-sensitivity analytical approaches for the structural characterization of glycoproteins. Chem Rev 113:2668–2732

Anandhakumar C et al (2015) Advancing small-molecule-based chemical biology with next-generation sequencing technologies. Chembiochem 16(1):20–38

Angel TE et al (2012) Mass spectrometry based proteomics: existing capabilities and future directions. Chem Soc Rev 41:3912–3928

Bird A et al (2002) DNA methylation patterns and epigenetic memory. Genes Dev 16:6–21

Brower-Toland B et al (2005) Specific contributions of histone tails and their acetylation to the mechanical stability of nucleosomes. J Mol Biol 346:135–146

Bussolati G et al (2011) Formalin fixation at low temperature better preserves nucleic acid integrity. PLoS One 6(6):e21043

Campos PF, Gilbert TM (2012) DNA extraction from formalin fixed material. Methods Mol Biol 840:81–85

Chen HC et al (2011) Quantitative analysis of multiple exocyclic DNA adducts in human salivary DNA by stable isotope dilution Nanoflow liquid chromatography–Nanospray ionization tandem mass spectrometry. Anal Chem 83:8543–8551

Cheow LF et al (2015) Multiplexed locus-specific analysis of DNA methylation in single cells. Nat Protoc 10:619–631

Dalgliesh GL et al (2010) Systematic sequencing of renal carcinoma reveals inactivation of histone modifying genes. Nature 463(7279):360

Deaton AM, Bird A (2011) CpG islands and the regulation of transcription. Genes Dev 25(10):1010–1022

deVos T, Tetzner R, Model F et al (2009) Circulating methylated SEPT9 DNA in plasma is a biomarker for colorectal cancer. Clin Chem 55(7):1337–1346

Dupont JM et al (2004) De novo quantitative bisulfite sequencing using the pyrosequencing technology. Anal Biochem 333:119–127

Eads CA et al (2000) MethyLight: a high-throughput assay to measure DNA methylation. Nucleic Acids Res 28(8):e32-00

Eads CA et al (2001) Epigenetic patterns in the progression of esophageal adenocarcinoma. Cancer Res 61:3410–3418

Egger G et al (2004) Epigenetics in human disease and prospects for epigenetic therapy. Nature 429(6990):457

El Messaoudi S et al (2013) Circulating cell free DNA: preanalytical considerations. Clin Chim Acta 424:222–230
Esteller M (2007) Cancer epigenomics: DNA methylomes and histone-modification maps. Nat Rev Genet 8(4):286
Esteller M, Garcia-Foncillas J, Andion E et al (2000) Inactivation of the DNA-repair gene MGMT and the clinical response of gliomas to alkylating agents. N Engl J Med 343(19):1350–1354
Esteller M et al (2007) Epigenetic gene silencing in cancer: the DNA hypermethylome. Hum Mol Genet 16(R1):R50–R59
Ficz G et al (2011) Dynamic regulation of 5-hydroxymethylcytosine in mouse ES cells and during differentiation. Nature 473(7347):398–402
Fraga MF, Esteller M (2002) DNA methylation: a profile of methods and applications. BioTechniques 33(3):632–649
Frankel A (2012) Formalin fixation in the '-omics' era: a primer for the surgeon-scientist. ANZ J Surg 82(6):395–402
Fu LJ et al (2015) Simultaneous quantification of methylated cytidine and adenosine in cellular and tissue RNA by Nano-flow liquid chromatography-tandem mass spectrometry coupled with the stable isotope-dilution method. Anal Chem 87:7653–7659
Gitan RS et al (2018) Methylation-specific oligonucleotide microarray: a new potential for high-throughput methylation analysis. Genome Res 12(1):158–164
Goto Y et al (2003) Differential patterns of histone lysine methylation and acetylation distinguish active and repressed alleles at X-linked genes. Cytogenet Genome Res 99:66–74
Gravina S et al (2015) Single-cell, locus-specific bisulfite sequencing (SLBS) for direct detection of epimutations in DNA methylation patterns. Nucleic Acids Res 43:1–7
Gregory R et al (2001) DNA methylation is linked to deacetylation of histone H3, but not H4, on the imprinted genes Snrpnand U2af1-rs1. Mol Cell Biol 21:5426–5436
Gun-Do K et al (2002) Co-operation and communication between the human maintenance and de novo DNA (cytosine-5) methyltransferases. EMBO J 21(15):4183–4195
Hacia JG et al (1999) Resequencing and mutational analysis using oligonucleotide microarrays. Nat Genet 21:42–47
Havik AB et al (2012) MGMT promoter methylation in gliomas-assessment by pyrosequencing and quantitative methylation-specific PCR. J Transl Med 10:36–49
Hayatsu H et al (1976) In: Cohen WE (ed) Progress in nucleic acid research and molecular biology, vol 16. Academic, New York, pp 75–124
Hegi ME, Diserens AC, Gorlia T et al (2005) MGMT gene silencing and benefit from temozolomide in glioblastoma. N Engl J Med 352(10):997–1003
Herman JG et al (1996a) Methylation-specific PCR: a novel PCR assay for methylation status of CpG islands. Proc Natl Acad Sci 93(18):9821–9826
Herman JG et al (1996b) Methylation-specific PCR: a novel PCR assay for methylation status of CpG islands. Proc Natl Acad Sci U S A 93:9821–9826
Holdenrieder S et al (2008) DNA integrity in plasma and serum of patients with malignant and benign diseases. Ann N Y Acad Sci 1137:162–170
House MG et al (2003) Molecular progression of promoter methylation in intraductal papillary mucinous neoplasms (IPMN) of the pancreas. Carcinogenesis 24:193–198
How Kit A et al (2012) DNA methylation based biomarkers: practical considerations and applications. Biochimie 94:2314–2337
Itzkowitz S, Brand R, Jandorf L et al (2008) A simplified, noninvasive stool DNA test for colorectal cancer detection. Am J Gastroenterol 103(11):2862–2870
Jahr S, Hentze H, Englisch S et al (2001) DNA fragments in the blood plasma of cancer patients: quantitations and evidence for their origin from apoptotic and necrotic cells. Cancer Res 61(4):1659–1665
Jason LJM et al (2002) Histone ubiquitination: a tagging tail unfolds? BioEssays 24(2):166–174
Jiang D et al (2014) Meta-analyses of methylation markers for prostate cancer. Tumor Biol 35:10449–10455
Jones PA (2012) Functions of DNA methylation: islands, start sites, gene bodies and beyond. Nat Rev Genet 13(7):484

Kanwal R, Gupta S (2012) Epigenetic modifications in cancer. Clin Genet 81(4):303–311

Kelly TK et al (2010) H2A. Z maintenance during mitosis reveals nucleosome shifting on mitotically silenced genes. Mol Cell 39(6):901–911

Kneip C, Schmidt B, Seegebarth A et al (2011) SHOX2 DNA methylation is a biomarker for the diagnosis of lung cancer in plasma. J Thorac Oncol 6(10):1632–1638

Laird PW et al (2010) Principles and challenges of genome-wide DNA methylation analysis. Nat Rev Genet 11(3):191–203

Lewis JM, Heineck DP, Heller MJ (2015) Detecting cancer biomarkers in blood: challenges for new molecular diagnostic and point-of-care tests using cell-free nucleic acids. Expert Rev Mol Diagn 15:1187–1200

Leygo C et al (2017) DNA methylation as a noninvasive epigenetic biomarker for the detection of cancer. Dis Markers 2017:3726595

Licchesi JD et al (2009) Methylation-specific PCR. Methods Mol Biol 507:305–323

Lin X-C et al (2015) Mass spectrometry based ultrasensitive dna methylation profiling using target fragmentation assay. Anal Chem 88:1083–1087. https://doi.org/10.1021/acs.analchem.5b04247

Liu S et al (2013a) Quantitative assessment of Tet-induced oxidation products of 5-methylcytosine in cellular and tissue DNA. Nucleic Acids Res 41:6421–6429

Liu J et al (2013b) Liquid chromatography tandem mass spectrometry for the measurement of global DNA methylation and Hydroxymethylation. J Proteomics Bioinform 01(S2):005

Liu S et al (2015) Mass spectrometry for the assessment of the occurrence and biological consequences of DNA adducts. Chem Soc Rev 44:7829–7839

Michaelson-Cohen et al (2011) Genome-wide de novo methylation in epithelial ovarian cancer. Int J Gynecol Cancer 21(2):269–279

Moss TJ, Wallrath LL (2007) Connections between epigenetic gene silencing and human disease. Mutat Res Fundam Mol Mech Mutagen 618(1):163–174

Mulero-Navarro S, Esteller M (2008) Epigenetic biomarkers for human cancer: the time is now. Crit Rev Oncol Hematol 68(1):1–11

Nakayama M, Bennett CJ, Hicks JL et al (2003) Hypermethylation of the human glutathione S-transferase-pi gene (GSTP1) CpG island is present in a subset of proliferative inflammatory atrophy lesions but not in normal or hyperplastic epithelium of the prostate: a detailed study using laser-capture microdissection. Am J Pathol 163(3):923–933

O'Driscoll L (2007) Extracellular nucleic acids and their potential as diagnostic, prognostic and predictive biomarkers. Anticancer Res 27(3A):1257–1265

Okano M et al (1999) DNA methyltransferases Dnmt3a and Dnmt3b are essential for de novo methylation and mammalian development. Cell 99(3):247–257

Olek A et al (1996) A modified and improved method for bisulphite based cytosine methylation analysis. Nucleic Acids Res 24:5064–5066

Orlando V et al (2000) Mapping chromosomal proteins in vivo by formaldehyde- crosslinked-chromatin-immunoprecipitation. Trends Biochem Sci 25:99–104

Pastor WA et al (2011) Genome-wide mapping of 5-hydroxymethylcytosine in embryonic stem cells. Nature 473(7347):394–397

Pogribny I (2010) Epigenetic events in tumorigenesis: putting the pieces together. Exp Oncol 32(3):132–136

Portela A, Esteller M (2010) Epigenetic modifications and human disease. Nat Biotechnol 28:1057–1068

Radman-Livaja M, Rando OJ (2010) Nucleosome positioning: how is it established, and why does it matter? Dev Biol 339(2):258–266

Robzyk K et al (2000) Rad6-dependent ubiquitination of histone H2B in yeast. Science 287(5452):501–504

Ronaghi M et al (1998) A sequencing method based on real-time pyrophosphate. Science 281:363–365

Ryan L et al (2009) Human DNA methylomes at base resolution show widespread epigenomic differences. Nature 462(7271):315

Salvi S et al (2016) Cell-free DNA as a diagnostic marker for cancer: current insights. OncoTargets Therapy 9:6549–6559

Satish S et al (2012) High resolution methylome map of rat indicates role of intragenic DNA methylation in identification of coding region. PLoS One 7(2):e31621

Sellar GC et al (2003) OPCML at 11q25 is epigenetically inactivated and has tumor-suppressor function in epithelial ovarian cancer. Nat Genet 34(3):337

Shapiro R et al (1974) An N-glycosidase from Escherichia coli that releases free uracil from DNA containing deaminated cytosine residues. J Biol Chem 248:4060–4064

Sharma S et al (2010) Epigenetics in cancer. Carcinogenesis 31(1):27–36

Shen L et al (2007) Genome-wide profiling of DNA methylation reveals a class of normally methylated CpG island promoters. PLoS Genet 3(10):e181

Shi H et al (2015) Association between P16(INK4a) promoter methylation and HNSCC: a meta-analysis of 21 published studies. PLoS One 10(4):e0122302

Strahl BD et al (2000) The language of covalent histone modifications. Nature 403:41–45

Su YH, Wang M, Brenner DE et al (2004) Human urine contains small, 150 to 250 nucleotide-sized, soluble DNA derived from the circulation and may be useful in the detection of colorectal cancer. J Mol Diagn 6(2):101–107

Swarup V, Rajeswari MR (2007) Circulating (cell-free) nucleic acids--a promising, non-invasive tool for early detection of several human diseases. FEBS Lett 581(5):795–799

Taby R et al (2010) Cancer epigenetics. CA Cancer J Clin 60(6):376–392

Tae Hoon K et al (2018) ChIP–quantitative polymerase chain reaction (ChIP-qPCR). Cold Spring Harb Protoc 2018(5):pdb.prot082628

Taiwo O et al (2012) Methylome analysis using MeDIP-seq with low DNA concentrations. Nat Protoc 7(4):617–636

Tang Y et al (2012) Cytosine DNA methylation is found in Drosophila melanogaster but absent in Saccharomyces cerevisiae, Schizosaccharomyces pombe, and other yeast species. Anal Chem 84:7249–7255

Tost J et al (2003) Analysis and quantification of multiple methylation variable positions in CpG islands by pyrosequencing. BioTechniques 35:152–156

Tost J et al (2006) Serial pyrosequencing for quantitative DNA methylation analysis. BioTechniques 40:721–722, 724, 726

Tost J et al (2007) DNA methylation analysis by pyrosequencing. Nat Protoc 2(9):2265–2275. https://doi.org/10.1038/nprot.2007.314

Tost J et al (2012) DNA methylation analysis by MALDI mass spectrometry. https://doi.org/10.1002/3527600906.mcb.201100025

van Engeland M et al (2003) Effects of dietary folate and alcohol intake on promoter methylation in sporadic colorectal cancer: the Netherlands cohort study on diet and cancer. Cancer Res 63:3133–3137

Van Neste L, Bigley J, Toll A et al (2012) A tissue biopsy-based epigenetic multiplex PCR assay for prostate cancer detection. BMC Urol 12:16

Wang J et al (2011) Quantification of oxidative DNA lesions in tissues of long-Evans cinnamon rats by capillary high-performance liquid chromatography-tandem mass spectrometry coupled with stable isotope-dilution method. Anal Chem 83:2201–2209

Widschwendter M et al (2017) The potential of circulating tumor DNA methylation analysis for the early detection and management of ovarian cancer. Genome Med 9:116

Wittenberger T et al (2014) DNA methylation markers for early detection of women's cancer: promise and challenges. Epigenomics 6:311–327

Xiong J et al (2014) Association between DAPK1 promoter methylation and cervical cancer: a meta-analysis. PLoS One 9(9):e107272

Yang X et al (2014) Gene body methylation can alter gene expression and is a therapeutic target in cancer. Cancer Cell 26(4):577–590

Yong W et al (2016) Profiling genome-wide DNA methylation. Epigenetics Chromatin 9:26

Part VI

Techniques Related to Disease Diagnosis and Therapeutic

Paper-Based Sensors for Biomedical Applications

15

Mohd Aurif Shergujri, Rabeuj Jaman, Arup Jyoti Baruah, Mrityunjoy Mahato, Davidson Pyngrope, L. Robindro Singh, and Manashjit Gogoi

Abstract

Paper-based sensor is a new analytical technique applicable in healthcare monitoring, environmental monitoring, food quality control, etc. Currently, these technologies are drawing great attention due to ease of fabrication, cost-effectiveness, and their simple, portable, and disposable nature. The inherent qualities of paper, such as availability in large quantities, cheap, lightweight, and biodegradability, are being exploited to develop sustainable devices. The application of paper as a substrate material in litmus paper was dated back to seventeenth century. The semiquantitative detection of glucose in urine and immune chromatographic paper test strips for pregnancy test kits are examples of paper-based sensing devices. Today, microfluidic paper-based sensors with advanced designs involving complex 3D geometrics for handling multiples analytes are under investigation. It is foreseen that advancements in the fabrication and analytical techniques will help in improving the accuracy and sensitivity of paper-based sensors. Different types of detection approaches used in microfluidic paper-based devices are colorimetry, luminescence, electrochemical detection, fluorescence, and surface-enhanced Raman scattering.

M. A. Shergujri · R. Jaman · D. Pyngrope · L. R. Singh
Department of Nanotechnology, North-Eastern Hill University, Shillong, Meghalaya, India

A. J. Baruah
Department of General Surgery, North Eastern Indira Gandhi Regional Institute of Health and Medical Sciences, Shillong, Meghalaya, India

M. Mahato
Physics Division, Department of Basic Sciences and Social Sciences, North-Eastern Hill University, Shillong, Meghalaya, India

M. Gogoi (✉)
Department of Biomedical Engineering, North-Eastern Hill University, Shillong, Meghalaya, India
https://orcid.org/0000-0002-1753-192X

S. Paul (ed.), *Biomedical Engineering and its Applications in Healthcare*,
https://doi.org/10.1007/978-981-13-3705-5_15

This book chapter focuses on different microfluidic paper-based devices used for healthcare applications. The coverage of this chapter is also extended to its current status and future prospects with elaborative and graphical examples.

Keywords
Paper-based sensor · Microfluidic · Lab-on-a-chip · Diagnostic · Analytical sensors

15.1 Introduction

Generally, paper is extensively used for writing, printing, drawing, and packaging. It has been, now, ventured in new applications due to the ease of tuning the physical properties like thickness, weight, and flexibility. Liquid can be transported easily within the cellulose fiber matrix of paper without any active pump or external power source. These properties allow paper to be used in sensor applications. The properties of cellulose like hydrophilicity, permeability, and reactivity could be customized as per requirement (Liana et al. 2012). Recently, paper-based sensors have gained a significant position in analytical fields, due to the high abundance, versatility, inexpensiveness, lightness, and biodegradability of paper. Apart from these, paper as a substrate for analytical devices enjoys the advantages (Von Lode 2005) like ease of patterning into discrete hydrophilic and hydrophobic zones, capability to transport fluids by capillary action, ease of transport, etc. Moreover, the paper-based analytical devices can be integrated as per requirement.

Paper-based analytical devices have been used since mid-seventeenth century, as Boyle introduced litmus paper for determining acid and base. In the year 1956, another paper-based device was developed for semiquantitative detection of glucose in urine (Comer 1956). Later, different lateral flow or assays were developed based on ELISA techniques (Von Lode 2005). These immunoassays composed of a paper strip containing a sample pad (source of sample), reagent pad (contain target antigen-specific antibodies conjugated to a signal indicator), and a test line to capture analytes (Liana et al. 2012). Since the year 2007, these assays or devices have been developed very rapidly due to the development of simple methods for creating microfluidic channels by patterning paper with hydrophobic barriers. The research on microfluidic paper-based analytical devices (μPADs) for environmental analysis has emerged along with the point-of-care diagnostic devices. Today, μPADs are recognized as potentially powerful analytical platforms.

In the recent years, research focus on paper-based sensor has been gradually shifted from basic design concepts to development of extreme fabrication and patterning techniques for better results. Whitesides and co-workers introduced a new procedure for fabricating microfluidic channels in paper for multiple analytes detection. Basically, the detection method was based on colorimetry, where color intensity was proportional to the concentration of the analyte (Martinez et al. 2007). Recent advances in paper-based devices along with fabrication processes help in exploring new detection techniques like optical, electrochemical, MEMS-based, and electrochemiluminescence methods. Each of these techniques has its own advantages as well as disadvantages. (Liana et al. 2012). In the recent years, paper-based devices have been used for performing large-scale and complicated laboratory tests in different areas including clinical, food, and environmental sectors. Applications of paper-based point-of-care (POC) devices could reduce the overall cost in healthcare sector (Liana et al. 2012). In this chapter, paper-based sensors and their fabrication process applications in biomedical engineering are discussed with illustrative examples and figures.

15.2 Fabrication Methods

There are several chemical and physical processes for modification of papers which help in improving the properties of paper as well as enabling them for direct usage in number of applications (Liana et al. 2012). The choice of techniques and materials depend on cost, simplicity, and production efficiency. The common processes for paper-based device fabrication are photolithography, screen printing, inkjet printing, flexography printing, laser treatment, wax printing, paper cutting, etc. These processes are elaborated in the following section.

15.2.1 Photolithography

Photolithography is one of the most common techniques used for microfabrication. In this technique, photomask-guided UV light is used to make a geometric pattern on light-sensitive chemical or photoresist on the particular substrate. Then the patterns are carved out by etching. Complex patterns can be fabricated using multiple steps of deposition and etching. SU-8 is generally used as a patterning photoresist for fabricating paper-based microfluidic devices. The photoresist is selectively polymerized by exposing it to photomask-guided UV light. The unexposed photoresist is then removed by etching to create hydrophilic zones (Li et al. 2012, Martinez et al. 2007).

15.2.2 Screen Printing

Screen printing is widely used for fabricating different electrodes for paper-based devices. In screen printing, a black and white image is created using design software where the black areas indicate hydrophobic features and white areas indicate hydrophilic features or vice versa. This image is transferred to a nylon screen on wooden frame. Then adequate amount of ink has to be added on the framed screen and transferred manually onto the paper (Whatman No #1 filter paper) to create hydrophobic features. The printed papers were dried at room temperature until the solvent evaporate completely (Sun et al. 2015). Wax, carbon ink, chitosan, and graphene are used as ink.

15.2.3 Wax Printing

Wax is a low-cost hydrophobic material and widely used in the fabrication of paper-based analytical devices (Sun et al. 2015). In this process, a solid wax printer prints the features of the sensor onto the paper. Then the wax printed paper sensor is heated using a hot plate or oven. Upon heating, the wax melts and penetrates the paper to form the hydrophobic features in the paper. The fabrication procedure can be completed within a very short period of time at a very low cost in comparison to other procedures. Wax is mostly used because it is cheaper than patterning agent polydimethylsiloxane, i.e. a UV curable photoresist (Govindasamy et al. 2012). (Fig. 15.1)

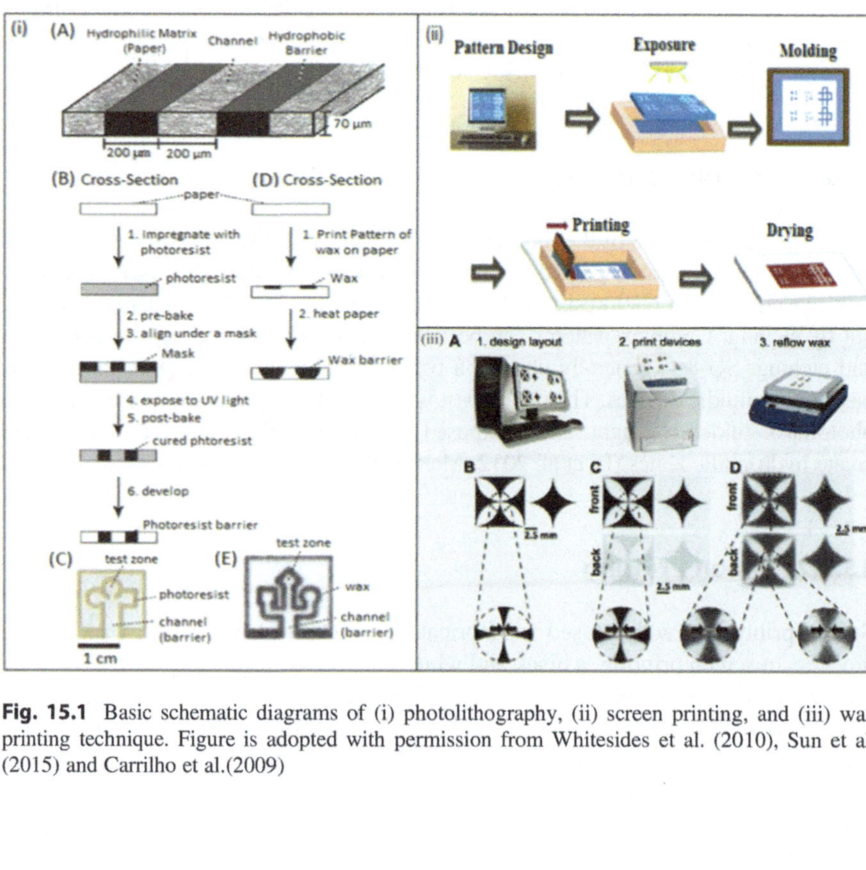

Fig. 15.1 Basic schematic diagrams of (i) photolithography, (ii) screen printing, and (iii) wax printing technique. Figure is adopted with permission from Whitesides et al. (2010), Sun et al. (2015) and Carrilho et al.(2009)

15.2.4 Inkjet Printing

Inkjet printing is increasingly gaining importance in industrial fabrication processes. Inkjet printing is used because it's a time saving, low-cost alternative to other fabrication techniques. Sensor parts can be directly printed on paper substrate instead of the multistep process of the other techniques. Furthermore, inkjet printing can be used for biological samples with minimal contamination and damage of the substrate. However, inject printing has been used for fabrication of electrochemical sensing devices only (Abe et al. 2008). (Fig. 15.2)

15.2.5 Flexographic Printing

In flexographic printing technique, mostly polystyrene is used as a patterning agent. Polystyrene is deposited on the paper to print the hydrophobic features/zones of sensor using the flexographic technique. Polystyrene penetrates into the paper and forms a hydrophobic wall, and the regions of the paper that do not contain polystyrene are hydrophilic (Yong et al. 2015). Patterning principle in flexographic technique is physical deposition of reagent on fiber surface.

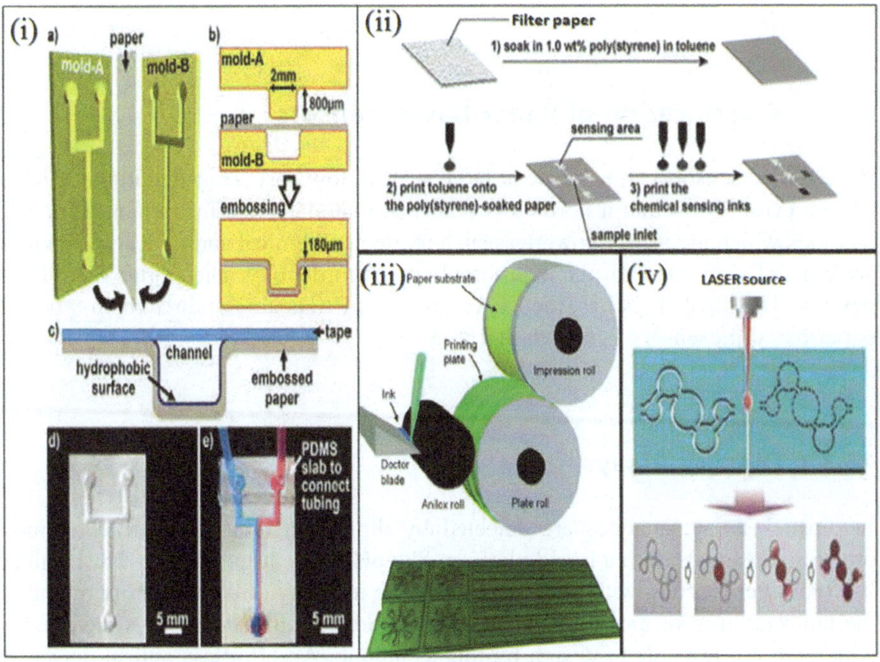

Fig. 15.2 Basic processes of (i) paper cutting fabrication, (ii) inkjet Printing, (iii) flexographic printing, and (iv) laser cutting/engraving. Figure is adopted with permission from Abe et al. (2008), Olkkonen et al. (2010), Nie et al. (2013), and Thuo et al. (2014)

15.2.6 Laser Treatment

Light amplification by stimulated emission of radiation (LASER) treatment is used for fabricating μPADs. Need of expensive equipment and careful processing limit the use of this method. Chitnis et al. (2011) fabricated a μPAD with parchment paper as substrate. Desired pattern was printed on specially treated parchment paper using laser beam in a very controlled manner. Later, the patterned areas were modified with silica microparticles. In another study, a μPAD was fabricated placing a nitrocellulose substrate within two thin polymeric layers. In this process, CO_2 laser cutting and ablative etching methods were used for constructing fluidic channels (Sun et al. 2015).

15.2.7 Paper Cutting

Paper cutting or cut and stack method is generally used for fabrication of 3D devices with varying depth of channel. Paper cutting method in combination with origami can be used for building complex fluid transport systems or devices. The cut and stack method allows one to build more robust devices in a reliable manner using simple tools (Thuo et al. 2014).

15.3 Classification of Paper-Based Sensors

Paper-based sensors can be made at large scale at low cost. Transduction method is chosen carefully so that it needs less reagents, materials, and instrumentation. The methods which are nearly powerless such as electrochemical and optical techniques are generally selected to maintain simplicity, affordability, and portability of the devices. The dipstick assays, the LFAs, and the μPADs are the three main types of paper-based biosensors as shown in Table 15.1.

15.3.1 Dipstick Assays

In dipstick assay, analytes are detected by dipping a blotting paper containing pre-stored reagents, into a liquid platform like pH paper strip. The dipstick changes its color after reacting with the analytes and the intensity of color is proportional to the concentration of analytes. In diagnosis, more sophisticated assays are needed which are not possible to design by this technique. An effortless optical detection technique is the dipstick assay which can be carried out through naked eye, such as pH strips, in the most cases.

Table 15.1 Paper-based sensors: Detection techniques, merits, and demerits (Paroloa and Merkoçi 2013)

Type of paper-based sensor	Possible detection techniques	Merits	Demerits
Dipstick	Optical	Design is easy	Single step
		Optimization is fast	Used for optical detection only
			No quantification mostly
LFA	Optical	Versatile	Optimization time is long
		Flow	
		Electrochemical detection	
		Quantification is possible	Fabrication is long
	Electrochemical	Versatile	Sample volume (around 100 μL)
μPADs	Optical	Flow	Optimization time is long
	Electrochemical	Different detection methods	
	Chemiluminescence	Quantification	
	MEMS	Small simple volume (less than 10 μL)	
		Massive production	

15.3.2 Lateral Flow Assay (LFA)

LFAs have all the reagents pre-stored in the strip and they allow the samples to flow through different zones of the strip. Different ELISA techniques are used for fabricating LFAs. Generally, LFA is composed of four distinct sections, i.e., conjugation pad, sample pad, absorbent pad, and detection pad. These sections play their own roles and help in detecting the analytes efficiently. By increasing the amount of sample, the sensitivity can be increased with the help of absorbent pad, that helps in draining the fluid through the membrane. Multiple and quantitative assays can be carried out using LFAs (Posthuma-Trumpie et al. 2009).

15.3.2.1 Optical LFA

LFAs are simple tests which depend on the attachment of target analyte with specific biomarkers on the designated regions of the test strips at the time of flow of fluid sample through the strip. LFAs are widely utilized for POC testing due to their easy operating procedure, affordability, and fast result turnaround time.

Conventional LFA test gives mainly qualitative results (e.g., "positive" or "negative"); attaching with cellphone sensing platforms enable them to produce quantitative results. Combination of LFA with mobile sensing platform increases the sensitivity and wide dynamic range. A mobile based rapid diagnostic test (RDT) reader as shown in Fig. 15.3a investigated for detection of diseases like TB, malaria, and HIV etc. using both Android and iPhone platforms (Mudanyali et al. 2012). This

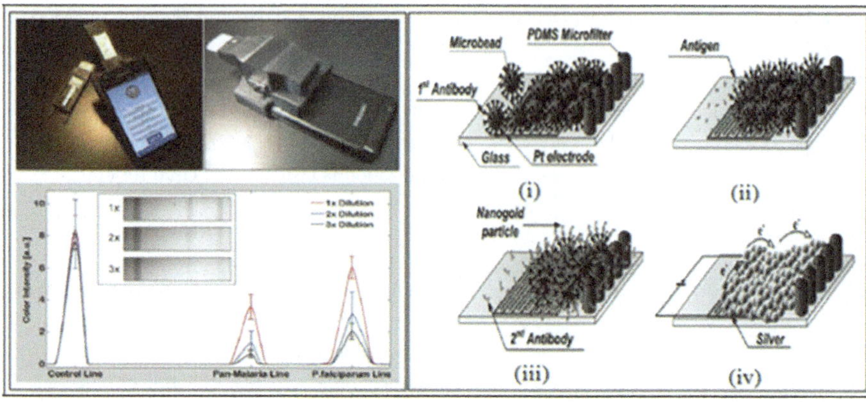

Fig. 15.3 (**a**) Cellphone-based rapid diagnostic test (RDT) reader that can assimilate different LFA tests such as TB, malaria, and HIV (Mudanyali et al. 2012). (**b**) Depiction of the electrical immunoassay. (i) Assuring the primary antibody immobilized microbeads. (ii) Reaction of first antibody with the antigen. (iii) Reaction with secondary antibody-conjugated AuNP. (iv) Amplification of electrical signal with the help of silver enhancer and the electrical signal detection produced from the immune reaction (Ko et al. 2008)

cellphone attachment uses two circuitous green LED arrangement accumulated in front of the LFA test strip for capturing reflection-based imaging. A third circuitous LED array was placed behind the test strip to capture image based on transmitted light, if needed. Later, this RDT reader technology was commercialized by Holomic LLC (2015). In another study, a mobile-based fluorescent LFA reader was demonstrated by Lee et al. (2013) using low-cost plastic film or colored glass filters and a light-emitting diode as the excitation light. Lin et al. (2015) also developed lateral e-flow assay (LeFA) by fusing it with fluorescence cell phone reader for detecting hepatitis C virus (HCV) antigens, namely, c100p, c22p, and c33c.

15.3.2.2 Electrochemical Detection

An amperometric immunosensor was produced by Yuan et al. (2009) for detecting carcinoembyronic antigen (CEA) biomarker. On the exposed portion of glassy electrode, nickel hexacyanoferrate nanoparticles decorated gold nanoparticles were immobilized by electrochemical reduction. In order to attach with anti-CEA, another layer of AuNPs is deposited. BSA was added to improve signal amplification and to stop non-specific sites. By using cyclic voltammetry (CV) and Electrochemical impedance spectroscopy (EIS), CEA was detected at low cost. This immunoassay was highly sensitivity and able to reproduce the results. Silica NPs (SiNPs) are electroactive and hence can also be used for detecting biomarkers.

A microchip-based multifold electro-immuno sensing system for the detecting multiple cancer biomarkers, viz., alpha-fetoprotein (AFP), CEA, and prostate-specific antigen (PSA) was developed by Ko et al. (2008). At first, polystyrene microbeads conjugated with antibodies were immobilized onto the electrode where they bound with their specific antigen and then secondary antibodies-conjugated with AuNPs were adjoined. In order to increase the electrical signal (current), silver was segmented at the surface of the AuNPs. An electrical signal generated by immune reactions was measured and monitored by PC-based system as shown in Fig. 15.3b.

15.3.3 Microfluidic Paper-Based Analytical Devices (μPAD)

The μPADs are analytical devices that assimilate the merits of paper with microfluidics. These microfluidic devices are capable of performing fluidic manipulations like transportation, sorting, mixing, and separation cost-effectively without any extrinsic energy source for transporting the fluid over the channel. These devices are capable of performing multiplexed and quantitative analysis with very small amount of sample (Rezk et al. 2012). Hydrophilic channels are created in hydrophobic paper using different fabrication techniques (Dungchai et al. 2011).

15.3.3.1 Colorimetric Detection

Paper-based colorimetric microfluidic techniques are used for qualitative or semi-quantitative detection. In colorimetric detection, change in color due to enzymatic or chemical interactions may be visualized by eye. Martinez et al. (2007) first proved paper-based microfluidic devices for detecting glucose and protein using colorimetric detection. In glucose assay, the color changes due to the enzymatic reaction of glucose with corresponding enzymes, which does not require additional instrumentation. This process is not accurate as UV-visible spectroscopy or atomic absorption spectroscopy as the color and intensity interpretation varies with different individual, ambient light condition, and the paper substrate (dry or wet) condition. With the help of camera phone, Martinez et al. (2008) measured visible color changes to enhance the accuracy and sensitivity of the method as shown in Fig. 15.4(a). The captured image was interpreted with imaging software depending on color intensity change, to determine the concentration of analytes.

15.3.3.2 Fluorescence Detection

Fluorescent signal detection occurs due to the intercourse of target molecules and fluorescent dyes or fluorophores (Pires et al. 2014, Terai and Nagano 2013). In this process, a light source at a predefined wavelength excites a fluorophore, then the emitted light is filtered and separated from excitation photons, and the emission photons are detected and converted to electrical signal. Paper-based device was developed using immobilized synthetic DNA oligonucleotides to detect DNA successfully (Ali et al. 2009). Paper-based devices can detect DNA at a lower cost. Poly (N-isopropylacrylamide) microgels coupled with an oligonucleotide was spotted onto a paper strip for DNA detection. In another study, lactoferrin was quantified in the human tear samples using fluorescence-based microfluidic sensor (Yamada et al. 2013).

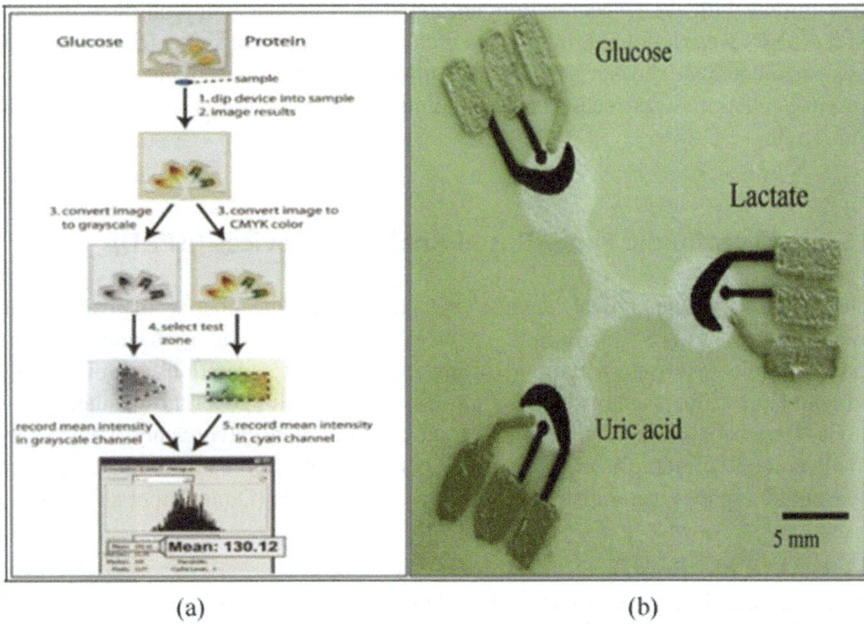

Fig. 15.4 (**a**) Quantification of the amount of glucose and protein in a urine sample (Martinez et al. 2008). (**b**) Paper-based microfluidic devices having three electrodes (Dungchai et al. 2009)

15.3.3.3 Electrochemical Detection

Electrochemical detection requires a three-electrode system, i.e., a counter electrode, reference electrode, and working electrode. In this, a three-electrode system is cloned on paper using conductive inks to replace the traditional solid electrodes. Carbon inks are generally used for fabricating the working and counter electrodes, while silver/silver chloride ink is utilized as the reference electrode (Dungchai et al. 2009; Nie et al. 2010a). The conductive electrode pads were screen-printed on the paper and enzymes were subsequently immobilized. The sample solution were added at the middle of the paper and allowed to flow through reaction sites containing the three-electrode system, as shown in Fig. 15.4a. The analysis was carried out with the help of chronoamperometry for producing hydrogen peroxide at the optimal potential. The electrical signal generated as a result of redox reactions between enzymes and substrates like glucose, uric acid, and lactate leads to estimation of substrate concentration. Similarly, gold and iron were detected using redox reactions (Apilux et al. 2010). Cyclic voltammetry and square wave voltammetry experiments were used for these electrochemical analyses.

Paper-based microfluidic platforms in combination with commercial handheld readers were used for detection of glucose and alcohol (Nie et al. 2010b). The paper device was fabricated on the paper substrate with the help of wax printing and screen printing. In this study, the electrodes and electronic wires were printed from graphite and silver ink, respectively. Upon introduction of the sample in the microfluidic sensing platform, the liquid sample moves towards the sensing region and the glucometer determines the analyte concentrations using amperometric measurement technique.

15.3.3.4 Chemiluminescent and Electrochemiluminescent Detection

In chemiluminescence (CL), chemical energy is converted into light as a result of movement of electrons from a high energy state to a low energy state. A host of compounds react with hydrogen peroxide or oxygen, resulting in the formation of intermediate complex as well as emission of light (Parveen et al. 2013), i.e., an organic compound luminol (5-amino-2, 3-dihydro-1, 4-phthalazinedione, or 3-aminophthalhydrazide) and hydrogen peroxide react to produce light. CL detection methods are less expensive and highly sensitive. A device was fabricated in order to detect uric acid on the basis of enzymatic reaction, which produces hydrogen peroxide during transition of the substrate. In order to get CL, peroxide and rhodamine derivatives produced react in acidic medium. The peroxide and rhodamine derivatives produced react in acidic medium to yield CL (Yu et al. 2011). The LOD was 1.9 mmol/L and the curve was straight over the range of 2.6–49.0 mmol/L. In another study, sandwich CL-ELISA was carried out on μPADs for determining different tumor markers simultaneously, i.e., AFP, CA125, and CEA (Wang et al. 2012). In order to detect these tumor markers, the detection limits are 0.06 ng/mL, 0.33 U/mL, and 0.05 ng/mL respectively.

Combination of chemiluminescence with electrochemical method is known as electrochemiluminescence. It gives better selectivity and enhances the dynamic concentration range (Delaney et al. 2011; Ge et al. 2012). Delaney et al. (2011) invented paper microfluidic devices on the basis of this analysis method where electrochemiluminescence emission was generated through the interactions between the oxidized ruthenium complex with its co-reactant 2-(dibutylamino)-ethanol or nicotinamide adenine dinucleotide (Deng et al. 2009). They added a cellphone for detecting electrochemiluminescence emission in which red pixel intensity is used for determining analyte concentration (shown in Fig. 15.5(a)).

15.3.3.5 Microelectromechanical Systems (MEMS)

Paper-based microelectromechanical systems have been drawing huge attention, in recent days, as cost-effective POC devices for detecting molecular biomarkers for different diseases. MEMS sensors offer quick and accurate method to detect *Mycobacterium tuberculosis*, whereas conventional detection methods require minimum of 4 h to maximum of 72 h to identify TB (Saranya et al. 2013; Sangeetha and Vimala 2014). The interactions of TB biomarkers with their corresponding antibodies on the surface of microcantilever lead to mechanical bending of microcantilever which can be detected as a change in capacitance, piezoresistance, or resonance frequency of the microcantilever material. The bending mechanism and antibody immobilization are shown in Fig. 15.5(b) (Gopinath et al. 2015; Murthy et al. 2016).

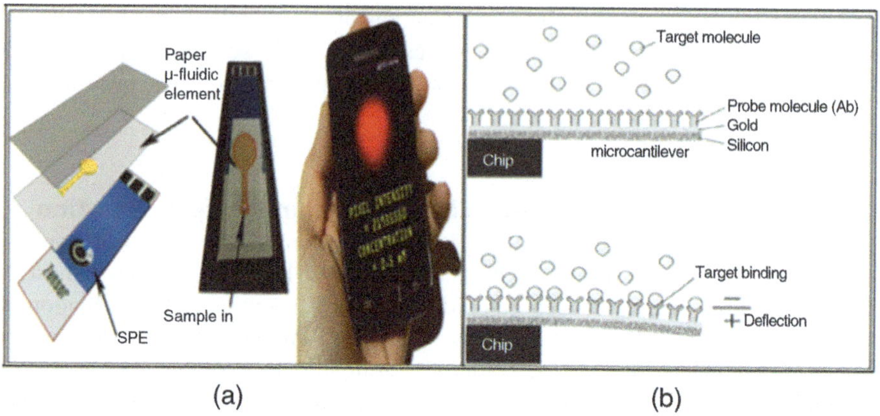

Fig. 15.5 (**a**) Procedure for analyzing paper-based microfluidic electrochemiluminescence sensors and their fabrication (Delaney et al. 2011). (**b**) Immobilization of antibodies on the cantilever surface and bending of cantilever after biomolecular recognition (Murthy et al. 2016)

15.4 Application of Paper-Based Sensors in Medical Diagnostics

In this chapter, we have described number of circumstances where a handheld device could be useful to declamate the real public health needs, demonstrating applicability of POC sensors in the field of diagnosis (Frank and Hargreaves 2003). In determining sexually transmitted infections (STIs) (Phan et al. 2009), renal injury biomarkers having acute kidney injury (AKI) and (Marrer and Dieterle 2010) cardiac biomarkers are necessary for diagnosis and treatment of number of lethal diseases (Blick 2014). Successful use of POC devices depends on their specificity, sensitivity, ease of use, reliability, multiplexing capability, cost-effectiveness, and low turnaround time (Vignali 2000).

15.4.1 Sexually Transmitted Infections

POC tests for early diagnosis of STIs are required because they facilitate treatment and counseling, and prevent their further transmission (Rompalo et al. 2013). Conventional rapid POC tests are not accurate enough to meet the clinician's need, moreover, they are very expensive and difficult to read. Wet mount test, a urine dipstick, and a rapid HIV test are the typical STI tests, and the turnaround time for these tests are 2–14 days (Hsieh et al. 2010). Newly developed tests are more sensitive, cost-effective, indirect, and easy to use and read, consume less time (5–20 min), and have better specificity (Rompalo et al. 2013; Hsieh et al. 2010). Gonorrhea and chlamydia are the two precedence areas for development of POC test for STI, which follows herpes simplex virus and seroconversion for HIV (Rompalo et al. 2013; Hsieh et al. 2010; Hsieh et al.

2011). Development of POC test probe for the top few STIs will clearly make a great impact on public health.

15.4.2 Acute Kidney Injury Biomarkers

The current tests involving measurement of blood urea nitrogen (BUN) and serum creatinine are not adequate for reducing or preventing AKI, because these biomarkers are generally not raised until 50% of kidney function is damaged (Hoffmann et al. 2010). Small amount of sample is required to determination of AKI biomarkers using the POC tests. In case of a biological attack, a POC device could be used to detect lipopolysaccharide (LPS), an endotoxin released by of gram-negative bacteria which can evoke an innate immune response (Amoto et al. 2012; Han et al. 2012). Exposure to LPS will leads to systemic vasodilation, reduced renal perfusion and subsequent AKI (Han et al. 2012). Moreover, exposure to the toxin resulting from inflammation and cytokine release can cause a nephrotoxic effect, which leads to raised levels of NGAL and KIM-1, which will denote early kidney damage (Amoto et al. 2012). A POC device is also useful in the developing world, where AKI is a major medical complication particularly with regard to sepsis, diarrheal illnesses, and infectious diseases. Low-resource medical tests like POCs are helpful for diagnosing AKI which follows crash injuries and congenital calamity such as earthquakes, which can adversely affect kidneys.

15.4.3 Cardiac Biomarkers

Study showed that among a host of biomarkers, cardiac troponin l (c -Tnl) alone is enough to diagnose chest pain and obviate myocardial infarction (Collinson et al. 2012). In POC tests, the use of c-Tnl provides accurately high diagnosis (100% sensitivity and almost 100% specificity) and the other two biomarkers may not give extra information, especially when patients suffer from renal dysfunction (Caragher et al. 2002; McCullough et al. 2002). In fact, even the current medical guidelines state the need of c-Tnl biomarker-based diagnosis results for treatment of myocardial infarction (Blick 2014). These three biomarkers, i.e., cardiac troponin l (c-Tnl), myoglobin, and the MB isoenzyme of creatine kinase (CK-MB), can be used for diagnosis of acute coronary syndrome in patients who are discharged from the hospital within 2 h (Macdonald and Nagree 2008), and use of CK-MB may be helpful diagnosing myocardial infarction (Haq et al. 2011).

15.4.4 Glucose Monitoring

Various colorimetric methods that use either glucose oxidase (GOx) or GOx with enzymes like horseradish peroxidase have been developed to monitor glucose. Due to its greatest role in diagnosing diabetes, glucose is considered as one of the significant clinical analytes. Zhu et al. (2014) used a tree-shaped μPADS with 2,4,6-tribromo-3-hydroxy benzoic acid and 4-aminoantipyrine for the detection of glucose. The system was used for estimation of glucose in serum. In another study, protein and glucose were measured in urine with the help of a three-dimensional μPAD as well as a commercial assay kit (Sechi et al. 2013). In this device, the standard concentration range detected for glucose and protein were 0.25–8 mg/dL and 5–20 mg/dL, respectively. Demirel and Babur (2014) developed another device for detecting albumin, uric acid, ALP, and alanine aminotransferase (ALT) including enzymatic detection of glucose as well. In the colorimetric reaction between albumin and tetra bromophenol blue, albumin was measured for quantifying total protein and with nitro-blue tetrazolium and 5-bromo-4-chloro-3′-indolyl phosphate, ALP was enzymatically detected. These results were validated with a commercially available enzymatic assay kit. In order to perform the liver fuction test, Pollock et al. (2012) prepared a 3D-μPAD for determining the level of ALP and aspartate amino transferase. The device was entirely covered in plastic, which increases the mechanical stability and also decreases the evaporation of the solution at the time of transport through the channels. For the detection of glucose, the color intensity was studied based on the type of paper substrate. Yetisen et al. (2014) produced a smart phone algorithm using paper sensors to increase glucose detection mobility. The accuracy of the measurement was enhanced by taking care of the space between the camera and paper substrate using the algorithm automatically.

A selective and sensitive platform was offered by μPADs coupled with electrochemical detection of glucose. Noiphung et al. (2013) narrated an electrochemical glucose detection in whole blood with the help of reusable, external screen-printed carbon electrode (SPCE) which is modified using a mediator. From the whole blood, blood plasma having glucose was isolated in a detection region to form hydrogen peroxide when it reacts with glucose oxidase and is detected electrochemically using amperometry. Santhiago and Kubota (2013) also weighed glucose with the help of GOx at graphite pencil electrodes with a p-aminophenylboronic acid as mediator. A delicate finding of D-glutamic acid, a neurotransmitter associated with brain damage, was reported by Ge et al. (2013), using GNP-coated cellulose fibers with electropolymerized molecular imprint polymer on the surface. A decrease in hexacyanoferrate oxidation was calculated with the help of differential pulse voltammetry (DPV) when D-glutamic acid is adsorbed by the electrode surface that complement with D-glutamic acid concentration (Ge et al. 2013). The results provided were matching with HPLC testing of real human samples, and this PAD device was competent for sub-nanomolar findings. By using thin-layer chromatography, co-migrating neurotransmitters paracetamol and ascorbic acid or dopamine and ascorbic acid were successfully detected by Dossi et al. (2013) with the help of pencil-drawn electrodes by flowing through PAD. With the help of configuration switching of an aptamer upon binding with the target analyte, Cunningham et al. (2014) cracked DNA and protein detection. There is a change in the position of an electrochemical label (methylene blue) out from the gold electroplated SPCE surface due to binding, which turn the signal "off" and provide limiting detection for both DNA and thrombin in the low nanomolar range.

15.4.5 Nanoparticle-Conjugated Immunoassays for TB (Tuberculosis) Patients

Nanomaterials have found widespread use as detection reagent because of their stability, higher extinction coefficients, and good sensitivity in case of target analytes (e.g., cancer antigens) (Posthuma-Trumpie et al. 2009; Sajid et al. 2014). Due to the capability of these NPs to attain lower detection limit and higher sensitivity, a good number of nanoparticle-conjugated immunoassays have been developed. Detection is often accomplished visually, which is highly competent for POC applications. Functionalized gold nanoparticles have been used for diagnosis of TB DNA (Tsai et al. 2013). Single-stranded DNA (ssDNA) and modified AuNPs were hybridized with correlative dsDNA from TB-positive patients. A change in color from red to blue was seen upon DNA hybridization with AuNPs, and the change in color intensity was measured using a mobile phone. With this method, *Mycobacterium tuberculosis* concentration up to a level of 2.6 nM could be detected and provide test result within 1 h.

15.4.6 Immunosensor for Monitoring Cancer Biomarkers

A colorimetric immune sensor for carcinoembryonic antigen (CEA) was created by Liu et al. (2014a, b) using $ZnFe_2O_4$-carbon nanotubes. Metal nanotubes attached with secondary antibodies to capture CEA. The primary and secondary antibody bind with the help of charge-transfer complex composed of a free -OH radical and 3,3′,5,5′-tetramethylbenzidine resulting in a change in blue-green color apparently. The detection limit of CEA was 2.6 pg/mL. Electrode was modified with Au NPs to produce multiplexed device made by Ge et al. (2013) comprising of an auxiliary electrode having four sample pads around it for detecting D-glutamate.

15.4.7 Cellular Activity Monitoring and Drug Screening

In tissue culture, cancer cells are useful for monitoring cellular activity and screening the therapeutic efficacy of drugs. Cancer cells cultured on 3D paper-based devices are used to monitor cellular events like apoptosis, glycan production, and release of hydrogen peroxide (Cunningham et al. 2014; Su et al. 2014; Liu et al. 2014a; Su et al. 2015). Drug-induced apoptosis and the release of hydrogen peroxide from living cells were detected non-enzymatically by Pt nanosphere-coated fibers (Liu et al. 2014b). Carbon paper electrodes were modified with the help of carbon nanotube, graphene oxide, and manganese oxide aerogel to develop another 3D PAD cell culture device for detecting H_2O_2 (Renault et al. 2013).

CL is affordable, highly sensitive, and thereby very attractive. A subset technique of CL, PL with the help of luminescence resonant energy transfer associated with up-converting phosphors was demonstrated by Zhou et al. (2014) with μPADs for point-of-care applications. Hybridized target DNA probe was detected using green light emitting phosphors containing paper-based device. For DNA hybridization measurement with the help of photoelectrochemiluminescence (PEL), Wang et al. (2013) improved a μPAD integrating Au-paper electrodes in order to get approximately femtomolar detection limits. PEL signal was amplified and detected using an integrated novel, paper-based super capacitor.

A power source is integrated directly on the sensor leading to the removal of bulk and expense incurred within electrochemical workstations (Yan et al. 2013; Yang et al. 2014). To improve on-device power with a relatively simple circuit, Zhang et al. (2013) developed an environment friendly battery by adding water for activation. The authors engaged an origami folding technique to keep the features aligned at the time of fabrication and to enhance fabrication simplicity. Without using for long time periods, the devices are stored as Fe(III) is more stable as compared to the typical Ag(I). After using for at least 250 s, the open circuit voltage was reported to be stable and did not change significantly for 15 consecutive days of storage. The successful power density attained was 0.52 ± 0.026 mJ cm^{-2} ($V_{oc} = 1.3$ V, $I_{sc} = 0.4 \pm 0.02$ mA cm^{-2}).

An established "color bar code" device for active pharmaceutical ingredient testing in anti-tuberculosis (TB) drugs was developed by Lieberman's group (Weaver et al. 2013). Similarly, a testing for the counterfeit anti-malarial drug artesunate was provided by Koesdjojo et al. (2014). For measuring blood hemoglobin, a paper-based sensor was also produced by Yang et al. (2014). HIV DNA was detected by Rohrman and Richards-Kortum (2012) using μPADs. In order to combine the enzyme storage, reaction component mixing, and recombinase polymerase amplification of HIV DNA steps on the paper, the enzymatic HIV DNA amplification was increased successfully to 10 copies within 15 min.

15.5 Challenges and Future Prospects

Paper-based sensors are gaining importance in recent years due to their advantages like cost-effectiveness, ease of fabricating disposable devices at large scale, and point of care applications. Combination of paper and microfluidic technology enables these devices for multiple analytes detection. Mostly, these devices are capable of giving qualitative information; quantitative analysis requires addition of mobile or detector-based platforms.

Improvement of paper-based analytical devices in the future depends on:

(i) The shift in focus toward personalized, preventive medicine, demands of home health monitoring, and self-testing procedures.
(ii) The evolution in printing technologies that enable rapid deposition of various types of materials on the paper including metals, nanoparticles, and biomolecules.

(iii) Widespread usage of smartphones or mobile communication devices capable of performing imaging and image analysis with the help of disposable devices even in low-resource countries.
(iv) Increased level of sustainability awareness that raises the use of bio-based eco-friendly renewable materials such as cellulose.

Paper-based analytical devices take the advantage of this friendly environment so that to overcome a number of challenges including improving the stability of paper-based devices in order to protect the biological activity at the time of transport and storage and maintaining the physical and structural properties of cellulose paper, and hence functionality of paper should not vary.

15.6 Conclusion

The importance of paper-based sensors has been increasing day by day in diagnostic field, especially in clinical analysis. Paper is an adaptable, easy to fabricate, affordable and, biocompatible material. It is environment friendly, degradable and disposable analytical custom-tailored miniaturized electrodes can be made from it. Miniaturized printed electrodes are easily conjugated to biomolecules, thereby increasing the analytical reading from the sensitivity and selectivity point of view. Paper-based POC assays can play important roles in disease management at an affordable price. But, the specificity, sensitivity, and reliability of these devices need to be improved prior to their commercialization and widespread uses in clinical operations.

Glossaries

AKI	acute kidney injury
ALP	alkaline phosphatase
ALT	alanine aminotransferase
APTES	(3-aminopropyl) triethoxysilane
AST	aspartate aminotransferase;
CEA	carcinoembryonic antigen
CL	chemiluminescence
CMOS	complementary metal oxide semiconductor
DNA	deoxyribonucleic acid
DPV	differential pulse voltammetry
dsDNA	double-stranded deoxyribonucleic acid
ECL	electrochemiluminescence
GOx	glucose oxidase
hCG	human chrionic gonadotropin
HCV	hepatitis C virus
HIV	human immunodeficiency virus

HPLC	high-pressure liquid chromatography
HRP	horseradish peroxidase
LeFA	lateral e-flow assay
LFA	lateral flow assay
MEMS	microelectromechanical systems
MGs	microgels
μPADS	microfluidic paper-based analytical devices
PEL	photoelectrochemiluminescence
PL	photochemiluminescence
PSA	prostate-specific antigen
POC	point-of-care
RCA	rolling circle amplification
SPCE	screen-printed carbon electrode
ssDNA	single-stranded deoxyribonucleic acid
STD	sexually transmitted diseases
STI	sexually transmitted infections

References

Abe K, Suzuki K, Citterio D (2008) Inkjet-printed microfluidic multianalyte chemical sensing paper. Anal Chem 80(18):6928–6934

Ali MM, Aguirre SD, Xu Y, Filipe CDM, Pelton R, Li Y (2009) Detection of DNA using bioactive paper strips. Chem Commun 43:6640–6642

Amoto AA, Adamson JW, Achermann JC, Aboulhosn J, Abbruzzese JL, Harrison TR, Loscalzo J, Hauser SL, Fauci AS, Jameson JL (2012) Harrison's principles of internal medicine, 18th edn. McGraw Hill Financial, Inc, New York

Apilux A, Dungchai W, Siangproh W, Praphairaksit N, Henry CS, Chailapakul O (2010) Lab-on-paper with dual electrochemical/colorimetric detection for simultaneous determination of gold and iron. Anal Chem 82:1727–1732

Blick KE (2014) The benefits of a rapid, point-of-care "Tnl-only" zero and 2-hour protocol for the evaluation of chest pain patients in the emergency department. Clin Lab Med 34:75–85

Caragher TE, Fernandez BB, Jacobs FL, Barr LA (2002) Evaluation of quantitative cardiac biomarker point-of-care testing in the emergency department. J Emerg Med 22:1–7

Carrilho E, Martinez AW, Whitesides GM (2009) Understanding wax printing: a simple micropatterning process for paper-based microfluidics. Anal Chem 81(16):7091–7095

Chitnis G, Ding Z, Chang C-L, Savran CA, Ziaie B (2011) Laser-treated hydrophobic paper: an inexpensive microfluidic platform. Lab Chip 11:1161–1165

Collinson P, Goodacre S, Gaze D, Gray A, Arrowsmith C, Barth J, Benger J, Bradburn M, Capewell S, Chater T (2012) Very early diagnosis of chest pain by point-of-care testing: comparison of the diagnostic efficiency of a panel of cardiac biomarkers compared with troponin measurement alone in the RATPAC trial. Heart 98:312–318

Comer JP (1956) Semiquantitative specific test paper for glucose in urine. Anal Chem 28:1748–1750

Cunningham JC, Brenes NJ, Crooks RM (2014) Paper electrochemical device for detection of DNA and thrombin by target-induced conformational switching. Anal Chem 86(12):6166–6170

Delaney JL, Hogan CF, Tian JF, Shen W (2011) Electrogenerated chemiluminescence detection in paper-based microfluidic sensors. Anal Chem 83:1300–1306

Demirel G, Babur E (2014) Vapor-phase deposition of polymers as a simple and versatile technique to generate paper-based microfluidic platforms for bioassay applications. Analyst 139:2326–2331

Deng L, Zhang L, Shang L, Guo S, Wen D, Wang F, Dong S (2009) Electrochemiluminescence detection of NADH and ethanol based on partial sulfonation of sol–gel network with gold nanoparticles. Biosens Bioelectron 24(7):2273–2276

Dossi N, Toniolo R, Piccin E, Susmel S, Pizzariello A, Bontempelli G (2013) Pencil-drawn dual electrode detectors to discriminate between analytes comigrating on paper-based fluidic devices but undergoing electrochemical processes with different reversibility. Electroanalysis 25:2515–2522

Dungchai W, Chailapakul O, Henry CS (2009) Electrochemical detection for paper-based microfluidics. Anal Chem 81:5821–5826

Dungchai W, Chailapakul O, Henry CS (2011) A low-cost, simple, and rapid fabrication method for paper-based microfluidics using wax screen-printing. Analyst 136:77–82

Frank R, Hargreaves R (2003) Clinical biomarkers in drug discovery and development. Nat Rev Drug Discov 2:566–580

Ge L, Yan J, Song X, Yan M, Ge S, Yu J (2012) Three-dimensional paper-based electro chemiluminescence immune device for multiplexed measurement of biomarkers and point of care testing. Biomaterials 33:1024–1031

Ge L, Wang S, Yu J, Li N, Ge S, Yan M (2013) Molecularly imprinted polymer grafted porous Au-paper electrode for an microfluidic electro-analytical origami device. Adv Funct Mater 23:3115–3123

Gopinath PG, Anitha VR, Aruna Mastani S (2015) Microcantilever based biosensor for disease detection applications. J Med Bioeng 4(4):307–311

Govindasamy K, Potgieter S, Land K, Muzenda E (2012) Fabrication of paper based microfluidic devices. In: Proceedings of the world congress on engineering- III (WCE 2012), July 4–6, 2012, London

Han M, Li Y, Liu M, Li Y, Cong B (2012) Renal neutrophil gelatinase associated lipocalin expression in lipopolysaccharide-induced acute kidney injury in the rat. BMC Nephrol 13(25):1–10. https://doi.org/10.1186/1471-2369-13-25

Haq SA, Tavakol M, Silber S, Bernstein L, Kneifati-Hayek J, Schleffer M, Banko LT, Heitner JF, Sacchi TJ, Puma JA (2011) Enhancing the diagnostic performance of troponins in the acute care setting. J Emerg Med 40:367–373

Hoffmann D, Fuchs TC, Henzler T, Matheis KA, Herget T, Dekant W, Hewitt P, Mally A (2010) A evaluation of a urinary kidney biomarker panel in rat models of acute and subchronic nephrotoxicity. Toxicology 277(1–3):49–58

Holomic LLC (2015) Available: http://www.holomic.com, https://doi.org/10.1021/nn500614kviewedon01/12/2018

Hsieh YH, Hogan MT, Barnes M, Jett-Goheen M, Huppert J, Rompalo AM, Gaydos C (2010) A perceptions of an ideal point-of-care test for sexually transmitted infections – a qualitative study of focus group discussions with medical providers. PLoS One 5:e14144

Hsieh YH, Gaydos C, Hogan T, Uy O, Jackman J, Jett-Goheen MR (2011) A perceptions on point-of-care tests for sexually transmitted infections-disconnect between frontline clinicians and professionals in industry. Sex Transm Infect 87:A82–A83

Ko YJ, Maeng JH, Ahn Y, Hwang SY, Cho NG, Lee SH (2008) Microchip-based multiplex electro-immunosensing system for the detection of cancer biomarkers. Electrophoresis 29(16):3466–3476

Koesdjojo MT, Wu Y, Boonloed A, Dunfield EM, Remcho VT (2014) Low-cost, high-speed identification of counterfeit antimalarial drugs on paper. Talanta 130:122–127

Lee LG, Nordman ES, Johnson MD, Oldham MF (2013) A low-cost, high-performance system for fluorescence lateral flow assays. Biosensors 3:360–373

Li X, Ballerini DR, Shen W (2012) A perspective on paper-based microfluidics: current status and future trends. Biomicrofluidics 6(1):11301–1130113

Liana DD, Raguse B, Gooding JJ, Chow E (2012) Recent advances in paper-based sensors. Sensors 12(9):11505–11526
Lin R, Skandarajah A, Gerver RE, Neira HD, Fletcher DA, Herr AE (2015) A lateral electrophoretic flow diagnostic assay. Lab Chip 15(6):1488–1496
Liu W, Yang H, Ding Y, Ge S, Yu J, Yan M, Song X (2014a) Paper-based colorimetric immunosensor for visual detection of carcinoembryonic antigen based on the high peroxidase-like catalytic performance of $ZnFe_2O_4$–multiwalled carbon nanotubes. Analyst 139:251–258
Liu F, Ge S, Yu J, Yan M, Song X (2014b) Electrochemical device based on a Pt nanosphere-paper working electrode for *in situ* and real-time determination of the flux of H_2O_2 releasing from SK-BR-3 cancer cells. Chem Commun 50:10315–10318
Macdonald SP, Nagree Y (2008) Rapid risk stratification in suspected acute coronary syndrome using serial multiple cardiac biomarkers: a pilot study. Emerg Med Australas 20:403–409
Marrer E, Dieterle F (2010) Impact of biomarker development on drug safety assessment. Toxicol Appl Pharmacol 243:167–179
Martinez AW, Phillips ST, Butte MJ, Whitesides GM (2007) Patterned paper as a platform forinexpensive, low-volume, portable bioassays. Angew Chem Int Ed Eng 46:1318–1320
Martinez AW, Phillips ST, Carrilho E, Thomas SW, Sindi H, Whitesides GM (2008) Simple telemedicine for developing regions: camera phones and paper-based microfluidic devices for real-time, off-site diagnosis. Anal Chem 80:3699–3707
McCullough PA, Nowak RM, Foreback C, Tokarski G, Tomlanovich MC, Khoury NE, Weaver WD, Sandberg KR, McCord J (2002) Performance of multiple cardiac biomarkers measured in the emergency department in patients with chronic kidney disease and chest pain. Acad Emerg Med 9:1389–1396
Mudanyali O, Dimitrov S, Sikora U, Padmanabhan S, Navruz I, Ozcan A (2012) Integrated rapid-diagnostic-test reader platform on a cellphone. Lab Chip 12:2678–2686
Murthy KSN, Prasad GRK, Saikiran NLNV, Manoj TVS (2016) Design and simulation of MEMS biosensor for the detection of tuberculosis. Indian J Sci Technol 9(31). https://doi.org/10.17485/ijst/2016/v9i31/90638
Nie ZH, Nijhuis CA, Gong JL, Chen X, Kumachev A, Martinez AW, Narovlyansky M, Whitesides GM (2010a) Electrochemical sensing in paper-based microfluidic devices. Lab Chip 10:477–483
Nie ZH, Deiss F, Liu XY, Akbulut O, Whitesides GM (2010b) Integration of paper-based microfluidic devices with commercial electrochemical readers. Lab Chip 10:3163–3169
Nie J, Liang Y, Zhang Y, Le S, Li D, Zhang S (2013) One-step patterning of hollow microstructures in paper by laser cutting to create microfluidic analytical devices. Analyst 138:671–676
Noiphung J, Songjaroen T, Dungchai W, Henry CS, Chailapakul O, Laiwattanapaisal W (2013) Electrochemical detection of glucose from whole blood using paper-based microfluidic devices. Anal Chim Acta 788:39–45
Olkkonen J, Lehtinen K, Erho T (2010) Flexographically printed fluidic structures in paper. Anal Chem 82:10246–10250
Paroloa C, Merkoçi A (2013) Paper-based nanobiosensors for diagnostics. Chem Soc Rev 42:450–457
Parveen S, Aslam MS, Hu L et al (2013) Electrogenerated chemiluminescence: protocols and applications. Springer, Berlin/Heidelberg
Phan JH, Moffitt RA, Stokes TH, Liu J, Young AN, Nie SM, Wang MD (2009) Convergence of biomarkers, bioinformatics and nanotechnology for individualized cancer treatment. Trends Biotechnol 27:350–358
Pires NMM, Dong T, Hanke U, Hoivik N (2014) Recent developments in optical detection technologies in lab-on-a-chip devices for biosensing applications. Sensors 14(8):15458–15479
Pollock NR, Rolland JP, Kumar S, Beattie PD, Jain S, Noubary F, Wong VL, Pohlmann RA, Ryan US, Whitesides GM (2012) A paper-based multiplexed transaminase test for low-cost, point-of-care liver function testing. Sci Transl Med 4(152):152ra129

Posthuma-Trumpie GA, Korf J, van Amerongen A (2009) Lateral flow (immuno)assay: its strengths, weaknesses, opportunities and threats: a literature survey. Anal Bioanal Chem 393:569–582

Renault C, Li X, Fosdick SE, Crooks RM (2013) Hollow-channel paper analytical devices. Anal Chem 85:7976–7979

Rezk AR, Qi A, Friend JR, Li WH, Yeo LY (2012) Uniform mixing in paper-based microfluidic systems using surface acoustic waves. Lab Chip 12:773–779

Rohrman BA, Richards-Kortum RR (2012) A paper and plastic device for performing recombinase polymerase amplification of HIV DNA. Lab Chip 12:3082–3088

Rompalo AM, Hsieh YH, Hogan T, Barnes M, Jett-Goheen M, Huppert JS, Gaydos CA (2013) Point-of-care tests for sexually transmissible infections: what do 'end users' want? Sex Health 10:541–545

Sajid M, Kawde A-N, Daud M (2014) Designs, formats and applications of lateral flow assay: a literature review. J Saudi Chem Soc 19(6):689–705

Sangeetha P, Vimala JA (2014) Simulation and analysis of micro cantilever sensor for enhanced biosensing of disease causing pathogens. Conference proceeding (NCETSE-2014), Int J Eng Dev Res (IJEDRCP 1403027), pp 131–136

Santhiago M, Kubota LT (2013) A new approach for paper-based analytical devices with electrochemical detection based on graphite pencil electrodes. Sensors Actuators B Chem 177:224–230

Saranya R, Saranya K, Ceemati D, Chandra DK, Meenakshi SN (2013) Design of MEMS based micro cantilever for tuberculosis detection. COMSOL conference, Bangalore

Sechi D, Greer B, Johnson J, Hashemi N (2013) Three-dimensional paper-based microfluidic device for assays of protein and glucose in urine. Anal Chem 85:10733–10737

Su M, Ge L, Ge S, Li N, Yu J, Yan M, Huang J (2014) Paper-based electrochemical cyto-device for sensitive detection of cancer cells and in situ anticancer drug screening. Anal Chim Acta 847:1–9

Su M, Ge L, Kong Q, Zheng X, Ge S, Li N, Yu J, Yan M (2015) Cyto-sensing in electrochemical lab-on-paper cyto-device for in-situ evaluation of multi-glycan expressions on cancer cells. Biosens Bioelectron 63:232–239

Sun JY, Cheng CM, Liao YC (2015) Screen printed paper-based diagnostic devices with polymeric inks. Anal Sci 31(3):145–151

Terai T, Nagano T (2013) Small-molecule fluorophores and fluorescentprobes for bioimaging. Pflugers Arch 465(3):347–359

Thuo MM, Martinez RV, Lan W-J, Liu X, Barber J, Atkinson MBJ, Bandarage D, Bloch J-F, Whitesides GM (2014) Fabrication of low cost paper based microfluidic devices by embossing or cut-and-stack methods. Chem Mater 26(14):4230–4237

Tsai TT, Shen SW, Cheng CM, Chen CF (2013) Paper-based tuberculosis diagnostic devices with colorimetric gold nanoparticles. Sci Technol Adv Mater 14:044404

Vignali DA (2000) A multiplexed particle-based flow cytometric assays. J Immunol Methods 243:243–255

Von Lode P (2005) Point-of-care immunotesting: approaching the analytical performance of central laboratory methods. Clin Biochem 38:591–606

Wang S, Ge L, Song X, Yu J, Ge S, Huang J, Zeng F (2012) Paper-based chemiluminescence ELISA: lab-on-paper based on chitosan modified paper device and waxscreen-printing. Biosens Bioelectron 31(1):212–218

Wang Y, Ge L, Wang P, Yan M, Ge S, Li N, Yu J, Huang J (2013) Photoelectrochemical lab-on-paper device equipped with a porous Au-paper electrode and fluidic delay-switch for sensitive detection of DNA hybridization. Lab Chip 13:3945–3955

Weaver AA, Reiser H, Barstis T, Benvenuti M, Ghosh D, Hunckler M, Joy B, Koenig L, Raddell K, Lieberman M (2013) Paper analytical devices for fast field screening of beta lactam antibiotics and antituberculosis pharmaceuticals. Anal Chem 85:6453–6460

Whitesides GM, Martinez AW, Phillips ST (2010) Diagnostics for the developing world: microfluidic paper-based analytical devices. Anal Chem 82(1):3–10

Yamada K, Takaki S, Suzuki K (2013) Microfluidic paper-based analytical device for fluorescence detection of lactoferrin in tear fluid. In: Chemical and biological microsystems society. 17th international conference on miniaturized systems for chemistry and life sciences (MicroTAS 2013). Chemical and Biological Microsystems Society, Freiburg, Germany

Yan J, Yan M, Ge L, Yu J, Ge S, Huang J (2013) A microfluidic origami electrochemiluminescence aptamer-device based on a porous au-paper electrode and a phenyleneethynylene derivative. Chem Commun 49:1383–1385

Yang H, Kong Q, Wang S, Xu J, Bian Z, Zheng X, Ma C, Ge S, Yu J (2014) Hand-drawn & written pen-on-paper electrochemiluminescence immunodevice powered by rechargeable battery for low-cost point-of-care testing. Biosens Bioelectron 61:21–27

Yetisen AK, Martinez-Hurtado JL, Garcia-Melendrez A, da Cruz Vasconcellos F, Lowe CR (2014) A smartphone algorithm with inter-phone repeatability for the analysis of colorimetric tests. Sensors Actuators B Chem 196:156–160

Yong H, Wu Y, Fu J-Z, Wu W-B (2015) Fabrication of paper-based microfluidic analysis devices: a review. RSC Adv 5:78109–78127

Yu J, Wang S, Ge L, Ge S (2011) A novel chemiluminescence paper microfluidic biosensor based on enzymatic reaction for uric acid determination. Biosens Bioelectron 26(7):3284–3289

Yuan YR, Yuan R, Chai YQ, Zhuo Y, Miao XM (2009) Electrochemical amperometric immunoassay for carcinoembryonic antigen based on bi-layer nano-au and nickel hexacyanoferrates nanoparticles modified glassy carbon electrode. Electroanal Chem 626:6–13

Zhang X, Li J, Chen C, Lou B, Zhang L, Wang E (2013) A self-powered microfluidic origami electrochemiluminescence biosensing platform. Chem Commun 49:3866–3868

Zhou M, Yang M, Zhou F (2014) Paper based colorimetric biosensing platform utilizing cross-linked siloxane as probe. Biosens Bioelectron 55:39–43

Zhu W-J, Feng D-Q, Chen M, Chen Z-D, Zhu R, Fang H-L, Wang W (2014) Bienzyme colorimetric detection of glucose with self-calibration based on tree-shaped paper strip. Sensors Actuators B Chem 190:414–418

Emerging Point-of-Care Diagnostic Methods for Disease Detection

16

Smriti Singh, Pranav Tripathi, and Seema Nara

Abstract

Diagnosis being an integrated part of healthcare industry majorly focuses on sensitivity, specificity and cost-effectiveness to ensure maximum projection of population. These parameters are interdependent and tend to change according to the disease to be diagnosed. Being an inseparable part of diagnostics, point-of-care diagnostics (POCD) holds a significant place and captures a giant market segment too. Despite tangible progresses in POC diagnostics, issues regarding consistency, data acquisition, correlation, data annotation and data relocation require novel strategies for addressing. In this context, this chapter discusses the existing as well as the emerging technologies where an optimal sequential advancement can be studied in reference to POC diagnostics. The chapter discusses the change in recent approaches due to which admirable reports have illuminated novel technologies.

Keywords

POCD · Lateral flow immunoassay · Microfluidics · Surface plasmon resonance · Point-of-care testing

16.1 Introduction

For an accomplished treatment and cure of disease, knowledge about the cause and symptoms are very important. The treatment solely depends upon the identification and confirmation of the reason which is directly or indirectly responsible for infection or diseased state which we call as diagnosis. The consequences of

S. Singh · P. Tripathi · S. Nara (✉)
Department of Biotechnology, Motilal Nehru National Institute of Technology, Allahabad/Prayagraj, Uttar Pradesh, India
e-mail: seemanara@mnnit.ac.in

S. Paul (ed.), *Biomedical Engineering and its Applications in Healthcare*,
https://doi.org/10.1007/978-981-13-3705-5_16

inaccurate or delayed diagnosis may result in increased morbidity or mortality. An effective and specific treatment of a disease depends on how early a disease can be diagnosed. Conventional diagnostic methods like culture-based methods, biochemical tests, histological studies, etc. are limited to the well-equipped laboratory settings, costly, time-consuming, and for various diseases do not allow timely diagnosis. Some limitations of conventional diagnostic methods include:

- More time for sample analysis.
- High costs.
- Low sensitivity and specificity.
- Need for specialized and trained persons to conduct the tests.
- Patient (samples) clinical tests cannot be done at home; sometimes due to distance, it becomes fatal.
- Sometimes invasive and painful for patient, e.g. biopsy, hysteroscopy.
- Most amount of the sample remains unused, and only a single test can be performed with sample; rest of the components are wasted.

To address the above-said limitations of conventional methods, a major focus of the diagnostic sector is now on developing point-of-care diagnostic devices. Point-of-care diagnostics (POCD) allows brisk detection of disease near the subject and facilitates improved disease diagnosis, prognosis and management. It enables swift medical decisions, as the diseases can be diagnosed at a basic stage, leading to better health outcomes for patients by enabling a timely treatment. A point-of-care testing (POCT) device is simple to use, cost-effective, and easy to evaluate, doesn't require any toxic component and has robust storage and usage of reagents and consumables (Loonen et al. 2012). POCT devices can offer various benefits such as less clinical visits, trimmed hospital stay, optimized cure, appropriate use of drugs and improved quality of life. The global POCT market is expected to rise from US$23.16 billion (2016) to US$ 36.96 billion in 2021 (Point-of-Care/Rapid Diagnostics Market by Testing 2016).

WHO has issued ASSURED guidelines for POC devices as elaborated below. It is difficult for a single POC device to meet all ASSURED criteria; however, an effective POC device shall meet maximum possible requirements.

- **A**ffordable – for those at risk of infection
- **S**ensitive – minimal false negatives
- **S**pecific – minimal false positives
- **U**ser-friendly – minimal steps to carry out test
- **R**apid and **r**obust – short turnaround time and no need for refrigerated storage
- **E**quipment-free – no complex equipment
- **D**elivered – to end users

Some of these devices are also known as lab-on-a-chip devices as they translate a simple chemical lab in a small portable and disposable format by integrating measuring and sensing components on a single platform (Harpaz et al. 2017).

Such miniaturized POCT devices like microfluidics or nanofluidics have minimal sample/chemical requirements, are rapid and may perform multiple steps in one process (Price et al. 2010).

POCT industry has traversed from simple dipsticks to lateral flow immunoassays, paper-based fluidics and microfluidic devices. Recent advancements are witnessing the upsurge of nanofluidics platform, integration of smartphone or wireless technology with POCT devices and amalgamation of nanotechnology with POCT (Vashist et al. 2015). This chapter intends to acquaint the readers with basic concepts of some existing POCT technologies and then traverse them further along the recent advancements happening in the area of POCT devices for disease detection.

16.2 Existing POCT Technologies

Numerous POCT technologies are being presently used not only at patient's bedside but also in critical care units, nursing homes, physician's office, emergency vehicles, pharmacies, etc. Such technologies include POCT analysers which determine a broad range of analytes using spectrometry, turbidimetry and potentiometry analysis. Other such technologies include biosensor devices like glucose meters and dipsticks, e.g. urine analysis, lab-on-chips and lateral flow devices (e.g. pregnancy testing). This section presents some of these commercially successful POCT technologies.

16.2.1 Dipsticks

It is the simplest form of POCT devices usually made of paper or membrane which is impregnated with reagents and provide qualitative or semi-quantitative estimation. A test reagent which can be an antibody or any other biochemical is impregnated on the test strip and allowed to interact with the analyte present in the test sample. Upon interaction with the analyte, test reagent generates a detectable signal by usually bringing a change in colour at the test spot (Price and Thorpe 1999). Urine analysis using dipstick is most popularly used for overall urine analysis and indicates the presence of proteins, cell (Ohly and Teece 2003), bile pigments and glucose (Oneson and Groschel 1985). Apart from urine dipstick, some other commercially available dipstick assays are listed in Table 16.1. Dipsticks are ready to use, cost-effective, and have simple framework, rapid processing and semi-quantitative outputs. There are some limitations like less sensitivity, limited detection applications and low throughput.

16.2.2 Lateral Flow Assay

Lateral flow assays (LFAs) are performed on membrane strip with pre-immobilized reagents, which become activated by flow of liquid sample. It combines the

Table 16.1 Some commercially available dipstick-based assays

Product name	Company	Analyte	Disease	References
LEPTOCHECK®-WB	Tulip Diagnostics	IgM	*Leptospirosis*	http://www.tulipgroup.com/
SLIM HIV ™	Tulip Diagnostics	HIV-1/HIV-2 antibodies	AIDS	
Gravi-Chek®	Tulip Diagnostics	hCG	Pregnancy	
Dengue check-™ NS1	Zephyr Biomedicals	Dengue NS1 antigen	Dengue	
Leish check	Zephyr Biomedicals	Visceral *leishmaniasis* K39 antibody	*Leishmaniasis*	
HIV 1/2 STAT-PAK® DIPSTICK Assay	Chembio	HIV-1 or HIV-2 antibodies	AIDS	http://chembio.com
UTRiPLEX ™	Mologic Ltd	Leucocytes, blood, nitrite	UTI	https://mologic.co.uk
Keto-Diastix	Bayer Diagnostics	Glucose and ketone	Diabetes	https://www.allegromedical.com

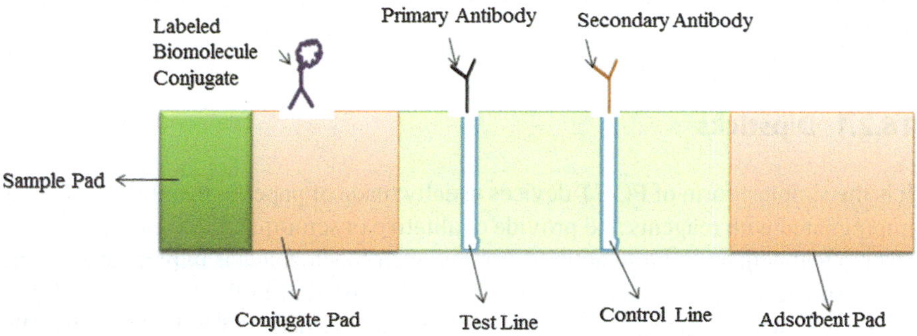

Fig. 16.1 Schema of lateral flow assay

advantage of bio-recognition molecule and chromatography. They are remarkably designed for single use at point-of-care/need, i.e. outside the laboratory.

These tests are also known as immunochromatographic tests, because the movement of molecule across the membrane is solely governed by capillary force. LFA consists of various overlapping layers like sample application pad, test line, absorbent pad, control line and conjugate pad (Fig. 16.1).

The basic principle, working and detailed architecture of LFIA can be studied through various reviews (Singh et al. 2015; Tripathi et al. 2017). On the basis of capture/bio-recognition molecule used, LFIA is mainly of two types, antibody-based lateral flow immunoassay (ALFIA) and nucleic acid lateral flow immunoassay (NALFIA) (Gonzalez et al. 2011; Liu et al. 2015; Chen et al. 2016). For better

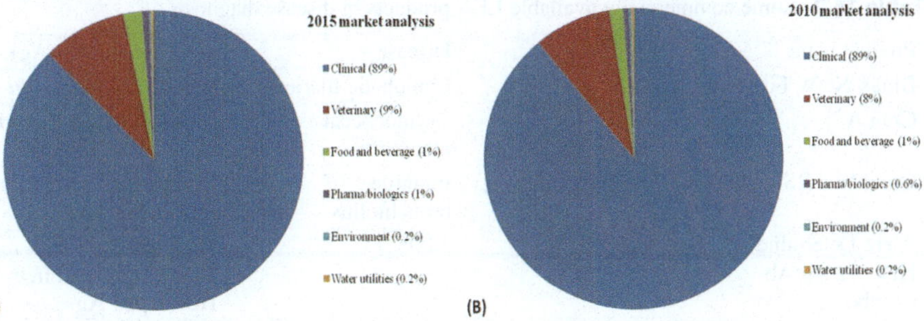

Fig. 16.2 World lateral flow test sales revenue (US$ Million) of the year (**a**) 2015 (**b**) 2010 (O'Farrell 2013)

signalling, these recognition molecules are tagged with some fluorophore or nanoparticles, which produce detectable signal for analysis of assay.

The global market for lateral flow-based tests is estimated to be at US$ 6.0 billion in 2018, and with an increment of 7.7% compounded annual growth rate (CAGR), it is supposed to touch $8.7 billion mark in 2023 (Lateral Flow Assay Market by Application 2018). Figure 16.2 depicts the global sales revenue of lateral flow tests in 2010 and 2015 along with their compound annual growth rate (CAGR). Table 16.2 enlists various commercially available LFA tests used in diagnostics in various fields.

LFAs are rapid, portable, user-friendly and reproducible, have less interference and have long shelf life. However, their limitations include low signal intensity, mostly qualitative and semi-quantitative, and sometimes needing pre-treatment of sample.

16.2.3 Microfluidics

A recent report from Yole Développment estimates that the market for microfluidic chips and microfluidic-based tests for point-of-need (PoN) testing applications should rise from $4 billion in 2017 to 13.2 billion in 2023 with 23% of growth rate per year (Asma and Sebastian 2018). Microfluidic devices can provide a fully integrated multiplexed point-of-care testing (MPOCT) device for fluid handling, sample processing and signal generation (Kozel and Bumham-Marusich 2017; Song et al. 2015). Microfluidic-based devices use channels to transfer small amounts of fluid by actuation forces. Microfluidics possess a great potential in an array of biological relevances (Edington et al. 2011; Sun et al. 2012), enzymatic assays (Cooksey et al. 2011; Sareenoi et al. 2012), immune hybridization reactions (Tavares et al. 2012, Seefeld et al. 2011) and PCR (Zhang et al. 2012; Yu et al. 2012). Some FDA-approved microfluidic-based POCT devices and their applications are mentioned in Table 16.3.

Table 16.2 Some commercially available LFA products in disease diagnosis

Product name	Company	Disease	Analyte/antigen (Ag)
Binax NOW Filariasis	Alere	Lymphatic filariasis	*Wuchereria bancrofti*
Crag A	IMMY	Cryptococcal meningitis	*C. neoformans, C. gattii*
QuickVue RSV Test	Quidel Corp.	Infantile bronchiolitis	Respiratory syncytial virus (RSV) Ag
Alere Determine HIV-1/2 Ag/Ab Combo	Alere	AIDS	HIV-1/HIV-2 antibodies and free HIV-1 p24 Ag
Panbio Dengue Duo Cassette	Alere	Dengue fever	IgM and elevated IgG
Alere™ Influenza A & B Test	Alere	Influenza	Influenza A and B nucleoprotein Ag
Alere Determine TB LAM	Alere	Tuberculosis (TB) inHIV-positive patients	Lipoarabinomannan (LAM) Ag
Uni-Gold™ Legionella Urinary Antigen PLUS	Trinity Biotech	Legionnaire's disease	*Legionella pneumophila* serogroup 1 Ag
Binax NOW	Alere	Malaria	*Plasmodium Ag*
Binax NOW	Alere	Pneumonia	*S. pneumoniae* Ag
On Site PSA Rapid Test	CTK Biotech	Prostate cancer	Prostate-specific antigen (PSA)
Alere NMP22 BladderChek	Alere	Bladder cancer	Nuclear matrix protein (NMP22)
Clearview iFOBT	Alere	Colon cancer	Faecal occult blood
Oncoe6 Cervical Test	Arbor Vita Corp.	Cervical cancer	E6 oncoproteins
CA125 rapid test kit	Quicking Biotech Co., Ltd	Ovarian cancer	CA125 Ag
AFP Test	Innovation Biotech	Hepatocellular cancer	Alpha-fetoprotein Ag
Status first CHF NT-probnp	Lifesign	Congestive heart failure	NT-probnp
Rapid Response CK-MB Test	BTNX Inc.	Myocardial infarction (MI)	Creatine kinase MB (CKMB)
Ichroma™ CK-MB Test	Boditech Med Inc.	Myocardial infarction (MI)	Creatine kinase MB (CKMB)
Ramp myoglobin test	Response Biomedical Corp.	Acute myocardial infarction (AMI)	Myoglobin
Meritas® Troponin I	Trinity Biotech	Myocardial infarction (MI)	Troponin I
Rapid response d-dimer test	BTNX Inc.	Venous Thromboembolism (VTE)	D-Dimer
Instant-View Troponin I	American Screening Corp.	Acute myocardial infarction (AMI)	Troponin I
Kala Azar test	Oscar Medicare Pvt. Ltd.	Leishmania	Anti K-39

Adopted from Sharma et al. (2015) under CC BY 4.0

Table 16.3 List of FDA-approved microfluidic-based POC tests

Test name	Company	Tested analyte/ parameter	Required component training
The Piccolo®	Abaxis	Blood chemistries	Disposable test discs and portable analyser
Alere Triage® MeterPro	Biosite (Alere)	Blood/urine chemistries including "waived" Lipid and liver panels	Disposable cartridges and metre
The epoc® Blood Analysis System	Epocal (Alere)	Blood chemistries	Test cards and wireless card reader with a host mobile computer
Simplexa	Focus Dx (Quest)	Flu A/B and RSV	PCR platform (3M™ Integrated Cycler) and amplification discs with sample wells
BD MAXTM GBS Assay IDI-Strep B Assay	HandyLab (BD)	Group B streptococcus (GBS)	Cepheid Smart Cycler® RT-PCR system, disposable cartridge and computer system
i-STAT Analyser	i-STAT Corp (Abbot)	Blood chemistries; coagulation; cardiac markers	Handheld analyser and self-contained cartridges
FilmArray	Idaho Technologies	RP panel of respiratory pathogens	Film array instrument and freeze-dried reagents in pouches
Liat™ Influenza A/B Assay	I Quum	H1N1 influenza viral RNA	Benchtop analyser disposable assay tubes
ABORhCard	Micronics (Sony)	ABO and Rh blood typing	Disposable cards that with anti-A, B and D antibodies printed into discrete microfluidic channels
Medical Proxima	Sphere	Blood chemistry	Disposable multi-parameter microanalyser, silicon chips
TearLab	TearLab	Osmolarity System	Disposable microchip works in conjunction with a pen that conveys data to the osmolarity reader

Adopted from Sharma et al. (2015) under CC BY 4.0

On the basis of fabrication substrate of the chip, microfluidic devices are glass based, silicon based, polymer based and paper based. Microfluidic-based assay has some outstanding features in comparison with other POCT methods like simultaneous processing of samples, rapid sampling, accuracy and control of multiple sample handling, low power utilization and flexible format for inclusion of various recognition schemes, thereby leading to augmented sensitivity. This technique also has few limitations like having limited multiplexed detection, sometimes poor separation of sample.

16.3 Emerging Technologies in POC Diagnostics

POCT devices are evolving each and every year in terms of their sensitivity, specificity and technology integration. Nowadays, POCT devices are an outcome of integrating existing technologies with upcoming ones like plasmonic in microfluidics, modified NPs in microfluidics or plasmonic. This section of the chapter presents such emerging technology in POCT which are being used for disease detection.

16.3.1 Plasmonic Resonance Technologies

Generally, the optical biosensor devices are based on monitoring of refractive index change, absorption and spectroscopic parameters. These devices comprise technologies like photonic crystal fibres (PCF), nano/microring resonator structures (NRS), interferometric, plasmonic nano/microarrays and surface plasmon resonance (Fan et al. 2008; Qavi et al. 2009).

Plasmonics can be divided into two categories:

(i) Propagating surface plasmon resonance (SPR) of electromagnetic forms along a metallic or dielectric interface. Surface plasmons are the delocalized electron oscillations that exist at the interface between two media, viz. metal-air interfaces, that are triggered by an electromagnetic wave, hence delocalizing electron oscillations and generating local electric fields.
(ii) Localized surface plasmon resonance (LSPR) in nanoscopic metallic structures impounds light to metal nanostructures that are smaller than the wavelength of light, thereby enabling the measurement of the dielectric changes (Li et al. 2015).

SERS effect of nanostructures was employed for developing rapid point-of-care diagnostic assays. Fu et al. (2016) have used malachite green isothiocyanate functionalized gold nanoparticles (MGITC-AuNPs) as SERS nanotags in lateral flow assay for detection of HIV-1 DNA marker. These Raman reporter AuNPs give strong Raman enhancement effect against conventional AuNPs as shown in Fig. 16.3. Mevold et al. (2015) have used graphene-poly(diallyldimethylammonium chloride) [PDDA] nanosheets for absorption of AuNPs onto it which are then used for creating strong SERS signal. Their detection ability is demonstrated for detection of *Staphylococcus aureus* and small molecules (adenine).

A shift in the LSPR peak of gold nanostructures has been used for developing label-free immunoassays. Yang et al. (2016) have reported a plasmonic ELISA-based sensitive detection of acute myocardial infarction (AMI) using myoglobin biomarker. In this sandwich ELISA, second antibody is labelled with glucose oxidase which catalyses the breakdown of glucose to generate H_2O_2. H_2O_2 under a specific HRP concentration etch the surface of gold nanorods (GNRs) causing a significant blue shift in their LSPR. This group has also developed a smartphone-

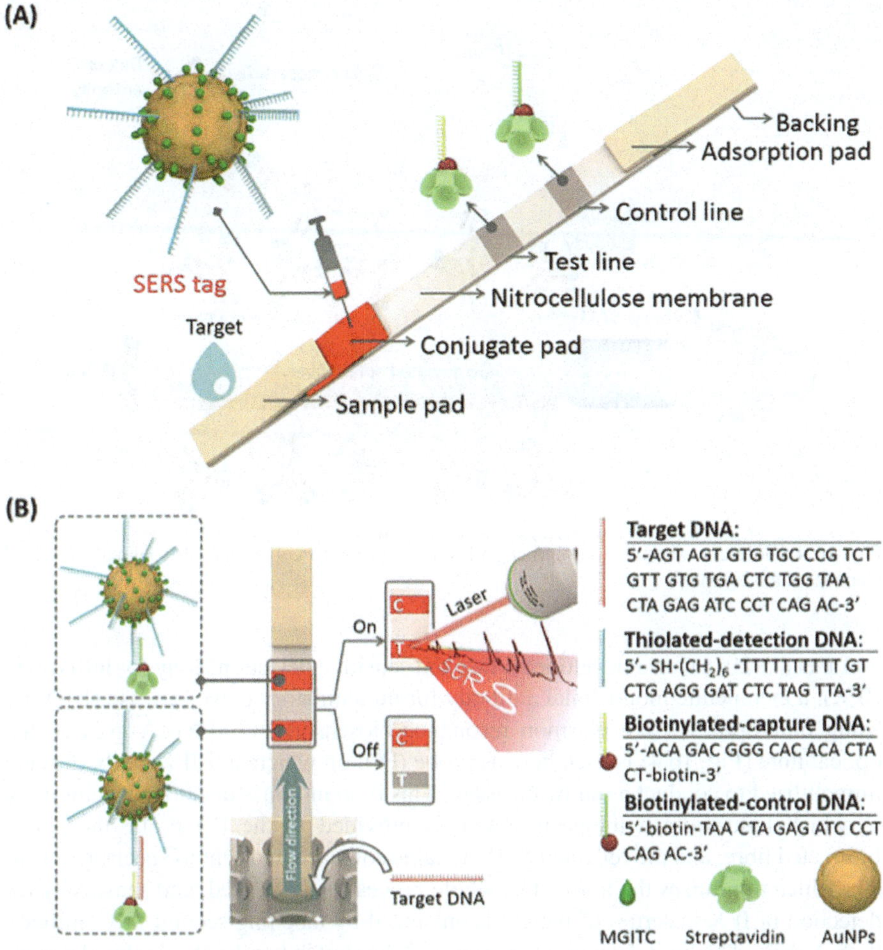

Fig. 16.3 Raman reporter-based enhancement of SERS signal. (**a**) General schema. (**b**) Raman reporter-based amplification of both control and test lines. (Adapted from Fu et al. 2016)

based plasmonic immunoassay reader which is used to quantitatively analyse the blue shift in LSPR of GNRs. The limit of detection is 0.057 ng mL^{-1}. Transition from disaggregated to aggregated phase brings a change in LSPR peak of gold nanorods, and this principle has been used for developing a rapid label-free detection assay for galectin-1 cancer biomarker (Zhao et al. 2016). GNRs are surface modified with lactose (Lac-GNRs) which is specifically recognized by galectin-1 cancer biomarker. Galectin-1 in the test sample interacts with Lac-GNRs through lactose and causes their aggregation resulting in the shift of LSPR peak and appearance of a new peak at ~ 900 nm. The developed assay is highly specific in aqueous solutions or in the complex and heterogeneous serum specimens.

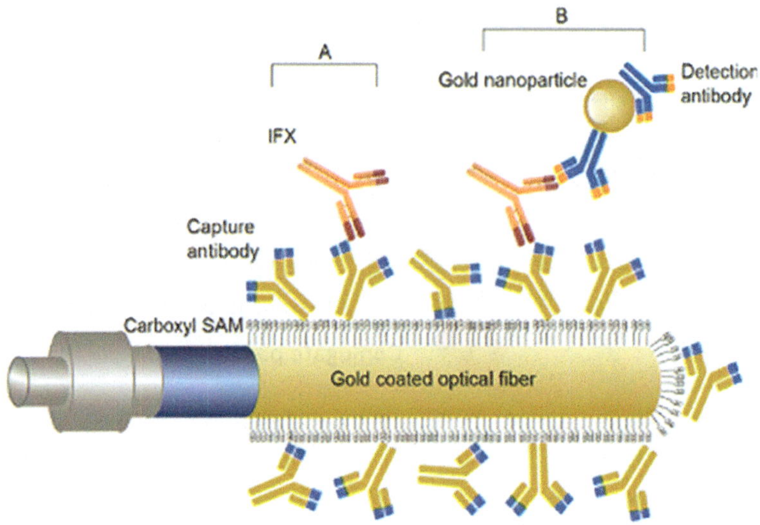

Fig. 16.4 Gold-coated optical comprising plasmonic probe to collect the SPR signal. (Adapted from Lu et al. 2016)

Lu et al. (2016) have developed a fast and sensitive detection assay for infliximab (IFX), a therapeutic monoclonal antibody for inflammatory bowel disease (IBD) by using fibre-optic surface plasmon resonance biosensor (FO-SPR). A gold-coated optical fibre (Fig. 16.4) is used here as probe (FO) on which anti-IFX antibodies are immobilized to conduct a sandwich assay. This is an in-house developed technology in which light from a tungsten source is provided to the FO probe through a bifurcated fibre, and the reflected SPR signal is collected and sent to spectrophotometer which minimizes the noise. The whole process is automated, and sensitivity for detection of IFX in serum of patient is enhanced by labelling second IFX antibody with AuNPs. The limit detection achieved is 2.2 ng/ml in hundredfold diluted serum.

Monteiro et al. (2016) have developed a plasmonic microfluidic device, for direct detection of analyte (HER2) without inference of any labelled molecules. The device relies on the sensitive refractive index changes of a nanohole array developed on a gold film. Interaction of HER2 analyte with its specific antibody immobilized on substrate greatly reduces the refractive index of the medium and cuts down the intensity of light transmitted through nanohole arrays. The light is channelled to a CCD camera of a microscope which captured the magnified images that are finally analysed by ImageJ software. The assay is highly sensitive with limit of detection of 3 ng mL^{-1} known as HER2 antigen.

The above-mentioned reports regarding plasmonics technology show a distinct trend that the development is majorly focussed on two phases: first before the signal capture and second after the signal capture. There is an upcoming saturation in the signal amplification section due to scarcity of novel reports, so the emerging technologies focus on the array of systems used for capturing the signal.

16.3.2 Nanostructures in POC Diagnostics

According to European Commission, nanostructures are defined as "A material having 50% or more particles of size in range 1–100 nm and these particles can be natural, incidental or manufactured material containing particles, in an unbound state or as an aggregate or as an agglomerate" on 18 October 2011. With extraordinary properties of reduced size, high electron mobility, and specific surface area, quantum effect with thermal and electrical conductivity, nanostructures suffice for a suitable candidate for development of sensors (Pingarron et al. 2008; Kuila et al. 2011). Tripathi et al. (2017) have discussed the use of nanostructures as labels in POC diagnostic assays. A number of nanostructures like metallic nanoparticles (Au, Ag and Fe), quantum dots, graphene and carbon nanotubes have been incorporated as labels or signal enhancers in POC diagnostics giving rise to a new phase of diagnostics called as nano-diagnostics. For the detection of infectious diseases, nano-diagnostic platforms achieve rapid and reliable conclusions by using various patient matrices. Another review from Quesada-González and Merkoçi (2018) presents various nanomaterial-based diagnostic devices that were used for point-of-care applications.

Gold nanoparticle-graphene oxide nanohybrids are recently being used for developing various point-of-care assays because of facile bioconjugation on graphene sheets and enhanced SPR effects. Saeed et al. (2017) have developed a diagnostic platform for early-stage diagnosis of breast cancer marker using capture/reporter oligonucleotides and gold nanoparticle tailored graphene oxide (AuNPs-GO) surface. A conjugate of gold nanoparticles-graphene oxide used for signal amplification is drop casted over a glassy carbon electrode and used as a novel platform for adsorption of thiolated capture oligonucleotide which can specifically hybridize with a terminal end of the target DNA biomarker of breast cancer. Another reporter oligonucleotide conjugated with horseradish peroxidase hybridizes with another end of the target DNA. Upon reaction, cyclic voltammetry analysis of redox-based conversion of TMB by HRP is carried out. The response shows detection limit of 1.6×10^{-10} M and 2.3×10^{-10} M with sensitivity 219 nA/nM and 378 nA/nM for the target CD24 and ERBB2, respectively. Chiu et al. (2018) have demonstrated that adsorption of AuNPs on graphene oxide sheet causes enhanced surface plasmon absorption characteristics of AuNPs. Such AuNPs-GO nanocomposites result in highly sensitive detection of anti-BSA (limit of detection = 145 fM).

De la Escosura-Muñiz et al. (2016) have reported the use of AuNPs and magnetic beads (MBs) for electrochemical recognition of isothermal amplified *Leishmania* DNA. It helps in easy purification of amplified target DNA (through MBs) and amplified signal generation (catalytic action of AuNPs to convert H^+ to H_2). The technology incorporates dual-tagged DNA primers for amplification of analyte from genomic DNA extracted from blood. The assay exhibits a good reproducibility and sensitivity; up to 0.8 parasites per mL of blood (8×10^{-3} parasites per DNA amplification reaction) and magnetic beads provide easy separation. In addition to the advantages of single-step recognition and uncomplicatedness of the amplified product separation, the performance of the proposed proof of concept is superior to various other point-of-care tests for *Leishmania*.

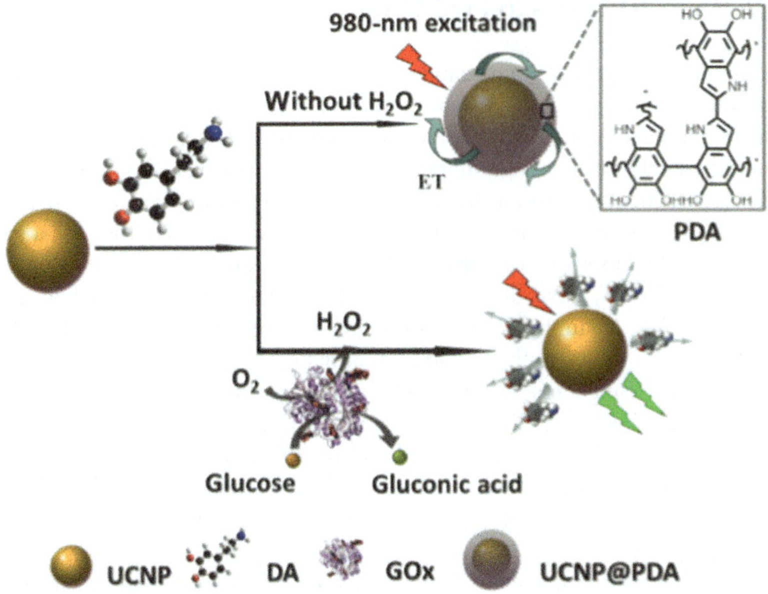

Fig. 16.5 Use of upconversion nanoparticles for glucose detection through polydopamine quenching mitigation. (Adapted from Yan Liu et al. 2018)

Liu et al. (2018) have used upconversion nanoparticles polydopamine (UCNPs-PDA) nanosystem for the accurate detection of glucose in human serum. Polymerization of dopamine on the lanthanide surface of ligand-free UCNPs remarkably quenches their upconversion luminescence (97.4%) under 980-nm excitation. PDA acts as an effective quencher either due to FRET or inner filter effect. H_2O_2 in the solution can inhibit this quenching, and the luminescence of UCNPs can be retained (Fig. 16.5). This principle can be employed for developing label-free point-of-care diagnostic assays of disease-specific biomarkers.

The emerging trend in the use of nanostructures for point-of-care applications is more lineated towards using hybrid nanostructures as reporters or probe so as to amalgamate various properties of nanostructures in a single probe. Further, the focus is more on developing nanostructure-based label-free immunoassays, wherein their intrinsic catalytic properties are being explored a lot.

16.3.3 Smartphone-Integrated Diagnostic Platforms

Smartphone technology has been a reform and develops day by day (Eisenstein 2012). A common smartphone with operating systems like Android or iOS is enriched with high quality of microphone and high-resolution camera, which can be useful in detection of biological signals. These smartphones possess a powerful processor and memory to analyse, annotate and curate the data also (Lane et al. 2010; Mertz 2012). Earlier reports documented the use of smartphone for microscopy and

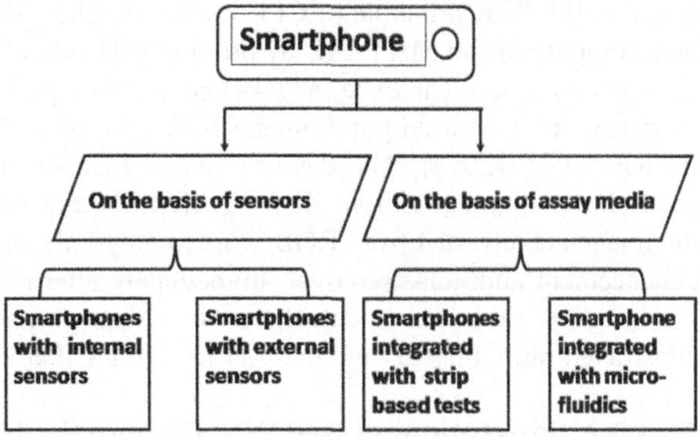

Fig. 16.6 Different types of smartphone integrations with POC diagnostics

optoelectronic sensing, in the form of smartphone-based microscopy and smartphone-based cytometry (Ozcan 2019). Current smartphones are enabled with about 40,000 mobile health applications that are made available on application downloading platforms (Gold 2012). Smartphones may contribute towards POCT devices by making a diagnostic assay quantitative, portable and cost-effective. The most recent report of Coleman et al. (2019) has established a new concept of using video recording feature present in smartphones to monitor the pixel intensity. It is demonstrated for detection of Zika virus on a 96-well plate where a distinct region of interest was identified on each well (1 region/well). From the obtained region of interest, mean pixel intensity is calculated and correlated with absorbance values. The assay results correlate well with ELISA-based kit. Smartphone-based diagnostics can be classified as shown in Fig. 16.6.

16.3.3.1 Smartphone-Based POC Diagnostics Integrated with Internal Sensors

Smartphones have various inbuilt features such as advanced sensors, microphone and camera (the resolution and quality of camera) which can be used in POC diagnostics. These features have seen a continuous improvement within past decades, e.g. images of eyes, fingertips and skin or video captured through smartphones can be post-processed with various pre-available algorithms. Most common image processing algorithms consist of Fourier transformation, colour signal (RGB) analysis, pattern recognition and segmentation of region (Jonathan and Leahy 2010; Pal et al. 2013; Scully et al. 2012; Kim et al. 2004; Wadhawan et al. 2012).

Ra et al. (2018) have established a proof of concept of correlating the data of a Donut-Shaped Nearness Urine Tester (DONUT) array with image acquired from a smartphone. The DONUT array is an array to qualitatively confirm the presence of ten different parameters of urine. The report perfectly correlates the data by defining a distinct region of interest which is further subjected to calculation of a matching factor from the available colour codes. Though this method does not provide an efficient solution in urine analysis, it may act as a platform to examine specific parameters of other diseases for a better prognosis. Barnes et al. (2018) have reported a smartphone-based "Bacticount" application for analysis of qPCR data. The process relies on RT-PCR principle of CFU counting, where the analysis is generally done on a computer using MATLAB by providing Tt values. These values when given to the app render it to calculate the CFU counts with proper correlation.

A smartphone-based pupillometer is developed to determine the pupil diameter by Kim et al. (2013). The devised sensor focuses on image acquisition from a smartphone coupled with an IR filter, IR LED and white LED. After acquisition, the image is converted from RGB scale to greyscale and further subjected to coarse enhancement and noise removal. Immediately after the morphological processing, the image is subjected to calculation of object pixels followed by calculation of constricted ratio, thus giving an exact idea of the diameter.

16.3.3.2 Smartphone-Based POC Diagnostics Integrated with External Sensors

External sensor systems have been formulated and employed to smartphone for extracting diagnostic information. Health information are extracted by using external sensors and processed or transmitted using a smartphone in the form of one-dimensional (e.g. body temperature and pulse rate), two-dimensional (e.g. ultrasound image) or three-dimensional (e.g. time-sequence ultrasound images) signals. With considerable advancements in smartphone's high-definition display, high resolution and processing capabilities, smartphone-based medical imaging has emerged as an area with great potential (Xie and Liu 2010).

Akraa et al. (2018) have reported a smartphone-based sensor for detection of proteinuria in urine of chronic kidney disease subjects. The study focuses on the correlation with data produced by BSPOTPE (an environmentally stable and synthetic fluorescent probe for detection of albumin in urine) reaction with urine or artificial urine. Smartphone is aided with external UV filters and UV LEDs to optimize the image. Upon capture, the image is subjected to analysis of HSL (hue, saturation and lightness) values for a distinct cognitive data and luminance was further measured. Hou et al. (2017) have reported a smartphone-based qualitative integration for lateral flow immunochromatographic strips. The array consists of an external light source coupled with angled mirrors for adjustment of image parameters to optimum (Fig. 16.7). Various operators are incorporated into the algorithm to enhance sensitivity and specificity by increasing the distinguishing parameters of test and control lines.

Liao et al. (2016) have reported a simple, minimally instrumented, cost-effective, easy-to-use device named as smart cup equipped with a smartphone for molecular detection of human simplex virus (HSV). Smart cup makes use of loop-mediated isothermal amplification of DNA to analyse the pathogens. The cup works by a thermal battery without any external electrical power supply. Generation of heat is a factor of reacting magnesium with water. The flashlight of phone is utilized to excite the fluorescence probe tagged to PCR amplified fragments, and inbuilt camera is used to capture and analyse the image. It is recently shown that an ultrasonographic probe plugged into a smartphone can acquire and display ultrasonographic results on the smartphone. It can take the images of muscular architecture of mouth floor and suprahyoid airway. This image can be transmitted to diagnostic centre for further interpretation (Wojtczak and Bonadonna 2013).

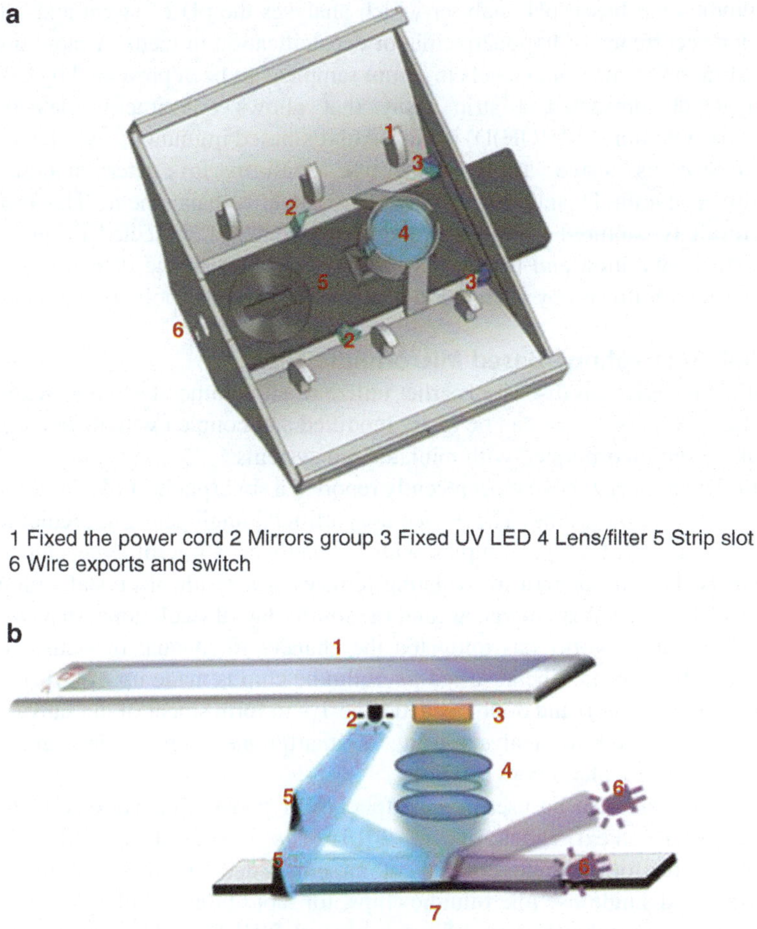

Fig. 16.7 Signal amplification in LFIA strips. (**a**) The external attachment, (**b**) image acquisition. (Adopted from Yafei Hou et al. 2017 under CC BY 4.0)

16.3.3.3 Smartphone-Based Strips, Tubes and Specimen-Based Tests

The commonly used diagnostic tests for various diseases are performed using strips and tubes and are colorimetric or based on counting of analytes (Hu et al. 2013). With integration of smartphone for POC, an image of strip or a test tube is acquired first and then analysed post-processing for counting targets.

Zhu et al. (2013) have also reported a compact smartphone-based microscopic platform for the measurement of the WBCs and haemoglobin concentration through image processing. The image is basically subjected to cropping the region of interest and extraction of saturation channel. The extracted saturation channel is used to generate bitmask image and blob detection which directly correlates with the concentration of cells. The specificity test of the strip is done with sweat, saliva and water. Onsescu et al. (2013) have demonstrated an integrated smartphone-based sensor for diagnosing the changes in pH of sweat and saliva. This assay is basically a microfluidic chip-based pH analyser which analyses the pH of sweat and saliva to check for occurrence of hyponatraemia or decalcification in teeth. A rapid quantification system for vitamin D levels in serum samples has been presented by Lee et al. (2014) which presented a strip assay that allows colorimetric detection of 25-hydroxyvitamin D (25(OH)D) using a AuNPs-based immunoassay. The reported immunoassay is tested as a competitive platform for determination using 25-hydroxyvitamin D antibody followed by silver enhancement. The proposed assay result is captured using an iPhone, and image is subjected to analysis of HGB (Hue, saturation and brightness) parameters between the detection area and reference area followed by calculation using a second-order polynomial equation.

16.3.3.4 Smartphone-Based Microfluidic Tests

Microfluidic devices as discussed earlier utilize small volume of sample (Wang et al. 2014; Lee et al. 2011) and can be easily modified and coupled with different pumps and valves and incorporated with miniaturized systems.

Potluri et al. (2019) have most recently reported a smartphone-linked microfluidic device for ovulation testing. The device uses artificial intelligence to predict ferning patterns in air-dried saliva samples with the help of a microfluidic chip in just 31 seconds. Ferning pattern in ovulating females is a result of crystal structure of salivary electrolytes that emerge around the fourth day of ovulation. Analysis using artificial neural networks has mitigated the chances of manual or statistical (low sample number) errors. The air-dried microfluidic chip is made up of PMMA (Poly (methyl methacrylate)) and double-sided adhesive to form smear of the saliva, which is further subjected to analysis using a smartphone-based optical attachment containing a 9 mm plano-convex lens.

Micropads can also be used in smartphone for performing various diagnostics assays. In this context, Lillehoj et al. (2013) have recently developed a compact smartphone platform which consists of an embedded circuit for digital signal processing and single-use microfluidic chips, for rapid biomolecular detection. The assay comprises sandwich format for detection of *Pf*HRP2 (a biomarker of malaria parasite) with the help of HRP-labelled secondary antibodies. The data from potentiostat is sent to a microcontroller and further transferred to the smartphone via a universal serial bus. The data is further analysed using the app-based graphical user interface.

16.3.4 POC Technologies with Wireless Integration

Wireless networks are another feature that is present with smartphones as well as accessible spectrum to transfer data at different permissible rates. Secure and immediate access to records is the most advanced advantage of this technology as this could help in error management as well as enabling crosschecking of data at a faster rate for a better diagnosis. A wide range of commercial sensors, such as electroencephalography (EEG), electromyography (EMG), electrocardiography (ECG), blood pressure, pulse oximetry, blood sugar and temperature sensors, have

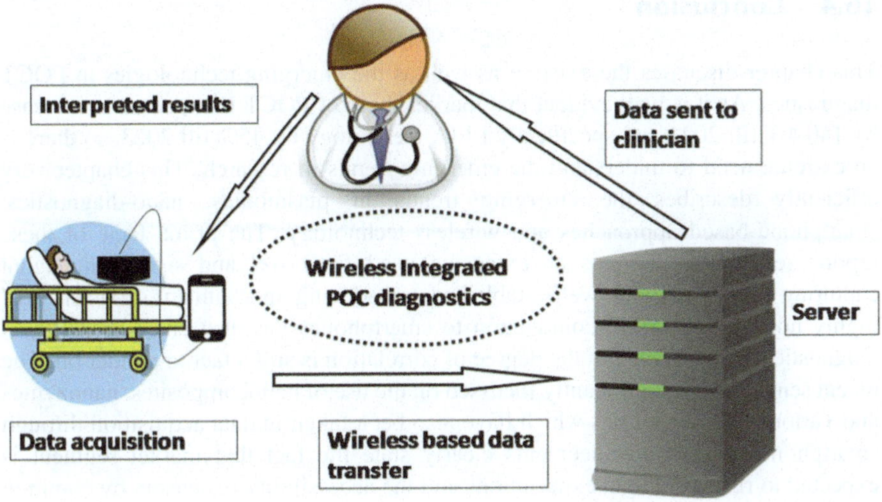

Fig. 16.8 Integration of wireless technology to attain highly accurate, cross-checked, reliable and fast results of tests

secured their place in market. There are basically four working blocks of a wireless technology-based sensor, i.e. the power unit, radio system, central processing unit and sensor module. The sensor or detection module detects the physiological condition, converts to machine language (digitization) and transmits to a data acquisition centre. The central processing unit controls the data annotation and curation portion and is responsible for communication with data servers. The radio system and power unit are responsible for data transfer and power management for increasing the shelf life of sensor, respectively.

Wireless technology has immensely changed the diagnostics due to the bridging the gap between data acquisition centres (which is patient itself in case of POC testing) and a physician. Figure 16.8 shows a simple diagram of communication between a patient, data acquisition centre and physician.

Fan et al. (2017) have recently reported a wireless POCT system comprising μPADs, electrochemical detector and smartphone to accomplish DPV measurements. The design of the portable electrochemical sensor on the printed circuit board is a conglomeration of thionine, gold and graphene-based nanocomposites on the working electrode. Results are transmitted to the smartphone using built-in Bluetooth from electrochemical sensor and are retrieved in the smartphone screen in real time.

Another report from Darwish and Hassanien (2011) has beautifully mentioned various types of wireless sensors including the wearable as well as the implantable ones. The need for POC has been enhanced greatly by integration of wireless technology as it renders fluency, accuracy and reliability in interpretations on such assays that are out of the zone for POC diagnostics. Wireless technology has also decreased the cost of skilled manpower ensuring efficiency.

16.4 Conclusion

This chapter discusses the existing as well as the emerging technologies in POCT diagnostics. As it is well evident that market share of POCT is expected to increase by 140.4% till 2021 and specifically LFIA technology by 45% till 2023, so there is an extreme need to understand the emerging trends in research. This chapter very efficiently describes the emerging trends in plasmonics, nano-diagnostics, smartphone-based approaches and wireless technology. The prima facie of these reports reveal that there is an extensive need for novel and significant signal capturing strategies with well-established transducing mechanisms. Existence of highly notable sensors in conjugation to smartphones has also emerged as novel diagnostic assay readers but the degree of correlation is still a fact to ponder on. The recent sensors have significantly focussed on the use of nanocomposites, nanozymes and various new deviations which have also been taken in data acquisition through smartphones. These advancements clearly state the fact that market segment is expected to rise more than expectations and the accessibility of sensors by common people is going to be realistically attainable.

Acknowledgements We greatly acknowledge the Department of Biotechnology, MNNIT, Allahabad, for providing all necessary support towards writing this chapter.

References

Akraa S, Anh PTT, Shen H et al (2018) A smartphone-based point-of-care quantitative urinalysis device for chronic kidney disease patients. J Netw Comput Appl 115:59–69

Asma S, Sébastien C (2018) Positive dynamics of microfluidic-based point-of-need. In: Point-of-need testing: application of microfluidic technologies report, Yole Dévelopement. http://www.yole.fr/iso_upload/News/2018/PR_POINT_of_NEED_TESTING_MarketGrowth_YOLE_February2018.pdf. Accessed on 29 Dec 2018

Barnes L, Heithoff DM, Mahan SP et al (2018) Smartphone-based pathogen diagnosis in urinary sepsis patients. EBio Med 36:73–82

Chen Y, Chen Q, Han M et al (2016) Near-infrared fluorescence-based multiplex lateral flow immunoassay for the simultaneous detection of four antibiotic residue families in milk. Biosens Bioelectron 79:430–434

Chiu NF, Chen CC, Yang CD et al (2018) Enhanced plasmonic biosensors of hybrid gold nanoparticle-graphene oxide-based label-free immunoassay. Nanoscale Res Lett 13:152

Cooksey GA, Elliott JT, Plant AL (2011) Reproducibility of a real-time microfluidic cell toxicity assay. Anal Chem 83(10):3890–3896

Darwish A, Hassanien AE (2011) Wearable and implantable wireless sensor network solutions for healthcare monitoring. Sensors (Basel) 11(6):5561–5595

de la Escosura-Muñiz A, Baptista-Pires L, Serrano L, Altet L, Francino O, Sánchez A, Merkoçi A (2016) Electrocatalytic detection: magnetic bead/gold nanoparticle double-Labeled primers for electrochemical detection of isothermal amplified DNA (small 2/2016). Small 12(2):204–204

Dengue Check Dipstick assay (2018). http://www.tulipgroup.com/Zephyr_New/qrg/Dengucheck%20NS1%20QRG.pdf. Accessed on 29/12/2018

Eisenstein M (2012) Miniature wireless sensors presage smart phone medicine. Nat Biotechnol 30(11):1013–1014

Fan XD, White IM, Shopova SI et al (2008) Sensitive optical biosensors for unlabeled targets: a review. Anal Chim Acta 620:8

Fan Y, Liu J, Wang Y et al (2017) A wireless point-of-care testing system for the detection of neuron-specific enolase with microfluidic paper-based analytical devices. Biosens Bioelectron 95:60–66

Fu X, Cheng Z, Yu J et al (2016) A SERS-based lateral flow assay biosensor for highly sensitive detection of HIV-1 DNA. Biosens Bioelectron 78:530–537

Gold J (2012) FDA regulators face daunting task as health apps multiply. In: USA Today [Online]. Available: http://usatoday30.usatoday.com/news/health/story/2012-06-22/health-apps-regulation/55766260/1

Gonzalez JM, Foley MW, Bieber NM et al (2011) Development of an ultrasensitive immunochromatography test to detect nicotine metabolites in oral fluids. Anal Bioanal Chem 400:3655–3664

Gravi Chek Dipstick for Pregancy Test (2018). http://www.tulipgroup.com/Qualpro_New/html/pack_inserts/Gravichek%20dipstick.pdf. Accessed on 29/12/2018

Harpaz D, Eltzov E, Seet R, Marks RS, Tok AI (2017) Point-of-care-testing in acute stroke management: an unmet need ripe for technological harvest. Biosensors 7:30

HIV-1/2 STAT PAK Dipstick assay (2018). http://chembio.com/products/human-diagnostics/hiv-12-stat-pak-dipstick-assay/. Accessed on 29/12/2018

Hou Y, Wang K, Xiao K et al (2017) Smartphone-based dual-modality imaging system for quantitative detection of color or fluorescent lateral flow immunochromatographic strips. Nanoscale Res Lett 12:291

Hu J, Wang L, Li F et al (2013) Oligonucleotide linked gold nanoparticle aggregates for enhanced sensitivity in lateral flow assays. Lab Chip 13(22):4352–4357

Jonathan E, Leahy M (2010) Investigating a smartphone imaging unit for photoplethysmography. Physiol Meas 31(11):N79–N83

Keto-diastix reagent strips for urine analysis (2018). https://www.allegromedical.com/diabetic-supplies-c520/keto-diastix-reagent-strips-p172163.html. Accessed on 29/12/2018

Kim TS, Yoon G, Lee J et al (2004) Method of extracting region of interest from tongue image and health monitoring method and apparatus using the tongue image. Europe EP1450287A2

Kim U, Ghanbari S, Ravikumar A et al (2013) Rapid, affordable, point-of-care water monitoring via a microfluidic DNA sensor and a mobile interface for global health. IEEE Trans Eng Heal Med 2013:1

Kozel TR, Burnham-Marusich AR (2017) Point-of-care testing for infectious diseases: past, present, and future. J Clin Microbiol 55:2313–2320

Kuila T, Bose S, Khanra P et al (2011) Recent advances in graphene-based biosensors. Biosens Bioelectron 26(12):4637–4648

Lane ND, Miluzzo E, Lu H et al (2010) A survey of mobile phone sensing. IEEE Commun Mag 48:140–150

Lateral Flow Assay Market by Application (Clinical Testing (Pregnancy, Infectious Disease (Mosquito, Influenza, STI, Hepatitis, TB), Cardiac Marker, Lipid Test) Veterinary, Food Safety) Product (Reader, Kits) Technique, End User – Global Forecast to 2023 (2018). https://www.marketsandmarkets.com/Market-Reports/lateral-flow-assay-market-167205133.html.Online. Accessed on 27/12/2018

Lee WG, Kim YG, Chung BG et al (2011) Nano/ microfluidics for diagnosis of infectious diseases in developing countries. Adv Durg Deliv Rev 62:449–457

Lee S, Oncescu V, Mancuso M et al (2014) A smartphone platform for the quantification of vitamin D levels. Lab Chip 14(8):1437–1442

Leishmaniasis Dipstick assay (2018). http://www.tulipgroup.com/Zephyr_New/qrg/leishcheck.pdf. Accessed on 29/12/2018

Leptochek Dipstick assay (2018). http://www.tulipgroup.com/Zephyr_New/html/pack_inserts/Leptochek.pdf. Accessed on 29/12/2018

Li WB, Zhang L, Zhou JH et al (2015) Well-designed metal nanostructured arrays for label-free plasmonic biosensing. J Mater Chem C3:6479
Lillehoj PB, Huang M, Truong N et al (2013) Rapid electrochemical detection on a mobile phone. Lab Chip 13:2950–2955
Liu Y, Wu A, Hu J et al (2015) Detection of 3-phenoxybenzoic acid in river water with a colloidal gold-based lateral flow immunoassay. Anal Biochem 483:7–11
Liu Y, Tu D, Zheng W et al (2018) A strategy for accurate detection of glucose in human serum and whole blood based on an upconversion nanoparticles-polydopamine nanosystem. Nano Res 11(6):3164–3174
Loonen AJ, Schuurman R, Van Den Brule AJ (2012) Highlights from the 7th European meeting on molecular diagnostics. Expert Rev Mol Diagn 12(1):17–19
Lu JD, Van Stappen T, Spasic D et al (2016) Fiber optic-SPR platform for fast and sensitiveinfliximab detection in serum of inflammatory bowel disease patients. Biosens Bioelectron 79:173–179
Mertz L (2012) Ultrasound? Fetal monitoring? Spectrometer? There's an app for that! IEEE Pulse 3(2):16–21
Mevold AHH, Hsu WW, Hardiansyah A et al (2015) Fabrication of gold nanoparticles/graphene-PDDA nanohybrids for bio-detection by SERS nanotechnology. Nanoscale Res Lett 10:397
Monteiro JP, de JH O, Radovanovic E et al (2016) Microfluidic plasmonic biosensor for breast cancer antigen detection. Plasmonics 11:45–51
O'Farrell B (2013) Lateral flow immunoassay systems: evolution from the current state of the art to the next generation of highly sensitive, quantitative rapid assays. In: Wild D (ed) The immunoassay handbook: theory and applications of ligand binding, ELISA and related techniques, 4th edn. Newnes, Boston, pp 89–107
Ohly N, Teece S (2003) Accuracy of negative dipstick urine analysis in ruling out urinarytract infection in adults. Emerg Med J20:362–363
Oncescu V, O'Dell D, Erickson D (2013) Smartphone based health accessory for colorimetric detection of biomarkers in sweat and saliva. Lab Chip 13:3232–3238
Oneson R, Groschel DH (1985) Leukocyte esterase activity and nitrite test as a rapid screen for significant bacteriuria. Am J Clin Pathol 83:84–87
Ozcan A (2019) Mobile phones democratize and cultivate next-generation imaging, diagnostics and measurement tools. Lab Chip 20:14
Pal A, Sinha A, Choudhury AD et al (2013) A robust heart rate detection using smart-phone video categories and subject descriptors. Proc MobileHealth 13:43–48
Pingarron JM, Yanez-Sedeno P, Gonzalez-Cortes A (2008) Gold nanoparticle-based electrochemical biosensors. Electrochim Acta 53:5848–5866
Point-of-Care/Rapid Diagnostics Market by Testing (Glucose, Lipids, HbA1c, HCV, HIV, Influenza, Urinalysis, Hematology, Cancer, Pregnancy, PT/INR), Platform (Lateral Flow, Immunoassay), Mode (Prescription, OTC), End-User – Global Forecast to 2022 (2016). Available online: https://www.marketsandmarkets.com/Market-Reports/point-of-care-diagnostic-market-106829185.html. Accessed on 27 Dec 2018
Potluri V, Kathiresan PS, Kandula H et al (2019) An inexpensive smartphone-based device for point-of-care ovulation testing. Lab Chip 19:59
Price CP, Thorpe G (1999) Disposable analytical devices in point-of care testing. In: Point-of-Care Testing (ed) CP Price and JM Hicks. AACC Press, Washington, pp 17–40
Price C, St John A, Kricka L (2010) Point-of-care testing. Needs, opportunities and innovation, 3rd edn. AACC Press, Washington
Qavi AJ, Washburn AL, Byeon JY et al (2009) Label-free technologies for quantitative multiparameter biological analysis. Anal Bioanal Chem 394:121
Quesada-González D, Merkoçi A (2018) Nanomaterial-based devices for point-of-care diagnostic applications. Chem Soc Rev 47(13):4697–4709
Ra M, Muhammad MS, Lim C et al (2018) Smartphone-based point-of-care urinalysis under variable illumination. IEEE J Transl Eng Health Med 6:2168–2372

Saeed AA, Sánchez JLA, O'Sullivan CK et al (2017) DNA biosensors based on gold nanoparticles-modified graphene oxide for the detection of breast cancer biomarkers for early diagnosis. Bioelectrochemistry 118:91–99

Sareenoi Y, Koehler K, Shapiro J et al (2012) Microfluidic electrochemical sensor for on-line monitoring of aerosol oxidative activity. J Am Chem Soc 134(25):10562–10568

Scully CG, Lee J, Meyer J (2012) Physiological parameter monitoring from optical recordings with a mobile phone. IEEE Trans Biomed Eng 59(2):303–306

Seefeld TH, Zhou WJ, Corn RM (2011) Rapid microarray detection of DNA and proteins in microliter volumes with surface Plasmon resonance imaging measurements. Langmuir 27(10):6534–6540

Sharma S, Zapatero-Rodríguez J, Estrela P et al (2015) Point-of-care diagnostics in low resource settings: present status and future role of microfluidics. Biosensors 5(3):577–601

Singh J, Sharma S, Nara S (2015) Evaluation of gold nanoparticle based lateral flow assays for diagnosis of enterobacteriaceae members in food and water. Food Chem 170:470–483

SLIM HIV Dipstick Assay (2018). http://www.tulipgroup.com/Qualpro_New/html/pack_inserts/SLIM_HIV_Dipstick_packinsert.pdf. Accessed on 29/12/2018

Song Q, Gao Y, Zhu Q et al (2015) A nanoliter self-priming compartmentalizaton chip for point of-care digital PCR analysis. Biomed Microdevices 17:64

Sun J, Gao Y, Isaacs RJ et al (2012) Simultaneous on-chip DC dielectrophoretic cell separation and quantitative separation performance characterization. Anal Chem 84(4):2017–2024

Tavares AJ, Noor MO, Vannoy CH et al (2012) On-chip transduction of nucleic acid hybridization using spatial profiles of immobilized quantum dots and fluorescence resonance energy transfer. Anal Chem 84(1):312–319

Tripathi P, Upadhyay N, Nara S (2017) Recent advancements in lateral flow immunoassays: a journey for toxin detection in food. Crit Rev Food Sci Nutr 58(10):1715–1734

Vashist SK, Luppa PB, Yeo LY et al (2015) Emerging technologies for next-generation point-of-care testing. Trends Biotechnol 33:692–705

Wadhawan T, Situ N, Rui H et al (2012) Implementation of the 7-point checklist for melanoma detection on smart handheld devices. Proc IEEE Eng Med Biol Soc 2011:3180–3183

Wang S, Tasoglu S, Chen PZ et al (2014) Micro-a-fluidics ELISA for rapid CD4 cell count at the point-of-care. Sci Rep 4:3796

Wojtczak J, Bonadonna P (2013) Pocket mobile smartphone system for the point-of-care submandibular ultrasonography. Am J Emerg Med 31(3):573–577

Xie QM, Liu J (2010) Mobile phone based biomedical imaging technology: a newly emerging area. Recent Pat Biomed Eng 3(1):41–53

Yang Q, Cai R, Xiao W et al (2016) Plasmonic ELISA for sensitive detection of disease biomarkers with a smart phone based reader. Nanoscale Res Lett 13:397

Yu Y, Baker CA, Zhang X et al (2012) Quantitative polymerase chain reaction using infrared heating on a microfluidic chip. Anal Chem 84(6):2825–2829

Zhang H, Jenkins G, Zou Y et al (2012) Massively parallel single-molecule and single-cell emulsion reverse transcription polymerase chain reaction using agarose droplet microfluidics. Anal Chem 84(8):3599–3606

Zhao YT, Tong LP, Li Y et al (2016) Lactose-functionalized gold nanorods for sensitive and rapid serological diagnosis of cancer. ACS Appl Mater Interfaces 8(9):5813–5820

Zhu H, Sencan I, Wong J et al (2013) Cost-effective and rapid blood analysis on a cell-phone. Lab Chip 13(7):1282–1288

Computer Aided Design and Synthesis for Marker Proteins of HT Carcinoma Cells: A Study

17

Shruti Jain and Durg Singh Chauhan

Abstract

Computer-aided diagnosis (CAD) system is one of the leading research topics in diagnostic radiology and medical imaging. The elementary idea of CAD is to provide a computer output to assist radiologists and other healthcare experts in reading and understanding the image and to differentiate between various anomalies. This is considered to be the "second opinion" tool in assessing the extent of disease, detecting lesions, and making diagnostic decisions for improving the interpretation component of medical imaging. The CAD system consists of input medical images, segmentation module, or region of interest (ROI) extraction module, feature extraction module, and the classification module. Various tests (parametric, nonparametric, regression analysis, clustering analysis, etc.) were discussed for different types of data variables (categorical or continuous) for different marker proteins.

Keywords

Computer aided design · Parametric tests · Nonparametric tests · Regression analysis · Data preprocessing

17.1 Introduction

Numerous imaging modalities like ultrasound (US), X-ray, magnetic resonance imaging (MRI), computerized tomography (CT), etc. can be inferred for diagnosing any abnormalities. In the past several years, ultrasonography has become a very popular tool for imaging physiological systems in the body. Widely available US

S. Jain (✉) · D. S. Chauhan
Jaypee University of Information Technology, Waknagaht, Solan, Himachal Pradesh, India

GLA University, Mathura, Uttar Pradesh, India

S. Paul (ed.), *Biomedical Engineering and its Applications in Healthcare*,
https://doi.org/10.1007/978-981-13-3705-5_17

imaging modality is the first choice for the diagnosis of diseases, and it helps in diagnosing the diseases of soft tissue organs such as kidney, brain, lung, etc. because of its real-time, inexpensive, nonradioactive, and noninvasive properties (Rana et al. 2016a; Bhusri et al. 2016a; Sa Sharma and Jain 2017). With the growth of computer technology, medical image processing algorithms have given the abundant opportunity for researchers to investigate the potential of computer-aided diagnosis (CAD) systems for characterization of radiological images. The main goal of emerging a computerized characterization system is to provide additional diagnostic information about the underlying tissue which cannot be captured by visual examination of medical images, and this information is useful for improving the diagnostic accuracy of sonograms. The discrimination between different tissue echogenicity and texture patterns by visual analysis usually depends upon the experience of radiologists. In the human body, the development of tissue masses is the result of uncontrollable growth of cells called tumors. Tumors are considered to be of two types noncancerous (benign tumors) and cancerous (malignant tumors) which can be primary malignant and secondary malignant. The process of spreading of cancerous cells from primary cancerous tumors of the other body parts is termed as metastasis or secondary cancer. Nowadays, it is considered as one of the major problems and is mostly found in people over the age of 60. There are more than 200 types of cancer but prostate, colorectal, breast, and lungs are the four common cancers.

17.1.1 Related Work

Alvarenga et al. (2010) investigated seven morphological features to discriminate between the malignant and the benign breast tumors on ultrasound images. The accuracy achieved is 84 %, and the classification rate (A_z) is 0.8. Wan et al. (2011) formulated that by detecting and choosing essential features, the low-rank matrix based on feature selection method can improve the classification outcomes. Lio group (2011) established a new set of features for differentiating benign tumors from the malignant tumors. Sonograms of 321 pathologically proven breast lesions were analyzed and evaluated using SVM classifier. Gomez et al. (2010) proposed a CAD system for the breast ultrasound cases in which a differential evolution technique was used to optimize the structure of a radial basis function neural network. Shi and his team (2010) developed a CAD system, in which the classifier used is fuzzy SVM classifier. Gómez et al. (2012) show the accuracy of 92.83% which is achieved from 5500 images of prostate cancer by combining the histogram, GLCM, gray-level run length matrix (GLRLM) to differentiate the different types. Huang et al. (2008) achieved an A_z of 0.909 morphological feature extraction using a support vector machine (SVM) classifier from 118 breast ultrasound images. Wu et al. (2008) combined autocovariance morphological features and texture feature to discriminate breast tumors with an accuracy of 92.86% using SVM classifier from 210 breast ultrasound images. In a later work (Wu et al. 2012) using the SVM genetic classifier on the same database, an accuracy of 96.14% is achieved. Alvarenga et al. (2012) use fisher linear discriminant analysis (FLDA) classification on 246 ultrasound images and achieved 85.37% accuracy. The study in (Raja et al. 2008) proposed an approach for automatic judgment and classification of kidney ultrasound images into normal (Nor), medical renal disease (MRD), and a cyst with Gabor feature, statistical, power spectral, and algebraic moment invariant (MI). Raja et al. (2008) use a total of 36 features which were extracted using various kinds of support classifiers named as hybrid fuzzy neural network system and enhanced multilayer backpropagation network. The study in (Raja et al. 2007a) reported classification of three kidney classes for the analysis of power spectral features. The study in (Susan Jose et al. 2012) uses spatial gray label-dependent matrix (SGLDM) feature with Bayesian classifier for classification task with various feature extraction techniques. The study in (Subramanya et al. 2014) uses statistical features like first-order statistics (FOS) features, RLM, gradient-based features, GLCM, MI, and law features which are extracted from AOIs' using SVM classifier. Depending on the overall classification accuracy (OCA), only some feature sets are considered for the classification task such as differential evolution feature selection (DEFS). Authors in (Raja et al. 2007b) reported classification with statistical analysis methods like FOS, MI, and power spectral features. In this study, 28 features are extracted using artificial neural network (ANN) with OCA of 90.4% for Nor, 86.6% for MRD, and 85.7% for Cyst.

17.1.2 Contribution

There are different industrial sectors in India out of which healthcare sector is a very different sector. It is one of the high-priority sectors where people are expecting the highest level of care and services. This sector consumes a large amount of percentage of the budget. An interpretation of medical data is done by a medical expert of that field. In this paper, we discuss various steps involved in CAD design. CAD nowadays increases its popularity as a research topic among various researchers. CAD helps the doctor as a second opinion tool. Image acquisition and image interpretation depend on accurate diagnoses of disease. Image acquisition devices have improved substantially over the recent few years with much higher resolution, while image interpretation has been done by deep learning and machine learning processes. Deep learning is providing exciting solutions with good accuracy for medical imaging, real-world application, and a key method for future applications in the health sector. Different regression analyses (RA), cross-validation (CV), parametric tests (PT), and nonparametric tests (NPT) were also explained.

17.2 Methodology

A review has been done for the CAD design. The different steps shown in Fig. 17.1 are taken into consideration: data collection, data preprocessing, feature extraction, partitioning, and classification/clustering (Bhusri and Jain 2017; Rana et al. 2016b).

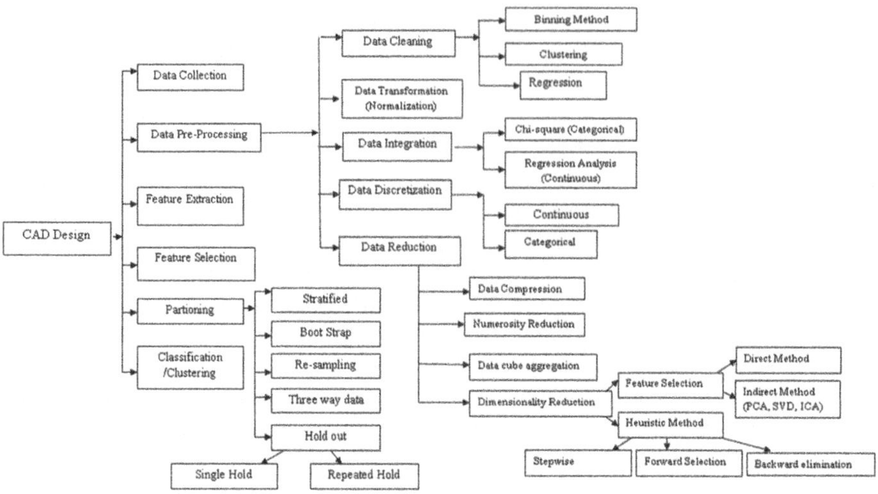

Fig. 17.1 Various steps for designing CAD

17.2.1 Data Collection

Data is collected for different marker proteins. We can say that if these proteins are absent, then it leads to cell death but if these proteins are present, then there is a cell survival. These proteins are due to the combination of three input proteins (EGF, insulin, and TNF) and four different outputs (membrane permeability (MP), phosphatidylserine exposure (PE), caspase substrate cleavage (CCK), and nuclear fragmentation (NF)). The data is collected for all marker proteins for ten different concentrations of input proteins (in ng/ml). For output values, the average of four values is considered.

17.2.2 Data Preprocessing

It is the step, we process the data by data integration, data reduction, data cleaning, data discretization, and data transformation.

1. *Data Cleaning*: This involves the smoothening of noisy data, filling the missing values, identifying or removing outliers, etc. Noisy data can be solved in different ways:
   (a) *Binning method*: It consists of two types of partitioning: *Equal width/distance partitioning* where the range is divided into N equal size intervals. If we have A and B as the lowest and highest values of feature, then the width of the interval is $W = (B\text{-}A) / N$. This type of the partitioning does not handle skewed data. Second is *equal depth/ frequency partitioning* where the range is divided into N intervals with the same number of samples. Through this

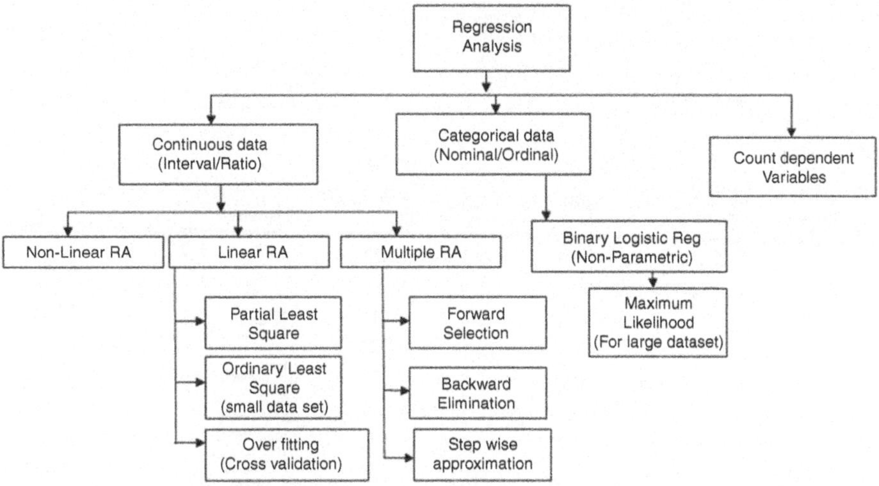

Fig. 17.2 Regression analysis with different types of data

method, we obtain good data scaling. Categorical attributes managed is a little bit tricky in this partitioning.

(b) *Clustering*: In this process, we can detect and remove outliers. The clusters are categorised as into two parts: unsupervised linear clusters and unsupervised nonlinear clusters which is further divided into Gaussian, hierarchal, fuzzy c-mean, quality threshold, etc., which are the parts of linear clustering, while kernel k-mean, density based, and MST based are the parts of nonlinear clustering.

(c) *Regression*: In this process, the smoothening of data is done by fitting into regression function/analysis (RA). Various types of RA are shown in Fig. 17.2. Continuous data, categorical data, and count variable data are the different types of data.

Linear RA, nonlinear RA, and multiple RA are the various types of continuous data for which we use parametric tests (PT) which are used for small data sets. Binary logistic regression is used for categorical data in which the maximum likelihood approach was used. Nonparametric tests (NPT)/distribution-free tests are used for categorical data analysis. In this test, large data set is used. Different PT and NPT tests are shown in Fig. 17.3. PT is used where we get information about population parameters while no information about population parameters in NPT. For PT, we use the mean of the population; where for NPT, the median is used. PT is sensitive and power efficient. These tests are used when the sample size is more.

But if the size is greater than 100, the central limit theorem can be applied. Correlation, regression, ANOVA, etc. are some statistical procedures which are known as PT. These tests are used on the data which are normally distributed. NPT statistics give exact probabilities regardless of the shape of the distribution

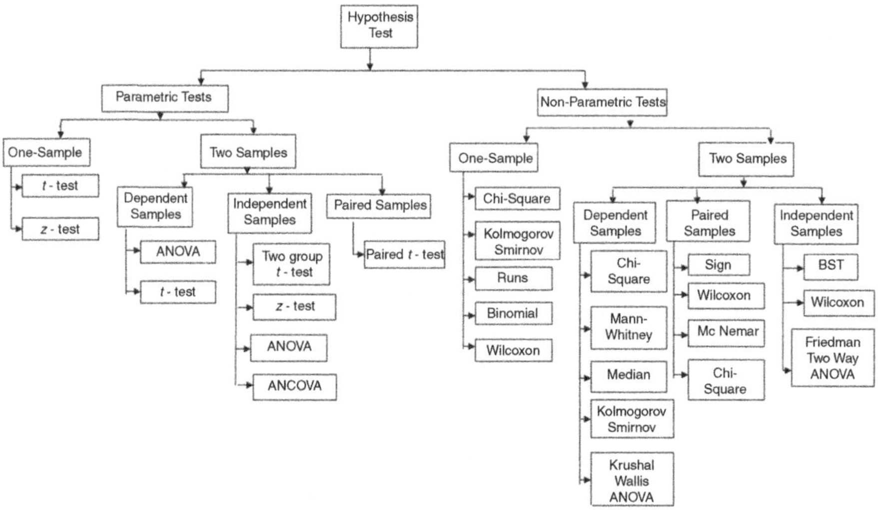

Fig. 17.3 Different parametric and nonparametric tests for different samples

from which the random samples were drawn. Figure 17.4 shows the difference between PT and NPT.

There are different types of tests where different PT and NPT tests are used: *tests of central tendency, tests of dispersion, tests of distribution,* and *measures of association.*

2. *Data Integration*: As the name suggests, different data is collected and integrated from multiple databases. This also includes removing duplicates and redundant data. Detecting and resolving data involve conflicts. Chi-square test was done for getting correlation values. We can get covariance values also. Chi-square test is used for categorical data while RA is done for continuous data.
3. *Data Transformation*: This step involves normalization, smoothening, and aggregation of data. Normalization is of different types: z-normalization, min-max approach, and normalization by decimal scaling. Smoothening means removal of the noise from the data while aggregation as the name suggests to summarize or to aggregate the data.
4. *Data Reduction*: It means to remove unimportant attributes. It consists of data reduction strategies, regression and log-linear model, aggregation and clustering, sampling, data compression, and histograms.

Data reduction strategies consist of data compression, numerosity reduction, data cube aggregation, and dimensionality reduction. Figure 17.5 explains the different types of feature selection process. Dimensionality reduction is divided into two steps: feature selection and heuristic methods. Feature selection is of two types: the first is a direct method in which selection of a minimum set of feature (attribute) is done which is sufficient for data mining, while the second is indirect method

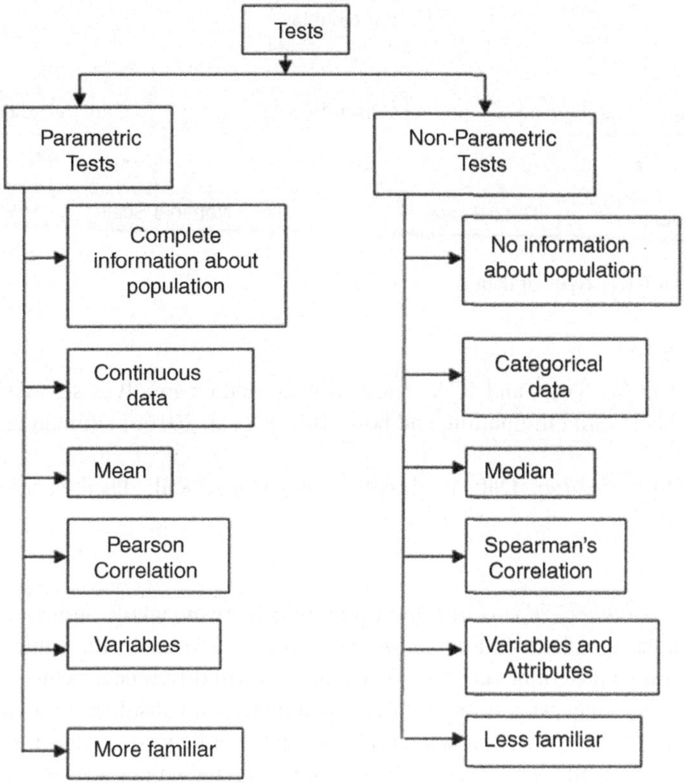

Fig. 17.4 Difference b/w PT and NPT

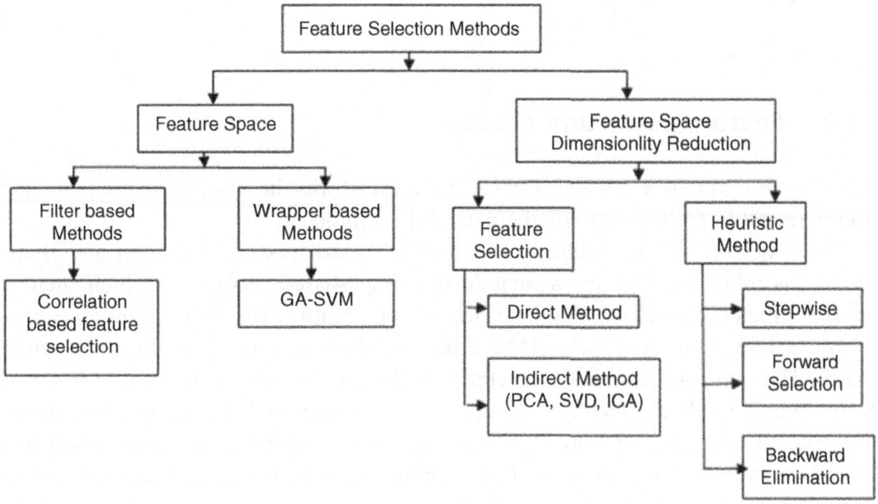

Fig. 17.5 Different feature selection techniques

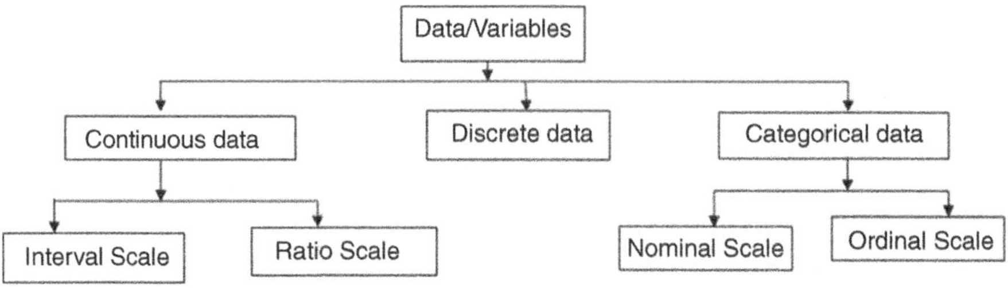

Fig. 17.6 Different types of data

consists of PCA, SVD, and ICA. The heuristic method involves stepwise forward selection, backward elimination, and both (Bhusri et al. 2016b; Dhiman et al. 2016).

5. *Data Discretization*: Data is classified into categorical, discrete, and continuous variables as shown in Fig. 17.6.

In *discrete numerical variable*, the values which are the whole number (counts) of those variables are discrete. For *continuous numerical variable,* the values which lie within the range are count variables. The magnitude of differences is the same which is of two types: interval and ratio scale. When there is no absolute zero value, equal magnitude, and rank, then that is interval scale data, while vice versa is for ratio scale (absolute zero), while for *categorical variables* when the values are in a small group, then data is categorical data. If the categories are ordered, then those are ordinal data, while if categories are not ordered (data can be in tabular or graphical form), then those are called nominal data.

17.2.3 Feature Extraction Module

The feature extraction is the process used to transform the visually extractable and non-extractable features into mathematical descriptors.

These descriptors are intensity distribution based (textural features) and shape based (morphological features include the shape-based properties which include area, perimeter, convexity, diameter, major axis, minor axis, Euler number, eccentricity, extent, hole area ratio (HAR), and solidity are calculated over the entire class). The intensity distribution methods include (*a*) statistical methods, (*b*) transform domain methods, and (*c*) signal processing-based methods. Figure 17.7 shows the different feature extraction methods. Statistical methods can be classified into FOS, second-order statistics (GLCM), higher-order statistics (GLRL), and other statistical features, which are edge features (mean and variances absolute gradient are calculated), NGTDM, SFM, and GLDS. Laws' mask texture analysis is used as

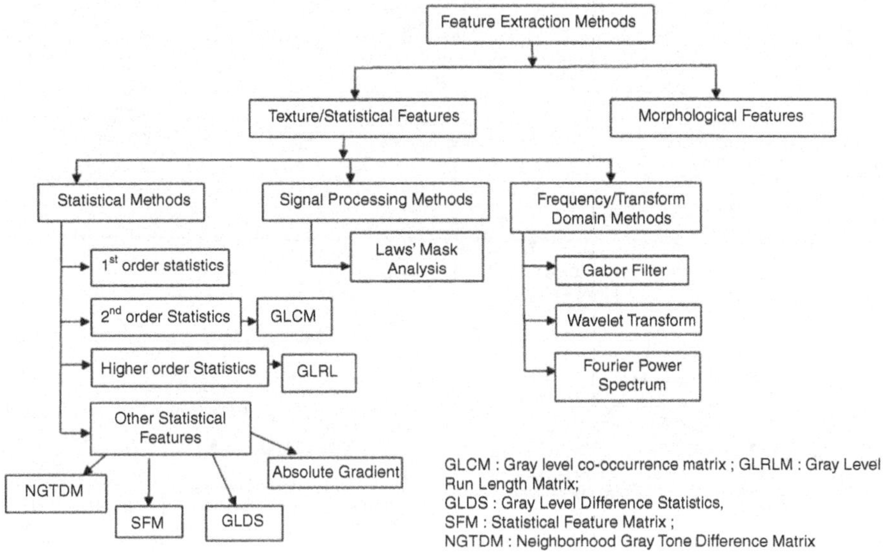

Fig. 17.7 Different feature extraction methods

signal processing-based methods. Fourier power spectrum (FPS), Gabor wavelet transform (GWT), and wavelet transform are used as frequency-domain methods.

17.2.4 Partitioning

For validation of data, it is divided into two parts: one is known as testing, while another is known as training. Data partitioning is done by different methods: holdout, stratified sampling, bootstrap, resampling, or three-way data splits. In general, holdout approach is used. In this process testing to training ratio is divided as ⅔ to ⅓ ratio where ⅔ for training and ⅓ for testing. Holdout can be the single holdout or repeated holdout. Holdout is a part of cross-validation (CV). For *k*-fold CV, we can train *k*-1 partitioned data rest use for testing, while in leave one out CV train $k = n$ data. Figure 17.8 shows the different cross-validation techniques.

17.2.5 Classification

Classification is one of the major techniques used in data mining shown in Fig. 17.9 while other being the clustering. Classification and prediction are sometimes used as synonyms, but in actual, there is a difference between the two. Classification refers to the prediction of categorical values; however, prediction models predict continuous values as well. In classification, the class labels are already known. Classification

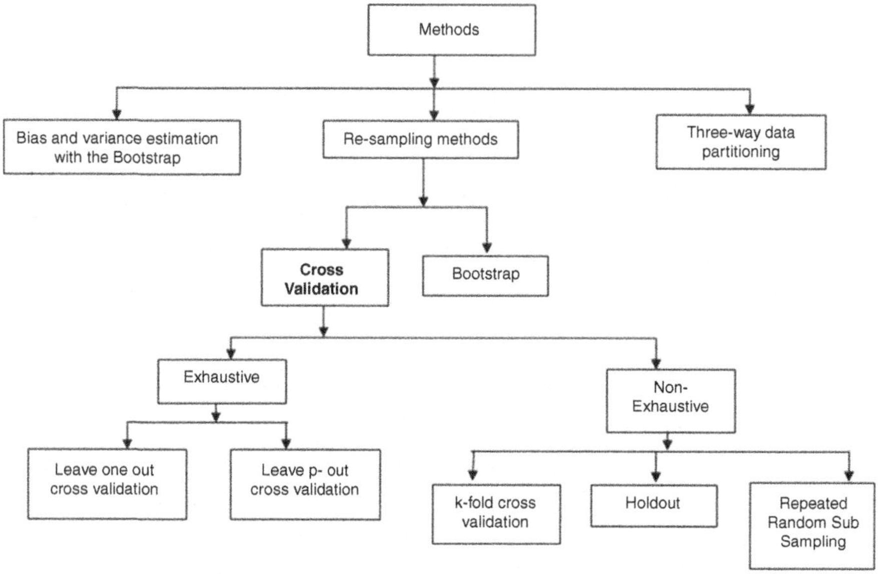

Fig. 17.8 Different partitioning techniques

algorithms include k-nearest neighbors (*k*-NN), naïve Bayes, ID.3, classification and regression tree (CART), chi-square automatic interaction detector (CHAID), and MARS (shown in Fig 17.9) which extends the decision trees in order to handle numerical data more precisely.

17.3 Results and Discussions

The data was collected from Gaudet et al. (2005) and Jain et al. (2012). There are different marker proteins, FKHR, MEK, MK2, ERK, pAkT, AkT, JNK, IRS, IKK, ptAkT, and EGFR for the HT carcinoma cells which lead to cell death/survival. Each image containing a sub-13 parts shows the time from 0 to 24 h which is further divided into 0, 5, 15, 30, 60, and 90 min and 2, 4, 8, 12, 16, 20, and 24 h (shown in Fig. 17.10).

For tests of central tendency, PT used for *one sample* can be solved by *z*-test (in this test, we should know the standard deviation and mean of data) and *t*-test (in this test, we must know the mean of data). *Two or more independent samples* (samples should comprise different samples with different mean values) can be solved by *t*-test, one-way ANOVA (in a group of three, five, or ten at least two samples should have different mean values). For a planned comparison, we can use multiple *t*-tests, while for unplanned comparison, Fisher's LSD test can be used, analysis of covariance (ANCOVA). For *two or more dependent samples*, *t*-test or ANOVA can be used. NPT uses median tests for *one sample* which can be solved by Wilcoxon signed-rank test (WSR) (median should be known). T*wo or more independent samples* can be solved by Mann-Whitney U Test (samples should have

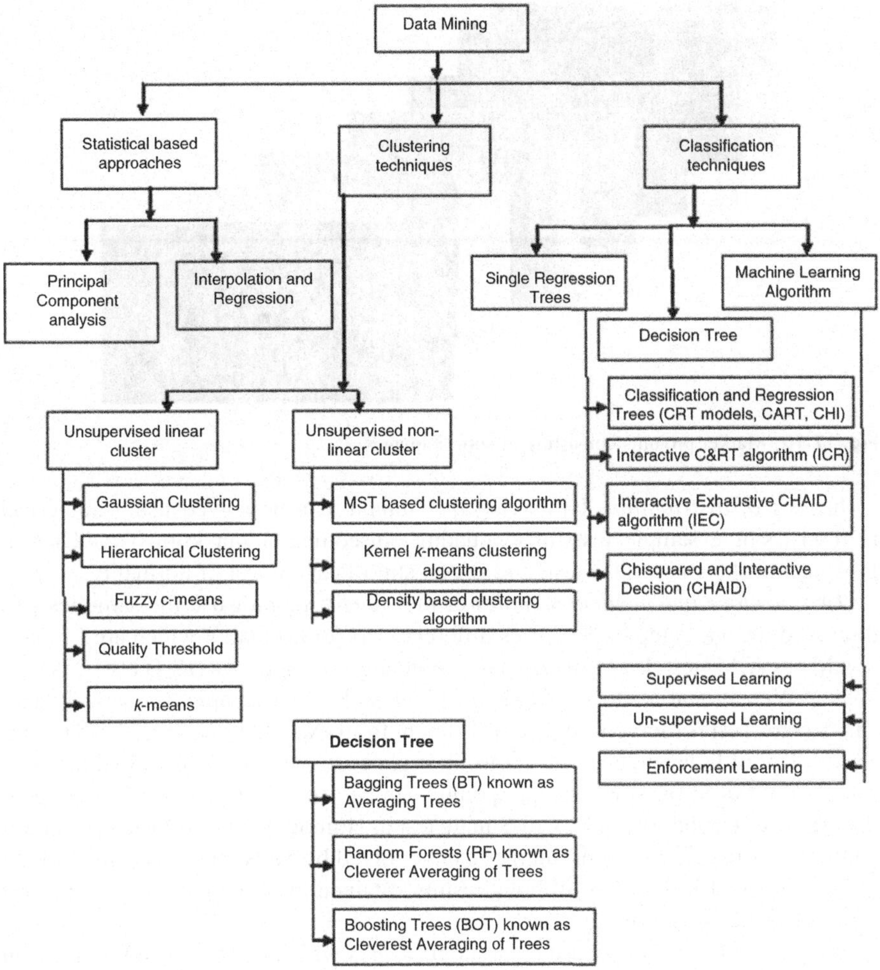

Fig. 17.9 Different data mining steps

different median values) and Kruskal-Wallis ANOVA (in a group of three, five, or ten, at least two samples should have different median values). Two *or more dependent samples* can be solved by WSR, binomial sign test (BST), or Friedman two-way ANOVA (group tests).

In *tests of dispersion*, PT uses variance method in which *one sample* can be solved by single chi-square test, *two or more independent samples* can be solved by Hartley's *F*(max) test for homogeneity of variance, and *two or more dependent samples* can be solved by the *t*-test for homogeneity of variance. NPT can be done for categorical data. But no method has been discussed in literature. NPT for two or more independent samples which can be solved by the Siegel-Tukey test for equal variability (samples with different variances) and Moses test for equal variability.

Fig. 17.10 Marker proteins consisting of sub-13 images

In *tests of distribution*: PT having *one sample* can be solved using numerical methods: single sample test for evaluating Anscombe-Glynn kurtosis (AG) test, D'Agostino skewness (DS) test, and D'Agostino-Pearson test of normality.

Two or more independent samples or *dependent samples* use the same tests as discussed for one sample. NPT uses different tests for normality which are Cramer-von Mises (CM) test, Lilliefors (LF) test, Kolmogorov-Smirnov (K-S) test, Shapiro-Wilk (SW) test, and Anderson-Darling (AD) test. For one sample categorical data : Chi-Square Test (observed frequency different from expected frequency), BST, Chi-Square Test of Independence (relationship between two variables measured for same samples, not used for small sample points) can be used. The two or more independent samples can be solved by Chi-Square test of Homogeneity and Fisher Exact test while two or more dependent samples can be solved by McNemar Test and Cochran Q Test. Figure 17.11 shows different results obtained from the K-S test and LF test using different marker proteins.

Measures of association use bivariate measures which can be solved by Pearson product-moment correlation coefficient if there is any linear relationship between two variables and multivariate measures which can be solved by multiple correlation coefficient and partial correlation coefficient (compute r^2 values). NPT uses bivariate measures which can be solved by Spearman's rank-order correlation coefficient and multivariate measures which can be solved by Kendall's coefficient of concordance. NPT for categorical data uses chi-square test for homogeneity which includes the contingency coefficient (applied for more than two variables), Yule's Q (only for two variables), Cramer's phi coefficient, the phi coefficient, and the odds ratio. Different correlation graphs are shown in Fig. 17.12.

In this type, division of range for the continuous features was done into intervals because some data mining algorithms only accept categorical attributes. This can be done by binning method or entropy-based method, clustering analysis (shown in Fig. 17.13), and histogram analysis (shown in Fig. 17.14).

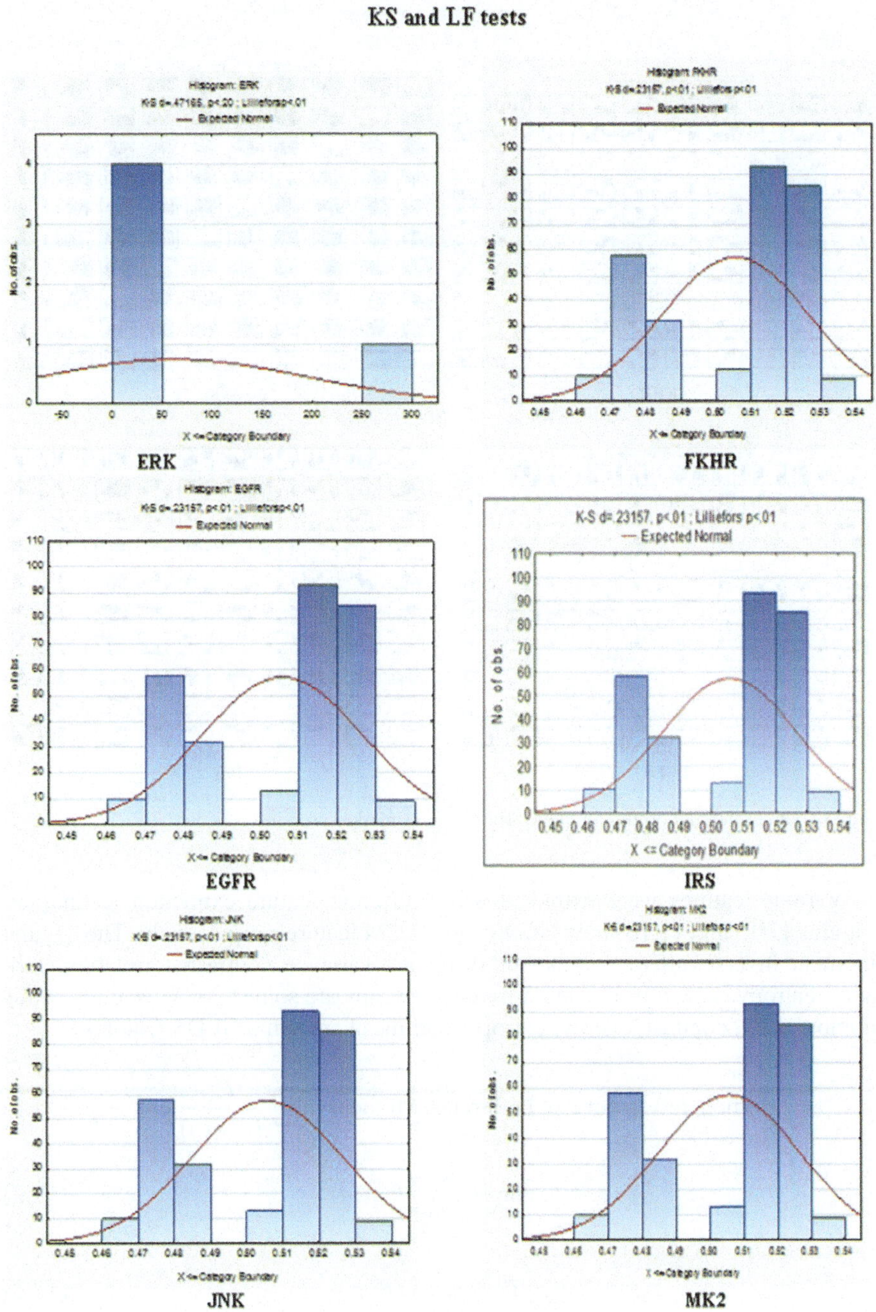

Fig. 17.11 Different K-S tests and LF tests for different marker proteins

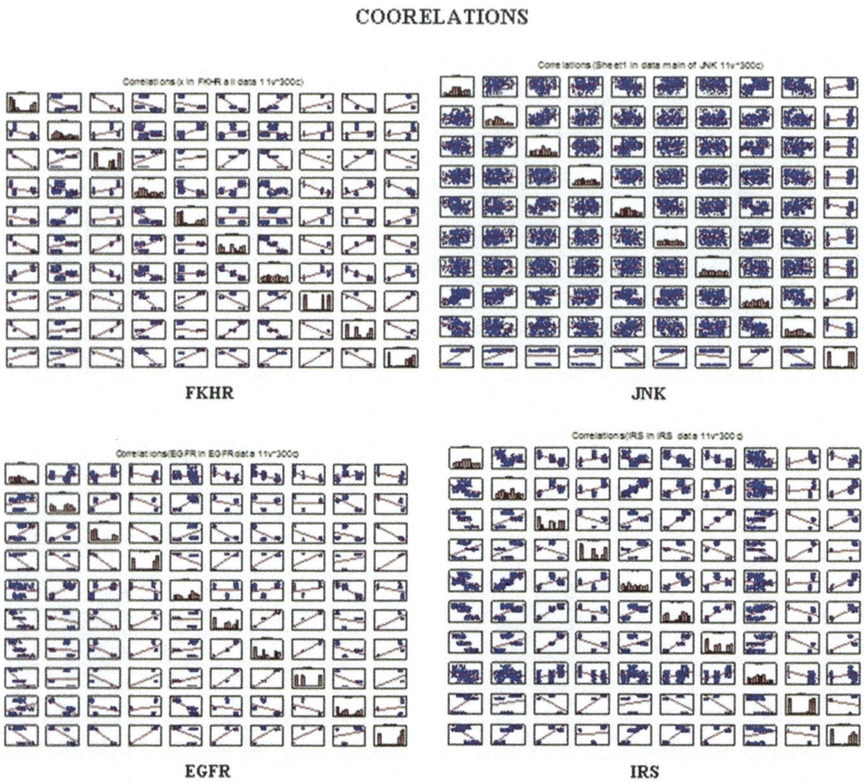

Fig. 17.12 Different correlation values for different marker proteins

Various features were extracted using different feature extraction techniques. Figures 17.15 and 17.16 show GLDS and GLCM features, respectively. The angular moment, inverse difference moment, difference variance, contrast, correlation, variance, entropy, sum entropy, and difference entropy are some GLCM features, while homogeneity, contrast, energy, entropy, and mean are some GLDS features.

$$\text{Homogenity or Inverse Difference Moment}: \sum_{i,j} \frac{P_{i,j}}{1 + (i\text{-}j)^2} \tag{17.1}$$

$$\text{Contrast}: \sum_{i,j} P_{i,j}(i\text{-}j)^2 \tag{17.2}$$

$$\text{Energy}: \sqrt{P_{i,j}^2} \tag{17.3}$$

$$\text{Entropy}: \quad -\sum_{i,j} P_{i,j} \log\left(P_{i,j}\right) \tag{17.4}$$

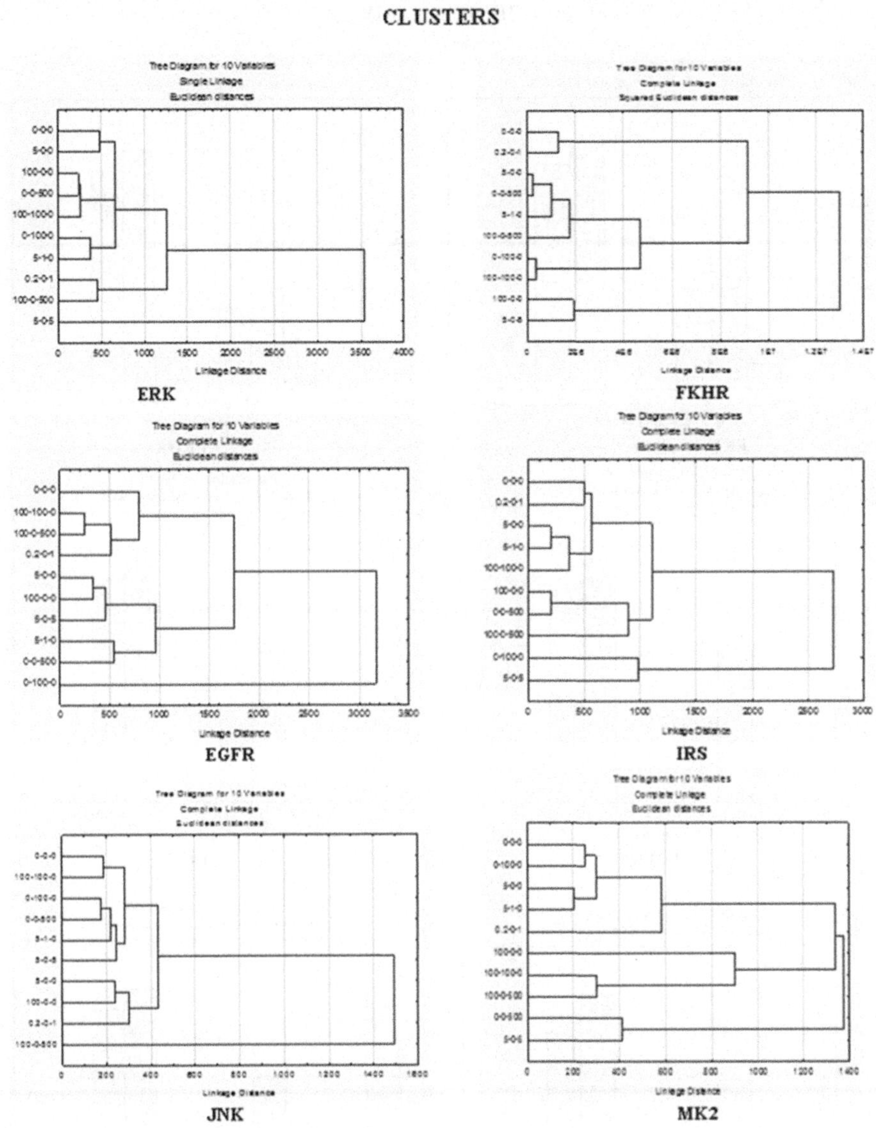

Fig. 17.13 Different cluster figures for different marker proteins

$$\text{Mean} = \frac{1}{m}\sum_{i,j} i.P_{i,j} \tag{17.5}$$

$$\text{Angular Moment} = \sum_{i,j} P_{i,j}^2 \tag{17.6}$$

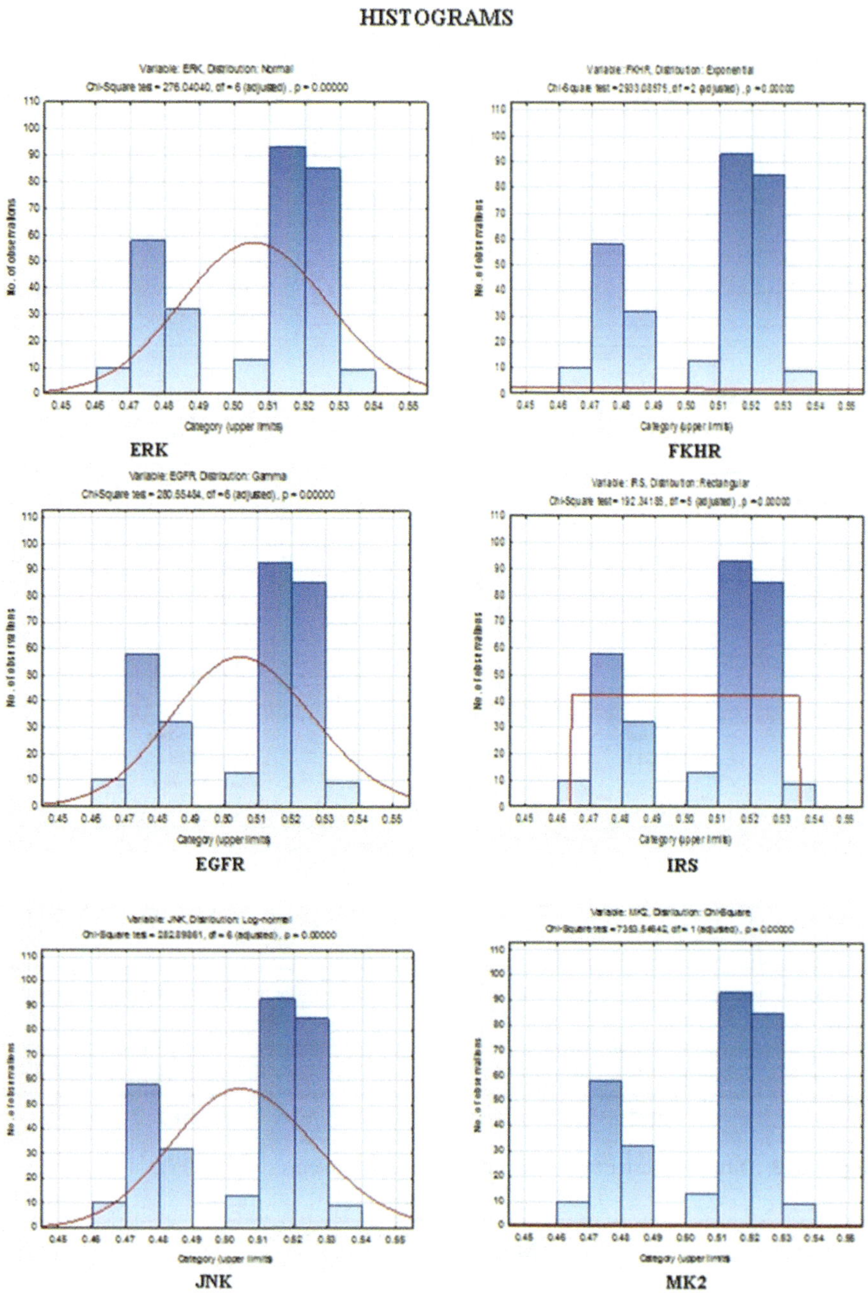

Fig. 17.14 Different histograms for different frequency distribution functions for different marker proteins

	A	B	C	D	E	F
1	Image Name	Homogeneity	Contrast	Energy	Entropy	Mean
2	training1.jpg	0.854987202	113.2381544	0.652933633	0.917039293	2.272441317
3	training2.jpg	0.847357643	131.7127817	0.64349056	0.976306029	2.230833836
4	training3.jpg	0.853626963	58.79003009	0.642359228	0.897115732	1.575407522
5	training4.jpg	0.876544645	68.76001769	0.689967145	0.811290516	1.38938121
6	training5.jpg	0.895257071	86.25917661	0.743151232	0.702375735	1.763189907
7	training6.jpg	0.841037151	97.15740904	0.63250689	0.974279108	1.983091402
8	training7.jpg	0.870878384	10.01150494	0.675819682	0.801835284	0.682674555
9	training8.jpg	0.877617321	27.10220569	0.678342701	0.764936527	0.835425999
10	training9.jpg	0.897779625	60.28578813	0.739976531	0.704064059	1.331711702
11	training10.jpg	0.880112832	67.38399868	0.702833229	0.790918361	1.53115434
12	training11.jpg	0.874703573	16.35436678	0.688712928	0.785249445	0.811738269
13	training12.jpg	0.858368037	25.75473928	0.663206927	0.887452265	1.093128829
14	training13.jpg	0.897126857	95.64714901	0.748572702	0.701132442	1.803323909
15	training14.jpg	0.893375222	110.603529	0.74868205	0.729537738	2.022123188
16	training15.jpg	0.86816662	87.49889848	0.673174881	0.85475627	1.575461224
17	training16.jpg	0.858618521	48.41669017	0.654469506	0.881123474	1.276601546
18	training17.jpg	0.90314625	67.82614905	0.770392178	0.65072029	1.479560083
19	training18.jpg	0.886861278	93.43135622	0.720160031	0.767720634	1.847457593
20	training19.jpg	0.865935053	69.79375574	0.669000325	0.853160863	1.485280251
21	training20.jpg	0.872286602	33.16310084	0.684546452	0.820957401	1.144531113

Fig. 17.15 GLDS features

	A	B	C	D	E	F	G	H	I	J	K	L	M	N
1	Image Name	ASM-M	Cont-M	Corr-M	SOSV-M	IDM-M	SuMAv-M	SumVar-M	SumEnt-M	Ent-M	DiffVar-M	DiffEnt-M	InfMCorr1-M	InfMCorr2-M
2	Training1.jpg	0.21668272	110.6379869	0.350477917	85.63480426	0.788795671	0.788795671	292.9514833	231.9012301	2.113752881	2.474573262	103.3833777	1.102668134	-0.503816844
3	Training2.jpg	0.199242332	101.1299376	0.344406289	83.83790814	0.766340689	0.766340689	292.5938465	234.221695	2.193306676	2.584148397	93.50778064	1.17351443	-0.502660193
4	Training3.jpg	0.113248102	65.33643014	0.689107697	104.796385	0.751212648	0.751212648	258.5192219	353.8491099	2.759852676	3.176091433	59.20823593	1.243519403	-0.610721181
5	Training4.jpg	0.081614268	86.22365177	0.877029828	352.1270385	0.763766302	0.763766302	227.6738346	1322.284502	3.230206604	3.629825315	78.73904119	1.289237596	-0.68236978
6	Training5.jpg	0.143655595	333.157413	0.912775306	1906.748692	0.805280497	0.805280497	235.4743067	7293.837357	2.791980175	3.071798072	307.9350745	1.14291351	-0.666258628
7	Training6.jpg	0.147632133	208.9373863	0.846386148	680.1232967	0.74554805	0.74554805	256.1609822	2511.5558	2.844121195	3.275742544	192.0095898	1.327411002	-0.583503958
8	Training7.jpg	0.048416802	58.7467229	0.887246209	261.7898458	0.743144795	0.743144795	184.3402697	988.4126604	3.439129572	3.903748902	49.30211676	1.368459762	-0.702936792
9	Training8.jpg	0.052629237	157.748132	0.927381581	1086.023232	0.750450777	0.750450777	159.5706401	4186.344796	3.458335868	3.850520298	143.3253004	1.336182467	-0.692172816
10	Training9.jpg	0.075726394	319.5221248	0.921879185	2045.714882	0.807609657	0.807609657	207.89851	7863.337404	3.238385486	3.518842071	288.9144325	1.20107481	-0.736525935
11	Training10.jpg	0.08158175	219.6874729	0.89777278	1072.981213	0.781040271	0.781040271	252.8850961	4072.23738	3.143563506	3.464947267	199.7308001	1.246550459	-0.704254807
12	Training11.jpg	0.064842344	117.6293207	0.885280318	513.0156981	0.772335784	0.772335784	185.7236965	1934.433472	3.370223827	3.723918783	106.5187261	1.266706474	-0.718699809
13	Training12.jpg	0.074298549	117.0443581	0.92150392	745.4749866	0.751571894	0.751571894	187.8911278	2864.855588	3.185045371	3.605097197	103.6116017	1.371787681	-0.659321242
14	Training13.jpg	0.213175428	119.0250112	0.606203178	150.742981	0.82164843	0.82164843	287.4316473	483.9469129	2.272864951	2.530092287	109.5583328	1.03870545	-0.614725067
15	Training14.jpg	0.263009165	111.1292918	0.415925193	94.79962732	0.813804386	0.813804386	293.2351977	268.0692174	2.108899353	2.382336226	103.6749235	1.03249979	-0.588217924
16	Training15.jpg	0.061999714	224.3284968	0.905403081	1185.424625	0.75934867	0.75934867	205.544335	4517.370002	3.296577873	3.715114471	197.7812542	1.376126108	-0.689694903
17	Training16.jpg	0.055558874	172.0148322	0.925376866	1155.631894	0.751917009	0.751917009	176.9846631	4450.512743	3.517107429	3.969923127	154.2976291	1.392139906	-0.707389263
18	Training17.jpg	0.139252882	173.273741	0.894557634	820.2541555	0.817801951	0.817801951	256.2895837	3107.742881	2.620945569	2.862731311	158.8390829	1.037257994	-0.690700188
19	Training18.jpg	0.189991171	109.8112214	0.626150817	149.5137413	0.797869465	0.797869465	286.9807552	488.2437436	2.407570001	2.750363414	100.5733484	1.158313223	-0.595159062
20	Training19.jpg	0.096945221	180.0026126	0.855503942	621.7826775	0.777879497	0.777879497	217.6966175	2307.128097	3.042109232	3.379592775	167.2069057	1.22217923	-0.675439241
21	Training20.jpg	0.079850203	190.4519406	0.92117127	1207.509915	0.781369826	0.781369826	183.2370778	4639.587721	3.214323292	3.533955636	176.8765542	1.219674264	-0.71674138
22	Training21.jpg	0.062531087	344.4168474	0.914964092	2024.838232	0.791253244	0.791253244	213.7912235	7754.93608	3.488420197	3.777742837	312.2217909	1.269741639	-0.74545385
23	Training22.jpg	0.141715905	314.8236211	0.910821592	1761.823464	0.79356396	0.79356396	247.6111697	6732.470235	2.786640659	3.068324004	288.7293439	1.220408962	-0.654417568
24	Training23.jpg	0.06232528	228.8938368	0.890640677	1044.572145	0.753456156	0.753456156	215.2404696	3949.394744	3.358502469	3.742364905	211.2633962	1.313571097	-0.686837192
25	Training24.jpg	0.049261322	126.5110836	0.912547622	726.8480178	0.755740086	0.755740086	144.9492291	2780.880988	3.506053446	3.924720227	107.3697871	1.39740374	-0.687185581

Fig. 17.16 GLCM features

$$\text{Correlation}: \sum_{i,j} P_{i,j}\left[\frac{(i\text{-}\mu_i)(j\text{-}\mu_j)}{\sigma_i \sigma_j}\right] \tag{17.7}$$

$$\text{Variance}: \sum_{i,j} P_{i,j}(i\text{-}\mu_i)^2 \tag{17.8}$$

After feature extraction process, classification is done. Figure 17.17 shows a different tool for the neural network in MATLAB. A different classification algorithm divides the data set into two portions which are called training set and test set.

These sets are usually divided into the ratio of 70:30, 70% being the training set and 30% being the test set. Then the algorithm uses the training set to train the model for accurate prediction. To check the accuracy of the developed model, it is then applied to the test set, and a confusion matrix is created which shows how many records belonging to a particular attribute/field have been correctly predicted. If the training sets are defined earlier, then the classification is known as supervised; otherwise, it is unsupervised classification. There are different classifiers like *k*-NN, PNN, SVM, and smooth support vector machine (SSVM) to classify the unknown testing data of different classes based on the training data.

17.4 Conclusion

Computer-aided methods are developed to facilitate better diagnosis and testing. In this paper, a review work has been carried out on benchmark database images for the analysis of different lesions. In the present work, ROIs from different images are extracted manually. The presentation of the proposed procedures remains to be tested on images of different resolutions. In this paper, different techniques for preprocessing (regression analysis, parametric tests, nonparametric tests, correlation, clustering, histograms, and different data variables), feature extraction (morphological or statistical features), and classification techniques or data mining techniques were explained. In the future, we can make an algorithm for automatic ROI extraction which can be developed by employing various pattern recognition concepts to identify the center of the lesions and then extract an ROI of some specified size automatically. We can use a real-time database to evaluate the performance of the computer-aided methods.

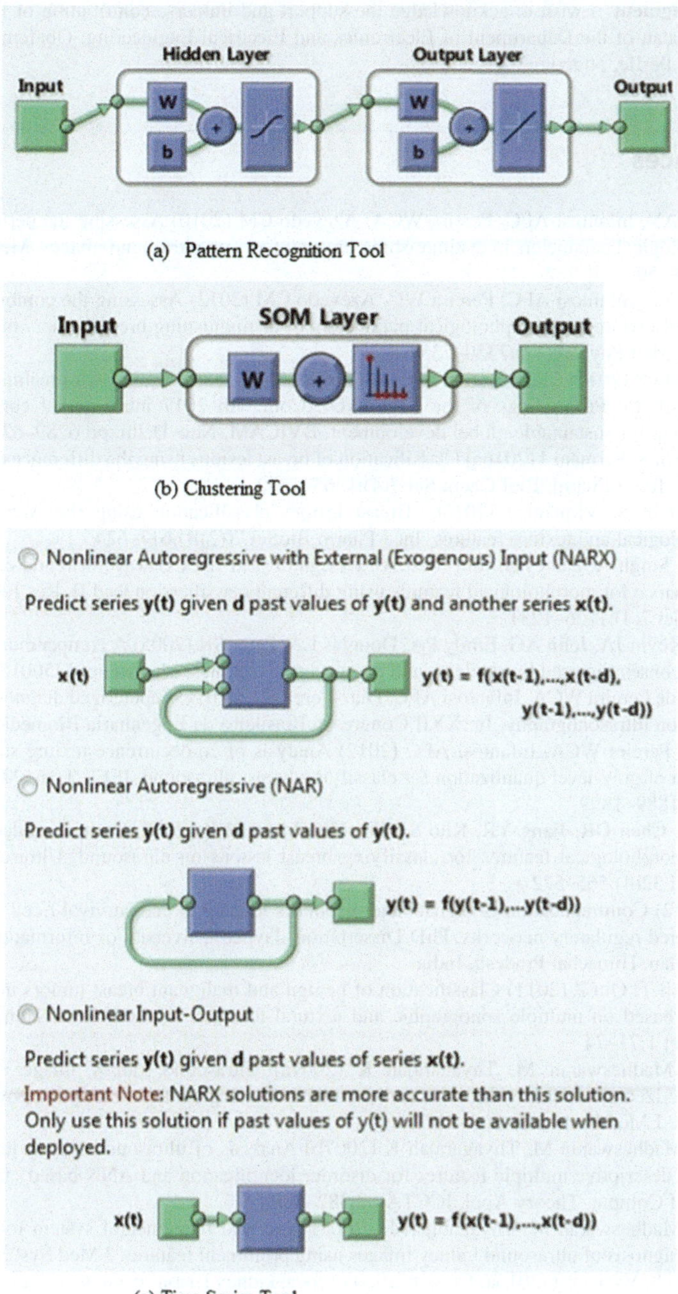

Fig. 17.17 Neural network tool of MATLAB

Acknowledgment I wish to acknowledge the support and immerse contribution of Dr. Ayodeji Olalekan Salau of the Department of Electronics and Electrical Engineering, Obafemi Awolowo University, Ile-Ife, Nigeria.

References

Alvarenga AV, Infantosi AFC, Pereira WCA, Azevedo CM (2010) Assessing the performance of morphological parameters in distinguishing breast tumors on ultrasound images. Med Eng Phys 32(1):49–56

Alvarenga AV, Infantosi AFC, Pereira WC, Azevedo CM (2012) Assessing the combined performance of texture and morphological parameters in distinguishing breast tumors in ultrasound images. Med Phys 39(12):7350–7358

Bhusri S, Jain S (2017) Classification of breast lesions using texture ratio vector technique, March 1st – 3rd, In: Proceedings of the 11th INDIACom: 4th 2017 international conference on computing for sustainable global development, BVICAM, New Delhi, pp 6289–6293

Bhusri S, Jain S, Virmani J (2016a) Classification of breast lesions using the difference of statistical features. Res J Pharm, Biol Chem Sci 7(4):1365–1372

Bhusri S, Jain S, Virmani J (2016b) Breast lesions classification using the Amalagation of morphological and texture features. Int J Pharm BioSci 7((2)B):617–624

Dhiman A, Singh A, Dubey S, Jain S (2016) Design of lead II ECG waveform and classification performance for morphological features using different classifiers on lead II. Res J Pharm, Biol Chem Sci 7(4):1226–1231

Gaudet S, Kevin JA, John AG, Emily PA, Douglas LA, Peter SK (2005) A compendium of signals and responses trigerred by prodeath and prosurvival cytokines, Manuscript M500158-MCP200

Gomez W, de Pereira WCA, Infantosi AFC, Diaz-Perez A (2010) Computerized diagnosis of breast lesions on ultrasonography. In: XXII Congresso Brasileiro de Engenharia Biomedica

Gómez W, Pereira WCA, Infantosi AFC (2012) Analysis of co-occurrence texture statistics as a function of gray-level quantization for classifying breast ultrasound. IEEE Trans Med Imaging 31(10):1889–1899

Huang YL, Chen DR, Jiang YR, Kuo SJ, Wu HK, Moon WK (2008) Computeraided diagnosis using morphological features for classifying breast lesions on ultrasound. Ultrasound Obstet Gynecol 32(4):565–572

Jain S (2012) Communication of signals and responses leading to cell survival / cell death using engineered regulatory networks. PhD Dissertation, Jaypee University of Information Technology, Solan, Himachal Pradesh, India

Liao R, Wan T, Qin Z (2011) Classification of benign and malignant breast tumors in ultrasound images based on multiple sonographic and textural features. Int Conf Intell Hum-Mach Sys Cybernet 1:71–74

Raja BK, Madheswaran M, Thyagarajah K (2007a) Ultrasound kidney image analysis for computerized disorder identification and classification using content descriptive power spectral features. J Med Syst 31(5):307–317

Raja BK, Madheswaran M, Thyagarajah K (2007b) Analysis of ultrasound kidney images using content descriptive multiple features for disorder identification and ANN based classification. Int Conf Comput, Theory Appl, ICCTA 07:382–388

Raja BK, Madheswaran M, Thyagarajah K (2008) A hybrid fuzzy-neural system for computer-aided diagnosis of ultrasound kidney images using prominent features. J Med Syst 32(1):65–83

Rana S, Jain S, Virmani J (2016a) Classification of focal kidney lesions using wavelet-based texture descriptors. Int J Pharm Bio Sci 7((3) B):646–652

Rana S, Jain S, Virmani J (2016b) SVM-based characterization of focal kidney lesions from B-mode ultrasound images. Res J Pharm, Biol Chem Sci (RJPBCS) 7(4):837–846

Sa Sharma S, Jain SB (2017) Two class classification of breast lesions using statistical and transform domain features. J Global Pharm Technol 9(7):18–24

Shi X, Cheng HD, Hu L, Ju W, Tian J (2010) Detection and classification of masses in breast ultrasound images. Digital Signal Process 20(3):824–836

Subramanya MB, Kumar V, Mukherjee S, Saini M (2014) SVM-based CAC system for B-mode kidney ultrasound images. J Digit Imaging 28:448. https://doi.org/10.1007/s10278-014-9754-4, Society for Imaging Informatics in Medicine

Susan Jose J, Sivakami R, Uma Maheswari N, Venkatesh R (2012) An efficient diagnosis of kidney images using association rules. Int J Comput Technol Electron Eng (IJCTEE) 2(2):14

Wan T, Liao R, Qin Z (2011) A robust feature selection approach using low rank matrices for breast tumors in ultrasonic images. In: 18th IEEE international conference on image processing

Wu WJ, Moon WK (2008) Ultrasound breast tumor image computer-aided diagnosis with texture and morphological features. Acad Radiol 15(7):873–880

Wu WJ, Lin SW, Moon WK (2012) Combining support vector machine with genetic algorithm to classify ultrasound breast tumor images. Comput Med Imaging Graph 36(8):627–633

Part VII

Biosignals and Its Significance

Biomedical Signal Analysis and Its Usage in Healthcare

18

Abdulhamit Subasi

Abstract

Biomedical signals are collected from a body that can be at the organ level, cell level, or molecular level. There are different biomedical signals including the electroencephalogram (EEG), which is the electrical activity from the brain; the electrocardiogram (ECG), which is the electrical activity from the heart; the electromyogram (EMG), which is the electrical activity from the muscle sound signals; the electroneurogram; the electroretinogram from the eye; and so on (Muthuswamy, Biomedical signal analysis. In: Myer Kutz (ed) Standard handbook of biomedical engineering and design, vol 14. McGraw-Hill Education, New York, pp 1–18. 2004). Biomedical signals are primarily used to diagnose or detect specific pathological or physiological conditions. Additionally, these signals are employed to analyze biological systems in the healthcare. The aims of signal processing are signal denoising, precise recognition of signal model through analysis, feature extraction and dimension reduction for decisive function or dysfunction, and prediction of future functional or pathological events by employing machine learning techniques. The objective of this chapter is to present how biomedical signals are used in the healthcare and what are the steps of biomedical signal analysis.

Keywords

Electrocardiograms (ECG) · Electroencephalograms (EEG) · Electromyograms (EMG)

A. Subasi (✉)
Effat University, College of Engineering, Jeddah, Saudi Arabia
e-mail: absubasi@effatuniversity.edu.sa

S. Paul (ed.), *Biomedical Engineering and its Applications in Healthcare*,
https://doi.org/10.1007/978-981-13-3705-5_18

18.1 Introduction

Biomedical signals are acquired from a medical or biological source which can be at the cell level, molecular level, or organ level. Several biomedical signals are generally employed in the research laboratory, clinic, and occasionally even at home. The electroencephalogram (EEG), or electrical activity from the brain or electrical responses of the brain to specific peripheral stimulation; the electrocardiogram (ECG), or electrical activity from the heart; the electromyogram (EMG), or electrical activity from the muscle; the electroretinogram from the eye; the electroneurogram, or field potentials from local regions in the brain; and so on are the widely known examples of the biomedical signals. In clinics, biomedical signals are mainly recorded to detect definite physiological or pathological conditions and diagnose and evaluate the therapy. Biomedical signal analysis is employed to remove the noise, create accurate signal model and analyze its components, extract features for decisive function or dysfunction, and predict future pathological or functional events in brain, heart, or muscle (Muthuswamy 2004). Biomedical signals contain information which is beneficial for understanding of the complex pathophysiologic mechanisms and behavior of living systems. However, such information may not be obtainable directly from the raw signals which might be disguised by other biomedical signals sensed or suppressed in additive noise. Because of these reasons, biomedical signal processing is generally needed to improve the related information and to describe the level of pathology for routine clinical diagnosis, rehabilitation, or therapy. Numerous signal processing approaches, which are sometimes termed as preprocessing techniques, can be employed for these purposes such as denoising, averaging, filtering, spectral estimation, and feature extraction. Biomedical signals are collected from sensors, electrodes, or transducers and then transmitted, stored, and treated (Mainardi et al. 2006).

Biomedical signals collected from the body deliver information related to the organs. Their spectral and temporal specifications might be associated with pathological or normal conditions. According to temporal variations in the function of the organs, the biomedical signals may show nonstationary as well as time-varying characteristics. Time-frequency analysis methods, such as wavelet, are more suitable for biomedical signals analysis. These methods are used for the analysis of time-varying and transient events in biomedical signals including cardiac and neurological signals. Biomedical signal processing utilizes complex mathematical techniques to achieve information hidden in the signals recorded from sensors. In biomedical engineering, these sensors and electrodes collect signals from biological tissue to check their well-being and health in clinical environment. Hence biomedical signal processing techniques needs employing appropriate signal modelling to extract features which is important for diagnosis. Since most of the biomedical signals are time-varying, it is essential to capture transient phenomena in both abnormal and healthy states. A crucial characteristic of several biomedical signals is the frequency domain feature. Similarly, biomedical signal pattern represents the transition from simple normative to unhealthy states of an organism occasionally undertaking severe variations that can be easily distinguished utilizing time-frequency methods such as wavelet transform. Biomedical signals are generally spread out over wide range of the frequency spectrum, the frequency content of a biomedical signal varies quickly as in the case of the heartbeat fibrillating in an ECG and seizure spikes in epilepsy. Wavelet transform, which has wideband representation of signals, is a usual choice in biomedical signal processing (Thakor et al. 2000).

The focus of this chapter is to support the biomedical researcher in order to choose the suitable representation or study of the biomedical signal from the existing models and then guide the engineer toward an ideal approach for enumeration. Hence, this chapter discusses the usage of basic biomedical signals (ECG, EEG, and EMG) in healthcare and techniques for analyzing them employing fundamental signal-processing and classification methods which find widespread application in biomedical signal analysis.

18.2 Biomedical Signals

18.2.1 The Electrocardiogram (ECG)

The electrocardiogram (ECG) signals are electrical activities originated from heart on the body surface so that isopotential surfaces can be calculated and analyzed over time. The contemporary ECG device is entirely combined with an analog front end, an analog-to-digital (A/D) converter, dedicated input-output (I/O) processors, and a microcomputer. The better hospital-based system can collect these changes and keep an ECG database which includes the permutation of parameters, e.g., all females, elder than age 30, with an inferior congenital heart disease. There are hundreds of demonstrative approaches where a particular diagnosis is completed for every ECG, but there are only five or six main classification sets for which the ECG is employed. The initial step in ECG analysis needs computation of the rhythm and rate for the atria and ventricles. This includes whichever conduction instability either in the connection among the different chambers or within the chambers themselves. Then feature identification that would be connected to the absence or existence of damaging because of the myocardial infarction would be done (Berbari 2000; Subasi 2019).

The heart consists of four chambers; the lower two chambers are named as ventricles and the upper two chambers are named as atria. The atria collect blood from the venous circulation. Positioned in the upper right atrium are a group of cells that operate as the main pacemaker of the heart. The nature of the body surface waves is dependent on the total of tissue activating at one time and the relative speed and direction of the activation wave front. Thus, the pacemaker potentials that are produced by a tiny tissue mass are not visible on the ECG. As the activation wave front faces the amplified mass of atrial muscle, the beginning of electrical activity is noticed on the body surface, and the initial ECG wave of the cardiac cycle is visible (Berbari 2000). The heart is one of the main organs of the human body, crucial to our existence. It is basically a huge pump, with only function is to keep blood circulation and preserve organs alive (Begg et al. 2008; Subasi 2019).

The QRS complex is an electrical ventricular system and the most well-known waveform showing electrical activity inside the heart. It is the basis for automatic recognition of heart rate and also as an access point for classification schemes. The QRS complex morphology describes the mechanical action of the heart offering a view into how each chamber is functioning. The waves of depolarization extending all the way through the heart via each cardiac cycle produce electrical impulses. These impulses travel via a variety of body fluids, e.g., blood, up to the body's surface where they can be recorded using surface electrodes. These signals are then sent to an electrocardiograph. The main characteristics of the QRS wave that describe significant data related to cardiac health are as follows (Begg et al. 2008; Subasi 2019).

(a) P wave
(b) QRS complex
(c) T wave
(d) QRS intervals.

There are different heart arrhythmias. In this chapter, four major heart arrhythmias are selected because these are most frequent arrhythmias. These are premature ventricular contraction (PVC), atrial premature contraction (APC), right bundle branch block (RBBB), and left bundle branch block (LBBB). In the following text, it will be given short description of these four arrhythmias (Jones 2008).

Many researchers studied on biomedical signal analysis and machine learning techniques for computer aided diagnosis (CAD) using ECG signals. Earlier researches done in this field suggest that biomedical signals taken from complex systems under healthy conditions may have a fractal temporal structure (Bassingthwaighte et al. 2013). Many researches (Muller et al. 1992; Lai and Chan 1998; Esgiar and Chakravorty 2004) presented that ECG signals are proper models as self-affined fractal sets and it is realizable to perform accurate classification by using fractal dimension. Recently, variety of detection algorithms have been proposed. Thakor et al. (1990) proposed the sequential hypothesis testing; the threshold-crossing intervals, the auto-correlation function and the VF-filter were suggested by Clayton et al. (1993), and algorithms based on neural-networks were suggested by Yang et al. (1994).

ECG based CAD systems contains two functional parts: feature extraction and classification. There are different ways for feature extraction. In (de Chazal and Reilly 2003; Hu et al. 1997; Moraes et al. 2002), feature extraction was done in time domain. In Acharya et al. (2004) and Minami et al. (1999), feature extraction was done in frequency domain. Al-Fahoum and Howitt (1999) extracted features by using multiscale decomposition. Osowski and Linh (2001) used statistical measures for feature extraction. A lot of work has also been dedicated to the improvement of classification techniques for these feature sets, such as linear discrimination, decision trees, neural networks, and the combination of experts systems (Yu and Chou 2007).

There are also numerous works done reporting the use of multifractal model for cardiac signal analysis (Ivanov et al. 1999; Wang et al. 2007). Raghav and Mishra

(2008) tried to use local fractal dimension based on nearest neighbor classification algorithm for ECG signal-based heart disorders, and results achieved were promising (A. K. Mishra and Raghav 2010). Qin et al. (2005) used radial basis function neural network (RBFNN) to classify ECG signals. Different neural network techniques were used such as techniques suggested in (Prasad and Sahambi 2003; Yu and Chou 2008; Yong et al. 2009; Nadal and de Bossan 1993) were used for ECG signal classification. Classification has also been performed using support vector machines (Acır 2006; Asl et al. 2008; Besrour et al. 2008; Melgani and Bazi 2008). Other works were conducted using genetic algorithms (Kutlu and Kuntalp 2011). Özbay et al. (2006) suggested fuzzy logic classification tool. k-nearest neighbor (k-NN) classification tool was used in Arif et al. (2009), Karimifard et al. (2006), Christov et al. (2005), and Arif and Akram (2010).

During ECG recording, several sources can add noise to the acquired signal. Imperative noise sources can be electrical power line interference, noise due to instrumentation or electronic devices, impedance changes at the skin/electrode edge, movement artifacts such as electrode movement, baseline drifts caused by respiration, and electrosurgical noise. Thus, precise preprocessing of the ECG signals is crucial to reduce the variety of noise components and enhance the signal-to-noise ratio (SNR) meaningfully. Consequently, a single band-pass filter having 10–25 Hz pass-band is utilized for the ECG signal filtering (Begg et al. 2008). Since ECG signal denoising and feature extraction present an indication about several cardiac abnormalities for diagnosis of cardiovascular disorders, it received lot of attention from the medical societies. Alickovic and Subasi (2015) denoised ECG signals using multiscale PCA (MSPCA) for arrhythmia detection. To achieve a better classification performance, noise should be removed. The proposed framework demonstrated that MSPCA denoising increases the classification accuracy. Moreover, Alickovic and Subasi (2016) proposed another framework where MSPCA was utilized for denoising, DWT for feature extraction, and random forest classifier for ECG signal classification, and they achieved higher accuracy. Usta and Yildiz (2017) employed random forests (RF) classifier to classify heart arrhythmia using ECG signals.

Afkhami et al. (2016) used MIT-BIH arrhythmia database including several forms of common arrhythmias. Dual tree complex wavelet transform based feature extraction technique is proposed by Thomas et al. (2015) for automatic classification of cardiac arrhythmias. The results showed that dual tree complex wavelet transform (DTCWT)-based feature extraction technique achieved better accuracy than discrete wavelet transform (DWT) for five types of ECG beats of MIT-BIH arrhythmia database. Li and Zhou (2016) utilized statistical features extracted from DWT, ICA for dimension reduction, and PNN, k-NN, DT, and SVM for classification. Li and Zhou (2016) proposed a method to classify ECG signals using wavelet packet entropy and random forest. Cruz et al. (2016) employed DWT in ECG signal classification and compared support vector machine (SVM) and adaptive neuro-fuzzy inference system (ANFIS). The experimental result showed that SVM achieved better performance in terms of sensitivity, specificity, accuracy, and training time, while ANFIS had the fastest evaluating time. H. Li et al. (2016) proposed a new framework for classification ECG signals by combining WPD and ApEn for feature extraction and LIBSVM for classification. The algorithm utilizes the PSO to determine the best parameters. Ai et al. (2015) used GND-ICA feature-fusion method based on a multilearning subspace-learning algorithm for ECG heartbeat classification by utilizing MIT-BIH arrhythmia database in all experiments (Subasi 2019).

KalaiSelvi et al. (n.d.) utilized DTCWT-based feature set for classifying ECG beats. The performance of the developed technique is compared with DWT feature extraction and it is realized that the proposed feature set was achieved higher recognition performance than the DWT based feature set. Kiranyaz et al. (2015) discussed the classification of ECG heartbeat is used CNN to record and detect the heart problem. The approaches are given maintaining a robust fast, and patient-specific scheme with a superior classification performance of the heart problem. Qurraie and Afkhami (2017) developed a novel algorithm based on the TF to get the features and the decision tree for the arrhythmia classification (Subasi 2019).

18.2.2 The Electroencephalogram (EEG)

Electrical signals produced by the brain characterize the brain function and the status of the whole body. This delivers the motivation to utilize biomedical signal processing techniques to the electroencephalogram (EEG) signals acquired from the brain of a human subject. The physiological characteristics of brain activities have numerous issues regarding to the characteristics of the original sources and medium and their actual patterns. The medium describes the path from the neurons, that are signal sources, to the electrodes, that are the sensors in which some form of mixtures of the sources are collected. Understanding of neurophysiological properties and neuronal functions of the brain together with the working principle underlying the signal generation and acquisition is useful when dealing with these signals for recognition, analysis, and treatment of brain disorders. EEG presents the way of diagnosis of several neurological disorders and abnormalities in the human body (Sörnmo and Laguna 2005; Subasi 2019).

Electroencephalograms (EEGs) are recordings of the electrical potentials created by the brain, usually less than 300 μV. EEG has been conducted mostly in research facilities and medical settings with the aim of detecting pathologies and epilepsies. An electroencephalographer, an individual trained to qualitatively differentiate normal and abnormal EEG activity within pretty long EEG records, was for many years the only person qualified for visual interpretation of the EEG. Therefore, researchers and clinicians were left and covered up in a bunch of EEG paper records. However, the arrival of modern powerful computers and related technologies opened a whole new door of possibilities for applying various methods to quantify EEG signals (Bronzino 1999; Subasi 2019).

Even though the EEG has lost portion of its supremacy in medical routine due to these modalities, it still remains a very powerful tool for the analysis of many diseases like epilepsy, sleep disorders and dementia. Moreover, the EEG signal is essential for real-time monitoring of progress of patients with encephalopathies or the ones in a coma. In these applications, the temporal resolution of the EEG is unmatched by the imaging. Furthermore, the overall cost related with recording equipment and skilled workers required for managing the instrumentation is significantly lower than the cost related with neuroimaging. For a simple recording system, the technical demands on instrumentation are quite modest. They are limited to a set of electrodes, a signal amplifier and a PC for data storage, signal analysis and graphical demonstration (Sörnmo and Laguna 2005; Subasi 2019).

The EEG signal from the scalp has time duration of 0.01–2 s and amplitude of around 100 μV (Kerem and Geva 2017). The frequency components of the EEG signals are generally employed for the analysis taking into account of the following frequency bands: Delta (up to 4 Hz), Theta (4–8 Hz), Alpha (8–12 Hz), Beta (12–26 Hz), and Gamma (26–100 Hz). Waveform activities differs from each other according to the brain function related to the mental and physical tasks. For instance, low-frequency waves (delta and theta) dominate during sleep times, while an EEG signal acquired during awake times includes a higher percentage of high-frequency waves (alpha and beta). Also, transition from an eyes open state to an eyes closed state changes the EEG frequency spectrum; the state with closed eyes has a diverse peak in the spectrum around 10 Hz (Felzer and Freisieben 2003). The EEG signals are semi-stationary time-dependent and non-stationarity in the waveforms. Hence, these characteristics cannot be detected easily. Power spectrum achieves a quantitative measure of the frequency distribution of the EEG at a cost of the amplitude distribution and information related to the existence of EEG patterns. Even though, these primary efforts are unsatisfactory, they allow the use of frequency analysis in the analysis of brain wave activity. Therefore, time-frequency methods such as wavelets are applied for feature extraction from the EEG signal (Bigan 1998). Furthermore time-frequency methods such as discrete wavelet transform (DWT), wavelet packed transform (WPT), tunable Q wavelet transform (TQWT), dual tree complex wavelet transform (DTCWT), empirical wavelet transform (EWT), and empirical mode decomposition (EMD) are essential to explain the different behavior of the EEG to express it in both the time and frequency domain. It should also be highlighted that the wavelet is appropriate for the analysis of nonstationary signals such as EEG, ECG, and EMG. Therefore, the wavelet is appropriate for detecting transient events, such as spikes occur during epileptic seizures (Bronzino 2000; Adeli et al. 2007; Subasi 2007; Subasi and Gursoy 2010).

The lack of clear difference in EEG activities makes the visual detection of different disorders from EEG signals challenging (Bigan 1998). Therefore, computer aided decision support systems were developed to allow more precise and quicker detection of disorders from the EEG recordings. ANN-based classifiers' performance was compared in Pang et al. (2003). They trained ANN with features selected from a raw EEG signal. A system for the automatic analysis and detection of epileptic seizures using wavelet transform to extract features from EEG signals was developed by Bigan (1998). An ANN model was used for the automated analysis of the EEG recordings. A discrete wavelet transform (DWT) and a mixture of expert model were employed for EEG signal classification in (Subasi 2007).

In Subasi and Gursoy (2010), DWT was used for feature extraction; PCA, ICA, and LDA were used for dimension reduction; and support vector machines (SVM) is used for the classification of EEG signals. Subasi et al. (2017) proposed an epileptic seizure detection model to determine the optimum parameters of support vector machines (SVMs) by employing particle swarm optimization (PSO) and genetic algorithm (GA). Alickovic et al. (2018) compared wavelet packet decomposition (WPD), discrete wavelet transform (DWT), and empirical mode decomposition (EMD) for epileptic seizure detection and seizure prediction. Soleimani et al. (2012) proposed a robust technique to elaborate and evolve a neuro-fuzzy model which works as an online adaptive method with a patient-independent parameter for a seizure prediction. Furthermore, prediction is improved by using multiple features to detect the preictal patterns. Liu et al. (2012) proposed wavelet decomposition of multichannel intracranial EEG (iEEG) within five scales by selecting three frequency bands. They extracted effective features such as relative amplitude, relative energy, and coefficient of variation and fluctuation index at particular scales. SVM is employed for classification and achieved low false detection rate and high sensitivity for seizure detection in long-term iEEG. Williamson et al. (2012) combined multivariate EEG features with patient-specific machine learning for seizure prediction. The proposed algorithm calculates the covariance matrices and eigenspectral of space-delay correlation from EEG data and classifies the data using support vector machine. Aarabi and He (2012) implemented a rule-based patient-specific seizure prediction framework for focal epilepsy. They used five univariate measures, including largest Lyapunov exponent, Lempel-Ziv complexity, correlation entropy, correlation dimension, and noise level as well as one bivariate measure, nonlinear interdependence which are extracted from electrodes implanted deep in the brain.

18.2.3 The Electromyogram (EMG)

The duty of human skeletal muscular system is to achieve the forces required to carry out a several activities. Such system composed of the nervous system and the muscular system, together creating the neuromuscular system (Begg et al. 2008). Motion and arrangement of limbs are administered by electrical signals propagating back and forth among the muscles and the peripheral and central nervous system (Bronzino 1999). The central nervous system (CNS) administers via nerve signals and muscles excitation. The skeletal-muscular system is composed of muscle sets connected to bones via tendons and a motion is done once nerve signals produce muscle contractions and relaxations that either attract or repel the bone (Begg et al. 2008).

Neuromuscular disorders consist of abnormalities initially appearing in the nervous system, in the neuromuscular junctions, and in the muscle fibers. These abnormalities have distinct degrees of severity going from negligible damages of muscle to amputation caused by neuron or muscle death. In more severe disorder like amyotrophic lateral sclerosis (ALS), decease is generally assured. In the majority of cases, clinical testing is insufficient to detect and prevent disorder from spreading

(Preston and Shapiro 2012) since a lot of dissimilar abnormalities could be results a specific symptom. Correct diagnosis of the disorder is, for that reason, of supreme significance so as result more purposeful healing can be done employing electromyography (EMG) that was first used as a method of evaluating neuromuscular states established on cell action capabilities throughout muscle activity. The understanding of EMG is as a rule done by trained and expert neurologists who besides examining EMG waveforms employ methods such as needle conduction researches and muscle acoustics as well. Trouble appears once there are too hardly any specialists to assemble the demand of patients and, consequently, it is imperative to build up automated diagnostic systems established on EMG readings. This offers range for the usage of machine learning methods for the discovery and classification of neuromuscular irregularities based on EMG processing. These smart systems will help doctors in discovery of abnormalities in the neuromuscular system. The goal of smart diagnostic and artificially administered neuromuscular systems is to primary preprocess the raw EMG signal and after that take out characteristic data or features that may be taken out comprising of time and frequency domain data. This data may after that be employed as input data for classifiers that may classify neuromuscular disorders. The challenge in this case is to invent signal-processing method that protect or confine significant discriminatory data so as to give a good quality set of features for classification. Neuromuscular disorders are anomalies related to the peripheral nervous system. They can be classified based on the location and reason of the disorders. Two main disorders are neuropathy and myopathy. Neuropathy is a disease about nerves that cause the pain and some disability. The causes of neuropathic disorders are injury, alcohol abuse, infection, diabetes, and cancer chemotherapy. It can be categorized as mononeuropathy and polyneuropathy. Myopathy is a disorder generally related with the skeletal muscle that caused by injury of a muscle group or some genetic mutation. The patient suffering with myopathic diseases has week muscles and depending on severity of disorder, has problem with the performing regular task or impossible to make any movement by using effected muscles (Begg et al. 2008; Subasi 2019).

A number of researches have been done to design an accurate automated diagnosis system by employing different classification algorithm to classify EMG signals (Richfield et al. 1981; Subasi et al. 2006). It is possible to find commercial version of some algorithm on the market, but almost none of them are used at clinics to diagnosis of the neuromuscular disorders (Bozkurt et al. 2016). Autoregressive (AR) analysis and time domain analysis were used together to classify the EMG signals by Pattichis and Elia (1999). De Michele et al. (2003) applied the wavelet cross-correlation analysis on the two different muscle and find out that it is possible to make detailed classification. MUAP parameters were used as input to the parametric sequential pattern recognition classifiers by Pattichis et al. (1995). Loudon et al. (1992) used eight different MUAP parameters as input to the statistical pattern recognition techniques to classify the EMG data. Hassoun et al. (1994a, b) applied the time domain parameters as input to a tree-layered ANN and used "Pseudo Unsupervised" algorithm as classifier in their study. Two different classification methods were used at the same time to classify the EMG signals by

Christodoulou and Pattichis (1999). They proposed to use unsupervised machine learning algorithms, including self-organized feature maps, learning vector, and Euclidian distance. Genetic algorithms were used to classify EMG signals by Schizas and Pattichis (1997). Recent years, wavelet neural network (WNN) was used to analyze the EMG signals. Subasi et al. (2006) used AR signal processing with WNN to classify EMG data. They stated that they classify the EMG signals with the accuracy of 90.07% and it is possible to develop a simple, accurate, and reliable enough automated classification system for routine clinical usage. Katsis et al. (2006) used SVM to classify the EMG data. The features of EMG signals were extracted by wavelet method and used as input to ANN algorithm and learning vector quantization (LVQ) by Guo et al. (2006). Jiang et al. (2006) performed the wavelet transformation of EMG signals and then used the statistical characteristic of wavelet coefficients as input to an ANN. By using similar techniques, Cai et al. (1999) extracted feature vectors by using wavelet transform and used them as input to an ANN which was trained by a standard backpropagation algorithm to classify the EMG signals.

EMG presents a comprehensive information to describe neuromuscular activity and muscular morphology. The EMG signals must be decomposed, classified, and analyzed in order to describe a muscle using quantitative EMG (QEMG) data. In order to diagnose neuromuscular disorders, EMG signal must be classified for the detection of abnormalities (Yousefi and Hamilton-Wright 2014). Recently, Rasheed et al. (2008) developed a model to distinguish individual MUP waveforms from a raw EMG signal to extract relevant features, and classify the MUPs. The adaptive fuzzy k-NN classifier with time domain features and with wavelet domain features are presented. EMG signal are segmented utilizing threshold technique to identify possible MUAPs. Statistical pattern recognition technique is utilized for clustering the identified MUAPs. After employing autoregressive (AR) method for feature extraction, MUAPs are classified utilizing binary support vector machine (SVM) classifier (Kaur et al. 2010).

Subasi (2012a) utilized several feature extraction methods and machine learning techniques such as multilayer perceptron neural networks (MLPNN), dynamic fuzzy neural network (DFNN), and adaptive neuro-fuzzy inference system (ANFIS) to classify the EMG signals and compared them according to their accuracy. In the proposed framework, neuro-fuzzy classification methods were capable to classify the EMG signals with the high accuracy. Moreover, Subasi (2012b) developed an effective combination of classifier and features to classify the EMG signals. LDA, RBFN, MLPNN, C4.5 DT, SVM, and Fuzzy SVM classifiers are used with statistical features extracted from DWT sub-bands. It is reported that the FSVM and the statistical features extracted from DWT sub-bands utilizing the internal cross validation method achieved a better performance than other classifiers. Subasi (2013) proposed a framework in which DWT is utilized to decompose the EMG signals into the frequency sub-bands and then a set of statistical features were extracted from these sub-bands. Substantial improvements in terms of classification accuracy was realized by the developed PSO-SVM classification system. Furthermore, Subasi (2015) used an evolutionary approach to classify EMG signals utilizing SVM classifier and the frequency sub-bands of DWT. A comparison research was done between combined neural network (CNN) and feedforward error backpropagation ANN (FEBANN) classifiers by Bozkurt et al. (2016).

Gokgoz and Subasi (2014) studied the effect of multiscale principal component analysis (MSPCA) denoising in EMG signal classification. Multiple signal classification (MUSIC) processing technique was utilized for feature extraction to classify EMG signals into normal, ALS, or myopathic. It was realized that MSPCA denoising was improved the classification accuracy. After denoising EMG signals with MSPCA, the classification accuracy was 92.55% for SVM, 90.02% for ANN, and 82.11% for k-NN. The same researchers (Gokgoz and Subasi 2015) presented a framework for classification of EMG signals utilizing decision tree algorithms for classification, DWT for feature extraction, and MSPCA for denoising. Parsaei and Stashuk (2013) employed the k-means clustering, and the supervised classification is realized by utilizing a certainty-based algorithm. Dobrowolski et al. (2012) used the wavelet index to classify myopathic, neuropathic, or normal EMG signals. Kamali et al. (2013) proposed a scheme which utilizes both time and time-frequency features of a MUAP along with an ensemble of SVM classifiers. Time-frequency features are DWT coefficients of the MUAP. Time domain features consist of peak to peak amplitude, turn, area, duration and phase of the MUAP. Kamali et al. (2014) employed ensemble of support vector machines (SVMs) classifiers to determine the class label (myopathic, neuropathic, or normal) using both time domain and time-frequency domain features of the EMG signal.

Artameeyanant et al. (2016) proposed a normalized weight vertical visibility algorithm as a feature extraction method for ALS and myopathy detection. In the proposed method, the features are extracted by utilizing selective statistical mechanics and measurements, and the extracted features are used as a feature matrix for classifier input. Finally, powerful classifiers, such as multilayer perceptron neural network, support vector machine, and k-nearest neighbor classifiers are employed to categorize signals into healthy, ALS, and myopathy. Naik et al. (2016) presented a classification technique for neuromuscular disorders (myopathic, and ALS) which utilized a single-channel EMG sensor. The single-channel EMG signals are decomposed by employing ensemble empirical mode decomposition algorithm, then the FastICA algorithm is used for dimension reduction. A reduced set of features are classified utilizing the linear discriminant analysis, and the classification results are fine-tuned with a majority voting scheme. Khan et al. (2016) proposed a framework that utilizes both time domain and time-frequency domain features extracted from the EMG signals. Several classification approaches including single classifier and multiple classifiers with time domain and time-frequency domain features were examined. SVM and k-NN classifiers are employed to predict class label (Normal, Neuropathic, or Myopathic). Sengur et al. (2017) proposed a deep learning based classifier for effectively categorization of normal and ALS EMG signals. They used different time-frequency methods combined with convolutional neural network for EMG signal classification. Mishra et al. (2017) employed improved empirical mode decomposition (IEMD) in combination with the least squares support vector machine (LS-SVM) classifier is used for the analysis of

ALS and normal EMG signals. The proposed technique is achieved 96.33% accuracy. Hazarika et al. (2018) presented a real-time feature extraction and fusion model for an automated classification of electromyogram signals with amyotrophic lateral sclerosis (ALS), myopathy (MYO) and normal (NOR) using Discrete wavelet transform and canonical correlation analysis (CCA).

18.3 Biomedical Signal Analysis Framework

The general framework for the biomedical signal analysis is shown in Fig. 18.1. This framework includes three main modules: (1) signal preprocessing/denoising, (2) feature extraction/dimension reduction, and (3) classification. In this section, comprehensive explanation of each module will be provided.

18.3.1 Biomedical Signal Denoising

The biomedical signals contain several types of artifacts including internal or external interfering noises. These artifacts can be eliminated by employing signal denoising techniques to filter out most of the artifacts and noise (Sanei 2013). The biomedical signal analysis and processing are implemented in three main steps: preprocessing/denoising, feature extraction/dimension reduction, and detection/classification. The main goal of preprocessing is to simplify succeeding procedures without losing related information and to enhance the signal quality by increasing the signal-to-noise ratio (SNR). Filters and transformations such as ICA, PCA, KPCA, and MSPCA are often used during preprocessing. Researchers employ these methods to eliminate or at least reduce the unwanted signal components by transforming the signals. The capacity of PCA can be combined with the ability of wavelet analysis to form multiscale principal component analysis (MSPCA) in order

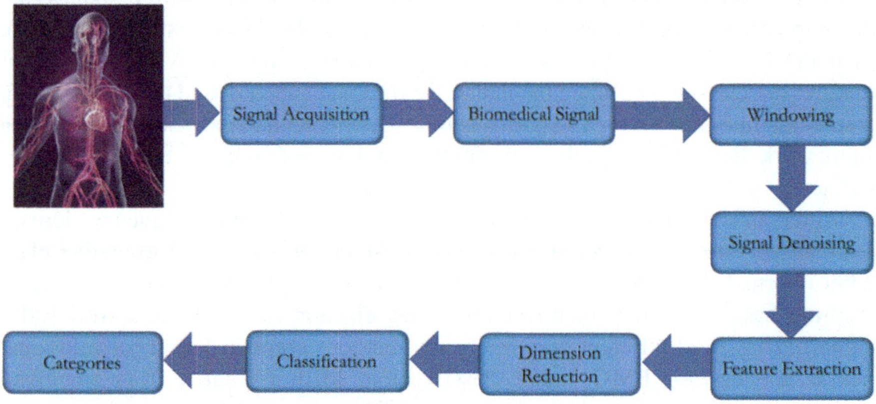

Fig. 18.1 General biomedical signal analysis framework

to eliminate the relationship among the variables with the ability of wavelet transform to extract features and to remove the relationship between auto-correlated measurements. The PCA of the wavelet coefficients at every scale is calculated by MSPCA with integrating the results at relevant scales. MSPCA is efficient since it includes contributions of events of which behaviors become different over time and frequency (Bakshi 1998).

The MSPCA denoising technique can be realized in three main steps. All signals from X_{nxm} is decomposed using wavelet in the initial step. Then, for each wavelet decomposition level, PCA denoising algorithm is applied separately, and wavelet coefficients that have certain threshold value are kept. In the last step, the PCA application for all levels are combined, to get a denoised input signal matrix $\dot{X}_{nxm}$ (Bakshi 1998). The MSPCA shows better denoising performance than PCA algorithm (Kevric and Subasi 2017).

18.3.2 Feature Extraction

One of the crucial steps in the biomedical signal analysis is the feature extraction. Therefore, the biomedical signals composed of several data points, and informative features can be extracted by using different feature extraction methods. These distinctive and informative parameters describe the behavior of the signal waveform which specify a precise action. The biomedical signal patterns can be represented by frequencies and amplitudes. These features can be extracted using different feature extraction algorithms which is another step in signal processing to simplify the succeeding stage for classification (Graimann et al. 2009). The biomedical signals can be decomposed using time-frequency (TF) methods which can detect changes in both time and frequency (Sanei 2013). It is important to deal with a smaller number of values that characterize proper features of the signals to accomplish better performance. Features are generally collected into a feature vector by transforming signals into a relevant feature vector known as feature extraction. Distinctive features of a signal are analyzed by a signal classification framework, and depending on those distinctive features, class of the signal is decided (Siuly et al. 2016). Time-frequency methods, such as Wigner-Ville transform, short-time Fourier transform (STFT), wavelet transform (WT), discrete wavelet transform (DWT), wavelet packet transform (WPT), tunable Q-factor wavelet transform (TQWT), dual tree complex wavelet transform (DTCWT), empirical mode decomposition (EMD), and ensemble EMD, decompose signals in both time and frequency domain.

The wavelet transform (WT) is a time-frequency signal decomposition algorithm on a set of orthogonal basis functions taken by contractions, dilations, and shifts of a mother wavelet. WT, which achieves a good time resolution for high-frequency components and good frequency resolution for low-frequency components, has been employed broadly for biomedical signal processing (Muthuswamy 2004). The continuous wavelet transform is denoted as

$$W_x(u,s) = \frac{1}{\sqrt{S}} \int_{-\infty}^{+\infty} x(t)\psi^*((t-u)/s)\,dt \tag{18.1}$$

The orthogonal basis functions can be taken by scaling and shifting a *mother wavelet* function ψ(t)

$$\psi_{mk}(t) = 2^{-m/2}\psi(2^{-m}t - k) \tag{18.2}$$

In the discrete TFRs both time and scale variations are discrete. Scaling for the DWT includes sampling rate changes. A larger time band is enclosed for a larger scale for a given number of samples. Naturally, a binary or dyadic scaling structure is used so that given a discrete wavelet function, $\psi(x)$, is scaled by values that are binary. Hence

$$\psi 2^j(t) = 2^j\psi(2^j t) \tag{18.3}$$

where j is the scaling index and $j = 0, -1, -2, \ldots$. In a dyadic scheme, subsampling is always decimation in-time by a power of 2. Translations in time will be correspondingly larger as well as for a more sizable scale. Once the scale is enlarged, resolution is lowered. Resolution mainly corresponds to the frequency. Signals are decomposed into a series of orthogonal wavelets in a way that every orthogonal vector space represent signal components with varying levels of scale and resolution. Mallat (1989) called this algorithm *multiresolution signal decomposition*. In every step of the algorithm, wavelets are generated with successively finer depictions of signal content. In order to produce an orthogonal wavelet depiction, a given wavelet function is first dilated by the scale coefficient 2^j, then translated by 2^{-jn} (Thakor et al. 2000). This process is shown in Fig. 18.2.

The application of DWT as a feature extraction from the biomedical signals will be given in sect. 18.4.

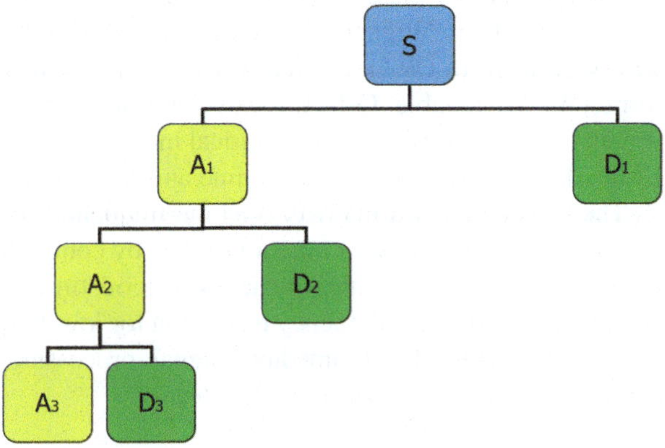

Fig. 18.2 DWT for scale level 3

18.3.3 Dimension Reduction

Dimension reduction is a process to decrease the dimension of the original feature vector, while keeping the most distinctive information and removing the unrelated information (Phinyomark et al. 2013). Most of the feature extraction methods yield redundant features. Actually, in order to improve the performance of a classifier and achieve minimum classification error, some types of feature selection/reduction methods that produce a new set of features must be applied. Several methods employed for dimension reduction and feature selection to achieve better a classification accuracy (Wołczowski and Zdunek 2017).

The dimension of biomedical signals should be reduced to analyze the data for achieving more accurate results. Small number of parameters are employed to reduce the dimension of the biomedical signals through different ways. Furthermore, the features or dimensions must be minimized for achieving better classification accuracy. For example, the DWT produces wavelet coefficients to describe the signal energy distribution in both time and frequency domains and they describe the biomedical signals with set of wavelet coefficients. Meanwhile, wavelet-based feature extraction methods yield the feature vector that are too big in size to be employed as an input to a classifier, a dimension reduction method should be utilized to achieve a smaller number of features from wavelet coefficients. Recently various dimension reduction methods such as Lyapunov exponents, higher order statistics, and entropies have been employed for dimension reduction. First, second, third and fourth order statistics of the sub-bands of the wavelet decomposition can be employed for reducing dimension. The six statistical features are utilized for the biomedical signal classification which are:

1. Mean absolute values of the signal coefficients in every sub-band,
2. Average power of the signal coefficients in every sub-band,
3. Standard deviation of the signal coefficients in every sub-band,
4. Ratio of the absolute mean values of signal coefficients of adjacent sub-bands,
5. Skewness of the signal coefficients in every sub-band,
6. Kurtosis of the signal coefficients in every sub-band.

18.3.4 Machine Learning Methods

Machine learning algorithms utilized to optimize a performance criterion using historical data or learned experience. The model defined with system parameters designed and controlled with hyperparameters, and learning is the optimization of the parameters by the execution of a machine learning algorithm to search the optimal parameters using the training data, which are historical data or previously acquired experience. The main learning goal can be predictive to make forecasts from the labeled data such as classification and regression models. Machine learning algorithms mainly utilize the theory of statistics in designing models, because the aim is to describe the samples or make an inference from the samples. In order to develop machine learning models, in training, you need to consider the performance by means of accuracy to find a solution for the optimization problem. Moreover, once a model is trained, its representation and soft computing solution for learning inference need to be efficient by means of space and time complexity as well. In certain applications, efficiency has also great importance, for example, efficiency is often as much as important as accuracy for the data mining applications. In medicine, learning programs are used for disease identification and diagnosis (Alpaydin 2014).

18.4 Biomedical Signal Analysis Applications

In order to get a consistent assessment of the quality of the target approximation characterized by the model, the assessment of the classification model is utilized. Depending on the model's intended application, diverse performance measures can be used. Since the model is generated based on a training data set, it is crucial for model's quality to check generalization ability. Hence, it important to distinguish between the value of a specific dataset, value of training set and its anticipated performance on the whole domain (Cichosz 2014).

In order to obtain the dataset performance of a model, the value of one or more selected performance metrics on a specific dataset with true class labels existing is calculated. Dataset performance represents the matching degree of the model and the target concept on this dataset. Training performance of the model is determined by the evaluation of the model on the training set that was employed to build the model. While this performance is beneficial for better understanding of the model, it is not of significant interest as the classification of the training data is not the purpose of classification models. Expected performance of the model, on the whole domain, represents its true performance. True performance of the model shows its ability to classify new examples from the given domain correctly. In order to assess the true performance, i.e., to consistently evaluate the unknown values of the accepted performance metrics on the whole domain, comprising generally previously unseen examples, appropriate assessment measures are needed (Cichosz 2014).

k-fold cross-validation is a sophisticated assessment process which handles the tradeoff of bias vs. variance. It randomly divides the existing dataset into k subsets of the same size and then iterates over these subsets. When all k iterations are accomplished, the model built without specific instance in the training set is employed to produce a predicted class label for every instance in the dataset. The resultant vector of predictions is compared to the true class labels utilizing one or more chosen performance measures. The k-fold cross-validation process successfully virtualizes the training and validation or test sets. All existing instances from the set are employed for both model creation and evaluation, but not at the same time (Cichosz 2014).

18.4.1 Performance Measures

Performance measures of model are created by comparing the true class labels of the instances from dataset and the predictions produced by the classifier on the same dataset. For binary classification, a confusion matrix can be characterized with TP as true positive, TN as true negative, FN as false negative, and FP as false positive counts. A new learning problem presented focusses on its domain but ignores a comprehensive analysis. This brings the most employed measure, accuracy, unable to differentiate between the number of correct labels of different classes (Sokolova et al. 2006):

$$Accuracy = \frac{TP + TN}{TP + FP + FN + TN}$$

In some applications, the number of instances in one class is frequently considerably lower than the overall number of instances. The experimental setting is represented as follows: there is a class of special interest (usually positive) within the set of classes. The rest of the classes are either left, as is in the case of multiclass classification, or combined into one, as in binary classification. The measures of selection are taken for the positive class (Sokolova et al. 2006). The ratio between true positives and false positives is represented by precision.

$$Precision = \frac{TP}{TP + FP}$$

A relation between correctly classified instances and misclassified instances is called recall.

$$Recall = \frac{TP}{TP + FN}$$

$$F - Measure = 2 * \frac{Precision * Recall}{Precision + Recall}$$

The F-measure is the harmonic mean of the precision and recall indicators. Recall is a function of its true positives and its false negatives. Precision is a function of true positives and false positives (Sokolova et al. 2006).

Receiver operating characteristic (ROC) employs Cartesian coordinate system in which its y-axis characterizes the true-positive rate, while the x-axis characterizes the false-positive rate. Single point on the ROC plane which visualizes the underlying tradeoff between true positives and false positives characterizes the performance of the classifier. Performance of a classifier which is based only on its scoring function element can be graphically designated utilizing ROC curve. It is essential to determine all possible operating points of a scoring classifier to generate the ROC curve. Occasionally a simple assessment measure is desired even in such more complex cases. Such usually employed measure is the area under the ROC curve (AUC).

During the comparison of models, the model which has greater AUC value is considered superior with respect to its overall predictive performance potential (Cichosz 2014).

Kappa statistic is another performance measure which takes expected figure into account by taking it from the predictor's successes. It represents the result as a proportion of the total for a perfect predictor. The measurement of the consensus between observed and predicted categories of a dataset, and correcting the agreement which happens by chance is achieved by the Kappa statistic (Hall et al. 2011).

18.4.2 ECG Signals Analysis in Diagnosis of Heart Arrhythmia

Cardiovascular disorders (CVDs) are one of the main mortality reasons in world-wide. The creation of accurate and fast techniques for automated ECG heartbeat signal classification is vital for clinical diagnosis of different CVDs (Thaler 2017), e.g., an arrhythmia. Notion arrhythmia is employed to represent a group of circumstances in which irregular electrical activities coming from heart and are characterized by the ECG beats or patterns (Pan and Tompkins 1985; De Chazal et al. 2004). ECG is an effective, simple, noninvasive tool for heart disease recognition. Medical doctors investigate several waveforms based on their characteristics (amplitude, polarity, etc.) and diagnose and treat based on this investigation (Subasi 2019).

Human heart is the most crucial muscle in human body, which together with blood vessels forms cardiovascular system and pumps the blood into each cell of the body. There is no precise heart failure diagnosis tool for detecting the heart failure. The electrocardiogram (ECG) is a noninvasive instrument that acquires electrical activity of the heart and demonstrates the heartbeat irregularities. It shows the possible arrhythmias of the heart or the heartbeat irregularities (Passanisi 2004). Therefore, ECG is an imperative tool for defining the healthiness and the function of the cardiovascular system. Furthermore, it is substantial to describe precise and appropriate diagnosis of physicians to circumvent more damage and to define appropriate approaches and techniques (Son et al. 2012). Still, the problem arises when there is inadequate number of doctors to encounter the requests of patients. Hence, it is essential to implement an ECG based efficient and automated diagnostic systems, together with the machine learning techniques to classify heart beats. These diagnostic schemes support physicians in distinguishing the cardiovascular anomalies (Masetic and Subasi 2016; Subasi 2019).

Different types of arrhythmias exist and each of them is related to a pattern to classify and identify its type. The arrhythmias can be classified into two main groups. The first group composed of arrhythmias formed by a single irregular heartbeat and the other group composed of arrhythmias formed by a set of irregular heartbeats. The classification and identification of arrhythmias can be very difficult since it needs the analysis of each heartbeat of the ECG records, recorded by a Holter monitor, during hours, or even days. Furthermore, physicians can make a mistake during the ECG analysis, due to exhaustion. The employment of computational methods for automated classification of heartbeats is an alternative. Fully automated framework for arrhythmia detection from the ECG can be divided in three stages: (1) ECG signal preprocessing; (2) feature extraction/dimension reduction; and (3) detection/classification (Luz et al. 2016).

MIT-BIH arrhythmia database records taken from the Beth Israel Hospital Arrhythmia Laboratory[1] is used. This database includes 48 two-lead ECG records recorded from 47 different patients, and duration of each of these records is about 30 minutes. Sampling frequency is 360 Hz. Five different heartbeat categories are selected for this study: normal (N), left bundle branch blocks (LBBB), right bundle branch blocks (RBBB), atrial premature contractions (APC), and premature ventricular contractions (PVC) heartbeats (Alickovic and Subasi 2016). The results of classification are shown in Table 18.1.

Table 18.1 presents values for four different evaluation criteria employed in heart arrhythmia classification for nine machine learning techniques. Features have been extracted using discrete wavelet transform from raw ECG signals. The least effective method was LAD Tree with average accuracy of only 87.7%. The best result was achieved with k-NN classifier reaching total accuracy of 98.1%. SVM and ANN also achieved good accuracy. Random forest is the best for F-measure and AUC and k-NN is the best for Kappa statistics.

18.4.3 EEG Signal Analysis in Epileptic Seizure Detection and Prediction

Electroencephalographic (EEG) signals are generally examined by spectrum analysis methods, separating the EEG signal into different frequency bands (α, β, θ, δ). Straightforward spectrum analysis techniques are beneficial when these events are slowly unfolding. But, once transient events such as epileptic seizures happen, bursting series of events or sharp spikes in the recorded signal is seen. The discrete wavelet transform can be used to detect the beginning of the seizure burst. Moreover, it can be used for the onset seizure detection and the termination of seizures (Thakor et al. 2000).

Machine learning techniques are employed to solve biomedical engineering problems and, especially, in biomedical signals analysis. They can accomplish to identify and diagnose in real time. EEG analysis has improved considerably with the extensive use of mathematical modelling and machine learning tools. Machine learning tools have also enabled the classification of patterns within the EEG to enhance the recognition, making EEG signals valuable for recognition of brain disorders and primary pathologies. Therefore, several studies on characteristics of the EEG signals related to neurological diseases have been carried out (Begg et al. 2008).

[1]http://physionet.ph.biu.ac.il/physiobank/database/html/mitdbdir/mitdbdir.htm

Table 18.1 Classification of different ECG heartbeat signals

	Normal	APC	PVC	RBBB	LBBB	Average	F-measure	AUC	Kappa
SVM	0.993	0.947	0.93	0.94	0.997	0.961	0.961	0.976	0.9517
k-NN	1	0.963	0.963	0.983	0.997	0.981	0.981	0.997	0.9767
ANN	1	0.953	0.963	0.977	0.98	0.975	0.975	0.999	0.9683
Random forest	0.953	0.893	0.863	0.883	0.893	0.897	0.998	1	0.9683
CART	0.947	0.85	0.863	0.867	0.92	0.889	0.951	0.973	0.8617
C4.5	0.96	0.907	0.847	0.887	0.917	0.903	0.97	0.982	0.8792
REP tree	0.933	0.833	0.847	0.87	0.92	0.881	0.881	0.953	0.8508
Random tree	0.94	0.883	0.877	0.81	0.91	0.884	0.884	0.928	0.855
LAD tree	0.947	0.827	0.83	0.87	0.91	0.877	0.877	0.981	0.8458

There are around 2 million epilepsy patients in the United States alone, responsive therapeutic intervention facilitated by seizure detection algorithm which increases the efficiency of the method. Raghunathan et al. (2011) devised a two-stage cascaded seizure detection solution, with full detection efficiency. The proposed solution is based on the usage of features that results in unique patterns during the seizure. The proposed technique shows high sensitivity rate and low detection duration. Yuan et al. (2012) proposed a new method for multichannel long-term EEG. Novel nonlinear features of EEG signals are derived from the fractal geometry, as the linear feature comes from the relative fluctuation index. The vector of the feature is then merged into an extreme learning machine for classification. For more stable results, post processing techniques are employed such as smoothing and channel fusion and they are tested on 21 subjects with the segment-based and event-based analysis.

Epilepsy is a severe disease characterized by temporary changes in the bioelectrical functioning of the brain. These fluctuations cause irregular neuronal synchronization and seizures that affect awareness, sensation or movement. Epileptic seizures represent unexpected bursts of wild electrical activity in a group of neurons of the cerebral cortex. Due to the location of the focus (origin) of the electrical activity and sequential enrolment of various brain regions, epileptic seizures may be manifested in numerous ways. For instance, some sort of auditive or visual sensation follows seizures which focus is located in the sensory regions of the cortex. A group of neurons with reduced functionality is referred to as the epileptic focus (Sörnmo and Laguna 2005). Seizure EEG signals contain characteristic patterns that health professionals use to distinguish them from normal (nonseizure) EEG signals. Therefore, their detection may be used to respond to a forthcoming or ongoing seizure. Also, automated recognition techniques have been tested to decrease the amount of data and enable quicker and more accurate detection of pathological EEG waveforms which characterize epileptic seizures. In addition, a few techniques have been suggested to identify spikes in the EEG to predict epileptic events (Begg et al. 2008).

During normal conditions, there is a stable relationship between inhibitory and excitatory signals. The former signals prevent neurons from firing and thus decrease the brain's electrical activity, while the latter signals force neurons to fire. Nonetheless, an important cause of epilepsy lies in impaired balance between these two actions. The reason for producing this imbalance or unstable condition hides inside the neurotransmitters which are responsible for chemical transfer of the signals in the synapse. Consequently, bursts of wild electrical activity will arise when the inhibitory neurotransmitters are being inactive, or the excitatory ones are being excessively active. This neurotransmitter imbalance can be improved by increasing the inhibitory activity or decreasing the excitatory activity, which is the main challenge of antiepileptic drugs (Sörnmo and Laguna 2005).

Epileptic seizures, hardly causing long-lasting injuries or death, may result in loss of consciousness or cause slight mental confusion only. Seizures possess an extremely variable time duration and rate of occurrence. Their duration may range from several seconds to several minutes. There are epileptic patients who experience just a few seizures during the entire life, while some of them have several seizures during a single day. Therefore, (Niedermayer 1999) built a scheme to classify seizures into groups based on the EEG characteristics. Groups are formed based on the epileptic seizure focus (origin): primary generalized seizures include the whole brain, while partial seizures start in a limited brain area. The latter group is associated with a single epileptic focus, which cannot be said for the former group. Therefore, some partial seizures by eliminating a small part of the cortex during surgery can be treated. To make sure that the location of the epileptic focus is correctly bordered, a series of very systematic and in-depth studies and examinations must be carried out before the surgery. A partial seizure may sometimes progress to other brain areas. Such seizure is denoted as a partial seizure with secondary generalization (Sörnmo and Laguna 2005).

Health specialists use distinctive patterns within ictal (seizure) EEG waveforms to differentiate them from interictal (nonseizure) EEG waveforms. Any form of long-term EEG monitoring creates huge amounts of data as a result. This data requires a lot of time to be properly analyzed. In order to reduce the amount of data and allow quicker and better detection of abnormal EEG signals related to epileptic seizures, automatic detection systems have been tested. Moreover, a few methods have been suggested to predict epileptic seizures by discovering spikes in the EEG. Several signal processing techniques considering mathematical representation of interictal and ictal data are essential for the design of these detection algorithms (Y. Khan and Gotman 2003). Noise and artifact elimination are an additional factor that should be taken into consideration. An effective seizure prediction algorithm may warn a patient wearing an ambulatory recording device to consider proper security actions before the seizure occurs (Sörnmo and Laguna 2005).

A medical application designed to control seizures comprises of two systems (Winterhalder et al. 2003): (1) a seizure prediction algorithm raising an alarm when it senses an upcoming seizure and (2) a system acting to control a seizure. Moreover, one simple alert of upcoming seizure may be enough for a patient to leave dangerous situations or actions, like climbing stairs or playing sports. Seizure prediction methods are based on the extraction of EEG features, calculated over a short time window of a few seconds to a few minutes. There exist univariate measures, calculated separately for each EEG channel, and bivariate (or multivariate) measures, which quantify some relationship between two or more EEG channels (Subasi 2019).

Table 18.2 presents values for four different evaluation criteria employed in epileptic seizure detection and prediction for nine machine learning techniques. Features have been extracted using discrete wavelet transform from raw ECG signals. The least effective method was Random Tree with average accuracy of only 94.3%. The best result was achieved with SVM classifier reaching total accuracy of 98.7%. Random forest is the second best among the classifiers with a total accuracy of 98.5%. F-measure of the classifiers is almost the same as total accuracy results. Random forest is the best for AUC and Kappa statistics with a value of 0.999.

Table 18.2 Classification of different EEG signals for epileptic seizure prediction and detection

	Interictal	Preictal	Ictal	Average	F-measure	AUC	Kappa
SVM	0.997	0.973	0.99	0.987	0.987	0.992	0.98
k-NN	0.987	0.918	0.967	0.957	0.957	0.968	0.936
ANN	0.981	0.95	0.955	0.962	0.962	0.987	0.943
Random forest	0.997	0.969	0.988	0.985	0.985	0.999	0.999
CART	0.977	0.936	0.942	0.952	0.952	0.965	0.9275
C4.5	0.978	0.946	0.961	0.962	0.962	0.974	0.9425
REP tree	0.977	0.937	0.953	0.956	0.956	0.986	0.9335
Random tree	0.962	0.922	0.945	0.943	0.943	0.957	0.9145
LAD tree	0.973	0.91	0.958	0.947	0.947	0.992	0.9205

Table 18.3 Classification of different EMG signals for diagnosis of neuromuscular disorders

	Control	Myopathy	ALS	Average	F-measure	AUC	Kappa
SVM	0.989	0.986	0.999	0.991	0.991	0.995	0.9871
k-NN	0.972	0.976	0.997	0.981	0.981	0.986	0.9721
ANN	0.98	0.975	0.998	0.984	0.984	0.998	0.9767
Random forest	0.974	0.984	0.999	0.986	0.986	1	0.9788
CART	0.933	0.949	0.993	0.958	0.958	0.976	0.9375
C4.5	0.947	0.952	0.989	0.963	0.962	0.975	0.9437
REP tree	0.937	0.953	0.985	0.958	0.958	0.985	0.9375
Random tree	0.913	0.928	0.968	0.936	0.936	0.952	0.9042
LAD tree	0.941	0.955	0.985	0.96	0.96	0.993	0.9404

18.4.4 EMG Signal Analysis in Diagnosis of Neuromuscular Disorders

The needle EMG is the usual clinical recording method employed for diagnosis of the neuromuscular pathology. Once a patient goes to a doctor for muscle weakness, recording of the needle EMG during contraction of specific muscles will be done. The morphology of single MUAP waveforms gives necessary clinical data regarding the muscle's skill to answer to the central nervous system. This data may assist to identify irregular activity happening in situations like muscles irritation, injury to nerves in the arms and legs, pinched nerves, and muscular dystrophy. The needle EMG is also investigated together with nerve wound and can be employed to find out if the wound restores and go back to normal with complete muscle re-activity, for instance, by analyzing alterations in motor unit accomplishment over a definite time period. The diagnostic EMG comprises of investigation of unplanned motor action that can be throughout muscle relaxation. In ordinary situations, the muscle is electrically quiet during relaxation period; on the other hand, irregular unplanned waveforms and waveform patterns can be produced that are connected with spontaneous muscular activities and seizures (Sörnmo and Laguna 2005; Subasi 2019).

Table 18.3 presents values for four different evaluation criteria employed in diagnosis of neuromuscular disorders for nine machine learning techniques. Features have been extracted using discrete wavelet transform from raw ECG signals. The least effective method was Random Tree with average accuracy of only 93.6%. The best result was achieved with SVM classifier reaching total accuracy of 99.1% and F-measure of 0.991. Random forest is the second best among the classifiers with a total accuracy of 98.6%. F-measure of the classifiers is almost the same as total accuracy results. Random forest is the best for AUC with a value of 1. SVM is the best for the Kappa statistics with a value of 0.9871.

18.5 Discussion and Conclusions

Biomedical signals are principally used to diagnose or detect specific pathological or physiological conditions. Additionally, these signals are employed to analyze biological systems in the healthcare. Biomedical signals are utilized in the research laboratory, clinic, and even at home. The ECG, the EEG, and the EMG are the widely used examples of the biomedical signals. In healthcare, biomedical signals are utilized to detect physiological or pathological conditions and diagnose different disorders. Biomedical signal analysis is utilized to remove the noise, create accurate signal model and analyze its components, and predict pathological events in the brain, heart or muscle (Muthuswamy 2004). Biomedical signals include information to recognize the complex pathophysiologic mechanisms. However, such information may not be attainable directly from the raw signals suppressed in additive noise. Because of these reasons, biomedical signal processing is needed to enhance the related information and to designate the level of pathology for routine clinical diagnosis, rehabilitation or therapy. Several signal processing techniques can be used for denoising, filtering, spectral estimation, and feature extraction (Mainardi et al. 2006).

Time-frequency feature extraction techniques are used for the analysis and interpretation of biomedical signals with time-varying characteristics. For instance, the P-QRS-T characteristic of the ECG signals show localized low frequencies in the P- and the ST-segments and high frequencies in the QRS complex. The QRS segment can be localized by means of time-frequency analysis such as discrete wavelet transform. Moreover, neurological signals with potential applications in the analysis of EEG, and epileptic spikes and seizures can be analyzed by discrete wavelet transform. The need for an effective analysis of biomedical signals with time-frequency methods is apparent, and spectral variations can be well localized with them. Analyzing the signal at different scales can achieve meaningful information. The discrete wavelet transform seems to have strong theoretical features permitting innovative interpretation of biomedical signals (Thakor et al. 2000).

Machine learning algorithms offer a powerful tool for the biomedical signal analysis. This chapter has reviewed the biomedical signals such as ECG, EEG, and EMG and applications of machine learning methods in cardiology, neurology, and brain signals. Machine learning methods have been employed broadly in different fields including biomedical signal analysis. In addition to these applications, many studies is still going on to find optimal values for the parameters and optimal algorithms employed in these techniques (Micheli-Tzanakou 2000).

Three different biomedical signals, namely, ECG, EEG, and EMG signal analysis, and classification results are presented in this chapter as an example of biomedical signal usage in healthcare. For each type of biomedical signal dataset, the analysis results are given in Tables 18.1, 18.2, and 18.3. The SVM, ANN, k-NN, and random forest classifiers achieved better results in most of the cases.

References

Aarabi A, He B (2012) A rule-based seizure prediction method for focal neocortical epilepsy. Clin Neurophysiol 123(6):1111–1122

Acharya R, Kumar A, Bhat P, Lim C, Kannathal N, Krishnan S (2004) Classification of cardiac abnormalities using heart rate signals. Med Biol Eng Comput 42(3):288–293

Acır N (2006) A support vector machine classifier algorithm based on a perturbation method and its application to ECG beat recognition systems. Expert Syst Appl 31(1):150–158

Adeli H, Ghosh-Dastidar S, Dadmehr N (2007) A wavelet-chaos methodology for analysis of EEGs and EEG subbands to detect seizure and epilepsy. IEEE Trans Biomed Eng 54(2):205–211

Afkhami RG, Azarnia G, Tinati MA (2016) Cardiac arrhythmia classification using statistical and mixture modeling features of ECG signals. Pattern Recogn Lett 70:45–51

Ai D, Yang J, Wang Z, Fan J, Ai C, Wang Y (2015) Fast multi-scale feature fusion for ECG heartbeat classification. EURASIP J Adv Signal Process 2015(1):46

Al-Fahoum A, Howitt I (1999) Combined wavelet transformation and radial basis neural networks for classifying life-threatening cardiac arrhythmias. Med Biol Eng Comput 37(5):566–573

Alickovic E, Subasi A (2015) Effect of multiscale PCA de-noising in ECG beat classification for diagnosis of cardiovascular diseases. Circuits, Syst, Signal Process 34(2):513–533

Alickovic E, Subasi A (2016) Medical decision support system for diagnosis of heart arrhythmia using DWT and random forests classifier. J Med Syst 40(4):108

Alickovic E, Kevric J, Subasi A (2018) Performance evaluation of empirical mode decomposition, discrete wavelet transform, and wavelet packed decomposition for automated epileptic seizure detection and prediction. Biomed Signal Process Control 39:94–102

Alpaydin E (2014) Introduction to machine learning. MIT Press, Cambridge, MA

Arif M, Akram MU (2010) Pruned fuzzy K-nearest neighbor classifier for beat classification. J Biomed Sci Eng 3(04):380

Arif M, Akram MU, Afsar FA (2009) Arrhythmia beat classification using pruned fuzzy k-nearest neighbor classifier. In: Presented at the soft computing and pattern recognition, 2009. SOCPAR'09, International Conference of IEEE, pp 37–42

Artameeyanant P, Sultornsanee S, Chamnongthai K (2016) An EMG-based feature extraction method using a normalized weight vertical visibility algorithm for myopathy and neuropathy detection. Springerplus 5(1):2101

Asl BM, Setarehdan SK, Mohebbi M (2008) Support vector machine-based arrhythmia classification using reduced features of heart rate variability signal. Artif Intell Med 44(1):51–64

Bakshi BR (1998) Multiscale PCA with application to multivariate statistical process monitoring. AICHE J 44(7):1596–1610

Bassingthwaighte JB, Liebovitch LS, West BJ (2013) Fractal physiology. Springer, New York

Begg R, Lai DT, Palaniswami M (2008) Computational intelligence in biomedical engineering. CRC Press, Boca Raton

Berbari EJ (2000) Principles of electrocardiography. The biomedical engineering handbook, vol 1. CRC Press, Boca Raton, pp 13–11

Besrour R, Lachiri Z, Ellouze N (2008) ECG beat classifier using support vector machine. In: Presented at the information and communication technologies: from theory to applications, 2008. ICTTA 2008, 3rd International Conference on, IEEE, pp 1–5

Bigan C (1998) A system for neural networks detection and automatic identification of EEG epileptic events. IEE Colloquium on Intelligent Methods in Healthcare and Medical Applications (Digest No. 1998/514), York/UK, pp 13/1–13/4. https://doi.org/10.1049/ic:19981044

Bozkurt MR, Subasi A, Koklukaya E, Yilmaz M (2016) Comparison of AR parametric methods with subspace-based methods for EMG signal classification using stand-alone and merged neural network models. Turk J Electr Eng Comput Sci 24(3):1547–1559

Bronzino JD (1999) Biomedical engineering handbook, vol 2. CRC Press, Boca Raton

Bronzino JD (2000) Principles of electroencephalography. The biomedical engineering handbook, 2nd edn. CRC Press LLC, Boca Raton, p 1

Cai L, Wang Z, Zhang H (1999) An EMG classification method based on wavelet transform [and ANN]. In: Presented at the [Engineering in medicine and biology, 1999. 21st Annual Conference and the 1999 Annual Fall Meeting of the Biomedical Engineering Society] BMES/EMBS Conference, 1999, vol 1. Proceedings of the First Joint, IEEE, pp 565-vol

Christodoulou CI, Pattichis CS (1999) Unsupervised pattern recognition for the classification of EMG signals. IEEE Trans Biomed Eng 46(2):169–178

Christov I, Jekova I, Bortolan G (2005) Premature ventricular contraction classification by the kth nearest-neighbours rule. Physiol Meas 26(1):123

Cichosz P (2014) Data mining algorithms: explained using R. John Wiley & Sons, Malden

Clayton RH, Murray A, Campbell RW (1993) Comparison of four techniques for recognition of ventricular fibrillation from the surface ECG. Med Biol Eng Comput 31(2):111–117

Cruz CIM, Marasigan JP, Perez APG, Pillejera JEV, Veron NP, dela Cruz AR (2016) A comparative study between DWT-ANFIS and DWT-SVM in ECG classification. In: Region 10 Conference (TENCON), 2016 IEEE, IEEE, pp 2052–2055

de Chazal P, Reilly RB (2003) Automatic classification of ECG beats using waveform shape and heart beat interval features. In: Presented at the acoustics, speech, and signal processing, 2003. Proceedings. (ICASSP'03), vol 2. 2003 IEEE International Conference on, IEEE, pp II–269

De Chazal P, O'Dwyer M, Reilly RB (2004) Automatic classification of heartbeats using ECG morphology and heartbeat interval features. IEEE Trans Biomed Eng 51(7):1196–1206

De Michele G, Sello S, Carboncini MC, Rossi B, Strambi S-K (2003) Cross-correlation time-frequency analysis for multiple EMG signals in Parkinson's disease: a wavelet approach. Med Eng Phys 25(5):361–369

Dobrowolski AP, Wierzbowski M, Tomczykiewicz K (2012) Multiresolution MUAPs decomposition and SVM-based analysis in the classification of neuromuscular disorders. Comput Methods Prog Biomed 107(3):393–403

Esgiar A, Chakravorty P (2004) Electrocardiogram signal classification based on fractal features. In: Presented at the computers in cardiology, 2004, IEEE, pp 661–664

Felzer T, Freisieben B (2003) Analyzing EEG signals using the probability estimating guarded neural classifier. IEEE Trans Neural Sys Rehab Eng 11(4):361–371

Gokgoz E, Subasi A (2014) Effect of multiscale PCA de-noising on EMG signal classification for diagnosis of neuromuscular disorders. J Med Syst 38(4):31

Gokgoz E, Subasi A (2015) Comparison of decision tree algorithms for EMG signal classification using DWT. Biomed Signal Process Control 18:138–144

Graimann B, Allison B, Pfurtscheller G (2009) Brain–computer interfaces: a gentle introduction. In: Brain-computer interfaces. Springer, Heidelberg, pp 1–27

Guo X, Yang P, Chen L, Wang X, Li L (2006) Study of the control mechanism of robot-prosthesis based-on the EMG processed. Presented at the Intelligent Control and Automation, 2006. WCICA 2006, vol 2. The Sixth World Congress on, IEEE, pp 9490–9493

Hall M, Witten I, Frank E (2011) Data mining: practical machine learning tools and techniques. Kaufmann, Burlington

Hassoun MH, Wang C, Spitzer A (1994a) NNERVE: neural network extraction of repetitive vectors for electromyography. I. Algorithm. IEEE Trans Biomed Eng 41(11):1039–1052

Hassoun M, Wang C, Spitzer A (1994b) NNERVE: neural network extraction of repetitive vectors for electromyography. II. Performance analysis. IEEE Trans Biomed Eng 41(11):1053–1061
Hazarika A, Dutta L, Boro M, Barthakur M, Bhuyan M (2018) An automatic feature extraction and fusion model: application to electromyogram (EMG) signal classification. Int J Multimedia Inf Retr:1–14. https://doi.org/10.1007/s13735-018-0149-z
Hu YH, Palreddy S, Tompkins WJ (1997) A patient-adaptable ECG beat classifier using a mixture of experts approach. IEEE Trans Biomed Eng 44(9):891–900
Ivanov PC, Amaral LAN, Goldberger AL, Havlin S, Rosenblum MG, Struzik ZR, Stanley HE (1999) Multifractality in human heartbeat dynamics. Nature 399(6735):461
Jiang M, Wang R, Wang J, Jin D (2006) A method of recognizing finger motion using wavelet transform of surface EMG signal. In: Presented at the Engineering in Medicine and Biology Society, 2005. IEEE-EMBS 2005. 27th Annual International Conference of the, IEEE, pp 2672–2674
Jones SA (2008) ECG Success Exercises in ECG Interpretation. F. A. Davis Company, Philadelphia
KalaiSelvi A, Sasireka M, Senthilkumar A, Maheswari S (n.d.) Complex wavelet transform based cardiac arrhythmia classification. Int J Emerg Technol Comput Sci Electron 20(2):34–37
Kamali T, Boostani R, Parsaei H (2013) A hybrid classifier for characterizing motor unit action potentials in diagnosing neuromuscular disorders. J Biomed Phys Eng 3(4):145
Kamali T, Boostani R, Parsaei H (2014) A multi-classifier approach to MUAP classification for diagnosis of neuromuscular disorders. IEEE Trans Neural Sys Rehab Eng 22(1):191–200
Karimifard, S., Ahmadian, A., Khoshnevisan, M., & Nambakhsh, M. (2006). Morphological heart arrhythmia detection using hermitian basis functions and kNN classifier. In: Presented at the engineering in medicine and biology society, 2006. EMBS'06. 28th Annual International Conference of the IEEE, IEEE, pp 1367–1370
Katsis CD, Goletsis Y, Likas A, Fotiadis DI, Sarmas I (2006) A novel method for automated EMG decomposition and MUAP classification. Artif Intell Med 37(1):55–64
Kaur G, Arora A, Jain V (2010) EMG diagnosis via AR modeling and binary support vector machine classification. Int J Eng Sci Technol 2(6):1767–1772
Kerem DH, Geva AB (2017) Brain state identification and forecasting of acute pathology using unsupervised fuzzy clustering of EEG temporal patterns. In: Fuzzy and neuro-fuzzy systems in medicine. CRC Press, Boca Raton, pp 19–68
Kevric J, Subasi A (2017) Comparison of signal decomposition methods in classification of EEG signals for motor-imagery BCI system. Biomed Signal Process Control 31:398–406
Khan Y, Gotman J (2003) Wavelet based automatic seizure detection in intracerebral electroencephalogram. Clin Neurophysiol 114(5):898–908
Khan M, Singh J, Tiwari M (2016) A multi-classifier approach of EMG signal classification for diagnosis of neuromuscular disorders. Int J Comput Appl 133(4):13–18
Kiranyaz S, Ince T, Hamila R, Gabbouj M (2015) Convolutional neural networks for patient-specific ecg classification. In: Engineering in Medicine and Biology Society (EMBC), 2015 37th Annual International Conference of the IEEE. IEEE, pp 2608–2611
Kutlu Y, Kuntalp D (2011) A multi-stage automatic arrhythmia recognition and classification system. Comput Biol Med 41(1):37–45
Lai K, Chan K (1998) Real-time classification of electrocardiogram based on fractal and correlation analyses. In: Presented at the Engineering in Medicine and Biology Society, 1998. Proceedings of the 20th Annual International Conference of the IEEE, IEEE, pp 119–122
Li T, Zhou M (2016) ECG classification using wavelet packet entropy and random forests. Entropy 18(8):285
Li H, Feng X, Cao L, Li E, Liang H, Chen X (2016) A new ECG signal classification based on WPD and ApEn feature extraction. Circuits, Syst, Signal Process 35(1):339–352
Liu Y, Zhou W, Yuan Q, Chen S (2012) Automatic seizure detection using wavelet transform and SVM in long-term intracranial EEG. IEEE Trans Neural Sys Rehab Eng 20(6):749–755

Loudon G, Jones N, Sehmi A (1992) New signal processing techniques for the decomposition of EMG signals. Med Biol Eng Comput 30(6):591–599

Luz EJ, Schwartz WR, Cámara-Chávez G, Menotti D (2016) ECG-based heartbeat classification for arrhythmia detection: a survey. Comput Methods Prog Biomed 127:144–164

Mainardi LT, Bianchi AM, Cerutti S (2006) Digital biomedical signal acquisition and processing. In: Medical devices and systems. CRC Press, Boca Raton, pp 49–72

Mallat SG (1989) A theory for multiresolution signal decomposition: the wavelet representation. IEEE Trans Pattern Anal Mach Intell 11(7):674–693

Masetic Z, Subasi A (2016) Congestive heart failure detection using random forest classifier. Comput Methods Prog Biomed 130:54–64

Melgani F, Bazi Y (2008) Classification of electrocardiogram signals with support vector machines and particle swarm optimization. IEEE Trans Inf Technol Biomed 12(5):667–677

Micheli-Tzanakou E (2000) Neural networks in biomedical signal processing. In: Bronzino JD (ed) The biomedical engineering handbook, 2nd edn. CRC Press LLC, Boca Raton, pp 49–72

Minami K, Nakajima H, Toyoshima T (1999) Real-time discrimination of ventricular tachyarrhythmia with Fourier-transform neural network. IEEE Trans Biomed Eng 46(2):179–185

Mishra AK, Raghav S (2010) Local fractal dimension based ECG arrhythmia classification. Biomed Signal Process Control 5(2):114–123

Mishra VK, Bajaj V, Kumar A, Sharma D, Singh G (2017) An efficient method for analysis of EMG signals using improved empirical mode decomposition. AEU-Int J Electron Commun 72:200–209

Moraes J, Seixas M, Vilani F, Costa E (2002) A real time QRS complex classification method using Mahalanobis distance. In: Presented at the Computers in Cardiology, 2002, IEEE, pp 201–204

Muller P, Contento N, Rix H (1992) Fractal dimension on ECG. In: Presented at the Engineering in Medicine and Biology Society, 1992 14th Annual International Conference of the IEEE, IEEE, vol 3, pp 977–978

Muthuswamy J (2004) Biomedical signal analysis. In: Kutz M (ed) Standard handbook of biomedical engineering and design, vol 14. McGraw-Hill Education, New York, pp 1–18

Nadal J, de Bossan MC (1993) Classification of cardiac arrhythmias based on principal component analysis and feedforward neural networks. In: Presented at the computers in cardiology 1993, Proceedings, IEEE, pp 341–344

Naik GR, Selvan SE, Nguyen HT (2016) Single-channel EMG classification with ensemble-empirical-mode-decomposition-based ICA for diagnosing neuromuscular disorders. IEEE Trans Neural Sys Rehab Eng 24(7):734–743

Niedermayer E (1999) Abnormal EEG patterns: Epileptic and paroxysmal. In: Niedermayer E, Silva FL (eds) Electroencephalography. Basic principles, clinical applications and related fields. Williams & Wilkins, Baltimore, pp 235–260

Osowski S, Linh TH (2001) ECG beat recognition using fuzzy hybrid neural network. IEEE Trans Biomed Eng 48(11):1265–1271

Özbay Y, Ceylan R, Karlik B (2006) A fuzzy clustering neural network architecture for classification of ECG arrhythmias. Comput Biol Med 36(4):376–388

Pan J, Tompkins WJ (1985) A real-time QRS detection algorithm. IEEE Trans Biomed Eng 32(3):230–236

Pang CC, Upton AR, Shine G, Kamath MV (2003) A comparison of algorithms for detection of spikes in the electroencephalogram. IEEE Trans Biomed Eng 50(4):521–526

Parsaei H, Stashuk DW (2013) EMG signal decomposition using motor unit potential train validity. IEEE Trans Neural Sys Rehab Eng 21(2):265–274

Passanisi C (2004) Electrocardiography. Delmar Learning, New York

Pattichis CS, Elia AG (1999) Autoregressive and cepstral analyses of motor unit action potentials. Med Eng Phys 21(6):405–419

Pattichis CS, Schizas CN, Middleton LT (1995) Neural network models in EMG diagnosis. IEEE Trans Biomed Eng 42(5):486–496

Phinyomark A, Quaine F, Charbonnier S, Serviere C, Tarpin-Bernard F, Laurillau Y (2013) EMG feature evaluation for improving myoelectric pattern recognition robustness. Expert Syst Appl 40(12):4832–4840

Prasad GK, Sahambi J (2003) Classification of ECG arrhythmias using multi-resolution analysis and neural networks. In: Presented at the TENCON 2003. Conference on Convergent Technologies for the Asia-Pacific Region, IEEE, vol 1, pp 227–231

Preston DC, Shapiro BE (2012) Electromyography and neuromuscular disorders: clinical-electrophysiologic correlations. Elsevier Health Sciences, London

Qin Z, Chen J, Liu Y, Lu J (2005) Evolving RBF neural networks for pattern classification. In: Presented at the International Conference on Computational and Information Science, Springer, pp 957–964

Qurraie SS, Afkhami RG (2017) ECG arrhythmia classification using time frequency distribution techniques. Biomed Eng Lett 7(4):325–332

Raghav S, Mishra AK (2008) Fractal feature based ECG arrhythmia classification. In: Presented at the TENCON 2008-2008 IEEE Region 10 Conference, IEEE, pp 1–5

Raghunathan S, Jaitli A, Irazoqui PP (2011) Multistage seizure detection techniques optimized for low-power hardware platforms. Epilepsy Behav 22:S61–S68

Rasheed S, Stashuk D, Kamel M (2008) A software package for interactive motor unit potential classification using fuzzy k-NN classifier. Comput Methods Prog Biomed 89(1):56–71

Richfield EK, Cohen BA, Albers JW (1981) Review of quantitative and automated needle electromyographic analyses. IEEE Trans Biomed Eng 7:506–514

Sanei S (2013) Adaptive processing of brain signals. John Wiley & Sons, Chichester

Schizas CN, Pattichis CS (1997) Learning systems in biosignal analysis. Biosystems 41(2):105–125

Sengur A, Akbulut Y, Guo Y, Bajaj V (2017) Classification of amyotrophic lateral sclerosis disease based on convolutional neural network and reinforcement sample learning algorithm. Health Inf Sci Syst 5(1):9

Siuly S, Li Y, Zhang Y (2016) EEG signal analysis and classification. Springer, Cham

Sokolova M, Japkowicz N, Szpakowicz S (2006) Beyond accuracy, F-score and ROC: a family of discriminant measures for performance evaluation. In: Presented at the Australasian Joint Conference on Artificial Intelligence, Springer, pp 1015–1021

Soleimani-B H, Lucas C, Araabi BN, Schwabe L (2012) Adaptive prediction of epileptic seizures from intracranial recordings. Biomed Signal Process Control 7(5):456–464

Son C-S, Kim Y-N, Kim H-S, Park H-S, Kim M-S (2012) Decision-making model for early diagnosis of congestive heart failure using rough set and decision tree approaches. J Biomed Inform 45(5):999–1008

Sörnmo L, Laguna P (2005) Bioelectrical signal processing in cardiac and neurological applications, vol 8. Academic Press, San Diego

Subasi A (2007) EEG signal classification using wavelet feature extraction and a mixture of expert model. Expert Syst Appl 32(4):1084–1093

Subasi A (2012a) Classification of EMG signals using combined features and soft computing techniques. Appl Soft Comput 12(8):2188–2198

Subasi A (2012b) Medical decision support system for diagnosis of neuromuscular disorders using DWT and fuzzy support vector machines. Comput Biol Med 42(8):806–815

Subasi A (2013) Classification of EMG signals using PSO optimized SVM for diagnosis of neuromuscular disorders. Comput Biol Med 43(5):576–586

Subasi A (2015) A decision support system for diagnosis of neuromuscular disorders using DWT and evolutionary support vector machines. SIViP 9(2):399–408

Subasi A (2019) Practical guide for biomedical signal analysis using machine learning: a MATLAB based approach. Academic Press, Elsevier Inc, London, UK, ISBN 978-0-12-817444-9

Subasi A, Gursoy MI (2010) EEG signal classification using PCA, ICA, LDA and support vector machines. Expert Syst Appl 37(12):8659–8666

Subasi A, Yilmaz M, Ozcalik HR (2006) Classification of EMG signals using wavelet neural network. J Neurosci Methods 156(1):360–367

Subasi A, Kevric J, Canbaz MA (2017) Epileptic seizure detection using hybrid machine learning methods. Neural Comput & Applic:1–9

Thakor NV, Zhu Y-S, Pan K-Y (1990) Ventricular tachycardia and fibrillation detection by a sequential hypothesis testing algorithm. IEEE Trans Biomed Eng 37(9):837–843

Thakor NV, Gramatikov B, Sherman D (2000) Wavelet (time-scale) analysis in biomedical signal processing. In: Bronzino JD (ed) The biomedical engineering handbook, vol 56, 2nd edn. CRC Press LLC, Boca Raton, pp 1–56

Thaler M (2017) The only EKG book you'll ever need. Lippincott Williams & Wilkins, Philadelphia

Thomas M, Das MK, Ari S (2015) Automatic ECG arrhythmia classification using dual tree complex wavelet based features. AEU-Int J Electron Commun 69(4):715–721

Usta N, Yildiz M (2017) Classification of ECG arrhythmia with machine learning techniques. In: Machine Learning and Applications (ICMLA), 2017 16th IEEE International Conference on. IEEE, pp 546–549

Wang G, Huang H, Xie H, Wang Z, Hu X (2007) Multifractal analysis of ventricular fibrillation and ventricular tachycardia. Med Eng Phys 29(3):375–379

Williamson JR, Bliss DW, Browne DW, Narayanan JT (2012) Seizure prediction using EEG spatiotemporal correlation structure. Epilepsy Behav 25(2):230–238

Winterhalder M, Maiwald T, Voss H, Aschenbrenner-Scheibe R, Timmer J, Schulze-Bonhage A (2003) The seizure prediction characteristic: a general framework to assess and compare seizure prediction methods. Epilepsy Behav 4(3):318–325

Wołczowski A, Zdunek R (2017) Electromyography and mechanomyography signal recognition: Experimental analysis using multi-way array decomposition methods. Biocybernetics Biomed Eng 37(1):103–113. https://doi.org/10.1016/j.bbe.2016.09.004

Yang T-F, Devine B, Macfarlane P (1994) Artificial neural networks for the diagnosis of atrial fibrillation. Med Biol Eng Comput 32(6):615–619

Yong, Z., Wenxue, H., & Yonghong, X. (2009). ECG beats feature extraction based on geometric algebra computational intelligence and software engineering. In: Presented at the International Conference on Computational Intelligence and Software Engineering, vol 2009, pp 1–3

Yousefi J, Hamilton-Wright A (2014) Characterizing EMG data using machine-learning tools. Comput Biol Med 51:1–13

Yu S-N, Chou K-T (2007) A switchable scheme for ECG beat classification based on independent component analysis. Expert Syst Appl 33(4):824–829

Yu S-N, Chou K-T (2008) Integration of independent component analysis and neural networks for ECG beat classification. Expert Syst Appl 34(4):2841–2846

Yuan Q, Zhou W, Liu Y, Wang J (2012) Epileptic seizure detection with linear and nonlinear features. Epilepsy Behav 24(4):415–421

Importance of Bio-signal for Rehabilitative Engineering

19

Uvanesh Kasiviswanathan and Neeraj Sharma

Abstract

Use of Bio-signals in the field of Rehabilitative engineering places a vital role in enhancing the quality-of-life (QOL) of individuals with several levels of disabilities. Assistive devices like motorized wheelchairs for lower-limb disabled persons, smart canes for visually impaired persons, hearing aids, robotic hands (or bionic arm) for the amputee, etc. are invented in order to improve the QOL of individuals with several levels of disabilities. Rehabilitative aids are mainly intended to assist the disables, so it is essential to have accurate and adequately functioning aids/devices with user-friendly manner, which can be made possible by interfacing Human-computer/machine (HCI or HMI) with proper feature extraction and computation models. Moreover, combining electrical bio-signals with non-electrical bio-signal will gradually increase the level of accuracy, but it will load the work of computational algorithm.

Keywords

Bio-signals · Origin of bio-signals · Importance of bio-signals · Rehabilitative engineering

19.1 Introduction

Signal refers to *information about/of something*, which can be a behavioral information or else information about some phenomenon, which may be varied over a certain period of time (Oppenheim 1981; Proakis 2001). Generally, signals are classified broadly into (1) analog signal and (2) digital signal. An analog signal is

U. Kasiviswanathan (✉) · N. Sharma
School of Biomedical Engineering, Indian Institute of Technology (Banaras Hindu University), Varanasi, Uttar Pradesh, India

S. Paul (ed.), *Biomedical Engineering and its Applications in Healthcare*,
https://doi.org/10.1007/978-981-13-3705-5_19

a continuous signal which is represented by time-varying features or quantity, whereas a digital signal has a discrete or discontinuous or digital value to represent information over the varying time period, and this can also be termed as "continuous-time signals and digital-time signals," respectively (Oppenheim 1981).

Biosignals can be classified into electrical and non-electrical bio-signals based on source and type of production of occurrence (Uvanesh 2015; Naït-Ali 2009).

19.2 Electrical Bio-signals

Under certain physiological state/condition, the biological system (such as cell, tissue, and/or organ) individually or in a group will produce some measurable change in electric potential (voltage/current signal), which is termed as electrical bio-signals or bioelectric signal or bioelectric potential (Uvanesh 2015; Uvanesh et al. 2016a). Electrical biosignals are generally fallen under time-variant signals, and it can easily be described using its magnitude and frequency, and also sometimes using phase (Uvanesh 2015). Some general electrical bio-signal are listed below:

- Electroencephalography (EEG) (Sanei and Chambers 2007; Webster 2009; Khandpur 2005)

The origin of this signal is due to the movement of ionic current between neuronal cells (i.e., the flow of ions from the post-synaptic junction of one neuronal cell to next pre-synaptic junction of other neuronal cells attached to it or just firing of neurons) present in human brain. The ionic current gets accumulated in the cerebral cortex and propagates through the skull to the scalp as an EEG signal with a frequency range of about 0.1–100 Hz, with a magnitude of about 2–200 μV. It can be recorded by placing electrodes over the scalp).

Neuromuscular disorders like amyotrophic lateral sclerosis (ALS), myasthenia gravis, multiple sclerosis, spinal muscular atrophy, etc. are some severe diseases developed due to the progressive degeneration of nerve cells, both in the spinal cord and brain (Kasiviswanathan et al. 2019). These sort of diseases lead to lack of control on the coordination of human body parts (especially leg and hand movement and coordination). The instrument used to record of electroencephalography is called as "Electroencephalogram."

- Electrooculography (EOG) (Webster 2009; Khandpur 2005)

Table 19.1 EEG signal's wave pattern types and its corresponding frequency range

EEG wave pattern	Frequency range (Hz)
Delta wave	0.5–3
Theta wave	3.5–7.5
Alpha wave	7.5–13
Beta wave	14–22
Gamma wave	22–100

The origin of this signal is due to the potential developed in corneo-retinal muscle (due to the movement of eye-ball). The EOG signal generally occurs in the frequency range of about 0.1–10 Hz with a peak magnitude of 0.04–3.5 mV. It can be measured by placing (surface) electrodes across the horizontal and vertical axis on the sides (above and below) of the human eye to collect the information the eye-ball movement or rotation with one reference electrode at center forehead. The instrument used to record of electrooculography is called as "Electrooculogram."

- Electrocardiography (ECG) (Webster 2009; Khandpur 2005)

The origin of this signal is due to the electrical activity associated with the functioning of the heart. Generally, an electrical signal is generated from the Sino-Atrial (SA) node (which is located in the right atrium (RA)) to stimulate the atria to get a contract. Then, these signals move to the atrioventricular (AV) node (located in the interatrial septum). Later with a delay, these signal gets divided into multiple signal and pass through the (left and right) bundle of Hiss to the respective Purkinje fibers (located each side of the heart) and also to the endocardium (located at the apex of the heart), to reach the ventricular epicardium, which causes the heart wall to get contract and relax. By placing (surface) electrodes, a tiny change in electrical potential over the surface of the human body (skin) can be acquired at each heartbeat due to the depolarization effect of the heart muscle (as shown in Fig. 19.7). The EEG signal generally occurs in the frequency range of about 0.01–300 Hz with a magnitude of about 0.1–5 μV. The instrument used to record of electrocardiography is called as "Electrocardiogram."

- Electromyography (EMG) (Webster 2009; Khandpur 2005)

The origin of these signals is due to the electrical activity associated with the contraction and relaxation of the muscle fibers (mainly due to a process of the calcium (Ca^{2+}) ions influx and efflux across the actin-myosin binding sites). The EMG signal generally occurs in the frequency range of about 5–3000 Hz with a magnitude of about 0.1–10 mV. By placing (surface) electrodes, a tiny change in electrical potential over the surface of the human body (across the muscle) can be acquired, but the generated EMG signal at the surface of the muscle will have the electrical activity of more than one motor units of the muscle. This is the major disadvantage of the EMG signal, has it is prone to the signal contamination, but due to this reason, the EMG signal is widely used to understand the physiological states/ properties of motor neurons (collectively) and muscle, and also found it variety of application in biomedical engineering (especially in Rehabilitative aspects) (Uvanesh et al. 2016b). The instrument used to record of electromyography is called as "Electromyogram."

- Galvanic skin response (GSR) (Webster 2009)

GSR is nothing but the measuring changes in the electrical characteristics of the skin caused due to the production of the sweating across the human body.

The importance of electrical bio-signal for rehabilitative engineering will be discussed in the later section with examples.

19.3 Non-electrical Bio-signals

Through any mechanism or principle of operation, the biological entity is being monitored such a mechanism comes under non-electrical bio-signal (Kasiviswanathan et al. 2019). Simply to define, it is nothing but an indirect way or procedure to measure or monitor the biological behavior using any device or sensor. Examples of such are mechanical signals (e.g., the mechanomyogram or MMG), acoustic signals (e.g., phonetic and nonphonetic utterances, breathing), chemical signals (e.g., pH, oxygenation), and optical signals (e.g., movements). This non-electrical bio-signals can be used as an additional parameter to build a rehabilitative aids (which will be discussed in the later section with examples).

19.4 Background

19.4.1 Origin of Bio-signal or Bioelectric Signal or Bioelectric Potential (da Silva et al. 2014; Kaniusas 2012)

A bio-signal is any signal in living beings that can be continually measured and monitored. The term bio-signal is often used to refer to bioelectrical signals/ potentials, but it may refer to both electrical and non-electrical signals. The usual understanding is to refer only to time-varying signals, although spatial parameter variations (e.g., the nucleotide sequence determining the genetic code) are sometimes subsumed as well (Uvanesh 2015).

Typically if we consider the human body as a source (power generating unit), then the source of that potential developed from/across the cellular region will be only due to ionic concentration across the cell membrane. All the cell membrane has the envelop (selectively permeable or semipermeable) that regulates, the in and out of the ions (such as sodium (Na^+), potassium (K^+), calcium (Ca^{2+}), magnesium (Mg^{2+}), chloride (Cl^-), etc.) in order to maintain the cell potential (Khandpur 2005).

Upon excitation, the nerve and/or muscle gets a contraction; this is due to the development of action potential across the cell membrane. The development of action potential serves as the base or primary requirement for the development of bioelectric signal in-order to avail the physiological information about that particular area (nerve, muscle, organ, etc.), and importantly the development of action potential is due to change in ionic concentration across that particular region or part, which is developed as the result of development of difference in potential difference.

Briefly, cell membrane has cytoplasm enclosed within it, which has "n" number of proteins and nucleic acids (i.e., biomolecules), and this membrane also serves as a protective layer by isolating/separating the cell from the surrounding environment.

The cell membrane is generally made up of double layered phospholipids (which are considered to be amphiphilic). The term "amphiphilic" defines that the molecule will have both lipophilic structures/group (i.e., hydrocarbon) and hydrophilic polar functional groups (which may be either ionic or neural). Typically, the amphiphilic molecule is arranged or positioned their polar groups outside (in order to separate the surrounding aqueous medium from) and lipophilic chains toward inside of the polar group resulting formation of the bilayer (i.e., phospholipids bilayer). Thus, the phospholipid bilayer within this membrane helps/prevents the movement of ions/molecules in and out of the cell (by forming channels and the pumps selective toward the movement of particular ion or molecule). The membrane surface also has some receptor proteins that allows, the cell to monitor/respond to external signaling molecules (such as hormones).

As mentioned earlier, the bioelectric signals/potentials are mainly generated at the cellular level by muscles and nerves due to change in concentration of ions (such as sodium (Na^+), potassium (K^+) and chloride (Cl^-), as shown in Fig. 19.1) in and out of the cell, which creates/generates the potential differences (i.e., bioelectric signals/potentials). The potential difference can also be generated/developed through electrochemical changes, which is associated with the conduction phenomenon developed in the form of neurotransmitter and receiver by the neuron in and out from the brain. The developed/generated signals together with artifacts/noise will be in the order of very few microvolts (generally −90 mV to +20 mV) (Webster 2009).

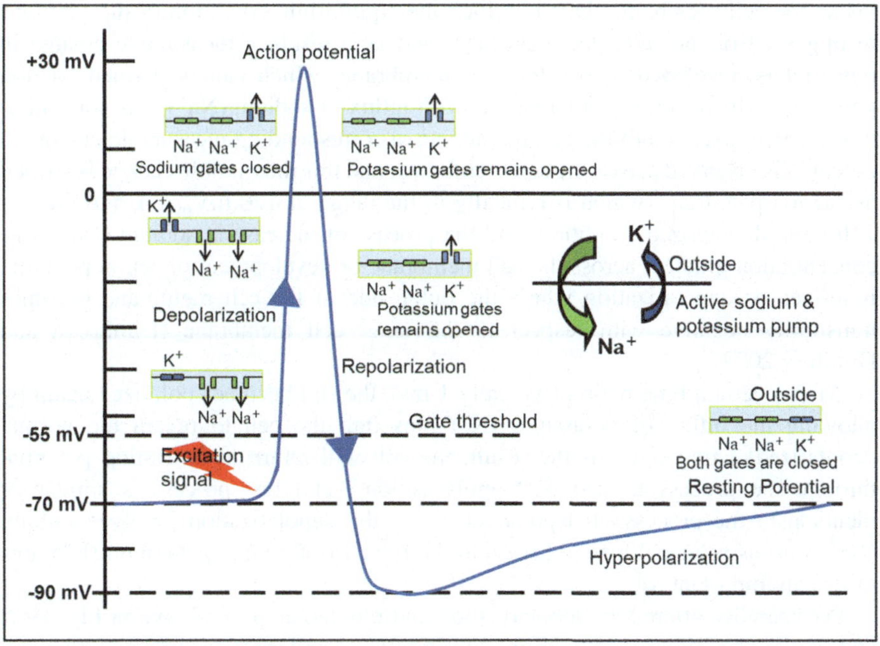

Fig. 19.1 Sodium (Na^+) and potassium (K^+) pump mechanism overview

The cells are separated from the surrounding ionic body fluids by a semipermeable membrane. Through these ionic body fluids only, the bioelectric signal (especially neuronal/synaptic information) will be transceived between one cell to another cell or a group of cells for a specific function to be performed or functionalized. While exciting a particular cell, the cell membrane of that particular cell voluntarily allows the influx of potassium (K^+) and chloride (Cl^-) ions into the cell, while restricting the entry of sodium (Na^+) ion into the cell though the presence of highly concentrated sodium (Na^+) ion gradient across its cell membrane, then that of the inside. Due to presence of excess sodium (Na^+) ion (positively charged ion) outside the cell membrane and negative charged ion along the inner surface of the cell membrane, then the cell will be in its "resting state" or "polarized condition," and the unequal charge distribution developed is caused by certain electrochemical reactions and processes within the biological cell, and the phenomenon of measuring the potential is termed as "resting potential" (which will be around −60 mV to −100 mV) (Uvanesh 2015; da Silva et al. 2014).

When a particular cell is being excited by supplying ionic current or with external stimuli or with any trigger input, the cell membrane's characteristics gets effected/changed with respect to the corresponding trigger. Upon the trigger input, the cell membrane begins to allow the entry or influx of sodium (Na^+) ions into the cell, and this effect is termed as "avalanche effect." When more sodium (Na^+) ions rush into the cell, the potassium (K^+) ions will start to leave the cell slowly, as the potassium (K^+) ion concentration gradient were much higher inside the cell during its resting state. Now there will be an increased in an accumulation of sodium (Na^+) ions (positive charge) inside the cell membrane, which results in a change in the polarity inside the cell membrane. Due to which the equilibrium condition of the previous resting state of the cells gets disturbed, and as a result, a measurable change in potential is developed across the cell membrane, which can be termed "action potential." The action potential begins with influx of sodium (Na^+) ions concentration gradient excess into the cell membrane, a corresponding measurable change in potential is observed across the cell membrane, and this measurable change is termed as "action potential" (which is typically in the range of +20 mV_{peak}), and the cell will be in depolarized condition, and the process of increase in sodium (Na^+) ions concentration gradient across the cell membrane or development of action potential is termed as depolarization where the outer side of the cell membrane becomes transitorily negative with respect to the inner cell membrane (Dimitrova and Dimitrov 2003).

After a certain time period (typically 1 ms), the cell becomes polarized again by allowing the influx of potassium (K^+) ions into the cell to regain the normal depolarized/resting state, as the result, the cell will return to its resting potential through the process termed as "repolarization," and this process is similar or identical to the process of depolarization, as the depolarization involves sodium (Na^+) ions as principal ions, whereas in the repolarization, the potassium (K^+) ions places an important role.

Permeability offered by depolarization and repolarization is shown in Fig. 19.2, and it can be simply defined by, during the depolarization process, the cell

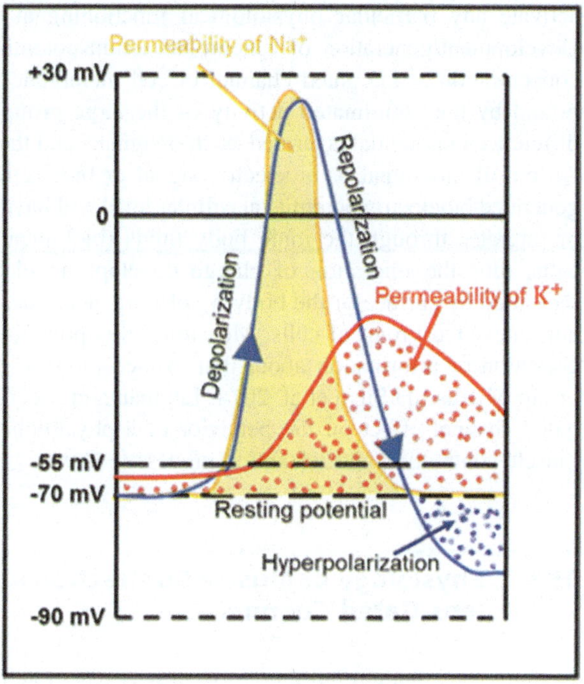

Fig. 19.2 Permeability of sodium (Na^+) and potassium (K^+) ions with respect to action potential developed

membrane will offer more permeability toward sodium (Na^+) ions and then that of potassium (K^+) ions. Later, the cell membrane's permeability gets changed immediately toward potassium (K^+) ions influx (as soon as possible) to regain its normal resting potential (due to repolarization) as shown.

Generally, during repolarization, the cell membrane's permeability toward the influx of potassium (K^+) ions is more, and it continues to some period. It was observed that the influx of potassium (K^+) ions into the cell is more than that of the normal resting state and leads to a new state called "hyperpolarization" state where the potassium (K^+) ions inside the cell will be more as compared to the normal resting state. After completing the action potential or after returning to its normal resting state, there will be some sodium (Na^+) ions still inside the cell for some time to suppress the activation of a new action potential, and this action is termed as "refractory period" (Webster 2009; Khandpur 2005).

19.5 Why It Shows Significant Importance to Use Bioelectric Signal or Bioelectric Potential

Upon excitation, the nerve and/or muscle gets a contraction; this is due to the development of neuronal action potential across the neurons (or cells) (Dimitrova and Dimitrov 2003). The development of action potential serves as the base or primary requirement for the development of any physiological action in order to activate any particular physiological functioning at nerve, muscle, organ level. Development/generation of a neuronal action potential is due to change in ionic concentration at the gated channel of cell membrane across that particular region or/and by the coordinated activity of the large group of cells. These measurable difference in potential is termed as "bio-signal," and the process can be re-defined as "origin of bio-signal or bioelectric signal or bioelectric potential." The produced/generated bioelectric potential at cellular level will have a charged that tends to move or migrates through the ionic body fluids (biological medium) nearby unexcited cells, thus the migration of charge develops an electric current, which can be measured by surface of the body by placing electrodes across the region of excited and unexcited group of cells. Measuring this potential difference will divulge the condition or information about that excited physiological status of that particular group of cells (da Silva et al. 2014; Lehman and McGill 1999). An example for ion gated channel effect on the behavior of a physiological function is described by muscle contraction and relaxation process (below).

19.6 Physiology of Muscle Contraction and Relaxation Through Ions Gated Channels

In short, an excitation pulse or wave or message or signal is made to travel from the nervous system of the brain to activate a particular muscle (Lehman and McGill 1999). Due to this, the sodium (Na^+) and potassium (K^+) ion's influx and efflux process take place (as mention above) across the motor neurons (or nerve cells) in-order to pass the trigger input to that particular muscle via nerves through neurotransmitters (i.e., acetylcholine). The propagation of that particular information across one cell to another in the form of a neuronal signal via ionic body fluids by changing the charge through a chemical reaction. Upon on reaching the target (particular muscle cell), the excitation signal/message will propagate/released via neuromuscular junction by the motor neuron. Acetylcholine, the neurotransmitter (is an organic chemical) binds to receptors present at the outer surface of the muscle fiber (muscle fiber is a cell present in skeletal muscle tissue) membrane and opens the membrane channel. Opening of the muscle fiber's membrane channels leads to depolarization (influx or entry) of positively charged sodium (Na^+) ions into the cytoplasm of the muscle fiber and triggers the action potential to depolarize the rest of the membrane along the T-tubules in order to release of stored calcium (Ca^{2+}) ions present at sarcoplasmic reticulum. Due to this phenomena, the calcium (Ca^{2+}) ions present get diffused into the troponin, which arranges the proteins present in actin binding sites unshielded to pull the actin strands by myosin to form shortening of the muscle or contraction of the muscle (the flow is shown in Fig. 19.3).

Later, when the nervous system stops sending the signal, the sarcolemma and T-tubules gets repolarized, to close the influx of calcium (Ca^{2+}) ions through the

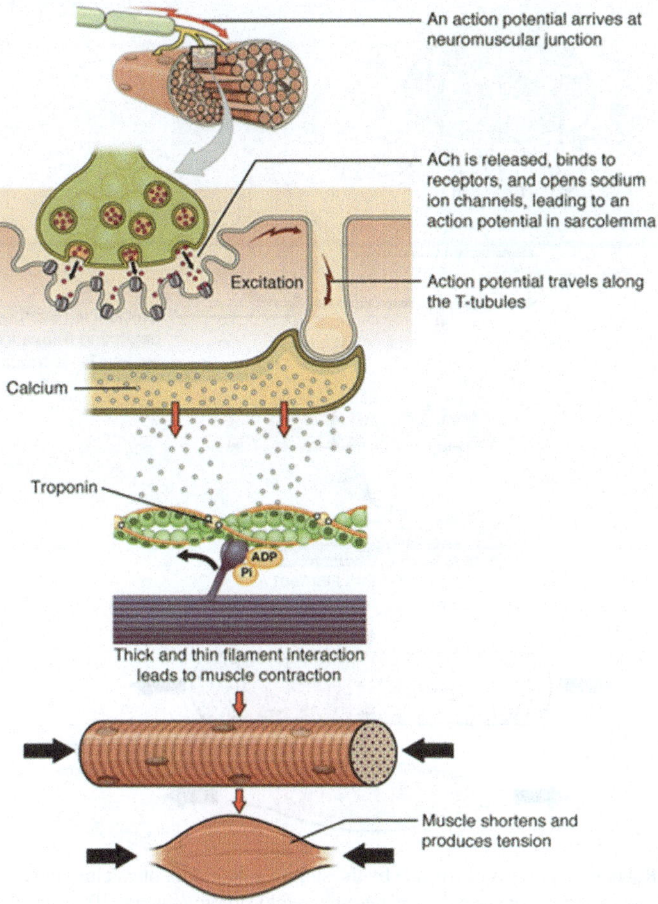

Fig. 19.3 Contraction of Muscle through the release of calcium ions into sarcolemma and T-tubules. (Photo courtesy: https://opentextbc.ca/anatomyandphysiology/chapter/10-3-muscle-fiber-contraction-and-relaxation)

channels. Then, the calcium (Ca^{2+}) ions are pumped/efflux out into the sarcolemma to creates inadequacy of calcium (Ca^{2+}) ions to tropomyosin to re-shield the unshielded the actin binding sites. As a result, the muscle becomes fatigue and relaxes that excited muscle (Fowles et al. 2002) (Fig. 19.4).

As the result of every contraction across the muscle, there will be corresponding bioelectric signals or potentials generated with respect to the unexcited muscle. This signals or potentials can be used as the source to divulge the information about that particular muscle.

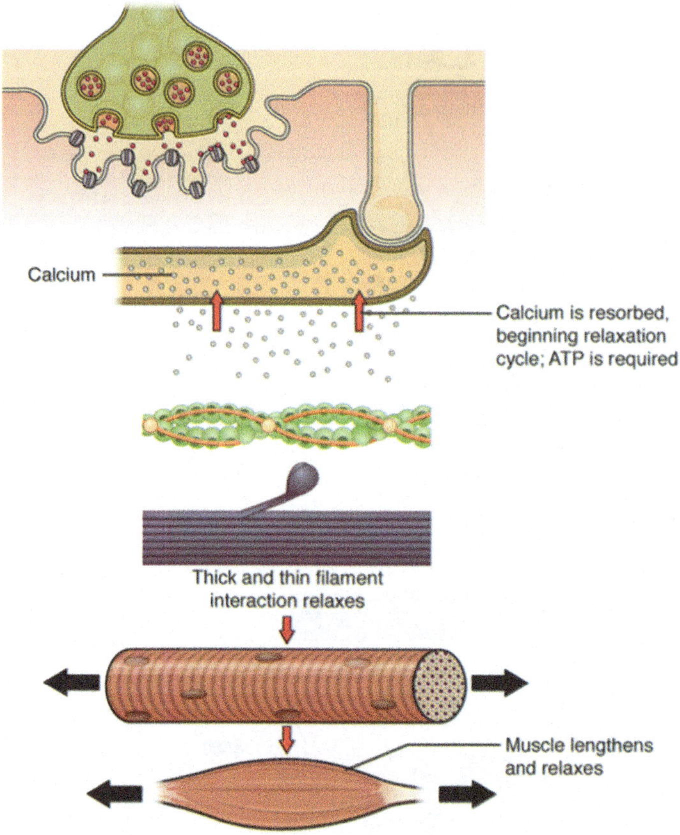

Fig. 19.4 Relaxation process of Muscle by the stopping the influx of calcium ions to tropomyosin. (Photo courtesy: https://opentextbc.ca/anatomyandphysiology/chapter/10-3-muscle-fiber-contraction-and-relaxation)

19.7 Use of Bio-signals in the Field of Rehabilitative Engineering

Rehabilitation engineering deals with the use of systematic application of engineering principles to enhance the quality-of-life (QOL) of individuals with several levels of disabilities (Uvanesh 2015, 2016a, b; Kasiviswanathan et al. 2019). Researchers used to develop assistive devices like motorized wheelchairs for lower-limb disabled persons, smart canes for visually impaired persons, hearing aids, robotic hands (or bionic arm) for an amputee, etc. to improve the quality-of-life in any way. Rehabilitative aids are mainly intended to assist the disables, so it is essential to have accurate and adequately functioning aids/devices with user-friendly manner (i.e., should not create more burden to operate). Moreover, nowadays the bio-signal driven rehabilitative aids have attracted the researchers, as it combines both human biological signal and computer with a machine (for decision-making and for performing the task) and it also provides mental support that the assistive aids/device became a part of them. Human-computer/machine interface (HCI or HMI) deals with interfacing a human biological system with a rule-driven intelligent computer to command/control the machine for performing a specific task (which was explicitly intended to do). We have listed an example for each biosignal-driven rehabilitative aids.

19.8 EEG Signal-Driven Rehabilitative Aids

Due to the movement of ionic current between the neuronal cells present in the brain, EEG is developed. This ionic current gets accumulated in cortex and propagates through the skull to the surface of the scalp with an amplitude and frequency of about 2–200 μV and 0.1–100 Hz. So, by placing proper surface electrodes (as shown in Fig. 19.5) across the scalp at different position respected to what type of wave needed, the EEG signal can be recorded. Generally, researchers and clinicians use, Ag/AgCl (or gold plated) surface electrode to acquire the EEG signals and to avoid the problem of fixation, the hair in the scalp is removed. The acquired raw EEG signal is very small to analysis, so in-order to analysis the signal, the acquired raw EEG signal is amplified by using proper amplification (which may be single or multiple) stages to boost the signal to a certain level (≈few mV). Later, to avoid signal contamination or the unwanted signals (source of signal contamination might be either human biological signals or powerline interface or other), a proper filtering technique need to be used. Through the interfacing channel(s), the filtered EEG

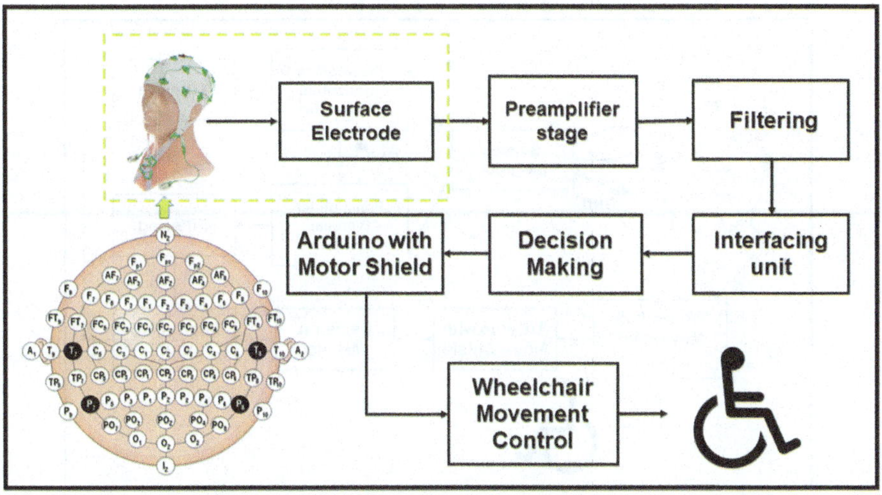

Fig. 19.5 EEG signal-driven rehabilitative aids

signals were feed to the controller (computer). The controller will take or make a decision based on the set of predefined rule, and the corresponding command will be transferred to the machine to perform that specific task. For example, if a motorized wheelchair or a robotic arm is connected to the controller, then depending upon the EEG signal (sensory raw EEG signal), a control signal carrying the command will be generated to control that assistive device (Garrett et al. 2003).

19.9 EOG Signal-Driven Rehabilitative Aids

Potential developed due to the movement of eye-ball in the corneo-retinal muscle is called as EOG signal. The EOG signal has a frequency range of about 0.1–10 Hz with a peak magnitude of 0.04–3.5 mV. So, it can be measured by placing surface electrodes across the horizontal and vertical axis on the sides (above & below) of a human eye with one reference electrode at center forehead or elsewhere (as shown in Fig. 19.6). To avoid loss in the information (while processing any signal), a preamplification stage is used for both horizontal and vertical direction signals. Later, to prevent unwanted signals (source of signal contamination might be either human biological signals or powerline interface or other), a proper filtering technique need to be used. Through the interfacing channel(s), the filtered EOG (containing information about the horizontal movement and vertical movement signals) signals were feed to the controller (computer). The controller will send the filtered EOG signals into a feature extraction process, to extract the useful information's (spectral information), then based on that information, a decision will be taken by considering the predefined rule and corresponding command will be transferred to the machine to perform that specific task. For example, if a motorized wheelchair is connected to the controller, the movement of the motorized

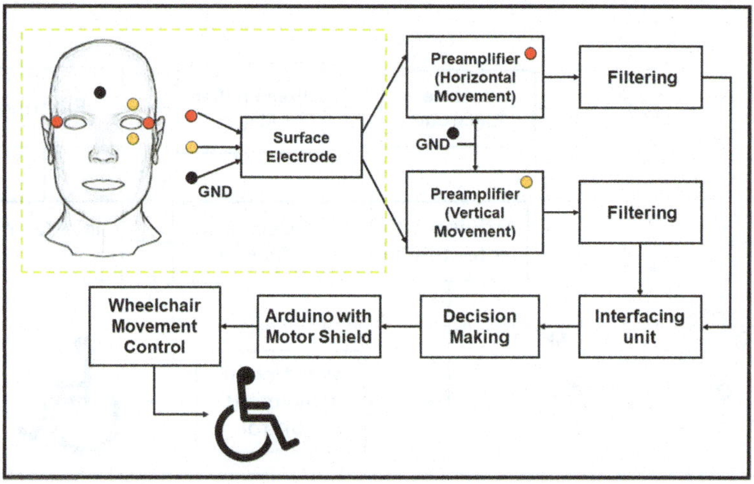

Fig. 19.6 EOG signal-driven rehabilitative aids

wheelchair will depend on the EOG signal (sensory raw EOG signal), and the corresponding control signal generated through the computation methods, which carrying the command will control that assistive device (Ferreira et al. 2007).

19.10 Use of ECG Signal in Rehabilitative Engineering

Potential developed due to the electrical activity associated with functioning of the heart and the functioning of the heart is controlled by the electrical impulse (repeating periodical, synchronized rhythmic signal) generated from the sino-atrial (SA) node (which is located in the right atrium (RA)) to stimulate the atria to get contract. Then, these signals move to the atrioventricular (AV) node (located in the interatrial septum). Later with a delay, this signal gets divided into multiple signal and pass through the (left and right) bundle of Hiss to the respective Purkinje fibers (located each side of the heart) and also to the endocardium (located at the apex of the heart), to reach the ventricular epicardium, which causes the heart wall to get contract and relax (Pumprla et al. 2002). By placing surface electrodes, a tiny change in electrical potential over the surface of the human body (across the chest) can be acquired at each heartbeat due to the depolarization effect of the heart muscle (as shown in Fig. 19.7).

Due to the presence of any obstacles (normally presence of cloth or due to the presence of damaged tissue) at any direction of flow of electrical impulse from SA node to endo-epicardium will create shift (disturbance) in the functioning of the heart muscle, which leads to failure of the heart functioning. Thus to rectify the functioning of the heart to its normal function, an artificial pacemaker (to provide electrical impulse, when the heart's pumping mechanism gets disturbed) is used (Webster 2009; Khandpur 2005).

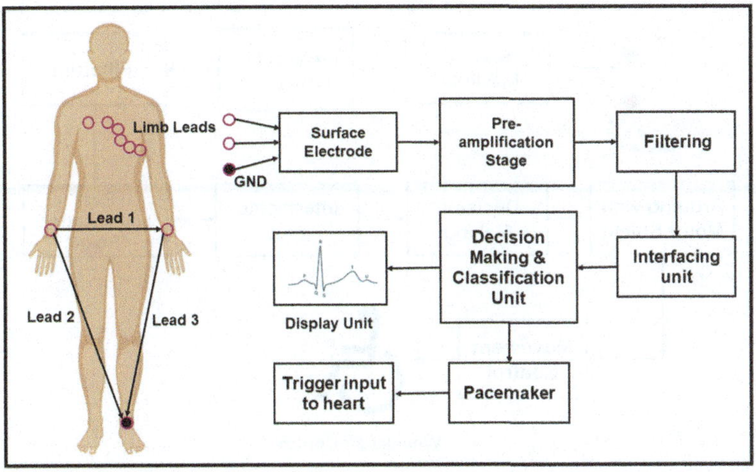

Fig. 19.7 Use of ECG signal in rehabilitation engineering

The Pacemaker, first analysis of the functioning of the normal heart function through acquiring the ECG signal. When it finds any disturbance in the function of the heart, the pacemaker immediately generates an electrical impulse to excite at functioning immediately in synchronization with normal heart function. A typical overview is shown in Fig. 19.7.

19.11 EMG Signal-Driven Rehabilitative Aids

Generated due to the electrical activity associated with the contraction and relaxation of the muscle fibers (especially due to the process of the calcium (Ca^{2+}) ions influx and efflux across the actin-myosin binding sites) (Sadoyama et al. 1983). The EMG signal as the frequency range of about 5–3000 Hz with a magnitude of about 0.1–10 mV. Thus, by placing surface electrodes across a respective muscle, the change in electrical potential manifested over the surface of that muscle can be acquired. So, the acquired raw EMG signal will undergo preamplification and filtering stages, respectively, in order to avoid loss and signal contamination present in the signal. The filtered EMG signal was feed to the controller (computer) connected to an assistive device(s). The controller will send the filtered EOG signals into a feature extraction process, to extract the useful information's (spectral information), then based on that information, a decision will be taken by considering the predefined rule and corresponding command will be transferred to the machine to perform that specific task. For example, if a motorized wheelchair is connected to the controller wirelessly, then, after filtration process, a postamplification stage with a rectification mechanism is needed to shift the negative axis EMG signal to the positive axis to avoid loss in the information. This process has been illustrated in Fig. 19.8, and as usual, the movement of the motorized wheelchair will be controlled

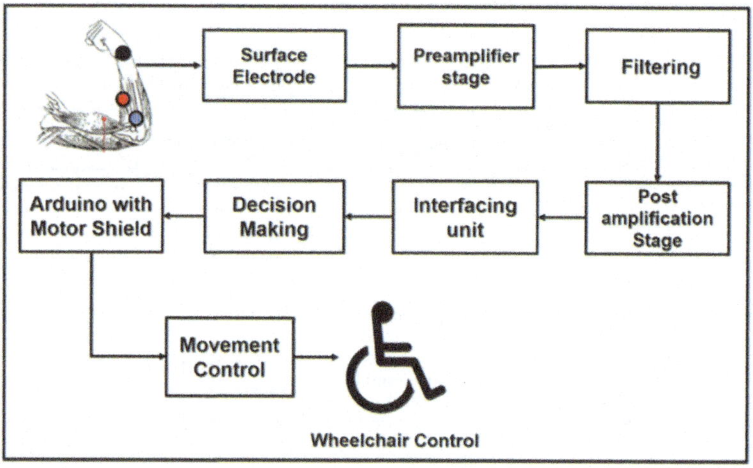

Fig. 19.8 EMG signal-driven rehabilitative aids

by generating the corresponding control signal with proper computation techniques (feature extraction and classification method), which carries the command will control that assistive device (Uvanesh 2015, 2016a, b).

19.12 Use of Non-electrical Bio-signals in Rehabilitative Engineering

The non-electrical bio-signals (mechanical signals, acoustic signals, chemical signals, and optical signals) are generated due to monitoring the behavior of some biological entity (based on the need). The human speech signal is one good example for non-electrical bio-signal, which can play a vital role in developing rehabilitative aid (Khandpur 2005; Kasiviswanathan et al. 2019; Kushwaha et al. 2017). Here, the source is human speech signal but it is acquired through a mic, and later the signal is amplified, and post-processed (feature extraction and classification) can be done to map the human speech signal to a corresponding command signal for an assistive device (Kasiviswanathan et al. 2019; Kushwaha et al. 2017). So, now the non-electrical bio-signal can also be used in assisting the person with the lower-limb disability, and it can be compared with electrical biosignal as shown in Fig. 19.9.

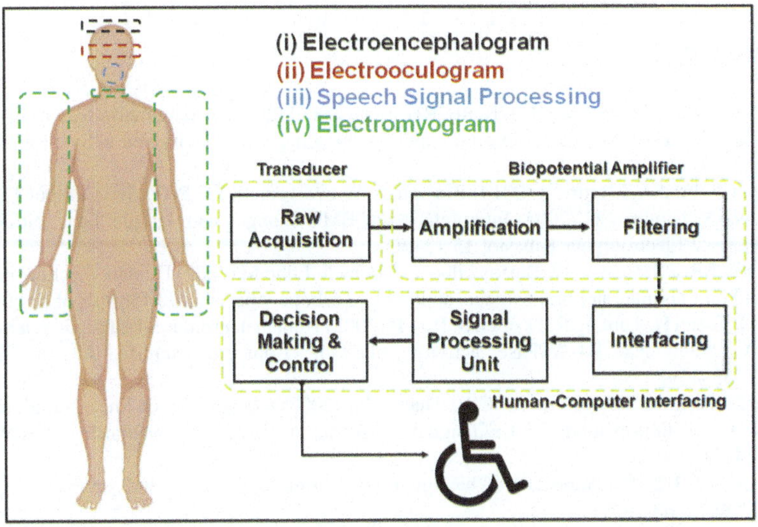

Fig. 19.9 Non-electrical bio-signal and electrical bio-signal in rehabilitative engineering

19.13 Need of Feature Extraction with Computation Methods

For generating the control signal, researchers have developed and used various types of computational models (artificial neural network, genetic algorithm, K-means, etc.) with optimization algorithm (Antlion optimization, Particle swarm optimization, Artificial bee colony optimization, etc.) to classify the bio-signals to a specific function or task. The computational models get optimized based on the features which are extracted from the source (bio-signals). Feature extraction method relies on the estimation of spectral analysis of a signal for a short period. Spectral analysis is used to estimate frequency components of a non-stationary or continuous time-varying signal, and it is determined by the energy or power spectrum of the signal (Trier et al. 1996; Rafiee et al. 2011; Burke et al. 2005).

19.14 Conclusion

Bio-signals have shown their importance in the field of applied (rehabilitation) engineering. The need is to fulfil and enhancing the quality-of-life (QOL) of individuals with several levels of disabilities by providing highly efficient assistive device which is being driven by their own biological system. The assistive devices driven by the (user's) bio-signal provide them a mental peace and mind-set or realization of thinking that the assistive devices to be their own original part. Use of different computational methods showed an increase level of accuracy in order to process (feature extraction) the bio-signals and classifying them to a corresponding action or task to fulfil the purpose of being generation of biosignals.

References

Burke DP, Kelly SP, de Chazal P, Reilly RB, Finucane C (2005) A parametric feature extraction and classification strategy for brain-computer interfacing. IEEE Trans Neural Syst Rehabil Eng 13:12–17

da Silva HP, Fred A, Martins R (2014) Biosignals for everyone. IEEE Pervasive Comput 13:64–71

Dimitrova N, Dimitrov G (2003) Interpretation of EMG changes with fatigue: facts, pitfalls, and fallacies. J Electromyogr Kinesiol 13:13–36

Ferreira A, Silva R, Celeste W, Bastos Filho T, Sarcinelli Filho M (2007) Human–machine interface based on muscular and brain signals applied to a robotic wheelchair. J Phys Conf Ser:012094

Fowles J, Green H, Tupling R, O'brien S, Roy B (2002) Human neuromuscular fatigue is associated with altered Na+-K+-ATPase activity following isometric exercise. J Appl Physiol 92:1585–1593

Garrett D, Peterson DA, Anderson CW, Thaut MH (2003) Comparison of linear, nonlinear, and feature selection methods for EEG signal classification. IEEE Trans Neural Syst Rehabil Eng 11:141–144

Kaniusas E (2012) Fundamentals of biosignals. In: Biomedical signals and sensors I. Springer, Heidelberg, pp 1–26

Kasiviswanathan U, Kushwaha A, Sharma S (2019) Development of human speech signal-based intelligent human-computer Interface for driving a wheelchair in enhancing the quality-of-life of the persons. In: Intelligent Systems for Healthcare Management and Delivery. IGI Global, Hershey, pp 21–60

Khandpur RS (2005) Biomedical instrumentation: technology and applications. Mcgraw-hill New York, New York

Kushwaha A, Kasiviswanathan U, Sharma S (2017) Feature extraction and classification of speech signal using hidden Markov-Gaussian mixture model (HM-GMM) for driving the rehabilitative aids. In: 2017 14th IEEE India council international conference (INDICON), pp 1–6

Lehman GJ, McGill SM (1999) The importance of normalization in the interpretation of surface electromyography: a proof of principle. J Manip Physiol Ther 22:444–446

Naït-Ali A (2009) Advanced biosignal processing. Springer Science & Business Media, Berlin

Oppenheim AV (1981) Digital signal processing. Massachusetts Inst of Tech Cambridge, Cambridge, MA

Proakis JG (2001) Digital signal processing: principles algorithms and applications. Pearson Education India, New Delhi

Pumprla J, Howorka K, Groves D, Chester M, Nolan J (2002) Functional assessment of heart rate variability: physiological basis and practical applications. Int J Cardiol 84:1–14

Rafiee J, Rafiee M, Yavari F, Schoen M (2011) Feature extraction of forearm EMG signals for prosthetics. Expert Syst Appl 38:4058–4067

Sadoyama T, Masuda T, Miyano H (1983) Relationships between muscle fibre conduction velocity and frequency parameters of surface EMG during sustained contraction. Eur J Appl Physiol Occup Physiol 51:247–256

Sanei S, Chambers JA (2007) EEG signal processing. Wiley, Chichester

Trier OD, Jain AK, Taxt T (1996) Feature extraction methods for character recognition-a survey. Pattern Recogn 29:641–662

K. Uvanesh (2015) Linear and non-linear classification of EMG signals for probable applications in designing control system for assistive aids

Uvanesh K, Nayak SK, Champaty B, Thakur G, Mohapatra B, Tibarewala D et al (2016a) Classification of surface electromyogram signals acquired from the forearm of a healthy volunteer. In: Classification and clustering in biomedical signal processing. IGI Global, Hershey, pp 315–333

Uvanesh K, Nayak SK, Champaty B, Thakur G, Mohapatra B, Tibarewala D et al (2016b) Development of a surface EMG-based control system for controlling assistive devices: a study on robotic vehicle. In: Classification and clustering in biomedical signal processing. IGI Global, Hershey, pp 335–355

Webster JG (2009) Medical instrumentation application and design. Wiley, Hoboken

Part VIII

Medical Imaging and Image Processing

Elements of Medical Image Processing

20

T. Emami, S. S. Janney, and S. Chakravarty

Abstract

The advent of technology has taken us so far in our lives that we cannot imagine any field without technology or devices. Name any area today, for example, business, education, media and communication, aerospace, etc. There are no surprises that health care has become one of the most advanced prospectives for technologies and its application to be used. Currently we are in the era where medical professionals are using applications to speed up diagnosis, treatment, surgical procedures, recovery, etc., to provide better services to the public. One of the most interesting aspects is the medical image processing which has come a long way from requiring human intervention to current day scenario where application accurately predicts the cause and location of tumor or abnormalities from ultrasound, MRI, PET scan, CT scan, X-ray data, etc. Buzz is going on in the medical arena that in the near future technologies will replace some of the health-care professional jobs. Until then let us start by understanding the current state of affair between technology in biomedical image processing field and its applications.

20.1 Medical Imaging

20.1.1 Introduction

Medical imaging is like a portal to view internal parts of bodies which are used for assisting medical diagnosis and analysis. This creates a visual representation for medical professionals to get an understanding of the current functioning or previous ailment of organs, tissues, or any interior section of the body. Biomedical images provide a wide range of patterns and designs like bones, muscles, etc. to diagnose diseases.

T. Emami · S. S. Janney · S. Chakravarty (✉)
Department of Electrical Engineering, Kennesaw State University, Kennesaw, Georgia

S. Paul (ed.), *Biomedical Engineering and its Applications in Healthcare*,
https://doi.org/10.1007/978-981-13-3705-5_20

20.1.2 History

Prevalence of medical imaging started after the discovery of X-ray by Wilhelm Conrad Roentgen, a professor in Wuerzburg University in Germany, in 1895. While working in the lab, he discovered that only few rays showed internal parts of human body where the rays could penetrate for other its was opaque. He received a Nobel Prize in 1901 (Bradley 2008). The discovery of X-ray opened up new avenues in research for understanding human body structure. Interest was in the area of surgery and medical diagnosis. Few months after the discovery, experts in Europe and the United States started to use radiographs as a guide for medical professionals specially used in wars to locate bullets in wounded soldiers (Fig. 20.1).

X-ray tomography started becoming popular in 1900. Slices of anatomy information also called as "tomograms" were taken into consideration to view the internal body pattern. Other topographies that are frequently used are computerized axial tomography (CAT) scanning or computed tomography (CT) scanning that came around the 1970s and magnetic resonance imaging (MRI) that does not X-ray (Bradley 2008; Toennies 2012) (Fig. 20.2).

In the 1950s, nuclear medicine became famous. Today positron emission tomography or "PET" scanning is a well-known nuclear medicine. Instead of emitting gamma rays, it emits positrons (Bradley 2008). Depending on the focus of scanning, contrasting material (agent) injections are given to patients to enhance the visibility of certain tissues or blood vessels.

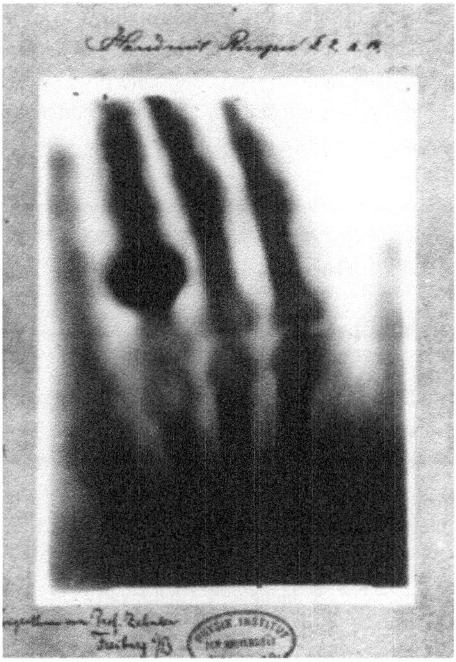

Fig. 20.1 Wilhelm Roentgen's first medical X-ray of his wife's hand (Kevles 1996; Sample 2007)

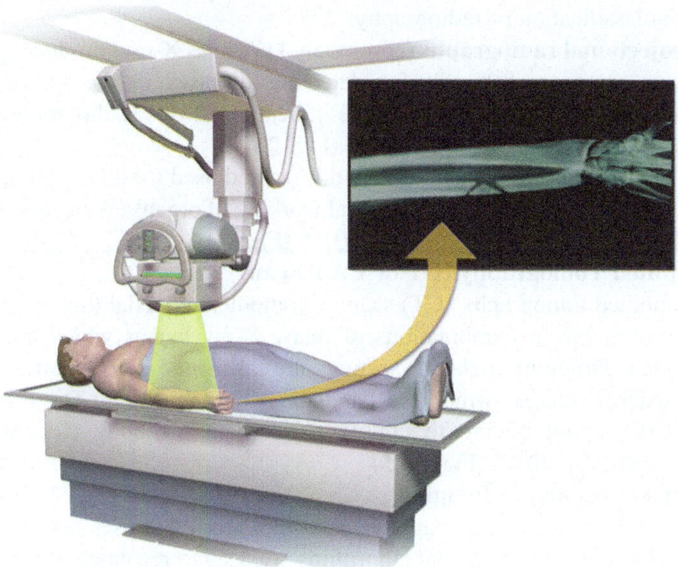

Fig. 20.2 X-ray machine showing crack in the bone structure (http://www.imaginis.com/faq/history-of-medical-diagnosis-and-diagnostic-imaging)

The benefits of these techniques are (Toennies 2012; http://www.imaginis.com/faq/history-of-medical-diagnosis-and-diagnostic-imaging):

- Small quantity of X-ray is required.
- Improved quality of images.
- Easy to store the medical data as images for analysis.
- Improve diagnosis by proper analyzing techniques.

20.1.3 Types of Medical Imaging Modalities

Technology has made it easy to obtain medical images without having invasively extracting information. There is a huge range of image modalities that could be considered to classify the types of medical images or techniques that are considered in biomedical image analysis.

(a) **Radiography**

This imaging modality uses various rays like X-rays, gamma rays, and other radiations to view internal structures (Carroll 2014). X-ray beams are projected on the object, and the object would absorb the radiation to display structural design, composition, and density.

The study of anatomy using radiographic information is known commonly as radiographic anatomy (James and Dasarathy 2014).

Sub-classification of radiography:

- **Projectional radiographs** (commonly known as X-ray) produce 2D images. They are normally used for detecting diseases in the lungs, stomach, intestines, etc. (Radiology – acute indications 2017; Radiographic Standard Operating Protocols (PDF) 2015) (Fig. 20.3).
- **Fluoroscopy** – X-ray used at low dosage and used for image-guide surgeries for getting visual display of internal working of organs (Wang and Blackburn 2000; Last Image Hold Feature 2010) (Fig. 20.4).

(b) **Computed Tomography (CT or CAT Scan)**

Computed tomography (CT) scan or computerized axial tomography (CAT) scan makes use of combinations of many X-ray beams which are computer processed. Different angles or sections are taken into consideration to get the final scanned images which are handled by the computer. These images are obtained without having to cut open the patient's body (CT Scan 2018; Shrimpton et al. 2011) (Fig. 20.5).

(c) **Magnetic Resonance Imaging**

MRI scan is a technique applied by radiologists that uses magnetism of the huge magnetic coil drum, and algorithms are used to reconstruct image of body structures (Bradley 2008). The MRI scanner is a circular drum that contains a magnet and a sliding table as shown in Fig. 20.6. The patient is placed on a moveable bed that is inserted into the magnet. The magnet creates strong magnetic field/signals which are later used by Fourier transformation using

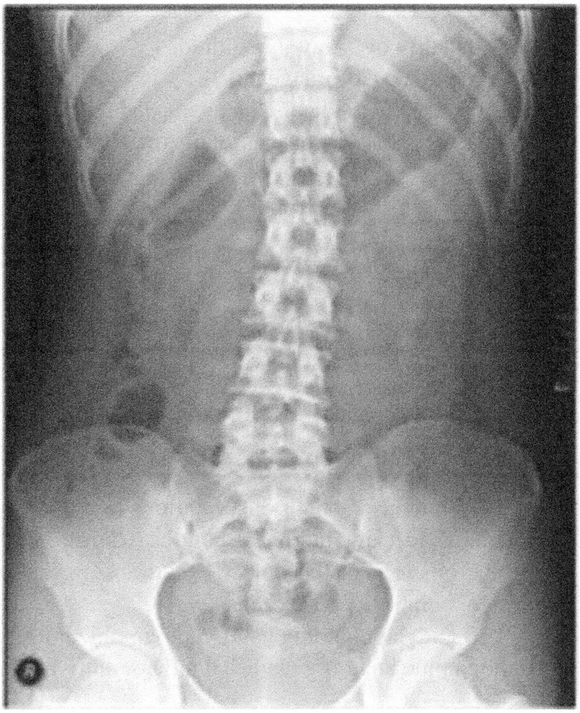

Fig. 20.3 Abdominal radiographs (https://en.wikipedia.org/wiki/Projectional_radiography)

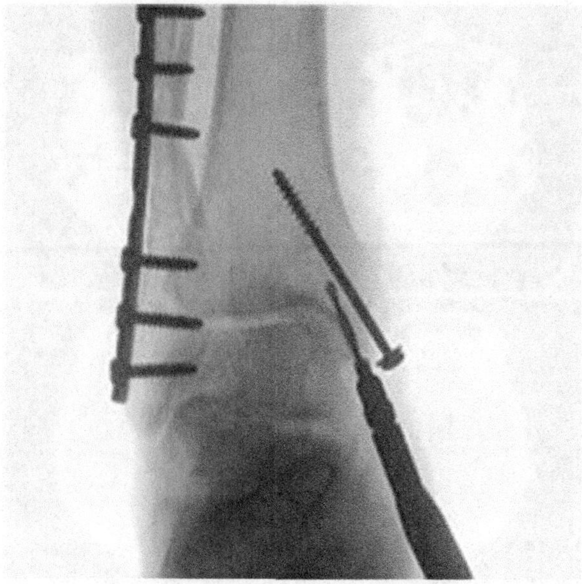

Fig. 20.4 A fluoroscopy X-ray image shown during implant surgery (https://en.wikipedia.org/wiki/Fluoroscopy)

compressed sensing concepts to produce the final image, what we all know as MRI (Zhu 2003).

(d) **Nuclear Molecular Imaging**

Nuclear medicine when used in diagnostic imaging is commonly referred to as molecular imaging which uses properties of particles that are emitted from radioactive material to diagnose or treat various diseases. Figure 20.7 shows the nuclear medicine image of the whole body which is used to diagnose bone-related diseases such as fracture, infections, abnormalities, etc. (https://en.wikipedia.org/wiki/Nuclear_medicine).

They are sub-classified as:

- SPECT – 3D images taken from gamma-based cameras from different projected angles are combined to reconstruct this image (https://en.wikipedia.org/wiki/Nuclear_medicine).
- PET – positron emitting radionuclide (a form of gamma rays) is captured after it reflects from the human tissue molecules to form PET image (Bailey et al. 2005). It is commonly used in neurology, cardiology, muscular-skeletal imaging, etc. (Carlson 2012) (Fig. 20.8).

(e) **Ultrasound**

Ultrasound is a sound wave-based application which is considered to be a diagnostic imaging technique. These sound waves are known to have very high frequencies which are way higher than human hearing ability (Novelline 1997). Famous application is in obstetrics scanning which means ultrasound scanning of pregnant women to view the development of the fetus. It is also used to view internal body structures such as organs, bones, muscles, etc. (DistanceDoc and MedRecorder 2011; Ultrasound Imaging of the Pelvis 2008) (Fig. 20.9).

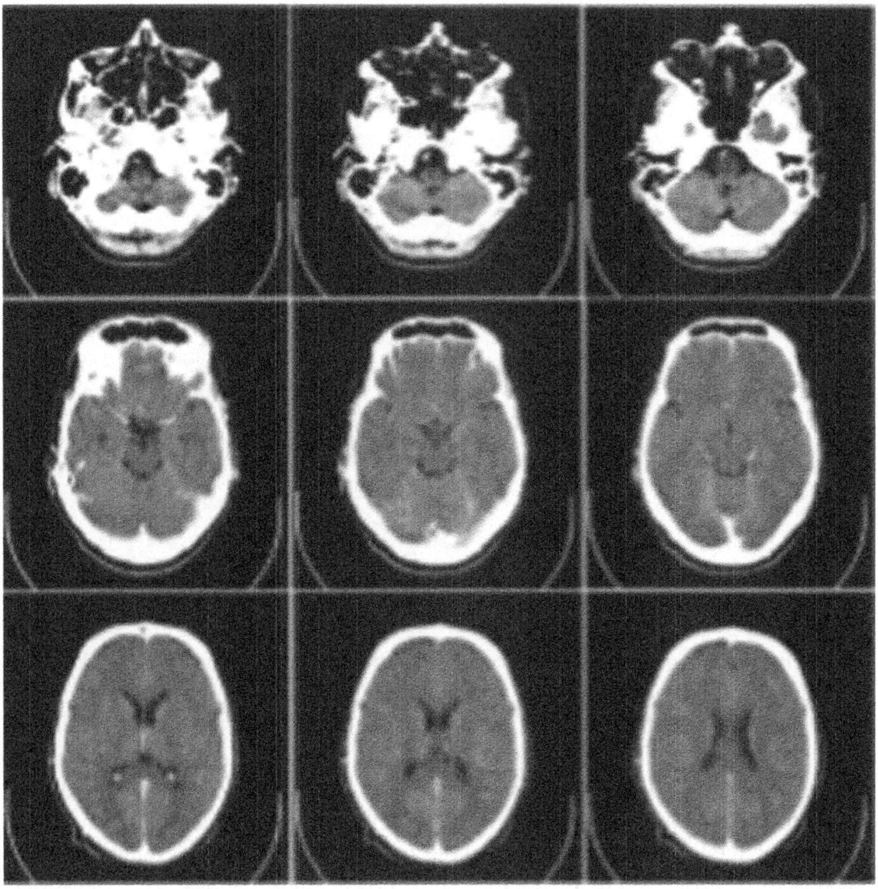

Fig. 20.5 CT image of the skull showing different cross-sections (http://pleasantonimaging.com/services/computed-tomography-ct/)

(f) **Functional Near-Infrared Spectroscopy**

Functional near-infrared spectroscopy is an optical imaging technique that is non-invasive. This technique used low-level light to see the internal working of the brain and its activity through the movement of blood flow in them (http://researchimaging.pitt.edu/content/near-infrared-spectroscopy-nirs-brain-imaging-laboratory; Coyle et al. 2007) (Fig. 20.10).

(g) **Magnetic Particle Imaging**

Magnetic particle imaging is an imaging technique used for tracking superparamagnetic iron oxide nanoparticles. It is highly sensitive, and depth of the structure can be analyzed (https://en.wikipedia.org/wiki/Magnetic_particle_imaging). This technique has been used for researches in areas such as cardiovascular performance, neuroperfusion, and cell movement tracking (Weizenecker et al. 2009; Yu et al. 2017) (Fig. 20.11).

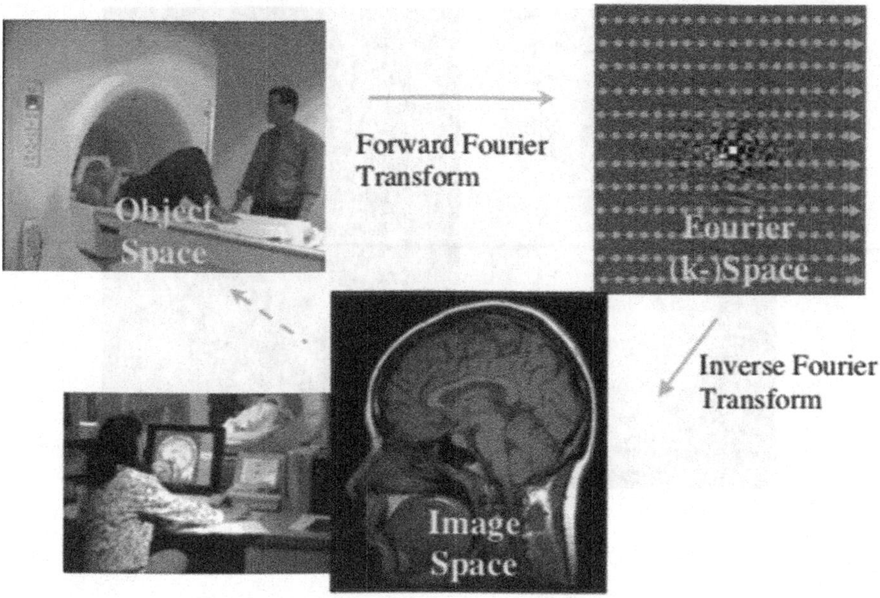

Fig. 20.6 Brain MR image reconstruction process (Zhu 2003)

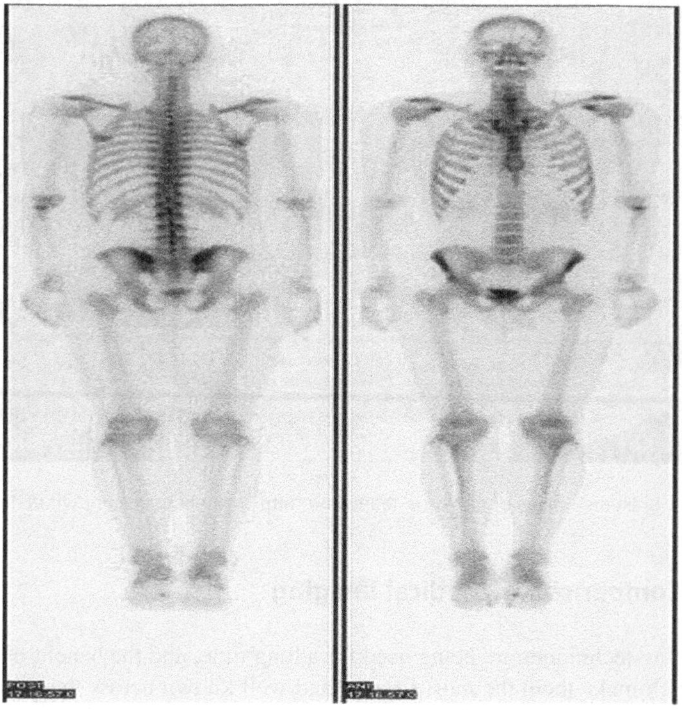

Fig. 20.7 Nuclear molecular image of a whole body (https://en.wikipedia.org/wiki/Nuclear_medicine)

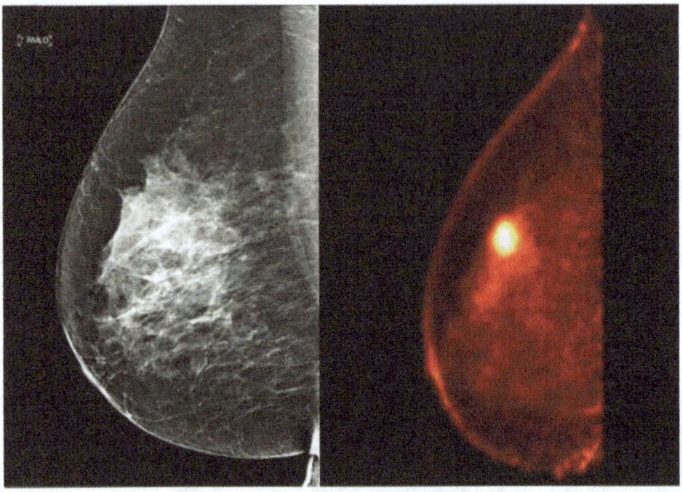

Fig. 20.8 PET scan image of the breast (left) and cancer cell detection shown using a tracer (right) (Images courtesy of Dr. Kathy Schilling, Medical Director of Lynn Women's Health and Wellness Institute)

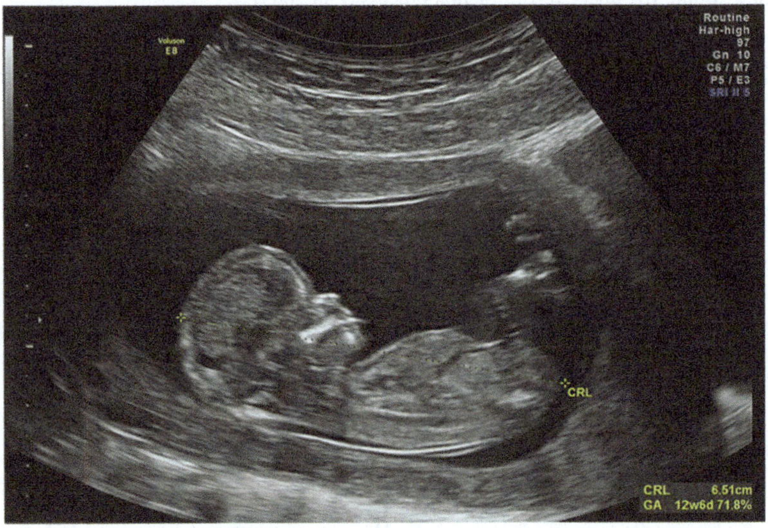

Fig. 20.9 Ultrasound image of a fetus in the womb (https://en.wikipedia.org/wiki/Ultrasound)

20.1.4 Comparison of Medical Imaging

Few imaging techniques are being used for a long time, and the benefit of them still exists which make them the most famous and well known across the medical field. Table 20.1 shows some of these techniques and their comparison as regards the common properties that make these images feasible.

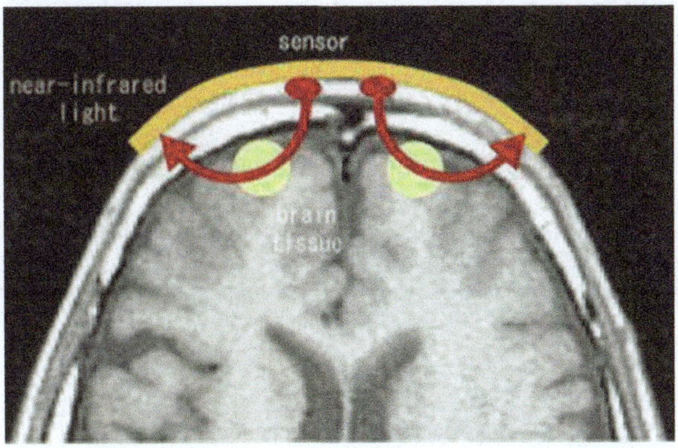

Fig. 20.10 Brain monitoring with near-infrared spectroscopy (http://www.aorticdissection.com/DISEASES%20OF%20AORTA.htm)

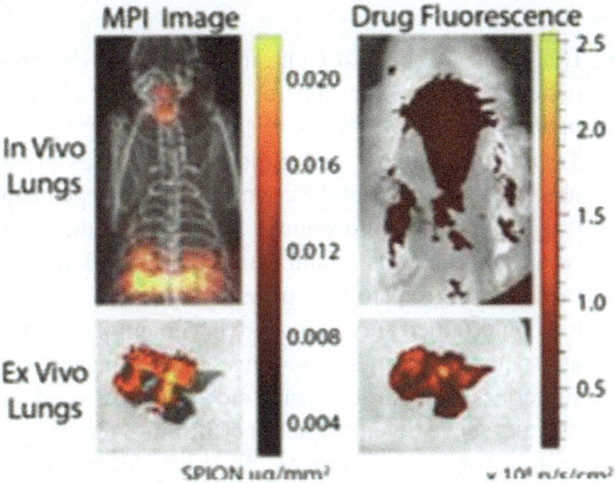

Fig. 20.11 Magnetic particle imaging of lungs tracking drug deposition (Dolovich and Labiris 2004)

20.2 Medical Image Analysis

20.2.1 Introduction

Analysis of medical images has been the integral part of any diagnoses, treatments, procedures, etc. These analyses could be carried out by medical professionals to help predict or take action with regard to the patients' health. Since these images are obtained non-invasively and can be stored, they serve to act one important aspect in electronic health record (EHR) for future references (Toennies 2012).

Reasons for carrying out medical image analysis are as follows (https://www.doc.ic.ac.uk/~jce317/history-medical-imaging.html):

- Clinical study – to detect patterns or structure in images that could describe or prove hypotheses of the study. These are used for scientific analyses and future case study analyses for educational institutions for training budding medical professionals.

Table 20.1 Comparing medical images (Goel et al. 2016)

	X-ray	Ultrasound	MRI	CT scan
Cost	Low	Moderate	Relatively high	High
Availability	Maximum	Maximum	Less than CT	Less difficult
Technique	Ionizing radiation	Non-ionizing radiation	No	Ionizing radiation
Speed	Short	Depends on machine handler	Long	Moderate
Data acquisition	Low	Low	High	High
Image resolution	Normal	Depends on selection of transducer	Best	Moderate

- Diagnosis – to diagnose chronicle illness or diseases by detecting tumor or other patterns. Doctors or experts in their field identify the medical conditions of patients.
- Treatment planning – after diagnosing comes the treatment to illness. Analyses need to be done about the course of action to be taken for diseases which could be drugs or medical procedures. Planning for the treatment needs serious research regarding previous health conditions or allergies. This can be obtained from medical image history of patients.
- Computer-aided surgeries – advancement in technology has given an automated assistance to doctors in various areas of health care from diagnoses, treatment, surgeries, post-surgery care, etc. They are used as guided tools for surgeries. Doctors have even started performing remote operations which could save millions of lives (Toennies 2012; http://www.imaginis.com/faq/history-of-medical-diagnosis-and-diagnostic-imaging).

Modern radiologists have various tasks to be performed during the diagnosis process. Medical image data is not only about reading the image, but other aspects contribute toward the analysis which are as shown in Fig. 20.12.

Image processing has a wide range of applications especially in medical area. Visual images have been contributing to various medical analyses (Goel et al. 2016; Rao and Rao n.d.). Few of the applications are mentioned below:

- Tumor detection
- Fracture detection
- Structural disabilities
- Cancer detection
- Heart defects and diseases
- Tuberculosis
- Birth defects
- Neurological functioning

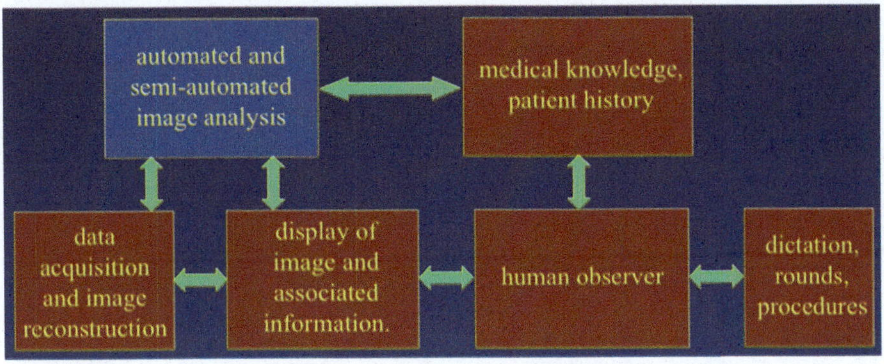

Fig. 20.12 Flow chart of image analysis by a radiologist. (Image from J. Galeotti, class material from "Methods in Medical Image Analysis", Carnegie Mellon University 2018)

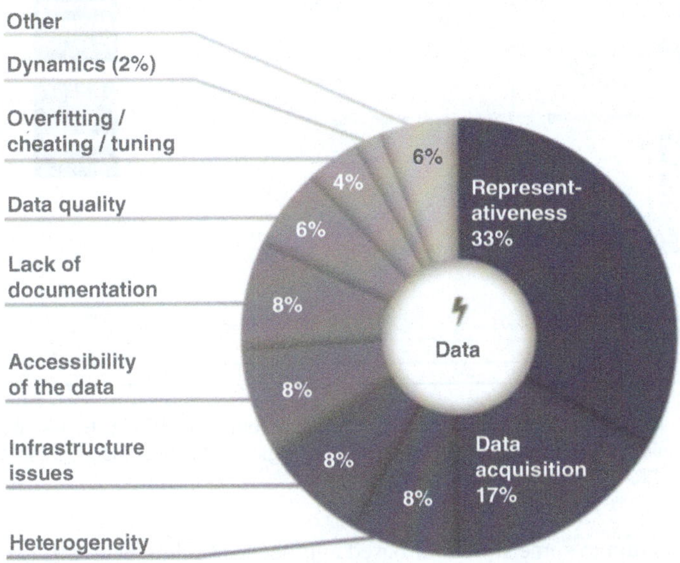

Fig. 20.13 Data-related problems (Maier-Hein et al. n.d.)

20.2.2 Image Pre-processing

The medical field majorly deals with data problems like understanding, acquiring, accessing, denoising, cleaning, and analysis of data as shown in Fig. 20.13 (Image from J. Galeotti, class material from "Methods in Medical Image Analysis", Carnegie Mellon University 2018).

Image data experts have been trying to extract information based on content and textual description, but image feature extraction has been the key point (Scholl et al. 2011; Deserno et al. 2009). Feature analyses range from the entire image to specific

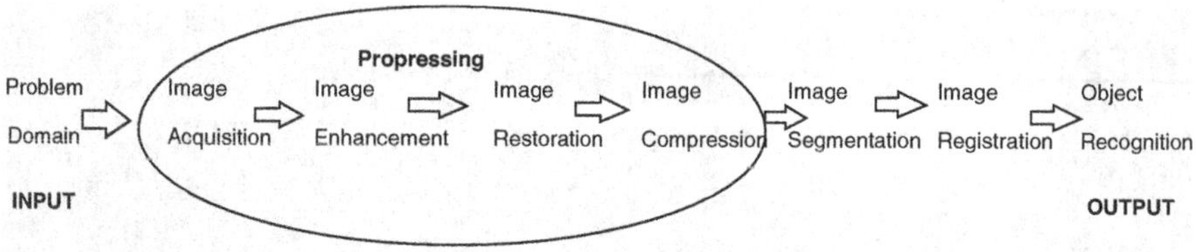

Fig. 20.14 Steps in image pre-processing (Goel et al. 2016)

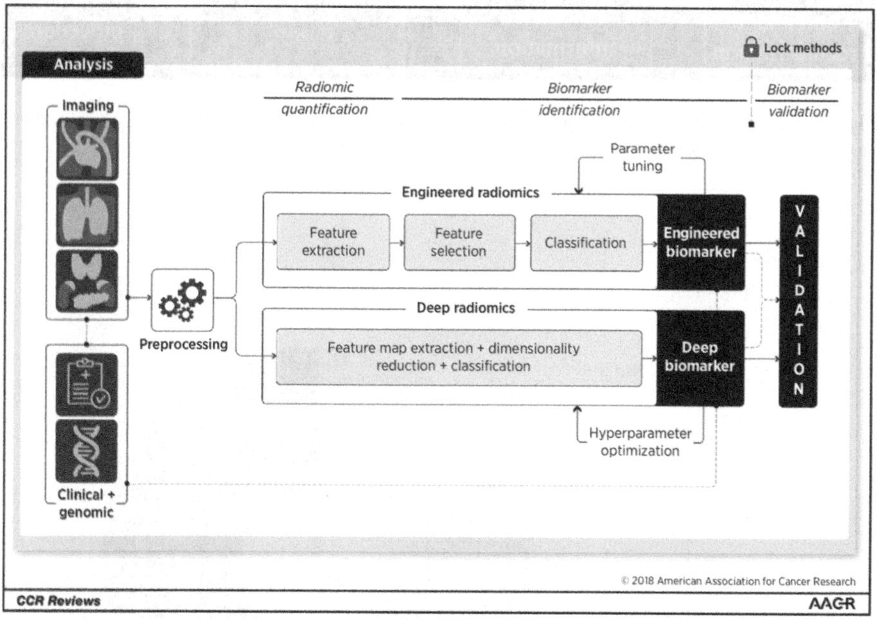

Fig. 20.15 Flow of medical image processing (Image from C. Parmar, D.B. Joseph, A. Hosny, J. Quackenbush, H.J.W.L. Aerts, "Data Analysis Strategies in Medical Imaging", DOI: 10.1158/1078-0432.CCR-18-0385 Published August 2018)

localized section to some structural-based approaches (Scholl et al. 2011; Long et al. 2009; Tagare et al. 1997).

Medical image pre-processing has a series of steps that need to be taken into consideration as shown in Fig. 20.14 (Goel et al. 2016).

20.2.3 Challenges in Medical Image Analysis

There are a number of specific challenges in medical image processing (Thirumaran and Shylaja 2014) (Fig. 20.15). They are:

- Pre-processing of image using image enhancement and restoration for best quality of image data

- Automated and accurate image segmentation of features of interest (region of interest)
- Automated and accurate image registration and fusion of multiple images
- Classification of image features or properties
- Simulation software that can be used to rehearse and plan procedures, evaluate access strategies, and plan treatments.
- Latest being is visualization of the environment in which image-guided procedures or reconstruction of working of human body in 3D.

Medical image analysis has key tasks, which will be explained in detail in subsequent sections of this document such as:

- Classification
- Segmentation
- Registration
- Deep learning (DL)-based analysis

20.2.4 Conclusion

Images play an important role in health care. Technology advancement in medical image has helped doctors to get an insight into the human body without having to cut open the body (Goel et al. 2016; Binh 2010) and to achieve the best possible diagnosis, treatment, and other surgical procedures via image analysis obtained after noise removal and high-quality resolution (Tsui et al. 2012).

20.3 Medical Image Classification

20.3.1 Introduction

Recently, rapid development in the combination of machine learning (ML) and medical field has become a popular and active topic in research area. Thus, medical image classification plays a significant role in computer-aided diagnosis (Lai and Deng 2018). The main concern for researchers in this area is how to extract features from medical image and classify them into the same model to achieve an accurate result for identifying the parts of a patient's body which are affected by the specific disease (Aberle et al. 2010).

The main purpose of image classification in the medical field is specifying the affected parts of the human body by disease, instead of gaining the high accuracy result; therefore, in this chapter we discuss the various medical image classification techniques in detail (Miranda et al. 2016).

20.3.2 Overview of Image Classification Techniques

Image classification process is divided into three stages, namely, pre-processing, feature extraction and feature selection, and classification. After the pre-processing level, by using feature extraction methods, analyze the images to extract the most appropriate features from input data for classification process, and then using feature selection methods, select the most correlated features to reduce the dimension of data which can be effective in improving performance of classification methods (Miranda et al. 2016; Lashari and Ibrahim 2013) (Fig. 20.16).

Some of the main feature selection techniques are:

- **Genetic algorithms-based optimization**

  The generic algorithms are one of the powerful methods which are based on natural selection (Kaushik et al. 2013). This technique has some disadvantages, which cause deflection in medical image segmentation (Cao et al. 2017).
- **Linear discriminant analysis**

  It is one of the dimensionality reduction techniques which goal is to preserve most of the features without eliminating any data in order to separate different classes as much as possible (Dhawan 2008; Sharma 2015).
- **Principal component analysis (PCA)**

  Another method of dimensionality reduction is PCA which is used for transformation methods to reduce a number of correlated variables to a smaller number of variables in a new subspace (Ashour and Salem 2015).

  Therefore, PCA is an applicable technique in medical image processing that can be used in feature extraction, feature selection, image segmentation, and image registration (Ashour and Salem 2015). PCA cannot be efficient in selecting features, if input images include noises (Dhawan 2008) (Fig. 20.17).

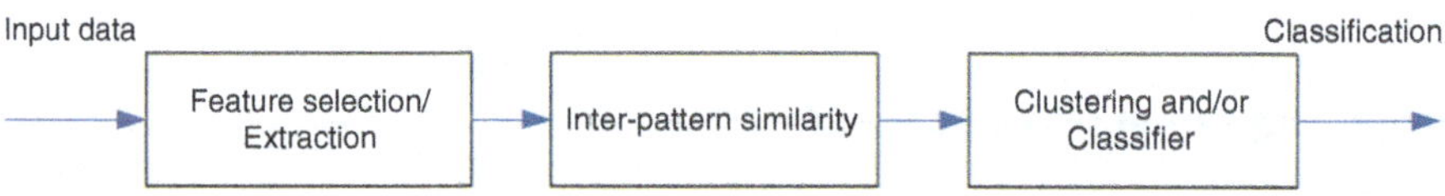

Fig. 20.16 A classification process (Miranda et al. 2016)

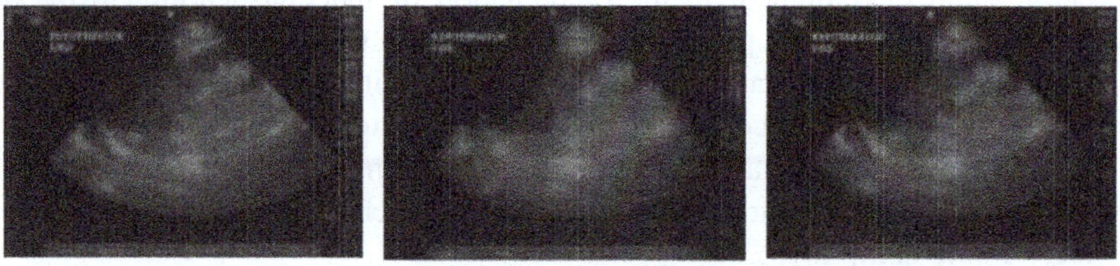

(g)Original image (h)PCA based compressed image after keeping 10 eigenvectors (i)PCA based compressed image after keeping 20 eigenvectors

Fig. 20.17 Performance of PCA on ultrasound images (Cao et al. 2017)

After completing the feature extraction and feature selection steps, start classifying images.

This part provides a brief explanation of some of the classification techniques which are more applied to classify and detect abnormalities in medical images (Lashari and Ibrahim 2013).

(a) **Neural Network Classification**

Neural Network (NN) is a computational model that has an important impact on classification by using supervised and unsupervised learning techniques. Neural Network models have some advantages; the first advantage is they are non-linear models; therefore, they are flexible in performing any complicated real-world application models. The second advantage of Neural Network is universal functional approximations which enable to approximate any function with ideal accuracy results, and the third advantage is Neural Network can regulate itself to the input data without any characteristics operational. In other words, it is a self-adaptive model (Lashari and Ibrahim 2013; https://pdfs.semanticscholar.org/1ba9/d67c80b6a762c11b9d519367e9e13a9c5c4f.pdf).

(b) **Support Vector Machine (SVM)**

SVM is the machine learning model that uses different algorithms to analyze data for the purpose of reaching the efficient classification outcome (Sharma 2015). Furthermore, this model is a binary classifier which provides the maximum separation line between two classes. However, SVM has disadvantages as it needs longer time for training data and does not manage discrete features (Lashari and Ibrahim 2013; Ehteshami 2017).

(c) **Statistical Classification Methods**

These models are based on supervised and unsupervised approach. Supervised learning method can be accomplished by using Bayesian decision theory because it is based on statistical classification and probabilistic methods. For instance, nearest neighbor and Bayesian model are the most practical classifiers. In addition, for performing supervised methods, besides training data and test data, it requires label data as well (Table 20.2).

The unsupervised learning technique classifies the data by separating the feature space, like K-means (Miranda et al. 2016; https://pdfs.semanticscholar.org/1ba9/d67c80b6a762c11b9d519367e9e13a9c5c4f.pdf; Dhawan and Dai 2008).

20.3.3 Medical Image Classification Challenges

- The first challenge is the variety of features in medical images which make challenges for training dataset; as a result, the classification outcome will be decreased.

Table 20.2 List of medical image classification methods (Lashari and Ibrahim 2013)

Author Name	Year	Methodology		Pros and Cons
		Method	Imaging Modalities	
Nguyen, Long D and Lin (Nguyen et al. 2018)	2018	Convolutional Neural Network	PAP-smear	Concatenated features with better performance at higher computation cost
Berahim, Mazniha and Samsudin (Berahim et al. 2018)	2018	Convolutional Neural Network	CT, MRI, X-ray	This is review paper which surveys some of the well-known methods
Jin, Kyong Hwan and McCann (Jin et al. 2017)	2017	Deep Convolutional Neural Network	CT, MRI	Super-resolution requiring multiple images
Wang, Lei and Pedersen (Wang et al. 2017)	2017	SVM	Diabetic Foot Ulcer Color Images	Multiple SVM with high computational requirements
A. Kumar et al. Erickson, Bradley J (Erickson et al. 2017)	2017	Neural Network k-Nearest Neighbors SVM Decision Tree Naïve Bayes Deep Learning	CT and MRI	This is review paper which surveys some of the well-known methods
A. A. A. Setio et al. (Setio et al. 2016)	2016	Multi-View Convolutional Networks (ConvNets)	Pulmonary CT	False positive reduction
				The CAD sensitivity performance should be enhanced
D. Mittal and A. Rani (Mittal and Rani 2016)	2016	SVM	Ultrasound image	High accuracy
				Each classifier being trained on only two out of N classes
G. Van Tulder and M. De Bruijne (Van Tulder and De Bruijne 2016)	2016	Convolutional classification Restricted Boltzmann Machine (RBM).	Lung CT	High mean classification accuracy
				Suitable for smaller representations learning with smaller filters or hidden nodes
J. Hong el al (Hong et al. 2016)	2016	Principal Nested Spheres (PNS), Distance Weighted Discrimination (DWD)	MRI	AUC > 0.600. Apply PNS separately

(continued)

Table 20.2 (continued)

Author Name	Year	Methodology		Pros and Cons
		Method	Imaging Modalities	
K. Seetharaman and S. Sathiamootthy (Seetharaman and Sathiamoorthy 2016)	2016	Adaptive Binary Tree Based Support Vector Machine (ABTSVM)	CT, MRI, Microscopy, Mammogram, Ultrasound, X-ray and Endoscopy images	Low computational and storage cost The relevance judgments are performed using the ground truth and the subjectivity of the individual user
K. Sirinukunwattana et al (Sirinukunwattana et al. 2016)	2016	Neighboring Ensemble Predictor (NEP) + Convolutional Neural Network (CNN)	Histopathology images	Accurately predict The Weighted Average FI score and Multiclass AUC result not considerably different with soffmax CNN + SSPP
M. Anthimopoulos et al (Anthimopoulos et al. 2016)	2016	Convolutional Neural Network (CNN)	Lung CT Scan Drawback	High classification performance The training time becomes slower due to very large number of parameters
M. J. J. P. Van Grinsven et al (Van Grinsven et al. 2016)	2016	Convolutional neural networks (CNNs) + Selective Sampling (SeS)	Color fundus image	High performance. Uses the reference guide from a single expert
Q. Dou et al (Dou et al. 2016)	2016	3D Convolutional Neural Network (CNN)	Cerebral micro-bleeds (CMBs) MRI	High sensitivity 93:16%. The accuracy and detection speed are not balance
A. Masood and A. Aljumaily (Masood and Al-jumaily 2015)	2015	SVM	Biopsy samples	High accuracy The error rate of classification decreased about 16.5% for Histopathological and 6% for Dermoscopic images

(continued)

Table 20.2 (continued)

Author Name	Year	Methodology		Pros and Cons
		Method	Imaging Modalities	
F. Khalvati, A. Wong, and M. A. Haider (Khalvati et al. 2015)	2015	SVM classifier	Multi-parametric magnetic resonance imaging (MP-MRI)	High sensitivity and specificity (>80%) A limited number of datasets and the target of Gleason score is >= 7, the proposed model was not assessed by clinicians
K. Chung et al (Chung et al. 2015)	2015	Pre-Trained Convolutional Neural Networks (CNN)	CT scan	AUC = 0.868. Time-consuming since Peri-Fissural Nodules (PEN) characterization was subjective, it suggests the increment of the number of 2D views may give the higher accuracy of characterization
V. Gopalakrishnan, P. G. Menon, and S. Madan (Gopalakrishnan et al. 2015)	2015	Bayesian rule learning (BRL) methods	Cardiovascular Magnetic Resonance Imaging (cMRI)	High accuracy A limited number of datasets
Y. Iwahori et al (Iwahori et al. 2015)	2015	K-means++	Endoscope	The accuracy is higher because using the edge-based features The computational time was decreased if HOG features used to detect the polyp region
Y. Song et al (Song et al. 2015)	2015	Locality-constrained Subcluster Representation Ensemble (LSRE)	High Resolution Computed Tomography (HRCT)	High accuracy The Locality-constrained Linear Coding (LLC) did not use advanced distance function

- The second challenge is about the size of the medical images. Since, the medical images are very small, extracting and selecting enough valid information from dataset is not easy (Lai and Deng 2018).

20.3.4 Conclusion

Study in medical image classification can be beneficial for both computer-aided diagnosis and teaching purposes in medical fields. Recent research in this area could be helpful for analyzing and diagnosing diseases rapidly. This chapter has provided the overview of image classification methods and algorithms which are using medical images to identify the human body in order to distinguish images showing diseases from ones which do not (Miranda et al. 2016).

20.4 Medical Image Segmentation

20.4.1 Introduction

Today, with growing usage of computed topography (CT) and magnetic resonance (MR), X-ray image, digital mammography, and other imaging modalities, analyses of these images manually are not possible; therefore, digital image processing and computer algorithms, such as image segmentation methods, play an important role in diagnosing diseases and progressing biomedical research areas especially in medical imaging applications (Petitjean and Dacher 2011; Pham et al. 2000). Image segmentation is a process that divides an image to many homogeneous sub-regions which have the same characteristics as color, depth, and intensity (Withey and Koles 2007).

For instance, MR images which provide high resolution of three dimensional (3D) are the most common applications that use image segmentation techniques. Image segmentation can analyze both 2D and 3D images, and the main difference between them is processing the pixels in 2D and voxels in 3D (Despotovi 2015) (Fig. 20.18).

In the following section, some important methods of medical image segmentation are reviewed in order to introduce the segmentation process and its importance in analyzing medical images accurately for diagnosing disease.

20.4.2 Review of Medical Image Segmentation Techniques

As it is mentioned before, segmentation is a technique that provides wide diagnostic insights in the medical field. Using this technique in medical images can be improved, detecting of image's boundaries, cell counting, scaling organs of human body and many other applications automatically (https://www5.cs.fau.de/research/groups/medical-image-segmentation/).

In this part, some of the medical image segmentation methods are provided:

(a) **Intensity-based segmentation method**

In this method, pixels in 2D images and voxels in 3D images are classified based on their intensity, for instance, brain MR images contain three tissue

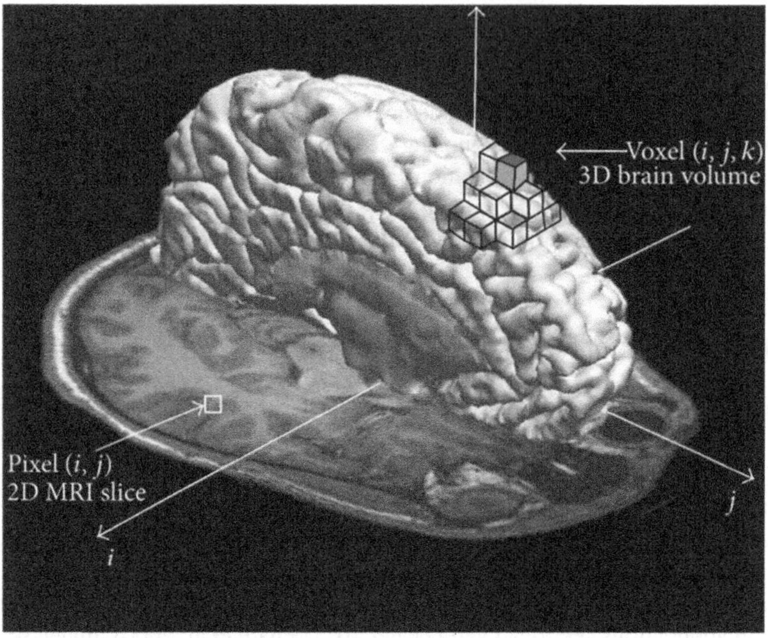

Fig. 20.18 Visualization of brain MR image. The square in the 2D MRI slice represents an image pixel (i, j), and the cube in 3D image illustrates an image voxel (i, j, k) (Despotovi 2015)

types, namely, cerebrospinal fluid (CSF), white matter (WM), and gray matter (GM), which can be identified based on intensity after using intensity-based segmentation method (Despotovi 2015) (Fig. 20.19).

(b) **Thresholding segmentation method**

This technique thresholding histogram gray scale images which it is one of the traditional and simplest method in medical image segmentation that it can be categorized in intensity-based methods. In other words, while thresholding segmentation is applied on medical images, intensity histogram is used in order to distinguish intensity value and separate different classes. The result of applying thresholding method on an abdomen CT image is illustrated in Fig. 20.20 (Despotovi 2015; Aggarwal 2010).

Thresholding method includes multiple groups, such as (Despotovi 2015):

- Local threshold which is dependent on the position in the image
- Adaptive thresholding
- Global or single thresholding
- Multi-thresholding

Thresholding is an efficient and fast technique; however, it has some limitations; first, choosing an appropriate value for threshold for different medical images causes many difficulties, and second, in low-contrast images, it processes the distributed class of pixels rather than connected areas, so it is required to use connectivity algorithm before thresholding process (Despotovi 2015; Sahoo et al. 1988).

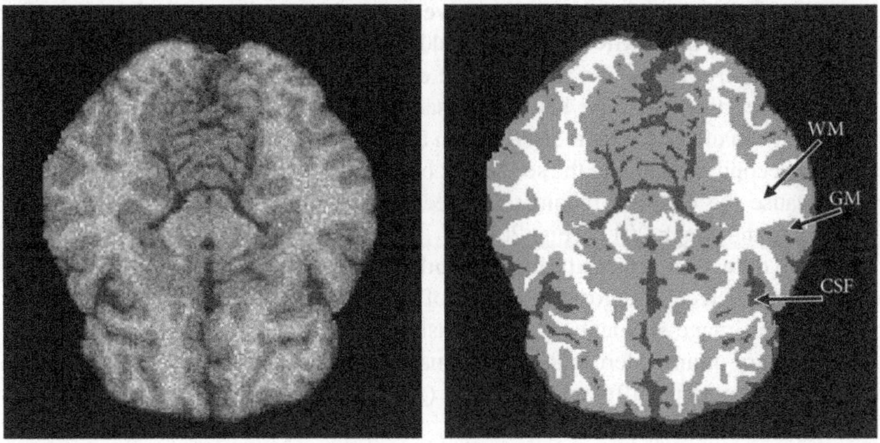

(a) (b)

Fig. 20.19 Illustration of an example of intensity-based segmentation method. (**a**) Original brain MR image and (**b**) related segmentation image. The segmentation image indicates three main tissue types (Lashari and Ibrahim 2013)

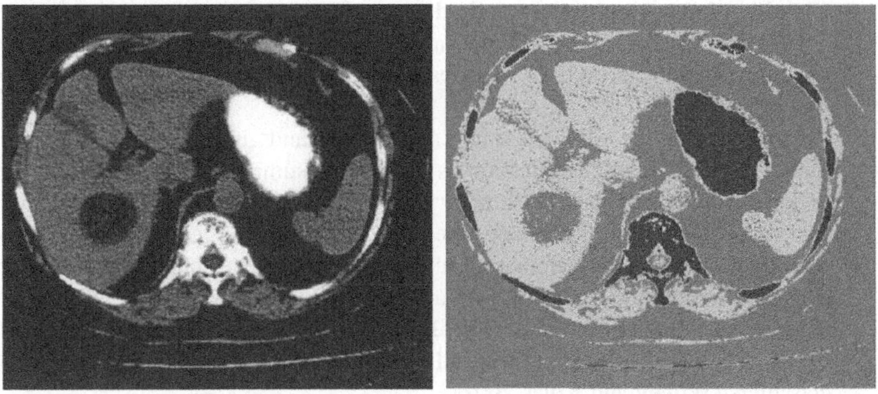

(a) (b)

Fig. 20.20 An example of thresholding segmentation. (**a**) Original abdomen CT image and (**b**) segmented abdomen CT image by thresholding technique (Aggarwal 2010)

(c) **Region growing segmentation method**

The major purpose of this method is to form a region for segmentation based on more homogeneity between pixels or voxels (Withey and Koles 2007). In addition, this method can be categorized in intensity-based segmentation as well. Initially, for processing the region growing techniques, require selecting a seed point by an operator manually, then after testing similarity of neighborhood pixels or voxels, region continue growing until to get to the heterogeneity pixels (Withey and Koles 2007; Despotovi 2015).

Region growing obtains impressive result to segment medical images. For example, in brain MR images, it could segment brain tumor and brain vessels successfully (Despotovi 2015; Passat et al. 2005; Haralick and Shapiro 1985). However, this technique has some disadvantages: first, it is sensitive to noise so that it can affect the segmented region and disconnect it from the related region, and second, initializing the seed point by human is difficult and time-consuming because the operator requires to select a seed point for each different region (Kaur and Singh 2011) (Fig. 20.21).

(d) **Edge detection segmentation method**

Edge detection algorithm is one of the common methods that are used in medical image segmentation. The basic idea of this technique is to detect the boundary or any interruption in the image (Upadhyay and Kashyap 2016). This method contains different algorithms (Aggarwal 2010):

- Hough transform based
- Edge relaxation
- Border detection method

Some limitations of this method are that it is sensitive to noise, segmentation can be completed by combination of edge detection and region growing techniques, and some lines appear that are not edge in outcome which effect on the final result (Aggarwal 2010) (Fig. 20.22).

(e) **Classification segmentation method**

Classifier methods are known as statistical pattern recognition which is a type of segmentation techniques that can divide images to feature space by labeling them.

This method is divided into supervised and unsupervised learning. Supervised learning classification is a time-consuming method because the image has to be segmented manually rather than using automatic segmentation for other images; therefore, supervised needs training and test dataset and also label dataset. Moreover, another disadvantage of supervised classification is that using the same training dataset for various images reduces the quality of result (Pham et al. 2000; Withey and Koles 2007; Anbeek et al. 2005). Some examples of supervised classification method are K-nearest neighbor and Bayesian classifiers (Withey and Koles 2007).

Unsupervised classification is a type of statistical clustering which uses expectation-maximum (Pham et al. 2000).

20.4.3 Conclusion

In this chapter, we discussed some segmentation techniques briefly and also limitations and disadvantages in medical images. In conclusion, image segmentation is one of the most challenging areas in image processing study which can be most efficient in computer-aided diagnosis and identify diseases in medical images

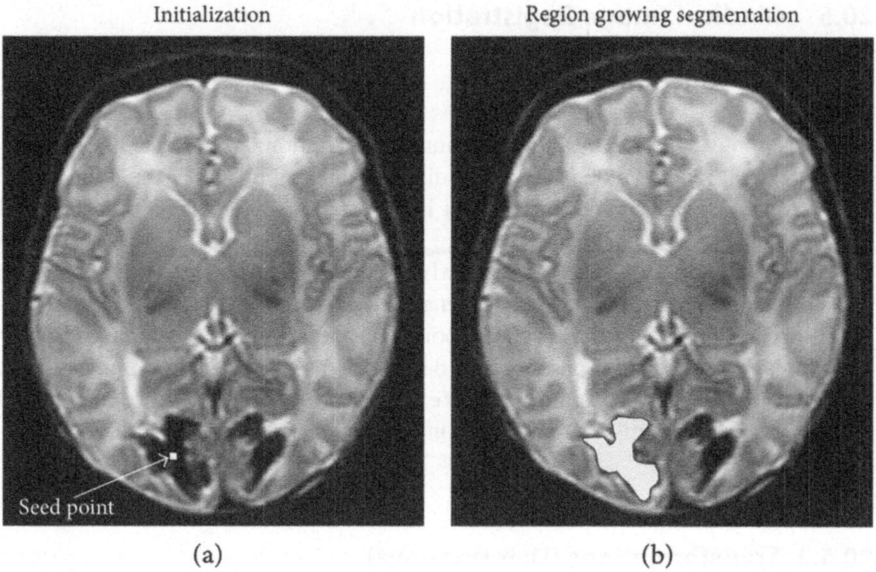

Fig. 20.21 Illustration of the result of applying region growing method on a brain MR image. (**a**) Original brain MR image with selected seed point manually and (**b**) segmented brain MR image by region growing method which indicates the final segmented region (Despotovi 2015)

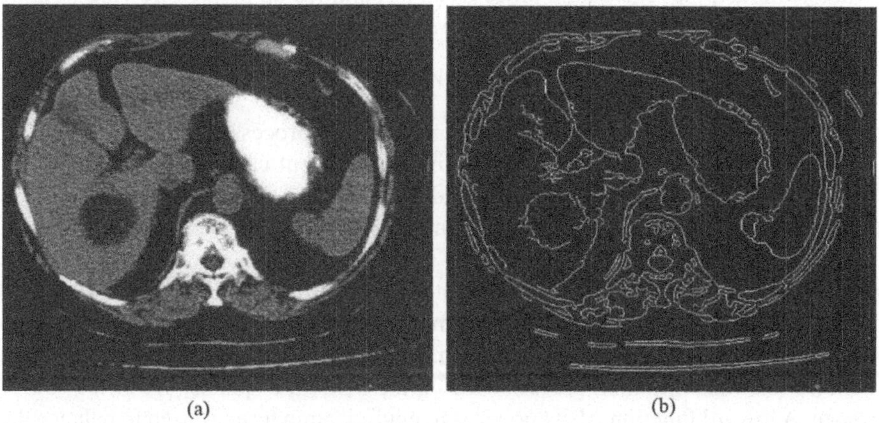

Fig. 20.22 An example of edge detection technique. (**a**) Original abdomen CT image and (**b**) result of applying edge detection technique on abdomen CT image (Aggarwal 2010)

especial 3D images. It is expected that these algorithms will become more practical in biomedical field to achieve faster and higher-quality diagnosis of diseases from 3D medical images reconstructing and visualizing the anatomical structures (Despotovi 2015).

20.5 Medical Image Registration

20.5.1 Introduction

In this section, we discuss the domain of image registration. We provide the elements of this field, current developments in this direction, the applications, and future scope of this topic. Medical image registration is one of the key elements in analysis of medical data. It is essential for mapping medical images and data to the correct physio-structural components of the body, for relating changes in captured data across spatio-temporal plane, and for generating an atlas of human structural biology. Registration consists of matching points of interest between source and destination images, transformation of source data to the target data, and optimization of the result to best suit the application. We next detail each of the abovementioned three aspects of the registration process and review the related literature.

20.5.2 Transformations (Deformations)

As previously mentioned, a typical registration process consists of three key steps: correspondence, transformation, and optimization. Although correspondence can be considered a first step of the process, it often is dictated by the transformation process. In fact, the registration process can be summed up in the following formula for the optimal transformation:

$$\Psi((\mathrm{D}, S\phi W), \mathrm{R}(\mathrm{W})) \tag{20.1}$$

Here, Ψ denotes the quality of the transformation process. This is often considered as a measure of the transformation in literature but can also be considered as qualitative measure when manual validation of the transformation process is performed. This is often the case where medical professionals exert their choice of the registration result to enhance automated transformation. Automated transformations mostly reply on quantitative transformation measure. D and S denote the destination and the source image, respectively, with the aim being to transform the source image to destination. The transformation is aided by ϕ, which denotes the transformation process used on the domain of pixels, W, of the source image. A reward function, also known as a regularization term, is used to balance the transformation process. We will focus on the transformation process next. The transformation process is often referred as the deformation process. Transformation processes have unique properties. Transformation properties are mostly ill posed according to Hadamard definition (Hadamard 1923). In perspective of real-time processing transformation parameters are as less as possible. Likewise, the constraint should ensure that as little as displacement is incurred in the transformation process.

We now discuss the variety of transformations available and presented in recent literature.

Transformation process can be categorized into physical model-based approaches, geometric-based methods, and knowledge-based paradigms. It is noted that this approach of categorization is not independent but has significant intersection. The aim of this categorization is to enforce logical categorization to the transformation field.

Physical-based models use plant models like elastic model by Navier-Cauchy is planar or hierarchical approach (Bajcsy and Kovačič 1989; Gee and Bajcsy 1999). The physical models can be linear (Leow et al. 2005; He and Christensen 2003) or non-linear (Rabbitt et al. 1995; Pennec et al. 2005) in nature. These include elastic as well as diffusion model.

The next approach is geometric-based transformations. They include radial basis, elastic body, and piecewise affine models. The radial basis functions use kernels to find the transformations of source to target image. The radial basis function falls under kernelized approach wherein the distance better source and target is optimally mapped via kernel thereby generating the transformation parameters. Typical kernelized formulation is expressed as:

$$\Gamma(x) = \sum_{i=0}^{N-1} \vartheta_i F(x, p_i) \tag{20.2}$$

where Γ () denotes the transformation operation, ϑ_i is the weight of the ith parameter of the kernel transformation, and p_i is the ith parameter of the kernel. F () is the kernel used. Kernels like radial basis and thin plate spline have been frequently used in literature (Zagorchev and Goshtasby 2006; Yang et al. 2011; Bookstein 1989; Bookstein 1991). Recent advancements in this field include Tensor-based deformation, wherein each dimension is deformed using Tensor transformations (Declerck et al. 1997; Rueckert et al. 1999). Another variation in this direction often perused in literature is the use of piecewise affine model (Hellier et al. 2001).

The final approach that is presented in literature is the knowledge-based approach. In many situations, further information is available on the registration process. This can be in terms of the statistical variability of the transformations (Cootes et al. 1995), availability of biophysical models (like tumor growth models) (Clatz et al. 2005; Hogea et al. 2007), and biomechanical models of human organs (Bharatha et al. 2001; Hensel et al. 2007).

20.5.3 Matching (Correspondence)

Matching process consists of two aspects, namely, location matching and interpolation. Interpolation is typically used within each iteration of the registration process wherein we find the most plausible value of the transformation based on the neighborhood kernel. The key aspect of matching can, however, be considered as location matching. Location matching between source and destination (or target) images is what we primarily consider as matching. If we register a functional image to a structural image, for example, matching a PET image or fMRI data (functional

MR image) to a CT Image, we need to map functional image to structural image. In such case, the functional image may be first mapped to a template image, which provides structural bearings. It is then related to the structural image.

Registration of two functional images in similar lines entails two such intermediate mapping, one for each functional image. Thus, for matching either intrinsic or extrinsic markers are used. Extrinsic marker methods rely on human intervention to relate the markers, while the intrinsic ones use various algorithmic approaches including feature and intensity measures. We now discuss some of the key works related to image location matching.

The most commonly used matching method is geometric in nature. Geometric methods relate two images by minimizing geometric criterion at landmark locations in the image. Geometric matching includes finding points of interests, creating correspondence between the suitable points, creating transformations with the suitable points instead of using correspondence, and finally joint use of correspondence and transformation. Key point detection procedures include the famous Harris point detectors (Triggs 2004; Mikolajczyk and Schmid 2004), invariant feature detectors (Mikolajczyk et al. 2005), multiscale approach (Kadir and Brady 2001), and histogram-oriented approach like SURF and SIFT (Mikolajczyk and Schmid 2005; Morel and Yu 2009). Using the feature points detected correspondence-based matching can be used. They include use of descriptor distance as well as incorporating geometric constraints (Cheung and Hamarneh 2009; Ni et al. 2008; Torresani et al. 2008). Use of Gaussian mixture model (Jian and Vemuri 2011) provides an approach for transformation matching. The above methods are cleverly weaved together by the famous iterative closest point algorithm (Besl and McKay 1992).

The geometric-based techniques are extended to spatio-temporal to incorporate matching within modal, multimodal, and temporal data. In this situation, care is taken to constraint the prior mentioned matching processes to suit the variability of the data. The key criterion in such situations is the cross-correlation coefficients using either intensity or attribute features (Kim and Fessler 2004). Information theoretic approach is also used in literature with mutual information being the primary measure (Viola and Wells III 1997).

20.5.4 Optimization

Optimization framework is the core of the registration process. Using the optimization framework, we can choose the best transformation parameters which generate the best measure of transformation for a minimum cost function. The cost function is typically considered as a difference between source and target images. An implicit factor which plays a key role in the optimization process is the computational efficiency as well as the rate of convergence of the optimization process. As more and more registration algorithms are required to be near real time, this implicit constraint becomes highly significant. Based on the nature of variables, the optimization process can be broken into continuous parameter optimization, discrete parameter optimization, or hybrid optimizations. We will discuss these algorithms next.

As mentioned previously, optimization process may be separated into continuous, discrete, and hybrid forms. Continuous approach assumes the optimization parameters to be real. This allows the objective function for the continuous optimizations to be differentiable. Such case therefore utilizes iterative update to real optimal value using delta increment. The typical update rule is provided in the following equation:

$$\theta_{t+1} = \theta_t + \alpha_t f(\alpha_t) \tag{20.3}$$

The above equation shows the update of the continuous parameter using step size alpha and the update function *f*, which is often the derivative function. The above framework has mutated into multiple forms. They include gradient descent approach (Klein et al. 2007; Moré and Thuente 1994) which is the closest to the original formulation. Conjugate gradient methods on the other hand have better converge rates that gradient descent. They try to exploit the knowledge from previous gradient direction to generate a new direction of descent based jointly on the previous gradient direction and the derivative (Fletcher and Reeves 1964; Polyak 1969; Hestenes and Stiefel 1952; Hager and Zhang 2006). Other similar approaches include Powell Conjugate, (Maes et al. 1997), Gauss-Newton (Ashburner and Friston 2011; Haber and Modersitzki 2007; Modersitzki 2008), Levenberg-Marquardt (L-M) (Kybic and Unser 2003; Wu et al. 2000; Gefen et al. 2003), and stochastic gradient descent.

Gauss-Newton (G-N) works by optimizing the sum of squared error term differential. This is referred to in the literature as the Jacobian, and the search direction is denoted by the following equation:

$$g = \left(J^T(\theta)J(\theta)\right)^{-1}\nabla(\theta) \tag{20.4}$$

where $J(\theta)$ denotes the Jacobian operation, $\nabla(\theta)$ denotes the delta derivative, and ^{T} is the transpose operator. The L-M technique modifies the G-N method by adding a weighting term to the Jacobian above. The modified formulation is shown by the following equation:

$$g = \left(J^T(\theta)J(\theta) + \zeta I\right)^{-1}\nabla(\theta) \tag{20.5}$$

All the previously described approaches relay on being able to compute the gradient, which can be very demanding due to the vastness of the data source. In such case, the gradient is approximated by a stochastic version.

The second set of methods assumes the optimal parameters belong to a set of discrete values. One such approach is to use Markov random fields, which are probabilistic graphical models. Graphical models consist of graphs consisting of vertices and edges ($G = \{v, e\}$). Max-flow min-cut principle is the key formulation (Ford and Fulkerson 1956) and is the fundamental approach for graph segmentation. Alpha-expansion approach was used by Bokov (Greig et al. 1989; Boykov et al. 2001) by using extensive label check for registration. Belief propagation (Frey and MacKay 1997; Murphy et al. 1999) is another technique wherein local message exchange happens between nodes and backtracked to recover the best solution. Linear programming (Komodakis and Tziritas 2007; Komodakis et al. 2008) has also been used in literature to solve discrete optimization problems.

The last set of approaches uses hybrid or miscellaneous approaches like Greedy learning (Liu et al. 2004; Xue et al. 2004) and neural algorithms and evolutionary methods (Hansen and Ostermeier 2001). We will discuss DL methods in a separate section. These hybrid methods are heuristic or meta-heuristic in nature. Greedy methods relay on choosing the best conditional solution in each iteration without enforcing combinatorial check on the optimality of the solution. Evolutionary techniques on the other hand use genetic-based techniques to create mutation of parameters and choose the mutation that has the best survivability.

20.5.5 Brain Registration

Brain registration occupies a special place in the domain of registration techniques. There are multiple reasons behind it. We know well that the brain is by far the most complicated organ with millions of neural connections and pathways. It also has volumetric multidimensional network connections and functional linkages. Due to large variation between populations further complicated by deformations due to disease like Alzheimer's disease. Several attempts have been made to generate brain template or atlas. They include Talairach atlas (Talairach and Tournoux 1988) developed from a physical brain. Digital brain atlas has been developed from physical models (Kruggel and Yves von Cramon 1999; Nowinski and Thirunavuukarasuu 2001; Roland and Zilles 1994; Roland et al. 1997) at Harvard and Montreal Neurology. To encompass diversity of intermodal and intra-modal brain variability, probabilistic atlases were developed by considering distributions of the brain landmarks and their intensity and other features. International Consortium of Brain mapping has been led in this direction (Mazziotta 2002; Mazziotta et al. 1995). Deformable brain atlas has been another direction of research wherein use of non-rigid registration has allowed the pliability of brain map between subjects. This approach is also suitable for longitudinal brain study (Thompson et al. 2000; Ganser et al. 2004; Woods 2003). The utility of this approach is however extremely dependent on the registration technique used.

20.5.6 Conclusion

In this review we visited the need of image registration, particularly in perspective of medical image analysis. We reviewed the different components of the registration process, namely, the correspondence problem, the transformation between source and target domains, and finally the validation of the transformation process via

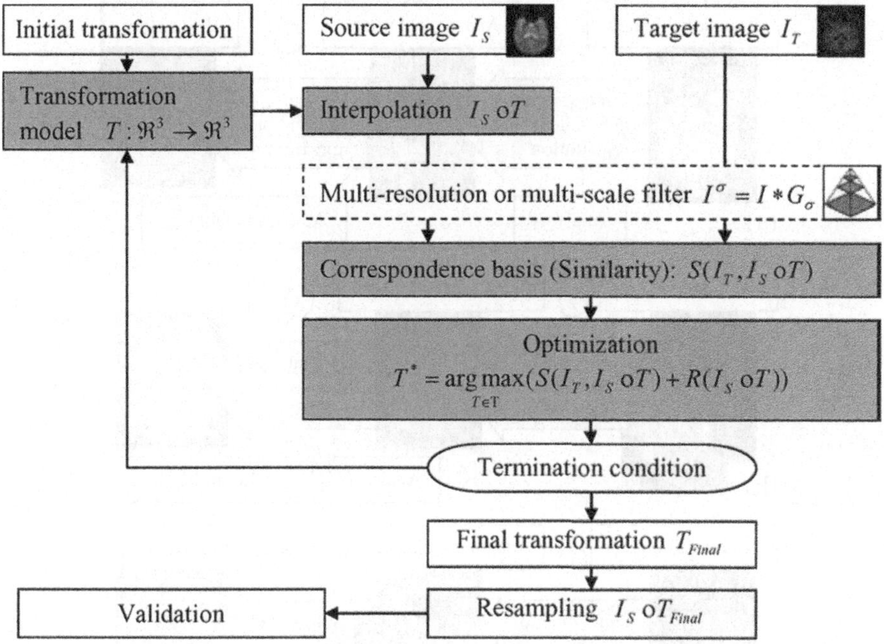

Fig. 20.23 Flowchart detail of the image registration process (Gholipur 2007)

optimization algorithms. We reviewed the cutting-edge techniques of each of the three components mentioned above and provided their pros and cons. Figure 20.23 (Gholipur 2007) shows the flowchart of the registration process detailing the source and target images, transformation, correspondence (and interpolation) and optimization processes, and the iterative loop.

We further elaborated the image registration process of the brain. Due to the uniqueness of the brain, registration process can happen between structural (anatomical) data and functional (image/data). Brain atlas has been a unique approach related to brain registration. We provided review of the cutting-edge works in this domain. Figure 20.24 (Gholipur 2007) provides comparative approaches available in this domain.

As detailed through the paper, most of the techniques in image registration require significant medical domain knowledge as well as signal processing expertise, thereby making the domain significantly challenging. A recent advancement in this domain is the use of deep learning and artificial deep networks to auto-train the registration process parameters. This is nascent and has a significant potential in future. Another aspect related to the registration process is the validation of its results. Currently, manual validation by domain experts is the norm. Such techniques require domain expertise, are laborious, and in many cases have limited availability. In the future, DL is anticipated to provide support in this direction as well.

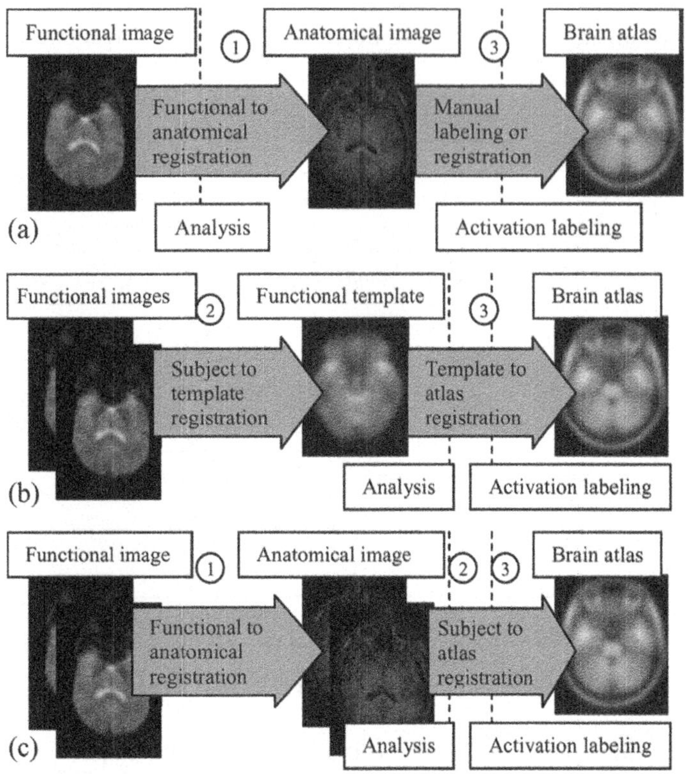

Fig. 20.24 Varieties of brain image registration, wherein the input data can be anatomical or functional (Gholipur 2007)

20.6 Deep Learning

20.6.1 Introduction

Deep learning is based on artificial neural network, which is built over basic biological system of working of brain network. Deep learning is a part of machine learning in artificial intelligence (AI) that learns patterns from unsupervised/ supervised data to improve the future prediction/recognition of new patterns using complex algorithms over different layers to achieve it.

Application of deep learning has become a trend in the current state of art in each and every field, for example, pattern recognition in images, speech, image and art restoration, language processing, news feed generating, and classification.

20.6.2 Concepts of Machine Learning, Deep Learning, and Artificial Intelligence

There is always a confusion about basic concepts between artificial intelligence (AI), machine learning, and deep learning because they are all inter-related and inter-connected. According to experts like Mr. Calum McClelland, Director of Big Data at Leverege:

- AI – "AI involves machines that can perform tasks that are characteristic of human intelligence."
- ML – "Machine learning is simply a way of achieving AI."
- DL – "Deep learning is one of many approaches to machine learning." (Fig. 20.25)

Machine learning algorithm can be classified as supervised, semi-supervised, or unsupervised learning:

- Supervised learning – (classification and regression problem) (https://machinelearningmastery.com/supervised-and-unsupervised-machine-learning-algorithms/) – label of the data is known.
- Semi-supervised learning – combination of supervised and unsupervised learning in the data.
- Unsupervised earning (clustering) – label of the data is unknown.

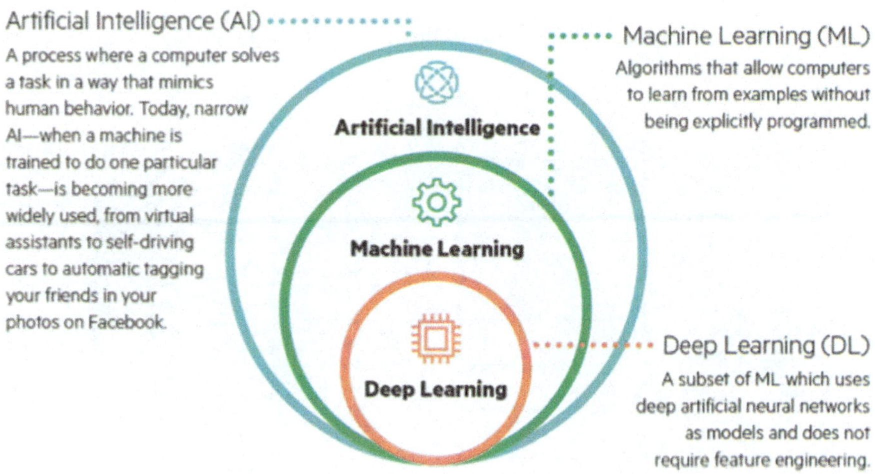

Fig. 20.25 Difference between AI, ML, and DL (Image by Curt Hopkins, Managing Editor, Hewlett Packard Labs)

20.6.3 Deep Learning Architectures

Neural Network was inspired by the biological working of brains. Multiple machine learning algorithms are combined together at different layers of the processes for specific purpose to solve on give complex data to find useful information/prediction is the overall concepts of neural network.

A deep neural network hierarchically combines multiple layers of neurons which contain important features. This network does memorize information from the training data and makes prediction on the test data or unseen data. Hence deep learning is very popular in computer vision and medical imaging area (LeCun 2013; Razzak et al. n.d.) (Fig. 20.26).

Some of the popular deep learning algorithms are compared in Table 20.3 such as Convolutional Neural Networks (LeCun 2013), Deep Neural Network, Deep Belief Network, Deep Autoencoder, Deep Boltzmann Machine, Recurrent Neural Network (Roell 2017), and Generative Adversarial Network.

20.6.4 Research Trends

Deep learning is trending with lot of research going on the health-care area. In terms of medical images, all the sections mentioned in the paper like classification, segmentation, registration, etc., can be done on a larger scale by using deep learning techniques. Some applications of deep learning in medical image are as mentioned below:

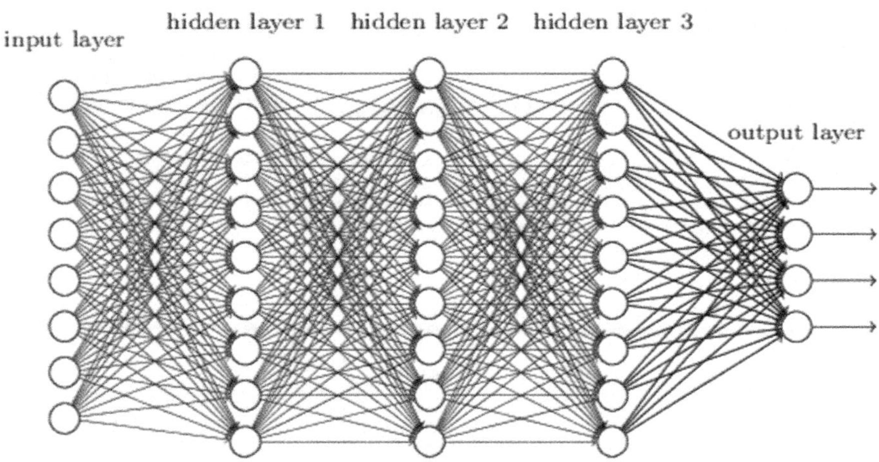

Fig. 20.26 Example of deep neural network (Image by M. Nielsen, http://neuralnetworksanddeeplearning.com/chap5.html)

Table 20.3 Comparison between few of the deep learning models (Lashari and Ibrahim 2013)

Type of Network	Description	Advantage	Disadvantage
Deep Neural Network (DNN)	There are more than two layers, which allow complex non-linear relationship. It is used for classification as well for regression	It is widely used with great accuracy	The training process is not trivial because the error is propagated back to the previous one layers, and they become very small. The learning process of the model is also too much slow
Convolution Neural Network (CNN) (LeCun, 2013)	This network is very good for 2D data. It consists of convolutional filters which transform 2D into 3D	very good performance, learning of model is fast	It needs a lot of labelled data for classification
Recurrent Neural Network (RNN) (Roell, 2017)	It has the capability of learning of sequences, the weights are sharing across all steps and neurons	Learn sequential events, can model time dependencies, there are many variations like LSTM, BLSTM, MDL- STM, HLSTM. These provide state of the art accuracies in speech recognition, character recognition and many other NLP related tasks	there many issues due to gradient vanishing and need of big datasets
Deep Boltz-Mann Machine (DBM)	This model is based on family of Boltzmann and it consists of unidirectional connections between all hidden layers	the top-down feed-back incorporates with ambiguous data for more robust inference	optimization of parameters is not possible for big dataset
Deep Auto-encoder (dA)	It is used in unsupervised learning and it is designed mainly to extraction and reduction of dimensionality of features. The number of inputs is equal to the number of output	Does not need labeled data. There is different variation like Sparse, De-nosing, etc. for robustness	It needs pre-training step. Its training may suffer from vanishing
Generative Adversarial Network (GAN) (Goodfellow et al. 2014)	It is used in unsupervised learning, technique can generate photographs that look at least superficially authentic to human observers, having many realistic characteristics	generate samples faster than fully visible belief nets Compared to variational autoencoders, GANs don't introduce any deterministic bias.	Training a GAN requires finding a Nash equilibrium of a game, hard to learn to generate discrete data

Table 20.4 List of few trending research papers in classification

Authors	Year	Paper name	Data type	Methods
Anthimopoulos et al. (Anthimopoulos et al. 2016)	2016	Lung pattern classification for interstitial lung disease using deep convolution neural network	CT images of Interstitial lung diseases (ILDs)	Deep convolutional neural networks
Sakamoto et al. (Sakamoto and Nakano 2016)	2016	Cascaded neural networks with selective classifiers and its evaluation using lung x-ray CT images	CT images of Lung nodules	Cascaded neural networks
Setio et al. (Setio et al. 2016)	2016	Pulmonary nodule detection in CT images: false positive reduction using multi-view convolutional networks	CT images of pulmonary nodules	Multi-view convolutional neural networks
Esteva et al (Esteva et al. 2017)	2017	Dermatologist-level classification of skin cancer with deep neural networks	Pixel based photo images of Skin	Deep convolutional neural networks
C. Ge et al. (Ge et al. 2018)	2018	3D multi-scale convolutional networks for Glioma grading using MR images	MRI brain images for Glioma	3D multi-scale convolutional network

(a) **Disorder classification**

Classification of disease is a basic requirement and needs accuracy. Table 20.4 shows few researches carried out for various parts of the body (Razzak et al. n.d.; Ker et al. 2017).

(b) **Tumor detection and segmentation**

Lesion/tumor detection and segmentation have been a very important research. Deep learning algorithms are now capable of handling diseases that could be missed by doctors. This provides a double assurance for their diagnosis. A few researches have been conducted in recent times (Razzak et al. n.d.; Ker et al. 2017) (Table 20.5).

(c) **Robotics surgery (autonomous)**

The Da Vinci robot revolutionized surgery avenues. This device acts as robotic limbs for surgeons. Accuracy is very important in these situations for fine detail and limited spaces are prone to human error which can be reduced by machine (Faggella 2018; https://www.youtube.com/watch?v=0XdC1HUp-rU) (Fig. 20.27).

Medical instruments for tracking, detecting, and performing surgery are common across the health-care arena. Hence a medical image obtained provides opportunity for using deep learning (Table 20.6).

Table 20.5 List of few research papers in detection and segmentation

Authors	Year	Paper name	Data type	Methods
H. R. Roth et al. (Roth et al. 2015)	2015	Deeporgan: Multi-level deep convolutional networks for automated pancreas segmentation	Pancreas	Multi-level deep convolutional networks (ConvNets)
W. Zhang et al. (Zhang et al. 2015)	2015	Deep convolutional neural networks for multi-modality isointense infant brain image segmentation	MRI of infant brain image for multi-modality	Deep convolutional neural networks
Pereira et al. (Pereira et al. 2016)	2016	Brain tumor segmentation using convolutional neural networks in MRI images	MRI of brain	Convolutional neural networks
R. Duggal et al. (Duggal et al. 2016)	2016	Overlapping cell nuclei segmentation in microscopic images using deep belief networks	Microscopic images of cells	Deep belief networks
Havaei et al. (Havaei et al. 2017)	2017	Brain tumor segmentation with deep neural networks	MRI of brain	2D convolutional neural networks
Q. Que et al. (Que et al. 2018)	2018	CardioXNet: automated detection for cardiomegaly based on deep learning	Presence of cardiomegaly on chest X-ray image	CardioXNet uses deep learning methods U-NET
Q. Zhu et al. (Zhu et al. 2018)	2018	A deep learning health data analysis approach: automatic 3D prostate MR segmentation with densely-connected volumetric ConvNets	MRI of prostate	Densely connected volumetric ConvNets

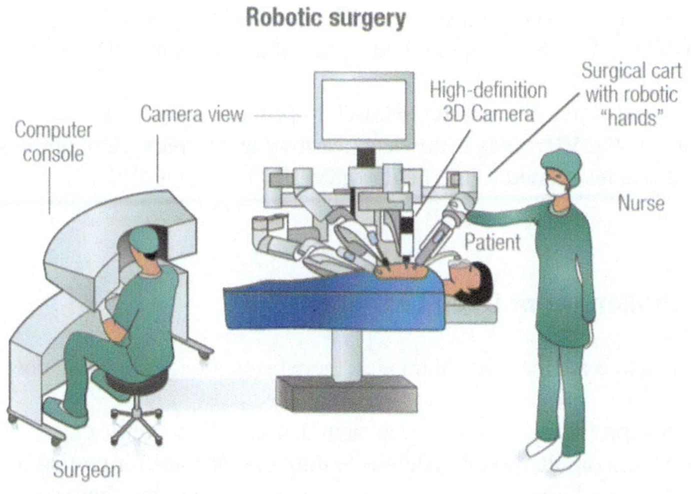

Fig. 20.27 da Vinci surgical system (https://www.materprivate.ie/dublin/centre-services/all-services/robotic-surgery/)

Table 20.6 List of few research papers in robotics surgery

Authors	Year	Paper name
Yamamoto et al. (Yamamoto et al. 2012)	2012	Augmented reality and haptic interfaces for robot-assisted surgery
R.F. Solodova et al. (Solodova et al. 2016)	2016	Instrumental tactile diagnostics in robot-assisted surgery
C. Varytimidis et al. (Varytimidis et al. 2016)	2016	Surgical video retrieval using deep neural networks
Sarikaya et al. (Sarikaya et al. 2017)	2017	Detection and localization of robotic tools in robot-assisted surgery videos using deep neural networks for region proposal and detection
S.A. Pedram et al. (Pedram et al. 2017)	2017	Autonomous suturing via surgical robot: an algorithm for optimal selection of needle diameter, shape, and path
S. Pestscharing et al. (Pestscharing and Schoffmann 2017)	2017	Learning laparoscopic video shot classification for gynecological surgery
Z. Zhao et al. (Zhao et al. 2017)	2017	Tracking-by-detection of surgical instruments in minimally invasive surgery via the convolutional neural network deep learning-based method
A. Ghose et al. (Ghose et al. 2018)	2018	New surgical robots on the horizon and the potential role of artificial intelligence
A. Shvets et al. (Shvets et al. 2018)	2018	Automatic instrument segmentation in robot-assisted surgery using deep learning

(d) **Virtual reality for visualization**

Major research companies have started exploring 3D technologies such as augmented reality and virtual reality (VR) visualization of human body for best equipping the doctors, medical professionals, and medical students to prepare them to provide the most personalized services. This provides a new avenue to get detailed understanding, practice to train and education them to deal with difficult situations (Fig. 20.28).

Recent trends have been in the research to provide innovative tools and technology to cater to the VR needs using deep learning techniques. Here are few research papers and articles related to the field (Table 20.7).

20.6.5 Challenges of Deep Learning

Apart from data issue for medical images, there lie challenges of using deep learning:

- Black-box problem – even though neural network is pretty clear, it is a huge collection or combination of machine learning algorithms for getting data, pattern recognition, building predictive models, and understanding the results. Selection of these algorithms for each dataset and problem statement varies.

Fig. 20.28 3D visualization of human body (https://www.mechdyne.com/healthcare-amp-medical.aspx)

Table 20.7 List of few research papers in VR for visualization

Authors	Year	Paper name
P. Richard et al. (Richard and Coiffet 1995)	1995	Human perceptual issues in virtual environments: sensory substitution and information redundancy
J. Liu et al. (Liu et al. 2007)	2007	Study and application of medical image visualization technology
Q. Lin et al. (Lin et al. 2013)	2013	Immersive virtual reality for visualization of abdominal CT
A.R. Lilja et al. (article) (Lilja et al. 2018)	2018	Design-led 3D visualization of nanomedicines in virtual reality

- Overfitting – there is a different training model created from the training data and testing on unseen test data. Performance of model varies. There will be lots of error that could create a bad prediction.
- Optimization of hyperparameters – choosing the right combination of hyperparameter to configuring is a critical point to get optimal solution model. These hyperparameters vary from model to model and dataset to dataset.
- High-performance hardware – deep learning surely requires a high-performance hardware to support the huge dataset and various algorithms are internally used. Sometimes even with high-performance system, it will take days.

20.6.6 Conclusion

Deep learning paves way for new avenues for doctors/medical professional to provide accurate, faster, and cheaper diagnosis, treatment, etc. Deep learning has more benefits even with tradeoff or challenges mentioned above. Future of medical images is tending toward deep learning.

References

Aberle D, El-Saden S, Abbona P, Gomez A, Motamedi K, Ragavendra N, Bassett L, Seeger L, Brown M, Brown K, Bui AAT, Kangarloo H (2010) A primer on imaging anatomy and physiology. In: Medical imaging informatics. Springer, New York, pp 17–53

Anbeek P, Vincken KL, van Bochove GS, van Osch MJP, van der Grond J (2005) Probabilistic segmentation of brain tissue in MR imaging. NeuroImage 27:795–804

Anthimopoulos M, Christodoulidis S, Ebner L, Christe A, Mougiakakou S (2016) Lung pattern classification for interstitial lung diseases using a deep convolutional neural network. IEEE Trans Med Imag 35(5):1207–1216

Ashburner J, Friston KJ (2011) Diffeomorphic registration using geodesic shooting and Gauss-Newton optimisation. NeuroImage 55(3):954–967

Bailey DL, Townsend DW, Valk PE, Maisey MN (2005) Positron-emission tomography: basic sciences. Springer, Secaucus, NJ. ISBN 1-85233-798-2

Bajcsy R, Kovačič S (1989) Multiresolution elasticmatching. Comput Vis Graph Image Process 46(1):1–21

Berahim M, Samsudin NA, Nathan SS (2018) A review: image analysis techniques to improve labeling accuracy of medical image classification. Int Conf Soft Comput Data Min 2018:298–307

Besl PJ, McKay ND (Feb. 1992) Amethod for registration of 3-D shapes. IEEE Trans Pattern Anal Mach Intell 14(2):239–256

Bharatha A, Hirose M, Hata N, Warfield SK, Ferrant M, Zou KH, Suarez-Santana E, Ruiz-Alzola J, D'Amico A, Cormack RA, Kikinis R, Jolesz FA, Tempany CMC (2001) Evaluation of three-dimensional finite element-based deformable registration of pre and intraoperative prostate imaging. Med Phys 28(12):2551–2560

Binh NT, Khare A (2010) Adaptive complex wavelet technique for medical image denoising. In Proceedings of third international conference on development of biomedical engineering, 195–198, Vietnam, January 11–14, 2010.

Bookstein FL (Jun. 1989) Principal warps: thin-plate splines and the decomposition of deformations. IEEE Trans Pattern Anal Mach Intell 11(6):567–585

Bookstein FL (1991) Thin-plate splines and the atlas problem for biomedical images. Proc Int Conf Inf Process Med Imag:326–342

Boykov Y, Veksler O, Zabih R (Nov. 2001) Fast approximate energy minimization via graph cuts. IEEE Trans Pattern Anal Mach Intell 23(11):1222–1239

Bradley WG (2008) History of medical imaging. Proc Am Philos Soc 152(3):349–361

Cao X, Miao J, Xiao Y (2017) Medical image segmentation of improved genetic algorithm research based on dictionary learning. World J Eng Technol (5):90–96

Carlson N (2012) "Physiology of behavior", methods and strategies of research, 11th edn. Pearson, London, p 151. ISBN 0205239390

Carroll QB (2014) Radiography in the digital age, 2nd edn. Charles C Thomas, Springfield, p 9. ISBN 9780398080976

Cheung W, Hamarneh G (2009) n-SIFT: N-dimensional scale invariant feature transform. IEEE Trans Imag Process 18(9):2012–2021

Chung K, Scholten ET, Oudkerk M, De Jong PA, Prokop M, Van Ginneken B (2015) Automatic classification of pulmonary peri-fissural nodules in computed tomography using an ensemble of 2D views and a convolutional neural network out-of-the-box. Med Image Anal 26(1):195–202

Clatz O, Sermesant M, Bondiau P-Y, Delingette H, Warfield SK, Malandain G, Ayache N (Oct. 2005) Realistic simulation of the 3-d growth of brain tumors in MR images coupling diffusion with biomechanical deformation. IEEE Trans Med Imag 24(10):1334–1346

Cootes TF, Taylor CJ, Cooper DH, Graham J (1995) Active shape models – their training and application. Comput Vis Imag Understand 61(1):38–59

Coyle SM, Ward TSE, Markham CM (2007) Brain–computer interface using a simplified functional near-infrared spectroscopy system. J Neur Eng 4(3):219–226. https://doi.org/10.1088/1741-2560/4/3/007

CT Scan (CAT Scan, Computerized Tomography) Imaging Procedure (2018) MedicineNet. Retrieved 29 Nov 2018

Declerck J, Feldmar J, Goris ML, Betting F (Dec. 1997) Automatic registration and alignment on a template of cardiac stress and rest reoriented SPECT images. IEEE Trans Med Imag 16(6):727–737

Deserno TM, Antani S, Long R (2009) Ontology of gaps in content-based image retrieval. J Digit Imaging 22(2):202–215

Despotovi I, Goossens B, Philips W (2015) MRI segmentation of the human brain: challenges, methods, and applications. Comput Math Methods Med 2015

Dhawan AP (2008) Image segmentation and feature extraction. In: Principles and advanced methods in medical imaging and image analysis. World Scientific Publishing Co. Pte. Ltd, Singapore, pp 197–228

Dhawan AP, Dai S (2008) Clustering and pattern classification. In: Principles and advanced methods in medical imaging and image analysis. World Scientific Publishing Co. Pte. Ltd, Singapore, pp 229–265

DistanceDoc and MedRecorder: New Approach to Remote Ultrasound Imaging, Solutions, Epiphan Systems Archived 2011-02-14 at the Wayback Machine.. Epiphan.com. Retrieved on 2011-11-13.

Dolovich M, Labiris R (2004) Imaging drug delivery and drug responses in the lung. Proc Am Thorac Soc 1:329–337

Dou Q, Member S, Chen H, Member S, Yu L, Zhao L, Qin J (2016) Automatic detection of cerebral microbleeds from MR images via 3D convolutional neural networks. IEEE Trans Med Imaging 11(4):1–14

Duggal R, Gupta A, Gupta R, Wadhwa M, Ahuja C (2016) Overlapping cell nuclei segmentation in microscopic images using deep belief networks. In: Proceedings of the tenth Indian conference on computer vision, graphics and image processing. ACM, New York, p 82

Erickson BJ, Korfiatis P, Akkus Z, Kline TL (2017) Machine learning for medical imaging. Radiographics 37(2):505–515

Esteva A et al (2017) Dermatologist-level classification of skin cancer with deep neural networks. Nature 542(7639):115–118

Faggella D (2018) Machine learning healthcare applications – 2018 and beyond, article published in techemergence.com in Mar 2018

Fletcher R, Reeves CM (1964) Function minimization by conjugate gradients. Comput J 7(2):149–154

Ford LR, Fulkerson DR (1956) Maximal flow through a network (PDF). Can J Math 8:399–404

Frey BJ, MacKay DJC (1997) A revolution: belief propagation in graphs with cycles. Proc Conf Adv Neural Inf Process Syst:479–485

Ganser KA, Dickhaus H, Metzner R, Wirtz CR (2004) A deformable digital brain atlas system according to Talairach and Tournoux. Med Imag Anal 8:3–22

Ge C et al (2018) 3D multi-scale convolutional networks for Glioma grading using MR images. IEEE Int Conf Imag Process Proc:141–145

Gee JC, Bajcsy R (1999) Elastic matching: continuum mechanical and probabilistic analysis. Brain Warp:183–197

Gefen S, Tretiak O, Nissanov J (Nov. 2003) Elastic 3-D alignment of rat brain histological images. IEEE Trans Med Imag 22(11):1480–1489

Gholipur K, Briggs G (2007) Brain function Localization: a survey of Image Registration techniques. IEEE Trans Med Imag 26(4):427–451

Ghose A, Ghose A, Dasgupta P (2018) New surgical robots on the horizon and the potential role of artificial intelligence. J Invest Clin Urol. https://doi.org/10.4111/icu.2018.59.4.221

Goel N, Yadav A, Singh BM (2016) Medical image processing: a review, IEEE CIPECH, Nov 2016

Goodfellow I, Pouget-Abadie J, Mirza M, Xu B, Warde-Farley D, Ozair S, Courville A, Bengio Y (2014) Generative adversarial networks. arXiv:1406.2661

Gopalakrishnan V, Menon PG, Madan S (2015) cMRI-BED: a novel informatics framework for cardiac MRI biomarker extraction and discovery applied to pediatric cardiomyopathy classification. BioMed Eng 14(suppl 2):1–16
Greig DM, Porteous BT, Seheult AH (1989) Exact maximuma posteriori estimation for binary images. J R Stat Soc Ser B (Methodol) 51(2):271–279
Haber E, Modersitzki J (2007) Image registration with guaranteed displacement regularity. Int J Comput Vis 71(3):361–372
Hadamard J (1923) Lectures on the Cauchy's Problem in Linear Partial Differential Equations. Yale Univ. Press, New Haven, CT
Hager WW, Zhang H (2006) A survey of nonlinear conjugate gradient methods. Pacific J Optimizat 2(1):35–58
Hansen N, Ostermeier A (2001) Completely derandomized self-adaptation in evolution strategies. Evolution Comput 9(2):159–195
Haralick RM, Shapiro LG (1985) Image segmentation techniques. Comput Vis Grap Imag Process 29(1):100–132
Havaei M et al (2017) Brain tumor segmentation with deep neural networks. Med Imag Anal 35:18–31
He J, Christensen GE (2003) Large deformation inverse consistent elastic image registration. Int Conf Inf Process Med Imag:438–449
Hellier P, Barillot C, Mémin É, Pérez P (2001) Hierarchical estimation of a dense deformation field for 3-D robust registration. IEEE Trans Med Imag 20(5):388–402
Hensel JM, Ménard C, Chung PW, Milosevic MF, Kirilova A, Moseley JL, Haider MA, Brock KK (2007) Development of multiorgan finite element-based prostate deformation model enabling registration of endorectal coil magnetic resonance imaging for radiotherapy planning. Int J Radiat Oncol Biol Phys 68(5):1522–1528
Hestenes MR, Stiefel E (1952) Methods of conjugate gradients for solving linear systems. J Res Nat Bureau Stand 49(6):409–436
Hogea C, Biros G, Abraham F, Davatzikos C (2007) A robust framework for soft tissue simulations with application to modeling brain tumor mass effect in 3d MR images. Phys Med Biol (23):6893–6908
Hong J, Vicory J, Schulz J, Styner M, Marron JS, Pizer SM (2016) Non-Euclidean classification of medically imaged objects via s-reps R. Med Image Anal 31:37–45
Iwahori Y, Hattori A, Adachi Y, Bhuyan MK, Robert J (2015) Automatic detection of polyp using hessian filter and HOG features. In: Procedia computer science international conference in knowledge based and intelligent information and engineering systems - KES2015, pp 730–739
James AP, Dasarathy BV (2014) Medical image fusion: a survey of state of the art. Inf Fusion 19:4–19. arXiv:1401.0166
Jian B, Vemuri B (2011) Robust point set registration using Gaussian mixture models. IEEE Trans Pattern Anal Mach Intell 33(8):1633–1645
Jin KH, McCann MT, Froustey E, Unser M (2017) Deep convolutional neural network for inverse problems in imaging. IEEE Trans Image Process 26(9):4509–4522
Kadir T, Brady M (2001) Saliency, scale and image description. Int J Comput Vis 45(2):83–105
Kaur G, Singh B (2011) Intensity based image segmentation using wavelet analysis and clustering techniques. Indian J Comput Sci Eng 2(3)
Kaushik D, Singh U, Singhal P, Singh V (2013) Medical image segmentation using genetic algorithm. Int J Comput Appl 81(18)
Ker J, Wang L, Rao J, Lim T (2017) Deep learning applications in medical image analysis, Special section on soft computing techniques for image analysis in the medical industry current trends, challenges and solutions, IEEE Access, Dec 2017
Kevles BH (1996) Naked to the bone medical imaging in the twentieth century. Rutgers University Press, Camden, NJ, pp 19–22. ISBN 978-0-8135-2358-3
Khalvati F, Wong A, Haider MA (2015) Automated prostate cancer detection via comprehensive multi-parametric magnetic resonance imaging texture feature models. BMC Med Imag:1–14

Kim J, Fessler JA (Nov. 2004) Intensity-based image registration using robust correlation coefficients. IEEE Trans Med Imag 23(11):1430–1444
Klein S, Staring M, Pluim JPW (Dec. 2007) Evaluation of optimization methods for nonrigid medical image registration using mutual information and B-splines. IEEE Trans Image Process 16(12):2879–2890
Komodakis N, Tziritas G (2007) Approximate labeling via graph cuts based on linear programming. IEEE Trans Pattern Anal Mach Intell 29(8):1436–1453
Komodakis N, Tziritas G, Paragios N (2008) Performance vs computational efficiency for optimizing single and dynamic MRFs: setting the state of the art with primal-dual strategies. Comput Vis Imag Understand 112(1):14–29
Kruggel F, Yves von Cramon D (Jun. 1999) Alignment of magnetic-resonance brain datasets with the stereotactical coordinate system. Med Imag Anal 3:175–185
Kybic J, Unser M (Nov. 2003) Fast parametric elastic image registration. IEEE Trans Imag Process 12(11):1427–1442
Lai ZF, Deng HF (2018) Medical image classification based on deep features extracted by deep model and statistic feature fusion with multilayer perceptron. Comp Intel Neurosci 2018
Lashari SA, Ibrahim R (2013) A framework for medical images classification using soft set. Proc Technol 11:548–556
Last Image Hold Feature (2010) Fluoroscopic Radiation Management. Walter L. Robinson & Associates. Retrieved April 3, 2010
LeCun Y (2013) LeNet-5, convolutional neural networks. Retrieved 16 Nov 2013
Leow A, Huang S-C, Geng A, Becker J, Davis S, Toga A, Thompson P (2005) Inverse consistent mapping in 3D deformable image registration: its construction and statistical properties. Int Conf Inf Process Med Imag:493–503
Lilja AR, Strong CW, Bailey BJ, Thurecht KJ, Houston ZH, Fletcher NL, McGhee JB (2018) Design-led 3D visualization of nanomedicines in virtual reality, VRST, Proceeding of the 24th ACM symposium on Virtual Reality Software and Technology Article No. 48
Lin Q, Xu Z, Li B, Baucom R, Poulose B, Landman BA, Bodenheimera RE (2013) Immersive virtual reality for visualization of abdominal CT. Proc SPIE 28:8673. https://doi.org/10.1117/12.2008050
Litjens, Geert, Kooi, Thijs, Bejnordi, Babak Ehteshami, Setio, Arnaud Arindra Adiyoso, Ciompi, Francesco, Ghafoorian, Mohsen, Van Der Laak, Jeroen Awm, Van Ginneken, Clara I. Sánchez. (2017). A survey on deep learning in medical image analysis, Med Image Anal, vol. 42, pp. 60–88
Liu T, Shen D, Davatzikos C (2004) Deformable registration of cortical structures via hybrid volumetric and surface warping. NeuroImage 22(4):1790–1801
Liu J, Ma W, Liu F, Hu Y, Yang J, Xu X (2007) Study and application of medical image visualization technology, ICDHM 2007: Digital Human Modeling, 668–677
Long LR, Antani S, Deserno TM, Thoma GR (2009) Contentbased image retrieval in medicine retrospective assessment, state of the art, and future directions. Int J Health Inform Syst Informat 4(1):1–16
Maes F, Collignon A, Vandermeulen D, Marchal G, Suetens P (1997) Multimodality image registration by maximization of mutual information. IEEE Trans Med Imag 16(2):187–198
Maier-Hein L, Eisenmann M, Reinke A, Onogur S, Stankovic M, Scholz P, Arbel T, Bogunovic H, Bradley AP, Carass A, Feldmann C, Frangi AF, Full PM, van Ginneken B, Hanbury A, Honauer K, Kozubek M, Landman BA, März K, Maier O, MaierHein K, Menze BH, Müller H, Neher PF, Niessen W, Rajpoot N, Sharp GC, Sirinukunwattanal K, Speidel S, Stock C, Stoyanov D, Taha AA, van der Sommen F, Wang C-W, Weber M-A, Zheng G, Jannin P, Kopp-Schneider A (n.d.) Is the winner really the best? A critical analysis of common research practice in biomedical image analysis competitions, https://arxiv.org/pdf/1806.02051.pdf
Masood A, Al-jumaily A (2015) Semi advised SVM with adaptive differential evolution based feature selection for skin cancer diagnosis. J Comput Comm 3:184–190

Mazziotta J (2002) The international consortium for brain mapping: a probabilistic atlas and reference system for the human brain. In: Toga AW, Mazziotta JC (eds) Brain mapping: the methods. Academic, New York, pp 727–755
Mazziotta JC, Toga AW, Evans A, Fox P, Lancaster J (1995) A probabilistic atlas of the human brain: theory and rationale for its development. NeuroImage (2):89–101
Mikolajczyk K, Schmid C (2004) Scale & affine invariant interest point detectors. Int J Comput Vis 60(1):63–86
Mikolajczyk K, Schmid C (Oct. 2005) A performance evaluation of local descriptors. IEEE Trans Pattern Anal Mach Intell 27(10):1615–1630
Mikolajczyk K, Tuytelaars T, Schmid C, Zisserman A, Matas J, Schaffalitzky F, Kadir T, Gool LV (2005) A comparison of affine region detectors. Int J Comput Vis 65(1–2):43–72
Miranda E, Aryuni M, Irwansyah E (2016) A survey of medical image classification techniques. Int Conf Inf Manag Technol (ICIMTech):56–61, 2016
Mittal D, Rani A (2016) Detection and classification of focal liver lesions using support vector machine classifiers. J Biomed Eng Med Imaging 3(1):21–34
Modersitzki J (2008) Flirt with rigidity-image registration with a local nonrigidity penalty. Int J Comput Vis 76(2):153–163
Moré JJ, Thuente DJ (1994) Line search algorithms with guaranteed sufficient decrease. ACM Trans Math Software 20(3):286–307
Morel J-M, Yu G (2009) Asift: a new framework for fully affine invariant image comparison. SIAM J Imag Sci 2(2):438–469
Murphy KP, Weiss Y, Jordan MI (1999) Loopy belief propagation for approximate inference: an empirical study. Proc Conf Uncert Artif Intell:467–475
Nandi D, Ashour AS, Samanta S, Chakraborty S, Salem MAM, Dey N (2015) Principal component analysis in medical image processing: a study. Int J Image Mining 1(1):65–86
Nguyen LD, Lin D, Lin Z, Cao J (2018) Deep CNNs for microscopic image classification by exploiting transfer learning and feature concatenation. In: Circuits and Systems (ISCAS), 2018 IEEE international symposium, pp 1–5
Ni D, Qu Y, Yang X, Chui Y, Wong T-T, Ho S, Heng P (2008) Volumetric ultrasound panorama based on 3d sift. Proc Int Conf Med Image Comput Assist Intervent:52–60
Novelline R (1997) Squire's fundamentals of radiology, 5th edn. Harvard University Press, Cambridge, MA, pp 34–35. ISBN 0-674-83339-2
Nowinski WL, Thirunavuukarasuu A (2001) Atlas-assisted localization analysis of functional images. Med Imag Anal 5:207–220
Passat N, Ronse C, Baruthio J, Armspach J-P, Maillot C, Jahn C (2005) Region-growing segmentation of brain vessels: an atlas-based automatic approach. J Magn Res Imag 21(6):715–725
Pedram SA, Ferguson P, Ma J, Dutson E, Rosen J (2017) Autonomous suturing via surgical robot: an algorithm for optimal selection of needle diameter, shape, and path. In: Proceedings of IEEE international conference on robotics and automation. IEEE, Singapore
Pennec X, Stefanescu R, Arsigny V, Fillard P, Ayache N (2005) Riemannian elasticity: a statistical regularization framework for non-linear registration. In International conference Medical Image Computing and Computer-Assisted Intervention, 943–950
Pereira S, Pinto A, Alves V, Silva CA (2016) Brain tumor segmentation using convolutional neural networks in MRI images. IEEE Trans Med Imag 35(5):1240–1251
Pestscharing S, Schoffmann K (2017) Learning laparoscopic video shot classification for gynecological surgery. Multimed Tools Appl 77:8061–8079. https://doi.org/10.1007/s11042-017-4699-5
Petitjean C, Dacher J-N (2011) A review of segmentation methods in short axis cardiac MR images. Med Imag Anal 15(2):169–184
Pham DL, Xu C, Prince JL (2000) Current methods in medical image segmentation. Ann Rev Biomed Eng 2(1):315–337
Polyak BT (1969) The conjugate gradient method in extremal problems. USSR Computat Math Math Phys 9(4):94–112

Que Q, Tang Z, Wang R, Zeng Z, Wang J, Chua M, Gee TS, Yang X, Veeravalli B (2018) CardioXNet: automated detection for cardiomegaly based on deep learning. IEEE EMBC. https://doi.org/10.1109/EMBC.2018.8512374

Rabbitt RD, Weiss JA, Christensen GE, Miller MI (1995) Mapping of hyperelastic deformable templates using the finite element method. In: Proceedings of SPIE Visual Geometry, pp 252–265

Radiographic Standard Operating Protocols (PDF) (2015) HEFT Radiology Directorate. Heart of England NHS Foundation Trust. Retrieved 27 Jan 2016

Radiology – acute indications (2017) Royal Children's Hospital, Melbourne. Retrieved 23 July 2017

Rao KMM, Rao VDP Medical image processing

Razzak MI, Naz S, Zaib A (n.d.) Deep learning for medical image processing: overview, challenges and future, https://arxiv.org/pdf/1704.06825.pdf

Richard P, Coiffet P (1995) Human perceptual issues in virtual environments: sensory substitution and information redundancy. In: Proceedings of IEEE international workshop on robot and human communication. IEEE, Tokyo

Roell J (2017) Understanding recurrent neural networks: the preferred neural network for time-series data, Article in towards data science, Jun 26, 2017

Roland PE, Zilles K (Nov. 1994) Brain atlases – a new research tool. Trends Neurosci 17:458–467

Roland PE, Geyer S, Amunts K, Schormann T, Schleicher A, Malikovic A, Zilles K (1997) Cytoarchitectural maps of the human brain in standard anatomical space. Hum Brain Mapp 5:222–227

Roth HR et al (2015) Deeporgan: multi-level deep convolutional networks for automated pancreas segmentation. Proc Int Conf Med Imag Comput Assist Intervent 2015:556–564

Rueckert D, Sonoda LI, Hayes C, Hill DLG, Leach MO, Hawkes DJ (1999) Nonrigid registration using free-form deformations: application to breast MR images. IEEE Trans Med Imag 18(8):712–721

Sahoo PK, Soltani S, Wong AKC (1988) A survey of thresholding techniques. Comput Vis Graph Image Proc 41:233–260

Sakamoto M, Nakano H (2016) Cascaded neural networks with selective classifiers and its evaluation using lung x-ray ct images. arXiv preprint arXiv:1611.07136

Sample S (2007-03-27) X-Rays. The electromagnetic spectrum. NASA. Retrieved 3 Dec 2007 https://en.wikipedia.org/wiki/X-ray

Sarikaya D, Corso JJ, Guru KA (2017) Detection and localization of robotic tools in robot-assisted surgery videos using deep neural networks for region proposal and detection. IEEE Trans Med Imag. https://doi.org/10.1109/TMI.2017.2665671

Scholl I, Aach T, Deserno TM, Kuhlen T (2011) Challenges of medical image processing. Comput Sci Res Dev 26:5–13. https://doi.org/10.1007/s00450-010-0146-9

Seetharaman K, Sathiamoorthy S (2016) A unified learning framework for content based medical image retrieval using a statistical model. J King Saud Univ Comput Inf Sci 28(1):110–124

Setio AAA, Ciompi F, Litjens G, Gerke P, Jacobs C, Van Riel SJ, Wille MW, Naqibullah M, Clara IS, Van Ginneken B (2016) Pulmonary nodule detection in CT images: false positive reductionusing multi-view convolutional networks. IEEE Trans Med Imag 35(5):1160–1169

Sharma A (2015) A refinement: better classification of images using LDA in contrast with SURF and SVM for CBIR system. Int J Comput App 117(16)

Sharma N, Aggarwal LM (2010) Automated medical image segmentation techniques. J Med Phys Assoc Med Phys India 35(1):3

Shrimpton PC, Miller HC, Lewis MA, Dunn M (2011) Doses from Computed Tomography (CT) examinations in the UK – 2003 Review Archived 2011-09-22 at the Wayback Machine.

Shvets A, Rakhlin A, Kalinin AA, Iglovikov V (2018) Automatic instrument segmentation in robot-assisted surgery using deep learning. bioRxiv. https://doi.org/10.1101/275867

Sirinukunwattana K, Raza SEA, Tsang Y, Snead DRJ, Cree IA, Rajpoot NM (2016) Locality sensitive deep learning for detection and classification of nuclei in routine colon cancer histology images. IEEE Trans Med Imaging 35(5):1–12

Solodova RF, Galatenko VV, Nakashidze ER, Andreytsev IL, Galatenko AV, Senchik DK et al (2016) Instrumental tactile diagnostics in robot-assisted surgery. Med Devices Evid Res 9:377–382. https://doi.org/10.2147/MDER.S116525

Song Y, Cai W, Huang H, Zhou Y, Wang Y, Feng DD (2015) Locality-constrained subcluster representation ensemble for lung image classification. Med Image Anal 22(1):102–113

Tagare HD, Jaffe CC, Duncan J (1997) Medical image databases: a content-base

Talairach J, Tournoux P (1988) Co-planar stereotaxic atlas of the human brain. Thieme, New York

Thirumaran J, Shylaja S (2014) Medical image processing – an introduction, IJSR, ISSN (Online): 2319-7064

Thompson PM, Woods RP, Mega MS, Toga AW (2000) Mathematical/computational challenges in creating deformable and probabilistic atlases of the brain. Hum Brain Mapp 9:81–92

Toennies KD (2012) Guide to medical image analysis. Springer Adv Patt Recogn. https://doi.org/10.1007/978-1-4471-2751-2

Torresani L, Kolmogorov V, Rother C (2008) Feature correspondence via graph matching: models and global optimization. Proc Eur Conf Comput Vis:596–609

Triggs B (2004) Detecting keypoints with stable position, orientation, and scale under illumination changes. Proc Eur Conf Comput Vis:100–113

Tsui P-H, Yeh CK, Huang C-C (2012) Noise-assisted correlation algorithm for suppressing noise-induced artifacts in ultrasonic Nakagami images. IEEE Trans Infor Technol Biomed 16(3)

Ultrasound Imaging of the Pelvis. radiologyinfo.org. Archived from the original on 2008-06-25. Retrieved 2008-06-21

Upadhyay A, Kashyap R (2016) Fast segmentation methods for medical images. Int J Comput Appl 156(3):18–23

Van Grinsven MJJP, Van Ginneken B, Hoyng CB, Theelen T, Clara IS (2016) Fast convolutional neural network training using selective data sampling: application to hemorrhage detection in color fundus images. IEEE Trans Med Imaging 35(5):1273–1284

Van Tulder G, De Bruijne M (2016) Combining generative and discriminative representation learning for lung CT analysis with convolutional restricted boltzmann machines. IEEE Trans Med Imaging 35(5):1262–1272

Varytimidis C, Rapantzikos K, Loukas C, Kolias S (2016) Surgical video retrieval using deep neural networks. In: Proceedings of workshop and challenges on modeling and monitoring of computer assisted interventions. MICCAI, Athens

Viola P, Wells WM III (1997) Alignment by maximization of mutual information. Int J Comput Vis 24(2):137–154

Wang J, Blackburn TJ (2000) The AAPM/RSNA physics tutorial for residents: X-ray image intensifiers for fluoroscopy. Radiographics 20(5): 1471–1477. doi:https://doi.org/10.1148/radiographics.20.5.g00se181471. ISSN 0271-5333. PMID 10992034.

Wang L, Pedersen PC, Agu E, Strong DM, Tulu B (2017) Area determination of diabetic foot ulcer images using a cascaded two-stage SVM-based classification. IEEE Trans Biomed Eng 64(9):2098–2109

Weizenecker J, Gleich B, Rahmer J, Dahnke H, Borgert J (2009) Three-dimensional real-time in vivo magnetic particle imaging. Phys Med Biol 54(5):L1–L10. https://doi.org/10.1088/0031-9155/54/5/L01

Withey DJ, Koles, ZJ (2007) Medical image segmentation: methods and software, 140–143

Woods RP (2003) Characterizing volume and surface deformations in an atlas framework: theory, applications, and implementation. Neuroimage 18:769–788

Wu Y-T, Kanade T, Li C-C, Cohn J (2000) Image registration using wavelet-based motion model. Int J Comput Vis 38(2):129–152

Xue Z, Shen D, Davatzikos C (Oct. 2004) Determining correspondence in 3-D MR brain images using attribute vectors as morphological signatures of voxels. IEEE Trans Med Imag 23(10):1276–1291

Yamamoto T, Abolhassani N, Jung S, Okamura AM, Judkins T (2012) Augmented reality and haptic interfaces for robot-assisted surgery. Int J Med Robotics Comput Assist Surg 8:45–56. https://doi.org/10.1002/rcs.421

Yang X, Xue Z, Liu X, Xiong D (2011) Topology preservation evaluation of compact-support radial basis functions for image registration. Pattern Recognit Lett 32(8):1162–1177

Yu YE, Bishop M, Zheng B, Ferguson RM, Khandhar AP, Kemp SJ, Krishnan KM, Goodwill PW, Conolly SM (2017) Magnetic particle imaging: a novel in vivo imaging platform for cancer detection. Nano Lett 17(3):1648–1654. https://doi.org/10.1021/acs.nanolett.6b04865

Zagorchev L, Goshtasby A (2006) A comparative study of transformation functions for nonrigid image registration. IEEE Trans Imag Process 15(3):529–538

Zhang W et al (2015) Deep convolutional neural networks for multi-modality isointense infant brain image segmentation. NeuroImage 108:214–224

Zhao Z, Voros S, Weng Y, Chang F, Li R (2017) Tracking-by-detection of surgical instruments in minimally invasive surgery via the convolutional neural network deep learning-based method. Comput Assist Surg 22:26–35. https://doi.org/10.1080/24699322.2017.1378777

Zhu H (2003) Medical image processing overview

Zhu Q, Du B, Wu J, Yan P (2018) A deep learning health data analysis approach: automatic 3D prostate MR segmentation with densely-connected volumetric convnets. IJCNN. https://doi.org/10.1109/IJCNN.2018.8489136

"Overview of X-Ray Tube Technology" 21

M. Anburajan and Jitendar Kumar Sharma

Abstract

X-ray is being used widely in medical and in other fields of science, engineering, and technology since its invention in the year 1895. Globally, the growing number of general population and the high prevalence of various critical diseases are augmenting the need for X-ray imaging equipment for an accurate diagnosis. Although X-ray tube technology (Coolidge tube) is about 106 years old, it has been still used widely today for medical diagnostic imaging. Various types of X-ray tubes are classified. The indicative list for the standards pertaining to performance and safety of medical electrical equipment and diagnostic X-ray tube assembly includes the following: IEC 60601-1-2/IS 13450-1-2, IEC 60601-1-3/IS 13450-1-3, IEC 60601-1-6, IEC 60601-1-8, IEC 60601-1-9, IEC 60601-2-28, IEC 60601-2-54, IEC 60522, IEC 60806, IEC 60336, IEC 61267, IEC 61674, IEC 61676, IEC 60613, IEC 60526, IEC 62304, IEC 62366, ISO 13485, ISO 14000, ISO 14971, and ISO 10993-1. The manufacturing processes of the X-ray tube assembly are listed as follows: (step 1) X-ray tube insert parts processing and cleaning; (step 2) X-ray tube insert parts (anode and cathode) assembly processing; (step 3) glass or metal-ceramic envelope processing; (step 4) degassing of X-ray tube insert; (step 5) seasoning of X-ray tube insert; (step 6) X-ray tube assembly processing; (step 7) final testing of X-ray tube assembly; and (step 8) quality control of X-ray tube assembly. As per the recommendations of Atomic Energy Regulatory Board (AERB), Government of India, the accuracy of the applied X-ray tube potential of the general radiography and fluoroscopy,

M. Anburajan (✉)
Andhra Pradesh MedTech Zone Limited (AMTZ), Visakhapatnam, Andhra Pradesh, India
e-mail: anburajan.m@amtz.in

J. K. Sharma
Andhra Pradesh MedTech Zone Limited (AMTZ), Visakhapatnam, Andhra Pradesh, India

Kalam Institute of Health Technology (KIHT), Visakhapatnam, Andhra Pradesh, India
e-mail: ceo@amtz.in

S. Paul (ed.), *Biomedical Engineering and its Applications in Healthcare*,
https://doi.org/10.1007/978-981-13-3705-5_21

computed tomography, mammography, and dental X-ray machine should be within ±5, ±2, ±1, and ±5, respectively, of the corresponding measured values. Linearity of the X-ray tube current (mA or mAs) loading stations (coefficient of linearity (CoL)) of the general radiography and fluoroscopy, computed tomography, mammography, and dental X-ray machine should be within <0.1, ±0.1, ±0.1, and <0.1, respectively. The most common causes for the X-ray tube failures include the following: (i) cathode filament burnout, (ii) target micro-cracking, (iii) tube arcing, (iv) slow leaks in the tube, and (v) bearings of rotating anode. Globally, more than 20,000 patents have been filed so far on the innovations of X-ray tube technology.

21.1 Introduction

Since the discovery of X-rays by Wilhelm Conrad Röntgen more than 123 years ago, significant developments have taken place in the X-ray tube to meet various requirements – these may be shorter exposure time, multiple repetitive exposures, capacity to accept heavy tube electrical load, enhanced tube life, etc. This has become possible by having multiple focal spots, faster-rotating anodes, and better anode disc materials or maybe by replacing glass envelope with metal alloy, etc., i.e., a lot of physical changes have taken place in the X-ray tube design, whereas principle of production of X-rays remains the same, and problem of high heat production remains still unresolved.

To produce X-rays, the following characteristic features are essential: (i) free electrons, (ii) high voltage to accelerate these electrons to greater speed, and (iii) an anode (target) to stop these high-speed electrons or to change the direction of these electrons at its surface. The basic principle of X-ray production is the kinetic energy of the fast-moving electrons which is converted into useful small amount of X-ray energy (1%), whereas the remaining kinetic energy is converted into the unwanted huge amount of heat energy (99%), due to the majority of electron interactions within the anode.

The aim of this review study was to overview and to discuss briefly on the following topics of interest, relevant to X-ray tube technology, particularly in the applications of medical diagnostic X-ray: (i) different types of X-ray tubes, (ii) regulation and standards of X-ray tubes, (iii) manufacturing processes of X-ray tube assembly, (iv) quality assurance of X-ray machine, (v) common causes of X-ray tube failures, and (vi) research and development, innovations, and patents in X-ray tube technology.

21.2 Different Types of X-Ray Tube

The X-ray tube is classified broadly based on the following: (i) oldest tube technology developed, (ii) type of cathode used, (iii) type of anode used, and (iv) applications (Fig. 21.1).

21.2.1 Oldest Tube Technology Developed (1875–1918)

X-ray tubes are classified based on its technology developed after the discovery of X-rays as follows: (i) Crookes tube, (ii) self-regulator tube, (iii) vacuum tube, (iv) Coolidge tube, and (v) shockproof dental X-ray unit.

i). **Crookes Tube (Gas Discharge or Ion Tube): 1875**

It is the "first"-generation "cold" cathode X-ray tube, invented by William Crookes in 1875. It is a partially evacuated glass bulb of sodium or cerium with a low atmospheric pressure of air ($7 \text{ x } 10^{-4}$ Torr to 4×10^{-5}). The device called "softener," placed at top of the tube, is used to control this gas pressure. It has three electrodes as follows: (i) anode, made up of platinum (atomic number Z = 78, melting point = 1768 °C); (ii) a concavely shaped cathode made up of aluminum (Z = 13, melting point = 660 °C); and (iii) anticathode made up of copper (Z = 29, melting point = 1083.5 °C). The anticathode is placed in line with the anode such that the anode is between the cathode and the anticathode. When high DC voltage (100 kV) is applied across anode and cathode, the gas atoms in the tube are ionized, and positive ions are produced. Then, it strikes the cathode and emits more electrons with greater speed. Afterward, these electrons strike on small focal spot of the anode (about 1 mm), which is inclined at an angle and produces X-rays as a "point" source. The X-ray production is by one of the two processes, either bremsstrahlung or X-ray fluorescence. A heat sink is used to dissipate the heat produced while the electrons strike the anode. This tube was used until the 1920s. The major demerits of the tube are given as follows: (i) relatively low intensity of X-ray produced (about 5 mA); (ii) unreliable and unstable as X-ray production depends on gas content inside the tube; (iii) the intensity and energy of the produced X-ray that cannot be controlled independently; and (iv) overheating of the tube following a series of X-ray exposures.

ii). **Self-Regulator Tube: 1902**

An automatic self-regulating and regenerative tube was designed by Sayen in 1896, and the tube was sold by Queen & Company. The anode and cathode parts of the tube are made up of platinum and aluminum, respectively. A large glass bulb with a high vacuum has a curved small cathode, which is connected to a small pear-shaped glass bulb (known as "regulating" bulb) that contains chemicals of caustic potash and potassium permanganate. The anode is kept in the center of the bulb at an

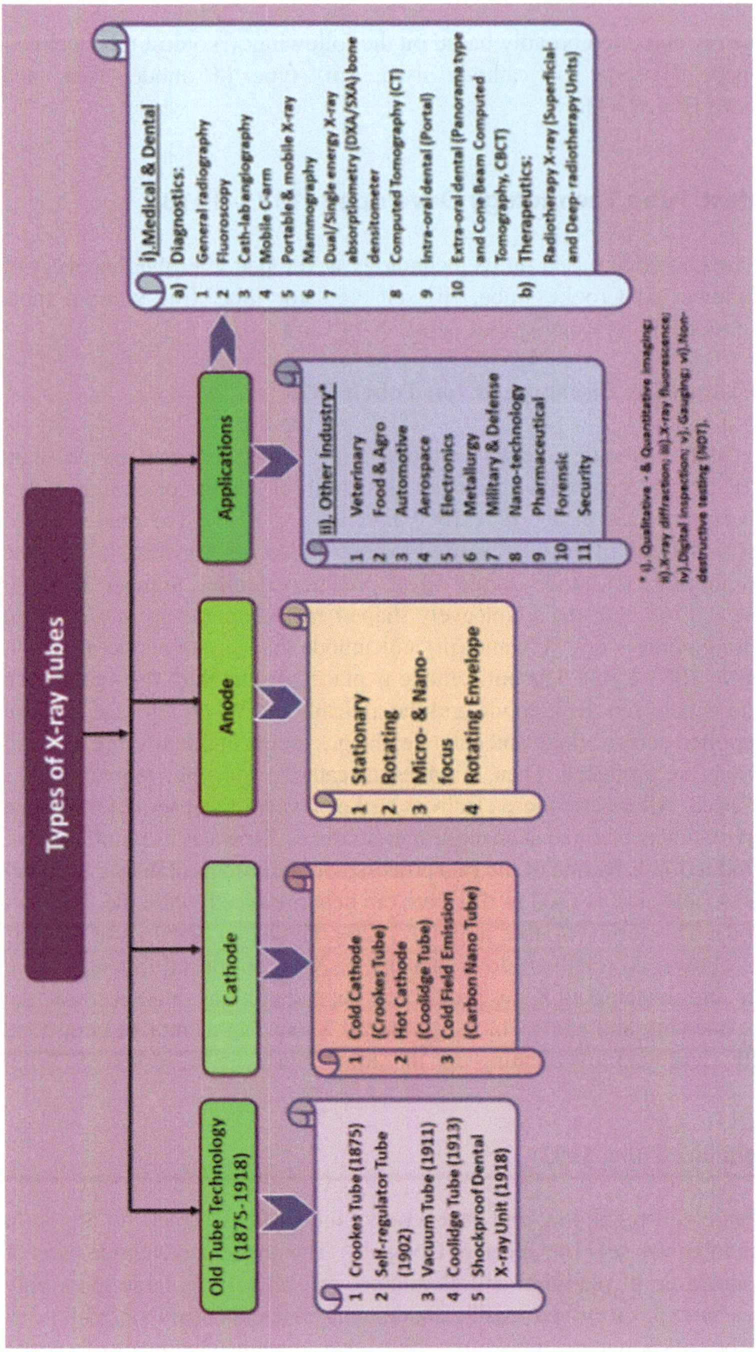

Fig. 21.1 Types of X-ray tube

angle of 45° off the long axis of the tube. It works on the principle that some chemicals (e.g., caustic potash and potassium permanganate) either release or absorb gasses upon heating or cooling, respectively. The cathode is electrically heated up through platinum wires that enter the ends of the glass arms. Due to electrical discharge, smaller bulb will be heated up. The gas pressure in the small bulb is adjusted by controlling the size of the air gap. When the vacuum in the large bulb becomes high, the resistance to the drain of electrons to the platinum target (anode) increases. Thus, it results in low emission of X-rays. On the other hand, when the resistance to the stream of electrons to the anode target is low, the cathode current developed will be high. It results in electrical heating of the caustic potash, which produces gas. Due to this, vacuum in the large bulb is lowered sufficiently to emit X-rays again. On the other hand, when the gas pressure in the main tube becomes high, its electrical resistance will decrease into normal value. Thus, it would help to increase the longevity of the X-ray tube.

iii). **Vacuum Tube: 1911**

Lilienfeld JE developed a vacuum tube in 1911, which works on the principle of field current principle to remove the gas and subsequently to stabilize the operation of X-ray tube. It uses a curved shape cathode. High potential is applied across the tube to extract electrons from the cathode. The process of such "cold" cathode tube is designated as "ticklish." The emitted electrons are lined up in the curved form and they will easily leak out (Lilienfeld Effect). To overcome this effect, a pointed shaped cathode is essential for increasing the drain of electrons from the cathode.

iv). **Coolidge Tube (Electron Tube): 1913**

The Crookes tube was improved by William Coolidge in 1913. It was the "second"-generation "hot" cathode X-ray tube. It is a spherical tempered borosilicate (pyrex) tube with high vacuum (10^{-6} Torr). It has two cylindrically shape arms of cathode and anode. The tungsten ($Z = 74$, melting point $= 3422$°C, evaporating point $= 5555$°C) embedded in copper is used as an anode. The shape of the anode may be either circular, square, or rectangular. The thickness and size of the anode ranged from 2 to 3 mm and 1.8 to 2.2 mm, respectively. The anode angle is ranged from 6–20° (typical angle $= 16.5$°). The tungsten filament is used as a cathode. The focusing cup, made of molybdenum, coated with nickel is used to streamline the electrons emitted from the cathode. Initially, the cathode filament is heated by an electric current. At red-hot condition (heated above 1000 °C), it emits free electrons, known as "thermionic emission." When a high voltage is applied between the cathode and the anode, these electrons are accelerated, and then strike the anode. Thus, it produces bremsstrahlung and characteristic X-rays. The power of the X-ray produced is ranged from 1 to 4 kW. It is the "prototype" of modern X-ray tubes being used today. It has the following technical limitations: (i) thermionic cathode generally has a slow temporal response time and high-power consumption and (ii) high temperature would lead to the failure of the cathode metal filament and thereby to reduce the lifetime of the X-ray tube.

The following are the two designs of the X-ray tube: (i) end-window tube and (ii) side-window tube. In an end-window tube, a thin "transmission-type anode" is used to allow emitted X-rays to pass through it. In this type, common cathode filament is around the "annular" or "ring-shaped anode," and the emitted cathode electrons follow a curved path. In a side-window tube, an electrostatic lens is used to focus the emitted cathode electrons onto a very small spot on the anode. Due to this intensely focused barrage of electrons, the anode requires any one of the following special designs to dissipate the heat and wear produced: (i) "rotating anode" to increase the area heated by the electrons and (ii) circulating coolant to cool the anode. The power of the tube ranges from 0.1 to 18 kW.

v). **Shockproof Dental X-ray Machine: 1918**

Coolidge and General Electric Corporation announced the Victor CDX model shockproof dental X-ray machine in 1918–1919. It removes the exposed high-electrical tension wires. Both the Coolidge hot cathode X-ray tube and high-voltage components are in-house in an oil-filled grounded metal compartment, which acted as an electrical insulator, coolant, and radiation shield. The length of the anode and the X-ray tube are reduced; thus, it facilitates to remove the excess heat generated faster. The advantage of this tube is that both the electrical and fire hazards are eliminated.

21.2.2 Based on the Type of Cathode Used

The X-ray tube is classified based on its type of cathode used as follows: (i) cold cathode tube (Crookes tube), (ii) hot cathode tube (Coolidge tube), (iii) cold field emission tube (carbon nanotube (CNT)).

i) **Cold Cathode Tube (Crookes Tube):** It is discussed earlier in Sect. 21.2.1(i).
ii) **Hot Cathode Tube (Coolidge Tube):** It is discussed earlier in Sect. 21.2.1(iv).
iii) **Cold Field Electron Emitter**

The carbon nanotube (CNT) has a large number of tall and thin sharp tip tubes arranged vertically on a conductive substrate (cathode). As it has both a high aspect ratio and physical and chemical inertness, it is used as the electron emitter for the X-ray tube. A gate mesh structure (anode) is coupled to an electrode inside the tube wall, and the same is positioned at a slight distance above the sharp tips of the CNTs. When an electric field is given at room temperature between the grid mesh and the substrate, it generates a greater number of pulsed electron beams at the tips of the CNTs. These electrons are accelerated toward the anode to produce X-rays. This tube has overcome some of the limitations of the conventional hot filament X-ray tube. Its merits are listed as follows: (i) miniaturized X-ray source, (ii) pulsed and shaped X-ray source, (iii) nanoengineered and (iv) controlled electron beam distribution, (v) low turn-on voltages, (vi) high temporal resolution, and (vii) negligible cathode sputtering.

21.2.3 Based on the Type of Anode Used

The X-ray tube is classified based on its type of anode used as follows: (i) stationary anode X-ray tube, (ii) rotating anode X-ray tube (conventional radiography tube, mammography tube, grid control X-ray tube), (iii) rotating envelope X-ray tube, and (iv) micro- and nano-focus X-ray tube.

i). **Stationary (Fixed) Anode X-Ray Tube**

This evacuated tube has one end fixed copper block and stem of circular/square/rectangular shape. At the stem end, a thin target (tungsten/potassium-doped tungsten (WVM)/tungsten-rhenium alloy) of thickness 2–3 mm and focal spot area of 1.0 × 1.0 mm are embedded with an angle 15–20°, which faced cathode (tungsten filament) just opposite to it. The target is the area of the anode, struck by electrons emitted from the cathode. Due to the small target area, its heat dissipation rate and the tube current are limited. Hence, it produces low-power X-rays. So, it is used to take an X-ray image, which consumes low tube current or low power X-rays. And this kind of tube is used in the following X-ray machines: (i) dental, (ii) portable/mobile, and (iii) portable fluoroscopy.

ii). **Rotating Anode X-Ray Tube**

It is a highly evacuated tube, which has the following components: (i) rotating target; (ii) anode stem, (iii) the rotor of induction coil, (iv) the stator of induction coil, (v) ball bearings, and (vi) safety circuit. It consists of a rotating molybdenum disc, backed with graphite, and the disc is connected to a molybdenum stem. On the disc, thin tungsten-rhenium alloy target is positioned with a bevel angle of 6–20°. The stem is connected to the copper bars with ball bearings of the rotor part of an induction motor. The stator, the other part of an induction motor is a series of electromagnets, which surround the rotor outside the X-ray tube envelope. When an alternating current pass through the stator windings, it induces an electrical current in the rotor copper bars. Due to this, a rotating magnetic field will be developed that will push the rotor and will cause the anode disc to rotate with a speed. Based on the electric power of single-phase (60 Hz) and three-phase (180 Hz) alternating current supplied to the induction motor, the anode rotates with a speed (revolutions per minute (RPM)) ranged from 3000 to 3600 and 9000 to 10000, respectively. The huge ball bearing contact surface and metal lubricant offer an operative method for transmission of heat, produced along with X-ray from the anode. This type of tube is used most commonly for almost all diagnostic X-ray applications, mainly because of their greater amount of heat loading and consequent higher X-ray output capabilities.

iii). **Micro- and Nano-focus X-Ray Tube**

An X-ray tube, which can generate very small focal spot size of diameter lesser than 50 μm, then the tube is called a microfocus X-ray tube. It is classified as follows: (i) solid metal anode tube and (ii) liquid metal jet alloy anode tube.

a) *Solid metal anode microfocus X-ray tube*: It works on the principle of Coolidge tube, and it has a solid metal anode. It operates at very low power of the order of 0.4–0.8 W/μm depending on the type of anode material used. It produces fine focus spot generally in the range 5–20 μm, but it can produce fine focus spot even smaller than 1 μm. The demerit of this tube is its low operating power. To overcome anode melting, the electron beam power density must be below a maximum value.
b) *Liquid metal jet alloy anode microfocus X-ray tube*: The anode is made up of a liquid metal jet alloy of lithium (95%) and bismuth or lanthanum (5%). The lithium is used to produce X-rays, whereas the bismuth/lanthanum is used as a coolant. The emitted electron beam bombards the target of the anode with a smaller focal spot 5 μm, which is about 400 times smaller than in conventional X-ray tube. It produces X-rays of high-power density ranged from 3 to 6 W/μm. These high-power X-rays are used to obtain a high-resolution image and to acquire the image faster. In a nano-focus X-ray tube, the spot size is very small in the order of 150 nm and it operates at very low power.

iv). **Rotating Envelope Tube (RET)**

It is one of the most advanced technologies used in computed tomography (CT). Here, the entire tube is rotated (maximum 150 Hz) with respect to its anode axis. It consists of the following four sub-systems: (i) tube envelope, (ii) electron emission cathode, (iii) magnetic deflection, and (iv) cooling. Nonmagnetic stainless steel is used as tube envelope, and it is attached to anode disc directly. It has an annular/circular window of thickness 0.2 mm. The anode disc is made of tungsten (90%) and rhenium with tungsten-zirconium-molybdenum (TZM) body alloy (10%) (boiling point = 4612 °C and melting point = 2600 °C). The electron emission cathode system consists of a tungsten filament (thickness = 100 μm and diameter = 5 μm) with a circular flat focusing cup. Magnetic deflection system consists of three coils: (i) R-coil, to deflect the electron beam radial direction onto a focal spot of the anode; (ii) Q-coil, to focus the electron beam to determine the size; and (iii) Phi-coil, to deflect flying focal spot of the anode in the tangential direction. In the cooling system (convective), the anode disc comes in direct contact with cooling oil. Turbine flow is used for mineral oil rotation, in which its flow rate (liter min^{-1}) during exposure and pump are 25 and 8, respectively. The merits of RET are listed as follows: (i) multiple focal spot sizes, (ii) better heat dissipation, (iii) high temporal resolution, (iv) useful in high kV and high mA technique for the prolonged duration, and (v) longer tube life.

21.2.4 Based on Applications

Although the primary area of application of X-rays remains in medical diagnosis and therapy, the X-ray source (produced using a customized X-ray tube) of an optimum power is used in almost all areas of science, engineering, and technology. The major applications of X-rays are given as follows:

i). Qualitative and quantitative imaging
ii). X-ray diffraction
iii). X-ray fluorescence
iv). Digital inspection
v). Gauging
vi). Nondestructive testing (NDT)

The industries which use a dedicated X-ray machine for the abovementioned applications are listed as follows: (i) medical and dental, (ii) veterinary, (iii) food and agro, (iv) automotive, (v) aerospace, (vi) electronics, (vii) metallurgy, (viii) military and defense, (ix) nanotechnology, (x) pharmaceutical, (xi) security, and (xii) forensic.

In the medical field, the X-ray machine (and its tube) is classified based on its clinical applications as follows:

1. General radiography
2. Fluoroscopy
3. Cath-lab angiography
4. Mobile C-arm
5. Portable and mobile X-ray
6. Mammography
7. Dual/single energy X-ray absorptiometry bone densitometer (DXA/SXA)
8. CT
9. Intraoral dental (portal)
10. Extraoral dental (panorama type and cone beam computed tomography (CBCT))
11. Radiotherapy X-ray tube (superficial and deep radiotherapy types)

21.3 Regulations and Standards

Based on the severity of risk associated with it, the US Food and Drug Administration (FDA) classified the diagnostic X-ray tube housing assembly as "Class-I device" (lowest risk), whereas it classified the therapeutic X-ray tube housing assembly as "Class-II device" (medium risk). Also, both the diagnostic and therapeutic X-ray equipment mentioned earlier are classified as "Class-II devices" (medium risk).

The European Union (EU) Directives on medical devices classified both the diagnostic and therapeutic X-ray equipment mentioned earlier as "Class-IIB devices" (elevated risk). On the other hand, the Union Ministry of Health & Family Welfare, Government of India, classified both the diagnostic and therapeutic X-ray equipment as "Class-C devices" (moderate-high risk) as per the Medical Devices Rules 2016 (under the Drugs and Cosmetics Act).

a) **Diagnostic X-Ray Tube**

The International Electrotechnical Commission (IEC) developed the international standards for the X-ray tube and its equipment. Also, the Ministry of Consumer Affairs, Food & Public Distribution, Government of India developed adopted Bureau of Indian Standard (IS) for the same. The standards for the diagnostic X-ray tube given below are indicative only and are not exhaustive (Table 21.1):

1). General Standard:

- IEC 60601-1/IS 13450-1: Basic safety and essential performance

a). Collateral Standard:

1. IEC 60601-1-2/IS 13450-1-3: Electromagnetic compatibility
2. IEC 60601-1-3/IS 13450-1-3: Radiation protection in diagnostic X-ray equipment
3. IEC 60601-1-6: Usability
4. IEC 60601-1-8: Alarm systems
5. IEC 60601-1-9: Environmentally conscious design

2). Particular Standard:

1. IEC 60601-2-28: X-ray tube assemblies for medical diagnosis
2. IEC 60601-2-54: Radiography and radioscopy

3). Normative Reference Standard:

i. IEC 60522: Determination of the permanent filtration of X-ray tube assemblies
ii. IEC 60806: Determination of the maximum symmetrical radiation field from a rotating anode X-ray tube for medical diagnosis
iii. IEC 60336: Characteristics of focal spots
iv. IEC 61267: Radiation conditions for use in the determination of characteristics
v. IEC 61674: Dosimeters with ionization chambers and/or semiconductor detectors as used in X-ray diagnostic imaging

Table 21.1 Regulation and standards of medical X-ray tube assembly

Regulation & standard (IEC/IS)	X-ray tube housing assembly			
	(I). Diagnostics		(II). Therapeutics	
(i). Regulation				
FDA	Class-I device (lowest risk)		Class-II device (medium risk)	
(ii). Standard (IEC/IS)				
(i). General	IEC 60601-1/ IS 13450-1	Basic safety and essential performance	IEC 60601-1/IS 13450-1	Basic safety and essential performance
(i.a). Collateral	(i). IEC 60601-1-2/ IS 13450-1-2	Electromagnetic compatibility		
	(ii). IEC 60601-1-3/ IS 13450-1-3	Radiation protection	IEC 60601-1-4	Programmable electrical medical systems
	(iii). IEC 60601-1-6	Usability		
	(iv). IEC 60601-1-8	Alarm systems		
	(v). IEC 60601-1-9	Environmentally conscious design		
(ii). Particular	(i). IEC 60601-2-28	X-ray tube assemblies for medical diagnosis	IEC 60601-2-8	Basic safety and essential performance
	(ii). IEC 60601-2-54	Radiography and radioscopy		
(iii). Normative	(i). IEC 60522	Permanent tube filtration	(i). IEC 62083	Safety of radiotherapy treatment planning systems
	(ii). IEC 60806	Maximum symmetrical radiation field from a rotating anode X-ray tube	(ii). IEC 62274	Safety of radiotherapy record and verify systems
	(iii). IEC 60336	Characteristics of focal spots	(iii). IEC 61674	Dosimeters with ionization chambers and/or semiconductor detectors as used

(continued)

Table 21.1 (continued)

Regulation & standard (IEC/IS)	X-ray tube housing assembly			
	(I). Diagnostics		(II). Therapeutics	
	(iv). IEC 61267	Radiation conditions for use in the determination of characteristics	(iv). IEC 60731	Dosimeters with ionization chambers as used
	(v). IEC 61674	Dosimeters with ionization chambers and/or semi-conductor detectors as used		
	(vi). IEC 61676	Dosimetric instruments used for non-invasive measurement of X-ray tube voltage		
	(vii). IEC 60613	Electrical and loading characteristics		
	(viii). IEC 60526	High-voltage cable plug and socket connections		
	(ix). IEC 62304	Software life cycle processes		
	(x). IEC 62366	Usability engineering		
(iv). Others: Generic management system standard	(i). ISO 13485	Managing quality systems		
	(ii). ISO 14000	An environmental management systems and risk analysis		
	(iii). ISO 14971	Risk management		
	(iv). ISO 10993-1	Biological risk management process		

vi. IEC 61676: Dosimetric instruments used for noninvasive measurement of X-ray tube voltage in diagnostic radiology
vii. IEC 60613: Electrical and loading characteristics of X-ray tube assemblies for medical diagnosis
viii. IEC 60526: High-voltage cable plug and socket connections for medical X-ray equipment
ix. IEC 62304: Software life cycle processes
x. IEC 62366: Usability engineering

4). Others: Generic Management System Standard

i. ISO 13485: Quality management systems
ii. ISO 14000: Environmental management systems and risk analysis
iii. ISO 14971: Risk management
iv. ISO 10993-1: Biological risk management process

b) **Therapeutic X-Ray**

For the therapeutic X-ray, the general standard is IEC 60601-1/IS 13450-1: Medical electrical equipment (Part 1) – General requirements for basic safety and essential performance. The IEC 60601-2-8 is the particular standard for the safety of the equipment. The collateral standard is IEC 60601-1-4: Programmable electrical medical systems. The normative reference standards are listed as follows: (1) IEC 62083, Safety of radiotherapy treatment planning systems; (2) IEC 62274, Safety of radiotherapy record and verify systems; (3) IEC 61674, Dosimeters with ionization chambers and/or semiconductor detectors as used; and (4) IEC 60731, Dosimeters with ionization chambers as used.

The summary of the regulation and the IEC standards and the adopted Indian Standards (IS) for the safety and essential performance of both the diagnostic and therapeutic X-ray equipment given in the Table 21.2 are indicative only and are not exhaustive.

21.4 Manufacturing Processes of X-Ray Tube Assembly

The components of X-ray tube assembly are classified majorly as follows: (1) an inner structure, X-ray tube insert assembly, and (2) an outer structure, X-ray tube housing assembly. The X-ray tube insert assembly has major parts, such as (1) anode structure, (2) cathode structure, and (3) envelope, glass or metal-ceramic type of material, whereas the X-ray tube housing assembly has parts, like (1) lead shielding, (2) two high-tension voltage cable sockets, (3) cathode filament circuit, (4) electrical insulator (oil), (5) stator electromagnetic coils, (6) insert cooler, (7) beam pre-collimation, (8) X-ray radiation port and window, (9) radiation control timer circuit, (10) bellow, and (11) heat exchanger parts such as (a) natural or forced

Table 21.2 Regulation and standards of medical X-ray equipment

Regulation and Standard (IEC/IS)	X-ray equipment			
	(I). Diagnostics		(II). Therapeutics	
(i). Regulation				
(a). FDA	Class-II devices (medium risk)		Class-II device (medium risk)	
(b). EU	Class-II B devices (elevated risk)		Class-II B devices (elevated risk)	
(c). Indian	Class-C devices (moderate-high risk)		Class-C devices (moderate-high risk)	
(ii). Standard (IEC/IS)				
Particular	IEC 60601-2-28	X-ray tube assemblies for medical diagnosis	IEC 60601-2-8	Therapeutic X-ray equipment (X-ray energy: 10 kV to 1 MV)
	IEC 60601-2-54/ IS 13450-2-54	Radiography and radioscopy	IEC 60601-2-68	X-ray-based image-guided radiotherapy
	IEC 60601-2-43/ IS 13450-2-43	Interventional X-ray	IEC 60601-2-29	Radiotherapy simulators
	IEC 60601-2-44/ IS 13450-2-44	Computed tomography (CT)		
	IEC 60601-2-45/ IS 13450-2-45	Mammography		
	IEC 60601-2-63/ IS 13450-2-63	Dental extra-oral X-ray		
	IEC 60601-2-65/ IS 13450-2-65	Dental intra-oral X-ray		
	IEC 60601-2-32/ IS 13450-2-32	Associated equipment of X-ray equipment		
	IEC 60601-2-7/ IS 13450-2-7	Diagnostic X-ray generator		

convection, (b) air blower or water jacket, (c) oil/water plates heat exchanger, and (d) oil/air heat exchanger attached or remote.

The major raw materials (metals and glass) required for manufacturing X-ray tube are listed as follows: (1) nickel, (2) copper, (3) tungsten, (4) aluminum, (5) lead, (6) glass, (7) stainless steel (SS), (8) mild steel (MS), and (9) graphite.

The manufacturing processing of the X-ray tube assembly involves the following steps (Fig. 21.2):

Step 1: X-ray tube insert parts processing and cleaning
Step 2: X-ray tube insert parts (anode and cathode) assembly processing
Step 3: Glass or metal-ceramic envelope processing
Step 4: Degassing of X-ray tube insert
Step 5: Seasoning of X-ray tube insert
Step 6: X-ray tube assembly processing
Step 7: Final testing of X-ray tube assembly
Step 8: Quality control of X-ray tube assembly

i). X-ray Tube Insert Parts Processing and Cleaning

X-ray tube requires a high vacuum to be maintained inside the tube; hence X-ray tube insert parts are to be washed deeply with pressure and cleaned perfectly to remove the dust, rust, oil, particulate, product residual, carbonaceous surface contamination, etc. if any present; after that, these parts should be dried appropriately to make them water-free. For this purpose, an automatic heavy-duty washing system is employed which uses cleaning agent like detergent and cleaning processes such as ultrasound vibration, isopropyl alcohol (IPA) washing, acid etching, vacuum degassing, and nitrogen drying.

ii). X-ray Tube Insert Parts Assembly Processing

In the clean room, the cleaned anode copper disc with a tungsten target is connected with rotor stem. In the rotor, surface-treated specific designed bearing balls are used, in which nonvolatile lubricant is applied. Further, the cleaned arm of the cathode assembly is connected with an electrostatic focusing cup. It is assembled with a cleaned nickel tube. It has both small- and large-sized cathode tungsten filaments and is spot-welded in its position with micro-precision. After assembling, both the anode and cathode structures are sealed off.

iii). Glass or Metal-Ceramic Envelope Processing

The anode and cathode structures are aligned at the correct distance between them. These are then assembled and positioned within the cleaned envelope (glass or

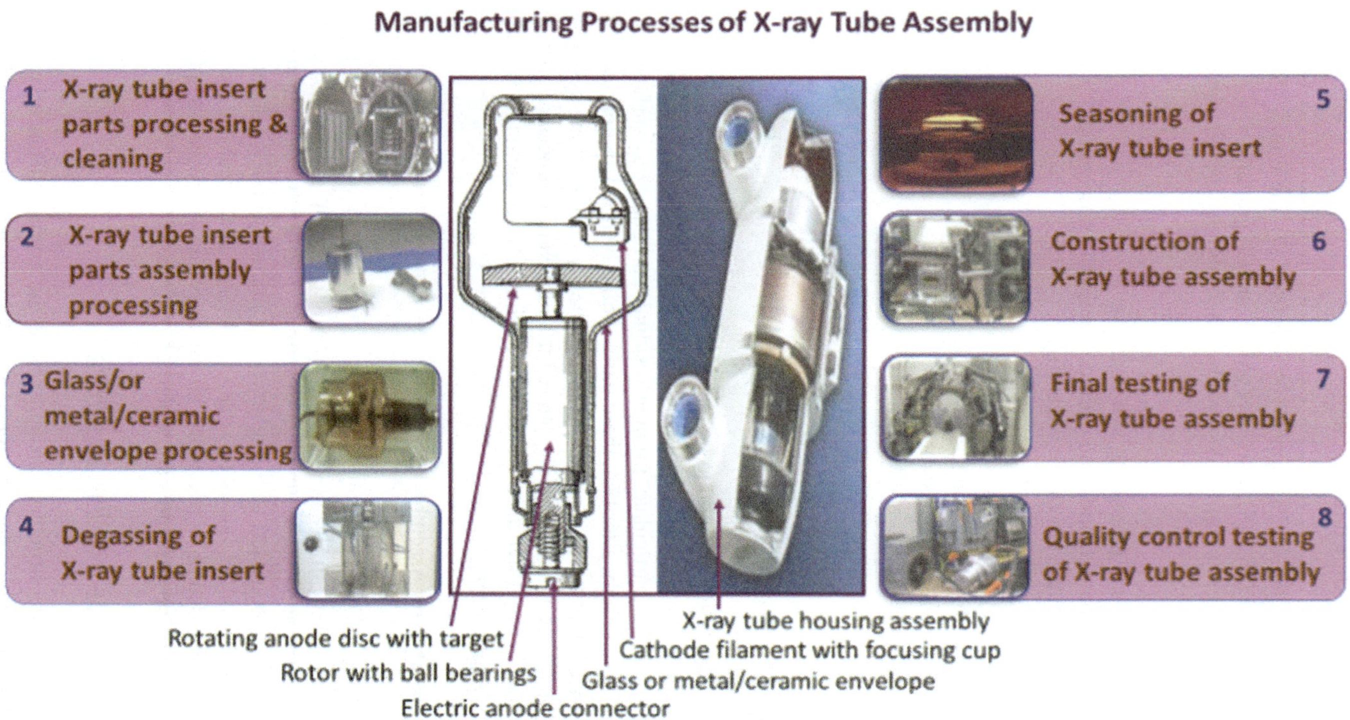

Fig. 21.2 Block diagram of manufacturing processes of X-ray tube assembly

metal-ceramic type). A strain in the glass or metal-ceramic envelope is removed appropriately.

iv). Degassing of X-Ray Tube Insert

In the manufactured and assembled X-ray tube insert, all gasses are evacuated out using a high vacuum turbo-molecular pump. Further, the tube insert is subjected to a higher temperature than its normal operating condition to remove all gasses (cleaning) and baked. The envelope is then sealed to maintain a vacuum inside the X-ray tube. The getter material (evaporable or non-evaporable), placed inside the X-ray tube insert, is generally activated either by activating (flashing) or actuating (raising its temperature).

v). Seasoning of X-Ray Tube Insert

Further, the manufactured and assembled X-ray tube insert is tested at higher voltages using dedicated equipment to assure its optimal performance with high accuracy and reproducibility. In the tube insert, the tube current and voltage are carefully raised to reduce any residual gas in the tube insert before the tube is operated at full output. Seasoning also minimizes the uneven distribution of potential/electric field on the tube glass. Following the recommended seasoning, schedule will help prolong the life of the X-ray tube and prevent tube arcing that can potentially cause irreversible damage to the X-ray tube.

vi). X-ray Tube Assembly

X-ray tube insert is loaded into the X-ray tube housing assembly. The following subassemblies and other major accessories are assembled between them with high-quality and safety standards:

a) Anode end horn assembly
b) Cathode end horn assembly
c) Centre frame assembly
d) Integration of anode end horn assembly, cathode end horn assembly, and center frame assembly
e) Stator assembly
f) High-tension cable subassembly
g) Snubber ring assembly
h) Window assembly
i) Envelope miscellaneous assembly
j) Coolant oil filling

vii). Final Testing of X-Ray Tube Assembly

The manufactured X-ray tube assembly is inspected visually and checked to verify its performances. It is subjected to various testing cycles of both normal and heavy load working conditions and checked whether all the obtained test results are within the limit. The following are the major testing and measurements: focal spot, noise, and vacuum level.

viii). Quality Control of X-Ray Tube Assembly

Further, the following quality measurements are carried out in the manufactured X-ray tube assembly to ensure its optimum performance: (1) bench test, to study voltage and current characteristics of the X-ray tube assembly; (2) leakage X-ray radiation, to certify that there is no leakage X-ray radiation from the tube assembly; (3) noise level testing, to make sure that noise level in both tube inserts and housing assemblies is well below the acceptance levels; (4) vibration level, to check and verify that vibration measurement of tube housing in full rotator anode speed condition is within the limit; and (5) focal spot size and centering, to ensure that the focal spot of the anode structure lies in the correct position and focalization.

21.5 Quality Assurance in Diagnostic X-Ray

The Atomic Energy Regulatory Board (AERB), Government of India, has recommended an accepted tolerance values for the various technical variables measured from different diagnostic X-ray machines, and the same are summarized in the Table 21.3. As per the recommendations, the accuracy of the applied X-ray tube potential of the general radiography and fluoroscopy, computed tomography, mammography, and dental X-ray machine should be within ± 5, ± 2, ± 1, and ± 5, respectively, of the corresponding measured values. Linearity of the X-ray tube current (mA or mAs) loading stations (coefficient of linearity (CoL)) of the general radiography and fluoroscopy, computed tomography, mammography, and dental X-ray machine should be within <0.1, ± 0.1, ± 0.1, and <0.1, respectively.

21.6 Common Causes of X-Ray Tube Failures

The common causes of X-ray tube failures are broadly classified as follows: (a) normal aging and (b) deficiency in manufacturing. The normal aging of the X-ray tube includes the following causes: (1) cathode filament burnouts, (2) target micro-cracking, (3) tube arcing, (4) slow leaks in the tube, (5) glass crazing, (6) bearings of rotating anode, (7) inactivity, and (8) accidental damage. On the other hand, the deficiencies in manufacturing X-ray tube are listed as follows:

Table 21.3 QA of diagnostic X-ray equipment

QA of diagnostic X-ray equipment						
			Accepted tolerance value			
S. No.	Testing X-ray tube parameters		Dental	General radiography and fluoroscopy	Computed tomography (CT)	Mammography
1	Congruence of X-ray and optical fields		NA	(1) Shift in the edges of radiation field within 2% of TFD; (2) differences in dimensions of radiation and optical field within 3% of TFD; (3) differences of sum of lengths and width of radiation and optical field within 4% of TFD	NA	NA
2	Central beam alignment (°)		NA	<1.5	NA	NA
3	Accuracy of applied X-ray tube potential (kV)		±5	±5	±2	±1
4	Accuracy of timer (sec)		<10%	NA	<10%	<10%
5	Linearity of timer (sec) loading stations (coefficient of linearity (CoL))		<0.1	<0.1	NA	NA
6	Linearity of tube current (mA or mAs) loading stations (coefficient of linearity (CoL))		<0.1	<0.1	±0.1	±0.1
7	Minimum total filtration (mm Al):	(1) For tube voltage <70 kVp	1.5	1.5	1.5	(1) First HVL at 30 kVp: $\geq$0.3
		(2) For tube voltage 70 kV< kVp $\leq$100	2.0	2.0	2.0	(2) First HVL at 40 kVp: $\geq$0.4

(continued)

Table 21.3 (continued)

QA of diagnostic X-ray equipment

S. No.	Testing X-ray tube parameters		Accepted tolerance value: Dental	General radiography and fluoroscopy	Computed tomography (CT)	Mammography
		(3) For tube voltage (kVp) ≥ 100	2.5	2.5	2.5	(3) First HVL at 50 kVp: ≥ 0.5
8	Reproducibility of radiation output (coefficient of variation (CoV))		≤0.05	≤0.05	≤0.05	≤0.05
9	Leakage radiation through tube housing (mGy $hour^{-1}$) at maximum rated tube potential (kVp) and current at that kVp		(1) For intraoral dental: <0.25	<1	<1	<0.02
			(2) For extraoral dental <1.0			
			(3) For CBCT, CT dose index (CTDI) = ±20% of the quoted value			
10	For fluoroscopy only: radiation exposure rate at tabletop (cGy min^{-1}):	(1) With automatic exposure control (AEC) mode	NA	≤10	NA	NA
		(2) Without automatic exposure control (AEC) mode		≤ 5		
11	For CT only: image slice thickness (mm)	(1) <1 mm	NA	NA	0.5	NA

		(2) 1–2 mm			±50%	
		(3) >2 mm			±1.0 mm	
12	Image resolution	(1) Low contrast	3.0 mm hole pattern should be visible		5.0 mm at 1% contrast	
		(2) Spatial	1.5 lp/mm should be visible		0.5 lp/cm at 10% contrast	

NA: Not applicable
Source: AERB, Govt. of India

(a) Immediate failures: (1) weed out by test, (2) hold period, (3) improper materials, and (4) process failures
(b) Latent failures: (1) process optimization, (2) marginal/poorly understood processes, and (3) failure analysis/untraceable causes

i). Cathode Filament Burnout: The length to diameter ratio of the cathode tungsten helix filament is in the range of 3–6. At high filament current heating (high temperature), the tungsten evaporates from its surface in a non-uniform way. The filament region with higher evaporation rate forms a "hot spot (visible as a notch)" at its crystal grain boundary. Due to this, the filament becomes thin in this hot spot region and ultimately open its burning. If the diameter of the filament decreases by 5–6% (about 10% reduction in filament mass), then it is considered as its end of life by many manufacturers.
ii). Target Micro-cracking: When the electron beam strikes the tungsten target, its temperature rises fast. It advances minute cracking at the surface and will grow up over a period of time. Due to this, when the incident electron beam falls into these cracks, it emits harder X-rays with reduced intensity. Further, it reduces heat transfer which increases the temperature of the focal spot of the target. Thus, it increases tungsten evaporation onto the glass envelope. For rotating anode, the occurrence of micro-cracking would be severe, and its side effects mentioned earlier are therefore greater.
iii). X-ray Tube Arcing (Internal Electrical Discharge): It is one of the most common causes of X-ray tube failure. If there is an internal ionization in the X-ray tube, it will reduce its vacuum level; hence an internal discharge (arcing spark) will occur between the anode and cathode of the X-ray tube or anyone electrode (usually cathode) and the tube envelope. Due to this, there is a temporary loss of X-ray output, and this will lead to a localized serious artifact (due to the repetition of X-ray exposures and increased patient radiation dose). More importantly, the arc will develop high current, which passes through the tube electrodes, and further through the rectifier and the high-tension transformer of the X-ray generator. If the tube arcing is very fast and strong, the safety circuit of the X-ray machine would fail; subsequently, it would damage both the X-ray tube and more significantly the X-ray generator. It is a phenomenon usually associated with tube aging, but it also presents in a new or unused X-ray tube. It cannot be observed directly, but an X-ray imaging technologist may notice a "hissing" noise (or pops) from the X-ray tube during X-ray exposure. Even a routine quality control (QC) test fails to detect this problem as it is with random manifestation.
iv). Slow Leaks in X-ray Tube: If there is a small leak in the glass-to-metal seal and metallic brazed joints of the X-ray tube, an external gas will enter into the tube and will decrease its vacuum state. Hence, anode and cathode materials will evaporate, and high voltage arc-over will occur.
v). Glass Crazing or Etching: Based upon tube factors and time used, tungsten of anode and cathode filament evaporates and deposits onto the glass

(an insulator) surface or metal surface causing an electric arc-over and tube failure.

vi). Ball Bearings of Rotating Anode: High temperature and high speed will reduce mostly the bearing life of the rotating anode. With the operation, the lubricant (which is usually silver or lead metal) wears off of the ball and race surfaces leaving steel-to-steel contact which leads to binding or jamming. With conservative use bearings usually outlast other failure mechanisms.

vii). Inactivity: If the X-ray tube is not operated for some time, gasses may be built up within the tube vacuum and will migrate along surfaces. When the cathode filament is energized, and the higher tube voltage is applied, an electric arc-over will occur. It is recommended that a warm-up procedure is to be followed depending on the inactive time period.

viii). Accidental Damage: It is caused by not following the recommended protocols during installation and operation of the X-ray tube. It will happen due to the following: (a) immediate failure and (b) latent or unpredictable failures. The failure analysis method is used by the manufacturers to find both the abovementioned types of X-ray tube failures. Sometimes, the cause for the failure is known, and other times additional testing and analysis are required to find out a root cause.

a) Immediate failure: In spite of standard procedures followed during the manufacturing process, no X-ray tube is perfectly identical. But even that minor changes in its design should not alter the functions of the tube.

i). Weed out by test: Once the tube is manufactured, it is operated to higher voltages typically in excess of 15% than its normal value. Such processing removes gasses and particles present if any in the tube. The tube is then exposed to various tests for checking its performance. A tube, failed in the performance test, is rejected/scrapped, but it is examined further to find the root cause for the failure, so improvements can be incorporated in the tube manufacturing process.

ii). Hold period: Occasionally a tube, passed in the performance test, may fail miserably in high tube operating voltage condition if the tube is held for 2–4 weeks. The cause may be a tiny leak formed in the tube, which will allow gas to enter and will reduce its vacuum level.

iii). Poor-quality materials: The raw materials used (tungsten alloy, copper, nickel, graphite, glass, ceramic, lead, and silver) in manufacturing X-ray tube assembly should ensure a high level of material quality for its best performance. The material, which is not up to standard, may result in failure in the tube.

iv). Process failures: A failure in the processes/equipment employed may result in marginal or reject X-ray tube.

b) Latent or unpredictable failures: It may occur in any time without a known cause and is often unanticipated.

i). Process optimization: Although processes used in manufacturing X-ray tube are optimized over many years and through hands-on experiences, sometime tube may not perform well due to unknown consequence.

ii). Marginal or poorly understood processes: Some tube failures are caused by effects that are not well known or for which side effects of various processes are not known.

21.7 Research and Developments, Innovations, and Patents in X-Ray Tube Technology

Today, X-ray tube has the following major limitations, like (1) limited life of X-ray tube, (2) high voltage instability, and (3) failures of X-ray tube processing. The challenges in designing of next-generation X-ray tube are listed as follows: (1) cathode filament assembly needs micron precision, (2) high vacuum and high voltage conditioning processes are very critical for performance, and (3) manufacturing tube design should ensure consistency in the processes. The research and development (R&D) processing and innovations involved in the X-ray tube technology are listed as follows, but are not limited (1) to carry out performance improvement in the X-ray tube; (2) to explore and upgrade manufacturing process improvement in the X-ray tube; (3) to incorporate next-generation smart product/material in the X-ray tube; and (4) to investigate the new design of the X-ray tube and its frame assembly, which would accommodate efficiently all its complex parts and components for its optimum performance and its maximum safety (Fig. 21.3).

Global-based published patent search on "X-ray AND tube" in the title and the year 1896 till January 21, 2019, in the publication period was entered in the European Patent Office website (Espacenet). The obtained results are tabulated (Table 21.4) according to a decade of publications, from 1895 to January 2019.

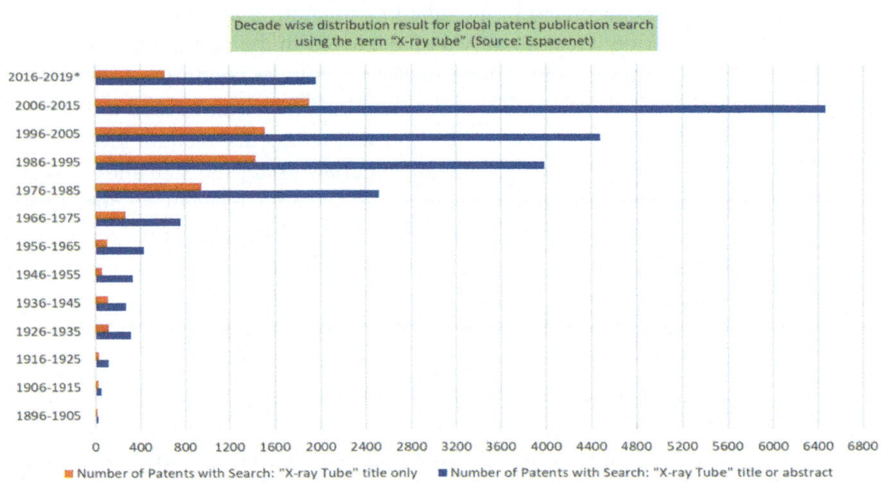

Fig. 21.3 Global patent publication search on X-ray tube

Table 21.4 Magnitude and rate of increase in number of globally published patents on X-ray tube from the year 1896 to 2019∗ (Search source: Espacenet)

	Total number of globally published patents on X-ray tube					
	Search: X-ray tube title or abstract			Search: X-ray tube title only		
		Decadal increase in total number of published patents			Decadal increase in total number of published patents	
Publication years	Total number of published patents	Absolute	Percent (%)	Total number of published patents	Absolute	Percent (%)
1896–1905	21	—	—	13	—	—
1906–1915	46	25	54.3	26	13	50
1916–1925	111	65	58.6	33	7	21.2
1926–1935	315	204	64.8	117	84	71.8
1936–1945	274	–(41)	–(13.0)	114	–(IAE, n.d.)	–(2.6)
1946–1955	334	60	18.0	57	–(57)	–(50.0)
1956–1965	430	96	22.3	109	52	47.7
1966–1975	759	329	43.3	270	161	59.6
1976–1985	2512	1753	69.8	950	680	71.6
1986–1995	3984	1472	36.9	1433	483	33.7
1996–2005	4476	492	11.0	1515	82	5.4
2006–2015	6464	1988	30.8	1902	387	20.3
2016–2019∗	1956	—	—	619	—	—
TOTAL	**21,682**			**7158**		

∗Till January 21, 2019

It was found that there were 21,682 and 7158 total number of patents published globally on X-ray tube with title/or abstract and title only respectively and the numbers of patent publication are still increasing. A general observation that the total number of globally published patents on X-ray tube are increasing in every decade from 1896 to the current year-2019, except one decade from 1936 to 1945. During these years, the percentage of a total number of patent publications on X-ray tube with title/abstract and title only was decreased by 13% and 2.6%, respectively, when compared to the corresponding value of the same published in the preceding decade. This may be due to the fact that the Second World War during these years would hamper the inventions and the subsequent filing of the patents. During the decades 1926–1935 and 1976–1985, the percentage increase in the total number of patents published was found to be greater, when compared to other decades since X-ray invention in 1985. Some of the latest globally published X-ray tube patents are listed in the Table 21.5.

Table 21.5 Recent globally published patents on X-ray tube (Search source: Espacenet)

Sl. No.	Patent publication number (date)	Inventor (applicant)	Patent abstract
1	US 2018/ 376574 A1 (2018-12-27)	Rogers Carey Shawn, Desrosiers Andrew J, Karesh Matthew (GE, USA)	**X-ray tube casing** The casing is made up metal matrix, which included metal and filler metal. It is manufactured using an additive manufacturing process to allow for tight tolerances. It includes a central frame having internal passages to supply a cooling fluid directly to the casing without the need for an external dedicated heat exchanger
2	US 2018/ 315577 A1 (2018-11-01)	William Ferneau (Thermo Scientific Portable Analytical Instruments Inc., USA)	**Target geometry for small spot X-ray tube** The electron gun is coupled to a side of the outer cylinder and a rod centrally positioned within the outer cylinder. The rod comprises a concave geometry, which is configured to position the target surface to have a focal spot size (2–6 μm) of an electron beam from the emission orifice
3	CN 108447755 A (2018-08-24)	Yang xiaohu, Xu xinlin, Liu jing, Rao Wei (Technical Inst Physics & Chemistry, China)	**Heat spreading cooled X-ray bulb tube based on liquid metal** It provides a heat spreading cooled X-ray bulb tube based on a liquid metal. It includes an anode metal target and a heat spreading module in contact with the anode metal target. The high-speed rotating flowing liquid metal is utilized to replace a traditional high-speed rotating solid-state anode target to achieve heat spreading
4	US 2018/ 204703 A1 (2018-07-19)	Andrews Gregory (Varex Imaging Corp, USA)	**Large angle anode target for an X-ray tube and orthogonal cathode structure** It provides a steep angle of a focal track of an anode of an X-ray tube. The focal track is positioned between the window-facing surface and the bearing-facing surface, wherein the focal track is angled with respect to the

(continued)

Table 21.5 (continued)

Sl. No.	Patent publication number (date)	Inventor (applicant)	Patent abstract
			window-facing surface and the angle between the focal track and the window-facing surface is between 45° and 89°
5	CN 207441650 U (2018-06-01)	Chen Zhiqiang, Tang Huaping, Li Yuanjing, et al. (Nuctech Co. Ltd., NuRay Tech Co. Ltd.)	**Multifocus X-ray tube and casing** The negative pole is located the casing and is produced the electronic beam current and the positive pole is located the casing and is produced the multi-beam X-ray. A plurality of target spots of anodal of casing align with setting up on the casing at the multifocus X-ray tube
6	JP 2018/073771 A (2018-05-10)	Demura Yasuhiro (Shimadzu Corp.)	**Rotating anode X-ray tube device and rotating anode driver** **Problem to be solved:** It includes a DC power supply, an inverter circuit including multiple switching elements and connected with the DC power supply, generating an AC voltage from the DC voltage of the DC power supply and outputting to a stator coil generating a revolving magnetic field of an X-ray tube
7	JP 2018/067529 A (2018-04-26)	John James Mccabe; Michael Scott Hebert; Hunt Ian Strider; et al. (GE)	**System and method for reducing relative bearing shaft deflection in X-ray tube** The X-ray tube includes a bearing configured to couple to an anode. The bearing includes a stationary member, a rotary member configured to rotate with respect to the stationary member during operation of the X-ray tube
8	WO 2018079946 A1 (2018-05-03)	Lee Donghoon, Kim Sanghyo, Kim Eunmin, et al. (Sunje Hi-Tek Co. Ltd., Korea)	**X-ray tube for improving electron focusing** It comprises a non-conductive tubular pipe forming a body and having a hollow; the X-ray irradiation window. It allows an X-ray to be irradiated therethrough; a stem part formed so as to shield the lower end part of

(continued)

Table 21.5 (continued)

Sl. No.	Patent publication number (date)	Inventor (applicant)	Patent abstract
			the tubular pipe; a plurality of metal wires which extend toward the inside of the tubular pipe from the outside of the stem part and to which a predetermined negative high voltage is applied
9	KR 101824135 B1 (2018-02-01)	Chae Young Hun (Kyungpook National University Industry Academic Cooperation Foundation, Korea)	**Thermal damage preventing rotating anode type X-ray tube**
			It comprises a cathode assembly disposed inside the case and colliding with electrons to emit X-rays; a bearing assembly supporting the cathode assembly to be able to rotate; and an anode assembly assembled in an upper part of the housing to emit the electrons. A plurality of heat transferring grooves are formed on the surface of the cooling member facing the cathode assembly, thereby enhancing cooling efficiency
10	CN 107546089 A (2018-01-05)	Peng Huaming, Hu junchao, Liao Youlin, Chen jinlu (Shanghai Chengming Electronic Tech Co. Ltd., China)	**High-power X-ray bulb tube**
			The high-power X-ray bulb tube comprises a tube shell with a hollow cavity, wherein ceramic flanges are arranged at two ends of the tube shell
			Cooling liquid is conveyed into the interiors of the stators through the cooling medium inlet pipelines. Heat generated by an anode can be quickly transmitted through the cooling liquid
11	EP 3264440 A1 (2018-01-03)	Ishihara Tomonari, Anno Hidero (Toshiba Electron Tubes & Device)	**X-ray tube device**
			It comprises a target surface generating an X-ray when the electron emitted from the cathode collides with the target surface, a vacuum envelope which accommodates the cathode and the anode target and is sealed in a vacuum-tight manner

21.8 Conclusion

The X-ray tube technology is continuously improving on its various limitations. As there is no alternate for X-ray source, many research studies are being carried out for the optimum performance of the X-ray tube. In the future, there may be a portable/foldable lightweight X-ray source for a tabletop X-ray imaging at field clinical applications. This would eventually result in more X-ray tube market growth opportunities anticipated in the coming years.

References

Bushberg JT, Seibert JA, Leidholdt EM Jr, Boone JM (2012) The Essential Physics of Medical Imaging, 3rd edn. Lippincott Williams & Wilkins, Philadelphia, PA, pp 97–140

U.S. Department of Health and Human Services (2019 May 8) Food and Drug Administration (FDA), Center for Devices and Radiological Health, "Medical X-Ray Imaging Devices Conformance with IEC Standards: Guidance for Industry and Food and Drug Administration Staff". https://www.fda.gov/media/99466/download

Spellman High Voltage Corporation, Common X-ray tube failure modes, Published online. Available: https://www.spellmanhv.com/en/Technical-Resources/Application-Notes-X-Ray-Generators/AN-02. Accessed 8 Mar 2019

IAE., Manufacturing of X-ray tube, Published online. Available: https://www.iae.it/manufacturing/

Atomic Energy Regulatory Board (AERB)., Quality Assurance (QA) of medical diagnostic x-ray equipment. Published online. Available: https://aerb.gov.in/images/PDF/DiagnosticRadiology/QUALITY-ASUURANCE-OF-DIAGNOSTIC-X-RAY-EQUIPMENT.compressed.pdf

Prospects of Avalanche Transit Time Terahertz Radiation Source in Biomedical Imaging: Application Feasibility in Health Engineering

22

Moumita Mukherjee

22.1 Introduction

The THz domain (0.1–10 THz) has attracted the attention of scientists and researchers in last decades for its huge application possibilities in the diversified fields of biomedical imaging, space science, spectroscopy, broadband communication, and remote sensing (Dhillon et al. 2017). This domain of EM spectrum throws a big challenge as it lies in between electronic and photonic technology gap. Electronic devices, viz., nanoscale transistors, Gunn and ATT (Avalanche Transit Time) diodes, and resonant tunneling diode (RTD), have been investigated extensively for lower THz domain, while photonic devices such as QCL (quantum cascade laser) and APD (avalanche photodiode) have been investigated for upper THz applications. Most of the available THz sources are complex and bulky and have low temperature. Also radiation power and detection sensitivity of such devices are not promising (Menikh 2010).

Therefore researchers are focusing on the development of suitable THz sources to bridge the so-called Tera Gap.

Development of potential THz sources will made this previously inaccessible region of EM spectrum accessible, for useful application in medical science. THz quanta are less energetic than those of X-ray radiation and therefore harmless for biological tissues. The reason of THz radiation's potential to differentiate between malignant and nonmalignant tissues is the variation of water absorption and density gradient. The shorter wavelengths of THz domain allow much greater spatial resolution than X-ray or microwave region. Therefore, in this paper, the authors have proposed a THz source and imaging system for nonhazardous biomedical applications.

M. Mukherjee (✉)
Adamas University, Kolkata, India

S. Paul (ed.), *Biomedical Engineering and its Applications in Healthcare*,
https://doi.org/10.1007/978-981-13-3705-5_22

Breast malignancy is the second leading cause of death from cancer among women population. The most common techniques of breast screening include X-ray mammography along with MRI and ultrasound; however, serious limitations such as false positive/negative results, failure to detect early-stage/in situ malignant tumor, and discomfort to patients are associated with the available diagonostic methods (Zhao et al. 2015). This paper is first time proposing a T-ray hyperthermia imaging technique for detection of early-stage breast tumors without causing damage to the healthy surrounding tissues.

The chapter is primarily divided into two sections:

(I) Design and analysis of low-noise, fast, and powerful room temperature T-ray radiation source.
(II) In silico T-ray imaging system design and development for snapshots/thermographs of human breast tissues under healthy and malignant conditions.

In Sect. 22.1 the authors will report a p^{++}-n^{-}-n-n^{++}-type two-terminal ATT device with GaN/AlGaN superlattice in the central active region. The carrier generation is contributed by both avalanche multiplication and inter-band tunneling phenomenon. The resultant device will operate in MITATT (Mixed Impact-ionization Tunneling ATT) mode, and the corresponding power generation will be in THz frequency regime. The proposed structure is a hexagonal Wz-GaN/AlGaN superlattice of periodicity four with asymmetrical doping and width distribution in the active region of the device. Superlattice configuration with asymmetrical doping density in active region results in spatial separation of mobile charges (electrons and holes) within the active region of MITATT device. This improves the electrical properties such as carrier lifetime and mobility significantly. The inter-sub-band transition and drifting of carriers through the active region of the device induce a current pulse in the external circuit, and the said transition generates an oscillation frequency in THz region. The authors have earlier developed a self-consistent, generalized large-signal simulator for the realistic modeling and analysis of *pin*/ATT devices, published elsewhere (Kundu et al. 2018a, b). The present study has used that indigenously developed simulator with some important modifications considering the various quantum aspects of carrier transition in asymmetrical superlattice structure. Quantum-Corrected Nonlinear Drift-Diffusion (QCNLCDD) model is used for solving Poisson and current continuity equations subject to appropriate boundary conditions (Tripathy et al. 2014). The validity of the model is established by comparing the simulated data with experimental findings (Kundu et al. 2018a, 2018b; Tripathy et al. 2014, Mukherjee and Roy 2009).

Substantial research works have already been done with wide bandgap (WBG) semiconductor-based ATT devices in recent years (Mukherjee et al. 2007, 2008). Most of the research works are focused on IMPATT mode of operation with flat doping profile. The said studies have established the superiority of WBG, GaN, SiC, and Si/SiC materials for generating THz power with a moderate to low efficiency (Mukherjee et al. 2007, 2008). However, to the best of authors' knowledge, asymmetrical superlattice ATT devices are not available in current literature.

High-frequency oscillation from a power device needs high mobility of carrier in transit. Specially designed superlattice structure is quite promising from this aspect; this has made the authors prompted to choose such exotic doping profile for designing a room temperature and efficient power source at Terahertz region.

Wide bandgap materials (III–V and IV–IV compound semiconductors) are promising for developing high-power efficient ATT devices. Power output from an ATT device depends upon the saturation velocity and critical electric field at breakdown of the base semiconductor. GaN and AlGaN, having saturation velocity $\sim 2 \times 10^5$ m/s, critical breakdown field $\sim 2 \times 10^8$ V/m, are expected to be a potential pair for developing a superlattice structure. The inherent mobility of AlGaN/GaN reduces transit time of carriers through the active region of the device. This makes the device suitable for oscillation at THz (0.1–10 THz) frequency region. Moreover the lattice mismatch in between sapphire substrate and epilayer AlGaN/GaN is minimum compared to flat GaN epilayer (Cheng et al. 2016). Thus the authors have chosen AlGaN/GaN superlattice for designing the high-power, high-frequency ATT device.

Section 22.2 will deal with the in silico development of T-ray imaging thermographs using COMSOL Multiphysics software coupled with the newly developed QCNLDD simulator. Maxwell's equations are solved, and results are used for getting the solution of bioheat equations (described in Sect. 22.2). The authors have compared the microwave and terahertz thermographs. In order to make the analysis more accurate, experimentally obtained electrical and thermal properties of the skin, breast, and tumor at microwave (11 GHz) and THz (0.15 THz and 0.5 THz) frequencies are incorporated in the simulation study (Xie et al. 2008). Radiation from the source is allowed to incident on healthy and malignant breast tissues in a breast like close structure. The incident radiation will increase the temperature in the malignant tumor above 40–45 °C while keeping normal temperature in the surrounding healthy tissues. Relatively high thermal conductivity of malignant tissues will increase the local heating potential, and thus high thermal gradient is expected to be observed in thermographs. However, healthy skin and glandular tissues also possess high thermal conductivity which results in secondary hotspots in thermographs. The unwanted hotspots cause significant side effects, viz., burn and pain. On the other hand, hotspots may degrade the screening efficiency by generating a false positive result. Therefore, to locate the tumors accurately, a sharp temperature gradient in thermographic images is required. This paper will through some light on this issue by comparing the relative thermographs at three different incident radiation frequencies.

22.2 Simulation Methodology

The work flow diagram of the developed in silico T-ray scanning and imaging system is shown in Fig. 22.1. Figure 22.2 denotes the structure of the device under test (DUT). The following subsections will deal with numerical modeling details.

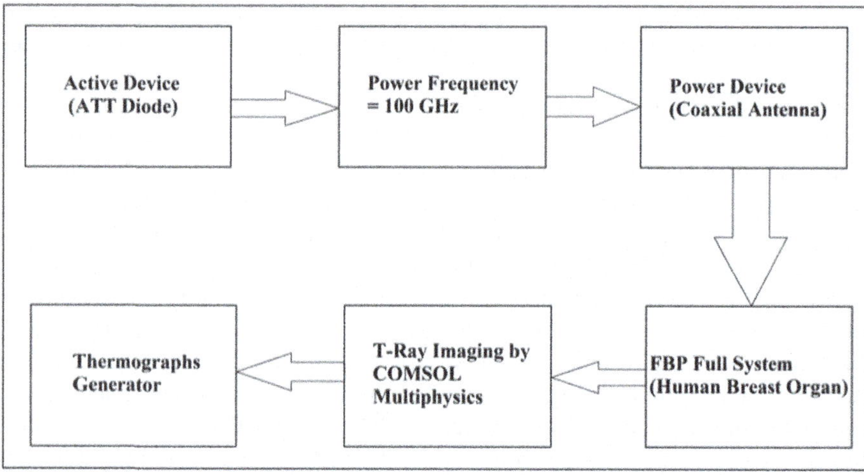

Fig. 22.1 Work flow diagram of T-ray imaging and detection system (in silico)

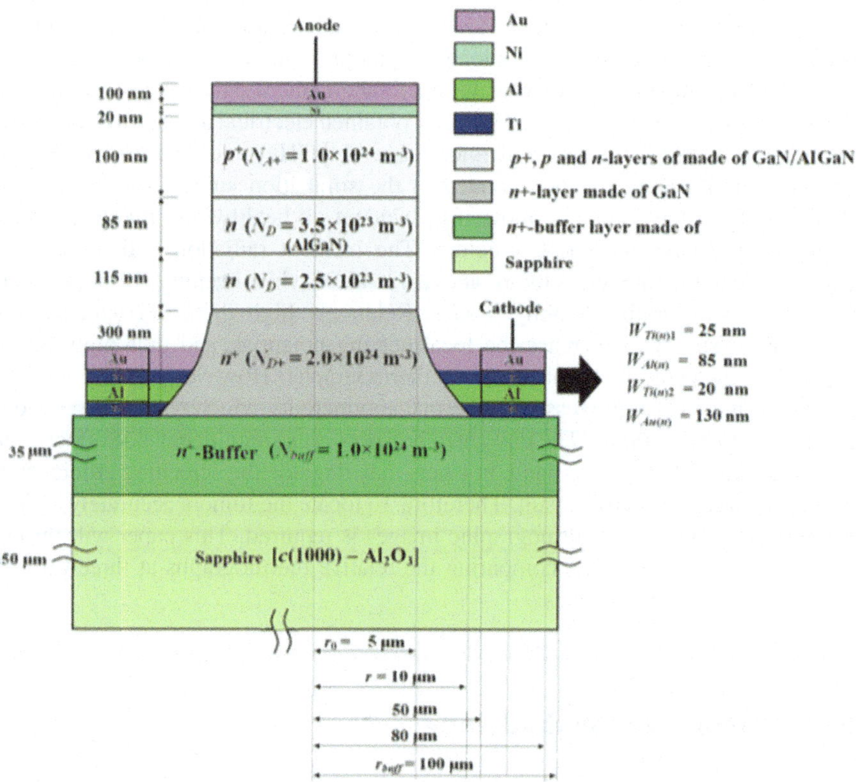

Fig. 22.2 Schematic view of the proposed T-ray semiconductor diode source

22.2.1 Quantum-Corrected Nonlinear Drift-Diffusion (QCNLDD) Model

The quasi 3D vertical and asymmetrically doped AlGaN/GaN-ATT (p^{++}-n^{-}-n^{+}-n^{++} doping profile) T-ray source and detector have been designed and analyzed in the paper. The physical/electrical/thermal properties of AlGaN/GaN material system along the symmetric axis of the device are taken into account for the design (Electronic Archive: www.ioffe.ru/SVA/NSM/Semicond/). The authors have made nonlinear large-signal (L-S) simulation in order to get realistic view of the device characteristics under various operating conditions. At each instant of time, the physical properties such as electric field, electron and hole current components, and recombination current are obtained by solving the nonlinear field and carrier transport equations, i.e., Poisson's equation and combined current continuity equations for various modulation factors at the edges of the active region, subject to satisfaction of appropriate boundary conditions. The authors have considered the effect of introducing an n-bump layer of appropriate doping concentration in between the substrate and epilayer. The following equations are solved in the QCNLDD model:

$$\frac{\partial^2}{\partial x^2}V(x,t) = -\frac{q}{\varepsilon}\left[N_d(x,t) - N_a(x,t) + C_p(x,t) - C_n(x,t)\right] \tag{22.1}$$

$$\frac{\partial}{\partial x}p(x,t) = -\left(\frac{1}{q}\right)\frac{\partial}{\partial x}J_p(x,t)^p + G_p(x,t) - R_p(x,t) \tag{22.2}$$

$$\frac{\partial}{\partial x}n(x,t) = \left(\frac{1}{q}\right)\frac{\partial}{\partial x}J_n(x,t)^n + G_n(x,t) - R_n(x,t) \tag{22.3}$$

$$J_p(x,t) = -q\mu_p\left[C_p(x,t)\frac{\partial}{\partial x}V(x,t) + \left(\frac{K_B T_j}{q}\right)\frac{d}{\partial x}C_p(x,t)\right] \tag{22.4}$$

$$J_n(x,t) = -q\mu_n\left[C_n(x,t)\frac{\partial}{\partial x}V(x,t) - \left(\frac{K_B T_j}{q}\right)\frac{d}{\partial x}C_n(x,t)\right] \tag{22.5}$$

$$J_t(x,t) = J_n(x,t) + J_p(x,t) \tag{22.6}$$

where $J_{p,n}(x,t)$ is the electron and hole current density $V(x,t)$ is the electric potential, $J_t(x,t)$ is the total current density, $C_{p,n}(x,t)$ is the charge carrier concentration, $G_{p,n}(x,t)$ is the carrier generation rate, $R_{p,n}(x,t)$ is the carrier recombination rates, $N_a(x,t)$ and $N_d(x,t)$ are the electron and hole current density, and $\mu_{p,n}, \varepsilon$, $T_j$are the mobility of charge carriers, permittivity, and junction temperature, respectively.

The carrier generation is due to the avalanche phenomenon and band to band tunneling of electron and hole. It can be written as.

$$G_{p,n}(x,t) = G_{A_{p,n}}(x,t) + G_{T_{p,n}}(x,t) + G_{ph_{p,n}}(x,t) \tag{22.7}$$

In the above equations, $G_{A_{n,p}}(x,t)$, $G_{T_{p,n}}(x,t)$, and $G_{ph_{p,n}}(x,t)$ represent the avalanche and tunneling carrier generation rates and photo-generation rate, respectively. The avalanche carrier generation rates for electron and hole can be expressed as

$$G_{A_p}(x,t) = G_{A_n}(x,t) = \alpha_p(x,t)v_p(x,t)C_p(x,t) = \alpha_n(x,t)v_n(x,t)C_n(x,t)$$

where $\alpha_{p,n}$ and $v_{p,n}$ are the ionization rate and drift velocities of the charge carriers, respectively.

The electron tunneling generation in GaN/AlGaN is expressed as

$$G_{T_n}(x,t) = a_T E^2(x,t)\exp\left[1 - \frac{b_T}{E(x,t)}\right]$$

where $E(x,t)$ represents the electric field. The coefficients a_T and b_T can be determined by.

$$a_T = \frac{q}{8\pi\hbar^2}\left(\frac{m_n^*}{E_g}\right)^{\frac{1}{2}}, b_T = \frac{1}{2q\hbar}\left(\frac{m_n^* E_g}{2}\right)^{\frac{1}{2}}$$

where E_g is the bandgap energy introduced in AlGaN/GaN superlattice, m_n^* is the effective mass of electron, $\hbar\left(\frac{h}{2\pi}\right)$ is the normalized Planck's constant, q (1.6×10^{-19} C) is charge of the electron, and h (6.625×10^{-34}) is the Planck's constant. The tunnel induced hole generation rate can be expressed as $G_{T_p}(x,t) = G_{T_n}(x',t)$. The tunnel induced hole generation rate at x is the function of electron generation rate due to tunneling at x', where $(x - x')$ is the spatial separation in between valance and conduction band at the same energy level. It can be obtained from the energy band diagram of p^{++} -n^{-} - n^{+} - n^{++} device.

22.2.2 T-Ray Thermograph Modeling

In this study, the authors have adopted the commercial software COMSOL to do the thermograph model. First of all, the relevant Maxwell's equations are solved by setting appropriate boundary conditions, and then computation is performed for determining the electromagnetic specific absorption rate (SAR). Bioheat (Zuar and Roetzel 1997) equations are further solved to obtain the temperature distribution in the breast model that consists of healthy and cancerous tissues. The radiation antenna is radiating T-ray/microwave on the subject under test.

The in silico breast model is developed, and detection is done by using an axisymmetric transverse magnetic (TM) formulation; Maxwell's equations can be simplified to a wave equation in H_ϕ

$$\nabla x\left[\left(\varepsilon_r - \frac{j\sigma_E}{\omega\varepsilon_0}\right)^{-1}\nabla xH \rightarrow_\phi\right] - \mu_r k_0^2 H \rightarrow_\phi = 0 \tag{22.8}$$

where σ_E denotes the electrical conductivity. The electric field $E \rightarrow$ is obtained once the magnetic field $H \rightarrow$ is solved from Eq. (22.8). Let $n \rightarrow$ denote the unit normal vector for a surface.

$$n \rightarrow xE \rightarrow = 0 \tag{22.9}$$

In COMSOL Multiphysics, a first-order boundary condition

$$n \rightarrow x\sqrt{\varepsilon}E \rightarrow -\sqrt{\mu}\quad H_\varnothing^{\rightarrow=-2\sqrt{\mu}H_{\varnothing 0}} \rightarrow \tag{22.10}$$

is used at outer boundaries of tissues. $\overrightarrow{H_{\varnothing 0}}$ is related to the input power $P \rightarrow_{avg}$ as shown in the following equations:

$$E \rightarrow = \frac{A}{r}e^{j(\omega t - kz)}e \rightarrow_r \tag{22.11}$$

$$H \rightarrow = \frac{A}{rz}e^{j(\omega t - kz)}e \rightarrow_\phi \tag{22.12}$$

where A, , and k are the amplitude, angular frequency, and wave number, respectively. The impedance of the wave is as follows:

$$z = \sqrt{\frac{\mu}{\varepsilon}} \tag{22.13}$$

where dielectric permeability is μ and permittivity is ε. The average power flow on the subject, along the z direction, is given by

$$P \rightarrow_{avg} = \int_{r_{inner}}^{r_{outer}} \mathrm{Re}\left(\frac{1}{2}E \rightarrow xH \rightarrow^*\right)2\pi r dr = \pi\frac{A^2}{Z}\ln\frac{r_{outer}}{r_{inner}}e \rightarrow_z \tag{22.14}$$

The temperature distribution inside a breast tissue during the coagulation can be obtained by solving the bioheat Eq. [22.13]

$$\rho c\frac{\partial T}{\partial t} = \nabla\cdot(\kappa\nabla T) + Q_{ext} - Q_{pfu} + Q_{met} \tag{22.15}$$

where ρ *denotes* the mass density, c denotes specific heat capacity, T is the absolute temperature, and κ is the thermal conductivity of the tissues, respectively. The first term on the right-hand side of Eq. (22.15) is the thermal conduction of tissue. The second term is the absorbed energy as a result of the resistive heat generation by the EM field in tissue, which is expressed as

$$Q_{ext} = \frac{1}{2}\mathrm{Re}\left[(\sigma - j\omega\varepsilon)\,E \rightarrow .\overset{\rightarrow *}{E}\right] \tag{22.16}$$

Specific absorption rate (SAR) is expressed as below:

$$\frac{Q_{ext}}{\rho} \tag{22.17}$$

The third term Q_{pfu} denotes the loss of heat due to microvascular blood perfusion

$$Q_{pfu} = \rho_b c_b \omega_b (T - T_b) \tag{22.18}$$

where $\rho_b$denotes the blood density, c_b specifies the blood specific heat capacity, ω_b represents the blood perfusion rate, and T_b represents the temperature of the blood. The metabolic heat generation term Q_{met} is neglected here, since its magnitude is substantially less than other terms in Eq. (22.15).

In this model the authors have assumed no heat flux across tissue surfaces, that is,

$$n \rightarrow \cdot \nabla T = 0 \tag{22.19}$$

Thermal, dielectric, and blood perfusion properties of healthy and malignant tissues are complex functions of EM wave, operating frequency, tissue contents, and temperature. The properties are summarized in Table 22.1. Mostly experimental data are used for the design and analysis (Xie et al. 2008).

22.3 Results

22.3.1 Nonlinear Characteristic Analysis of T-Ray Source

The authors have studied the THz characteristics of the designed device at 0.15 THz and 0.5 THz regime. A quasi 3D quantum-corrected nonlinear classical drift-diffusion (QCNLDD) simulator, as explained in Sect. 22.2, is used for this purpose. Temperature- and field-dependent material parameters of superlattice GaN/AlGaN are taken into account for making the analysis more accurate. The validity of the model is established elsewhere (Kundu et al. 2018a, b) for pin/ATT devices. Initially, the nonlinear static properties are studied for a voltage modulation factor ~ 50%. The corresponding electric field profile snapshots of 0.15 THz and 0.5 THz

Table 22.1 Dielectric and thermal parameters of healthy and malignant human breast tissue (Xie et al. 2008)

Parameters	Healthy tissues	Malignant tissues
Relative permittivity ε	2.53	50
Electrical conductivity σ (S/m)	2.06	7
Thermal conductivity k (W/mK)	0.21	0.545

devices are shown in Fig. 22.3(a–b). It is interesting to observe that the peak electric field at breakdown for 0.15 THz device is 3.25×10^8 V.m^{-1}, whereas the same increases to 4×10^8 V.m^{-1} for the 0.5 THz device. The breakdown voltage in case of W band (~0.14 THz) device is 102 V for an active region width of 450 nm, whereas the same decreases to 80 V in its 0.5 THz counterpart for an active region width of 200 nm. In Fig. 22.3(a) the effect of large signal on E(x) profile is shown for different phase angles ($0 < T < 2\pi$). It is observed that the magnitude of peak electric field is varying with the variation of phase angle within a complete time period.

Figure 22.4 depicts the normalized current density profiles of flat profile Si and superlattice GaN/AlGaN devices at W band. Due to the intrinsic material parameter limitations in Si, flat profile Si devices are not capable of generating significant power at around 0.5 THz. Therefore the authors have designed, studied, and compared the performance characteristics of Si-based ATT device with those of the GaN/AlGaN superlattice device at 0.15 THz only. Figure 22.5 shows the large-signal (50% voltage modulation) admittance characteristics of Si and GaN/AlGaN devices for different operating frequencies in lower THz region. It is interesting to observe that in case of Si device, the negative conductance at peak frequency (0.143 THz) is 27×10^6 S.m^{-2}, and on the other hand, due to the incorporation of superlattice-type doping profile in the active region of the ATT device, the peak negative conductance enhances to 43×10^6 S.m^{-2} at 0.145 THz. The device quality factor for 50% voltage modulations is found to be ~ 3.0 for Si ATT, and the same improves to 1.9 for the superlattice variant. Figure 22.6 shows the frequency negative conductance plots of superlattice GaN/AlGaN device for R_s (Parasitic Series Resistance) = 0 Ω and 4.68×10^{-10} Ω.m^2. The parasitic series resistance plays a dominant role in case of small-dimensional devices. Therefore, for the designing of 0.5 THz source, the authors have considered the effect of series resistance on the values of negative conductance and RF power output. The series resistance is estimated at oscillation threshold as described elsewhere (Mukherjee et al. 2011). The nonlinear analysis reveals that the magnitude of negative conductance decreases from 700×10^6 S.m^{-2} to 589×10^6 S.m^{-2} due to the effect of parasitic series resistance. Figure 22.6 also shows that the device is oscillating in between 0.5 THz and 0.55 THz.

The impedance characteristics of 0.15 THz and 0.5 THz devices are shown in Figs. 22.7 and 22.8 respectively. It is observed that the negative resistance at peak frequencies in case of 0.15 THz device is ~ 5×10^{-9} Ω.m^2 and ~ 1×10^{-9} Ω.m^2 for 0.5 THz device. With the increasing operating frequency, the dimension of the device (also mesa diameter) decreases, and this results in significant reduction of negative resistance values as observed in the present study. Though the negative resistance of the superlattice 0.5 THz ATT device decreases to 1×10^{-9} Ω.m^2, still it is higher than the corresponding parasitic series resistance value ~ 4.68×10^{-10} Ω.m^2. This ensures the oscillation condition.

The high-frequency properties of the superlattice devices within THz region are summarized in Table 22.2. The study reveals that the W band (0.14 THz) device is capable of developing a peak RF power ~ 6×10^{10} W.m^{-2}, whereas the 0.5 THz device will deliver ~ 56.0×10^{10} W.m^{-2}.

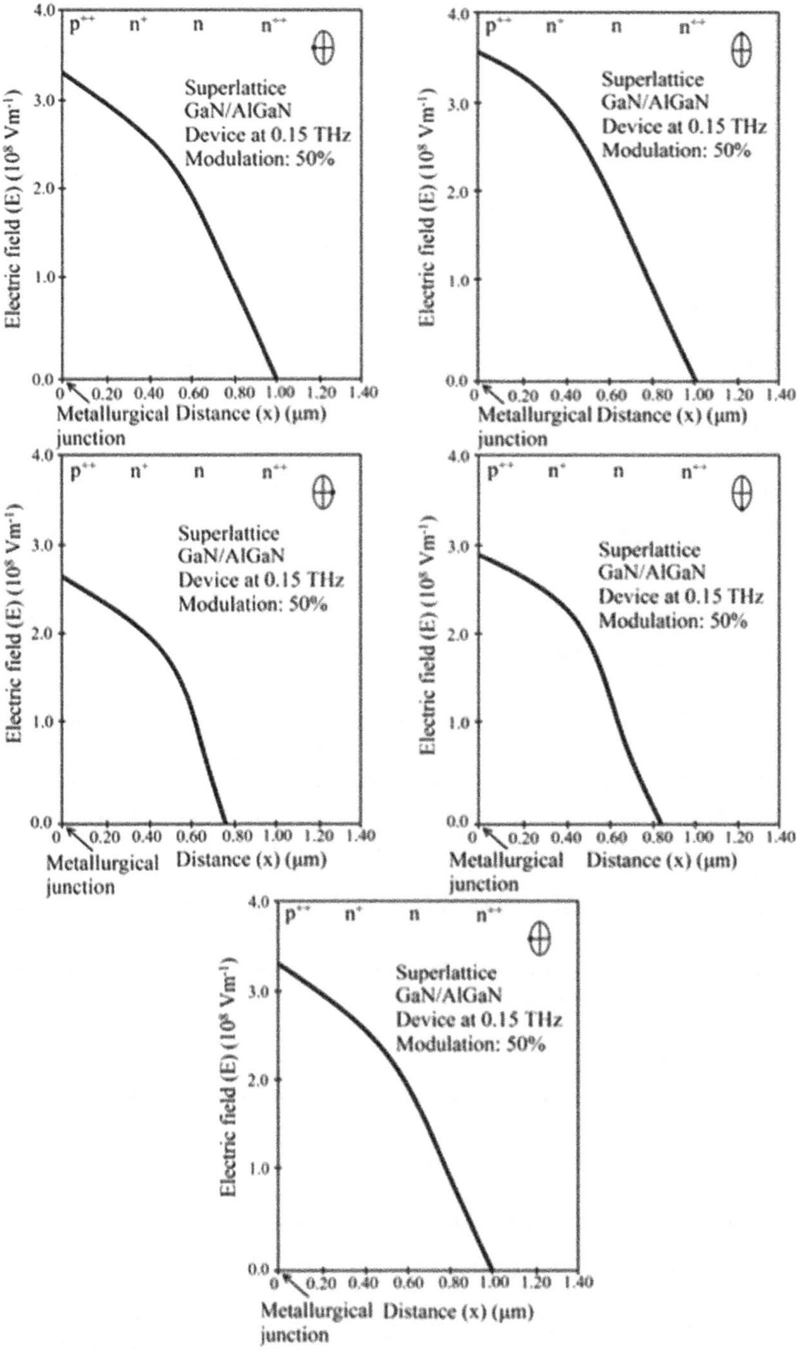

Fig. 22.3 (**a**) Electric field profiles of THz (0.15 THz) source under large-signal operating condition. (**b**) Electric field profile of superlattice ATT device operating at 0.5 THz

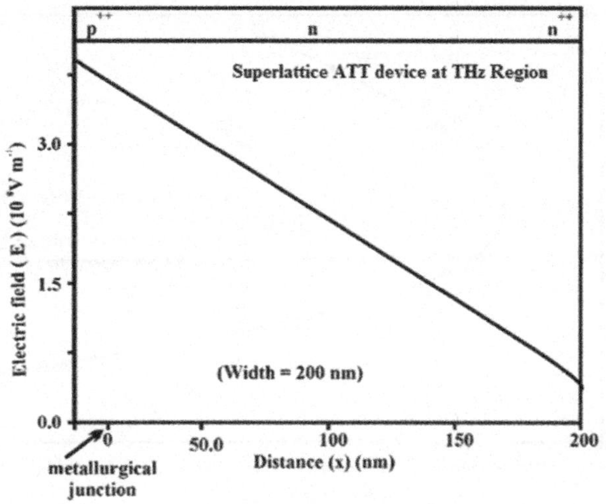

Fig. 22.3 (continued)

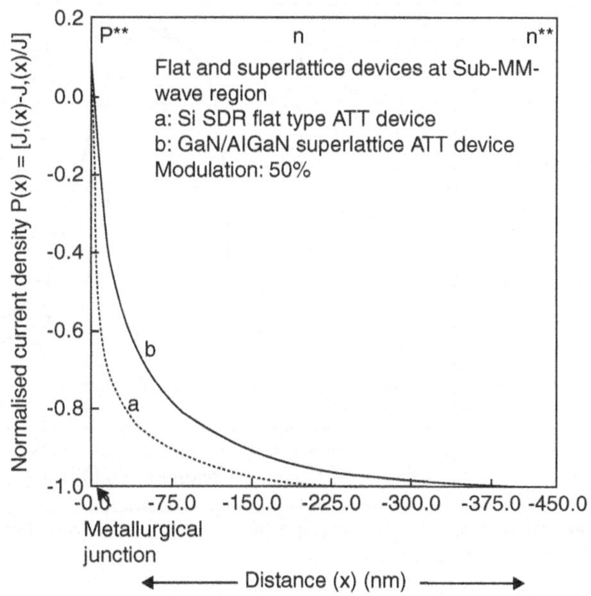

Fig. 22.4 Normalized current density profiles of Si and GaN/AlGaN superlattice ATT device at 0.15 THz

22.3.2 T-Ray Thermographs Analysis

The authors have developed an equivalent breast model using COMSOL Multiphysics 5.3a, which is taken in the form of a cylindrical cone of dimension 120 mm in length and 50 mm in diameter. The breast volume consists of breast skin, while the remaining

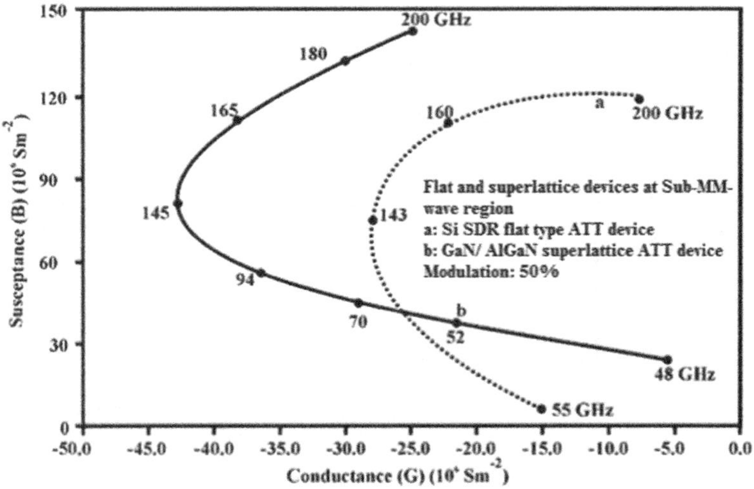

Fig. 22.5 Admittance characteristics of Si and GaN/AlGaN superlattice ATT device at 0.5 THz

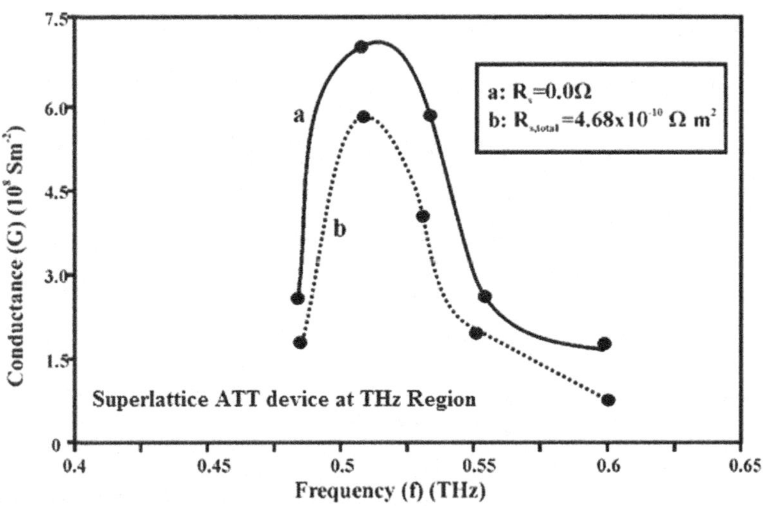

Fig. 22.6 Conductance-frequency plots of GaN/AlGaN superlattice ATT device at high-frequency (0.45–0.6 THz) operation

consists of breast fatty tissues. The skin thickness is not considered in the simulation studies. The tumor is considered to be 3–4 mm in diameter. The developed breast model is subjected to a series of incident radiations starting from low microwave region (11 GHz) to THz wave (0.15 THz and 0.05 THz) region. The corresponding thermographs are studied and reported in this section.

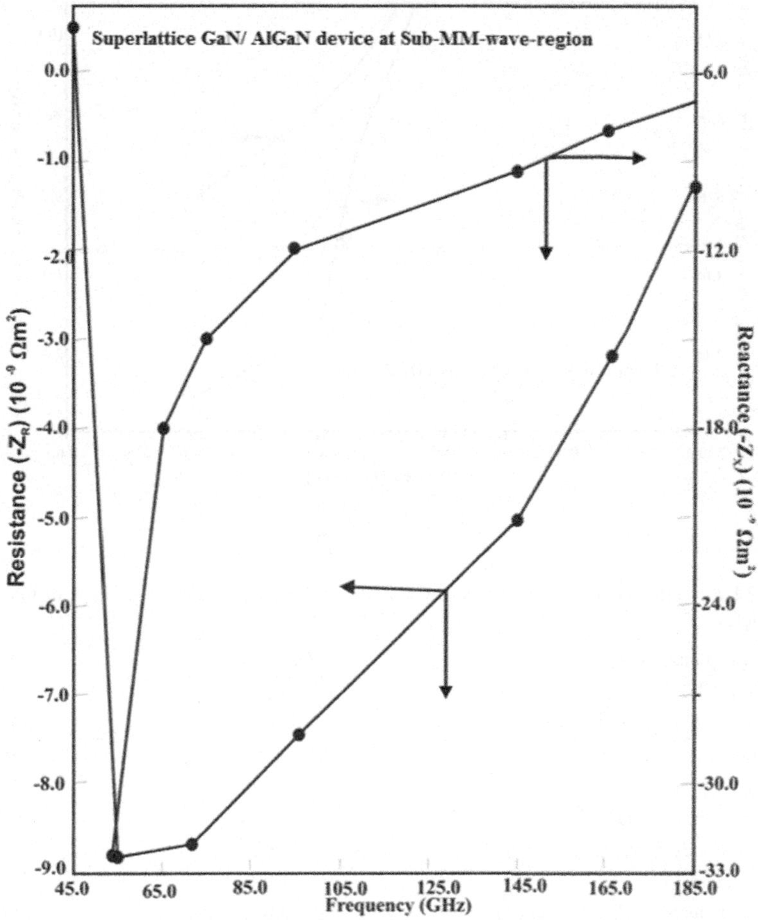

Fig. 22.7 Impedance characteristics of 0.15 THz ATT device

Before starting the analysis/modeling, several grid resolutions are tested. Two sets of grid meshes (default grid and refined grid) are shown in Figs. 22.9(a–b) and 22.10(a–b), respectively, for microwave and THz analysis.

Figure 22.11 (a–c) depicts the 3D/2D thermographs of normal fatty breast tissues and malignant breast tissues under THz and microwave radiation. Figure 22.11(a) depicts that due to the incident radiation at 0.15 THz frequency, there is no change in temperature in normal breast tissue thermograph. However, in case of malignant breast tissue, the temperature thermographs are showing a noticeable temperature gradient (rise in temperature from 37 °C to 550 °C) due to the presence of tumor in the system. With the increasing radiation frequency (0.5 THz), the temperature gradient is becoming more prominent as reflected in Fig. 22.11(b). Also, at higher

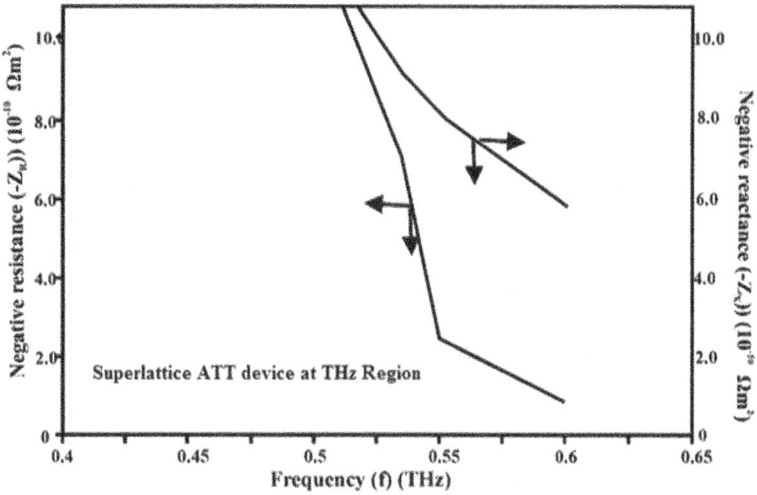

Fig. 22.8 Impedance characteristics of 0.5 THz ATT device

Table 22.2 High-frequency properties of GaN/AlGaN-based superlattice ATTs at different THz frequencies

Nonlinear parameters	$GaN|_{0.15THz}$	$GaN|_{0.5THz}$
$E_p(10^8 Vm^{-1})$	3.25	4.0
Breakdown voltage V_B(V)	102.0	80
Efficiency (η) (%)	23.50	13.5
Avalanche frequency f_a(THz)	0.048	0.48
Peak frequency f_p (THz)	0.145	0.5
Negative conductance $Gp(10^6\ Sm^{-2})$	43.00	700
Quality factor Q_p	1.90	2.5
Negative resistance $Z_{RP}(10^{-9}\Omega m^2)$	5.04	1.0
RF peak power $P_{rf}(10^{10}Wm^{-2})$	5.6	56.0

THz radiation frequency (Fig. 22.11(b)), normal breast tissues are showing no variation in temperature thermographs. On the other hand, the temperature gradient is more significant in malignant tissues (rise in temperature from 37 °C to 750 °C) compared to lower THz radiation. In order to make a comparison with microwave thermographs, the authors have studied the effect of microwave radiation on healthy and malignant breast tissues under similar operating conditions. Figure 22.11(c) clearly shows that compared to its THz counterpart, microwave thermographs are not so effective in detecting the position/location of tumors as the temperature gradient is not so sharp. Figure 22.12 shows the relative temperature thermographs of microwave and T-ray radiation. The comparative study reveals that T-ray nonionizing thermal imaging is an efficient technique in breast cancer screening. In this in silico imaging studies, the authors have also considered the issue of cell damage as a

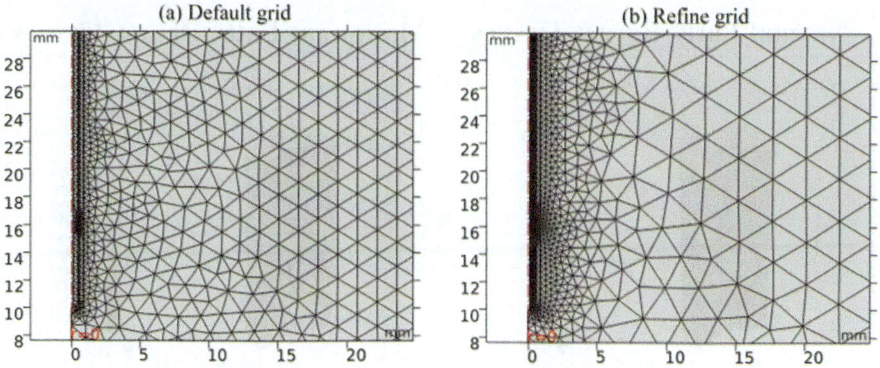

Fig. 22.9 Grid design for microwave imaging system development

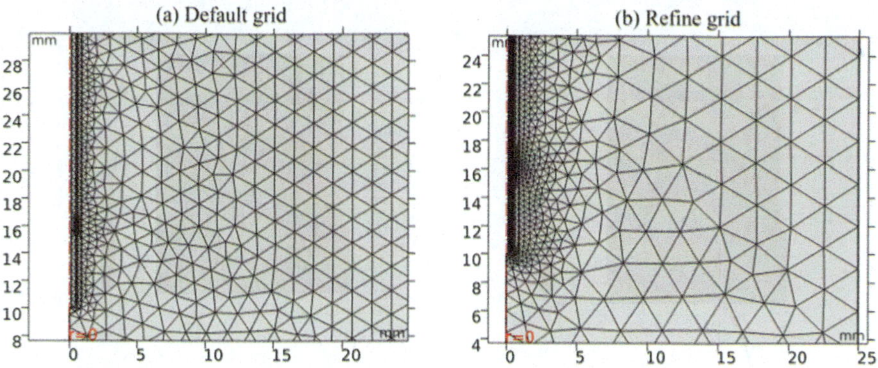

Fig. 22.10 Grid design for THz imaging system development

result of microwave/THz radiation. The study (Fig. 22.13) reveals that a significant fraction of healthy cell is getting damaged in case of microwave thermography diagnosis technique, whereas the cell damage issue is less significant in T-ray thermographs.

This increase in temperature is due to the presence of more water in cancer affected cell in breast organ. Also the cell damage is more significant near the T-ray radiation source and decreases gradually with distance. The dimension of the malignant tumor has been considered to be ~3 mm in diameter.

22.4 Conclusion

Superlattice GaN/AlGaN-based exotic ATT device oscillator is designed and studied for T-ray source applications. A generalized quantum-modified classical nonlinear drift-diffusion simulator is used for this purpose. The validity of the model is already

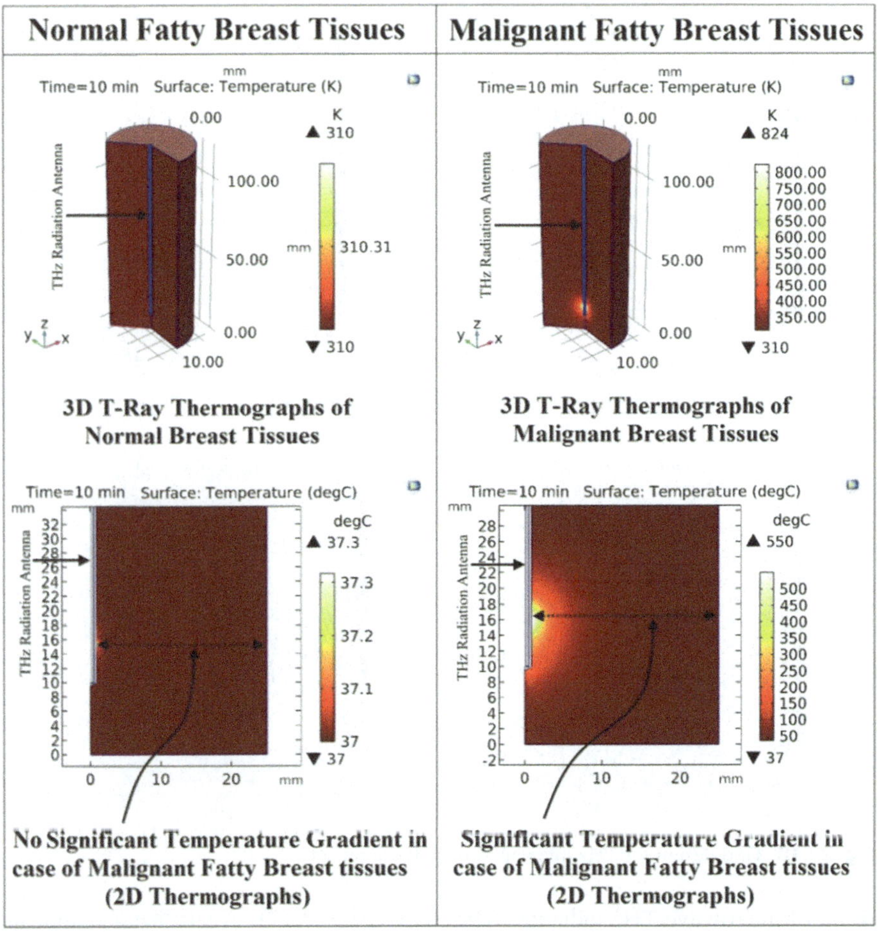

Fig. 22.11 (**a**) T-ray thermographs (0.15 THz) of healthy vs malignant breast tissues. (**b**) T-ray thermographs (0.5 THz) of healthy vs malignant breast tissues. (**c**) Microwave thermographs of malignant breast tissues

established. The necessity of incorporation of superlattice properties in conventional model is to improve the effective mobility of carriers in the central active region and, in turn, high-frequency electrical and thermal properties of the ATT device. GaN/AlGaN superlattice is found to be a good replacement of conventional flat profile GaN devices as far as improved admittance, electrical field profile, power output, and efficiencies are concerned. The authors have made the study much more realistic by incorporating experimentally obtained field-dependent material parameter data and also by incorporating the effects of series resistance on the power-frequency behavior of the designed device. The study has established the suitability of GaN/AlGaN material pair in developing high-power (~56.0×10^{10} W.m^{-2}) T-ray source.

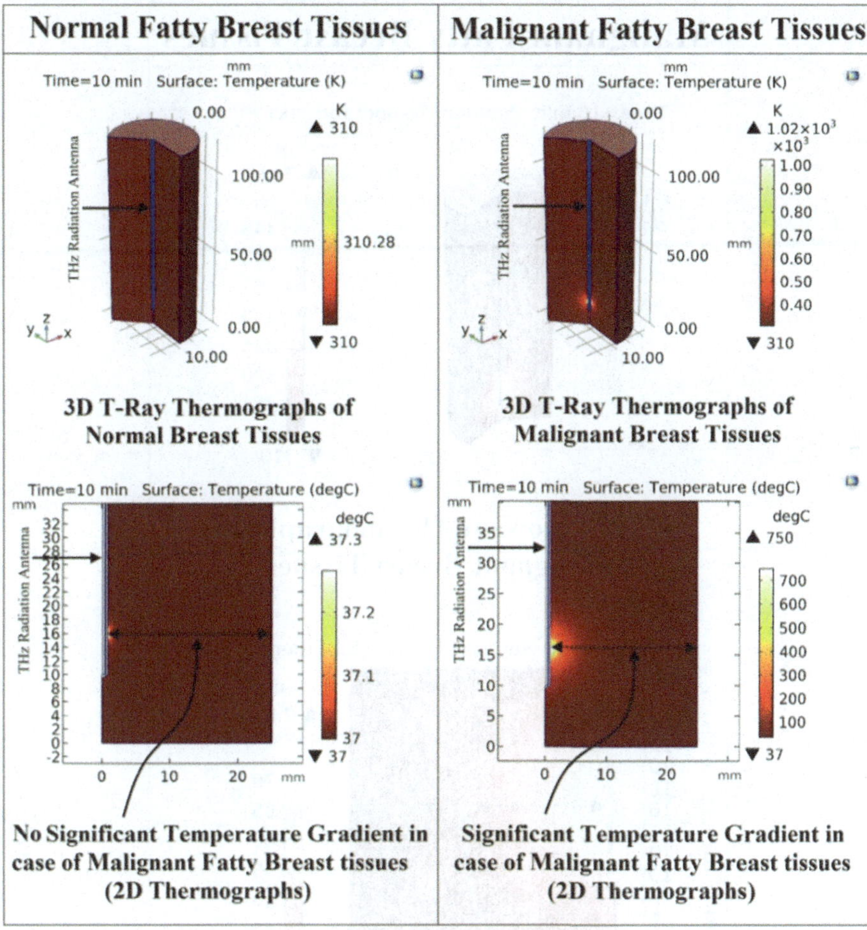

Fig. 22.11 (continued)

The authors have further studied the T-ray vs microwave radiation thermographs in detecting malignant breast tumor of ~ 3.0 mm in diameter. Through this comprehensive study, the authors have established that the nonionizing T-ray imaging technique could be an excellent alternative to ionizing X-ray mammograms in breast cancer screening. This is for better thermal gradient (as compared to microwave radiation) and very less amount of surrounding healthy tissue damage. The published literature, dealing with X-ray radiation, shows that malignant tumor of such a small dimension could not be predicted with such accuracy by simply adopting a cost-effective, room temperature, and easy technique (Zhao et al. 2015) as proposed here. To the best of authors' knowledge, this is the first study report on superlattice MITATT-based T-ray source and in silico radiation unit in noninvasive, low-cost, and accurate identification of breast cancer.

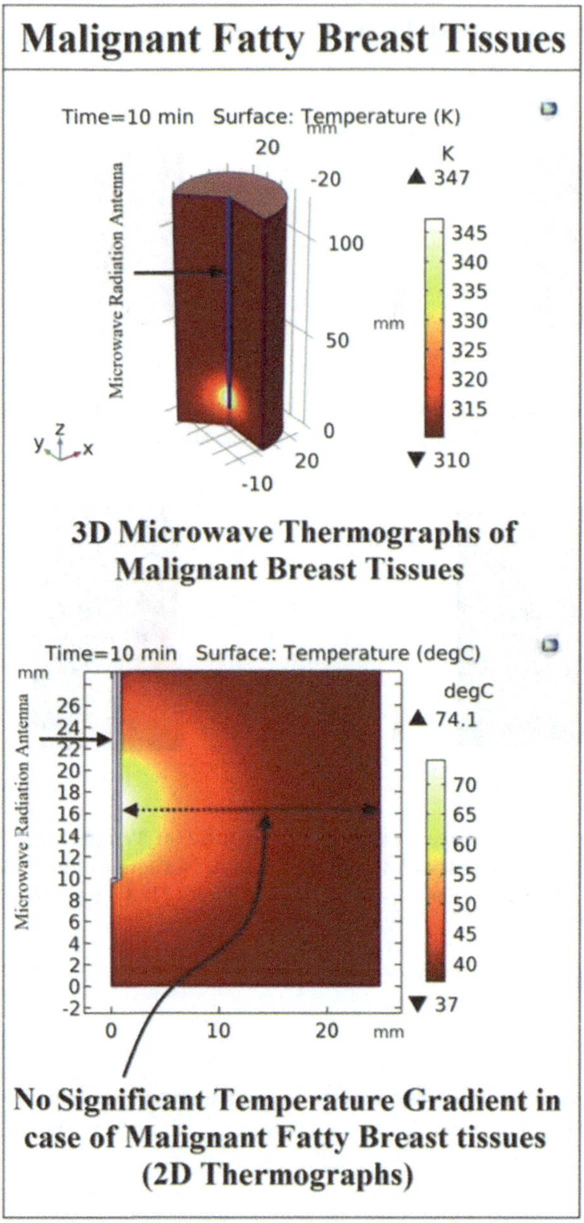

Fig. 22.11 (continued)

Fig. 22.12 Microwave vs T-ray temperature thermographs for malignant tumor detection

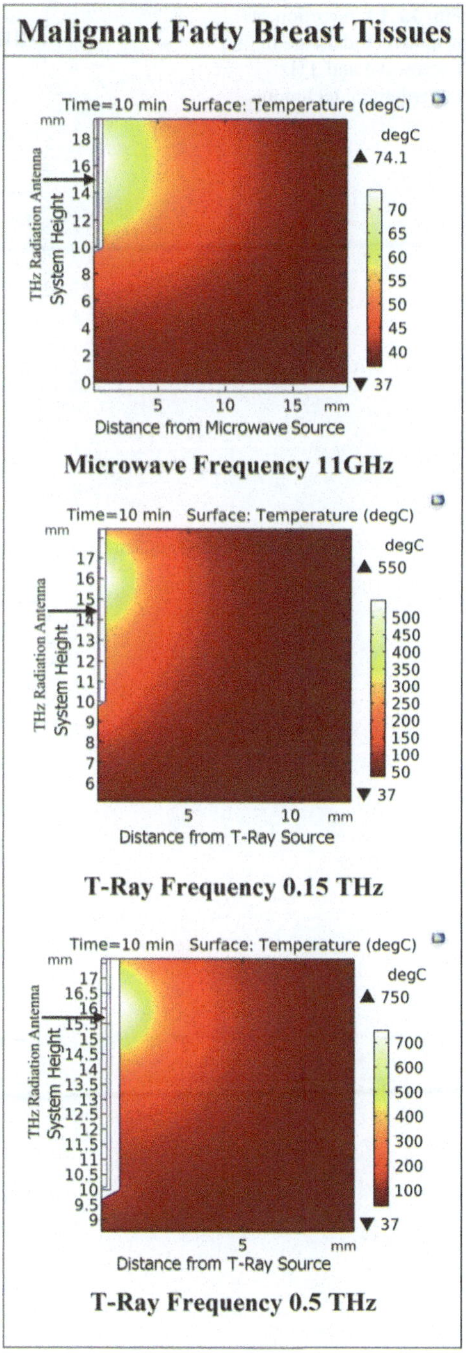

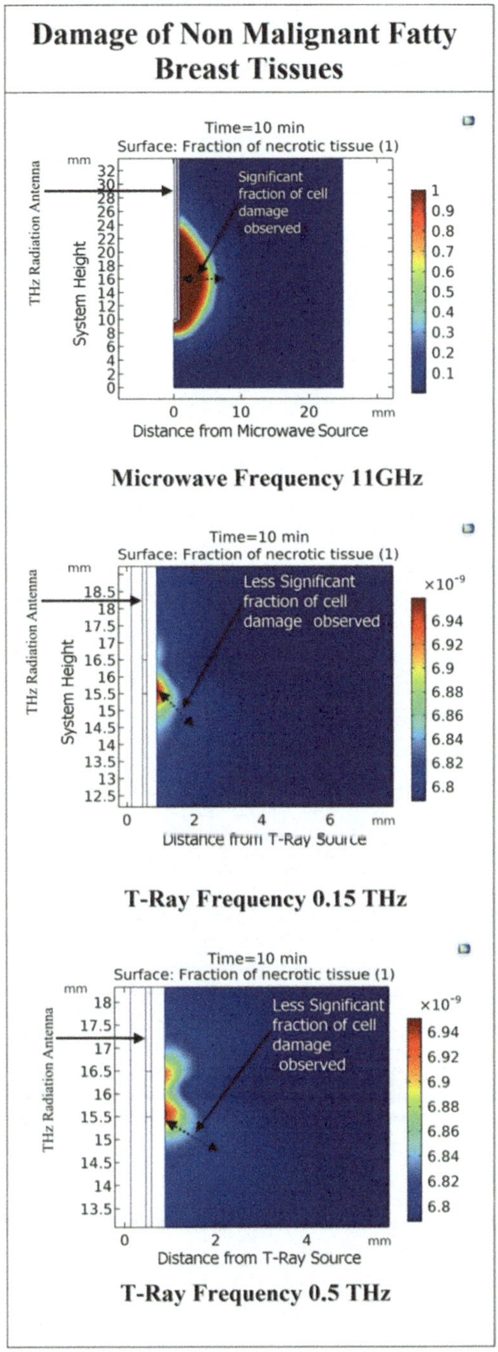

Fig. 22.13 Fraction of healthy cell damage in microwave and THz hyperthermia technique

References

Cheng J, Yang X, Sang L, Guo L, Zhang J, Wang J, He C, Zhang L, Wang M, Xu F, Tang N, Qin Z, Wang X, Shen B (2016) Growth of high quality and uniformity AlGaN/GaN heterostructures on Si substrates using a single AlGaN layer with low Al composition. Sci Rep 6(23020):1–7. https://doi.org/10.1038/srep23020

Dhillon SS, Vitiello MS, Linfield EH, Davies AG, Hoffmann CM, Booske J, Paoloni C, Gensch M, Weightman P, Williams GP (2017) The 2017 terahertz science and technology roadmap. J Phys D Appl Phys 50(4):1–49

Electronic Archive: www.ioffe.ru/SVA/NSM/Semicond/

Kundu A, Adhikari S, Das A, Kanjilal MR, Mukherjee M (2018a) Design and characterization of asymmetrical super-lattice Si/4H-SiC pin photo diode array: a potential opto-sensor for future applications in bio-medical domain. Microsyst Technol. https://doi.org/10.1007/s00542-018-4119-4

Kundu A, Kanjilal MR, Mukherjee M (2018b) III–V super-lattice SPST/SPMT pin switches for THz communication: theoretical reliability and experimental feasibility studies. Microsyst Technol 20. https://doi.org/10.1007/s00542-018-4053-5

Menikh A (2010) Terahertz-biosensing technology: progress, limitations and future outlook. Opt Guided-wave Chem Biosens 8:283–295. https://doi.org/10.1007/978-3-642-02827-4_11

Mukherjee M, Roy SK (2009) Optically modulated III–V nitride-based top-mounted and flip-chip IMPATT oscillators at terahertz regime: studies on the shift of avalanche transit time phase delay due to .... IEEE Trans Electron Devices 56(7):1411–1417

Mukherjee M, Mazumder N, Roy SK, Goswami K (2007) GaN IMPATT diode: a photo-sensitive high power terahertz source. Semicond Sci Technol 22(12):1258

Mukherjee M, Mazumder N, Roy SK (2008) Photosensitivity analysis of gallium nitride and silicon carbide terahertz IMPATT oscillators: comparison of theoretical reliability and study on experimental feasibility. IEEE Trans Device Mater Reliab 8(3):608–620

Mukherjee M, Mazumder N, Roy SK (2011) Prospects of photo-sensitive indium phosphide based top-mounted and flip-chip IMPATT oscillators for application in terahertz regime. Int J Electron 98(4):449–458

Tripathy PR, Mukherjee M, Choudhury SK, Pati SP (2014) Ge/Si based DDR IMPATTs: a potential millimeter-wave source at 140 GHz atmospheric window. Adv Sci Lett 20(3–4):846–849

Xie Y, Guo B, Li J, Ku G, Wang LV (2008) Adaptive and robust methods of reconstruction (ARMOR) for thermoacoustic tomography. IEEE Trans Biomed Eng 55(12):2741–2752

Zhao H, Zou L, Geng X, Zheng S (2015) Limitations of mammography in the diagnosis of breast disease compared with ultrasonography: a single-center retrospective analysis of 247 cases, Published online 2015 Apr 21. https://doi.org/10.1186/s40001-015-0140-6

Zuar Y, Roetzel W (1997) Bioheat equation of human thermal system. Chem Eng Technol 20(4):268–276. https://doi.org/10.1002/ceat.270200407

Part IX

Pathophysiology of Diseases Causing Physical Disability

Pathophysiology of the Disease Causing Physical Disability

23

Sachchida Nand Rai, Hareram Birla, Saumitra Sen Singh, Walia Zahra, Aaina Singh Rathore, Hagera Dilnashin, and Surya Pratap Singh

An "individual with a disability" has a physical impairment or record of such an impairment that substantially limits a "major life activity." Today, "person with a disability" is a more widely used term than "handicapped" as this phrasing reflects a positive approach by putting people first, not the disability. The disability term is used to refer an individual functioning including motor, sensory, cognitive, intellectual impairment, mental illness, and various types of chronic diseases (Oliver 1995; Crow 2008).

International Classification of Functioning, Disability and Health (ICF) is a classification of function and disability of the health components. On May 22, 2001, the World Health Assembly approved its abbreviation "ICF." This classification is first created in 1980 and called the International Classification of Impairments, Disabilities, and Handicaps (ICIDH) by the World Health Organization (Organization, W.H 1980).

A "physical disability" is the long-term loss or impairment of a person's body function, resulting in a limitation of physical performance and daily life activities. It can affect a person's ability to move about, to use arms and legs effectively, to swallow food, and to breathe unaided. It identifies and describes the more prevalent physical disabilities in society today (Oliver 1995).

A physical disability may have existed since birth or it could be result of an accident, illness, infection, disease, degeneration, and medical condition or the result of congenital factors. There are a wide variety of situations which people experience. Each person may have different causes, symptoms, and management strategies making it difficult to generalize physical disabilities.

S. N. Rai · H. Birla · S. S. Singh · W. Zahra · A. S. Rathore · H. Dilnashin · S. P. Singh (✉)
Department of Biochemistry, Institute of Science, Banaras Hindu University, Varanasi, Uttar Pradesh, India

S. Paul (ed.), *Biomedical Engineering and its Applications in Healthcare*,
https://doi.org/10.1007/978-981-13-3705-5_23

23.1 Causes of Physical Disability

Physical disabilities may either be congenital or acquired that is caused by a condition in later stage. On the basis of occurrence time, it is divided into:

- **Congenital disability:** The disability developed before or during the birth of a child. It is commonly related to the diseases acquired by mother during pregnancy. It is developed by the influence of external factors such as poisoning, alcohol, and drug consumption during pregnancy or using wrong medication. Genetic cause, that is, genetic incompatibility between parents or gene mutation, also contributes to congenital disability. Premature birth, head injury, and lack of oxygen due to respiratory tract obstruction might also cause the disability.
- **Acquired disability:** This type of disability is developed by external causes anytime in the life of a person. Accidents, serious infection, and other illnesses cause impairment after birth.

A person with a physical disability may require some assistance or the use of some sort of equipment (such as wheelchairs, walking sticks, and artificial limbs) to aid mobility. The two major categories of physical disabilities are (a) the musculo-skeletal disability, including muscular dystrophy and dwarfism, and (b) the neuro-muscular disability, including cerebral palsy, spina bifida, and spinal cord injury (SCI).

23.1.1 Cerebral Palsy

Cerebral Palsy (CP) is a group of non-progressive neurological disorders that affects normal movement in different parts of the body. CP is not contagious; necessarily it does not affect intelligence and cognitive ability (Minear 1956). The word "cerebral" refers to the brain's cerebrum, which regulates motor function, and "palsy" refers to paralysis which accounts for the lack of voluntary movement in certain parts of the body (Reddihough and Collins 2003). It is most common among young children, causing physical disability, and persists throughout the life span (Rosenbaum et al. 2007). It affects 2–3 out of 1000 children. In the United States, the major physical disability that affects the functional development of children is CP. Developmental disabilities affect only 17% of children.

CP is characterized by abnormal brain development or injury to the developing brain which results in inability to control motor functions. The brain damage occurs before, during, or within 5 years of birth that prevents the proper brain development. It has the potential to interfere with the overall development of a child accompanied by disturbances of sensation, cognition, learning ability, communication, and a seizure disorder (Jones et al. 2007). Its signs and symptoms vary from person-to-person. Some persons have difficulties in walking and sitting, while others have trouble in grasping objects. The symptoms can worsen over the time depending on the part of the brain affected.

23.1.2 History

CP was first reported in 1861 by **William John Little** and initially was called Little's disease named after Dr. Little (Rosenbaum et al. 2007; Bax et al. 2005). It was also referred to as "cerebral paralysis." Little correlates to lack of oxygen and brain damage during labor and delivery (Jones et al. 2007). It was originated from the German word *zerebrale Kinderlähmung* (cerebral child-paralysis). The term "cerebral palsy" was coined by **Sir William Osler** in the book entitled *Cerebral Palsies of Children* (Panteliadis et al. 2013). Later, in 1897, **Dr. Sigmund Freud** was first to state that CP might be caused by abnormal development before birth (Jones et al. 2007).

In 1950, **Leonard H. Goldenson** co-founded the United Cerebral Palsy Association (UCPA) to understand the disorder and promote awareness, when his daughter died with CP-associated complications. Today, UCPA is the fifth largest agency in the United States.

23.1.3 Etiology

Causes of CP are diverse including congenital, inflammatory, infectious, genetic, traumatic, and metabolic. In more than 30% children, there are no known risk factors. CP can be divided into prenatal, perinatal, and postnatal time periods. Around 70–80% of the cases are due to prenatal risk factors, with birth asphyxia cases being 10% (Sankar and Mundkur 2005; Torfs et al. 1990).

- **Prenatal:** These include prematurity, multiple births, hypoxia, genetic and metabolic disorders, maternal illness (such as thyroid disease, iodine deficiency), thrombophilic disorders including factor V Leiden mutations, teratogenic exposure, chorioamnionitis, and fetal brain malformation (Jones et al. 2007; Reid et al. 2006).
- **Perinatal:** These include birth asphyxia due to instrumental delivery, premature birth (<32 weeks or < 2500 g), abnormal fetal presentation, placental abruption, infection, blood incompatibility, rupture of the uterus or prolonged/obstructed labor, and postmaturity (Jones et al. 2007; Sankar and Mundkur 2005).
- **Postnatal:** These include seizure (within 48 h of birth), hyperbilirubinemia, sepsis, respiratory distress, cerebral infarction, intraventricular hemorrhage, periventricular leukomalacia, meningitis, and head injuries prior to 3 years (including child abuse and shaken baby syndrome).

In majority of cases, the risk factors are abnormal delivery, such as a breech birth (feet first), premature birth, low birth weight (less than 1500 g), poor maternal health, and being a twin or triplet (Jones et al. 2007; Torfs et al. 1990).

23.1.4 Classification

There are various parameters to classify CP that describe its role in influencing treatment. CP is divided on the basis of severity level, physiological (motor activity), topographical pattern of limb involvement, muscle tone impairment, and Gross Motor Function Classification System (GMFCS) (Minear 1956; Jones et al. 2007).

- **Severity:** CP is classified by severity level as mild, moderate, and severe. This classification is not specific and provides very little information, when compared to other means of classification. This is a very common and simple method and can be useful when accuracy is not necessary.
  1. **Mild:** In mild CP, daily activities are not limited and a person can move without assistance.
  2. **Moderate:** In moderate CP, a person needs braces, medications, and adaptive technology to accomplish daily activities.
  3. **Severe:** In case of severe CP, a person will need assistance support and require a wheelchair. They face a significant challenge in accomplishing daily activities (Jones et al. 2007).
- **Physiological (motor activity):** There are the (Minear 1956) two major physiological classifications indicating the brain areas affected by the motor activity disorders. These are spastic/pyramidal and non-spastic/extra-pyramidal.
  1. **Spastic/pyramidal CP:** It is an upper motor neuron damage that results from defects in the brain's corticospinal pathways. It accounts for nearly 70–80% of all CP cases. The predominant features observed in spastic CP are cognitive impairment (about 30%), increased muscle tone, persistent primitive reflexes, clonus, extensor Babinski response, and hyperreflexia (Jones et al. 2007). About 35–60% children with CP develops epilepsy. Quadriplegia and hemiplegia patients have higher tendency to develop epilepsy than diplegia and ataxic CP patients (Sankar and Mundkur 2005).
     - **Monoplegia:** It is a very rare condition, which involves defect in only one limb. Disability depends on the severity of the affected limb (Minear 1956).
     - **Hemiplegia:** It is also referred as unilateral spasticity. It affects one-half of the lateral side of the body. It occurs occasionally with a high rate of partial seizure. It is usually associated with strokes, vascular malformation, and unilateral intraventricular hemorrhage (IVH) or periventricular leukomalacia (PVL). Aphasias (impairment of language) more frequently occur in right than left hemiplegic patient, whereas it is more common in acquired than congenital CP (Minear 1956; Jones et al. 2007).
     - **Triplegia:** It is condition in which both legs and one arm are usually affected. Sometimes it represents hemiplegia (one-half lateral body side is affected) plus paraplegia (one limb is affected) or incomplete quadriplegia (both limbs and arms are affected). As the condition worsens, the affected arm will become shorter (Minear 1956).

- **Quadriplegia:** It is also referred as tetraplegia. It is the most severe form involving both legs and both arms. It accounts for 10–15% of all spastic CP cases, associated with severe asphyxia in all infants and severe IVH or PVL in premature babies. The upper limbs are more severely involved than the lower limbs with intrapartum asphyxia (metabolic academia measured at birth) and acute hypoxia. Approximately 50% of quadriplegic CP patients are at the risk of epilepsy, mental retardation, deafness, and severe vision impairment (Jones et al. 2007; Sankar and Mundkur 2005).
- **Diplegia:** It is also known as "bilateral paralysis," which occurs in 30–40% of all spastic CP cases and 50% was premature. It is associated with prematurity and low birth weight. Lower limbs are more severely affected than upper limbs in this condition. There is an impaired dorsiflexion of the feet with increased ankles tone in mild cases, whereas, in severe cases flexion of hips, knees, and elbows can be seen (Minear 1956; Sankar and Mundkur 2005).

2. **Non-spastic/extra-pyramidal:** It occurs due to the damage of nerve cells outside of the pyramidal tracts (basal ganglia) or the cerebellum. It accounts for 15–20% of all CP cases. This is a global disability with abnormal tone regulation coordination and postural control (Jones et al. 2007).
   - **Dyskinetic:** It accounts for 10–15% of all non-spastic CP patients. It causes damage to basal ganglia or deep motor neurons of thalamus. It is further divided into *athetoid* and *dystonic*. In athetoid CP, infants are hypotonic with random, rapid, jerky movement patterns. Dystonic infants have rigid posturing of neck and trunk. Dyskinetic patients suffer from speech problem such as *dysarthria (Minear* 1956; *Jones* et al. 2007).
   - **Ataxic:** It occurs only in 5% of all non-spastic CP cases. It causes damage to neurons of cerebellum with the impairment of voluntary movement and balance. They also suffer with tremors and poor head control (Jones et al. 2007).

- **Gross Motor Function Classification System (GMFCS)**: The first GMFCS was developed in 1997. It is the most reliable system to classify children with CP according to the movement ability across five age bands from I to V.
  - **I Group** (Before 2 years): Walks without limitations indoors or outdoors and climbs stairs without limitations. Speed, balance, and coordination are reduced.
  - **II Group** (Between 2 and 4 years): Walks with limitations indoors or outdoors and climbs stairs holding on to a rail. Experiences limitations walking on uneven surfaces and inclines, in crowds or confined spaces.
  - **III Group** (Between 4 and 6 years): Usually require assistance mobility device. They have independent floor mobility and can sit independently. Walks indoors or outdoors using a handheld mobility device and climbs stairs holding onto a railing.
  - **IV Group** (Between 6 and 12 years): Self-mobility with great limitations and may use powered mobility.

- **V Group** (Between 12 and 18 years): Physical impairments restrict voluntary control of movement and have no means of independent mobility. Transported in a manual wheelchair (Jones et al. 2007; Sankar and Mundkur 2005).

23.1.5 Pathophysiology

Depending on the etiology, it varies. The major signs that collectively lead to a CP are abnormal postural control, persistence of primitive reflexes, delayed motor activity, and also lesions in brain that result from prenatal, perinatal, and postnatal events. The premature neonatal brain is susceptible to two main pathologies that increase the risk of CP. These are **intraventricular hemorrhage (IVH)** and **periventricular leukomalacia (PVL)**. CP is more common in males (30%) than in females and is more likely caused by acquired injuries like infections, toxins, and environmental effects than genetic (Johnston and Hagberg 2007).

IVH is the result of bleeding into the ventricles of the brain from the subependymal matrix (origin of fetal brain cells). Severity of the IVH increases the risk of CP. Premature infants have underdeveloped periventricular blood vessels because the blood vessels around the ventricles develop late in the third trimester, predisposing them to increased risk of IVH. An anti-inflammatory drug *indomethacin* (also a prostaglandin inhibitor) reduces the incidence and severity of IVH by 50% in low birth weight infants. It also eliminates the parenchymal hemorrhage associated with males, but it has no significant effects on females (Johnston and Hagberg 2007).

PVL has a separate pathological process including IVH as a risk factor. PVL pathogenesis arises from two factors:

(a) **Ischemia/hypoxia**: Distal segments of adjacent cerebral arteries supply the periventricular white matter to the neonatal brain. If one artery is blocked due to thromboembolic stroke, the collateral blood flow from two arterial sources protects the area. Decreased cerebral blood flow in the overall brain (cerebral hypoperfusion) causes damage to this area. The premature infants have low cerebral blood flow; the periventricular white matter is susceptible to ischemic damage. The fetal brain is protected from hypoperfusion by autoregulation of cerebral blood flow as it is limited in premature infants due to immature vasoregulatory mechanisms and underdevelopment of arteriolar smooth muscles (Perlman 1997; Freeman and Nelson 1988).

(b) **Infection and inflammation**: Oligodendrocyte is a supportive brain cell that wraps around the neurons and form myelin sheath. It is also essential for white matter development. In the developing brain, when oligodendrocytes are damaged due to microglial (brain macrophage) cell activation and cytokine release. Cytokines such as interferon gamma (IFN-γ) and tumor necrosis factor alpha (TNF-α) are toxic to premyelinating oligodendrocytes. Cytokines are released in response to activation of fetal immune system during intrauterine infections. Fetal immune system activation also activates microglial cells, which releases free radicals. Premyelinating oligodendrocytes have immature defenses against ROS (reactive oxygen species) as there is low production of *glutathione,* an important antioxidant. IVH causes PVL because of the presence of iron-rich blood. Presence of iron in blood causes iron-mediated conversion of hydrogen peroxide to hydroxyl radical, contributing to oxidative damage (Yoon et al. 1997; Nelson and Willoughby 2000).

23.1.6 Parkinson's Disease

Parkinson's Disease (PD) is the second most common neurodegenerative disorder after Alzheimer's disease. It is a chronic and progressive disease affecting 1–2% of the population over 65 years. It is characterized by the loss of DAergic neurons in the substantia nigra pars compacta (SNpc) and the presence of intracytoplasmic inclusions called Lewy bodies or Lewy neurites (in neurons) (Cho et al. 2012). The loss of DAergic neurons reduces the level of dopamine (DA) and causes motor dysfunction including tremor at rest, rigidity, bradykinesia, and postural instability in PD cases. It also causes non-motor dysfunction like REM (rapid eye movement) sleep behavior disorder, autonomic dysfunction, and cognitive issues (Lim and Lang 2010; McPhee and Hammer 1995).

The cell body of the DAergic neurons is present in the substantia nigra pars compacta (SNpc) and extended to striatum. The cell body DAergic neurons produce dopamine (DA), which releases from the striatum. DA is a neurotransmitter responsible for movement coordination. Reduction in DA level due the DAergic neuronal loss causes characteristic motor symptoms of PD. The motor symptoms appear approximately after 60% loss of the DAergic neurons and 80–85% reduction of DA content in the striatum (Jankovic 2008). Lewy bodies (LBs) are mainly composed of misfolded alpha-synuclein (α-Syn). α-Syn aggregates into oligomers and form fibrils that are rich in β-sheet. Misfolded aggregates can move between the neurons (like prion) and act as template to promote normal α-Syn misfolding process. α-Syn and other protein accumulation occurs long before to the neuronal loss (McPhee and Hammer 1995).

23.1.7 Classification

The motor defect related to PD resulting in movement impairment is called "parkinsonism." There are several neurodegenerative disorders that show parkinsonism, and other features are "Parkinson plus," "atypical Parkinson," and "typical Parkinson."

Parkinsonism It is a syndrome of motor function disorder. It is chronic and progressive, characterized by tremors at rest, and worsens with emotional stress. Other features include rigidity and disordered gait and postural instability. It is caused by several neurodegenerative diseases including PD, which is the most important disease. Trauma, toxic agents, and DA antagonist's drugs are the other causes of parkinsonism. Externally, the brain is normal and atrophic. The hallmarks of the parkinsonism are the depigmentation of SN and locus coeruleus (nucleus in the pons of the brainstem) due to loss of neuromelanin pigment from neurons and accumulation of neuromelanin pigment in the glial cells. Some of the residual neurons in this region contains LBs (intracytoplasmic, eosinophilic, elongated inclusions) (Mohan 2005).

Parkinson Plus It is a syndrome characterized by the impairment of motor function (parkinsonism). In addition, it shows some other features not related to PD. There are five separate Parkinson plus syndrome. These syndromes are multiple system atrophy, progressive supranuclear palsy, parkinsonism-dementia-amyotrophic lateral sclerosis complex, corticobasal ganglionic degeneration, and dementia with Lewy bodies (Louis and Frucht 2007).

Atypical Parkinson It is characterized by Parkinsonian symptoms and other PD-related symptoms (involuntary, autonomous, GIT, etc.). It shows symmetrical response and no tremor at rest, no levodopa (L-DOPA), and no response to deep brain response (activation of SN through electrical shock) (Focke et al. 2011).

Typical Parkinson It is the Parkinson's disease characterized by tremor at rest. It is non-symmetrical and shows response to L-DOPA and also deep brain response (Louis and Frucht 2007; Focke et al. 2011).

23.1.8 History

Before the seventeenth century, **Galen** described it as "shaking palsy" in western medical literature. In 1817, **James Parkinson** published *An Essay on the Shaking Palsy* and established PD as a recognized medical condition. In the essay, Parkinson described six personally observed cases; three cases were observed on the street, termed "street watch methodology" by **professor Andrew Lees (Goetz** 2011). Later in 1877, Jean-Martin Charcot suggested the term Parkinson's disease instead of paralysis agitans or shaking palsy. The disease still remains a mystery. The symptoms are progressive and neurodegenerative, more common in older age (Pahwa and Lyons 2013).

In 1960s, chemical differences in PD patients were identified. The discovery of effective medical treatment of the disease was based on the evidence that the low level of dopamine in substantia nigra causes neurodegeneration. In the 1960s, levodopa was first administered to the patients to treat PD symptoms and became the "gold standard" in medication. In 1957, the Parkinson's Disease Foundation was established in America (Goetz 2011).

23.1.9 Etiology

A total of 90–95% PD cases are sporadic (environmental/genetic susceptibility), whereas only 5–10% are the familial PD cases, i.e., caused by gene mutation, which suggests that PD etiology is multifactorial (Pavlou and Outeiro 2017). Aging is the main risk factor for PD (Chandra et al., 2006). The exposure to environmental factors like pesticides, metals, and solvent could increase the PD development risk, whereas tobacco consumption provides protection against PD development (Xiao et al. 2011).

- **Genetic factors:** The known genetic cause of PD is mutations in six genes (SNCA, LRRK2, MAPT, PINK1, PARK2, and PARK7). The risk factors are polymorphism in three genes (MAPT, LRRK2, and SNCA) and loss-of-function mutations in GBA. The gene SNCA or PARK1 encodes α-Syn; LRRK2 or PARK8 encodes LRRK2 or dardarin; PARK7 encodes DJ1, a mitochondrial peroxidase; PARK6 or PINK1 encodes PTEN-induced putative kinase-1, mitochondrial kinase; and PARK2 encodes Parkin (Corti et al. 2011). PARK genes are inherited through an autosomal recessive, dominant, or X-linked mode. On early onset of PD, mutation in three genes (PINK1, PARK2 and PARK7) can cause autosomal recessive PD.

  SNCA gene is located on chromosome 4q21–23 and initially associated with familial PD. Even in sporadic PD, it is the largest risk factor. α-Syn regulates the dopamine release and transport (Corti et al. 2011). Six-point mutation in the SNCA gene leads to amino acid substitutions, duplication, and triplication of SNCA locus linked to autosomal dominant PD (Pavlou and Outeiro 2017). α-Syn is a presynaptic protein, which plays a role in synaptic vesicle pool regulation and release of neurotransmitter. Misfolded α-Syn causes impairment in protein care-taking systems such as ubiquitin-proteasome system (UPS) and autophagy-lysosome pathway (ALS) resulting in α-Syn accumulation. Further suppression in UPS and ALP function leads to neuronal death. Lmx1b is a LIM-homeodomain transcription factor, essential for normal ALP function and is required for the integrity and long-term DAergic neuron survival. DA quinone (enzymatic product of DA oxidation) causes mitochondrial dysfunction and protein modification including α-Syn, parkin, DJ-1, and UCH-L1. This protein dysfunction leads to PD.

  Multiple missense mutations in LRRK2 (leucine-rich repeat kinase 2) gene cause autosomal dominant form of PD and also associated with sporadic PD (Cho et al. 2012). PARK2 is located on chromosome 6q25. It has been identified in autosomal recessive juvenile parkinsonism. It is a ubiquitin E3 ligase that catalyzes the covalent ubiquitin addition to specific target protein for degradation. Mutation in PARK2 causes loss-of-function, which leads to disturbance in ubiquitin-mediated protein degradation.

Mutations in an enzyme glucocerebrosidase (GCase) account for sporadic PD (3%) and juvenile-onset PD cases (25%) and are involved in lysosomal processing. The activity of GCase is reduced in the SN of heterozygous patients (55%) and lower in sporadic PD patients (33%). Inhibition of GCase leads to α-Syn accumulation, which further leads to the inhibition of this enzyme.

- **Environmental factors:** Toxicants like pesticides, metals, and solvents cause alteration and dysregulation of the molecular mechanism related to PD development with the production of reactive oxygen species (ROS) as the main factor and lead to the selective loss of DAergic neurons in SNpc.

  Accumulation of heavy metals such as iron (Fe), lead (Pb), and manganese (Mn) and their combinations generate oxidative stress in substantia nigra (SN), associated with an increased risk of PD development. Pb significantly decreases the release of dopamine and the sensitivity of dopamine D1 receptor postsynaptically in rats (250 ppm of Pb treated for 3–6 weeks) (Kala and Jadhav 1995). Pb also increases lipid peroxidation and causes aggregation of α-Syn. Mn causes "manganism" or manganese-induced parkinsonism. It is a neurotoxic chemical, damages the globus pallidus, and also causes activation of microglia in the substantia nigra pars reticulata (SNpr) (Verina et al. 2011). Fe is an essential element transported into the brain by transferring receptor and divalent metal transporter 1 (DMT 1) (Zheng and Monnot 2012). SN has the highest level of Fe due to the neuromelanin in pigmented DAergic neurons of SNpc. Fe also reacts with ROS produced in the dopamine metabolism and further promotes highly toxic radical generation (Zecca et al. 2002).

  Pesticides like MPTP (1-methyl 4-phenyl 1,2,3,6-tetrahydropyridine) cause selective damage to DAergic neurons in the SN. Dieldrin can cross the blood-brain barrier (BBB) and remain in the brain. It can cause depletion of dopamine level in the brain, increase ROS level in DAergic neurons, reduce cellular ATP production by inhibiting mitochondrial oxidative phosphorylation, and alter mitochondrial membrane potential and cytochrome release (Wagner and Greene 1978; Purkerson-Parker et al. 2001). Quaternary nitrogen herbicide, paraquat, shows structural similarity with active metabolite of MPTP (MPP+). It causes DAergic neuron toxicity by oxidative stress (Rappold et al. 2011). Rotenone, an insecticide, induces a loss of striatal dopamine and α-Syn accumulation and poly-ubiquitin positive aggregates in remaining DAergic neurons (Betarbet et al. 2000).

  Solvents that are lipophilic in nature are easily absorbed by the CNS and PNS tissues associated with acute parkinsonism. Trichloroethylene (TCE) inhibits the activity of mitochondrial enzyme complex I, increases oxidative stress markers, and causes a significant loss of DAergic neurons in SNpc in a dose-dependent manner (Guehl et al. 1999).

  Exposure to chemicals also triggers epigenetic mechanism (including DNA methylation, acetylation of histone protein and miRNAs) that influence gene expression involved in neurodegenerative diseases (Pavlou and Outeiro 2017).

23.1.10 Pathophysiology

PD pathogenesis is based on the specific cardinal signs of the disease and levodopa response. It also involves several molecular mechanisms that lead to neuronal cell death including protein processing pathways, oxidative stress, mitochondrial dysfunction, microglial activation, and inflammation. These mechanisms are the result of reduced GSH levels, α-Syn aggregation, proteasome impairment, and autophagy dysfunction (Navarro-Yepes et al. 2014). PD has no cure; existing therapies give only brief relief from the motor symptoms of the disease by improving the dopamine deficit (Jankovic 2008). In PD, selective degeneration of monoamine-containing cell occurs in the brainstem and basal ganglia, particularly DAergic neurons of the SN. In addition, LBs along with parkin, synphilin, neurofilaments, and synaptic vesicle proteins are present. These protein aggregations are also found in basal ganglia, brainstem, spinal cord, and sympathetic ganglia. The gene identification also provides clues about the molecular pathogenesis mechanism (Corti et al. 2011).

The study of MPTP, a potent neurotoxin, provides an important clue about the PD pathogenesis. It is a by-product of a synthetic opioid derivative (meperidine) and selectively injures DAergic neurons that lead to PD-like symptoms. MPTP enters into the brain and converts into *N*-methyl-4-phenyl-dihydropyridine (MPDP +) by the enzyme monoamine oxidase B (MAOB). The enzyme MAOB is present in glial and serotonergic nerve terminals. MPDP + diffuses across glial membranes where it is converted into the active metabolite *N*- methyl-4-phenylpyridinium (MPP +) by non-enzymatic oxidation and reduction. MPP + interacts with complex I of the electron transport chain (ETC) inside the mitochondria and inhibits oxidative phosphorylation. This in turn inhibits ATP production and reduces molecular oxygen metabolism, which allows increase in reactive oxygen species (ROS) formation such as peroxide, hydroxyl radicals, and superoxide radicals. ROS react with lipids, proteins, and nucleic acids and cause cell injury (Navarro-Yepes et al. 2014).

Rotenone, an insecticide which inhibits mitochondrial complex I, leads to parkinsonism in animals providing evidence in support of a role for mitochondrial dysfunction and oxidative damage PD pathogenesis. Paraquat exposure, a common herbicide, inhibits complex I, which can lead to selective degeneration of DAergic neurons and α-Syn aggregation. Paraquat is structurally similar to MPP +. Complex I activity impairment has been observed in cell lines derived from PD patients, and NADH dehydrogenase 3 (a genetic variant in complex I) is associated with a reduced risk of PD among Caucasians. Thus, mitochondrial complex I activity alterations play an important role in the PD pathogenesis. Exogenous dopamine addition is toxic to neurons in culture; auto-oxidation of DA generates superoxide radicals. DA is metabolized by monoamine oxidase to generate hydrogen peroxide (H_2O_2). The conversion of superoxide to H_2O_2 is catalyzed by superoxide dismutase. Further, H_2O_2 is converted into water by glutathione peroxidase and catalase. H_2O_2 also reacts with ferrous iron to form highly reactive hydroxyl radicals. Thus, DA within DAergic neurons may provide a source of ROS, when coupled with reduced complex I function and may promote cell death.

Misfolded, damaged, or unassembled proteins undergo ubiquitin-mediated degradation. Ubiquitin is a 76-residue protein which marks proteins by a proteolytic complex (proteasome) for processing. A missense mutation in a component of the UPS, ubiquitin carboxyl terminal hydrolase L1, has been found to be associated with autosomal dominant PD in one family.

23.1.11 Multiple Sclerosis

Multiple Sclerosis (MS) means "many scars" and is also known as disseminated sclerosis. It is the most common demyelinating diseases of the central nervous system (CNS) with genetic and environmental effect (Geurts and Barkhof 2008; Wingerchuk and Carter 2014). It is a progressive disease, with the ratio of three women to two men. The onset of the disease occurs usually at the age of 20–40 years. MS involves an immune-mediated inflammatory process which attacks myelinated axon, destroying the myelin and the axon in variable degrees and producing significant physical disability (>30%). In MS, immune system also damages oligodendrocytes (myelin producing cells) and underlying nerve fibers (Goodin et al. 2002; Milo and Kahana 2010). MS is a long-lasting disease and causes problems in vision, balance, muscle control, and other body functions with the involvement of brain, spinal cord, and optic nerves (eyes). The hallmark of the MS is the presence of focal demyelinated plaques or lesions in discrete areas (Popescu et al. 2013). In worse condition, the disease presents as recurrent attacks called relapses.

23.2 Plaques

Plaques appear as grey-pink, swollen, and usually bilaterally symmetric areas and occur anywhere within the white matter of the CNS. Cerebellum, brainstem, spinal cord, and optic nerves are the most frequently affected sites. According to the age of the plaque, its features vary:

1. ***Active Enlarging Plaque (AEP)***: In AEP, there is accumulation of macrophages and lymphocytes at the plaque margin, where demyelination occurs. In addition, loss of oligodendrocytes and presence of reactive astrocytosis (abnormal increase in astrocytes number due to the destruction of nearby neurons) with numerous lipid-laden macrophages (microglia) are observed in the plaque. The axons are generally intact in the plaque (Popescu et al. 2013).
2. ***Old Inactive Plaque (OIP)***: In OIP no perivascular inflammatory cells infiltrate, and there is nearly total absence of oligodendrocytes and also some axonal loss. There is only limited regeneration of myelin in the plaque area, and as a result complete demyelination occurs. Well-developed gliosis (nonspecific reactive change of glial cells in response to damage) but less prominent astrocytes can be seen (Geurts and Barkhof 2008).

23.2.1 History

In 1868, **Jean-Martin Charcot** defined multiple sclerosis and gave its name (Orrell 2005). He defined and treated many neurological disorders and is also known for treatment of hysteria with hypnotism. During the1800s and 1900s, there are hundreds of therapies with no successful treatment of MS. Individuals with MS are once given ineffective and dangerous therapies including deadly nightshade (plant with poisonous fruits), arsenic, mercury, and injections of malarial parasites.

In 1951, MS relapses were first treated with a steroid called cortisone. It was found that it had no long-term effects on the disease, but it shortens the duration of relapse and to reduce its severity.

In 1993, the first drug for the effective long-term treatment of MS was approved (Orrell 2005). These are interferon beta-1a, interferon beta-1b, and glatiramer acetate (also known as "A-B-C" drugs because of their brand names) (Wingerchuk and Carter 2014; Goodin et al. 2002). At present day, 15 long-term treatments are approved for MS relapse. These long-term treatments are also referred as disease-modifying therapies (DMTs) (Wingerchuk and Carter 2014).

23.2.2 Etiology

The etiology of MS is still unknown, but some risk factors have been implicated in its pathogenesis including the role of both genetic and environmental risk factors. The etiology of MS is believed to be a viral infection (CNS), an autoimmune disorder, and a toxic encephalopathy (Watson and Killgore 2016).

- **Viral infection:** In the etiology of MS, two hypotheses were proposed – *poliomyelitis hygiene hypothesis* (most accepted) and *prevalence hypothesis*. According to the poliomyelitis hygiene hypothesis, the infection caused by the pathogen during early childhood provides protection from MS, whereas late contact with the same pathogen causes MS. Another theory, the prevalence hypothesis, is based on Faroe Islands epidemics. It states that in region of high MS frequency, the pathogens that are more common causes MS (Milo and Kahana 2010). **Warner** and **Carp** first proposed the potential role of Epstein-Barr virus (EBV). They emphasized that during teen years, infectious mononucleosis (IM) occurs shortly after which the incidence of MS peaks (Tselis 2011). EBV is a double-stranded virus (184 kb) and it is shed in saliva. They can cause asymptomatic persistent infection in young children. The virus mainly infects B-cells of the immune system and results into the activation and proliferation of B-cell. These infected B-cells are eliminated by CD8+ cytotoxic T-cells. (Tselis 2011; Steiner et al. 2001).
- **Environmental factors:** There are a number of environmental factors associated with MS including geographical region, sunlight, and ultraviolet radiation (UVR). Latitude is the most salient feature of MS cause and can be strongly correlated with the duration and intensity of sunlight. The higher the exposure of

sunlight (on average 2–3 h per day), the lesser the risk of MS and the mortality rate, whereas it increases the mortality rate of skin cancer (Milo and Kahana 2010; Ebers 2008).

The major source of vitamin D is sunlight exposure, but it is also supplied from the food. At moderate or high altitude, the intensity of sunlight decreases; at this stage, the only source of vitamin D is food. Ultraviolet B radiation (290–320 nm) coming along with the sunlight is absorbed by the cutaneous 7-hydroxycholesterol and converted into previtamin D3. Previtamin D3 isomerizes into vitamin D, which undergoes a series of hydroxylation in the liver and kidney and is converted into 25-hydroxyvitamin D3 (major circulating form) and 1,25-dihydroxyvitamin D3 (biologically active hormone). They play an important role in calcium and phosphate metabolism and maintain skeletal health. Deficiency of vitamin D in higher latitude, an important risk factor, leads to MS (Wacker and Holick 2013).

In childhood, the incidence of MS is low, whereas it reaches to the peak between the ages of 20 and 40 years. After that it will slowly declines and become rare after 50 years of age. Women are at a greater risk of MS than men. They are affected with the disease about 2 years earlier than men (Ebers 2008). There are certain other factors that reduce the incidence of MS including tetanus toxoid vaccination and some antibiotics and antihistamine (Milo and Kahana 2010).

23.2.3 Classification

Pain is the most common symptom associated with MS patients (70%). It can be originated from any point of the body during the course of the disease. There are certain drugs that accounts for nearly 30% of all symptomatic treatment. Pain is defined as the "highly unpleasant physical sensation caused by illness or tissue damage" (Merskey 1991). "Pain syndrome" is classified as **nociceptive pain** and **neuropathic pain**. Nociceptive pain is caused by primary afferents activation in somatic or visceral tissues. It occurs as an appropriate encoding of an actually or potentially tissue damaging event (noxious stimulus). Neuropathic pain is the dysfunction of peripheral nervous system (PNS) or central nervous system (Marchand 2008; Solaro et al. 2013). Pain associated with MS is classified into four categories:

1. **Continuous Central Neuropathic Pain:** The pain is related to the brain and spinal cord due to demyelinating lesions and axonal damage involved in pain. These MS lesions can cause CNS hyperexcitability and disruption of spinothalamic pathways that result in pain syndrome (O'connor et al. 2008). It occurs in nearly 50% of all MS patients and is the most common pain syndrome. The classical signs of pain are allodynia (painful response to non-painful stimuli) and hyperalgesia (increase pain sensation due to noxious stimuli). A constant burning sensation involving lower limbs has been reported in 40% patients. The

pain most frequently occurs distally than proximally. In 30% patients, pain is reported as deep or muscular aching (Solaro et al. 2013).

2. **Intermittent Central Neuropathic Pain:** Abnormalities in somatosensory pathways depending on the location of the lesions can result in the most common pain conditions in MS patients, extremity pain, trigeminal neuralgia, and Lhermitte's sign (Solaro et al. 2013; O'connor et al. 2008).

   Extremity pain: It is also referred as "dysesthetic" extremity pain (Clifford and Trotter 1984). MS lesion affects the inhibitory function of GABA interneurons in the spinal cord nociceptive pathways (O'connor et al. 2008). At the onset of the disease, this type of pain is reported in 11% MS patients, whereas in 23% patients it occurs during the course of MS (Indaco et al. 1994). It is a chronic pain in MS patients and described as bilateral continuous "burning pain" that affects legs and feet (Clifford and Trotter 1984; Moulin et al. 1988).

   Trigeminal neuralgia: It is an intense painful sensation in the side of the face. The prevalence of this type of pain range from 1.9% to 6.3% (Solaro et al. 2013). It is the result of demyelination of trigeminal sensory fibers within the nerve root or brainstem (Love and Coakham 2001). It occurs due to an electrical ectopic generation caused by a plaque or compression by blood vessels at the root entry of trigeminal neuralgia nerve (Love and Coakham 2001).

   Lhermitte's sign: It is quite common and reported in 40% MS patients. It is caused due to the lesions in the cervical cord. It is a painful short-lasting electric shock-like sensation related to the neck and moves from the back of the neck to the limbs traveling throughout the body (Solaro et al. 2013; O'connor et al. 2008; Al-Araji and Oger 2005).

3. **Musculoskeletal Pain:** Lesions affect motor neurons and cause damage to muscles, tendons, ligaments, or soft tissue that result in the painful muscle contraction (O'connor et al. 2008). Musculoskeletal pain in MS feels like the pain from common injuries unrelated to MS, such as a sprain or pulled muscle. It is not considered as neuropathic pain, although it is caused by demyelination. There is no effect on somatosensory pathways as no nerve damage appears in this type of pain. Hip and back pain are caused because of muscle spasticity and weakness and cause problems with balance. Back pain also arises due to the inability to stand or sit for long because of fatigue and walking difficulties (Solaro et al. 2013).

4. **Mixed Neuropathic/Non-neuropathic Pain:** It is the complex mixture of all types of pain including neuropathic, inflammatory, and musculoskeletal. Headache including migraine is the most common example of the mixed neuropathic pain (O'connor et al. 2008).

23.2.4 Pathophysiology

The pathophysiology of MS is usually related to the axonal conduction and the symptoms of MS (both positive and negative) during the course of the disease. The pattern of pathophysiology is revealed by magnetic resonance spectroscopy (MRI).

Understanding of the mechanism including demyelination, inflammation, and defect in synaptic transmission and circulating blocking factors play an important role MS pathophysiology (Frohman et al. 2006).

Axonal conductance blocked due to the axonal injury is the key feature of the disability in MS, which occur due to the demyelination. Axonal injury is associated with hallmarks of the MS including demyelination, inflammation, cellular infiltration, and defects in synaptic transmission and circulating blocking factors (Bitsch et al. 2000). Demyelinating lesion is the dominant feature of both peripheral nervous system (PNS) and central nervous system (CNS). The block of conductance specifically occurs at the demyelination site of the axonal region, whereas in the region that is not affected on the either side of the lesions, the conductance is not impaired. The blocking of the conductance at the demyelination site is related to the magnitude of myelin loss and also the time elapsed during the demyelination (Smith and McDonald 1999). Local inflammation, rise in the temperature of the body, and large number of impulse conductance also block the axonal conductance. In chronic active lesions, demyelination is associated with the immunoglobulin deposition and association of myelin to macrophages converted into droplets which undergo phagocytosis (Frohman et al. 2006).

Inflammation modifies the property of astroglia and microglia and contributes to neurological deficit. Astrocytes play an important role in regulation of potassium ion concentration in the brain. Astrocytic processes are also associated with neurons, dendrites, synapses, and central nodes of Ranvier. Disturbance in these processes or cells affects the normal function. There are certain cytokines that play an indirect role in conductance block like tumor necrosis factor alpha (TNF-α) and interferon gamma (IFN-γ) (Bitsch et al. 2000; Smith and McDonald 1999). These cytokines have direct role some ion channels including sodium ion channels and on the energy production in mitochondria. TNF-α and IFN-γ are the potent stimulators of the enzyme inducible nitric oxide synthase (iNOS), which produces nitric oxide in high concentration. Ultimately, nitric oxide blocks the axonal conductance within minutes. A recently introduced therapy for MS is the treatment with interferon beta (IFN-β) that particularly inhibits the iNOS activity (Bitsch et al. 2000). Inflammation, not restricted to the white matter, also occurs in the cerebral cortex. There are several factors associated with inflammation that disrupt the synaptic transmission in normal tissues. These factors are interleukin-1 (IL-1), interleukin-2 (IL-2), interferons (IFN), and nitric oxide.

23.2.5 Huntington's Disease

Huntington's disease (HD) is a fatal, progressive neurodegenerative illness characterized by the striatal neuron projections. It is an autosomal dominant disorder inherited through a gene (HTT) localized on the short arm of chromosome 4 (4p) (Reiner et al. 1988; Vonsattel et al. 1985). It is caused by the accumulation of huntingtin protein (Htt) containing abnormal polyglutamine [poly (Q)] expansion. A triplet (CAG) expansion mutation in the HTT gene causes Htt accumulation

(Steffan et al. 2004). At early onset of the disease, HD patients develop parkinsonian symptoms with tremor and stiffness, but at later onset, patients develop chorea (involuntary, rapid, jerky movement) and athetosis (slow writhing movement of the proximal limbs and trunks) caused by the GABAergic neuron (striatum) degeneration (McPhee and Hammer 1995). In addition to the movement disorder, HD is also characterized by cognitive defects and psychiatric disturbances as it also degenerates neurons in the deep layers of the cerebral cortex (early stage) and extends to the other brain regions, including the hippocampus and hypothalamus (later stage). Dopamine (DA) blocks remaining striatal neuronal inhibition and reduces the involuntary movements. HD is the only known human condition in which the homozygous mutant phenotype is identical to the heterozygous mutant phenotype (sometimes referred to as a "true dominant") (McPhee and Hammer 1995; Folstein et al. 1990).

The HTT gene is located on the short arm of chromosome 4 (4p) and encodes for a 3144-amino acid long protein, **huntingtin (Htt)**, that widely expresses and interacts with several proteins involved in intracellular trafficking and endocytosis, gene transcription, and intracellular signaling (Li and Li 2004). The protein contains a trinucleotide (CAG) repeat of 11–34 copies that encodes a [poly (Q)] domain and the expansion of [poly (Q)] causes the mutation in protein. The mutant protein is degraded, and the resulting fragments contain the glutamine repeats. The fragments form aggregates and are deposited into nuclear and cytoplasmic inclusions (Walker 2007). They may also bind abnormally to other proteins and interfere with normal protein processing or disrupt mitochondrial function. In the cerebral cortex, nuclear fragments may interfere with nuclear functions such as gene expression and the mutant protein reduces the brain-derived neurotrophic factor production by suppressing its transcription. In addition, normal protein blocks the procaspase-9 processing, thereby reducing programmed cell death (apoptosis) and thus is protective for cortical and striatal neurons. Therefore, both loss of neurotrophic support and enhanced caspase activity could promote striatal cell loss in HD (McPhee and Hammer 1995).

23.2.6 History

In 1842, the disease was first recorded by **Charles Oscar Waters** in his letter, published in *Practice of Medicine*. In 1860, Johan Christian Lund, a physician, also gave a description of the disease, high prevalence of dementia, along with jerky movement, particularly prevalent in Setesdalen area (Norway) (Bhattacharyya 2016). In 1872, **George Huntington** first described the condition of HD as an inherited disorder in his paper ***On Chorea.*** Later his paper was published in the *Medical and Surgical Reporter* of Philadelphia and the disease came to be known as Huntington's chorea. "Chorea" comes from the Latin and Greek words meaning *chorus* or a *group of dances* (Walker 2007; Vale and Cardoso 2015). Sir William Osler's interest in the disease combined with his influence in the medical field helped spread awareness of HD by the end of the nineteenth century.

In 1983, the result of US-Venezuela Huntington's Disease Collaborative Research Project discovered the approximate location of the gene involved in the disease. In 1993, exact location of the gene (4p16.3) was reported by the group. This was the first locus of an autosomal disease to be discovered by using linkage analysis.

In 1996, transgenic mouse studies and other animal models of the disease were developed and in 1997, mutated protein fragments (mHtt) were found to be misfolded and the nuclear inclusions caused by the protein were discovered (Bhattacharyya 2016).

23.2.7 Etiology

The genetic and environmental factors attribute to the CAG trinucleotide repeat by 40% and 60%, respectively (Mo et al. 2015).

- **Genetic factors:** It is caused by the single gene (HTT) defect or mutation on chromosome 4 which causes the many more repetition of a part of DNA. The repetition is called CAG trinucleotide repeat. In normal condition, the part of DNA is repeated 10–28 times, whereas it is repeated 36–120 times in HD patients. As the defect is in the HTT gene, it passes down to the next generation and the number of repeats increases. The higher the number of repeats in a person, greater will be the tendency of developing disease in early stage. Therefore, symptoms of the disease develop at younger ages as the gene passes along the next generation (van Dellen and Hannan 2004). It is an autosomal dominant disorder with 50% transmission to the next generation (Mo et al. 2015).
- **Synaptic and neuronal dysfunction:** The mutated HD gene (mHTT) induces selective dysregulation of neuronal gene expression, neurotransmitter receptors, and synaptic transduction pathways. Before the neuronal loss, there is decrease in the expression of these specific receptors that lead to onset of HD. In the striatum, cortex, and other regions of the brain, both pre-synaptic and post-synaptic functions are disrupted due to the alteration in the expression of the synaptic signaling-mediated molecules. Brain-derived neurotrophic factor, a key protein in the cortex, directly affects the synaptic plasticity by regulating the post-synaptic NMDA receptor (van Dellen and Hannan 2004; Hannan 2004).
- **Environmental factors:** The study of Venezuelan kindreds provides the first evidence for environmental factors in HD (Mo et al. 2015; Wexler 2004). Environmental enrichment effects show beneficial effects on neuronal survival particularly in the cortex and also in other regions of the brain. It delays the onset and progression of the HD. Dietary supplementation with essential fatty acids shows both slow disease progression and increased survival rate. Supplements with antioxidants (coenzyme Q10, polyphenols and flavonoids) have beneficial effects against the damage caused by aging and neurodegenerative diseases. Coenzyme Q10, scavenger of free radicals, is therefore a potential treatment of HD. Polyphenols, obtained from fruits and vegetables, reduce neurotoxicity and

improve motor activity. Physical activity enhances the hippocampal neurogenesis and brain-derived neurotrophic factor in adults (Mo et al. 2015; van Dellen and Hannan 2004; Hannan 2004).

23.2.8 Pathophysiology

In HD, the corpus striatum (caudate nucleus, putamen, and globus pallidus) is abnormal. The most striking features occur in neostriatum (caudate nucleus and putamen), which undergoes diffuse atrophy accompanied by selective loss of neurons with astrogliosis (Jimenez-Sanchez et al. 2016). The loss is more severe in caudate nucleus than putamen, and there is a marked neuronal loss in the deep layers of the cerebral cortex (Raymond et al. 2011). Other regions, including the globus pallidus, subthalamic regions, substantia nigra, medulla oblongata, spinal cord, amygdala, and cerebellum, show varying degrees of atrophy depending on the pathologic grade (Vonsattel et al. 1985).

Htt is a 348-kDa protein and ubiquitously expressed in neurons with high level. It is primarily located in the cytoplasm, but also found in the nucleus and organelles (transport vesicles, synaptic vesicles, microtubules, and mitochondria) (Jimenez-Sanchez et al. 2016). In addition to the glutamine stretch, the N-terminus also contains a polyproline region; both are involved in the pathophysiology. Htt plays a role in myriad normal processes (axonal and vesicular transport, endocytosis, post-synaptic signaling, and cell survival pathways).

In HD, a basis for grading the severity is the extent of gross striatal atrophy, neuronal loss, and gliosis and is divided into five grades from 0 to 4 (Wexler 2004; Raymond et al. 2011).

- In grade 0, a normal gross appearance is observed and has a positive family history suggesting HD occurrence. Neuronal loss in the caudate nucleus occurs 30–40%.
- Grade 1 cases can be microscopically detected and exhibit atrophy of the tail of the caudate with 50% neuronal loss.
- In grade 2, mild-moderate gross striatal atrophy of the caudate with more than 50% neuronal loss and a 16% increase in astrocytosis are observed.
- In grade 3, the caudate nucleus is flat and more severe striatal atrophy with more than 75% neuronal loss is observed.
- In grade 4, the medial surface of the caudate nucleus is concave, and striatal atrophy is most severe with greater than 95% neuronal loss (Wexler et al. 1987).

In HD, the genetic basis is the expansion of a triplet (CAG) repeat, i.e., cysteine-adenosine-guanine repeat encoding a polyglutamine stretch in the N-terminus of the protein product Htt. The association of Htt with the cytoplasmic surface of variety of organelles provides the possibility that might be relevant to neurodegeneration.

N-terminal fragments of mHtt accumulate and form inclusions in the brain's cell nucleus of HD patients. These inclusions are toxic and, hence, pathogenic (Walker 2007; Jimenez-Sanchez et al. 2016).

mHtt undergoes cleavage of the N-terminal region resulting in polyglutamine fragments, which can oligomerize and form both cytoplasmic and nuclear aggregates. These aggregations disrupt the intracellular processes including mitochondrial dysfunction, transcriptional dysregulation of various genes, altered axonal transport of critical factors, disrupted calcium signaling, alterations in proteasomal function, and autophagy (Jimenez-Sanchez et al. 2016).

The spiny GABAergic neurons of the striatum play an important role in Htt toxicity. In previous studies on HD cases, it was observed that indirect pathway is affected, which involves the GABAergic neurons of the striatum extended to the external globus pallidus. The glutamic acid decarboxylase (synthetic enzyme for GABA) is significantly decreased resulting in preferential loss of striatal projections to external globus pallidus and causes choreiform movements (Reiner et al. 1988). In this case, an overall inhibitory GABAergic neuron outflow to the thalamus is decreased. In later stage, direct pathway is affected, resulting in preferential loss of striatal projections to internal globus pallidus, and causes hypokinetic or akinetic state. An enhanced inhibitory GABAergic neuron outflow to the thalamus is seen in this case (McPhee and Hammer 1995; Reiner et al. 1988).

References

Al-Araji AH, Oger J (2005) Reappraisal of Lhermitte's sign in multiple sclerosis. Mult Scler J 11(4):398–402

Bax M et al (2005) Proposed definition and classification of cerebral palsy, April 2005. Dev Med Child Neurol 47(8):571–576

Betarbet R et al (2000) Chronic systemic pesticide exposure reproduces features of Parkinson's disease. Nat Neurosci 3(12):1301

Bhattacharyya KB (2016) The story of George Huntington and his disease. Ann Indian Acad Neurol 19(1):25

Bitsch A et al (2000) Acute axonal injury in multiple sclerosis: correlation with demyelination and inflammation. Brain 123(6):1174–1183

Chandra V et al (2006) Neurological disorders. In: Disease control priorities. World Bank, Washington, DC, p 21

Cho HJ et al (2012) MicroRNA-205 regulates the expression of Parkinson's disease-related leucine-rich repeat kinase 2 protein. Hum Mol Genet 22(3):608–620

Clifford DB, Trotter JL (1984) Pain in multiple sclerosis. Arch Neurol 41(12):1270–1272

Corti O, Lesage S, Brice A (2011) What genetics tells us about the causes and mechanisms of Parkinson's disease. Physiol Rev 91(4):1161–1218

Crow KL (2008) Four types of disabilities: their impact on online learning. TechTrends 52(1):51–55

Ebers GC (2008) Environmental factors and multiple sclerosis. Lancet Neurol 7(3):268–277

Focke NK et al (2011) Differentiation of typical and atypical Parkinson syndromes by quantitative MR imaging. Am J Neuroradiol 32(11):2087–2092

Folstein SE, Brandt J, Folstein MF (1990) Huntington's disease. Oxford University Press, New York

Freeman JM, Nelson KB (1988) Intrapartum asphyxia and cerebral palsy. Pediatrics 82(2):240–249
Frohman EM, Racke MK, Raine CS (2006) Multiple sclerosis—the plaque and its pathogenesis. N Engl J Med 354(9):942–955
Geurts JJ, Barkhof F (2008) Grey matter pathology in multiple sclerosis. Lancet Neurol 7(9):841–851
Goetz CG (2011) The history of Parkinson's disease: early clinical descriptions and neurological therapies. Cold Spring Harb Perspect Med 1(1):a008862
Goodin DS et al (2002) Disease modifying therapies in multiple sclerosis. Neurology 58(2):169–178
Guehl D et al (1999) Trichloroethylene and parkinsonism: a human and experimental observation. Eur J Neurol 6(5):609–611
Hannan AJ (2004) Molecular mediators, environmental modulators and experience-dependent synaptic dysfunction in Huntington's disease. Acta Biochim Pol Engl Ed 51:415–430
Indaco A et al (1994) Chronic and acute pain syndromes in patients with multiple sclerosis. Acta Neurol 16(3):97
Jankovic J (2008) Parkinson's disease: clinical features and diagnosis. J Neurol Neurosurg Psychiatry 79(4):368–376
Jimenez-Sanchez M et al (2016) Huntington's disease: mechanisms of pathogenesis and therapeutic strategies. Cold Spring Harb Perspect Med 7:a024240
Johnston MV, Hagberg H (2007) Sex and the pathogenesis of cerebral palsy. Dev Med Child Neurol 49(1):74–78
Jones MW et al (2007) Cerebral palsy: introduction and diagnosis (part I). J Pediatr Health Care 21(3):146–152
Kala SV, Jadhav AL (1995) Region-specific alterations in dopamine and serotonin metabolism in brains of rats exposed to low levels of lead. Neurotoxicology 16(2):297–308
Li S-H, Li X-J (2004) Huntingtin–protein interactions and the pathogenesis of Huntington's disease. Trends Genet 20(3):146–154
Lim SY, Lang AE (2010) The nonmotor symptoms of Parkinson's disease—an overview. Mov Disord 25(S1):S123–S130
Louis ED, Frucht SJ (2007) Prevalence of essential tremor in patients with Parkinson's disease vs. Parkinson-plus syndromes. Mov Disord 22(10):1402–1407
Love S, Coakham HB (2001) Trigeminal neuralgia: pathology and pathogenesis. Brain 124(12):2347–2360
Marchand S (2008) The physiology of pain mechanisms: from the periphery to the brain. Rheum Dis Clin N Am 34(2):285–309
McPhee SJ, Hammer GD (1995) Pathophysiology of disease: an introduction to clinical medicine. Lange Medical Books, New York
Merskey H (1991) The definition of pain. Eur Psychiatry 6(4):153–159
Milo R, Kahana E (2010) Multiple sclerosis: geoepidemiology, genetics and the environment. Autoimmun Rev 9(5):A387–A394
Minear W (1956) A classification of cerebral palsy. Pediatrics 18(5):841–852
Mo C, Hannan AJ, Renoir T (2015) Environmental factors as modulators of neurodegeneration: insights from gene–environment interactions in Huntington's disease. Neurosci Biobehav Rev 52:178–192
Mohan H (2005) Textbook of pathology. Jaypee Brothers Medical Publishers, New Delhi.
Moulin DE, Foley KM, Ebers GC (1988) Pain syndromes in multiple sclerosis. Neurology 38(12):1830–1830
Navarro-Yepes J et al (2014) Antioxidant gene therapy against neuronal cell death. Pharmacol Ther 142(2):206–230
Nelson KB, Willoughby RE (2000) Infection, inflammation and the risk of cerebral palsy. Curr Opin Neurol 13(2):133–139
O'connor AB et al (2008) Pain associated with multiple sclerosis: systematic review and proposed classification. Pain 137(1):96–111

Oliver M (1995) Understanding disability: from theory to practice. Macmillan International Higher Education, Basingstoke
Orrell RW (2005) Multiple sclerosis: the history of a disease. Demos Medical Publishing, New York
Pahwa R, Lyons KE (2013) Handbook of Parkinson's disease. CRC Press, Boca Raton
Panteliadis C, Panteliadis P, Vassilyadi F (2013) Hallmarks in the history of cerebral palsy: from antiquity to mid-20th century. Brain and Development 35(4):285–292
Pavlou MAS, Outeiro TF (2017) Epigenetics in Parkinson's disease. In: Neuroepigenomics in aging and disease. Springer, Cham, pp 363–390
Perlman JM (1997) Intrapartum hypoxic-ischemic cerebral injury and subsequent cerebral palsy: medicolegal issues. Pediatrics 99(6):851–859
Popescu BFG, Pirko I, Lucchinetti CF (2013) Pathology of multiple sclerosis: where do we stand? Continuum Lifelong Learn Neurol 19(4 Multiple Sclerosis):901
Purkerson-Parker S, McDaniel KL, Moser VC (2001) Dopamine transporter binding in the rat striatum is increased by gestational, perinatal, and adolescent exposure to heptachlor. Toxicol Sci 64(2):216–223
Rappold PM et al (2011) Paraquat neurotoxicity is mediated by the dopamine transporter and organic cation transporter-3. Proc Natl Acad Sci 108(51):20766–20771
Raymond LA et al (2011) Pathophysiology of Huntington's disease: time-dependent alterations in synaptic and receptor function. Neuroscience 198:252–273
Reddihough DS, Collins KJ (2003) The epidemiology and causes of cerebral palsy. Aust J Physiother 49(1):7–12
Reid S et al (2006) Factor V Leiden mutation: a contributory factor for cerebral palsy? Dev Med Child Neurol 48(1):14–19
Reiner A et al (1988) Differential loss of striatal projection neurons in Huntington disease. Proc Natl Acad Sci 85(15):5733–5737
Rosenbaum P et al (2007) A report: the definition and classification of cerebral palsy April 2006. Dev Med Child Neurol Suppl 109(Suppl 109):8–14
Sankar C, Mundkur N (2005) Cerebral palsy-definition, classification, etiology and early diagnosis. Indian J Pediatr 72(10):865–868
Smith KJ, McDonald W (1999) The pathophysiology of multiple sclerosis : the mechanisms underlying the production of symptoms and the natural history of the disease. Philos Trans R Soc Lond Ser B Biol Sci 354(1390):1649–1673
Solaro C, Trabucco E, Uccelli MM (2013) Pain and multiple sclerosis: pathophysiology and treatment. Curr Neurol Neurosci Rep 13(1):320
Steffan JS et al (2004) SUMO modification of huntingtin and Huntington's disease pathology. Science 304(5667):100–104
Steiner I, Nisipianu P, Wirguin I (2001) Infection and the etiology and pathogenesis of multiple sclerosis. Curr Neurol Neurosci Rep 1(3):271–276
Torfs CP et al (1990) Prenatal and perinatal factors in the etiology of cerebral palsy. J Pediatr 116(4):615–619
Tselis A (2011) Evidence for viral etiology of multiple sclerosis. In: Seminars in neurology. Thieme Medical Publishers, Inc., New York
Vale TC, Cardoso F (2015) Chorea: a journey through history. Tremor Other Hyperkinet Mov 5:pii: tre-5-296
van Dellen A, Hannan AJ (2004) Genetic and environmental factors in the pathogenesis of Huntington's disease. Neurogenetics 5(1):9–17
Verina T et al (2011) Manganese exposure induces microglia activation and dystrophy in the substantia nigra of non-human primates. Neurotoxicology 32(2):215–226
Vonsattel J-P et al (1985) Neuropathological classification of Huntington's disease. J Neuropathol Exp Neurol 44(6):559–577
Wacker M, Holick MF (2013) Sunlight and vitamin D: a global perspective for health. Derm Endocrinol 5(1):51–108

Wagner SR, Greene FE (1978) Dieldrin-induced alterations in biogenic amine content of rat brain. Toxicol Appl Pharmacol 43(1):45–55

Walker FO (2007) Huntington's disease. Lancet 369(9557):218–228

Watson RR, Killgore WD (2016) Nutrition and lifestyle in neurological autoimmune diseases: multiple sclerosis. Academic, Amsterdam

Wexler NS (2004) Venezuelan kindreds reveal that genetic and environmental factors modulate Huntington's disease age of onset. Proc Natl Acad Sci 101(10):3498–3503

Wexler NS et al (1987) Homozygotes for Huntington's disease. Nature 326(6109):194–197

Wingerchuk DM, Carter JL (2014) Multiple sclerosis: current and emerging disease-modifying therapies and treatment strategies. Mayo Clin Proc 89(2):225–240. Elsevier

World Health Organization (1980) International classification of impairments, disabilities, and handicaps: a manual of classification relating to the consequences of disease, published in accordance with resolution WHA29. 35 of the Twenty-ninth World Health Assembly, May 1976

Xiao W et al (2011) A genomic storm in critically injured humans. J Exp Med 208(13):2581–2590

Yoon BH et al (1997) Amniotic fluid inflammatory cytokines (interleukin-6, interleukin-1β, and tumor necrosis factor-α), neonatal brain white matter lesions, and cerebral palsy. Am J Obstet Gynecol 177(1):19–26

Zecca L et al (2002) The absolute concentration of nigral neuromelanin, assayed by a new sensitive method, increases throughout the life and is dramatically decreased in Parkinson's disease. FEBS Lett 510(3):216–220

Zheng W, Monnot AD (2012) Regulation of brain iron and copper homeostasis by brain barrier systems: implication in neurodegenerative diseases. Pharmacol Ther 133(2):177–188

Physical Impairments Associated with Diseases: A Pathophysiological Approach

24

Vignesh Kotian, Leela Mounica, Deepaneeta Sarmah, Harpreet Kaur, Geetesh Verma, Radhika Kesharwani, Pabbala Veeresh, Anupom Borah, Kiran Kalia, and Pallab Bhattacharya

Abstract

Today with the advent of chronic illnesses, disability associated with it has found its place as a major contributor to the global burden by impacting lives in all age groups in both developed and developing nations. "Disabilities" roots lie not just in terms of restriction in physical aspects of the patient but also in the context of social, emotional, and mental state too. With an ever-growing aging population prone to disabilities, it is quintessential today to understand the etiology and progression of diseases leading to disability. This chapter aims to focus on the pathogenesis of disease that is likely to cause disability as per the medical and social models of disability.

Keywords

Disability · Musculoskeletal disorder · Neuropsychiatric disorder · Cancer · Diabetes · Cardiovascular disorders

24.1 Introduction

Disability is an amalgamation of conditions involved in causing impairments and hindrance to a person's day-to-day normal activities. People with disabilities are found to have health concerns similar to non-disabled people like for immunization, cancer screening, etc. The current treatment interventions thus aim to fix the

V. Kotian · L. Mounica · D. Sarmah · H. Kaur · G. Verma · R. Kesharwani · P. Veeresh · K. Kalia
P. Bhattacharya (✉)
Department of Pharmacology and Toxicology, National Institute of Pharmaceutical Education and Research-Ahmedabad (NIPER-A), Gandhinagar, Gujarat, India
e-mail: pallab.bhattacharya@niperahm.ac.in

A. Borah
Department of Life Science and Bioinformatics, Assam University, Silchar, Assam, India

S. Paul (ed.), *Biomedical Engineering and its Applications in Healthcare*,
https://doi.org/10.1007/978-981-13-3705-5_24

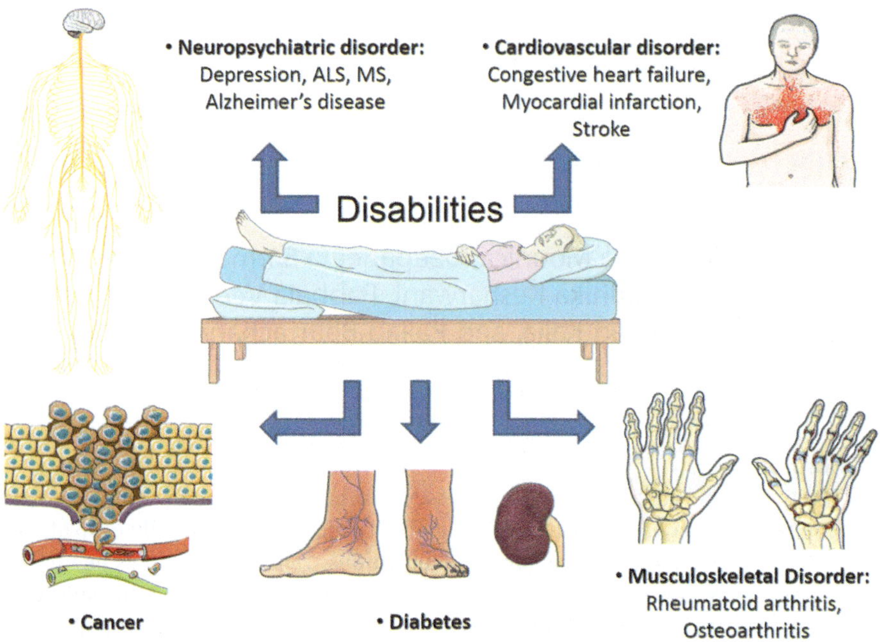

Fig. 24.1 Schematic representation of various diseases that cause disability

disability itself or at least offer help to adapt to disability for increasing the standard of living (Silvers 1998). The medical model of disability holds that a person's functional limitations lie at the crux of disadvantages experienced which can only be rectified by medical interventions (Morris 1996). However, in contrast, the social model of disability dictates it as a complex phenomenon that is interdependent on the interactions between features of a patient's body and society in which he/she resides (Goering 2015). As the population ages, a concomitant increase in the global burden of disability is eminent along with a poor quality of life. Thus to address such issues, the causative factors leading to disabilities need to be identified and worked upon, among which musculoskeletal disorder, cancer, neuropsychiatric disorder, diabetes, cardiovascular disorder, and stroke are found to be at the forefront (Fig. 24.1) (Klijs et al. 2011).

24.2 Musculoskeletal Disorders

Advancement in science and technology has decreased death rates globally. As a result, the life span of individuals has increased, but it has made them more susceptible to non-communicable diseases like musculoskeletal (MSK) disorders (Lozano et al. 2012). With a growing global prevalence and burden of MSK, it has been observed that they account for 21.3% of the total years lived with disability (YLDs) followed by neuropsychiatric disorder (23.2%) (Cross et al. 2014; Goldring

2008; Hoy et al. 2012). This is because MSK involves a large number of diagnoses shown to affect the basic locomotor system. Acute incidences like fractures, sprains, and strains to chronic conditions are associated with life-long pain and disability affecting all age groups. In the burden major conditions like osteoarthritis (OA), rheumatoid arthritis (RA), and others were identified (March et al. 2014).

24.2.1 Rheumatoid Arthritis

Rheumatoid arthritis (RA), a chronic progressive autoimmune disorder, is characterized by a functional loss and prolonged articular damage along with comorbidities in vascular, bone, metabolic, and psychological domains. The etiology of RA encompasses amalgamation of genetic and environmental factors that is found to propagate and perpetuate this disease (McInnes and Schett 2017). Genetic mutation in HLADR01/04 responsible for T-cell recognition of autoreactive peptide, in post-translational modifying enzymes like PADI, genes encoding for MHC-II molecules, in regulatory genes like PTPN22, TNFAIP3, and STAT3 involved in immune activation are observed in RA. To further ameliorate RA, perturbed microflora in the GIT, exposure to silica dust, vitamin D deficiency, and obesity may also contribute to long-term effect on immune regulation and maintenance of self-tolerance (Konig et al. 2016; Scher et al. 2013). It is initiated by alteration in immune reactivity and aberrant B- and T-cell cross-regulation resulting in the production of autoantibodies that recognize a range of post-translationally modified citrulline, carbamylated, and acetylene peptide residue mainly in mucosal tissues (Catrina et al. 2016; Källberg et al. 2011). However, this autoreactive process of conversion into chronic inflamed tissue is poorly elucidated. Localization of this insult is proposed to be a result of activation of a particular osteoclast by autoantibodies which target the citrullinated proteins and immune complexes that are causative factors for bone damage. This is accompanied by TNF-α and IL-8 release causing synovitis (Harre et al. 2012; Krishnamurthy et al. 2016). A complex cascade is said to be initiated which involves the cells and soluble immune mediators which intensifies with duration of RA. The lesion so formed in the synovium contains infiltrating T and B cells, plasma cells, mast cells, and macrophages within the extracellular matrix and synovial fibroblast. However, foregoing studies have only pinpointed potential causative reasons for seropositive disease and not seronegative RA (McInnes and Schett 2017). Thus, there exists a vast lacuna in terms of completely exploring all aspects of RA.

24.2.2 Osteoarthritis

This commonly occurring heterogeneous age-related musculoskeletal disorder is hallmarked by functional loss of cartilage in synovial joint in conjunction with joint pains, crepitus, synovitis, and subchondral bone changes (Dieppe and Lohmander 2005). In the pathology of osteoarthritis (OA), the most commonly affected area is the articular cartilage which undergoes rapid changes during disease progression. The matrix of cartilage cell is rich in collagen and proteoglycans conferring its functional properties. In OA, gradual degradation of the matrix as a whole contributes to the loss of cartilage volume, surface fibrillation, and cleft formation (Dieppe and Lohmander 2005; Sandy 2003). The cascade of events that occur in OA are less explored, but they include osteophyte development at the joint margin, alteration in vascularity, and changes in subchondrial bone leading to joint destruction (Burr 2003). Major environmental causes of OA are found to be mechanical like obesity, joint overload, joint instability, ligamentous instability, etc. (Sharma 2003). However there still is uncertainty whether events occurring in cartilage occur concomitantly or earlier in bone. Recent studies have found that chondrocyte is affected by the signaling molecules released by the synovium and cartilage (Dieppe and Lohmander 2005). Recent studies have also reported inflammatory pathways that contribute to OA progression (Dieppe and Lohmander 2005; Van den Berg et al. 2003).

24.3 Cancer

Worldwide, millions of lives are lost because of cancer. It is found that major contributors among cancers are prostate, breast, colorectal, and lung cancers to total disability-adjusted life-years (DALYs), adding to 18–50% of the total cancer burden (Soerjomataram et al. 2012). With over 200 types of cancers, each year 1.7 million new cases are registered. Some forms are non-invasive and treatable, whereas some are aggressive and require much aggressive treatment methods. Any type, grade, and stage of cancer is recognized as clearly disabling by the multitude effect on the patient's physiology and psychology apart from the effects of radiation, chemotherapy, or biological treatments. Thus there exists a dire need to treat this menace of cancer at the earliest.

Advances in cancer research have identified it as a genetically driven disease with a perturbed pathology involving multiple factors (Weinberg and Hanahan 2000). Today cancer is no longer limited to only genetics of visible tumor but encompasses areas of cancer stemness, its clonal expansion, and its surrounding microenvironment (Hanahan and Weinberg 2011). Carcinogenesis involves the interplay of genetic, epigenetic, and environmental factors. Genetic mutation in tumor suppressor genes, oncogenes, and stability genes is inculpated in most tumorigenic conditions. Chromosomal translocations by intragenic mutations are found responsible for either modulating a gene product activity or by a somatic mutation in an allele of oncogene that can lead to oncogene activation (Davies et al. 2002). In contrast, mutation of tumor suppressor gene leads to a decrease in activity by truncation in the proteins or its epigenetic silencing (Knudson 2002). Stability gene or caretaker genes are responsible for mismatch repair, base excision repairs, and nucleotide excision repairs. A mutation in the latter often leads to increasing the likelihood of another gene mutation (Friedberg 2003). In addition to the major mutation in cancer, there also exist other genes involved with no mutation yet with

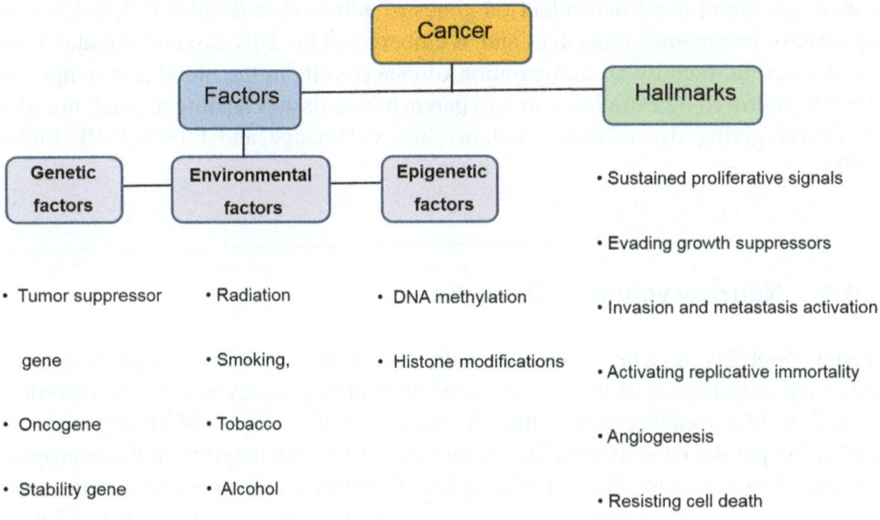

Fig. 24.2 Schematic representation of factors and hallmarks of cancer

an altered expression (Brown and Botstein 1999; Polyak and Riggins 2001). They are implicated with an epigenetic modification that is preserved in cancer cells (Jones and Baylin 2002). Unlike genetic changes, epigenetic changes do exist in normal cells at some stage of development, but in cancer it might lead to unwanted silencing of essential gene and might lead to progression of cancer or increase the likelihood of cancer (Fig. 24.2).

Although cancer is a very general term with each type having different ways of presenting itself, most cancer pathologies possess six major traits—sustained proliferative signals, evading growth suppressors, invasion and metastasis activation, activating replicative immortality, angiogenesis, and resisting death (Hanahan and Weinberg 2011). Sustained proliferative signals promote chronic proliferation of cancer by producing growth factor or activating the tumor stromal cells to supply growth factors or by deregulating receptor signaling (Cheng et al. 2008; Bhowmick et al. 2004). Apart from growth stimulatory signals, inactivated tumor suppressor genes are generally found in most types of animal or human cancers, and majorly the TP53 protein and RB (retinoblastoma-associated) are affected. They are responsible for the fate of cells to proliferate and to undergo senescence and apoptosis (Hanahan and Weinberg 2011). Cancer cells resist cell death by disrupting apoptosis-inducing stress signals by activating oncogene signaling and hyperproliferation due to DNA damage making it malignant and become resistant to therapy (Adams and Cory 2007). Replicative immortality is another striking feature of cancerous growth in contrast to most normal cells with a limited number of division cycles helping it to form tumor masses (Hanahan and Weinberg 2011). Neovascularization, by angiogenesis, fulfills the need for growth and proliferation of cancer cells. During tumor progression, "angiogenic switch" activation results in new vessels and sustains the cancer growth (Hanahan and Folkman 1996). Aggressive invasion and metastasis of cancer are found to be dependent on genes of adhesion molecules in some highly aggressive carcinomas (Hanahan and Weinberg 2011). This cascade initiates with local invasion, then by an intravasation of cancer cells in the blood and lymphatic vessels, followed by extravasation into parenchymal tissues leading to small nodules of cancer giving rise to cancer colonization (Talmadge and Fidler 2010; Fidler 2003).

24.4 Neuropsychiatric Disorder

Today, disability has a broad meaning with no clear definition to cover all its aspects. After the introduction of the social model of disability, today psychiatric disorders contribute to a sizeable chunk of the population with disabilities (W. H. Organization 2001). As per the Persons with Disabilities Act, 1995, the majority of the neuropsychiatric disorders cause a partial or complete disturbance in the person's intellect and emotional behavior resulting in a persistent inability or reduced functional ability to carry out day-to-day activities. Being a major contributor to the global burden of disease, the study of disability associated with neuropsychiatric disorder has become a matter of prime importance (Chaudhury et al. 2006). Figure 24.3 gives a schematic representation of disabilities associated with various neuropsychiatric disorders.

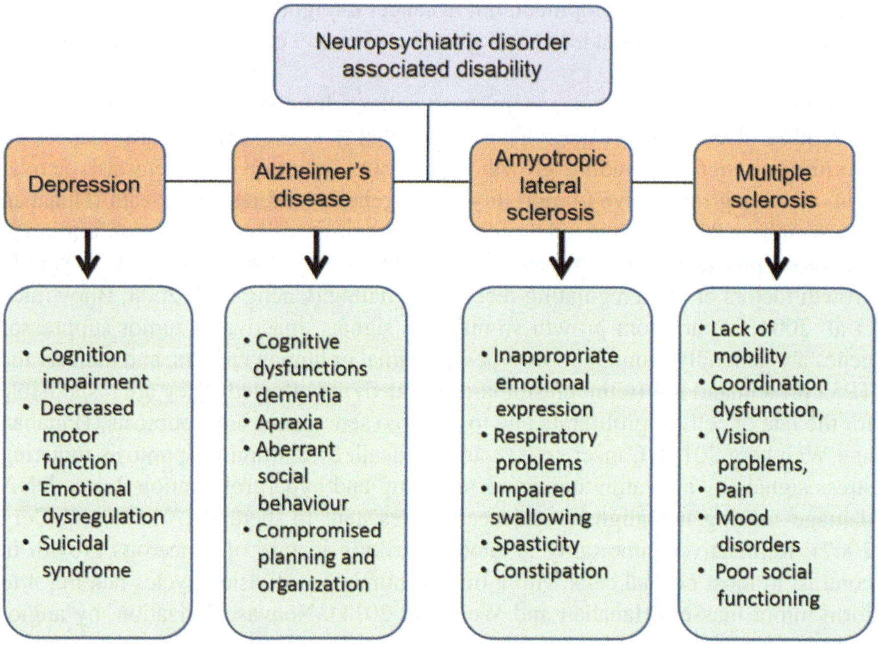

Fig. 24.3 Schematic representation of disabilities associated with various neuropsychiatric disorders

24.4.1 Depression

Depression is the most common and debilitating disorder with a still debatable etiology. It is hallmarked by cognition impairment, decreased motor function, and emotional dysregulation. Several hypotheses involving mechanisms like altered neurotransmitters, inflammatory signal activation, downregulated neurogenesis, and abnormalities in the HPA axis are potentially linked to depression. These findings are however not true in all cases, and the current treatments that target mechanisms are partially explored (Dean and Keshavan 2017). Monoamine hypothesis in depression is well-accepted which links depression to altered levels of monoamines such as serotonin (5-HT), norepinephrine (NE), and dopamine (DA). In the case of serotonergic involvement, an inhibitor of 5-HT has been shown to increase the levels of serotonin in the brain, whereas chronic antidepressant treatment has been shown to downregulate inhibitory presynaptic 5-HT1A somatodendritic autoreceptors (Richelson 2001). NE is said to be involved in mood regulation, and inhibitors of NE reuptake also are effective antidepressants (Leonard 2001). As NMDA antagonists are fast-acting antidepressants, this might suggest the active involvement of glutamatergic system in depression (Machado-Vieira et al. 2010). In recent decades, numerous abnormalities are associated with the HPA axis causing hyperactive response to stress in depression. The overall effect includes hypercortisolemia, hypersecretion of corticotropin-releasing factor (CRF) from the hypothalamus, enlarged adrenal glands, and impairment in negative feedback of the HPA axis (Pruessner et al. 2003). Increased Inflammatory levels of IL-2, IL-1β, TNF-α, IL-6, and PGE2 are also observed in depression (Felger and Lotrich 2013).

Considering the complexity of depression, ranging from moderate to severe intensity, depression leads to the precipitation of neurovegetative symptoms. In addition, it causes primary and secondary disability, making them susceptible to chronic medical illnesses and making them refuse medical assistance. Patients are found to function poorly at work, at school, and in the family. At its worst, depression leads to suicide, and close to 8 lakh of death is due to suicide every year, thus becoming the second leading cause of death among young adults. Barriers like lack of resources, untrained professionals, and social stigma about psychiatric disorders and inaccurate assessment have further contributed to its progression. Thus, WHO has ranked it as the main cause of disability worldwide with an average of 300 million people getting affected annually (Chaudhury et al. 2006).

24.4.2 Alzheimer's Disease

Alzheimer's disease (AD), leading among neurodegenerative disorders, claims significant mortality and morbidity among the elderly population (Duckett 2001), hallmarked by dementia and higher cognitive dysfunction caused by lesion confined to cortical and hippocampus regions. Pathologically indicators of AD are accumulation of interneuronal fibrillary tangles composed of hyperphosphorylated tau proteins, apoptotic neurons, and extracellular amyloid β protein (Aβ) plaques (Alzhelmer 1907). Although the symptomatic and pathological aspect of AD has been extensively studied, a thorough overview of its etiology is still in its infancy. There exist different hypotheses pertaining to the pathogenesis of this disease like cholinergic hypothesis, tau hypothesis, amyloid cascade hypothesis, inflammatory hypothesis, and mitochondrial hypothesis (Bartus et al. 1982; Gray et al. 1987; Hardy and Higgins 1992; Eikelenboom and Veerhuis 1999; Swerdlow and Khan 2004). Among these the amyloid hypothesis is the widely accepted hypothesis that provides direction for the development of current neurotherapeutics. According to this hypothesis, proteolytic cleavage of amyloid β precursor protein (APP) forms Aβ protein (Sun et al. 2012). This processing can be either amyloidogenic (pathological) or non-amyloidogenic pathway (physiological) (Bromley-Brits and Song 2012). Amyloidogenic pathway encompasses APP cleavage by β-secretase producing a secreted form of APP (sAPP β) along with a membrane-bound C-terminal fragment. γ-Secretase cleaves C fragment to produce Aβ ($A\beta_{1\text{-}40/42}$ are the most common isoforms) and CTFγ. The Aβ peptides initially oligomerize and end up with cross-β-sheet fibrils conformation called amyloid plaque (Ahmed et al. 2010; Kotler et al. 2015; Suzuki et al. 2013). These misfolded Aβ are implicated in disrupting cell membranes (Korshavn et al. 2016; Kotler et al. 2014) and impairment of the ion channels (Supnet and Bezprozvanny 2010). Aβ can also exert neurotoxic effects by binding to metal ions, membrane proteins, and free radical generation (Tõugu et al. 2011). This neurotoxic peptide also binds to the neuronal receptor for advanced glycation end product (RAGE) leading to oxidative stress and inflammatory signaling (Du Yan et al. 1996, 1997). The Aβ fibrils cause NMDA receptor excitation via integrins, resulting in aberrant Ca^{2+} influx and eventually cell death (Juhász et al. 2010). Till date there is no cure for AD; thus this progressive disease is said to worsen over time. Intellectual disability associated with AD comprises of apraxia, dementia, aberrant social behavior, and compromised planning, judgment, and organization along with loss of emotional control which are some of the factors hampering the quality of life in AD patients (Zanetti et al. 1993).

24.4.3 Amyotrophic Lateral Sclerosis

Amyotrophic lateral sclerosis (ALS) is a slow, painful, idiopathic, and fatal neurodegenerative disorder of motor nervous system. There seem to be multiple factors involved in the pathogenesis of ALS with a complex interplay between genes and molecular signals (Kiernan et al. 2011; Vucic and Kiernan 2009). ALS seems to be an adult manifestation of motor developmental disorder (Sutedja et al. 2009). Epidemiological studies have highlighted that involvement of many factors has increased the complexity of ALS (Sutedja et al. 2009). Current understanding of involvement of casual genes and various neurotransmitters in familial type of ALS have prompted research on exploring more in depth on the clinical problems and survival aspect of ALS (Vucic and Kiernan 2009). In ALS factors like early childhood infection during developmental stage, strenuous physical exercises in lifetime, smoking, and neurotoxins like β-amino-L-alanine have shown to be linked to ALS (Pasinelli and Brown 2006; Harwood et al. 2009; Kasarskis et al. 2009; Chio et al. 2005; Gallo et al. 2009; Cox and Sacks 2002). The main culprit associated with ALS is the SOD1 mutation; however, there is no clear understanding as to how it causes early neuronal death (Vucic and Kiernan 2009). SOD1 mutation causes misfolding of SOD1 peptide causing intracellular aggregates that result in disruption of axonal transport and hampering proteosomic function (Bruijn et al. 1997; Williamson and Cleveland 1999). Recent advances have also linked genetic mutation to toxic functioning of SOD1 enzyme leading to ROS generation causing neuronal death (Liu et al. 1998). Apart from this, glutamate-induced excitotoxicity is shown activating Ca^{+2}-dependent degrading enzymes and ROS generation resulting in neurodegeneration by damaging intracellular organelles and upregulation of inflammatory markers (Meldrum and Garthwaite 1990; Maher and Davis 1996; Hensley et al. 2006). However, this glutamate-induced excitotoxicity-mediated motor degeneration still remains unclear. Abnormalities in mitochondria, disrupted axonal transport systems, dysfunctioning of sodium/potassium ion pump, and autophagy have also been associated with ALS (Kiernan et al. 2011). Thus, considering possible pathogenesis of ALS, this motor neurodegenerative disease is said to cause a progressive weakness in upper and lower motor neuronal dysfunction which is found to have a profound effect on inappropriate emotional expression, respiratory problems, impaired swallowing, spasticity, constipation, and cognitive impairment accompanied by difficulties making decisions and grasping complex problem (Kasarskis et al. 2014; Cummings et al. 2006; Radunovic et al. 2009).

24.4.4 Multiple Sclerosis

Multiple sclerosis (MS), a chronic progressive CNS disorder, is associated with diffuse neurodegeneration and primary demyelination in spinal cord and brain regions (Lassmann et al. 2007). MS is hallmarked by brain atrophy and primary demyelination with partial preservation of axon (Marburg 1906; Bermel and Bakshi 2006). Axonal degeneration is found in acute MS lesion but does not contribute to neuronal disability as the human brain compensates for the axonal loss. However, this is not true in a chronic condition which is a characteristic feature of progressive MS (Bjartmar et al. 2000, 2003). Another feature of progressive MS is the cortical demyelination being the pathological substrate of cognitive disability in relapse cases also (Brownell and Hughes 1962; Peterson et al. 2001). Cortical lesion may be leucocortical identified by immune cells in white matter lesion of early MS patients (Peterson et al. 2001). Subpial lesion in a progressive MS patient extends across cortical layers associated with inflammation of meninges and microglial activation (Choi et al. 2012; Howell et al. 2011; Kutzelnigg et al. 2005). Gray matter demyelination is found in the cerebellar cortex, gray matter, hippocampus, and spinal cord contributing to MS. Progressive MS is hallmarked by diffuse pathologies in the normal brain. Diffuse axonal loss in white matter of the spinal cord corresponds to the degree of T-cell infiltration in meninges (Androdias et al. 2010). The anterograde and retrograde progression of MS can be indicated by the ballooning of neurons with an abnormal nucleus, and increased phosphorylated neurofilament in gray matter was also observed (Fischer et al. 2013). Failure in remyelination also leads to progression of MS; however, there exist remyelinated shadow plaques in all the stages of MS (Bramow et al. 2010).

Currently the major challenge is there are no models available that can replicate the disease progression along with its diffuse pathology (Mahad et al. 2015). Thus by acting in a multiple manner, MS causes rather bothersome disabling physical symptoms. Prominent among them are lack of mobility, coordination and cognitive dysfunction, vision problems, fatigue, and pain. Adding to them is ameliorated quality of life reduced by mood disorders and poor social functioning (Wu et al. 2007; Feinstein 2011). MS thus ranks second among the disabilities in young adults (Adelman et al. 2013).

24.5 Diabetes Mellitus

Diabetes mellitus is a chronic metabolic disorder which is characterized by hyperglycemia that ensues from insulin secretion or insulin action or both.

24.5.1 Types of Diabetes Mellitus

Type 1 Diabetes It is an immune-mediated diabetes, erstwhile known as insulin-dependent diabetes or juvenile or childhood-onset diabetes (Okur et al. 2017). This type of diabetes accounts for only 5–10% of those with diabetes (Maahs et al. 2010). The hallmark of type 1 diabetes is cellular-mediated autoimmune destruction of pancreatic β-cells which results in the gradual decrease of insulin levels. Immune destruction of markers of β-cells includes islet cell autoantibodies, autoantibodies to glutamic acid decarboxylase (GAD65), autoantibodies to insulin, and autoantibodies to the tyrosine phosphatases IA-2 and IA-2 (A. D. Association 2014). In type 1 diabetes mellitus (DM), the autoimmune trigger is the outcome of certain exposure of environmental factors in genetically susceptible individuals. This genetic susceptibility is strongly coupled with specific HLA genes that encode the major histocompatibility complex proteins (Gerstein 1994). These proteins play a crucial role in immune response regulation and self- versus non-self-cell identification. There are few HLA types associated with the development of type 1 DM in a great extent; among them HLA-DR3 and HLA-DR4 and HLA-DQ genes are frequently expressed in people with type 1 DM, while some other types (e.g., HLA-DR2) appear to be preventative against developing autoimmunity against β-cells (Zeitler 2009; Pociot et al. 2010).

Type 2 Diabetes It is the most typical form of DM formerly known as non-insulin-dependent diabetes or adult-onset diabetes. It accounts for about 90–95% of cases (A. D. Association 2014). It is caused preponderantly by severe insulin resistance and subsequent β-cell failure. In addition to the resistance of insulin, the enhanced demand for insulin could not be met by the β-cells of pancreas because of the defects in the function of these cells (Halban et al. 2014). On the contrary, due to the gradual destruction of pancreatic β-cells, secretion of insulin is reduced with the high demand for insulin (Druet et al. 2006). This loss of β-cells is multifactorial and includes glucolipotoxicity and amyloid deposition that ensued in β-cell apoptosis through oxidative and endoplasmic stress (Jurgens et al. 2011; Poitout and Robertson 2007). Obesity is the prominent cause behind insulin resistance, mainly accountable for causing type 2 diabetes (Ginsberg et al. 1975). Genes and the environment together are other determinants of insulin resistance and β-cell dysfunction. The most robust candidate variants reported were E23K variant in the KCNJ11 gene and the P12A variant in the peroxisome proliferator-activated receptor (PPARG) gene (Altshuler et al. 2000; Nielsen et al. 2003). Other genes include WFS1, HNF1B, GCK, and TCF7L2 (Hertel et al. 2013).

Cardiovascular disease is the major disabling disorder which causes mortality in DM (Nickerson and Dutta 2012). To a great extent, atherosclerosis is common in people with DM than others; in people aged 25–65, DM gradually increases the risk of heart attack five times more than in those without (Khoury et al. 2013). Approximately 12% of diabetic patients are affected by diabetic cardiomyopathy that independently develops from hypertension and coronary artery diseases which lead to overt heart failure, morbidity, and mortality (Lorenzo-Almoros et al. 2017; Papatheodorou et al. 2016). Some of the microvascular complications of diabetes include diabetic nephropathy, retinopathy, and neuropathy which are stimulated by hyperglycemia through initiation of oxidative stress, formation of pro-inflammatory microenvironment, and advanced glycation end products (AGEs) production causing death and disability (Chilelli et al. 2013; Nguyen et al. 2012). The diabetic population is more susceptible to develop these complications than the general population; diabetic retinopathy is the major cause of visual impairment and blindness in people with diabetes and can affect the macula densa, peripheral retina, or both (Afkarian et al. 2013). Around 50% of people with diabetes have few forms of peripheral neuropathy and monodiabetic or polydiabetic neuropathy (Dyck et al. 1993). Diabetic patients also often have cardiovascular autonomic dysfunction which is a type of autonomic neuropathy manifested as irregular heart rate and vascular disorder (Vinik et al. 2003). Another complication of diabetes is diabetic nephropathy which is characterized typically by microalbuminuria which gradually converts to albuminuria which causes renal dysfunction and finally leads to renal failure and is the prominent cause of end-stage renal disease (ESRD) (Brenner et al. 2001; Cade 2008; Drummond and Mauer 2002). Figure 24.4 gives a schematic representation of complications associated with diabetes mellitus.

24.6 Cardiovascular Diseases

Cardiovascular disease is the term for all types of disorders that affect the heart and blood vessels including congestive heart failure, coronary heart disease, hypertensive heart disease, atherosclerosis, arrhythmia, and stroke (Fig. 24.5).

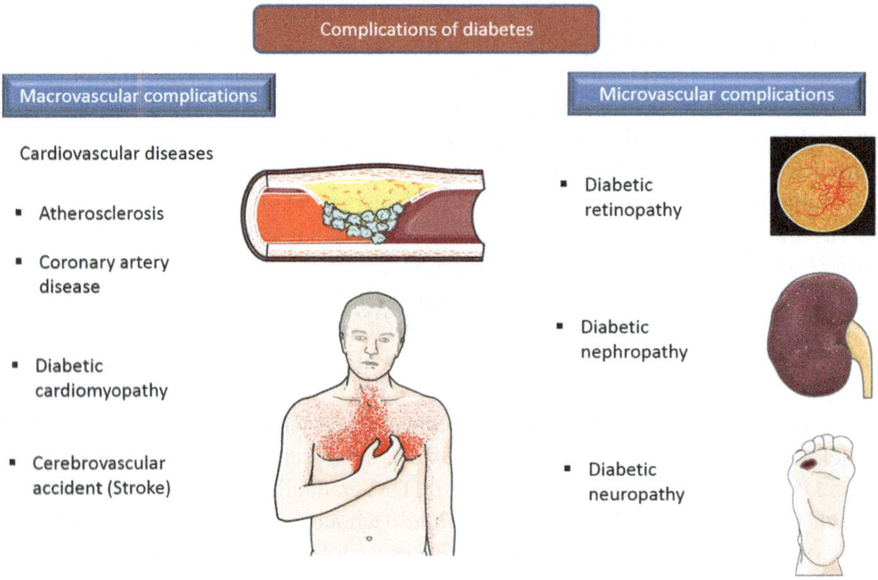

Fig. 24.4 Schematic representation of complications associated with diabetes mellitus

24.6.1 Myocardial Infarction

It is a multifactorial disorder characterized by sudden ischemic death of myocardial tissue. In general, myocardial infarction is the final phase in the continuum of cardiovascular disease (Jugdutt 2010; Lloyd-Jones et al. 2010). It is also termed as "heart attack." Myocardial infarction (MI) occurs due to the decreased O_2 supply as a result of thrombotic occlusion of the coronary arteries. MI is induced by vulnerable plaque rupture which therefore leads to necrotic cardiac cell death. In general, synthesis of cardiac energy relies predominantly on oxidative metabolism and is thus extremely sensitive to alterations in the intracellular O_2 levels. Occlusion of the coronary artery reduces the concentration of O_2 and upregulates anaerobic-dependent ATP synthesis, contributing to energy famishment and necrotic death of myocardiocytes (Brooks et al. 2005; Reimer and Jennings 1981; Ventura-Clapier et al. 2011). As a result, several deleterious processes swing into action like elevated cytosolic calcium (Ca^{2+}) concentrations and decreased pH, which culminates in the loss of integrity of cell membrane and an uncontrolled release of the intracellular content into the extracellular space which subsequently causes the death of myocardial cell (Orogo and Gustafsson 2013). Some of the risk factors for myocardial infarction are genetic factors, sedentary lifestyle, smoking, stress, dyslipidemia, type 2 diabetes mellitus, obesity, and metabolic syndrome (Gabriel-Costa 2018).

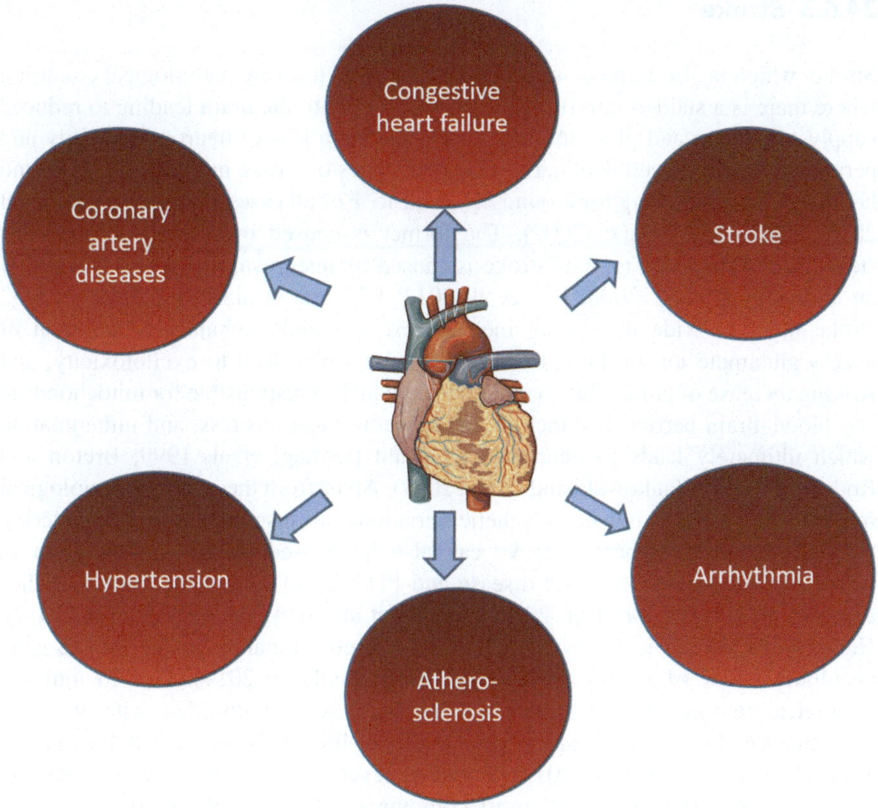

Fig. 24.5 Schematic representation of cardiovascular disorders associated with long-term disability

24.6.2 Congestive Heart Failure (CHF)

Congestive heart failure (CHF) is the most common cardiac disorder in elderly patients aged 65 years or more. It is a clinical syndrome that can result from any functional or structural defects in the myocardium that impairs the ventricle's ability to fill with or eject blood. CHF is characterized by various signs (peripheral edema, elevated jugular venous pressure, pulmonary crackles) and symptoms like ankle swelling, breathlessness, and fatigue (Figueroa and Peters 2006; Inamdar and Inamdar 2016). It is mainly caused by dysfunction of the left ventricle occurring primarily from hypertension, myocardial infarction, or both. The etiology of CHF encompasses ventricular remodeling, enhanced hemodynamic overload, immoderate stimulation of neuro-humoral transmission, aberrant myocyte, Ca^{2+} cycling accelerated apoptosis, and genetic mutations (Inamdar and Inamdar 2016; Dassanayaka and Jones 2015). Risk factors which may aggravate the cerebrovascular disease risk include elevated blood pressure, obesity, and physical inactivity (Heiland et al. 2018).

24.6.3 Stroke

Stroke, which is also termed as cerebral ischemia, is a severe pathological condition where there is a sudden interruption of blood supply to the brain leading to reduced supply of oxygen and glucose to the brain causing a loss of neuronal viability and permanent neuronal cell death. The two chief types of stroke includes ischemic and hemorrhagic, accounting for around 85% and 15% of all cases, respectively (Kanyal 2015; Hinkle and Guanci 2007). The former is caused by arterial occlusion or stenosis, and the latter type of stroke is caused by aneurysm rupture or leakage of an artery (Stoll et al. 2008; Lee et al. 2014; D'Souza et al. 2008). Etiologically, stroke involves wide interlacing mechanisms like ionic imbalance, liberation of excess glutamate ion in the extracellular spaces which lead to excitotoxicity, and striking increase of intracellular calcium level which is responsible for mitochondrial and blood–brain barrier dysfunction, oxidative/nitrosative stress, and inflammation which ultimately leads to neuronal cell death (Dirnagl et al. 1999; Bretón and Rodríguez 2012; Stankowski and Gupta 2011). Apart from these pathophysiological mechanisms, several molecular genetic variations are responsible for engendering cerebral ischemia. Ischemic stroke can also be caused by genetic mutation in HDAC9 related to large vessel disease and PITX2 and ZFHX3 related to cardio-embolic stroke (Traylor et al. 2012; Kilarski et al. 2014; Bellenguez et al. 2012). Hemorrhagic stroke is firmly related to the genetic mutation in the APOE gene peculiarly ε2 or ε4 alleles (Biffi et al. 2010; Lindgren 2014). Due to multiple etiological reasons behind stroke, patients may remain disabled with life-long consequences on functional ability and quality of life (Ahlsiö et al. 1984; Lindgren et al. 2008; Lekander et al. 2017). It causes severe lasting disability especially in elderly people. The major and most common disability resulting from stroke is hemiplegia (one-sided paralysis) and hemiparesis (one-sided weakness). Some patients with stroke have difficulty in swallowing, called dysphagia, and a disability called ataxia which affects coordination of muscle movement which is manifested by unsteady gait movements and speech changes. In addition stroke patients have problems with thinking and memory, impairment of language (aphasia), sensory disturbances including pain, and emotional disturbances which include depression, frustration, anxiety, and sense of grief due to their physical and mental deprivation. Thus, it is the third prominent reason of disability-adjusted life years (DALYs) worldwide, and according to WHO globally 10 million people suffer from stroke annually (Luengo-Fernandez et al. 2013; Murray et al. 2012; Mehta and Vemuganti 2014).

24.7 Conclusion

In summary, as a consequence of all the abovementioned factors, it is observed that there exists no disparity between population age and the exponential impact of conditions causing disability in both the developed and developing countries. The net result of which might turn out to be potentially detrimental to decimate national economies at this current rate. Moreover, large strata of this disabled population are found to be in their most productive years of life when it is most required. A better management of disability burden coupled with effective and affordable prevention and therapeutic strategies thus has become the need of the hour.

References

A. D. Association (2014) Diagnosis and classification of diabetes mellitus. Diabetes Care 37:S81–S90

Adams JM, Cory S (2007) The Bcl-2 apoptotic switch in cancer development and therapy. Oncogene 26:1324

Adelman G, Rane SG, Villa KF (2013) The cost burden of multiple sclerosis in the United States: a systematic review of the literature. J Med Econ 16:639–647

Afkarian M, Sachs MC, Kestenbaum B, Hirsch IB, Tuttle KR, Himmelfarb J et al (2013) Kidney disease and increased mortality risk in type 2 diabetes. J Am Soc Nephrol. https://doi.org/10.1681/ASN.2012070718

Ahlsiö B, Britton M, Murray V, Theorell T (1984) Disablement and quality of life after stroke. Stroke 15:886–890

Ahmed M, Davis J, Aucoin D, Sato T, Ahuja S, Aimoto S et al (2010) Structural conversion of neurotoxic amyloid-β 1–42 oligomers to fibrils. Nat Struct Mol Biol 17:561

Altshuler D, Hirschhorn JN, Klannemark M, Lindgren CM, Vohl M-C, Nemesh J et al (2000) The common PPARγ Pro12Ala polymorphism is associated with decreased risk of type 2 diabetes. Nat Genet 26:76

Alzhelmer A (1907) Uber eine eigenartige Erkrankung der Hirnrinde. Allg Z Psychiatr 64:146–148

Androdias G, Reynolds R, Chanal M, Ritleng C, Confavreux C, Nataf S (2010) Meningeal T cells associate with diffuse axonal loss in multiple sclerosis spinal cords. Ann Neurol 68:465–476

Bartus RT, Dean RR, Beer B, Lippa AS (1982) The cholinergic hypothesis of geriatric memory dysfunction. Science 217:408–414

Bellenguez C, Bevan S, Gschwendtner A, Spencer CC, Burgess AI, Pirinen M et al (2012) Genome-wide association study identifies a variant in HDAC9 associated with large vessel ischemic stroke. Nat Genet 44:328

Bermel RA, Bakshi R (2006) The measurement and clinical relevance of brain atrophy in multiple sclerosis. Lancet Neurol 5:158–170

Bhowmick NA, Neilson EG, Moses HL (2004) Stromal fibroblasts in cancer initiation and progression. Nature 432:332

Biffi A, Sonni A, Anderson CD, Kissela B, Jagiella JM, Schmidt H et al (2010) Variants at APOE influence risk of deep and lobar intracerebral hemorrhage. Ann Neurol 68:934–943

Bjartmar C, Kidd G, Mörk S, Rudick R, Trapp BD (2000) Neurological disability correlates with spinal cord axonal loss and reduced N-acetyl aspartate in chronic multiple sclerosis patients. Ann Neurol 48:893–901

Bjartmar C, Wujek J, Trapp B (2003) Axonal loss in the pathology of MS: consequences for understanding the progressive phase of the disease. J Neurol Sci 206:165–171

Bramow S, Frischer JM, Lassmann H, Koch-Henriksen N, Lucchinetti CF, Sørensen PS et al (2010) Demyelination versus remyelination in progressive multiple sclerosis. Brain 133:2983–2998

Brenner BM, Cooper ME, De Zeeuw D, Keane WF, Mitch WE, Parving H-H et al (2001) Effects of losartan on renal and cardiovascular outcomes in patients with type 2 diabetes and nephropathy. N Engl J Med 345:861–869

Bretón RR, Rodríguez JCG (2012) Excitotoxicity and oxidative stress in acute ischemic stroke. In: Acute ischemic stroke. InTech, Croatia

Bromley-Brits K, Song W (2012) The role of TMP21 in trafficking and amyloid-β precursor protein (APP) processing in Alzheimer's disease. Curr Alzheimer Res 9:411–424

Brooks G, Fahey T, Baldwin K (2005) "Exercise, disease, and disability," exercise physiology: human bioenergetics and its applications, 4th edn. McGraw Hill: Higher Education, New York, pp 649–680

Brown PO, Botstein D (1999) Exploring the new world of the genome with DNA microarrays. Nat Genet 21:33

Brownell B, Hughes JT (1962) The distribution of plaques in the cerebrum in multiple sclerosis. J Neurol, Neurosurg, Psychiatry 25:315

Bruijn L, Beal M, Becher M, Schulz J, Wong P, Price D et al (1997) Elevated free nitrotyrosine levels, but not protein-bound nitrotyrosine or hydroxyl radicals, throughout amyotrophic lateral sclerosis (ALS)-like disease implicate tyrosine nitration as an aberrant in vivo property of one familial ALS-linked superoxide dismutase 1 mutant. Proc Natl Acad Sci 94:7606–7611

Burr D (2003) Subchondral bone in the pathogenesis of osteoarthritis. Mechanical aspects. In: Osteoarthritis. Oxford University Press, Oxford, pp 125–133

Cade WT (2008) Diabetes-related microvascular and macrovascular diseases in the physical therapy setting. Phys Ther 88:1322–1335

Catrina AI, Joshua V, Klareskog L, Malmström V (2016) Mechanisms involved in triggering rheumatoid arthritis. Immunol Rev 269:162–174

Chaudhury PK, Deka K, Chetia D (2006) Disability associated with mental disorders. Indian J Psychiatr 48:95

Cheng N, Chytil A, Shyr Y, Joly A, Moses HL (2008) Transforming growth factor-β signaling–deficient fibroblasts enhance hepatocyte growth factor signaling in mammary carcinoma cells to promote scattering and invasion. Mol Cancer Res 6:1521–1533

Chilelli N, Burlina S, Lapolla A (2013) AGEs, rather than hyperglycemia, are responsible for microvascular complications in diabetes: a "glycoxidation-centric" point of view. Nutr Metab Cardiovasc Dis 23:913–919

Chio A, Benzi G, Dossena M, Mutani R, Mora G (2005) Severely increased risk of amyotrophic lateral sclerosis among Italian professional football players. Brain 128:472–476

Choi SR, Howell OW, Carassiti D, Magliozzi R, Gveric D, Muraro PA et al (2012) Meningeal inflammation plays a role in the pathology of primary progressive multiple sclerosis. Brain 135:2925–2937

Cox PA, Sacks OW (2002) Cycad neurotoxins, consumption of flying foxes, and ALS-PDC disease in Guam. Neurology 58:956–959

Cross M, Smith E, Hoy D, Nolte S, Ackerman I, Fransen M et al (2014) The global burden of hip and knee osteoarthritis: estimates from the global burden of disease 2010 study. Ann Rheum Dis 73:1323–1330

Cummings JL, Arciniegas DB, Brooks BR, Herndon RM, Lauterbach EC, Pioro EP et al (2006) Defining and diagnosing involuntary emotional expression disorder. CNS Spectr 11:1–11

D'Souza B, D'Souza V, Sowmya S, Seema G, Hemalatha C, Anu M et al (2008) A comparative study on oxidative stress and antioxidant status in ischemic stroke patients with and without diabetes. Indian J Clin Biochem 23:218

Dassanayaka S, Jones SP (2015) Recent developments in heart failure. Circ Res 117:e58–e63

Davies H, Bignell GR, Cox C, Stephens P, Edkins S, Clegg S et al (2002) Mutations of the BRAF gene in human cancer. Nature 417:949

Dean J, Keshavan M (2017) The neurobiology of depression: an integrated view. Asian J Psychiatr 27:101–111

Dieppe PA, Lohmander LS (2005) Pathogenesis and management of pain in osteoarthritis. Lancet 365:965–973

Dirnagl U, Iadecola C, Moskowitz MA (1999) Pathobiology of ischaemic stroke: an integrated view. Trends Neurosci 22:391–397

Druet C, Tubiana-Rufi N, Chevenne D, Rigal O, Polak M, Levy-Marchal C (2006) Characterization of insulin secretion and resistance in type 2 diabetes of adolescents. J Clin Endocrinol Metab 91:401–404

Drummond K, Mauer M (2002) The early natural history of nephropathy in type 1 diabetes: II. Early renal structural changes in type 1 diabetes. Diabetes 51:1580–1587

Du Yan S, Chen X, Fu J, Chen M, Zhu H, Roher A et al (1996) RAGE and amyloid-β peptide neurotoxicity in Alzheimer's disease. Nature 382:685
Du Yan S, Zhu H, Fu J, Yan SF, Roher A, Tourtellotte WW et al (1997) Amyloid-β peptide–receptor for advanced glycation endproduct interaction elicits neuronal expression of macrophage-colony stimulating factor: a proinflammatory pathway in Alzheimer disease. Proc Natl Acad Sci 94:5296–5301
Duckett L (2001) Alzheimer's dementia: morbidity and mortality. J Insur Med (New York, NY) 33:227–234
Dyck PJ, Kratz K, Karnes J, Litchy WJ, Klein R, Pach J et al (1993) The prevalence by staged severity of various types of diabetic neuropathy, retinopathy, and nephropathy in a population-based cohort: the Rochester Diabetic Neuropathy Study. Neurology 43:817–817
Eikelenboom P, Veerhuis R (1999) The importance of inflammatory mechanisms for the development of Alzheimer's disease. Exp Gerontol 34:453–461
Feinstein A (2011) Multiple sclerosis and depression. Mult Scler J 17:1276–1281
Felger JC, Lotrich FE (2013) Inflammatory cytokines in depression: neurobiological mechanisms and therapeutic implications. Neuroscience 246:199–229
Fidler IJ (2003) The pathogenesis of cancer metastasis: the 'seed and soil' hypothesis revisited. Nat Rev Cancer 3:453
Figueroa MS, Peters JI (2006) Congestive heart failure: diagnosis, pathophysiology, therapy, and implications for respiratory care. Respir Care 51:403–412
Fischer MT, Wimmer I, Höftberger R, Gerlach S, Haider L, Zrzavy T et al (2013) Disease-specific molecular events in cortical multiple sclerosis lesions. Brain 136:1799–1815
Friedberg EC (2003) DNA damage and repair. Nature 421:436
Gabriel-Costa D (2018) The pathophysiology of myocardial infarction-induced heart failure. Pathophysiology 25(4):277–284
Gallo V, Bueno-De-Mesquita HB, Vermeulen R, Andersen PM, Kyrozis A, Linseisen J et al (2009) Smoking and risk for amyotrophic lateral sclerosis: analysis of the EPIC cohort. Ann Neurol 65:378–385
Gerstein HC (1994) Cow's milk exposure and type I diabetes mellitus: a critical overview of the clinical literature. Diabetes Care 17:13–19
Ginsberg H, Kimmerling G, Olefsky JM, Reaven GM (1975) Demonstration of insulin resistance in untreated adult onset diabetic subjects with fasting hyperglycemia. J Clin Investig 55:454–461
Goering S (2015) Rethinking disability: the social model of disability and chronic disease. Curr Rev Musculoskelet Med 8:134–138
Goldring SR (2008) Targeting disease: musculoskeletal conditions. Pharm Res 25(12):2878–2880
Gray E, Paula-Barbosa M, Roher A (1987) Alzheimer's disease: paired helical filaments and cytomembranes. Neuropathol Appl Neurobiol 13:91–110
Halban PA, Polonsky KS, Bowden DW, Hawkins MA, Ling C, Mather KJ et al (2014) β-cell failure in type 2 diabetes: postulated mechanisms and prospects for prevention and treatment. J Clin Endocrinol Metab 99:1983–1992
Hanahan D, Folkman J (1996) Patterns and emerging mechanisms of the angiogenic switch during tumorigenesis. Cell 86:353–364
Hanahan D, Weinberg RA (2011) Hallmarks of cancer: the next generation. Cell 144:646–674
Hardy JA, Higgins GA (1992) Alzheimer's disease: the amyloid cascade hypothesis. Science 256:184
Harre U, Georgess D, Bang H, Bozec A, Axmann R, Ossipova E et al (2012) Induction of osteoclastogenesis and bone loss by human autoantibodies against citrullinated vimentin. J Clin Investig 122:1791–1802
Harwood CA, McDermott CJ, Shaw PJ (2009) Physical activity as an exogenous risk factor in motor neuron disease (MND): a review of the evidence. Amyotroph Lateral Scler 10:191–204
Heiland EG, Welmer A-K, Wang R, Santoni G, Fratiglioni L, Qiu C (2018) Cardiovascular risk factors and the risk of disability in older adults: variation by age and functional status. J Am Med Dir Assoc. https://doi.org/10.1016/j.jamda.2018.05.013

Hensley K, Mhatre M, Mou S, Pye QN, Stewart C, West M et al (2006) On the relation of oxidative stress to neuroinflammation: lessons learned from the G93A-SOD1 mouse model of amyotrophic lateral sclerosis. Antioxid Redox Signal 8:2075–2087

Hertel JKH, Johansson S, Midthjell K, Nygård O, Njølstad PR, Molven A (2013) Type 2 diabetes genes–Present status and data from Norwegian studies. Norsk Epidemiol 23. https://doi.org/10.5324/nje.v23i1.1597

Hinkle JL, Guanci MM (2007) Acute ischemic stroke review. J Neurosci Nurs 39:285–293

Howell OW, Reeves CA, Nicholas R, Carassiti D, Radotra B, Gentleman SM et al (2011) Meningeal inflammation is widespread and linked to cortical pathology in multiple sclerosis. Brain 134:2755–2771

Hoy D, Bain C, Williams G, March L, Brooks P, Blyth F et al (2012) A systematic review of the global prevalence of low back pain. Arthritis Rheum 64:2028–2037

Inamdar A, Inamdar A (2016) Heart failure: diagnosis, management and utilization. J Clin Med 5:62

Jones PA, Baylin SB (2002) The fundamental role of epigenetic events in cancer. Nat Rev Genet 3:415

Jugdutt BI (2010) Aging and heart failure: changing demographics and implications for therapy in the elderly. Heart Fail Rev 15:401–405

Juhász G, Barkóczi B, Vass G, Datki Z, Hunya Á, Fülöp L et al (2010) Fibrillar Aβ 1-42 enhances NMDA receptor sensitivity via the integrin signaling pathway. J Alzheimers Dis 19:1055–1067

Jurgens CA, Toukatly MN, Fligner CL, Udayasankar J, Subramanian SL, Zraika S et al (2011) β-cell loss and β-cell apoptosis in human type 2 diabetes are related to islet amyloid deposition. Am J Pathol 178:2632–2640

Källberg H, Ding B, Padyukov L, Bengtsson C, Rönnelid J, Klareskog L et al (2011) Smoking is a major preventable risk factor for rheumatoid arthritis: estimations of risks after various exposures to cigarette smoke. Ann Rheum Dis 70:508–511

Kanyal N (2015) The science of ischemic stroke: pathophysiology & pharmacological treatment. Int J Pharma Res Rev 4:65–84

Kasarskis EJ, Lindquist JH, Coffman CJ, Grambow SC, Feussner JR, Allen KD et al (2009) Clinical aspects of ALS in Gulf War veterans. Amyotroph Lateral Scler 10:35–41

Kasarskis EJ, Mendiondo MS, Matthews DE, Mitsumoto H, Tandan R, Simmons Z et al (2014) Estimating daily energy expenditure in individuals with amyotrophic lateral sclerosis. Am J Clin Nutr 99:792–803

Khoury JC, Kleindorfer D, Alwell K, Moomaw CJ, Woo D, Adeoye O et al (2013) Diabetes mellitus: a risk factor for ischemic stroke in a large biracial population. Stroke 44:1500–1504

Kiernan MC, Vucic S, Cheah BC, Turner MR, Eisen A, Hardiman O et al (2011) Amyotrophic lateral sclerosis. Lancet 377:942–955

Kilarski LL, Achterberg S, Devan WJ, Traylor M, Malik R, Lindgren A et al (2014) Meta-analysis in more than 17,900 cases of ischemic stroke reveals a novel association at 12q24. 12. Neurology 83:678–685

Klijs B, Nusselder WJ, Looman CW, Mackenbach JP (2011) Contribution of chronic disease to the burden of disability. PLoS One 6:e25325

Knudson AG (2002) Cancer genetics. Am J Med Genet 111:96–102

Konig MF, Abusleme L, Reinholdt J, Palmer RJ, Teles RP, Sampson K et al (2016) Aggregatibacter actinomycetemcomitans–induced hypercitrullination links periodontal infection to autoimmunity in rheumatoid arthritis. Sci Transl Med 8:369ra176–369ra176

Korshavn KJ, Bhunia A, Lim MH, Ramamoorthy A (2016) Amyloid-β adopts a conserved, partially folded structure upon binding to zwitterionic lipid bilayers prior to amyloid formation. Chem Commun 52:882–885

Kotler SA, Walsh P, Brender JR, Ramamoorthy A (2014) Differences between amyloid-β aggregation in solution and on the membrane: insights into elucidation of the mechanistic details of Alzheimer's disease. Chem Soc Rev 43:6692–6700

Kotler SA, Brender JR, Vivekanandan S, Suzuki Y, Yamamoto K, Monette M et al (2015) High-resolution NMR characterization of low abundance oligomers of amyloid-β without purification. Sci Rep 5:11811

Krishnamurthy A, Joshua V, Hensvold AH, Jin T, Sun M, Vivar N et al (2016) Identification of a novel chemokine-dependent molecular mechanism underlying rheumatoid arthritis-associated autoantibody-mediated bone loss. Ann Rheum Dis 75:721–729

Kutzelnigg A, Lucchinetti CF, Stadelmann C, Brück W, Rauschka H, Bergmann M et al (2005) Cortical demyelination and diffuse white matter injury in multiple sclerosis. Brain 128:2705–2712

Lassmann H, Brück W, Lucchinetti CF (2007) The immunopathology of multiple sclerosis: an overview. Brain Pathol 17:210–218

Lee Y, Lee S-R, Choi SS, Yeo H-G, Chang K-T, Lee HJ (2014) Therapeutically targeting neuroinflammation and microglia after acute ischemic stroke. BioMed Res Int 2014:297241

Lekander I, Willers C, Von Euler M, Lilja M, Sunnerhagen KS, Pessah-Rasmussen H et al (2017) Relationship between functional disability and costs one and two years post stroke. PloS One 12:e0174861

Leonard BE (2001) Stress, norepinephrine and depression. J Psychiatry Neurosci: JPN 26:S11

Lindgren A (2014) Stroke genetics: a review and update. J Stroke 16:114

Lindgren P, Glader E-L, Jönsson B (2008) Utility loss and indirect costs after stroke in Sweden. Eur J Cardiovasc Prev Rehabil 15:230–233

Liu R, Althaus JS, Ellerbrock BR, Becker DA, Gurney ME (1998) Enhanced oxygen radical production in a transgenic mouse model of familial amyotrophic lateral sclerosis. Ann Neurol 44:763–770

Lloyd-Jones D, Adams R, Brown T, Carnethon M, Dai S, De Simone G et al (2010) Heart disease and stroke statistics--2010 update: a report from the American Heart Association. Circulation 121:e46

Lorenzo-Almoros A, Tunon J, Orejas M, Cortés M, Egido J, Lorenzo Ó (2017) Diagnostic approaches for diabetic cardiomyopathy. Cardiovasc Diabetol 16:28

Lozano R, Naghavi M, Foreman K, Lim S, Shibuya K, Aboyans V et al (2012) Global and regional mortality from 235 causes of death for 20 age groups in 1990 and 2010: a systematic analysis for the Global Burden of Disease Study 2010. Lancet 380:2095–2128

Luengo-Fernandez R, Paul NL, Gray AM, Pendlebury ST, Bull LM, Welch SJ et al (2013) Population-based study of disability and institutionalization after transient ischemic attack and stroke: 10-year results of the Oxford Vascular Study. Stroke 44:2854–2861

Maahs DM, West NA, Lawrence JM, Mayer-Davis EJ (2010) Epidemiology of type 1 diabetes. Endocrinol Metab Clin 39:481–497

Machado-Vieira R, Salvadore G, DiazGranados N, Ibrahim L, Latov D, Wheeler-Castillo C et al (2010) New therapeutic targets for mood disorders. Sci World J 10:713–726

Mahad DH, Trapp BD, Lassmann H (2015) Pathological mechanisms in progressive multiple sclerosis. Lancet Neurol 14:183–193

Maher P, Davis JB (1996) The role of monoamine metabolism in oxidative glutamate toxicity. J Neurosci 16:6394–6401

Marburg O (1906) Die sogenannte akute multiple Sklerose: Рипол Классик.

March L, Smith EU, Hoy DG, Cross MJ, Sanchez-Riera L, Blyth F et al (2014) Burden of disability due to musculoskeletal (MSK) disorders. Best Pract Res Clin Rheumatol 28:353–366

McInnes IB, Schett G (2017) Pathogenetic insights from the treatment of rheumatoid arthritis. Lancet 389:2328–2337

Mehta S, Vemuganti R (2014) Mechanisms of stroke induced neuronal death: multiple therapeutic opportunities. Adv Anim Vet Sci 2:438–446

Meldrum B, Garthwaite J (1990) Excitatory amino acid neurotoxicity and neurodegenerative disease. Trends Pharmacol Sci 11:379–387

Morris J (1996) Encounters with strangers: feminism and disability. Womens Press Ltd, London

Murray CJ, Vos T, Lozano R, Naghavi M, Flaxman AD, Michaud C et al (2012) Disability-adjusted life years (DALYs) for 291 diseases and injuries in 21 regions, 1990–2010: a systematic analysis for the Global Burden of Disease Study 2010. Lancet 380:2197–2223

Nguyen DV, Shaw L, Grant M (2012) Inflammation in the pathogenesis of microvascular complications in diabetes. Front Endocrinol 3:170

Nickerson HD, Dutta S (2012) Diabetic complications: current challenges and opportunities. J Cardiovasc Transl Res 5:375–379

Nielsen E-MD, Hansen L, Carstensen B, Echwald SM, Drivsholm T, Glümer C et al (2003) The E23K variant of Kir6. 2 associates with impaired post-OGTT serum insulin response and increased risk of type 2 diabetes. Diabetes 52:573–577

Okur ME, Karantas ID, Siafaka PI (2017) Diabetes Mellitus: a review on pathophysiology, current status of oral pathophysiology, current status of oral medications and future perspectives. ACTA Pharma Sci 55:482–494

Orogo AM, Gustafsson ÅB (2013) Cell death in the myocardium: my heart won't go on. IUBMB Life 65:651–656

Papatheodorou K, Papanas N, Banach M, Papazoglou D, Edmonds M (2016) Complications of diabetes 2016. J Diabetes Res 2016. https://doi.org/10.1155/2016/6989453

Pasinelli P, Brown RH (2006) Molecular biology of amyotrophic lateral sclerosis: insights from genetics. Nat Rev Neurosci 7:710

Peterson JW, Bö L, Mörk S, Chang A, Trapp BD (2001) Transected neurites, apoptotic neurons, and reduced inflammation in cortical multiple sclerosis lesions. Ann Neurol Off J Am Neurol Assoc Child Neurol Soc 50:389–400

Pociot F, Akolkar B, Concannon P, Erlich HA, Julier C, Morahan G et al (2010) Genetics of type 1 diabetes: what's next? Diabetes 59:1561–1571

Poitout V, Robertson RP (2007) Glucolipotoxicity: fuel excess and β-cell dysfunction. Endocr Rev 29:351–366

Polyak K, Riggins GJ (2001) Gene discovery using the serial analysis of gene expression technique: implications for cancer research. J Clin Oncol 19:2948–2958

Pruessner M, Hellhammer DH, Pruessner JC, Lupien SJ (2003) Self-reported depressive symptoms and stress levels in healthy young men: associations with the cortisol response to awakening. Psychosom Med 65:92–99

Radunovic A, Annane D, Jewitt K, Mustfa N (2009) Mechanical ventilation for amyotrophic lateral sclerosis/motor neuron disease. Cochrane Database Syst Rev 4:CD004427

Reimer KA, Jennings RB (1981) Energy metabolism in the reversible and irreversible phases of severe myocardial ischemia. Acta Med Scand 210:19–27

Richelson E (2001) Pharmacology of antidepressants. Mayo Clin Proc 76:511–527

Sandy J (2003) Proteolytic degradation of normal and osteoarthritic cartilage matrix. Osteoarthritis 2003:82–92

Scher JU, Sczesnak A, Longman RS, Segata N, Ubeda C, Bielski C et al (2013) Expansion of intestinal Prevotella copri correlates with enhanced susceptibility to arthritis. elife 2:e01202

Sharma L (2003) Local mechanical factors in the natural history of knee osteoarthritis. Malalignment and joint laxity. Osteoarthritis 2:177–183

Silvers A (1998) A fatal attraction to normalizing: treating disabilities as deviations from ôspecies-typicalõ functioning. Enhanc Hum Traits 1998:95–123

Soerjomataram I, Lortet-Tieulent J, Parkin DM, Ferlay J, Mathers C, Forman D et al (2012) Global burden of cancer in 2008: a systematic analysis of disability-adjusted life-years in 12 world regions. Lancet 380:1840–1850

Stankowski JN, Gupta R (2011) Therapeutic targets for neuroprotection in acute ischemic stroke: lost in translation? Antioxid Redox Signal 14:1841–1851

Stoll G, Kleinschnitz C, Nieswandt B (2008) Molecular mechanisms of thrombus formation in ischemic stroke: novel insights and targets for treatment. Blood 112:3555–3562

Sun X, Bromley-Brits K, Song W (2012) Regulation of β-site APP-cleaving enzyme 1 gene expression and its role in Alzheimer's Disease. J Neurochem 120:62–70

Supnet C, Bezprozvanny I (2010) The dysregulation of intracellular calcium in Alzheimer disease. Cell Calcium 47:183–189

Sutedja NA, Veldink JH, Fischer K, Kromhout H, Heederik D, Huisman MH et al (2009) Exposure to chemicals and metals and risk of amyotrophic lateral sclerosis: a systematic review. Amyotroph Lateral Scler 10:302–309

Suzuki Y, Brender JR, Soper MT, Krishnamoorthy J, Zhou Y, Ruotolo BT et al (2013) Resolution of oligomeric species during the aggregation of Aβ1–40 using 19F NMR. Biochemistry 52:1903–1912

Swerdlow RH, Khan SM (2004) A "mitochondrial cascade hypothesis" for sporadic Alzheimer's disease. Med Hypotheses 63:8–20

Talmadge JE, Fidler IJ (2010) AACR centennial series: the biology of cancer metastasis: historical perspective. Cancer Res 70:0008–5472. CAN-10-1040

Tõugu V, Tiiman A, Palumaa P (2011) Interactions of Zn (II) and Cu (II) ions with Alzheimer's amyloid-beta peptide. Metal ion binding, contribution to fibrillization and toxicity. Metallomics 3:250–261

Traylor M, Farrall M, Holliday EG, Sudlow C, Hopewell JC, Cheng Y-C et al (2012) Genetic risk factors for ischaemic stroke and its subtypes (the METASTROKE collaboration): a meta-analysis of genome-wide association studies. Lancet Neurol 11:951–962

Van den Berg W, Van der Kraan P, Van Beuningen H (2003) Synovial mediators of cartilage damage and repair in osteoarthritis. In: Brandt KD, Doherty M, Lohmander LS (eds) Osteoarthritis, 2nd edn. Oxford University Press, Oxford, pp 147–155

Ventura-Clapier R, Garnier A, Veksler V, Joubert F (2011) Bioenergetics of the failing heart. Biochim Biophys Acta (BBA)-Mol Cell Res 1813:1360–1372

Vinik AI, Maser RE, Mitchell BD, Freeman R (2003) Diabetic autonomic neuropathy. Diabetes Care 26:1553–1579

Vucic S, Kiernan MC (2009) Pathophysiology of neurodegeneration in familial amyotrophic lateral sclerosis. Curr Mol Med 9:255–272

W. H. Organization (2001) International Classification of Impairment, Disability and Handicap (ICIDH-2 beta version). WHO Website, 1999, Available at (Accessed): http://www.who.int/icidh/. Date: June, 2001

Weinberg R, Hanahan D (2000) The hallmarks of cancer. Cell 100:57–70

Williamson TL, Cleveland DW (1999) Slowing of axonal transport is a very early event in the toxicity of ALS–linked SOD1 mutants to motor neurons. Nat Neurosci 2:50

Wu N, Minden SL, Hoaglin DC, Hadden L, Frankel D (2007) Quality of life in people with multiple sclerosis: data from the Sonya Slifka Longitudinal Multiple Sclerosis Study. J Health Hum Serv Adm 30:233–267

Zanetti O, Bianchetti A, Frisoni GB, Rozzini R, Trabucchi M (1993) Determinants of disability in Alzheimer's disease. Int J Geriatric Psychiatr 8:581–586

Zeitler P (2009) Update on nonautoimmune diabetes in children. J Clin Endocrinol Metab 94:2215–2220

Part X

Calibration, Repair and Safety Aspects

The Technical Support: Repair, Preventive Maintenance, and Inspection 25

Mana Sezdi

Abstract

Medical devices, which are an indispensable part of health services in hospitals, need technical support during their lifetime. In this process, medical devices need to be checked for performance and safety, and emergency intervention is needed in case of failure. All these applications applied to medical devices are carried out with the maintenance program. The maintenance studies performed by the biomedical staff of the hospital or the technical personnel of the device's manufacturer provide the medical devices to be more reliable, efficient, and safe.

25.1 Introduction

Medical equipment directly affect patient's life during all health services, such as diagnosis, treatment, and monitoring. Because of this, it is important to keep the medical devices available for all health applications. While they must be reliable and safe in working conditions, they must be activated in short time if they have breakdowns. This goal is achieved only by applying a maintenance program. Maintenance is defined as the work done in order to take precautions and to ensure the continuity of the system.

The maintenance program can be handled as two main headings (World Health Organization 2011): preventive maintenance, called as PM, and corrective maintenance (CM). Corrective maintenance may also be called as repair. The branching of the maintenance program can be seen in Fig. 25.1 as a diagram.

Preventive maintenance is performed to extend the device's life. Failure rates are reduced, and the device operates with full performance by applying preventive

M. Sezdi (✉)
Vocational School of Technical Sciences, Istanbul University-Cerrahpasa, Istanbul, Turkey
e-mail: mana@istanbul.edu.tr

S. Paul (ed.), *Biomedical Engineering and its Applications in Healthcare*,
https://doi.org/10.1007/978-981-13-3705-5_25

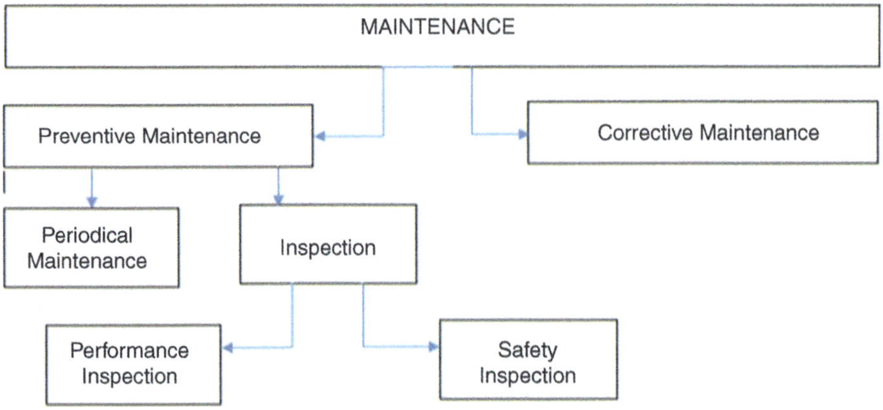

Fig. 25.1 Classification of maintenance program (World Health Organization 2011)

maintenance (World Health Organization 2011; Bahreini et al. 2018). It is a planned maintenance. Preventive maintenance consists of two important activities: inspection and periodical maintenance.

Inspection includes performance measurements and safety inspections of the medical devices. By applying the performance inspection, it is tested whether the device is working correctly. By applying safety inspection, electrical safety measurements are performed to ensure whether the device is safe for patient and user.

Periodical maintenance is performed in accordance with the instructions of the technical service and service manuals of each device to avoid any failure.

Corrective maintenance makes the device usable after failure. While preventive maintenance is a planned maintenance process, corrective maintenance is an unplanned maintenance process.

The staff responsible for maintenance and repair of the medical devices within the health facility is a clinical engineer or biomedical engineer. The responsibilities of these personnel in the process of maintenance are as follows:

- Making inventory of existing devices
- Managing the stocks of spare parts and materials
- Providing technical services needed for intensive care and emergency services 24 h a day
- Providing regular and continuous communication with the companies producing the devices, vendors, or their representatives
- Repairing the defective devices in their locations inside hospital or supplying the repair service from outside the hospital
- Providing and archiving documents such as handbooks, brochures, etc.
- Deciding the devices to be scrapped due to obsolescence or high repair costs

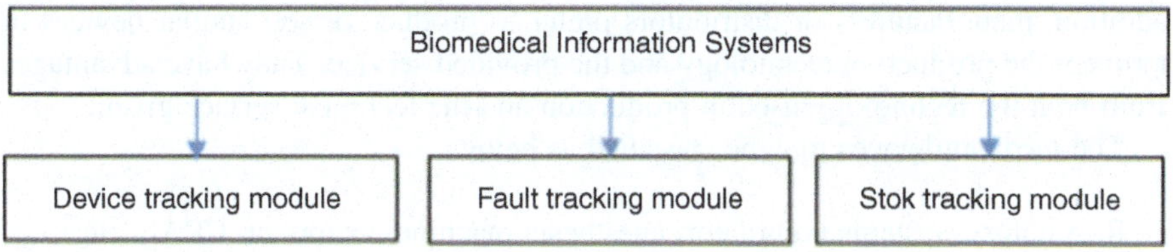

Fig. 25.2 Biomedical information systems

- Developing programs for regular inspection of devices
- Ensuring that periodical maintenance and inspections are carried out as scheduled
- Recording of the information of periodic maintenance and inspections

Whoever (the hospital's biomedical staff or the external company) performs the preventive maintenance and corrective maintenance, biomedical information systems have great importance in order to manage all maintenance process without any problems. Biomedical information systems must consist of three main modules. The modules can be seen in Fig. 25.2.

In order to manage maintenance program, firstly a current medical device inventory must be generated. The whole maintenance process can only be fulfilled and controlled by this way. The inventory studies are included in the "device tracking module." The inventory list must contain all of the device's information: brand model, serial number, biomedical number, location, year of purchase, periodical maintenance interval, period of performance inspection, etc. Inventory information should always be up-to-date. The information of the new devices should be added and the information of the scrapped devices should be deleted.

While storing the inventory information of medical devices, recording all information in the life cycle such as its faults and its inspections, provides the largest information infrastructure in clinical engineering applications. This type of information is stored in the "fault tracking module" and guides the technical staff in the event of malfunctions. It may give the information about the cause of the fault and the repair that may be done. In the first stage, biomedical staff takes information about the device, for example, "have the same malfunction been occurred before?" or "how has the similar problem been solved?" Additionally, the stock of the spare parts that are used during repair or periodical maintenance is managed by using "stock tracking module." These modules allow the biomedical service to accelerate, especially during breakdown repairs.

Grouping of medical devices in the maintenance, repair, and inspections increases the quality of the technical service. In particular, if the technical service is given by the hospital biomedical staff, the concentration of those staff on certain device groups increases the level of the qualified technical service. A biomedical staff cannot have the same high level of knowledge about all devices within the hospital. It is preferable to receive training and specialization on devices in certain groups. In addition, manufacturers or distributors prefer to produce or sell similar devices in terms of the production technology and the provided service. They have advantages from both the technology used in production and the technical service given.

The medical devices may be classified as below:

- Respiratory Systems (ventilator, anesthesia machine, vaporizer, CPAP, etc.)
- Electrosurgical Systems (electrosurgical unit, etc.)
- Electrotherapy Systems (defibrillator, pacemaker, TENS, etc.)
- Physiological Signal Monitoring Systems (patient monitor, ECG, EEG, etc.)
- Ultrasonic Imaging Systems (ultrasonic imaging device, echocardiography)
- Sterilization and Incubation Systems (autoclave, incubator, etc.)
- X-Ray Imaging Systems (angiography, fluoroscopy, etc.)
- Magnetic Resonance Imaging Systems (MR)
- Medical Light Systems (phototherapy device, laser, etc.)
- Analysis Systems (laboratory analysis devices, etc.)
- Microscopic Systems
- Audiometric Systems (audiometer, etc.)
- Endoscopic Imaging Systems (endoscopy, bronchoscopy, etc.)
- Dialysis Systems (hemodialysis, reverse osmosis systems, etc.)
- SI Unit Systems (infusion pump, flowmeter, blood storage cabinet, blanket, sphygmomanometer, etc.)

It is essential to store the information obtained from all maintenance-repair-inspection applications in the period from the purchasing of the device to the scrapping of the device, and to analyze them statistically. This information gives the history of the device. The statistical analysis of the knowledge provides great benefits to the employee, technical service, biomedical unit, and patient during all maintenance process.

By examining these statistical information, the following information about the device failure can be accessed:

- How many times has this device failed?
- What is the frequency of failures?
- What are failures?
- How did each fault fix?
- What parts were changed?
- Was the periodical maintenance applied to the device?
- Who applied the periodical maintenance?
- Are the performance measurements performed?

All this information provides information on how to plan preventive maintenance. When the device is faulted, it guides to repair.

25.2 Preventive Maintenance

Preventive maintenance is the activities to prevent malfunctions of medical devices before the problem is occurred. The goal is to keep a device usable with reliable and accurate results. The advantages of the preventive maintenance are (World Health Organization 2011; Bahreini et al. 2018):

- It enables the devices to be used more efficiently and faster.
- It provides a safer environment for patients and users.
- It extends the life of the devices.
- It protects the device from possible malfunctions.
- It enables devices to operate at the standard performance level reported by the manufacturers.
- It eliminates the delays in health services by removing the malfunctions quickly.

While carrying out the planned preventive maintenance activities, it is necessary to collect the systems according to their risk factors in three main groups: high-, medium- and low-risk systems. High-risk systems are intensive care, life-sustaining, diagnosis and treatment systems, and if they have a problem, they can cause serious problems in patient care. Moderately risky systems are those systems that cause problems in patient care, but do not seriously harm the patient or staff in the event of malfunction, misuse, or absence. Systems with low risk are the systems which do not harm the patient or personnel in case of malfunction, misuse, or absence and constitute the least problems in patient care.

Other factors affecting the frequency of maintenance can be listed as (Bahreini et al. 2018):

- Clinical practice
- Maintenance requests
- Manufacturer recommendations
- Problems in the past
- Loss of income due to device failure
- The high prices to be paid to the repair

Audits for preventive maintenance should be conducted throughout the year; past workforce, responsibilities for other units, and levels of patient intensity should be assessed during audits. Due to the change in the intensity of the patient according to the seasons, it is not possible to distribute the work-related intensity to the whole year equally. The priority should be given to the preventive maintenance in the summer period when the patient density is low.

It is important that the health services provided in health institutions are presented in a high-quality, effective, and reliable manner and the results of the examination and analysis are obtained in a correct, reliable, and shortest time. This is one of the important factors that directly affect the quality of service, the correct and timely diagnosis, and the correct application or orientation of the treatment. As medical devices directly affect patient health, it is also important to intervene without interruption to devices since they require an expensive and long procurement process. These interventions should be done within the scope of planned preventive maintenance.

Preventive maintenance is carried out at three levels:

- User-level maintenance
- Hospital-level maintenance
- Factory-level maintenance

User-Level Maintenance User-level maintenance is the basis of the entire maintenance system. The user is responsible for maintaining and calibrating all types of devices continuously, clean and in good working order. To do this, the user's job description is:

- Inspection and control of devices and materials
- Squeezing loose screws or parts
- Lubrication of some parts
- Cleaning of materials and parts
- Timely notice of malfunctions and problems

Hospital-Level Maintenance It is the maintenance performed by using testing equipment by the biomedical personnel responsible for the medical devices or nonhospital technical service personnel.

Factory-Level Maintenance It is the maintenance performed by the manufacturer. It is advanced maintenance. It also includes the factory settings.

Preventive maintenance services can be examined in two groups. These are:

- Periodical maintenance
- Inspection

25.2.1 Periodical Maintenance

Periodical maintenance is a preventive maintenance performed with specific time intervals, such as daily, monthly, or yearly. Periodical maintenance is performed in accordance with the instructions of the technical service and service manuals of each device without any fault by the user or technical service personnel.

Not performing periodical maintenance of the devices increases the risk of malfunction or incorrect evaluation. The methods and materials recommended by the manufacturer must be used during maintenance. Because the usage of unsuitable methods and consumables may cause malfunctions and problems in the devices, periodical maintenance should be carried out by trained personnel. Periodical maintenance must not be neglected to extend the life of the devices.

For all medical devices used, periodical maintenance intervals are determined depending on the characteristics of the devices and their location. These maintenance times are given in daily, weekly, monthly, quarterly, 6-month, or 1-year periods. Generally, daily and weekly maintenance are performed by the user and include precleaning and preuse maintenance. Each device manufacturer explains the detail of the periodical maintenance in the "User's Manual."

Periodical maintenance procedures must be applied to ensure that medical devices are able to work continuously and efficiently and to minimize downtime. The procedures can be listed as:

- Cleaning
- Lubrication
- Adjustment
- Software update
- Changing specific parts (sensors, filters, batteries, etc.)

Periodical maintenance service can be provided from the device manufacturer with various contracts. The firm has all kinds of maintenance, repair, and application rights.

25.2.2 Inspection

Inspection is performed to control the function and safety of medical devices. Inspection consists of scheduled activities performed at specific periods. The inspection controlling the function of the devices is called as performance inspection, while the inspection controlling the safety of the devices is called as safety inspection.

25.2.2.1 Performance Inspection

The performance inspection is to test the function of the device and to determine the accuracy of the device, by using calibrators or simulators. If there is a deviation in the measurement results, it is determined and reported (Sezdi 2012).

The performance inspection can be performed in two ways. Either by the biomedical staff working in the hospital or by outsourcing services from the neutral calibration companies. If the hospital's biomedical staff will carry out the measurements themselves, they must have the necessary training to perform measurements and the required calibrator and/or simulator pool for the measurements. If they take the performance inspection from outside the company, they should accompany the inspection, and they should supervise the inspection.

Performance measurements are performed in the following steps:

- Listing of the devices to be inspected
- Performing of measurements
- Labeling of devices according to measurement results
- Certification

The performance inspection is performed with the diagnostic or therapeutic devices having risks for both patient and user. These devices are also accepted as the high-risk group devices. To determine the devices in high-risk group, the method of "device management coefficient (DMC)" is used (Fennigkoh and Smith 1989). When calculating this coefficient, the function, risk, and preventive maintenance requirements of the device are taken into consideration, and each part is scored separately. Function score can be maximum of 10 points, risk score maximum of 5 points, and maintenance requirement score maximum of 5 points.

$$\text{DMC (20p)} = \text{Function Score} + \text{Risk Score} + \text{Maintenance Requirement Score}$$

If the device management coefficient is 12 or above, the device is included in the performance inspection planning, and it is performed every 12 months. If the coefficient is 17 and above, the inspection is performed every 6 months (Sezdi 2012). In addition to this coefficient planning, there are a number of additional factors to consider in determining the inspection period. Factors such as very intensive use of the device or frequent malfunction may require more frequent control of the device.

Performance measurements are carried out according to international standards and procedures. These are:

- ECRI (Emergency Care Research Institute) Benchmark procedures
- ISO (the International Organization for Standardization) standards
- IPEM (the Institute of Physics and Engineering in Medicine) procedures
- AAPM (the American Association of Physicists in Medicine) procedures

IPEM and AAPM procedures are used for imaging devices such as angiography, computerized tomography, and magnetic resonance imaging system. During inspection, the medical devices firstly are controlled physically. The results of these physical controls are reported as qualitative parameters in the certificate. Then, the device's functions are controlled. The obtained measurement results are reported as quantitative parameters in the certificate. The quantitative parameters that will be measured are different in each type of medical devices (Sezdi 2013; Emergency Care and Research Institute 2008). Energy is measured in the defibrillator, while temperature is measured in the newborn incubator.

The results of performance inspection performed according to international standards and procedures are interpreted according to the acceptance criteria set out in these standards and procedures (Emergency Care and Research Institute 2008). If the measurement results are within the specified acceptance criteria, the interpretation will be "the device is appropriate to the international standards." If not, it will be "the device is not appropriate to the international standards and it must not be used."

After performance inspection is performed and interpreted, color labels are attached to the medical devices according to the results of the measurement. The color of labels indicates whether the device can be usable or unusable. If the medical device conforms to international standards according to the performance measurement results, it is labeled with a green label, and if not, it is labeled with a red label. If one of the modules is defective in medical devices consisting of multiple modules, the device is labeled with a yellow color label. The user must not use the red labeled device when he/she can use the green labeled device. The red labeled device must be informed to the relevant biomedical unit or technical department. The yellow label means that the device is suitable for limited use. The user can use only the parts different from the problem module (Sezdi 2012).

The label must include the information about the serial number or biomedical number, certificate number, date of measurement, and date of next measurement (Republic of Turkey Ministry of Health, Clinical Engineering Management Department 2016). The sample labels pasted after performance inspection can be seen in Fig. 25.3.

The results of performance inspection must be reported as a calibration certificate. A calibration certificate must have a "cover page" and "measurement report pages." In addition to qualitative and quantitative measurement results, calibration uncertainties and calibration traceability information of the calibrators and simulators used in measurements must be available in the calibration certificates. The interpretation of the inspection results must be explained. If the device is not to be used, the interpretation must indicate what the problem is.

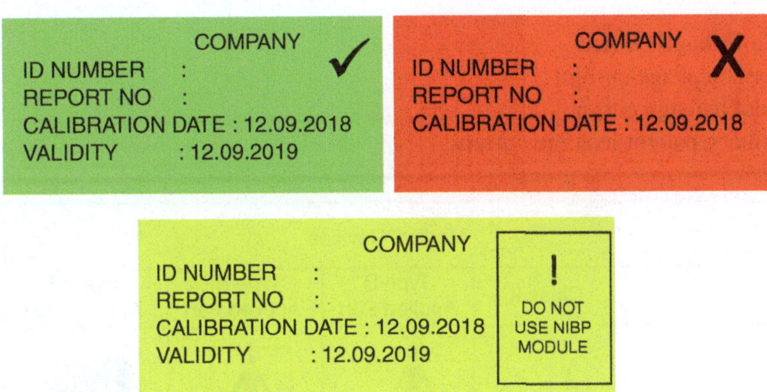

Fig. 25.3 Sample labels pasted after performance inspection (Republic of Turkey Ministry of Health, Clinical Engineering Management Department 2016)

25.2.2.2 Safety Inspection

Electrical safety inspection takes the first place in safety inspection. The leakage current is an important parameter to be checked because of shock of the patient. Electrical safety inspection must be performed every 12 months in order to detect leakage currents to protect both the patient and the user from electric shock. Electrical safety inspections are carried out according to TS EN 60601-1 (Medical electrical equipment Part 1: General requirements for basic safety and essential performance) and TS EN 62353 (Medical electrical equipment – Recurrent test and test after repair of medical electrical equipment) (International Electrotechnical Commission 2005, 2014). The electrical safety analyzer is used for safety inspection. TS EN 62353 standard provides a shorter time measurement method for electrical safety inspection than TS EN 60601 standard. TS EN 62353 standard provides service in hospital-level, while TS EN 60601 standard provides service in factory-level inspection (International Electrotechnical Commission 2005, 2014, Backes 2007a, 2007b).

The medical devices are classified according to their electrical safety specifications: Class I and Class II. Class I devices are the devices including basic insulation. The metal parts are connected to the earth. Class II devices have double insulation. They do not need any earthing conductor. The types of devices are identified on the device's label with the symbols. These symbols can be seen in Fig. 25.4.

Additionally, for the medical devices having patient application parts (electrodes, probes, etc.), the applied parts are classified as type B, type BF, or type CF. The types of classification indicated on the device with the symbols should be considered during the measurements and should be introduced to the electrical safety analyzer, and measurements should be made according to this classification.

Type BF is defined for devices in contact with the patient, while type CF applies to devices in contact with the heart of the patient. For example, the electrocardiogram is a medical device with CF type; the pulse oximeter is a BF type device.

Parameters measured using the electrical safety analyzer (Backes 2007a):

- Earth leakage current
- Chassis leakage current
- Patient leakage current
- Auxiliary patient leakage current

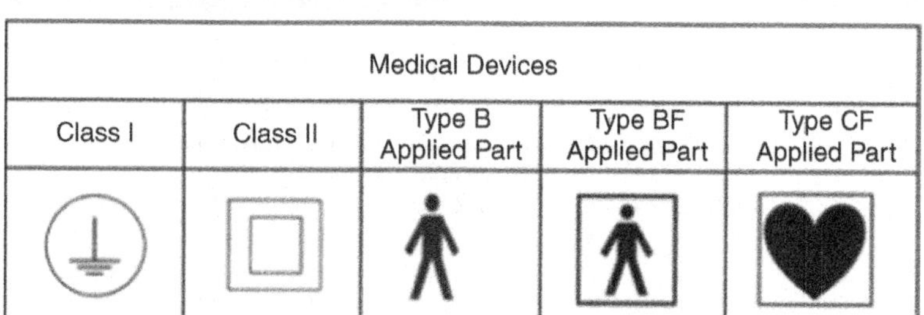

Fig. 25.4 Symbols indicating the class of medical devices (Backes 2007a)

All leakage measurements are performed under normal and single fault conditions. These conditions are (Backes 2007a, 2007b):

- Normal supply voltage
- Normal supply voltage, open neutral
- Normal supply voltage, open ground
- Reverse supply voltage
- Reverse supply voltage, open neutral
- Reverse supply voltage, open earth

The electrical safety analyzers compare the values they have measured with the standard acceptance criteria values installed on the system. The result is "there is leakage current" or "there is no leakage current."

25.3 Corrective Maintenance

The repair includes all activities performed to restore the normal functioning, safety, reliability, and efficiency of the device by eliminating the problems resulted from medical equipment failure or malfunctions. Within the scope of the repair, the circuit elements of the device can be changed or the device can be adjusted. The faults can be rectified directly by the biomedical personnel of the health institution to which the device belongs, or in cases where the biomedical personnel cannot make any attempts, the device is supported by the technical service of the manufacturer or the supplier of the device.

A few different ways can be followed in the repair workflow.

- The fault is detected by the hospital's biomedical staff and can be eliminated.
- The fault is determined by the hospital's biomedical staff. However, due to the material requirement, material is taken from the company and the fault is eliminated by the hospital staff.
- The fault cannot be detected by the hospital's biomedical staff. Service is taken from the related company for repair.

The workflow in the repair process can be summarized as shown in Fig. 25.5.

When the device broke down, the first question to be asked before interfering with the device should be "Is the device under warranty?" If the device is under warranty, it will be repaired or replaced by the related company. The quality of the repair service that will be provided by the company must be specified in the technical specifications prepared during the purchase process. For example

- The maximum time allowed for the first response to the fault and the penal sanction if the maximum time is over.

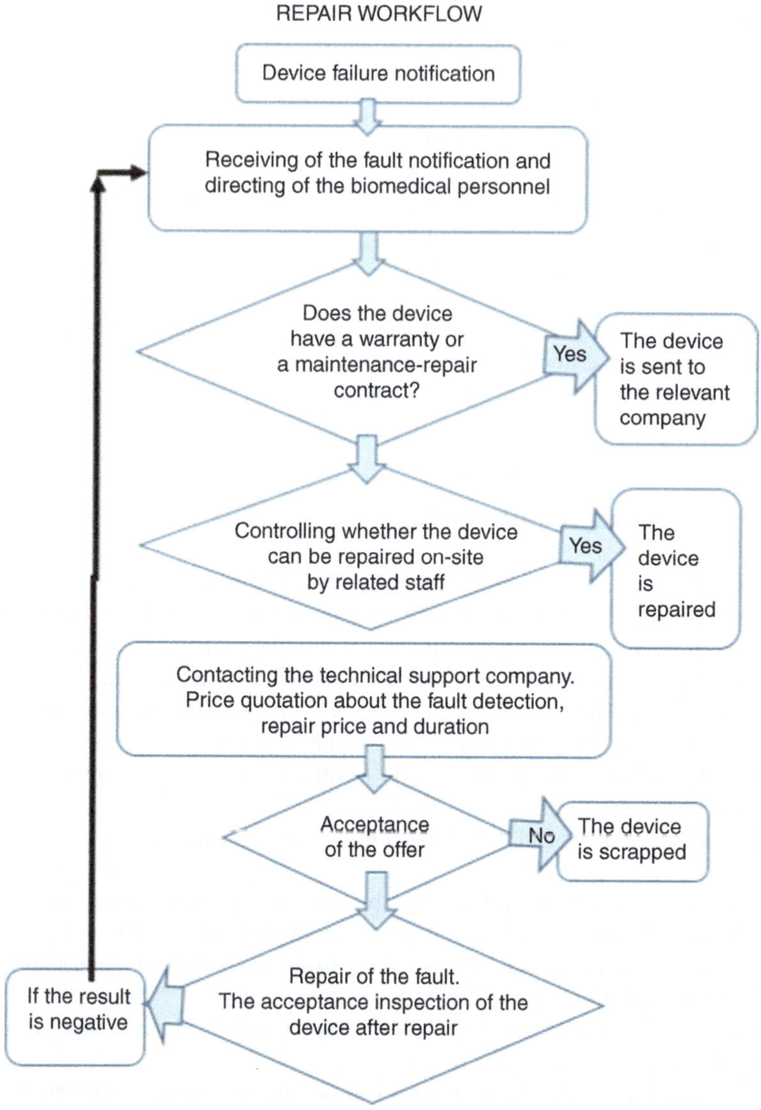

Fig. 25.5 Workflow of the repair process

- The maximum time allowed for the repair of the device and the penal sanction if the maximum time is over.
- Procurement of spare parts required for fault rectification.
- Providing of temporary equipment to be used in hospital if the repair is delayed because of the factors that cannot be controlled by the firm.

If the device is not under warranty, firstly the hospital biomedical staff attempt to repair the device. If they cannot repair, they contact the company.

Monitoring the device delivered to the company for repair is another important issue. The device is delivered to the company by signing the delivery form. If the repair of the device does not take place in a short time, the repair process of the device should be tracked by contacting the company. Sometimes, the device can be sent abroad for repair. In this case, it may take too long for the device to return to the hospital.

Corrective maintenance begins with a failure notification and ends when the device regains its old function. After failure notification, the biomedical personnel finds the origin of the failure and repairs it. During repair process, it may be sometimes required to change components of the electronic cards to change electronic board completely or all subsystems.

Even if the biomedical staff of the health facility has identified the malfunction, they cannot do anything if they do not have the spare parts that they must use for fault rectification, and they are still dependent on the technical service of the relevant distributor or manufacturer. In order to maintain the functionality of the medical devices, it is important to ensure that the consumables necessary for the devices to perform their functions are made available by the biomedical staff. Multi-use consumables that are required by the devices (ECG cables, SpO2 sensors, bacteria filter and oxygen sensors of ventilator devices, patient plates of electrocautery devices, carbon filters of reverse osmosis water treatment systems, hydrophobic filters of surgical aspirator devices) must be replaced with a new one. It is not possible to repair and use such consumables. For this reason, the period for which such consumables should be renewed is determined by the manufacturer, and the materials are determined on an annual basis. For this reason, biomedical personnel should take care to keep spare parts in stock. The following criteria should be taken into account when deciding which consumables to store in biomedical units:

- The most needed spare parts must be stocked.
- Spare parts of medical devices which may require urgent use on the patient should be stocked primarily, and spare parts of medical devices that do not adversely affect patient intervention or maintenance should be taken into the second plan.
- If there is too much of the same type of device within the health facility, spare parts for that device must be in stock.
- If the time when the defective device is not used is a significant loss of income for the healthcare facility, the stock of spare parts for such devices should be in the forefront.
- Spare parts which have long procurement period and which cannot be received for a long time after the order has been placed must be stocked in advance.
- Spare parts with low costs will not be too difficult for the budget of the hospital during the purchase process and will be able to be taken more in number and will provide a wide range of usage in case of need.
- Shelf life of the spare parts to be stored should be taken into consideration.
- When stocking the spare parts of too old or too problematic medical devices, the possibility of scrapping them at any time should be taken into account.

In addition to the spare parts that can be purchased and stocked to be used in the repair, a number of needs can be eliminated by installing a spare part recycling system from the medical devices that have been expired and disposed of within the health facility. This kind of implementation will provide a significant economic contribution to the healthcare provider and will provide an exemplary clinical engineering practice.

Scrap devices are devices that are withdrawn from use. Medical devices that complete the life cycle are withdrawn from use. The most important factor that starts this process is that the device cannot be repaired. In this case, either the device is too old or it is technically end-of-support.

An increase in the usage cost of the device as the device gets older can cause the device to be scrapped. The device may fail frequently and the cost of repair may be too high. The cost will be considerably higher if the service stops when the device fails. In such a case, the hospital administration may try to guarantee a more efficient service by receiving a new device. And, the problem device is scrapped. For the scrapping process, the device, which is planned to scrap, is removed from the inventory by creating a report. The report must clearly state why the device was disabled.

Due to its fault or technological life-span, the electronic materials of many devices that are scrapped can be used partially or completely. Upon analysis of the device failures in the hospitals, it is observed that approximately 50% of the faults are caused by the feeding units of the devices. The reasons for these failures are the imbalances in the supply voltage, the dust inside the device and the lack of sufficient air circulation. Most of which can be repaired with a defective component change. The remaining malfunctions are caused by the control units of the devices. 10–15% of all these faults can be made with spare parts removed from the scrap devices. Even if the numerical ratio of the faults made with the scraped devices is low, it provides a serious recycling and financial benefit to the institution.

The scope of malfunctions that can be made by the hospital's biomedical personnel is directly proportional to the scope of spare parts in stock and the technical training given in the relevant field.

A number of possible factors should be considered when intervening to the malfunction (Ridgway et al. 2009). These are:

- User errors
  - If the user does not know the device well
  - If the user does not use the device properly
  - If the user does not do the required preventive maintenance
- Electrical infrastructure problems
  - Leakage current
  - Power failure problem
  - Compensation problem between devices
- Air conditioning problems
- Plumbing problems
- Gas installation problems

User errors can be reduced easily by training. In the electrical infrastructure problems, the problem of leakage current can be solved by a good grounding system and the power problems can be solved by using generator and ups. Problems in the plumbing cause clogging of the filters of the devices using water, while problems in the gas system cause problems in the filters of gas-powered devices. If the air conditioning systems in the imaging departments such as MR, CT do not cool sufficiently, increasing the ambient temperature may cause the devices to malfunction. Continuous control of these systems will reduce the number of device faults. If it is detected that the fault is caused by these systems, intervening to the problematic system will prevent the occurrence of subsequent failures.

The success of corrective maintenance is determined by some criteria. These are (Tiwari and Tiwari 2014):

- The average time between faults
- The number of repeated faults
- The time between failure notification and start of repair
- The repair time
- The out-of-service time of the device

Extending the scope of repair that can be carried out by hospital biomedical staff is only possible with training. The greater the demand for technical service training in the purchase specification during the purchase of new equipment, the greater the intervention of the biomedical staff.

25.4 Conclusion

In this section, medical device maintenance applications are briefly mentioned and the points that will be considered during the applications have been explained.

Although the maintenance studies to be applied to medical devices is difficult in terms of technology and device diversity, it is important for patient and user safety. It should not be forgotten that the continuity and quality of the services provided by the medical devices will be ensured by the applied maintenance services.

References

Backes J (2007a) A practical guide to IEC 60601-1. Rigel Medical, Peterlee

Backes J (2007b) A practical guide to IEC 62353. Rigel Medical, Peterlee

Bahreini R, Doshmangir L, Imani A (2018) Factors affecting medical equipment maintenance management: a systematic review. J Clin Diagn Res 12(4):IC01–IC07

Emergency Care and Research Institute (2008) Electrical safety, biomedical benchmark – the technology support system, Philadelphia

Fennigkoh L, Smith B (1989) Clinical equipment management. JCAHO Plant Technol Safely Manag Ser 2:5–14

International Electrotechnical Commission (2005) IEC 60601-1:2005, Medical electrical equipment – part 1: General requirements for basic safety and essential performance
International Electrotechnical Commission (2014) IEC 62353:2014, Medical electrical equipment – recurrent test and test after repair of medical electrical equipment
Republic of Turkey Ministry of Health, Clinical Engineering Management Department (2016) Biyomedikal Metroloji Faaliyetleri Kılavuzu (The Guide for Biomedical Metrology Activities). Ankara, Turkey. (in Turkish)
Ridgway M, Atles LR, Subhan A (2009) Reducing equipment downtime. J Clin Eng:200–204
Sezdi M (2012) Medical technology management and patient safety. In: Kara S (ed) A Roadmap of biomedical engineers and milestones. InTech, Croatia, pp 183–208
Sezdi M (2013) Performance analysis for medical devices. Biomed Eng Res 2(3):139–146
Tiwari A, Tiwari A (2014) Performance evaluation of outsourced medical equipment maintenance service in a tertiary care hospital. Int J Sci Res Publ 4(9)
World Health Organization (2011) Medical equipment maintenance programme overview. In: WHO Medical device technical series. WHO, Geneva

Part XI

Advances in Biomedical Engineering

Recent Advances on Polymer Nanocomposite-Based Radiation Shielding Materials for Medical Science

26

Abhijit Nath, Aunggat Shah, Sanjeev Bhandari, Manashjit Gogoi, and Mrityunjoy Mahato

26.1 Introduction

In this era of technology we can't get rid of various types of natural (such as radon, cosmic rays) and manmade radiations used in communication and health care (such as microwave, X-ray, gamma ray, etc.). All these radiations are harmful depending on exposure time and doses. The unwanted exposure of these radiations can lead to health disorder (cancer, birth defect) and other kinds of accidents. Therefore, protection of our health and environment from the radiation is necessary, and development of radiation protection nanocomposite is now an emergent field of research. Usually, a high-density material lead is used to protect in general from any kind of radiation and in particular from nuclear radiation. However, due to high atomic weight and high toxicity of lead, it has limited applications, and researchers are encouraged to look for a lightweight, nontoxic, low-cost polymer nanocomposite material as an alternative for lead. In this chapter, the following points will be summarized such as radiation used in medical field, unwanted exposure of environmental radiation, existing radiation shielding materials, and recent advancement on polymer-based lightweight cum high-efficient composite shielding materials.

Polymer nanocomposites are the mixture of a polymer which is the matrix and nanofiller which is at least one dimension less than 100 nm (Mai and Yu 2006). The nanofillers are found as 1D, 2D, and 3D nanomaterials (Mai and Yu 2006). These

A. Nath · A. Shah · S. Bhandari · M. Mahato (✉)
Physics Division, Department of Basic Sciences and Social Sciences, School of Technology, North-Eastern Hill University, Shillong, Meghalaya, India

M. Gogoi
Department of Biomedical Engineering, North-Eastern Hill University, Shillong, Meghalaya, India

S. Paul (ed.), *Biomedical Engineering and its Applications in Healthcare*,
https://doi.org/10.1007/978-981-13-3705-5_26

nanomaterials have higher efficiency and unique properties than their bulk material counterparts. Polymer is a macromolecule which is composed of many repeating subunits known as monomers. Polymers are of four types such as thermoplastic, thermoset, elastomer, and natural/biodegradable polymer. Thermoplastics undergo molten state on heating and solidify on cooling, such as polyethylene and polypropylene. Thermosets are synthetic cross-linked polymer with irreversible chemical bond, and they do not melt upon heating, such as polyester resin, polyurethane, and bakelite. Elastomers are natural or synthetic polymers having weak and amorphous nature, such as polyisoprene and ethylene vinyl acetate. Natural polymers are biodegradable such as nucleic acid, proteins, natural rubber, and cotton fabric (Bhattacharya et al. 2008).

In polymer nanocomposites, the nanofiller and the polymer matrix are bound to each other by weak intermolecular forces such as hydrogen bond and van der Waals interaction (Mai and Yu 2006). Polymer nanocomposites generally show anisotropic behaviour due to inhomogeneous distribution of nanofiller within the matrix. The quality of polymer nanocomposites depend on the following parameters which demand for further research to be carried out (Jordon et al. 2005; Jeon and Baek 2010):

- Shape/size of nanofiller materials.
- Volume fraction of the nanofiller.
- Distribution of nanofiller within matrix.
- Cross-linking and mixing between different phases.
- Process used for fabrication of the nanocomposites.
- Morphology of the nanocomposites.

Polymer nanocomposites can be prepared by both chemical and physical processes, out of which mainly four processes have been established in literature such as (i) sol-gel method, (ii) in situ polymerization, (iii) intercalation method, and (iv) direct mixing method (Tanahashi 2010; Yang et al. 1998; Alexandre and Dubois 2000; Reddy 2010). In polymer nanocomposites, the nanofiller can easily interact with the polymer due to high surface-to-volume ratio. Polymer nanocomposites possess some special properties such as barrier properties, flame resistance, scratch resistance, magnetic properties, and electronic/optoelectronic properties. Due to their unconventional properties, polymer nanocomposites are used in a variety of application such as electronic devices, medical devices, consumer goods, adhesives, fire retardants, packaging materials, sensors, coatings, automobiles, aerospace, as well as radiation-shielding applications (Paul and Robeson 2008). To prepare an efficient polymer nanocomposite, it needs to be characterized by using some complementary techniques such as UV-visible spectroscopy to analyze size/band gap of the nanoparticle; Fourier transform infrared (FTIR) spectroscopy for identifying functional group involved in composite formation; X-ray diffraction (XRD) for determining particle size and crystallinity of the composite and scanning electron microscopy (SEM) for nanoparticle image, elemental analysis, and morphology (Mittal 2012).

The literatures on polymer nanocomposites for its application in radiation shielding have been summarized as follows. There is reported literature for *microwave shielding* by Manna et al. who prepared Fe_3O_4, carbon, and polyaniline trilaminar core-shell composites having shielding efficiency ~47 dB due to absorption (Manna and Srivastava 2017). Hosseini et al. prepared a composite using epoxy resin, MWCNT, $NiFeO_4$ and obtained a reflection loss value 19 GHz (Hosseini and Mahdavi 2018). Gupta et al. made a composite of polyurethane and modified MWCNT having shielding effectiveness value ~29 dB for the sample of 10 wt% MWCNT with thickness of 1.5 mm (Gupta et al. 2013). For *UV shielding*, Abidi et al. prepared cotton fabric with titania nanosol, and their UV protection factor (UPF) was 50 (Abidi et al. 2007). Camlibel et al. prepared a cotton fabric with polyacrylate polymer containing iron ores (goethite and hematite) and obtained a value of UPF 50 (Camlibel et al. 2018). Junqing et al. prepared PET-ZnO nanocomposite by in situ polymerization, and they did comparative study of the composites with various wt% of ZnO in PET. The result shows very less transmittance for 1.5 wt% of ZnO nanoparticles in PET, whereas pure PET shows 84.5% transmittance (He et al. 2009). Chen et al. prepared carboxymethyl chitosan/organic montmorillonite nanocomposite and obtained less transmittance percentage than ZnO and TiO_2 in both UVA and UVB regions (Chena et al. 2017). Sirohi et al. prepared multifunctional ZnO/polyester resin green composite films of thickness 0.75 mm. They obtained UPF as 100 for ZnO/polyester composite; however, for pure polyester, UPF was 50 (Sirohi et al. 2017). For *EMI shielding*, Aycik et al. prepared 10 mm-thick shielding material sample of PbO-reinforced polyethylene-based composite and obtained 8% attenuation at 1333 keV ionizing electromagnetic radiation (IEMR) (Aycik and Belgin 2018). Yim et al. prepared Ni-plated MWCNT/HDPE composite and shielding effectiveness obtained was up to 12 dB at 1.0 GHz (Yim et al. 2016). Saleh et al. prepared 1 mm-thick sample of 7.5 vol% MWCNT/PP composite and obtained an overall EMI SE of 36.4 dB at 12.4 GHz (Al-Saleh and Sundararaj 2009a). Sapurina et al. prepared PANI-wood composite and obtained transmittance coefficient of −10 dB (MHz range) and −40 dB (GHz range) (Sapurina et al. 2005). Jia et al. prepared calcium alginate/AgNW/polyurethane film and obtained an EMI SE of 20.7 dB (Jia et al. 2018). Zunfeng et al. prepared SWCNT/polyurethane composite and obtained EMI SE of ~17 dB at 20 wt% of filler (Liu et al. 2007). For *gamma shielding*, Singh et al. prepared composite of barium-borate-flyash glass and obtained a mass attenuation coefficient of 0.0523 cm^2/g at photon energy 1.33 MeV (Singh et al. 2008). More et al. studied gamma shielding efficiency of different types of nylon polymer and obtained mass attenuation coefficient of 0.0614 cm^2/g at 1.33 MeV photon energy (More et al. 2017). Harish et al. prepared PbO-filled isophthalate polyester resin composites and obtained mass attenuation coefficient of 0.0948 cm^2/g with 50 wt% of PbO NP at photon energy of 0.662 MeV from Cs-137 point source radiation (Harish et al. 2009). Badawy et al. prepared Fe_3O_4/polyvinyl alcohol nanocomposite and obtained nearly 59% shielding ability of lead, whereas pure polyvinyl alcohol has only 8% of that in terms of the linear attenuation coefficient. However, the magnetite nanocomposites are lighter and flexible than lead, and their density is only 16% compared to that of lead (Badawy 2015). Kim et al. prepared nanotungsten/polyethylene composite, and the attenuation coefficient for the composites was ~75% for Ba-133 (~0.3 MeV) (Kim et al. 2014). For *neutron shielding*, Zali et al. prepared thermoplastic natural rubber/nanosized boron carbide (B_4C) composite and obtained a half value layer 0.047 at 30 wt% of B_4C (Zali et al. 2018). Kaloshkin et al. prepared ultrahigh-molecular-weight polyethylene-based composite containing B_4C and tungsten (W) and obtained gamma ray attenuation coefficient of 3.43 cm^{-1} for 60 wt% of tungsten with photon energy of 1.02 MeV (Kaloshkin et al. 2012). For *alpha shielding*, the composites are simple such as HDPE, PET composites since alpha particles are very heavy, and their paths of attenuations are very less (Knoll 1999).

26.2 Radiations and Radiation Dose Measurements

Basically, radiation dose is the amount of radiation energy incident per kg of tissue. Before shielding of radiation, it is essential to know the activity, type of radiation, the energy, and the dose of radiation. Becquerel (Bq) is the unit to measure the intensity or activity of a radiation source. This unit is also given in terms of total activity such as for liquid Bq/L, for solid Bq/kg, and for gas Bq/m^3. For measuring radioactive pollution, Bq/m^2 is used. In addition, another essential task is to distinguish between α-rays, β-rays, neutrons, X-rays, γ-rays, and other radiations. The absorbed radiation energy dose is measured in terms of gray (Gy). In radiation therapy and various types of radiobiological experiments, correct dose measurement is important. In this context, measurement of doses also accounts for natural radiations. The unit of absorbed dose (gray) is defined as the total radiation energy absorbed (joules) per unit of mass (kg) of an exposed area of organ or tissue, i.e., 1 gray is 1 joule/kg. For γ-rays and β-rays, the effective dose which is expressed in Sievert (Sv) is equivalent to the absorbed dose. In case of α-radiation, the effective dose value is more, and it creates more damage to the body. For radiation measurements, we have four different but interrelated units (Maillie and Devid 2003):

- Radioactivity – It is the quantity of ionizing radiation released to the surrounding by a radioactive material.
- Exposure – It is the total amount of ionizing radiation traveling through the air.
- Absorbed dose – It is defined as the quantity of radiation absorbed by the target person or object.
- Effective dose – It is the combination of absorbed dose and the biological effects for that particular type of radiation.

Depending on the types of radiations, the detection or measurements of radiations are different such as photographic films for X-ray, Dosimetry, GM counter, scintillation detector are used for gamma ray. The principles of these techniques have been described as follows. In thermoluminescence dosimetry (TLD), a crystal is used (e.g., Mn-doped LiF). The impurity present in the crystal creates traps in the crystal lattice in which electrons are held when it is irradiated. Then, the crystal is heated and the trapped electron energy is emitted in the form of light. So, amount of emitted light is directly related to radiation dose encountered by the crystal lattice. When radiation of sufficient energy falls on any material, positive and negative ions are formed. This ionization is utilized for dose measurement as well as Bq measurements. To form an ion pair, some amount of energy is needed and dose is measured on the basis of knowledge of this energy. The well-known Geiger-Mueller tube, commonly known as GM counter, is based on the concept of measuring the electrical response pulses produced when new ion pair is formed. In the process of scintillation detection, various types of material compounds are used which will emit light when they are exposed to the radiation. Radiation exposure plays a key role in the intensity level of the emitted light, and the intensity of light is measured easily. For radiation detection and measurement, semiconductors are also used. When radiation falls on a semiconducting material, an electric current is generated and it can be measured. It also produces some free radicals inside the material. Free radicals are very reactive because they contain unpaired electrons. Then, also some solid materials trap them. The number of free radicals trapped inside reflects the amount of radiation dose. When the material is exposed to radiation, oxidation occurs by electron addition and reduction occurs by abstraction of electron; as a result, redox products are obtained. Initially, these changes are in the form of unstable free radicals, and then some chemical reactions occur, which ultimately result in stable reduction and oxidation products (Maillie and Devid 2003).

26.3 Environmental Radiation Health Hazards

Our environment contains various types of manmade and natural radiations such as microwave radiation (from cell phones, cell phone towers, cordless phones, broadcast antennas, radars, TVs, computers, microwave ovens, satellites, wireless, Internet etc.), X-ray radiation (from X-ray scanner, CT scan), gamma ray radiation (from radiotherapy, radio diagnosis, radioactive materials), as well as other natural internal radiations such as radon, cosmic rays, etc.

All types of radiation irrespective of their nature are hazardous depending upon the exposure dose, time of exposure, and length of exposure. In general, ionizing radiations are more dangerous. Alpha particles can penetrate into the body and releases the energy to a small area. But, high-energy ionizing radiations like gamma ray and X-ray penetrate and spread the energy in large volume, leading to damage in larger scale. Radiations damage tissue and cells which leads to cancer-like disease. Exposure to UV radiation, present in sunlight, also causes skin cancer. Nonionizing radiations do not create these types of problems in molecular level, but those can create electrical shocks or burns (Lippmann et al. 2003). As per the recommendation of International Commission on Radiological Protection (ICRP), the total effective dose of an individual should not exceed 50 mSv or 5 rem (www.icrp.org). Also, the equivalent dose should not exceed 150 mSv for eye, 500 mSv for skin, and 500 mSv for hands and feet.

26.4 Radiations Used in Medical Science

In medical science, radiations are used for many types of diagnostic examinations in some way. In the world, radiations exist in several forms. Some radiations are good for us, some cause harm, and some exist without causing any harm or good. These radiations are used in medical sciences. There are some classes of radiation, such as low-energy radiation (radio waves, microwaves and infrared), medium-energy radiation (visible light), and high-energy radiation (UV ray, X-rays, and gamma rays). The high-energy radiations are used for medical diagnostic purpose. There are two categories of this type of radiations, such as ionizing (X-ray, gamma ray) and nonionizing (wave used in MRI). Also, a mechanical wave ultrasound is used widely in medical science. X-ray is used to detect the injury/defect of bone or the presence of unwanted external objects in patients. CT scan gives us a computer-generated 3D image of the body cross sections. In nuclear medicine, a radioactive material is injected in a patient's body. This technique is used to locate tumours and assess the health of organs by using a detector outside.

X-rays are one of the main diagnostic tools in medicine since its discovery by Wilhelm Roentgen in 1895. Figure 26.1 shows the generation of X-rays by converting the kinetic energy (KE) of electrons into electromagnetic (EM) radiation. For production of X-ray, two electrons are placed inside an evacuated chamber of glass. A large potential difference is applied across the two electrodes. The cathode is the source of electron, and the target anode is kept at 45 degrees to the path of the electrons (Cierniak 2011). After the application of a very high electrical potential difference (~KeV), the electrons are released from the cathode and attain kinetic energy then start accelerating toward anode. Almost 99% of the gained KE is converted to heat due to collision-like interactions. Almost 0.5%–1% of gained KE is converted into X-rays due to strong Columbic interactions (Bremsstrahlung). When an e^- comes within the proximity of the charged nucleus present in the anode, the e^- are attracted and decelerated by Coulombic forces, resulting in a significant loss in KE, and cause changes in the path of electron. The

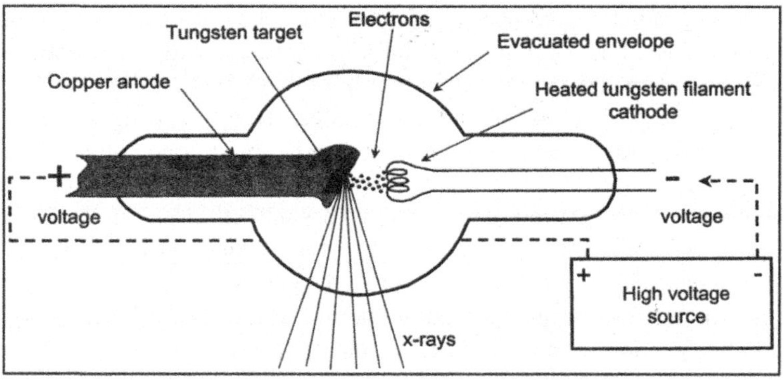

Fig. 26.1 Schematic diagram of X-ray production system as obtained from reference (Khan 2012)

difference in kinetic energy of the electron is converted to an X-ray photon with the same energy (conservation of energy) (Cierniak 2011). Figure 26.2 shows a digital X-ray machine used for medical purpose.

Computed tomography (CT) scanners also use X-ray radiations (Fig. 26.3). The X-rays are allowed to travel in a circular path around the patient's body which is captured at the other side of the transmitter. Then, the collected information are processed by the computer, and we get multiple images of the slices throughout the body. 3D images can also be obtained with the collected information. The radiation is switched on and off only to take the picture for these machines. So, the radiations are allowed to pass only through those body parts which are to be examined (Hathcock and Stickle 1993).

Gamma ray has the smallest wavelength (less than 10 pM) and greatest energy (few 100 KeV to few MeV) in the electromagnetic spectrum (Fig. 26.4).

The gamma decay process starts usually after alpha and beta decays. In the field of medical sciences, gamma decay is most commonly used as a source of gamma ray. Terrestrial thunderstorm is one natural phenomenon during which gamma rays are also produced. During thunderstorms, the electrons are accelerated due to highly energetic static electric fields, resulting in the production of gamma rays by the

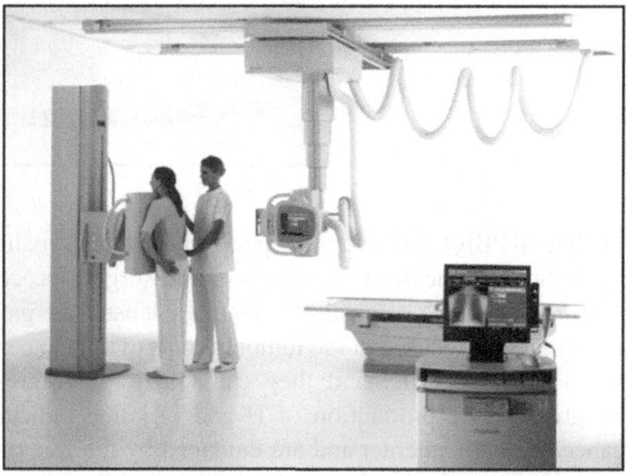

Fig. 26.2 An X-ray machine used in medical as obtained from reference (www.nibib.nih.gov)

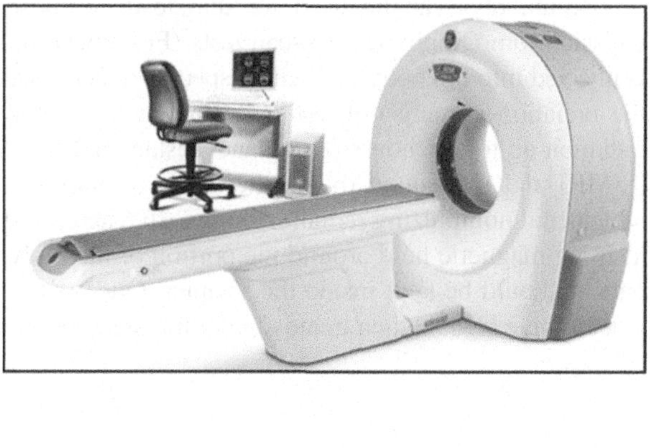

Fig. 26.3 A CT scan machine used in medical as obtained from reference (www.nibib.nih.gov)

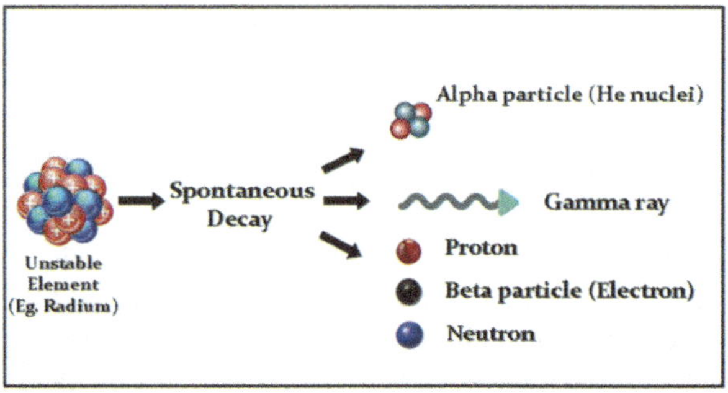

Fig. 26.4 Schematic diagram of spontaneous gamma ray emission process as obtained from reference (www.nibib.nih.gov)

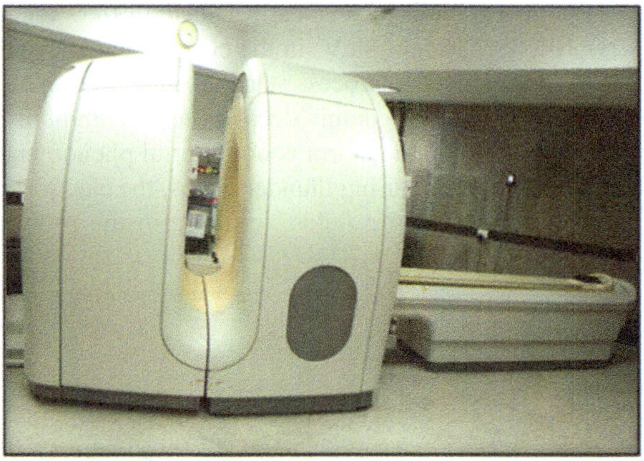

Fig. 26.5 A PET scan machine used in medical as obtained from reference (www.nibib.nih.gov)

method of "Bremsstrahlung". Specifically, to treat malignant and cancerous tumours in the area of oncology, in gamma knife surgery process, gamma rays are used. In this process, using some electromagnetic lenses, the gamma rays are concentrated to one point and focused at the tumours so as to kill cancerous cells. These gamma rays are highly energetic, and they can ionize water present in the cancerous cells, resulting in the production of H and OH free radicals. Chromosomes present in cancerous cells interact and are damaged by the free radicals because these radicals are highly reactive. Some radiations interact directly with the chromosomes present in the tumour without using free radicals and cause damages.

Gamma rays can be used for diagnostic purposes with the help of positron emission tomography (PET) scan tests (Fig. 26.5). In this test, a radioactive liquid is injected into the body and with a special camera placed outside the body to detect the radiation coming out. After the process, the body continues to send out the radiation up to few hours (Hogg and Testanera 2010; Saha 2010).

MRI and ultrasound diagnostic are two kinds of techniques used in medicals which emit nonionizing radiations (Figs. 26.6 and 26.7). The MRI scanner can create a strong magnetic field around the portion of the body to be studied. Therefore, the patient should be kept inside the scanner. Our tissues contain water molecules and react to give signal when treated under magnetic radiation. Initially, the energy from

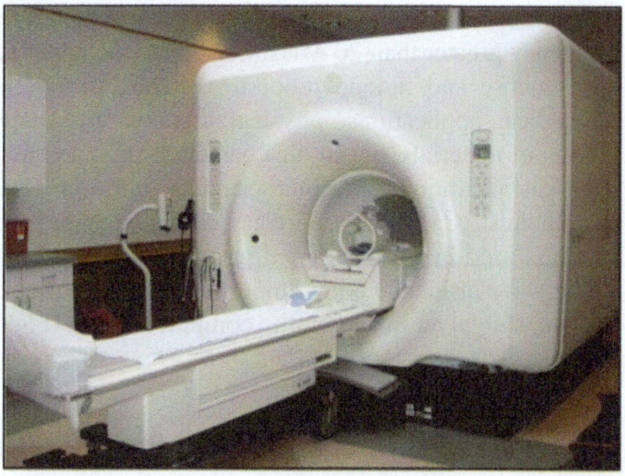

Fig. 26.6 An MRI machine used in medicine obtained from reference (www.nibib.nih.gov)

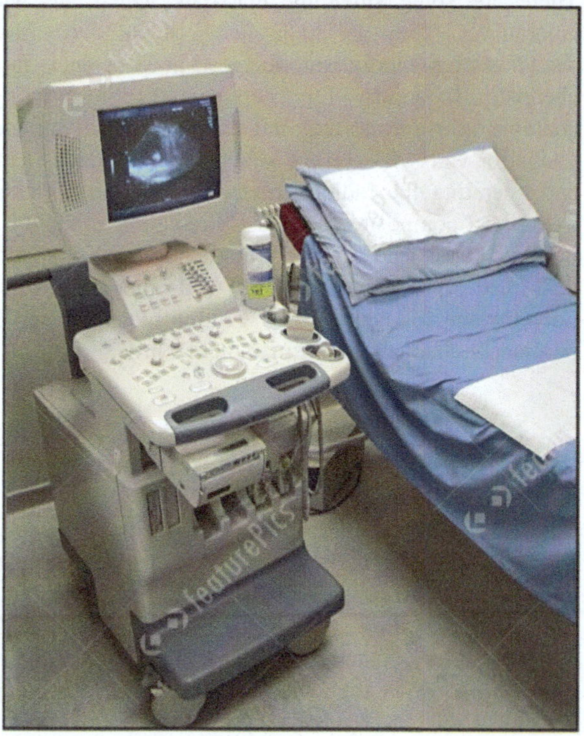

Fig. 26.7 An ultrasound diagnostic machine used in medical as obtained from reference (www.nibib.nih.gov)

a varying magnetic field is temporarily applied at resonance frequency to the person. Then, the protons present are excited and start emitting radio-frequency signals, which are received by coil. Then, the signal is processed by the computer and an image is obtained (McRobbie et al. 2006).

Ultrasound is one type of mechanical wave with nonionizing behaviour. It has frequency greater than 20 MHz, i.e., above the upper audible limit of sound. Due to high energy and unique properties, it has various applications in the fields of biomedical applications, physical therapy, ultrasonic treatment, processing and mixing of materials, ultrasonic characterization and manipulation of particles, ultrasonic disintegration, ultrasonic cleaning, ultrasonic humidifier, ultrasonic welding, sonochemistry, weapons, wireless communication, etc. In biomedical field, ultrasound is used in various directions. The ultrasound machine sends high-frequency sound wave inside the body. Impurities present in the body send back unusual wave, and we can get an image on the computer screen (Fig. 26.7) (Szabo 2004).

26.5 Existing Radiation Shielding Materials

There are some shielding materials which are widely used in medical and laboratories. They are high-energy shield decay drum with 0.5 inch lead for shielding high-energy isotopes, lead-lined cabinets in nuclear medicine and radiochemistry laboratory for keeping radiation emitting equipment, lead storage containers for transporting and storing radioactive materials, tungsten and tungsten alloys for vials and syringe shields which reduces hand exposure while preparing and operating the radiopharmaceuticals, lead curtains in PET and radiotherapy, nylon lead composite curtains for radiation shielding areas where permanent wall is not practical, lead glass for shielding use in X-ray laboratory and operation theatre, lead-lined plywood wall for PET and other diagnostic imaging room, boron-filled polyethylene for neutron and gamma blocking, and lead brick for temporary or permanent chamber (www.marshield.com).

26.6 Radiation Shielding Measurement

The ratio of transmitted power and incident power for a material is known as shielding efficiency (SE), and it is the sum of the reflected radiation from the material surface (SE_R), absorbed radiation (SE_A), and multiple internal reflections (SE_M) or transmitted radiation (Fig. 26.8). In this calculation, the source-to-shield distance should be greater than the free-space wavelength, so as to consider the measurements under far field. The total shielding efficiency can be expressed as (Chen et al. 2013; Cao et al. 2015a, 2015b; Gupta and Mool 2005; Li et al. 2015; Colaneri and Shacklette 1992).

$$SE_{total} = 10\log\frac{P_i}{P_0} = SE_R + SE_A + SE_M \qquad (26.1)$$

where P_i is the incident power and P_0 is the transmitted power. When $SE_{total} \geq 10$ dB, SE_M can be considered as negligible.

Fig. 26.8 Schematic diagram for shielding as obtained from reference (Wu et al. 2014)

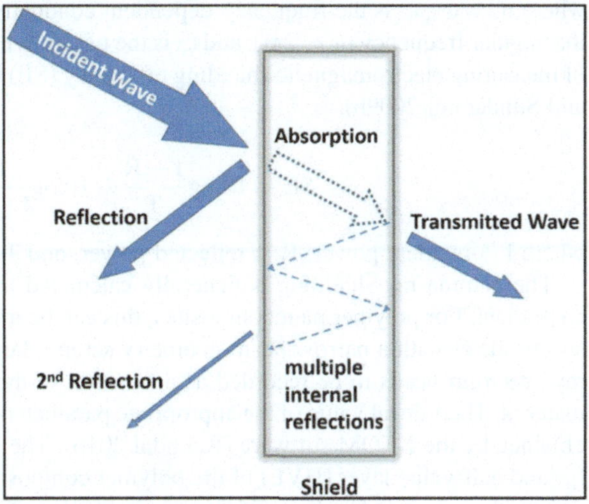

As per the transmission line theory (Naito and Suetake 1971), the input impedance (Z_{in}) at the material interface can be expressed as

$$Z_{in} = Z_0\sqrt{\frac{\mu_r}{\varepsilon_r}}\tanh\left(j\frac{2\pi fd}{c}\sqrt{\mu_r\varepsilon_r}\right) \tag{26.2}$$

where Z_0 is the free space impedance, μ_r is complex permeability ($\mu_r = \mu' - j\mu''$), ε_r is the complex permittivity ($\varepsilon_r = \varepsilon' - j\varepsilon''$), f is the frequency, d is the thickness of the material, and c is the speed of light. Therefore, the reflection loss (RL) can be expressed as

$$RL(dB) = 20\log\frac{|Z_{in} - Z_0|}{|Z_{in} + Z_0|} \text{ and } SE_A = -RL \tag{26.3}$$

For a material, the skin depth (δ) is the distance up to which the intensity of the electromagnetic wave decreases to 1/e of its original strength. It can be expressed as (Colaneri and Shacklette 1992; Singh et al. 2012)

$$\delta = \sqrt{\frac{2}{\mu_r\omega\sigma_s}} \tag{26.4}$$

According to electromagnetic theory, for electrically thick samples ($d > \delta$), the shielding efficiency due to reflection (SE_R) can be expressed as (Colaneri and Shacklette 1992; Singh et al. 2012)

$$SE_R(dB) = 10\log\left(\frac{\sigma_s}{16\omega\mu'\varepsilon_0}\right) \tag{26.5}$$

where $\sigma_s = \omega\varepsilon_o\varepsilon^{//}$ is the frequency-dependent conductivity (Ohlan et al. 2008), ω is the angular frequency (ω = 2πf), and ε_o is the permittivity of free space. Another way of measuring electromagnetic shielding efficiency (SE) is as per Eq. (26.6) (Al-Saleh and Sundararaj 2009b).

$$SE = 10\log\frac{I - R}{T} + 10\log\frac{I}{I - R} \tag{26.6}$$

where I = incident power, R = reflected power, and T = transmitted power.

The gamma ray shielding is generally calculated in terms of linear attenuation coefficient. For polymer nanocomposites, this can be measured by using gamma ray spectrometer with a narrow beam geometry setup (Harish et al. 2009). The gamma ray spectrum needs to be recorded as a function of the thickness of the composite material. Theoretical value of the appropriate parameters used for gamma rays can be obtained by the XCOM software (Kaundal 2016). The linear attenuation coefficient (μ) and half-value layer (HVL) of the polymer composite due to gamma ray irradiation can be calculated by Eqs. (26.7) and (26.8). However, the alpha shielding can be measured by measuring the path length traversed by the alpha particle on the chemically etched polymer composite sample.

$$\mu = \frac{1}{x}\ln\frac{N_0}{N_x} \tag{26.7}$$

where x = thickness of the absorbing material, N_0 = detector count without polymer composite target, and N_X = detector count with polymer composite target.

$$HVL = \frac{\ln 2}{\mu} = \frac{0.693}{\mu} \tag{26.8}$$

The above-mentioned Eqs. 26.1, 26.2, 26.3, 26.4, 26.5, 26.6, 26.7, 26.8 and 26.9 allow us to calculate the radiation shielding parameters for polymer composites for different types of radiations.

26.7 Polymer Nanocomposites as Shielding Materials

There are different preparation strategies for polymer nanocomposites such as sol-gel method, in situ polymerization, intercalation, and direct mixing method. The sol-gel is a bottom-up process, where sol is the colloidal state of the nanofiller and gel is the network of the polymer matrix. The colloidal nanofiller sol is dispersed in the polymer gel matrix by hydrolysis method. The polymer acts as a nucleating agent and helps in the growth of the layered crystals. As the crystals grow, the polymer is seeped between layers and thus nanocomposite is formed (Fig. 26.9).

Intercalation methods are of two types: chemical process and mechanical process. Chemical process is basically an in situ polymerization process. In the mechanical

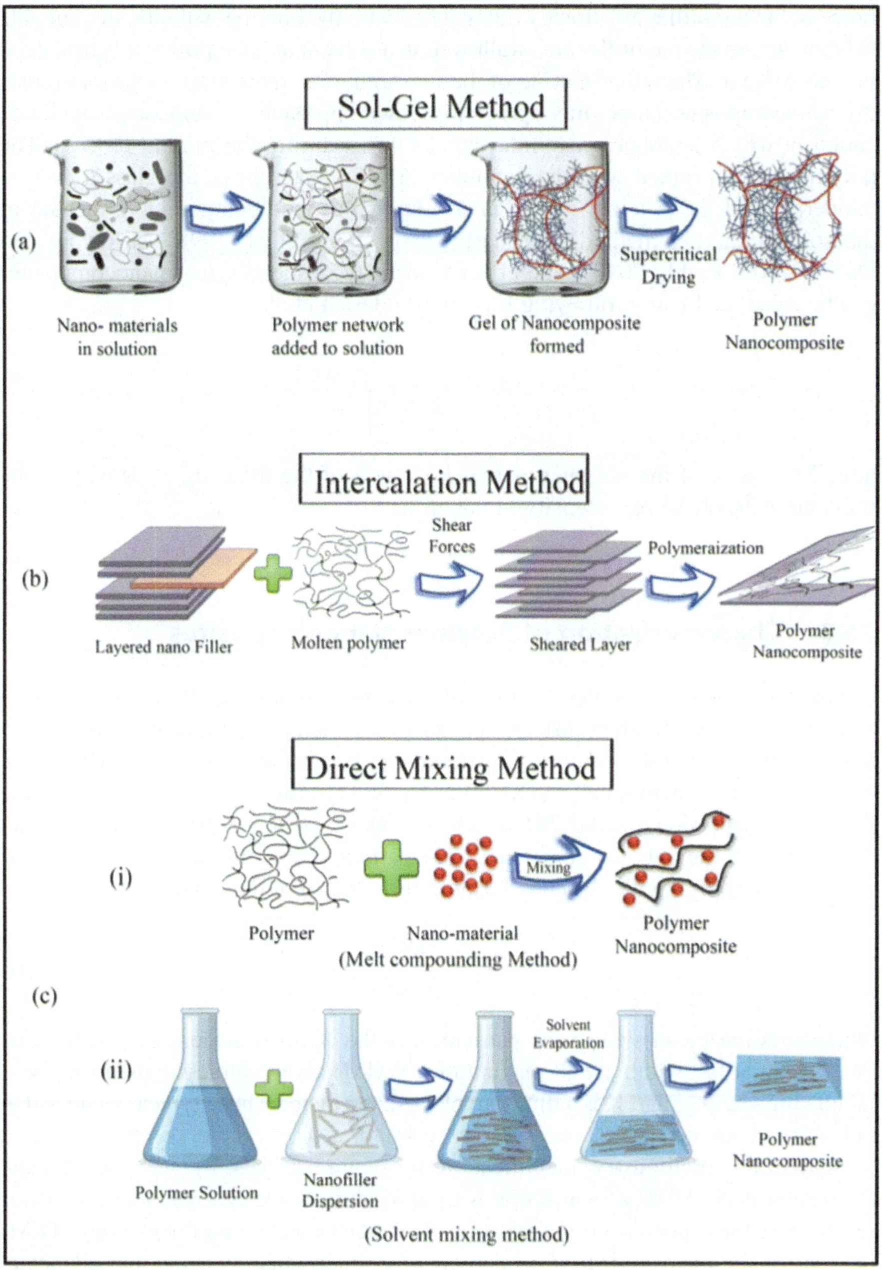

Fig. 26.9 Different methods for polymer nanocomposites preparation

process, the nanofiller are directly intercalated with the polymer solution in a suitable solvent, where the nanofiller are swallowed in a solvent and the polymer is dissolved in a co-solvent. Then after mixing of the two solutions, intercalation occurs to form the nanocomposite. Direct mixing is a top-down approach of nanocomposite fabrication in which aggregate nanofiller breaks down during the mixing process. The polymer can be mixed with the nanofiller in solid state above the glass transition temperature of the polymer, which is also known as melt compounding method or the solvent mixing method (Tanahashi 2010; Yang et al. 1998; Alexandre and Dubois 2000; Reddy 2010). The effective density of the polymer nanocomposites can be calculated using following Eq. (26.9) (Harish et al. 2009).

$$\rho_c = \frac{1}{\left[\frac{M}{\rho_m} + \frac{F}{\rho_f}\right]} \tag{26.9}$$

here, M = wt% of the polymer matrix, F = wt% of the filler, ρ_m = density of the polymer matrix, and ρ_f = density of nanofiller.

26.8 Characterization of Polymer Nanocomposites

Characterization of nanofillers and nanocomposites involves different investigation techniques to evaluate their different physical and chemical parameters. The dispersion of nanofiller in the composite material can be characterized using wide-angle X-ray diffraction technique (WAXRD) from its *d*-spacing values, which reflect the degree of dispersion (Mittal 2012). XRD is also useful for the composition and crystallinity of the polymer composites. The particle size of the nanomaterials can be obtained by applying the Scherrer formula (Eq. 26.10) in the XRD data.

$$L = \frac{K\lambda}{\beta\cos\theta} \tag{26.10}$$

where L is the mean size of the particle, λ is the X-ray wavelength, β is the line broadening at full width at half maximum (FWHM) after subtracting the instrumental line broadening, and K is a dimensionless shape factor whose typical value is 0.9, but varies with the actual shape of the crystallite. UV-visible spectroscopy is a technique for preliminary identification for nanomaterials by its characteristic absorption peak. FTIR spectroscopy is used to identify the type of chemical bonds responsible for nanocomposite formation. Transmission electron microscopy (TEM) is used for the characterization of particle size, interplanar spacing, and elemental analysis of the nanocomposites. The above-mentioned characterization techniques are essential to evaluate a polymer nanocomposite material after preparation.

26.9 Conclusion

In this chapter, effort has been made to summarize the importance of the radiation shielding material for medical purpose, existing shielding materials in the market, and the recent research on new type of shielding material such as polymer composite, etc. We have also discussed radiation-related hazards and how to combat it using polymer nanocomposites. Polymer nanocomposite-based shielding materials are much superior compared to existing lead-based shielding material in terms of lightweight and negligible toxicity. The production of polymer composite is also possible in large scale to make shielding product; however, there may be a limitation of having defects in polymer composites in large-scale production. In this chapter, the recent research on polymer composites as shielding materials have been discussed elaborately.

Acknowledgement Authors acknowledge DST-SERB, Government of India, for financial support through two SERB projects (No. EMR/2016/002634 and EMR/2016/ 004219) sanctioned to Dr. Mrityunjoy Mahato. Authors also would like to thank Department of BSSS, NEHU, Shillong, for providing laboratory working facility. AN and AS would like to acknowledge Non-NET NEHU PhD fellowship and SB for Non-NET SERB project fellowship to carry out research.

References

Abidi N, Hequet E, Tarimala S, Dai LL (2007) Cotton fabric surface modification for improved UV radiation protection using sol–gel process. J Appl Polym Sci 104:111–117

Alexandre M, Dubois P (2000) Polymer-layered silicate nanocomposites: preparation, properties and uses of a new class of materials. Mater Sci Eng 28(1–2):1–63

Al-Saleh MH, Sundararaj U (2009a) EMI shielding mechanisms of CNT/polymer composites. Carbon 47:1738–1746

Al-Saleh MH, Sundararaj U (2009b) Electromagnetic interference shielding mechanisms of CNT/polymer composites. Carbon 47:1738–1746

Aycik GA, Belgin EE (2018) Effect of polymer matrix type on EM radiation shielding performances of PbO reinforced/polyethylene, isophytalic polyester and bisphenol A vinyl ester based composites. Int J Chem Chem Eng Syst 3:1–4

Badawy SM (2015) Magnetite nanocomposite films for radiation shielding. Plastics Research Online, pp. 1–2

Bhattacharya SN, Kamal MR, Gupta RK (2008) Polymeric nanocomposites: theory and practice. s.l. Hanser Gardner Publications, Munich

Camlibel NO, Arik B, Avinc O, Yavas A (2018) Antibacterial, UV protection, flame retardancy and coloration properties of cotton fabrics coated with polyacrylate polymer containing various iron ores. J Text Inst 109:1–10

Cao M-S, Wang X-X, Caoa W-Q, Yuan J (2015a) Ultrathin graphene: electrical properties and highly efficient EMI shielding. J Mater Chem C 26:1–35

Cao M-S, Wang X-X, Caoab W-Q, Yuan J (2015b) Ultrathin graphene: electrical properties and highly efficient EMI shielding. J Mater Chem C 3:6589–6599

Chen Z, Xu C, Ma C, Ren W, Cheng H-M (2013) Lightweight and flexible graphene foam composites for high-performance EMI shielding. Adv Mater 25:1–5

Chena K, Guoa B, Luo J (2017) Quaternized carboxymethyl chitosan/organic montmorillonite nanocomposite as a novel cosmetic ingredient against skin aging. Carbohydr Polym 173:100–106
Cierniak R (2011) X-Ray Computed Tomography in Biomedical Engineering. s.l. Springer Verlag London Limited, London
Colaneri NF, Shacklette LW (1992) EM1 shielding measurements of conductive polymer Blends. IEEE Trans Instrum Meas 41(2):291–297
Gupta YY, Mool C (2005) Novel CNT-polystyrene foam composites for EMI shielding. Nano Lett 5(21):2131–2134
Gupta TK, Singh BP, Dhakate SR, Singh VN (2013) Improved nanoindentation and microwave shielding properties of modified MWCNTreinforced PU composite. J Mater Chem A 1:9138–9149
Harish V, Nagaiah N, Niranjana Prabhu T, Varughese KT (2009) Preparation and characterization of PbO filled unsaturated polyester based polymer composites for gamma radiation shielding applications. J Appl Polym Sci 112:1503–1508
Hathcock JT, Stickle RL (1993) Principles and concepts of computed tomography. Diagn Imaging 23(2):399–415
He J, Shao W, Zhang L, Deng C, Li C (2009) Crystallization behavior and UV-protection property of PET-ZnO nanocomposites prepared by in situ polymerization. J Appl Polym Sci 114:1303–1311
Hogg P, Testanera G (2010) Principles and practice of PET/CT. s.l.: EANM and SIEMENS. Brochures, Vienna
Hosseini H, Mahdavi H (2018) Nanocomposite based on epoxy and MWCNTs modified with NiFe2O4 nanoparticles as efficient microwave absorbing. Appl Organomet Chem 32:1–8
Jeon I, Baek J (2010) Nanocomposites derived from polymers and inorganic nanoparticles. Materials 3:3654–3674
Jia L-C, Yan D-X, Liu X, Ma R, Wu H-Y, Li Z-M (2018) Highly efficient and reliable transparent electromagnetic interference shielding film. ACS Appl Mater Interfaces 10:1–9
Jordon J, Jacob KI, Tannenbaum R, Sharaf MA, Jasiuk I (2005) Experimental trends in polymer nanocomposites-a review. Mater Sci Eng A 393:1–11
Kaloshkin SD, Tcherdyntsev VV, Gorshenkov MV, Gulbin VN, Kuznetsov SA (2012) Radiation-protective polymer-matrix nanostructured composites. J Alloys Compd 536:522–526
Kaundal RS (2016) Comparative Study of Radiation Shielding Parameters for Bismuth Borate Glasses. Mater Res 19:1–10
Khan FM (2012) The physics of radiation therapy. Lippincott Williams and Wilkins, New York
Kim J, Seo D, Lee BC, Seo YS, Miller WH (2014) Nano-W dispersed gamma radiation shielding materials. Adv Eng Mater 16:1–7
Knoll GE (1999) Radiation detection and measurement. Wiley, New York
Lippmann M, Cohen BS, Schlesinger R b (2003) Environmental health science. Oxford University Press, New York
Li Y, Pei X, Shen B, Zhai W, Zhang L, Zheng W (2015) Polyimide/graphene composite foam sheets with ultrahigh thermostability for EMI shielding. RSC Adv 5:24342–24351
Liu Z, Bai G, Huang Y, Ma Y, Du F, Li F, Guo T, Chen Y (2007) Reflection and absorption contributions to the EMI shielding of SWCNT/polyurethane composites. Carbon 45:821–827
Mai Y-W, Yu Z-Z (2006) Polymer nanocomposites. s.l. CRC Press/Woodhead Publishing, Boca Raton/Cambridge
Maillie TH, Devid H (2003) Radiation and health. Taylor and Francis, London New York
Manna K, Srivastava SK (2017) Fe3O4 carbon polyaniline trilaminar core-shell composites as superior microwave absorber in shielding of electromagnetic pollution. ACS Sustain Chem Eng 7b02682:1–38
McRobbie DW, Moore EA, Graves MJ, Prince MR (2006) MRI from picture to proton. Cambridge University Press, Cambridge
Mittal V (2012) Characterization techniques for polymer nanocomposites. Wiley, Weinheim

More CV, Bhosale RR, Pawar PP (2017) Detection of new polymer materials as gamma-ray-shielding materials. Radiat Eff Defects Solids 172:469–484
Naito Y, Suetake K (1971) Application of ferrite to electromagnetic wave absorber and its characteristics. IEEE Trans Microwave Theory Tech MTS-19:65–72
Ohlan A, Singh K, Chandra A, Dhawan SK (2008) Microwave absorption properties of conducting polymer composite with barium ferrite nanoparticles in 12.4–18 GHz. Appl Phys Lett 93:1–3
Paul DR, Robeson LM (2008) Polymer nanotechnology: nanocomposites. Polymer 49:3187–3204
Reddy RJ (2010) Preparation, characterization and properties of injection molded graphene nanocomposites. Mechanical Engineering, Wichita State University, Wichita. Kansas, USA: s.n. Master's Thesis
Saha GB (2010) Basics of PET imaging. Springer, New York
Sapurina I, Kazantseva NE, Ryvkina NG, Prokes J, Saha P, Stejskal J (2005) Electromagnetic radiation shielding by composites of conducting polymers and wood. J Appl Polym Sci 35:807–814
Singh S, Kumar A, Singh D, Singh Thind K, Mudahar GS (2008) Barium–borate–flyash glasses: As radiation shielding materials. Nucl Inst Methods Phys Res B 266:140–146
Singh AP, Garg P, Alam F, Singh K, Mathur RB, Tandon RP, Chandra A, Dhawan SK (2012) Phenolic resin-based composite sheets filled with mixtures of reduced graphene oxide, c-Fe2O3 and carbon fibers for excellent electromagnetic interference shielding in the X-band. Carbon 50:3868–3875
Sirohi S, Singh R, Jain N, Pani B, Dutt K, Nain R (2017) Synthesis and characterization of multifunctional ZnO/polyester green composite films. J Polym Res 24(193):1–10
Szabo TL (2004) Diagnostic ultrasound imaging: inside out. Elsevier Academic Press, California
Tanahashi M (2010) Development of fabrication methods of filler/polymer nanocomposites: with focus on simple melt-compounding based approach without surface modification of nanofillers. Materials 3(3):1593–1619
Wu F, Xu Z, Wang YWM (2014) Electromagnetic interference shielding properties of solid-state polymerization conducting polymer. RSC Adv 4:38797–38803
www.bbc.co.uk
www.ffden-2.phys.uaf.edu
www.icrp.org
www.labmate-online.com
www.marshield.com
www.nibib.nih.gov.
www.sciencing.com
Yang F, Ou Y, Zhongzhen YU (1998) Polyamide 6/silica nanocomposites prepared by in situ polymerization. J Appl Polym Sci 69:355–361
Yim Y-J, Rhee KY, Park S-J (2016) EMI shielding effectiveness of nickel-plated MWCNTs/HDPE composites. Composites Part B 98:120–125
Zali NM, Yazid H, Ahmad MHARM (2018) Neutron shielding behavior of thermoplastic natural rubber/B4C composites. IOP Conf Ser: Mater Sci Eng 298:1–6

27 Biodegradable Composite Scaffold for Bone Tissue Regeneration

Sandip Bag

Abstract

At micro-architectural viewpoint, human bone is composed of polymer ceramic composite having similar mechanical characteristics that can be tailored by synthetic composite materials. To mimic the properties of bone, research on bone substituted analogous biomaterials was initiated by reinforcing active biomolecules within the matrices of biocompatible polymers to formulate suitable bone analogous. The major advantages of the composites over conventional homogeneous materials like metals, ceramics, and polymers are superior mechanical, biological, and other physical properties that can be matched with the requirements of particular applications. Modern technology has not been able to provide a suitable bone substitute that replaces autogenous bone. The availability and suitability of conventional autogenous or homogeneous prosthetic elements to repair severe bone trauma or large defects caused by various bone diseases are critically limited; as a result, profound interest concentrated on application of man-made polymeric composite materials as biodegradable scaffold, which would provide support and a symptomatic, long-term function within the body or in contact with body fluid. In tissue engineering, biodegradable scaffolds play a crucial role, where matrix degradation and tissue in growth are of immense phenomenon for decisive performance of tissue-scaffold system during regenerative process.

Keywords

Bone · Bone graft · Bioactive composite · Biodegradable scaffold · Prosthesis · Tissue regeneration

S. Bag (✉)
JIS College of Engineering, Kalyani, West Bengal, India

S. Paul (ed.), *Biomedical Engineering and its Applications in Healthcare*,
https://doi.org/10.1007/978-981-13-3705-5_27

27.1 Introduction

The management of damaged and degenerative osseous tissues due to trauma or disease in the human body is more demanding and challenging one to reconstruct or regenerate via natural growth of the host tissue (Ami et al. 2012; Dumic-Cule et al. 2015). Treatment of such defects customarily focuses on transplanting tissue from the same patient (an autograft) or from different subject (an allograft). However these type treatment to meet the requirement have been radical and life saving but the main obstacle survive with these techniques includes extravagant, agonizing, embarrassed by anatomical limitations, combined with donor-site morbidity owing to infection (for autograft) and there are chance of repudiation by the patient's immune system & the opportunity for ventilating infection or septicity from the donor site to the patient (for allograft). Therefore, these defects must be restored with an artificial bridging material (Dinopoulos et al. 2012; Saska et al. 2015). In modern orthopedic surgery, biocompatible metals (such as SS-316 L, titanium alloys, etc.) and ceramics (like bio-inert alumina, zirconia and bioactive calcium phosphate ceramics, and bioglass ceramics) are very common (Ratner 2004; Mirza et al. 2010). But these materials are considerably heavier, denser, and stiffer than normal human cortical bone. The discrepancy of mechanical properties between an artificial implants and the host tissue can create bone absorption at the bone-implant articulation, which accelerates the prosthesis instability and failure (Sumner 2015; Chanlalit et al. 2012; Haase and Rouhi 2013).

The development of new bioinspired bone substitute materials considering the ultrastructure and mechanical behavior of natural bone is very crucial and important one (Ulrike et al. 2015). It is acknowledged that the suitable component for replacing a hard tissue is a substance whose properties are analogous to that of natural one (Blokhuis 2014). Advances in composite research have shown a new pathway of biocomposite development that imitate the structure and function and simulate the behavior of hard tissues. These novel materials may conquer the constraints of the traditional implantable materials (Maria Fátima Vaz et al. 2011). Additionally, the fields of tissue engineering aim toward the repair or reconstruction of the damaged tissues by evolving biological alternates that recover, retain, or promote tissue activity.

27.1.1 Bone: A Natural Composite

Due to its excellent all-round properties, bone is truly remarkable biological material suitable for structural and physiological support. Regardless of its light weight, it exhibits appropriate tensile and great compressive strength and quite high modulus of elasticity. Bone, a natural complex nanocomposite (Maria Fátima Vaz et al. 2011), comprises of an organic collagenous tissue and an inorganic mineral phase with distinct remodeling properties for modifying its morphology to

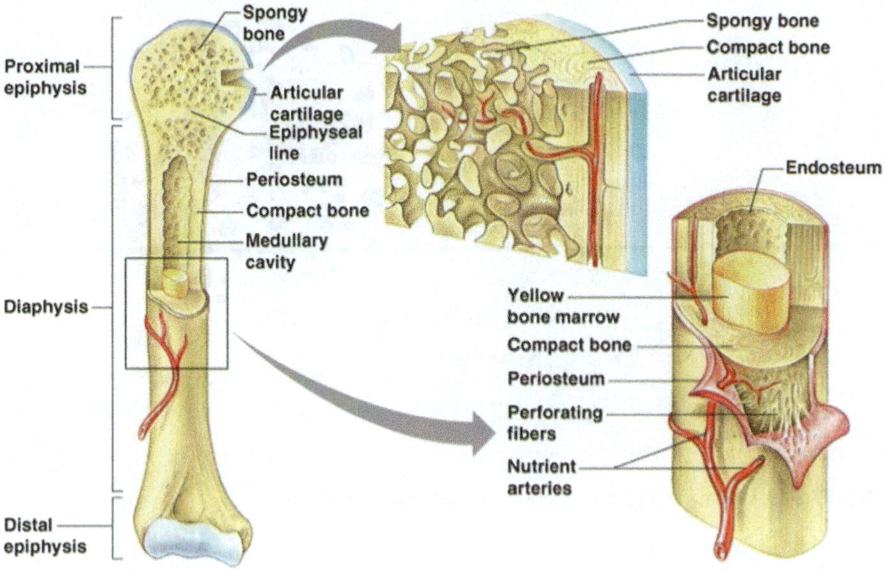

Fig. 27.1 Microstructure of a long bone

outermost mechanical loads (Laurin et al. 2011; Zeeshan et al. 2015; Currey 2002; Singh and Majumdar 2014; Goldberg et al. 2011). The distribution of the mineral phase within organic network varies in different parts of the bone and over time although the major ingredient of bone mineral is hydroxyapatite [$Ca_{10}(PO_4)_6(OH)_2$ (Bertazzo and Bertran 2006; Levrero et al. 2016). The complete structure of a long bone is shown in the following diagram (Fig. 27.1).

27.1.2 Scaffold

It is a structure that provides support and allows for body tissues to regenerate themselves. During bone fracture, the body will induce a soft callus around a broken part that allows cells to proliferate and heal the breach. So, the biodegradable polymeric scaffold will act as a template for tissue rehabilitation and reconstruction. Scaffolds made from various natural and synthetic materials are also applicable in tissues engineering and infused with cells to facilitate growth and healing.

Combination of tissue engineering and regenerative medicine can be useful to create "scaffolds" in the living body. These types of scaffolds are utilized to guide diseased or damaged organs and organ systems due to injury or trauma (Blokhuis 2014). Therefore, scaffolds are of great importance in modern regenerative medicine as these are helpful to rebuild organs and return normal function (Fig. 27.2).

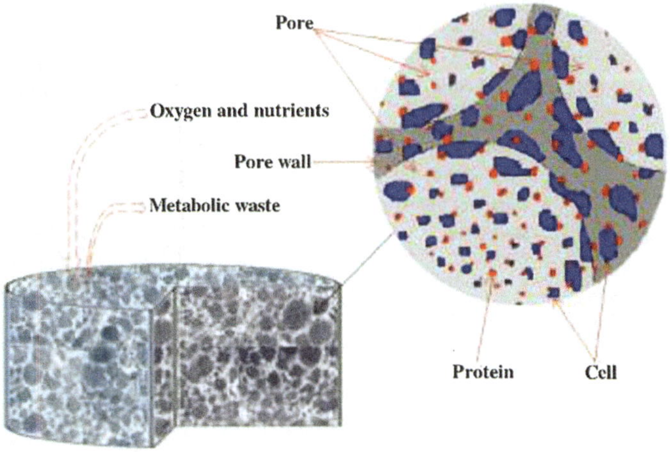

Fig. 27.2 A schematic diagram of a porous biodegradable scaffold

27.2 Scaffold Design

The most fundamental parameter in tissue engineering and regenerative medicine is the design of a scaffold that acts to compliment the regeneration process or to compensate the infected or damaged tissue (Hutmacher 2000). The scaffold is a makeshift/interim matrix that provides platform for cell attachment, procreation, and reproduction. Biodegradability of the scaffold is another essential factor in the design of scaffolds which falls in line with adequate mechanical properties of the scaffold (Bitar and Zakhem 2014). After implantation, the scaffold must degrade slowly with time to ensure proper refurbishing of the tissue. Highly porous scaffolds are crucial for cell infiltration especially when it comes to thick scaffolds where diffusion becomes a constraint. Porosity also plays a demanding role in providing greater surface area for cell attachment. An ideal scaffold is the result of a balance between all these factors and must have an excellent biocompatibility to ensure cell growth and minimal immune response after implantation (Raeisdasteh Hokmabad et al. 2017). Different biomaterials are available to synthesize the scaffolds, and different techniques are applicable to modulate the design of these scaffolds. Finally, physical and chemical alterations can be applied to the scaffolds to augment their bioactivity. Cell, scaffold, and growth factors are the three key parameters for tissue engineering (Chan and Leong 2008) (Fig. 27.3).

27.2.1 Scaffold Requirements

A large number of scaffolds originated from a wide range of biodegradable materials and are constructed using different fabrication methods to regenerate different

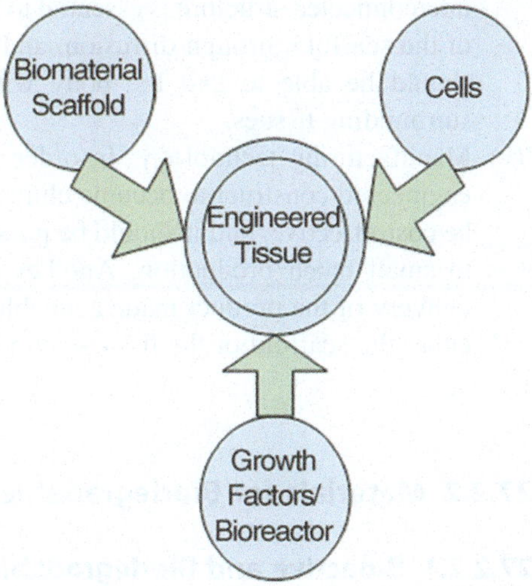

Fig. 27.3 Components of engineered tissue

tissues and organs in the body. Irrespective of the nature of tissue, a number of fundamental cogitations are important when designing or determining the efficacy of a scaffold for use in tissue engineering such as:

(i) Biocompatibility: The basic principle of any scaffold for tissue engineering is that it must be highly biocompatible, i.e., cells must attach, behave normally and immigrate onto the surface and finally through the scaffold, and begin to proliferate before laying down new matrix (Gunatillake and Adhikari 2003).
(ii) Biodegradability: The main goal of tissue engineering is to confess the body's own cells to replace the implanted scaffold or tissue-engineered construct over time. Therefore, the scaffold must be biodegradable so that the cells can easily harvest their own extracellular matrix (Haigang et al. 2010). The outgrowth of this degradation should also be nontoxic and able to exit the body without interfering with other organs.
(iii) Mechanical properties: Ideally, the scaffold should have good mechanical properties similar to the anatomical site into which it is to be implanted, and from a practical aspect, it must be strong enough to allow surgical handling during implantation (Sokolsky et al. 2007). Fabrication of scaffolds with adequate mechanical properties is one of the great challenges in attempting to engineer tissue like bone or cartilage. For such tissues, the implanted scaffold must have adequate useful integrity to operate throughout the remodeling process.
(iv) Scaffold architecture: The architecture of scaffolds used for tissue engineering is one of the most demanding and challenging tasks. Scaffolds should have an interconnected pore structure with high porosity to ensure cellular growth and sufficient diffusion of nutrients to cells within the construct and to the extracellular matrix formed by these cells (Ikada 2006). Furthermore, a porous

interconnected structure is essential to allow the excretion of waste products out of the scaffold through diffusion, and the by-products of scaffold degradation should be able to exit the body without hindrance with other organs and surrounding tissues.

(v) Manufacturing technology: In order to make a particular scaffold or tissue-engineered construct to become clinically and commercially feasible, it should be cost-effective, and it should be possible to scale up from research laboratory to small batch production. Another decisive component is determining the delivery of the product made available to the clinician. This will regulate how either the scaffold or the tissue-engineered construct will be stored.

27.2.2 Materials for Biodegradable Scaffold

27.2.2.1 Bioactive and Biodegradable Polymer

The choice of a suitable polymer for a specific biomedical application from large numbers of currently available synthetic polymers is confronting and difficult task. Like all other properties required for a particular medical application, clinical success of polymeric biomaterials depends on the biofriendly behavior of the polymer with surrounding tissues and biological fluids when interacting. From a wide range of polymers, some are considered as biocompatible polymers such as polyethylene (PE), polypropylene (PP), polyurethane (PU), polytetrafluoroethylene (PTFE), poly(vinyl chloride) (PVC), poly(methyl methacrylate) (PMMA), polyetheretherketone (PEEK), polysulfone (PSU) (Ambrose et al. 2015; Razak et al. 2012; Chandra and Rustgi 1998), etc. Their applications include from facial prostheses to tracheal tubes, from kidney and liver parts to heart components, and from dentures to hip and knee joints (Bret et al. 2011).

Recently many biodegradable polymers have been used to overcome the limitations inherent in the use of nondegradable polymers such as poly(L-lactic acid), poly(glycolic acid), poly{(DL-lactic-co-glycolic acid [PLGA])}, and collagen (Schaschke 2014; Lendlein and Sisson 2011). Innate and fabricated biodegradable polymers such as polyesters and polyamides are used as sutures, bone plates, and scaffolds (Huayu et al. 2012). Reconstituted collagen polymers are extensively set for restorations of arterial wall, heart valves, and the skin.

(a) **Natural Biodegradable Polymers**

These are distinctive types of polymers that degrade after specific use (Bastioli 2005). Biocompatibility, biodegradability, and thermo-processable nature make them attractive for tissue engineering (Iftikhar and Nazia 2016; Valappil et al. 2006; Zhao et al. 2012). They also facilitate an instinctive environment for cellular coupling, procreation, and differentiation and are designated as popular stuff for tissue engineering (Annalia and Luciana 2014; Wu et al. 2009; Wang et al. 2013).

(i) **Collagen**

Collagen is a tough, strong rope-like structural protein found mainly in the connective tissues like bones, skin, etc. (Buehler 2006; Brodsky and Persikov 2005) (Fig. 27.4). Collagen forms a complex chain of macromolecules that actuate the physical attributes of bone tissues, providing structural support, strength, and a degree of elasticity (in combination with elastin) (Bhattacharjee and Bansal 2005). The collagen has several different medical applications such as wound dressing, skin revitalization, etc. and can be used with a variety of medical devices like vascular prosthetics due to its biocompatible, biodegradable, and osteoconductive nature (Fabrizio et al. 2011; Khan and Khan 2013; Cunniffe and O'Brien 2011).

(ii) **Chitosan**

Chitosan is a naturally occurring biodegradable polymer and derived from chitin through deacetylation, composed of β-1,4-linked glucosamine and high degree of N-acetyl D-glucosamine (Fig. 27.5). It is an unbranched cationic

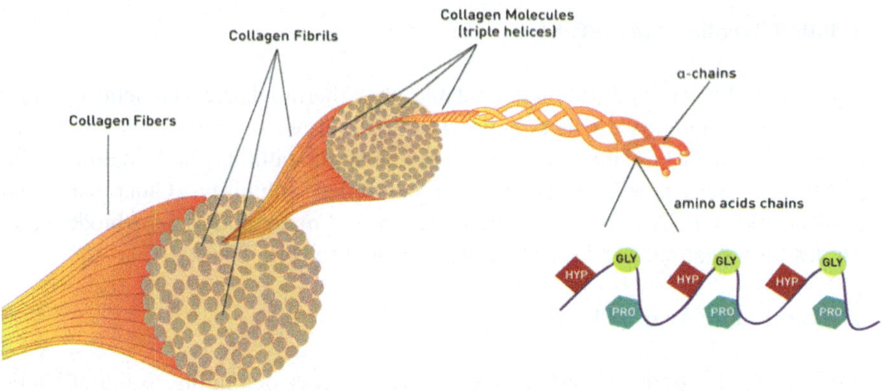

Fig. 27.4 Structure of collagen

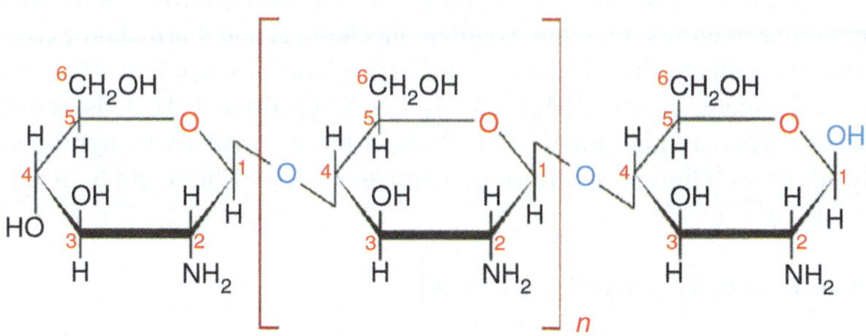

Fig. 27.5 Chemical structure of chitosan [β-(1,4)-D-glucosamine]

aminopolysaccharide with a high charge density in solution and carries a positive charge below pH 6.5. The enzyme chitosanase, papain, and lysozyme are responsible for in vitro degradation. The in vivo degeneration takes place due to the activity of lysozyme and is controlled by virtue of hydrolysis of the acetylated residues. This is very attractive biomaterial for its crystalline nature and excellent mechanical behavior. They are nontoxic and can be debased into normal body constituents (Lee et al. 2006).

(b) **Synthetic Biodegradable Polymers**

Two types of synthetic biopolymers are widely used, i.e., bulk biodegradable and surface reactive polymers (Asti and Gioglio 2014; Tian et al. 2012). Poly(lactic acid) (PLA), poly(glycolic acid) (PGA), poly(lactic-*co*-glycolic acid) (PLGA), and poly (ε-caprolactone) (PCL) belong to the group of saturated poly(α-hydroxy esters) and are the most frequently used synthetic biopolymers for three-dimensional scaffolds in tissue engineering (BaoLin and Ma 2014). They are considered as suitable biopolymers for hard tissue repair and bone regeneration for their exquisite biofriendly nature with the host tissue and biodegradable in the human body (Tan et al. 2013).

(i) **Poly(Glycolic Acid) [PGA]**

It is a highly crystalline, biodegradable, and thermoplastic synthetic polymer (45%–50% crystallinity). This polymer was first enlisted for clinical use as sutures and as biomedical implants due to its formidable crystallinity, high melting point (>200 °C), great tensile modulus, and controlled solubility (Gunatillake and Adhikari 2003). PGA is used as degradable scaffold material due to its biodegradation, less aggregation, and lack of cytotoxic response (Fig. 27.6).

(ii) **Poly(Lactic Acid) [PLA]**

It is a highly biodegradable and bioactive polymer belonging to the aliphatic thermoplastic polyester group with linear polymeric chains, and it undergoes in vivo biodegradation via enzymatic and hydrolytic pathways (Lim et al. 2008). It is primarily utilized for medical applications as sutures and rods for the treatment of mandibular fractures. PLA has excellent mechanical and thermal properties with excellent biocompatibility. Poly(lactic acid) is a kind of biopolymer that exists as two stereoisomeric forms: D-LA and L-LA. Crystalline L-PLA is resistant to hydrolysis, whereas amorphous D,L-PLA is more sensitive to hydrolysis and mostly used for clinical applications (Simamora and Chern 2006; Xiao et al. 2012) (Fig. 27.7).

(iii) **Poly(Lactide-co-Glycolide) [PLGA]**

This is the first synthetic biodegradable polymer that is used as resorbable material in incision. Copolymerization of PLGA in both the L- and D,L- forms is

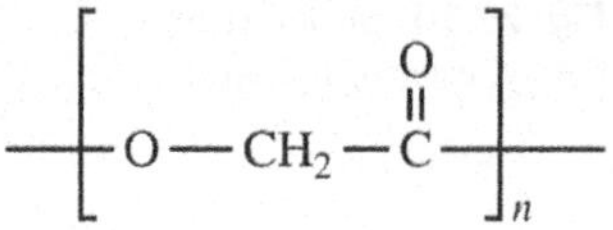

Fig. 27.6 Chemical structure of polyglycolic acid (PGA)

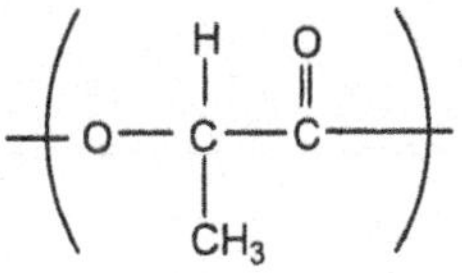

Fig. 27.7 Chemical structure of polylactic acid (PLA)

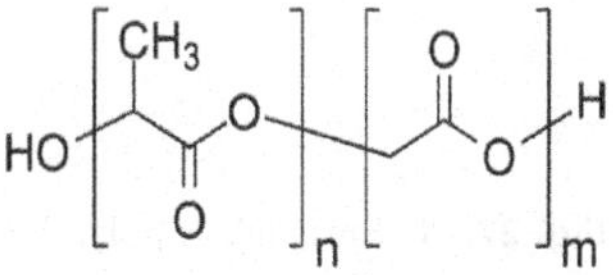

Fig. 27.8 Structure of poly (D,L-lactide-co-glycolide)

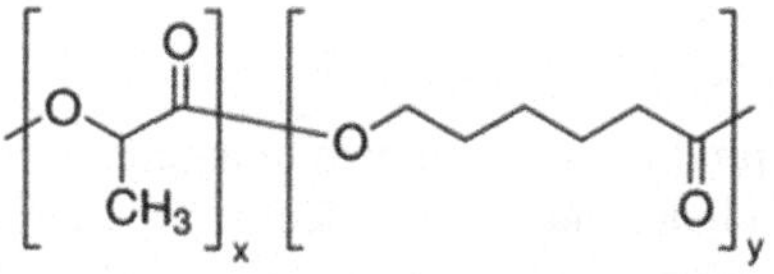

Fig. 27.9 Structure of poly (DL-lactide-co-caprolactone)

achieved by the combination of lactic and glycolic acid. It forms amorphous PLGA when compositional range is 25–75%. 50–50% PLGA has been shown to be hydrolytically unstable (Makadia and Siegel 2011). Although it has been extensively used in a variety of clinical applications, its use is limited in the field of orthopedics due the hydrophobic nature of PLGA (Fig. 27.8).

(iv) **Poly(ε-Caprolactone)**

PCL is an aliphatic biodegradable polyester with low melting temperature (60 °C) that is semicrystalline in nature (Fig. 27.9). Compared with other aliphatic degradable polymers, it shows very high thermal stability. It has been extensively used for long-term implants and controlled drug delivery systems but not effective for engineered tissue as PCL experiences with some drawbacks such as slow degradation rate, poor mechanical properties, and low cell adhesion property.

(v) **Benzyl Ester of Hyaluronic Acid**

Hyaluronic acid occupied an important position in the field of tissue engineering. Benzyl ester of hyaluronic acid (a hyaluronan derivative) known as HYAFF-11 is the most encouraging material for tissue engineering and regenerative medicine (Fig. 27.10). These materials exhibit good rate of degradation, and their degradation

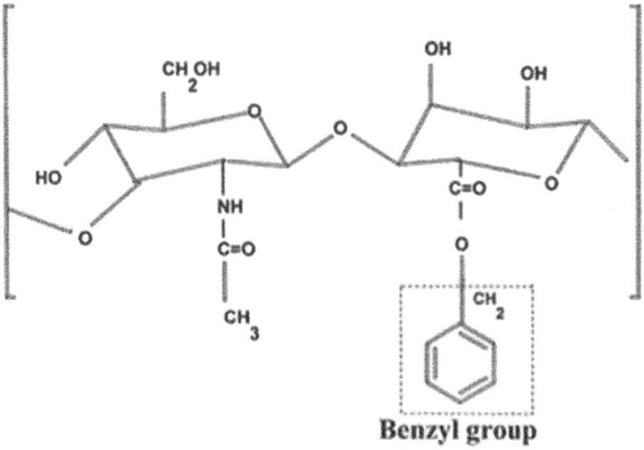

Fig. 27.10 Structure of a benzyl ester of hyaluronan

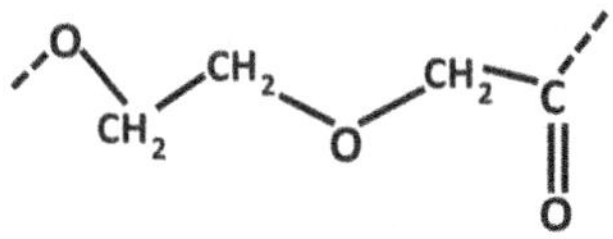

Fig. 27.11 Structure of poly-para-dioxanone

products are nontoxic in nature (Vindigni et al. 2009). The degradation of HYAFF-11 depends on the degree of esterification, and the esterified HYAFF-11 is more soluble and resembles the precursor of hyaluronic acid. Investigation of HYAFF-11 shows a promising candidate for bone tissue engineering and vascular graft preparation applications (Masina 2011).

(vi) **Poly-para-dioxanone [PDS]**

Poly-para-dioxanone is a synthetic, crystalline, biodegradable polymer of multiple repeating ether-ester units (Fig. 27.11). Due to its excellent biocompatibility, biodegradability, and flexibility, PDS has been greatly used in tissue regeneration and fracture repair applications particularly as internal fracture fixation (Si-Chong Chen et al. 2006). Other biomedical applications include maxillofacial and plastic surgery, orthopedic surgery, drug delivery, and tissue engineering (Middleton and Tipton 1998).

27.2.2.2 Bioactive Ceramic

Calcium phosphate includes hydroxyapatite and β-TCP, and various types of bioactive glass are well-known bioactive ceramic. When placed in bone tissue, these materials promote bone formation and bond to bone at various stages (Balani et al. 2015; Yamamuro 2012). Calcium phosphate ceramics are composed of several ingredients which differ not only in their chemical composition but also in their specific surface area, crystal structure, and macro- and microporosity. They are different due to variations in the calcium to phosphate ratio. For example, tricalcium phosphate, hydroxyapatite, and tetracalcium phosphate have Ca/P ratios of 1.5, 1.67,

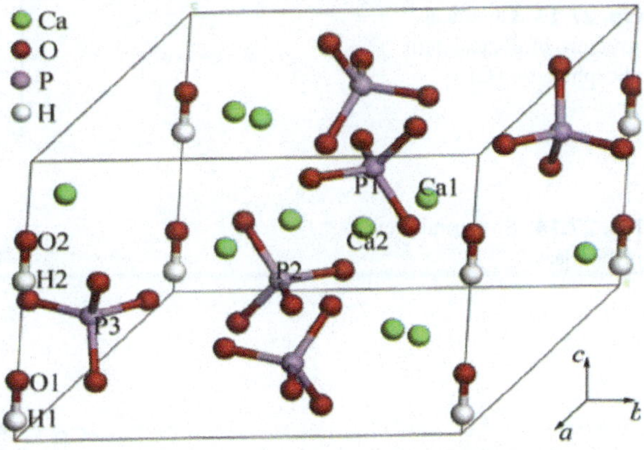

Fig. 27.12 Crystalline structure of hydroxyapatite

and 2, respectively (Raynaud et al. 2002; Osborn and Newesely 1980; Samavedi et al. 2013). Some ceramic materials have a direct role in cell activity. Bioactive glass for instance can cause stem cells to differentiate into osteoblasts Krishnan and Lakshmi 2013).

(a) **Hydroxyapatite**

The most popular bioactive ceramic materials used in medicine is hydroxyapatite (HA) (Oonishi 1991). The mineral (inorganic) constituent of bone is made up of biological apatites, which provide strength to the skeleton and act as storehouse for calcium, phosphorous, sodium, and magnesium (Fig. 27.12). These biological apatites are structurally similar to the mineral apatites hydroxyapatite [HAP] and brushite (B, $CaHPO_4{,}2H_2O$). The hexagonal structure of HA can be represented by the chemical formula $Ca_{10}(PO4)_6(OH)_2$ where the Ca/P ratio is 1.67 (Tamimi et al. 2012). It is highly stable in body fluid and in dry or moist air up to 1200 °C and does not decompose. These calcium phosphate ceramics show their bioactive nature due to its resorbable behavior. Hydroxyapatite seems to be the most suitable ceramic material for artificial teeth and bone due to its good tissue compatibility without immunological reaction or toxic reaction (Prakasam et al. 2015).

(b) **β-Tricalcium Phosphate**

β-Tricalcium phosphate is the most bioactive material in the calcium phosphate ceramics (Galois et al., 2002). It is represented by the chemical formula $[Ca_3(PO_4)_2]$ where the Ca/P ratio is 1.5. This type of calcium phosphate shows an X-ray pattern consistent with a pure hexagonal crystal structure, although the related α-TCP is monoclinic in nature (Fig. 27.13). Their high degree of solubility compared with hydroxyapatite has indicated their use in areas where high degree of bioactivity is desired such as dental, orthopedic, and reconstructive application (Liu and Lun, 2012; Miranda et al. 2006).

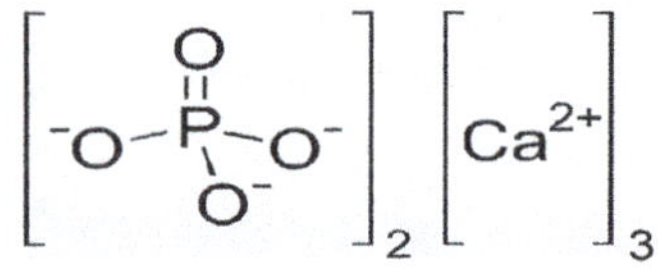

Fig. 27.13 Chemical structure of β-tricalcium phosphate (β-TCP)

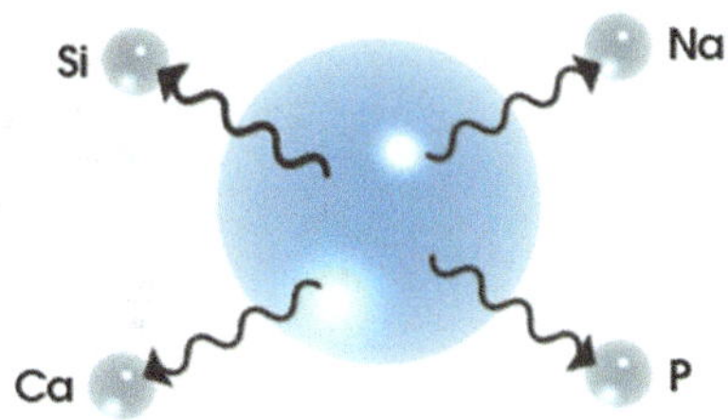

Fig. 27.14 Structural view of bioglass

(c) **Bioglass and Glass Ceramic**

Bioactive glass ceramics belong to Class A bioactive materials which are indicated by both osteoconductive (i.e., growth of bone at the implant surface) (Fig. 27.14) and osteoinductive (i.e., activation and recruitment of osteoprogenitor cells by the material itself stimulating bone growth on the surface of the material) behavior (PeitlFilho et al. 1996; Rahaman et al. 2011). The bioglasses are reinforced in a biopolymer support matrix to form prosthetics for hard tissues. Such prosthetics are biocompatible, show excellent mechanical properties, and are useful for orthopedic and dental prosthetics (Massera et al. 2012; Fu et al. 2011). Bioglasses are a subdivision of inorganic surface reactive biomaterials, competent toward the reaction with physiological fluids to form cohesive bond with the bone through the formation of bone-like hydroxyapatite layers and also the biological interaction of collagen with the material surface (Hench 2013; Rahaman et al. 2012; Vallittu et al. 2015).

27.3 Fabrication of Polymer Composite Scaffolds

By definition, a polymeric composite material consists of two or more distinct materials where polymers are used as matrix materials usually. These materials perform synchronously to produce the best properties of the scaffold, duplicating the same composite features of bone. In particular, polymer-based composite scaffolds may simulate the skeleton of natural bone which is mainly composed of HA and collagen (Polo-Corrales et al. 2014). Exhibiting better osteoconductivity, HA and bioglass can be considered as impeccable inorganic reinforcing phase of the composite scaffold, while polymers like PCL, PLLA, PGA, and PLGA serve as the organic matrix (Peck et al. 2011; Kurosch Rezwan et al. 2006).

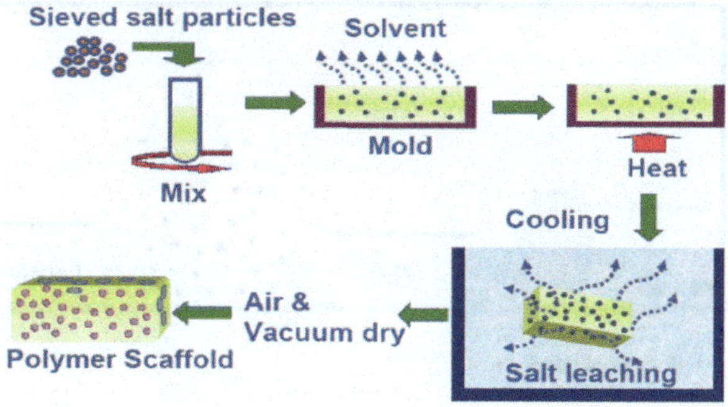

Fig. 27.15 The process diagram of solvent casting technique

Although a large number of polymer processing technique are available, solvent casting with and without particle leaching, thermally induced phase separation (TIPS) combined with freeze-drying and solid free-form method have been applied successfully for the fabrication of polymer-ceramic composite scaffolds (Jack et al. 2009).

27.3.1 Solvent Casting

This process is one the oldest technique for deposition of thin film over a substrate. Solvent casting technique includes the disunion of the polymer within organic solvent, incorporating with bioactive ceramic or glass particles and casting the solution into a predefined 3D mould (Ma 2004). After that, the solvent is subsequently conceded to evaporate. The major convenience of this processing technique is the ease of fabrication without any specialized equipment. The organic solvents used in this casting may reduce the activity of bioinductive molecules (e.g., protein). Comprehensive processing steps are shown in Fig. 27.15.

27.3.2 Solvent Casting/Particle Leaching and Microsphere Packing

Biocomposites can be molded by the blending of solvent casting, particle leaching, and microsphere packing methods. At first, polymer microspheres are formed through conventional water oil/water emulsion approach. Biodegradable composite scaffolds can then be constructed by blending solvent, salt or sugar particles (porogens), bioactive glass or ceramic granules, and pre-hardened microspheres (Chun-Jen et al. 2002). Hence, a 3D structure of controlled porosity is formed by this method combined with particle leaching and microsphere packing (Fig. 27.16). Unlike solvent casting technique, this method exhibits similar type of advantages and disadvantages.

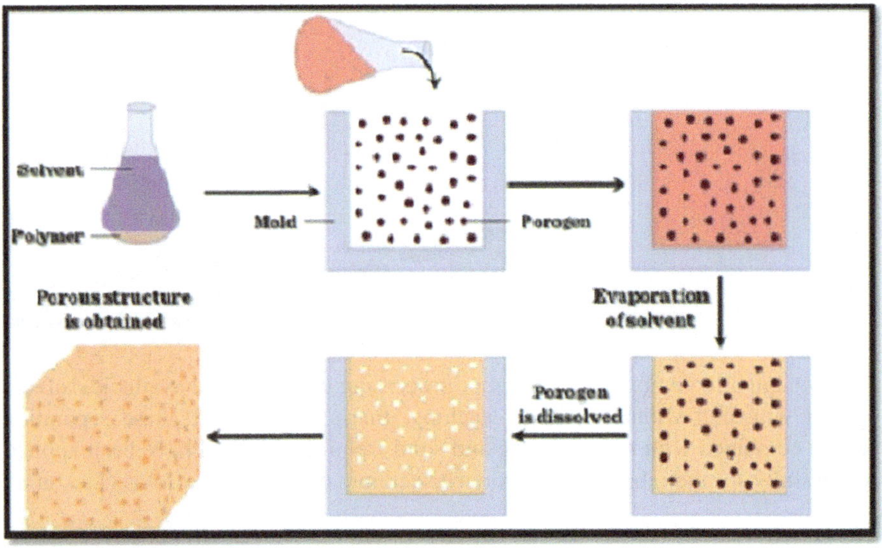

Fig. 27.16 Schematic diagram of solvent casting-particulate leaching scaffolding method

27.3.3 Thermally Induced Phase Separation (TIPS)/Freeze-Drying

It is an alternative technique that has been used successfully to fabricate highly porous scaffolds for bone tissue engineering through phase-separation and evaporation process (Zhao et al. 2011). In this process, dioxane exploits as a solvent and can be utilized to create a composite structure having interconnected pores. The solvent is solidified first imposing the polymer and ceramic mixture toward the interstitial spaces. Afterward the ice solvent evaporates from the frozen mixture through freeze dryer of a lyophilizer (Udeni Gunathilake et al. 2016; Francesca et al. 2013). The TIPS method (Fig. 27.17) can produce homogeneous and highly porous (~95%) scaffolds with greater anisotropic tubular morphology and extensive pore interconnectivity with pore sizes ranging from few microns to several hundred microns with improved mechanical properties (Boschetti et al. 2008; Akbarzadeh and Yousefi 2014).

27.3.4 Microsphere-Sintering

Microsphere-based tissue-engineered scaffolds proposed the advantage of shape-specific constitutes with excellent spatiotemporal authority over mechanical integrity and growth factor release and interconnected porous structures (Borden et al. 2004). The employment of the highly versatile scaffolds requires an approach to sinter the discrete microspheres close together into a cohesive network applying heat or organic solvents (Singh et al. 2010). In this mechanism, microspheres are formed (Fig. 27.18) by a polymer matrix, and the inclusion of bioglass or calcium phosphate

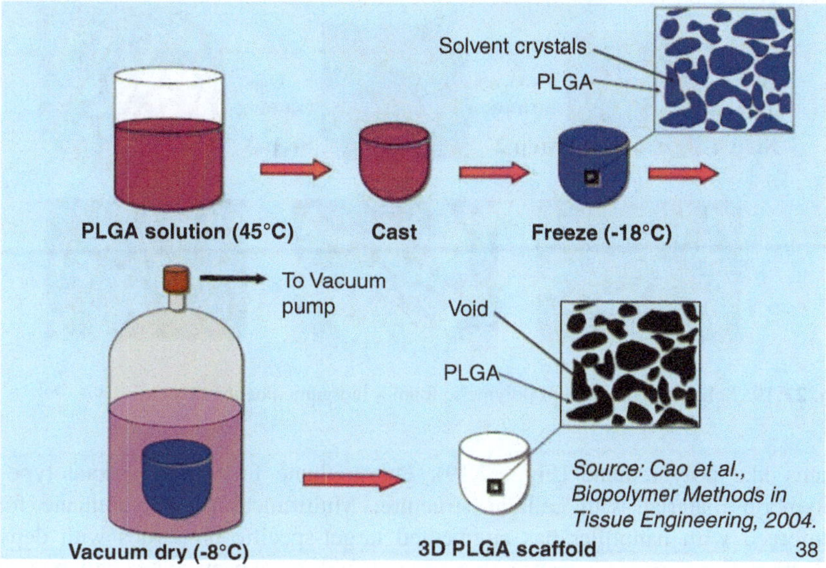

Fig. 27.17 Step details of thermally induced phase separation/freeze-drying

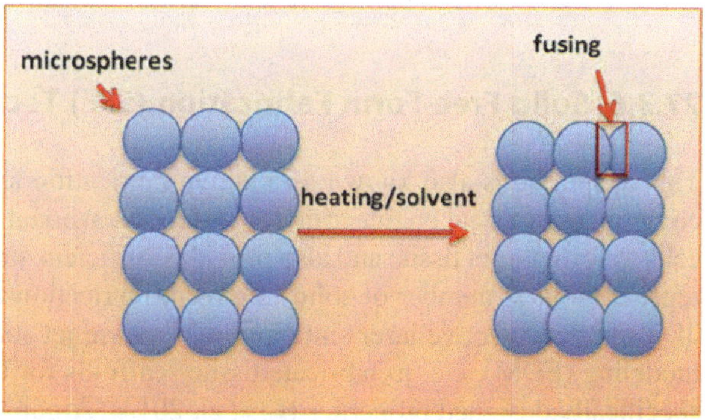

Fig. 27.18 Schematic of the generation of PLGA sintered microsphere scaffolds

ceramics was performed using a variety of scheme including the spraying of polymer solutions over bioglass or ceramic followed by non-solvent induced phase separation (NIPS). Once the composite microspheres have been synthesized, sintering is employed in 3D moulds to fabricate 3D porous composite scaffolds (Cox et al. 2015; Singh et al. 2010).

27.3.5 Polymeric Foam: Inorganic Coating

A novel way of encouraging the consolidation of biodegradable polymers and bioactive glass or ceramic materials is to coat the inorganic particles using polymeric

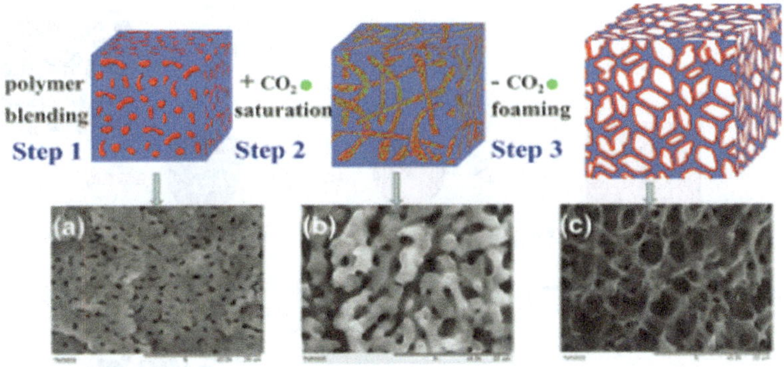

Fig. 27.19 Schematic diagram of polymeric foam – inorganic coating process

foams like polyurethane (Fig. 27.19). Polyurethane foam is a special type of polymeric material with cellular structure. Multifunctional polyurethane foam reinforced with nanofiller has augmented target-specific properties with density reduction. For example, porous polymeric scaffolds have been coated with bioactive glasses and other inorganic particles by slurry dipping or electrophoretic deposition method (Niu et al. 2012).

27.3.6 Solid Free-Form Fabrication (SFF) Techniques

This technique is also known as additive fabrication technology. It is an important component in tissue engineering for three-dimensional (3D) scaffold, which guides cells to form target tissue and also provides sufficient structural support during tissue regeneration. A number of solid free-form fabrication techniques are available like 3D printing, selective laser sintering, multiphase jet solidification, fused deposition modeling (FDM), etc. to fabricate tissue scaffolds for bone tissue engineering with specifically designed properties (Li et al. 2014). Among several SFF methods, FDM is most promising due to its ability to form 3D structures by layer-by-layer deposition. The scaffolds have a great degree of interconnectivity and the porosity that can be controlled by optimizing the processing parameters. This technique provides a unique opportunity to investigate the effect of micro-architecture of the scaffold upon cell proliferation and ECM generation. Furthermore, this process can be utilized to create scaffolds incorporating patient-specific information as well as an explicitly designed microenvironment (Lee et al. 2010a, b). Three-dimensional, porous, PCL-based composite scaffolds have also been prepared through the combination of a filament winding technique and salt leaching process incorporating β-tricalcium phosphate (β -TCP) particles and PLA fibers (Persson et al. 2014). The details of the process are shown in Fig. 27.20.

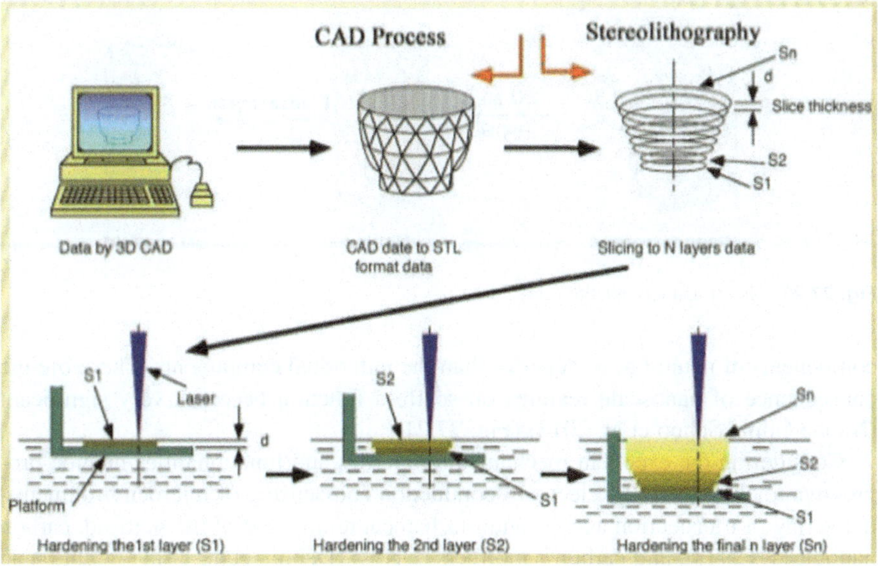

Fig. 27.20 Solid free-form fabrication (SFF) technique

Bioactive Nanocomposite Scaffold for Bone Tissue Regeneration
Since the natural bone is a compound of HA nanocrystal and collagen fiber, it is important to design composites containing either HA nanocrystals or an ingredient that induces the formation of HA nanocrystals, to mimic the bone composition and structure. For bone tissue reconstruction, it is imperative for the biomaterial to mimic the living bone tissue. As there is no such kind of material available to resemble the composition, structure, and properties of native bone, synthetic nanocomposites are the preferred choice for bone regeneration. These types of composites are able to afford the appropriate matrix environment, integrate desirable biological properties, and furnish controlled, sequential delivery of multiple growth factors for the different stages of bone tissue reformation. Bioactive polymer/hydroxyapatite nanocomposites are currently being exclusively explored as biomaterials for promotion of bone tissue rebirth and reclamation (Basile et al. 2015; Zhu et al. 2016a, b).

In recent years, there is a thriving interest on using nanomaterials in tissue engineering scaffolds to simulate the structure of natural bone tissue which possesses a nanocomposite structure knitted in a 3D matrix (Zhu et al. 2016a, b). The embodiment of nanoelements within the biopolymer matrix has the dual objective of improving the mechanical behavior along with the incorporation of nanotopographic features that mimic the nanostructure of natural hard tissue (Amin et al. 2014). Thus the function of the scaffolds is being extended from being a mere mechanical support to inclusion of intelligent surfaces capable of providing both chemical and physical signals to guide cell attachment, proliferation, and cell differentiation (Lee et al. 2010a, b). Nanocomposites can also mimic the

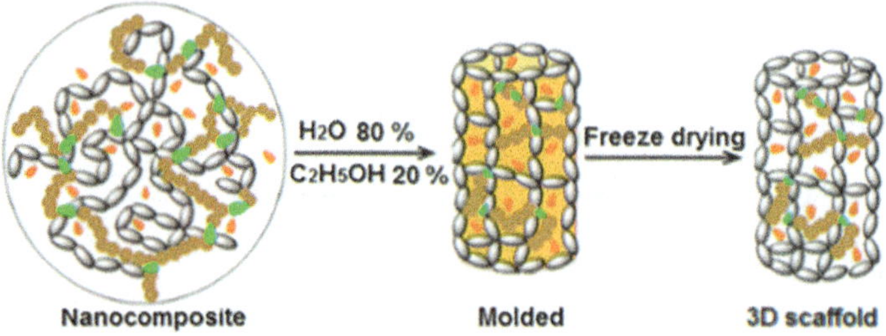

Fig. 27.21 Novel nanocomposite scaffold

components of natural bone superior than the individual components. Therefore the consequence of nanoscale features on scaffold function becomes very significant (Nanda Gopal Sahoo et al. 2013) (Fig. 27.21).

Considering the excellent mechanical properties and functionalities include progressive tracking of cells, electrical conduction and sensing of microenvironments, delivering of transfection agents, and mechanical reinforcing of the scaffold, carbon nanotubes (CNTs), and carbon nanofibers (CNFs) (Dahlin, et al. 2011) might be an interesting alternate strategy as reinforcing agents in polymer-based composite materials, especially in the framework for bone tissue engineering (Edwards et al. 2009; Mikael and Nukavarapu 2011). The incorporation of CNTs is also fruitful to tailor the (nano-)roughness and topography of matrix pore surfaces which have an immense response on early cell attachment behavior, subsequent cell adhesion, cytoskeletal organization, and gene expression (Oh and Islam 2015; Zanello et al. 2006; Lalwani et al. 2015).

Compared to other ceramic-based scaffolds for bone, the use of single-walled carbon nanotubes (SWCNTs) will allow the fabrication of light-weight scaffolds with very high strength due to their porous, flexible nature and a very high Young's modulus (Cardiel et al. 2014).

Another interesting route for developing multifunctional nanocomposite scaffolds in bone tissue engineering may be well-defined magnetic PCL/iron oxide scaffolds fabricated by 3D fiber deposition technique (Gloria et al. 2013). The application of suitably functionalized iron oxide nanoparticles with biocompatible coatings may lead to magnetic nanocomposite scaffolds that are ready to attract and adopt in vivo growth factors via magnetic driving, stem cells, or other bioagents bound to magnetic particles (Roberto De Santis et al. 2011).

27.4 Conclusion and Future Perspective

The application of biodegradable scaffold with bioactive composite in bone tissue engineering confess mimicking the complicated arrangement of natural bone tissue, providing an innovative and practical approach to the quality construction of biomaterials for bone tissue engineering. Fabricated or natural polymer matrices offer a wide range of mechanical properties and exhibit different biodegradation features, whereas various inorganic nanoparticle ceramics can maintain the excellent bioactivity and biostability. In addition, their integration with the bone tissue makes it possible to fabricate materials that resemble the structural and morphological organization of native bone. Considering all the aspects, an ideal design of scaffolds for tissue engineering and regenerative medicine is still a challenge. Different types of cells having a specific alignment are a constraint of the amalgamation of various scaffolds with precise structures that meet the native tissue. So, there is a great future to improve current biomaterials and develop advanced nanocomposite scaffold suitable for bone regeneration.

References

Akbarzadeh R, Yousefi AM (2014) Effects of processing parameters in thermally induced phase separation technique on porous architecture of scaffolds for bone tissue engineering. J Biomed Mater Res B Appl Biomater 102(6):1304–1315

Ambrose CG, Hartline BE, Clanton TO, Lowe WR, McGarvey WC (2015) Polymers in orthopaedic surgery. In: Advanced polymers in medicine. Springer, Berlin/Heidelberg, pp 129–145

Ami R, Amini, Laurencin CT, Nukavarapu SP (2012) Bone tissue engineering: recent advances and challenges. Crit Rev Biomed Eng 40(5):363–408

Amin Y, Wauthle R, Böttger AJ, Schrooten J, Weinans H, Zadpoor AA (2014) Crystal structure and nanotopographical features on the surface of heat-treated and anodized porous titanium biomaterials produced using selective laser melting. Appl Surf Sci 290:287–294

Annalia A, Luciana G (2014) Natural and synthetic biodegradable polymers: different scaffolds for cell expansion and tissue formation. Int J Artif Organs 37(3):187–205

Asti A, Gioglio L (2014) Natural and synthetic biodegradable polymers: different scaffolds for cell expansion and tissue formation. Int J Artif Organs 37(3):187–205

Balani K, Narayan R, Agarwal A (2015) Surface engineering and modification for biomedical applications. In Balani K, Verma V, Agaqrwal A, Narayan R (Eds) Biosurfaces: materials science and engineering perspective, John Wiley, pp 201–238

BaoLin G, Ma PX (2014) Synthetic biodegradable functional polymers for tissue engineering: a brief review. Sci China Chem 57(4):490–500

Basile MA, d'Ayala GG, Malinconico M, Laurienzo P, Coudane J, Nottelet B, Ragione FD, Oliva A (2015) Functionalized PCL/HA nanocomposites as microporous membranes for bone regeneration. Mater Sci Eng C Mater Biol Appl 48:457–468

Bastioli C (2005) Handbook of biodegradable polymers. Rapra Technology, Shawbury/Shrewsbury/Shropshire. ISBN 9781847350442

Bertazzo S, Bertran CA (2006) Morphological and dimensional characteristics of bone mineral crystals. Bioceramics 3(10):309–311

Bhattacharjee A, Bansal M (2005) Critical review collagen structure: the madras triple helix and the current scenario. IUBMB Life 57:161–172

Bitar KN, Zakhem E (2014) Design strategies of biodegradable scaffolds for tissue regeneration. Biomed Eng Comput Biol 6:13–20

Blokhuis TJ (2014) Bioresorbable bone graft substitutes. In: Bone substitute biomaterials, pp 80–92, Elsevier

Borden M, Attawia M, Khan Y, El-Amin SF, Laurencin CT (2004) Tissue-engineered bone formation in vivo using a novel sintered polymeric microsphere matrix. J Bone Joint Surg Br 86:1200–1208

Boschetti F, Tomei AA, Turri S, Swartz MA, Levi M (2008) Design, fabrication, and characterization of a composite scaffold for bone tissue engineering. Int J Artif Organs 31(8):697–707

Bret DU, Lakshmi SN, Cato TL (2011) Biomedical applications of biodegradable polymers. Biomedical applications of biodegradable polymers. J Polym Sci B Polym Phys 49(12):832–864

Brodsky B, Persikov AV (2005) Molecular structure of thecollagen triple helix. Adv Protein Chem 70:301–339

Buehler MJ (2006) Nature designs tough collagen: explaining the nanostructure of collagen fibrils. PNAS 103(33):12285–12290

Cardiel JJ, Zhao Y, Kim J-H, Chung J-H, Shen AQ, Shen AQ (2014) Electro-conductive porous scaffold with single-walled carbon nanotubes in wormlike micellar networks. Carbon 80:203–212

Chan BP, Leong KW (2008) Scaffolding in tissue engineering: general approaches and tissue-specific considerations. Eur Spine J 17(S4):467–479

Chandra R, Rustgi R (1998) Biodegradable polymers. Prog Polym Sci 23:1273–1335

Chanlalit C, Shukla DR, Fitzsimmons JS, An KN, O'Driscoll SW (2012) Stress shielding around radial head prostheses. J Hand Surg 37:2118–2125

Chun-Jen L, Chin-Fu C, Jui-Hsiang C, Shu-Fung C, Yu-Ju L, Ken-Yuan C (2002) Fabrication of porous biodegradable polymer scaffolds using a solvent merging/particulate leaching method. J Biomed Mater Res Part A 59(4):676–681

Cox SC, Thornby JA, Gibbons GJ, Williams MA, Mallick KK (2015) 3D printing of porous hydroxyapatite scaffolds intended for use in bone tissue engineering applications. Mater Sci Eng C 47:237–247

Cunniffe G, O'Brien F (2011) Collagen scaffolds for orthopedic regenerative medicine. J Miner Met Mater Soc 63(4):66–73

Currey JD (2002) The structure of bone tissue. In: Bones: structure and mechanics. Princeton University Press, Princeton, pp 12–14

Dahlin RL, Kasper FK, Mikos AG (2011) Polymeric nanofibers in tissue engineering. Tissue Eng B Rev 17:349–364

De Santis R, Gloria A, Russo T, Amora UD, Zeppetelli S, Tampieri A, Herrmannsdorfer T, Ambrosio L (2011) A route toward the development of 3D magnetic scaffolds with tailored mechanical and morphological properties for hard tissue regeneration: preliminary study. Virtual Phys Prototyping 6(4):189–195

Dinopoulos H, Dimitriou R, Giannoudis PV (2012) Bone graft substitutes: what are the options? Surgeon 10(4):230–239

Dumic-Cule I, Pecina M, Jelic M, Jankolija M, Popek I, Grgurevic L, Vukicevic S (2015) Biological aspects of segmental bone defects management. Int Orthop 39:1005–1011

Edwards SL, Werkmeister JA, Ramshaw JA (2009) Carbon nanotubes in scaffolds for tissue engineering. Expert Rev Med Devices 6(5):499–505

Fabrizio M, Lorenzo N, DianaChicon P, Massimo I (2011) New biomaterials for bone regeneration. Clin Cases Miner Bone Metab 8(1):21–24

Francesca G, Alessandro S, Giuseppe MP (2013) The biomaterialist's task: scaffold biomaterials and fabrication technologies. Joints 1(3):130–137

Fu Q, Saiz E, Rahaman MN, Tomsia AP (2011) Bioactive glass scaffolds for bone tissue engineering: state of the art and future perspectives. Mater Sci Eng C Mater Biol 31(7):1245–1256

Galois L, Mainard D, Delagoutte J (2002) Beta-tricalcium phosphate ceramic as a bone substitute in orthopaedic surgery. Int Orthop 26:109–115

Gloria A, Russo T, D'Amora U, Zeppetelli S, D'Alessandro T, Sandri M, Bañobre-López M, Piñeiro-Redondo Y, Uhlarz M, Tampieri A, Rivas J, Herrmannsdörfer T, Dediu VA, Ambrosio L, DeSantis R (2013) Magnetic poly(ε-caprolactone)/iron-doped hydroxyapatite nanocomposite substrates for advanced bone tissue engineering. J R Soc Interface 10(80):8–33

Goldberg M, Kulkarni AB, Young M, Boskey A (2011) Dentin: structure, composition and mineralization-the role of dentin ECM in dentin formation and mineralization. Front Biosci (Elite Ed) 3:711–735

Gunatillake PA, Adhikari R (2003) Biodegradable synthetic polymers for tissue engineering. Eur Cell Mater 5:1–16

Haase K, Rouhi G (2013) Prediction of stress shielding around an orthopedic screw. Using stress and strain energy density as mechanical stimuli. Comput Biol Med 43:1748–1757
Haigang GU, Zhilian Y, Bramasta N, Leong WS, Tan JP (2010) Control of *invitro* neural differentiation of mesenchymal stem cells in 3D macroporous, cellulosic hydrogels. Regen Med 5:245–253
Hench LL (2013) Chronology of bioactive glass development and clinical applications. Sci Res 3:67–73
Huayu Tian, Zhaohui Tang, Xiuli Zhuang, Xuesi Chen, Xiabin Jing (2012) Biodegradable synthetic polymers: preparation, functionalization and biomedical application. Prog Polym Sci 37(2):237–280
Hutmacher DW (2000) Scaffolds in tissue engineering bone and cartilage. Biomaterials 21(24):2529–2543
Iftikhar A, Nazia J (2016) Polyhydroxyalkanoates: current applications in the medical field. Front Biol 11(1):19–27
Ikada Y (2006) Challenges in tissue engineering. J R Soc Interface 3:589–601
Jack KS, Velayudhan S, Luckman P, Trau M, Grøndahl L, Cooper-White J (2009) The fabrication and characterization of biodegradable HA/PHBV nanoparticle-polymer composite scaffolds. Acta Biomater 5(7):2657–2667
Khan R, Khan MH (2013) Use of collagen as a biomaterial: an update. J Indian Soc Periodontol 17(4):539–542
Krishnan V, Lakshmi T (2013) Bioglass: a novel biocompatible innovation. J Adv Pharm Technol Res 4(2):78–83
Lalwani G, Gopalan A, D'Agati M, Sankaran JS, Judex S, Qin YX, Sitharaman B (2015) Porous three-dimensional carbon nanotube scaffolds for tissue engineering. J Biomed Mater Res 103 (10):3212–3225
Laurin M, Canoville A, Germain D (2011) Bone microanatomy and lifestyle: a descriptive approach. Comptes Rendus Palevol 10(5–6):381–402
Lee DW et al (2006) Strong adhesion and cohesion of chitosan in aqueous solutions. Langmuir 29(46):14222–14229
Lee JW, Kim JY, Kim JY, Cho D-W, Cho D-W (2010a) Solid free-form fabrication technology and its application to bone tissue engineering. Int J Stem Cells 3(2):85–95
Lee K-W, Wang S, Dadsetan M, Yaszemski MJ, Lu L (2010b) Enhanced cell ingrowth and proliferation through three-dimensional nanocomposite scaffolds with controlled pore structures. Biomacromolecules 11(3):682–689
Lendlein A, Sisson A (eds) (2011) Handbook of biodegradable polymers : synthesis, characterization and applications. Weinheim, Wiley-VCH
Levrero F, Margetts L et al (2016) Evaluating the macroscopic yield behaviour of trabecular bone using a nonlinear homogenisation approach. J Mech Behav Biomed Mater 61:384–396
Li X, Cui R, Sun L, Aifantis KE, Fan Y, Feng Q, Cui F, Watari F (2014) 3D-printed biopolymers for tissue engineering application. Int J Polym Sci 2014:1–13
Lim LT, Auras R, Rubino M (2008) Processing technologies for poly (lactic acid). Prog Polym Sci 33(8):820–852
Liu B, Lun DX (2012) Current application of β-tricalcium phosphate composites in orthopaedics. Orthop Surg 4:139–144
Ma PX (2004) Scaffolds for tissue fabrication. Mater Today 7(5):30–40
Makadia HK, Siegel SJ (2011) Poly lactic-*co*-glycolic acid (PLGA) as biodegradable controlled drug delivery carrier. Polymers (Basel) 3(3):1377–1397
Maria Fátima Vaz, Helena Canhão, João Eurico Fonseca (2011) Bone: a composite natural material. In: Pavla (ed) Advances in composite materials – analysis of natural and man-made materials. Intech, Croatia
Masina M (2011) Use of an absorbent non-woven fabric dressing based on benzyl ester of hyaluronic acid (Hyallofill®-F) in the treatment of difficult to heal ulcers of the lower extremities. Acta Vulcanol 9(4):173–181

Massera J, Fagerlund S, Hupa L, Hupa M (2012) Crystallization mechanism of bioactive glasses 45S5 and S53P4. J Am Ceram Soc 95(2):607–613

Middleton J, Tipton A (1998) Synthetic biodegradable polymers as medical devices. Med Plast Biomater Mag 5(2):30–39

Mikael PE, Nukavarapu SP (2011) Functionalized carbon nanotube composite scaffolds for bone tissue engineering: prospects and progress. J Biomater Tissue Eng 1(1):76–85

Miranda P, Saiz E, Gryn K, Tomsia AP (2006) Sintering and robocasting of β-tricalcium phosphate scaffolds for orthopaedic applications. Acta Biomater 2:457–466

Mirza SB, Dunlop DG, Panesar SS, Naqvi SG, Gangoo S, Salih S (2010) Basic science considerations in primary total hip replacement arthroplasty. Open Orthop J 4:169–180

Niu XF, Li XM, Liu HF (2012) Homogeneous chitosan/poly(L-lactide) composite scaffolds prepared by emulsion freeze-drying. J Biomater Sci Polym Ed 23:391–404

Oh Y, Islam MF (2015) Preformed Nanoporous carbon nanotube scaffold-based multifunctional polymer composites. ACS Nano 9(4):4103–4110

Oonishi H (1991) Orthopaedic applications of hydroxyapatite. Biomaterials 12(2):171–178

Osborn JF, Newesely H (1980) The material science of calcium phosphate ceramics. Biomaterials 1(2):108–111

Peck M, Dusserre N, McAllister TN, L'Heureux N (2011) Tissue engineering by self- assembly. Mater Today 14:218–224

PeitlFilho O, LaTorre GP, Hench LL (1996) Effect of crystallization on apatite-layer formation of bioactive glass 45S5. J Biomed Mater Res 30(4):509–514

Persson M, Lorite GS, Kokkonen HE, Cho SW, Lehenkari PP, Skrifvars M, Tuukkanen J (2014) Effect of bioactive extruded PLA/HA composite films on focal adhesion formation of preosteoblastic cells. Colloids Surf B Biointerfaces 121:409–416

Polo-Corrales L, Latorre-Esteves M, Ramirez-Vick JE (2014) Scaffold design for bone regeneration. J Nanosci Nanotechnol 14(1):15–56

Prakasam M, Locs J, Salma-Ancane K, Loca D, Largeteau A, Berzina-Cimdina L (2015) Fabrication, properties and applications of dense hydroxyapatite: a review. J Funct Biomater 6(4):1099–1140

Raeisdasteh Hokmabad V, Davaran S, Ramazani A, Salehi R (2017) Design and fabrication of porous biodegradable scaffolds: a strategy for tissue engineering. J Biomater Sci Polym Ed 28(16):1797–1825

Rahaman MN, Day DE, Bal BS, Fu Q, Jung SB (2011) Bioactive glass in tissue engineering. Acta Biomater 7(6):2355–2373

Rahaman MN, Liu X, Bal BS, Day DE, Bi L, Bonewald LF (2012) Bioactive glass in bone tissue engineering. Biomater Sci 237:73–82

Ratner BD (2004) Biomaterials science: an introduction to materials in medicine. Academic Press, Waltham

Raynaud S, Champion E, Bernache-Assollant D, Thomas P (2002) Calcium phosphate apatites with variable Ca/P atomic ratio I. Synthesis, characterisation and thermal stability of powders. Biomaterials 23(4):1065–1072

Razak SIA, Sharif N, Rahman W (2012) Biodegradable polymers and their bone applications: a review. Int J Basic Appl Sci 12:31–49

Rezwan K, Chen QZ, Blaker JJ, Boccaccini AR (2006) Biodegradable and bioactive porous polymer/inorganic composite scaffolds for bone tissue engineering. Biomaterials 27(18):3413–3431

Sahoo NG, Pan YZ, Li L, He CB (2013) Nanocomposites for bone tissue regeneration. Nanomedicine 8(4):639–653

Samavedi S, Whittington AR, Goldstein AS (2013) Calcium phosphate ceramics in bone tissue engineering: a review of properties and their influence on cell behavior. Acta Biomater 9(9):8037–8045

Saska S, Mendes LS, Gaspar AMM, de Oliveira Capote TS (2015) Bone substitute materials in implant dentistry. Implant Dent 2:158–167

Schaschke C, Audic JL (Editorial) (2014) Biodegradable materials. Int J Mol Sci 15:21468–21475

Si-Chong Chen, Zhi-Xuan Zhou, Yu-Zhong Wang, Xiu-Li Wang, Ke-Ke Yang (2006) A novel biodegradable poly(*p*-dioxanone)-grafted poly(vinyl alcohol) copolymer with a controllable in vitro degradation. Polymer 47(1):32–36

Simamora P, Chern W (2006) Poly-L-lactic acid: an overview. J Drugs Dermatol 5(5):436–440

Singh AB, Majumdar S (2014) The composite of hydroxyapatite with collagen as a bone grafting material. J Adv Med Dent Sci Res 2:53–55

Singh M, Sandhu B, Scurto A, Berkland C, Detamore MS (2010) Microsphere-based scaffolds for cartilage tissue engineering: using subcritical CO(2) as a sintering agent. Acta Biomater 6(1):137–143

Sokolsky PM, Agashi K, Olaye A, Shakesheff K, Domb AJ (2007) Polymer carriers for drug delivery in tissue engineering. Adv Drug Deliv Rev 59:187–206

Sumner DR (2015) Long-term implant fixation and stress-shielding in total hip replacement. J Biomech 48:797–800

Tamimi F, Sheikh Z, Barralet J (2012) Dicalcium phosphate cements: Brushite and monetite. Acta Biomater 8:474–487

Tan L, Yu X, Wan P, Yang K (2013) Biodegradable materials for bone repairs: a review. J Mater Sci Technol 29:503–513

Tian H, Tang Z, Zhuang X, Chen X, Jing X (2012) Biodegradable synthetic polymers: preparation, functionalization and biomedical application. Prog Polym Sci 37(2):237–280

Udeni Gunathilake TMS, Ching YC, Chuah CH, Sabariah JJ, Pai-Chen L (2016) Fabrication of porous materials from natural/synthetic biopolymers and their composites. Materials 9(12):991

Ulrike G, Wegst K, Bai H, Eduardo S, Antoni PT, Ritchie RO (2015) Bioinspired structural materials. Nat Mater 14:23–36

Valappil SP, Misra SK, Boccaccini AR, Roy I (2006) Biomedical applications of polyhydroxyalkanoates: an overview of animal testing and in vivo responses. Expert Rev Med Devices 3(6):853–868

Vallittu PK, Närhi TO, Hupa L (2015) Fiber glass–bioactive glass composite for bone replacing and bone anchoring implants. Dent Mater 31:371–381

Vindigni V, Cortivo R, Iacobellis L, Abatangelo G, Zavan B (2009) Hyaluronan benzyl ester as a scaffold for tissue engineering. Int J Mol Sci 10(7):2972–2985

Wang Y, Jiang XL, Peng SW, Guo XY, Shang GG, Chen JC, Wu Q, Chen GQ (2013) Induced apoptosis of osteoblasts proliferating on polyhydroxyalkanoates. Biomaterials 34(15):3737–3746

Wu Q, Wang Y, Chen GQ (2009) Medical application of microbial biopolyesters polyhydroxyalkanoates artificial cells. Blood Substitutes Biotechnol 37(1):1–12

Xiao L, Wang B, Yang G, Gauthier M (2012) Poly (lactic acid)-based biomaterials: synthesis, modification and applications. In: Biomedical science, engineering and technology, pp 247–282. InTech, Croatia

Yamamuro T (2012) Clinical applications of bioactive glass-ceramics. New Mater Technol Healthc 1:1–97

Zanello LP, Zhao B, Hu H, Haddon RC (2006) Bone cell proliferation on carbon nanotubes. Nano Lett 6(3):562–567

Zeeshan S, Shariq N, Zohaib K, Vivek V, Haroon R, Michael G (2015) Biodegradable materials for bone repair and tissue engineering applications. Materials (Basel) 8(9):5744–5794

Zhao J, Han W, Chen H (2011) Preparation, structure and crystallinity of chitosan nano-fibers by a solid-liquid phase separation technique. Carbohydr Polym 83:1541–1546

Zhao Q, Wang S, Kong M, Geng W, Li RK, Song C, Kong D (2012) Phase morphology, physical properties, and biodegradation behavior of novel PLA/PHBHHx blends. J Biomed Mater Res B Appl Biomater 100(1):23–31

Zhu S, Sun H, Geng H, Liu D, Zhang X, Cai Q, Yang X (2016a) Dual functional polylactide–hydroxyapatite nanocomposites for bone regeneration with nano-silver being loaded *via* reductive polydopamine. RSC Adv 6:91349–91360

Zhu W, Holmes B, Glazer RI, Zhang LG (2016b) 3D printed nanocomposite matrix for the study of breast cancer bone metastasis. Nanomedicine 12(1):69–79

Advancements and New Technologies in Drug Delivery System

28

Ajay Kumar Sahi, Pooja Verma, Pallawi, Kameshwarnath Singh, and Sanjeev Kumar Mahto

Abstract

Drug delivery is defined as administration of drug component inside the body, and the system adopted for the same is known as drug delivery system. Advancements in the drug delivery system are gaining more attention and popularity due to the use of nanoformulations that enables efficient, effective and specific targeting of the drug. Several drug carriers such as liposomes, aptamers, quantum dots, peptide, polymers, metals and magnetic nanoparticle-based delivery are categorised as advanced generation drug delivery systems. The structural complexity of nano-based drug delivery system, e.g. nanocapsules, dendrimers, nanosponges, nanocrystals, nanogels and nanocapsules, provides high surface area for precise targeting in the field of cancer management and several other life-threatening diseases.

Keywords

Drug delivery system · Nanoformulations · Nanotechnology · Aptamers · Quantum dots

A. K. Sahi · P. Verma
Tissue Engineering and Biomicrofluidics Laboratory, School of Biomedical Engineering, Indian Institute of Technology (Banaras Hindu University), Varanasi, Uttar Pradesh, India

Pallawi · K. Singh
Department of Rachana Sharir, Faculty of Ayurveda, Institute of Medical Sciences, Banaras Hindu University, Varanasi, Uttar Pradesh, India

S. K. Mahto (✉)
Tissue Engineering and Biomicrofluidics Laboratory, School of Biomedical Engineering, Indian Institute of Technology (Banaras Hindu University), Varanasi, Uttar Pradesh, India

Centre for Advanced Biomaterials and Tissue Engineering, Indian Institute of Technology (Banaras Hindu University), Varanasi, Uttar Pradesh, India
e-mail: skmahto.bme@iitbhu.ac.in

S. Paul (ed.), *Biomedical Engineering and its Applications in Healthcare*,
https://doi.org/10.1007/978-981-13-3705-5_28

28.1 Introduction

As per the US National Institute of Health (NIH), drug delivery system (DDS) is defined as 'Formulation of a component that enables the introduction of therapeutic substances into the body with improved efficiency as well as safety at controlled rate, time and place of release'. (Wanigasekara and Witharana 2016) A drug is one of the key components in clinical application, and it would not be valuable without an appropriate delivery system. The delivery of drug at an optimal rate inside body requires detailed information mainly about the physicochemical property of drugs, the type of delivering material and the release mechanisms of drug that has to be adapted (Park 2014). Drug delivery system is an extensive territory that comprises targeted delivery and restrained drug release (Jain 2008). The process of drug delivery can be predominantly partitioned into the following steps:

- Step1: The mode of administration for medication or remedial item can be classified either non-invasive (such as oral, topical/skin, nasal and inhalation routes) or invasive (include injection or nanoneedle array) mode of administration.
- Step2: The liberation of active component of delivered drug.
- Step3: Penetration and transport of drug active components across the biological membrane of cell or at the target site to perform particular action (Fig. 28.1).

Drug delivery system interface provides a link between the patient, the drug and the way of formulation of drug or device, resulting in successful delivery of drug to the particular site (Wanigasekara and Witharana 2016; Jabbari and Sadeghian 2012). Nowadays, advanced drugs comprising of nucleic acid, peptides and proteins are observed to possess higher surface area than the conventional drugs. Therefore, most of the drugs are delivered via parenteral route because of their vast size, constrained stability, short half-life and more sophistication in terms of targeted delivery (Yun et al. 2015) (Fig. 28.2)

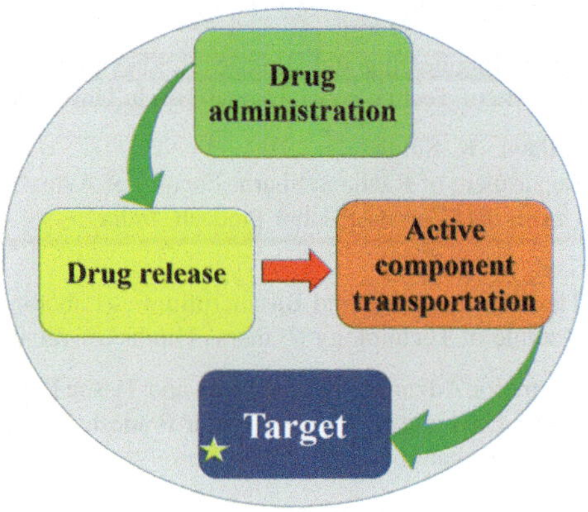

Fig. 28.1 Schematic depicting the key elements involved in drug delivery steps

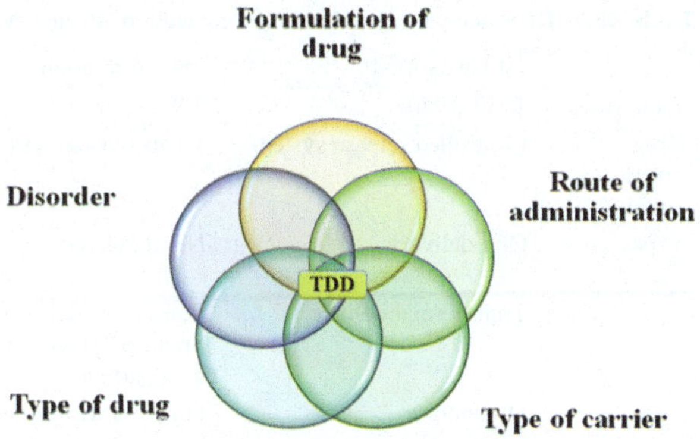

Fig. 28.2 Representation of correlation between the factors involved in targeted drug delivery (TDD)

Formulation of a drug includes many barriers such as poor bioavailability, limited loading capacity, fluctuations of drug in blood plasma, limited therapeutic effectiveness, low chances of in vivo stability, side effect and problem related to solubility, targeting to location like avascularised region in body and narrow range of drug action. The major objective of research in drug delivery systems is to develop a new formulation for clinical application to cure disease and to achieve these features. The research activity must consider the rate and target area of the drug release specifically (Jabbari and Sadeghian 2012).

28.2 History and Development

The most widely recognized type of administration of drug is via oral route. There is proof for the utilization of pills by ancestors of ancient Egypt in Ebers Papyrus. Their medicines were blended with various excipients or added substances, for example, sugars and starches to ensure and settle the detailing (Lavik et al. 2012). Since then oral route of delivery is the most commodious and adaptable method for administration of drug. However, many times it became difficult to formulate oral doses for a few drugs, which may have exceptionally weak tendency to dissolve in water or tremendously poor penetrability into cells. The whole journey of drug delivery is classified into three major generations on time scale, which is represented in Table 28.1 (Fig. 28.3).

28.3 Liposome-Mediated Drug Delivery/Lipid-Based DDS

Liposomes are homocentric structure composed of two molecular layers especially of amphipathic phospholipids. Liposomes are classified based on the number of bilayers as multilamellar (MLV), large unilamellar (LUVs) or small unilamellar (LUVs) with the range of 0.025–10 μm diameter. Liposomes show high effectiveness and low lethality, so it can be used to convey proteins, peptides, DNA, siRNA, antisense oligonucleotides and both lipid-based (hydrophobic) and water-soluble

Table 28.1 Represents the generation-by-generation advancement in drug delivery

	1st Generation	2nd Generation	3rd Generation
Time scale	1950–1980	1980–2000	2000 to present
Drug release system	Controlled release system	Smart release system	Modified release system
Types	Oral delivery	Nasal delivery	Long-term delivery
	Transdermal delivery	Pulmonary delivery Smart polymers and nanoparticles	Nucleic acid, peptide and protein delivery
Drug release mechanism	Dissolution	Diffusion-controlled system	DNA nanostructure mediated
	Diffusion		
	Osmosis	Chemically activated system	Quantum dot mediated
	Ion exchange	pH-sensitive system	
Duration	Once or twice a day	One or few weeks	6–12 months
Limitations	Physicochemical barriers (water solubility, cell permeability)	Physicochemical and biological barriers, in vitro, in vivo correlation	Formulation and biological barriers

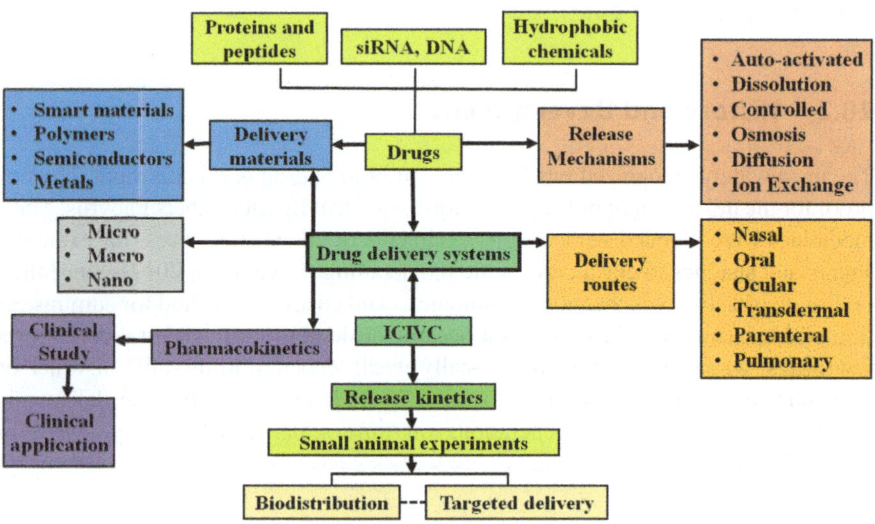

Fig. 28.3 Representation of correlation between the factors

(hydrophilic) chemotherapeutic medications. Drugs are linked into the liposomes by encapsulation strategy by varying size and physiochemical characteristic that results in preventing spillage from liposomes. Liposome composition, surrounding environment, osmotic potential, magnetic field, pH and radio-frequency control the medication discharge to the target site (Bamrungsap et al. 2012; Wilczewska et al. 2012; Çağdaş et al. 2014).Liposomes are used to deliver vaccine, drugs and genes to

Table 28.2 Examples of few liposome-mediated drugs and their treatment

Drug (liposome mediated)	Treatment	References
Amphotericin B	Fungal infections	Wanigasekara and Witharana (2016)
Vancomycin	Antibiotics drug	Wanigasekara and Witharana (2016)
Liposomal vincristine	Increased delivery to tumor site and lower acute lymphoblastic	Prasad et al. (2018)
	Systemic toxicity arising from side effect leukaemia (FDA approved 2012)	
Liposomal Irinotecan	Increased delivery to a tumor arising from side effect pancreatic cancer (FDA approved 2015)	Prasad et al. (2018)

treat a variety of disorders (Rao and Diwan 1997). As of now, accepted liposomal drug delivery system provides stable detailing, contributes enhanced pharmacokinetics and provides an idea about the level of 'passive' or 'physiological' targeting to tumor tissue (Bhat et al. 1995) (Table 28.2).

28.4 Micelle- and Dendrimer-Mediated Drug Delivery

Micelles are the circular supermolecular conglomeration of amphiphilic copolymer, where core of micelles house the hydrophobic drug while the hydrophilic brush-like structure remains in the outer region. This configuration of structure makes micelles more soluble in water and permits easy delivery of hydrophobic contents (Singh and Lillard 2009; De Jong and Borm 2008). On the other side, dendrimers are one of a kind globular, three-dimensional, nanopolymeric structure having features of water solubility, restricted polydispersity index, adjustable nanosize and molecular structure with hollow space inside and various useful functional groups attached to the border. These functional groups provide customized properties and enhance their flexibility. Furthermore, functional group at the terminal of structure acts as the stage for conjugation with drug that leads to drug targeting (Patri et al. 2002). With advancement in dendrimer and polymer chemistry, in the dendronized polymers novel groups of molecule are introduced which are arranged as linear polymers that have dendron as a repeating unit (De Jong and Borm 2008) (Table 28.3).

Table 28.3 List of drugs and therapeutic applications that are targeted using dendrimer

Drug (dendrimer mediated)	Application	References
Methotrexate	Antitumor	Wanigasekara and Witharana (2016)
Doxorubicin	Antitumor	Wanigasekara and Witharana (2016)
Ibuprofen	Anti-inflammatory drugs	Wanigasekara and Witharana (2016)
Piroxicam	Anti-inflammatory drugs	Wanigasekara and Witharana (2016)

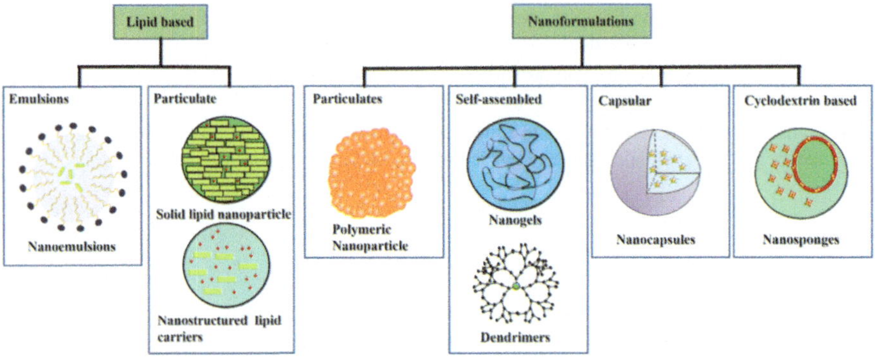

Fig. 28.4 Nanomaterials in therapeutic health care

28.5 Nanoparticles-Mediated Drug Delivery

In targeted drug delivery (TDD), the drugs are required to infuse directly into cell so that they can act as carrier itself; therefore, the particle size is compressed to nanoscale and surface is stabilized by using non-ionic surfactant and polymeric molecules. A nanoscale drug alters their interaction property to the target site due to increase in surface area of drugs. Nanoparticles can be used in different forms such as nanogels, nanoemlusions, nanosponges, nanocapsules, nanotubes, nanocrystals, dendrimers, nanocarbon tubes, nanolipids, nanoproteins and nano-nucleic acid complex structure (Fig. 28.4).

With the developing era of drug delivery, affixing drug molecules specifically to antibodies has been achieved. However, combining a bunch of drug molecule to an antibody essentially limits its targeting capacity because of chemical interaction with drug turns off the antibody activity. To overwhelm this limitation, various nanocarriers/devices are developed; a few examples of the nanocarriers with their application have been depicted in Table 28.4.

Table 28.4 List of nanocarriers/devices that are developed at preclinical and clinical stage

Nanocarriers/ devices	Application	Reference
Nanocapsules	Improvement in the cells therapeutic efficiency of drug in colon cancer	Feng et al. (2017)
Conjugated dendrimers	Enhanced efficacy of drug in pancreatic cancer	Öztürk et al. (2017)
Lipid nanocapsules	Effective in lung cancer therapy	Kim and Park (2017)
Chitosan-based nanoparticles	Chitosan nanoparticle and microparticle comparison for anticancer activity of propolis	Elbaz et al. (2016)
Nanographene oxide	In vivo targeting of metastatic breast cancer via tumor vasculature-specific nanographene oxide	Yang et al. (2016a)
Nano-Fe_3O_4/CA[a]	Breast cancer: Alkylating drug efficacy Nimustine >> Semustine > Chlormethine	He et al. (2017)
Aprepitant-containing nanocrystals	Antiemetic	Wanigasekara and Witharana (2016)

[a]*CA* citrate capped supermagnetic iron oxide

28.6 Metal and Magnetic Nanoparticle-Mediated Drug Delivery

Metal-based drug therapy shares a low percentage among other delivery systems, but the major advantages of metal-based therapeutic over organic based drug and its delivery system are due to its properties including its varying coordinate number, geometry, redox state and ability to form complex with organic molecules, which make it possible to combine with magnet-mediated delivery of drug. Numerous magnetic nanoparticles have been in process for early clinical trials and many have been accepted by FDA for clinical application (Sun et al. 2008; Kang et al. 2017). Furthermore, there is a wide range of utilization for magnetic nanoparticle and superparamagnetic iron oxide nanoparticle (SPION) in various forms for drug delivery (Veiseh et al. 2010; Mok and Zhang 2013), biomedical fields (Qiao et al. 2009; Oh and Park 2011), diagnostic imaging (Lee and Hyeon 2012; Amstad et al. 2009), biosensors (Wu et al. 2015), and hyperthermia treatments as represented in Table 28.5 (Kumar and Mohammad 2011).

28.7 Antibody-Mediated Drug Delivery

Antibodies or immunoglobulins are the Y-shaped glycoprotein composed of four polypeptide chains and are produced by immune cells in response to attack of foreign substances, i.e. antigens. Antibodies are most familiar targeting molecule after the development of monoclonal antibodies at clinical level for the treatment of

Table 28.5 List of drugs and therapeutic application that are targeted using dendrimer

Metal/magnetic nanoparticle	Indication	References
Iron sucrose Venofer® (Luitpold Pharm)	Iron deficiency in chronic kidney disease FDA approved 2000	Prasad et al. (2018)
Magnetite-based, silica-coated MNP prodrug + external magnetic field	Enhanced tumor extravasation	De Jong and Borm (2008)
Fe_3O_4 MNP	Enhanced penetration of blood-brain barrier (BBB), human ovarian tumor tissues and breast carcinoma xenograft models	De Jong and Borm (2008) and Klostergaard et al. (2007)
Fe_3O_4 poly e-caprolactone (PCL) + nanoparticle external magnetic field	Increased antitumor activity when loaded with gemcitabine in nude mice with subcutaneous (SC) xenografts of human pancreatic adeno-carcinoma cells	Klostergaard et al. (2007) and Klostergaard et al. (2010)
Iron oxide magnetic nanoparticle (IOMNPs) + external magnetic field	Magnetic targeting induced a fivefold increase in the total glioma exposure to the IOMNPs compared with nontargeted tumors	Chertok et al. (2008)

cancer (Stacker et al. 2002). Drug delivery through antibodies can be done in the form of radio-immunotherapy (Goldenberg 2007), immunoliposome mediated (Nielsen et al. 2002; Heitner et al. 2001; He et al. 2010), immunotoxines (Turturro 2007; Gattenlöhner et al. 2010) and antibody-drug conjugates (Ducry and Stump 2010; Alley et al. 2010). Few examples of the antibody-based drug delivery are represented in Table 28.6 and Fig. 28.5.

28.8 Aptamer-Mediated Drug Delivery

The introduction of SELEX (Systematic Evolution of Ligands by Exponential Enrichment) technology has grabbed the attention of researchers towards another targeting molecule, i.e. aptamer. It is small oligonucleotide of about 15–40 bases (Catuogno et al. 2016; Ellington and Szostak 1990; Tuerk and Gold 1990). Alike antibodies, aptamer represents an extraordinary feature towards its target and having high binding affinity because of its tendency to form a three-dimensional molecular complex (Levy-Nissenbaum et al. 2008). Aptamer offers several advantages over antibodies and monoclonal antibodies which include its low kD value (in nano–pico range), high ligand specificity, low immunogenic effect, simple production process and cost-effective and batch-to-batch fidelity. A few aptamer-related examples are discussed in Table 28.7.

28.9 Peptide- and Polymer-Mediated Drug Delivery

Although antibodies have very high target selectivity and binding affinities, they are potentially immunogenic. To overcome such limitation, antibodies could be engineered to acquire humanized or chimeric properties to avoid immune detection (De Jong and Borm 2008). In contrast to antibodies, peptides as targeting moieties offer several advantages as they lack immunogenicity and have low production cost (De Jong and Borm 2008).

Similar to peptides, polymeric materials contribute significantly in nanomedicine sector for pharmaceutical and biotechnology devices (Kamaly et al. 2012). Polymers deliver better option over metals and ceramics because of their optimal properties such as biodegradability, adaptable amalgamation techniques according to mechanical and physiological requirements and capability to get moulded into different morphologies including films, gels, fibre and particles (Ramakrishna and Rao 2011). PLGA has been widely studied for various medical applications. It has also been extensively reported for drug delivery for different drugs and diseases. Few other most commonly marketed formulations made of polymeric nanoparticles are

Table 28.6 Few examples of antibody-medicated drug delivery systems and their clinical status

Agent	Target	Indication	Clinical Status	Reference
IMGN388	αv integrin	Solid tumors	Phase I	Pathak and Benita (2012a)
AGS-16M18	AGS-16	Kidney and liver cancer	Phase I	Pathak and Benita (2012b)
Brentuximab Vedotin	CD30	Hodgkin lymphoma	Phase III	BrentuximabVedotin (ADCETRIS®)
Inotuvumab Ozogamicin	CD22	Follicular lymphoma	Phase I	Lundin et al. (2002)
SAR650984	CD38	Haematological malignancies	Phase I	Pathak and Benita (2012a)
Inotuvumab ozogamicin+ rituximab	CD22 + CD20	Follicular lymphoma, diffuse large B-cell lymphoma	Phase I/phase III	Fayad et al. (2008)
SGN-35	CD30	CD30 + hematologic malignancies	Phase I	Younes et al. (2008)

Fig. 28.5 Schematic depicts an antibody-drug conjugate

Table 28.7 Examples of aptamer-mediated delivery system for therapeutic drugs

Aptamer-mediated delivery systems	Ligands	Application	Reference
Aptamer drug (DOX) conjugate	Anti-CD38 aptamer	Multiple myeloma	Wen et al. (2016)
	Anti-prostate-specific antigen (PSMA) A10	Prostate cancer	Bagalkot et al. (2006)
Aptamer carbon nanomaterial	SWNTS	NRT light-controlled cancer cell delivery and killing	Yang et al. (2016b)
Block polymer nanoparticle	Tri-block copolymer (TCP)	Prostate cancer targeting	Gu et al. (2008)

Table 28.8 List of drugs carriers and devices that are targeted using dendrimer at laboratory or clinical level

Polymer base carrier/device	Application	References
Flash nanoprecipitation (FNP)	Multifunctional nanoparticle	Johnson et al. (2003)
ChemoRed	Ovarian cancer	Wang (2010) and Werner et al. (2011)
Polyurethane nanocarriers	Antitumor	Ding et al. (2013)
Polymer-protein conjugate (PEGylated IFN beta 1-a), Plegridy® (biogene)	Multiple sclerosis FDA approved 2014	Prasad et al. (2018)
Polymer-protein conjugate (PEGylation as porcine-like uricase) Krystexxa®	Chronic gout FDA approved 2010	Prasad et al. (2018)
Polymer-protein conjugate Adynovate (Baxalta)	Haemophilia FDA approved 2015	Prasad et al. (2018)

Decapeptyl®, Gonapeptyl Depot®, Enantone Depot® and Abraxane (Lherm et al. 1992; Cortesi et al. 2002). Other various examples for polymer-based carriers for drug delivery are discussed in Table 28.8.

28.10 DNA-Mediated Drug Delivery

The RNA interference (RNAi) and antibody-mediated delivery show intrinsic contradiction at cellular level in terms of poor solubility, difficulty in transportation through biological membrane, prone to enzymatic degradation, undesirable side effect and cellular toxicity. The exponential advancement in DNA nanotechnology since the last two decades holds the promising capability to conquer limitation in drug delivery (Table 28.9).

DNA nanotechnology-based drug delivery shows outstanding performance with respect to retina and brain cancer as well as due to its exceptional ability to pass through blood-brain barrier, which is the most challenging part of target drug.

DNA-based nanoparticle work in two successive steps:

- Primary targeting – it involves the accumulation of nanoparticle in particular organ where the target cells reside.
- Secondary targeting – it involves all the process after drug reaches the organ, and it includes direction of drug molecule on nanoparticle to the target cells and then to the subcellular site (Fig. 28.6).

28.11 Quantum Dot-Mediated Drug Delivery System

Quantum dot synthesis involves various semiconductor materials that deliver emission in wavelength from ultraviolet (UV) to the near infrared (NIR). Examples for few quantum dots are CdS, PbS, ZnS, CdTe and PbTe. Outstanding quantum confinement effects and unbeatable capacity of fine-tuning in quantum dots (QDs) allow a huge range for selection of quantum dots with its unique fluorescent emission based on their photoluminescence spectra range. A core-shell structure with wider band gap semiconductor shell component enhances fluorescent properties as well as acts as a barrier to prevent drug leach-out form core. This kind of core-shell QDs are generally used in biological applications.

In targeted drug delivery, quantum dots are affixed with DNA, proteins and peptides on its surface that provide bio-recognition (Medintz et al. 2005; Klostranec and Chan 2006; Susumu et al. 2007). They act as biosensor, bind to targeted molecule and facilitate cellular delivery. The table depicts the QDs with their ligand and their target sites (Medintz et al. 2005; Delehanty et al. 2009) (Table 28.10).

28.12 Conclusion

Drug delivery plays an indispensable character to reduce the noxious side effects and support the targeted drug delivery. Nanomaterial (NM)-based target drug delivery plays a vital role to deliver drugs at tumor sites. They also provide options to be

Table 28.9 List of ongoing researches in DNA-mediated targeted drug delivery

Carrier for delivery	Mediator	Device	Application	References
Small molecules	Fluorescent dyes	FRET (fluorescent resonance energy transfer), i.e. nanoflowers (NFs)	Multiplexed cellular imaging and traceable targeted drug delivery	Hu et al. (2014)
	Doxorubicin	DOX-loaded DNA origami nanostructures	Treated on human cancer breast cells	Zhao et al. (2012), Jiang et al. (2012) and Zhang et al. (2014)
		Multifunctional aptamer-based DNA assembly (AptNA)	Targeted cancer therapy	Wu et al. (2013)
		Controllable and targeted delivery of dox to cancer cells	Cancer	Zhang et al. (2015)
Oligonucleotides	CpGs	CpG nanoflower	Act as an immuno-stimulus, trigger the secretion of TNF-R, IL-6 and IL-12 that induce cancer cell apoptosis and necrosis	Keefe et al. (2010)
	siRNA	Self-assembled DNA tetrahedral nanostructure	Delivery of siRNAs in vivo to silence target genes in tumors	Lee et al. (2012) and Fakhoury et al. (2014)
		Spherical nucleic acid (SNA)	Neutralization of oncogene expression in glioblastoma multiforme (GBM)	Jensen et al. (2013), Dunn et al. (2012) and Wen and Kesari (2008)
	microRNA	(DNA star motif + miRNA) shuriken-like shape	Anti-proliferative effect on colorectal cancer cell line DLD-1, tumor suppression	Qian et al. (2017) and Nahar et al. (2017)
	CRISPR-Cas9	DNA Nanoclaw loaded with CAS9/sgRNA	Cell culture and tumor-bearing mice model	Sun et al. (2015) and Varkouhi et al. (2011)

(continued)

Table 28.9 (continued)

Carrier for delivery	Mediator	Device	Application	References
	Aptamers	AS1411	Cancer-targeting ligand	Kang et al. (2009), Kotula et al. (2012), Soundararajan et al. (2008) and Bates et al. (2009)
		Pyramidal DNA nanostructures + AS1411 aptamer loaded	Inhibited the growth of cancer cells in human cervical cancer cell line (HeLa)	Charoenphol and Bermudez (2014)

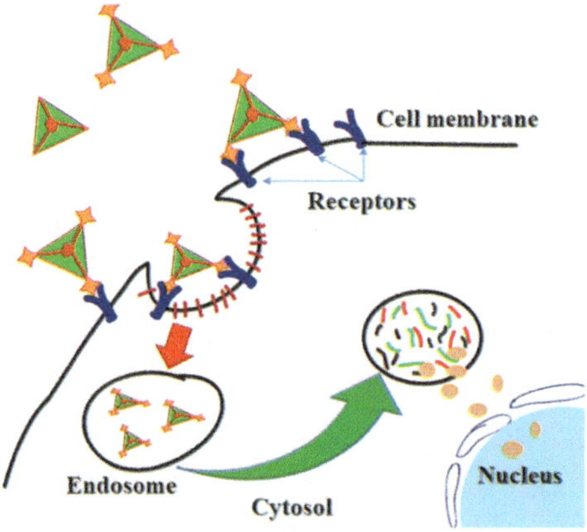

Fig. 28.6 DNA nanotechnology-enabled drug delivery system

modified as per the need and the target of interest. Furthermore, studies need to concentrate on the interaction of nanomaterials with their hosts in terms of genetic impairment, cellular structure damage, organ accumulation and bio-distribution. DNA-based, peptide-based and nanomaterial-based carriers are playing a unique and pivotal role in the drug delivery system. DNA-based carriers are very beneficial in the rare scenarios.

The use of non-biodegradable NMs such as carbon-based single- or multiwall carbon nanotubes, metallic nanocarriers and some inorganic oxides can also be explored in nanomedicine. Depending on their functionalization, biodegradable drug nanocarriers can take a number of paths within tissues. The pharmacokinetics (PK) and excretion routes of the various NMs demands exhaustive research to clear the path for their extensive human applications.

Table 28.10 List of targeted intracellular delivery of quantum dots (QDs)

Quantum dot functional group	Ligand	Target/cell lines	References
TAT peptide (<2 kDa)	Heparin sulfate, proteoglycans	Endosomes (HEK293T/17, 17/COS-1)	Delehanty et al. (2006)
Dopamine (<1 kDa)	Dopamine receptor, expressing dopamine receptors	Neural cells or those cytoplasm (transfected A9)	Zheng et al. (2006)
Epidermal growth factor	EGF receptor[ab] (EGF, 6 kDa)	Endosomes (CHO, A431)	Liu et al. (2008) and Lidke et al. (2004)
Anti-EGF receptor	EGF receptor[ab]	Membrane-localized/endosomes domain antibody (< 15 kDa) (SK-BR3, MDA-MB468)	Zaman et al. (2009)
RGD peptide (<1 kDa)	$a_v\beta_3$-integrins[a]	Membrane-localized (calvarial osteoblasts, SKOV-3)	Delehanty et al. (2009) and Lieleg et al. (2007)
Transferrin (~80 kDa)	Transferrin receptor	Endosomes (HeLa)	Zaman et al. (2009) and Chan and Nie (1998)
Cholera toxin B (12 kDa)	Ganglioside receptors	Endosomes/cytoplasmic vesicles (fibroblast)	Zhang et al. (2009)
Nerve growth factor (NGF, 30 kDa)	TrkA receptor	Neurons endosomes/along neural processes (neural PC12)	Cui et al. (2007) and Smith et al. (2008)
D-Galactosamine D-mannose (~180 Da)	Asialoglyco protein receptor	Liver cells, endosomes	Kikkeri et al. (2009)
WS2quantum dots (WQDs) nanoparticles	Coated with doxorubicin (DOX)-loaded periodic mesoporous organosilica	Anticancer activity	Liao et al. (2018)
Carbon-based quantum dot	Conjugated with protoporphyrin IX (PpIX)	Nucleus targeting and phototherapeutic property that lead to antitumor activity	Hua et al. (2018)

[a]Therapeutic target
[b]Upregulated or altered activity in certain cancers

28.13 Future Scope

Scaffolds based on second-generation technology is still lacking clinical products, and the clinical products based on third-generation technologies are still under development. Multiple factors need to be considered simultaneously for the development of novel drug delivery systems. It is essential to contemplate the complexity related to the drug delivery system advancement. Pharmacodynamics and pharmacokinetic analysis and incorporation of biology may help to overcome the limitations observed during earlier generations. Nanocarriers and nanoscale structures such as nanoparticles, quantum dots, aptamers, DNA, RNA-based carriers can define the next-generation drug delivery systems because of their unique advanced features and easy manipulation capability that make them fit for next-generation drug delivery system.

Acknowledgement This work was financially supported by a DST-INSPIRE (DST/INSPIRE/04/2013/000836) research grant from the Department of Science and Technology, Government of India. The authors would also like to thank the Institute Research Project (IRP) scheme for individual faculty provided by the Indian Institute of Technology (Banaras Hindu University) for the development of state-of-the-art facilities.

References

Alley SC, Okeley NM, Senter PD (2010) Antibody-drug conjugates: targeted drug delivery for cancer. Curr Opin Chem Biol 14:529–537

Amstad E et al (2009) Surface functionalization of single superparamagnetic iron oxide nanoparticles for targeted magnetic resonance imaging. Small Weinh Bergstr Ger 5:1334–1342

Bagalkot V, Farokhzad OC, Langer R, Jon S (2006) An aptamer-doxorubicin physical conjugate as a novel targeted drug-delivery platform. Angew Chem Int Ed Eng 45:8149–8152

Bamrungsap S et al (2012) Nanotechnology in therapeutics: a focus on nanoparticles as a drug delivery system. Nanomedicine 7:1253–1271

Bates PJ, Choi EW, Nayak LV (2009) G-rich oligonucleotides for cancer treatment. Methods Mol Biol Clifton NJ 542:379–392

Bhat M, Shenoy SD, Udupa N, Srinivas CR (1995) Optimization of delivery of betamethasone-dipropionate from skin preparation. Indian Drugs 32:211–214

BrentuximabVedotin (ADCETRIS®) Seattle genetics. Available at: http://www.seattlegenetics.com/pipeline/brentuximab-vedotin

Çağdaş M, Sezer AD, Bucak S (2014) Liposomes as potential drug carrier systems for drug delivery. Appl Nanotechnol Drug Deliv. https://doi.org/10.5772/58459

Catuogno S, Esposito CL, de Franciscis V (2016) Aptamer-mediated targeted delivery of therapeutics: an update. Pharm Basel Switz 9:69

Chan WCW, Nie S (1998) Quantum dot bioconjugates for ultrasensitive nonisotopic detection. Science 281:2016–2018

Charoenphol P, Bermudez H (2014) Aptamer-targeted DNA nanostructures for therapeutic delivery. Mol Pharm 11:1721–1725

Chertok B et al (2008) Iron oxide nanoparticles as a drug delivery vehicle for MRI monitored magnetic targeting of brain tumors. Biomaterials 29:487–496

Cortesi R, Esposito E, Luca G, Nastruzzi C (2002) Production of lipospheres as carriers for bioactive compounds. Biomaterials 23:2283–2294

Cui B et al (2007) One at a time, live tracking of NGF axonal transport using quantum dots. Proc Natl Acad Sci U S A 104:13666–13671

De Jong WH, Borm PJ (2008) Drug delivery and nanoparticles: applications and hazards. Int J Nanomedicine 3(2):133

Delehanty JB et al 2006 Self-assembled quantum dot-peptide bioconjugates for selective intracellular delivery. PubMed – NCBI. Available at: https://www.ncbi.nlm.nih.gov/pubmed/16848398. Accessed: 5th Jan 2019

Delehanty JB, Mattoussi H, Medintz IL (2009) Delivering quantum dots into cells: strategies, progress and remaining issues. Anal Bioanal Chem 393:1091–1105

Ding M et al (2013) Toward the next-generation nanomedicines: design of multifunctional multiblock polyurethanes for effective cancer treatment. ACS Nano 7:1918–1928

Ducry L, Stump B (2010) Antibody-drug conjugates: linking cytotoxic payloads to monoclonal antibodies. Bioconjug Chem 21:5–13

Dunn GP et al (2012) Emerging insights into the molecular and cellular basis of glioblastoma. Genes Dev 26:756–784

Elbaz NM, Khalil IA, Abd-Rabou AA, El-Sherbiny IM (2016) Chitosan-based nano-in-microparticle carriers for enhanced oral delivery and anticancer activity of propolis. Int J Biol Macromol 92:254–269

Ellington AD, Szostak JW (1990) In vitro selection of RNA molecules that bind specific ligands. Nature 346:818–822

Fakhoury JJ, McLaughlin CK, Edwardson TW, Conway JW, Sleiman HF (2014) Development and characterization of gene silencing DNA cages. Biomacromolecules 15:276–282

Fayad L et al (2008) Safety and clinical activity of the anti-CD22 Immunoconjugate InotuzumabOzogamicin (CMC-544) in combination with rituximab in follicular lymphoma or diffuse large B-cell lymphoma: preliminary report of a phase 1/2 study. Blood 112:266–266

Feng M, Zhong LX, Zhan ZY, Huang ZH, Xiong JP (2017) Enhanced antitumor efficacy of resveratrol-loaded nanocapsules in colon cancer cells: physicochemical and biological characterization. Eur Rev Med Pharmacol Sci 21(2):375–382

Gattenlöhner S et al (2010) A human recombinant autoantibody-based immunotoxin specific for the fetal acetylcholine receptor inhibits rhabdomyosarcoma growth in vitro and in a murine transplantation model. J Biomed Biotechnol 2010:187621–187621

Goldenberg DM (2007) Radiolabelled monoclonal antibodies in the treatment of metastatic cancer. Curr Oncol 14:39–42

Gu F et al (2008) Precise engineering of targeted nanoparticles by using self-assembled biointegrated block copolymers. Proc Natl Acad Sci U S A 105:2586–2591

He J et al (2010) Targeting prostate cancer cells in vivo using a rapidly internalizing novel human single-chain antibody fragment. J Nucl Med Off Publ Soc Nucl Med 51:427–432

He K et al (2017) The efficacy assessments of alkylating drugs induced by nano-Fe3O4/CA for curing breast and hepatic cancer. Spectrochim Acta A Mol Biomol Spectrosc 173:82–86

Heitner T et al (2001) Selection of cell binding and internalizing epidermal growth factor receptor antibodies from a phage display library. J Immunol Methods 248:17–30

Hu R et al (2014) DNA nanoflowers for multiplexed cellular imaging and traceable targeted drug delivery. Angew Chem Int Ed Eng 53:5821–5826

Hua X-W, Bao Y-W, Wu F-G (2018) Fluorescent carbon quantum dots with intrinsic nucleolus-targeting capability for nucleolus imaging and enhanced cytosolic and nuclear drug delivery. ACS Appl Mater Interfaces 10:10664–10677

Jabbari A, Sadeghian H (2012) Amphiphilic cyclodextrins, synthesis, utilities and application of molecular modeling in their design. Recent Adv Nov Drug Carr Syst. https://doi.org/10.5772/50220

Jain KK (ed) (2008) Drug delivery systems, vol 2. Humana press, Totowa

Jensen SA et al (2013) Spherical nucleic acid nanoparticle conjugates as an RNAi-based therapy for glioblastoma. Sci Transl Med 5:209ra152

Jiang Q et al (2012) DNA origami as a carrier for circumvention of drug resistance. J Am Chem Soc 134:13396–13403

Johnson BK, Prud'homme RK (2003) Flash nanoprecipitation of organic actives and block copolymers using a confined impinging jets mixer. Aust J Chem 56:1021–1024

Kamaly N, Xiao Z, Valencia PM, Radovic-Moreno AF, Farokhzad OC (2012) Targeted polymeric therapeutic nanoparticles: design, development and clinical translation. Chem Soc Rev 41:2971–3010

Kang WJ, Chae JR, Cho YL, Lee JD, Kim S (2009) Multiplex imaging of single tumor cells using quantum-dot-conjugated aptamers. Small Weinh Bergstr Ger 5:2519–2522

Kang T et al (2017) Surface design of magnetic nanoparticles for stimuli-responsive cancer imaging and therapy. Biomaterials 136:98–114

Keefe AD, Pai S, Ellington A (2010) Aptamers as therapeutics. Nat Rev Drug Discov 9:537–550

Kikkeri R, Lepenies B, Adibekian A, Laurino P, Seeberger PH (2009) In vitro imaging and in vivo liver targeting with carbohydrate capped quantum dots. J Am Chem Soc 131:2110–2112

Kim J-E, Park Y-J (2017) Paclitaxel-loaded hyaluronan solid nanoemulsions for enhanced treatment efficacy in ovarian cancer. Int J Nanomedicine 12:645. https://doi.org/10.2147/IJN.S124158

Klostergaard J et al (2007) Magnetic vectoring of magnetically responsive nanoparticles within the murine peritoneum. J Magn Magn Mater 311:330–335

Klostergaard J, Bankson J, Woodward W, Gibson D, Seeney C (2010) Magnetically-responsive nanoparticles for vectored delivery of cancer therapeutics. AIP Conf Proc 1311:382–387

Klostranec JM, Chan WCW (2006) Quantum dots in biological and biomedical research: recent progress and present challenges. Adv Mater 18:1953–1964

Kotula JW et al (2012) Aptamer-mediated delivery of splice-switching oligonucleotides to the nuclei of cancer cells. Nucleic Acid Ther 22:187–195

Kumar CSSR, Mohammad F (2011) Magnetic nanomaterials for hyperthermia-based therapy and controlled drug delivery. Adv Drug Deliv Rev 63:789–808

Lavik EB, Kuppermann BD, Humayun MS (2012) Drug delivery. In: Retina, 5th edn. Allen Institute for Artificial Intelligence, Seattle, pp 734–745. https://doi.org/10.1016/B978-1-4557-0737-9.00038-2

Lee N, Hyeon T (2012) Designed synthesis of uniformly sized iron oxide nanoparticles for efficient magnetic resonance imaging contrast agents. Chem Soc Rev 41:2575–2589

Lee H et al (2012) Molecularly self-assembled nucleic acid nanoparticles for targeted in vivo siRNA delivery. Nat Nanotechnol 7:389–393

Levy-Nissenbaum E, Radovic-Moreno AF, Wang AZ, Langer R, Farokhzad OC (2008) Nanotechnology and aptamers: applications in drug delivery. Trends Biotechnol 26:442–449

Lherm C, Müller RH, Puisieux F, Couvreur P (1992) Alkylcyanoacrylate drug carriers: II. Cytotoxicity of cyanoacrylate nanoparticles with different alkyl chain length. Int J Pharm 84:13–22

Liao W et al (2018) Fabrication of ultra-small WS2 quantum dots-coated periodic mesoporous organosilica nanoparticles for intracellular drug delivery and synergistic chemo-photothermal therapy. OncoTargetsTher 11:1949–1960

Lidke DS et al (2004) Quantum dot ligands provide new insights into erbB/HER receptor-mediated signal transduction. Nat Biotechnol 22:198–203

Lieleg O et al (2007) Specific integrin labeling in living cells using functionalized nanocrystals. Small Weinh Bergstr Ger 3:1560–1565

Liu W et al (2008) Compact biocompatible quantum dots functionalized for cellular imaging. J Am Chem Soc 130:1274–1284

Lundin J et al (2002) Phase II trial of subcutaneous anti-CD52 monoclonal antibody alemtuzumab (Campath-1H) as first-line treatment for patients with B-cell chronic lymphocytic leukemia (B-CLL). Blood 100:768–773

Medintz IL et al 2005 Quantum dot bioconjugates for imaging, labelling and sensing. PubMed – NCBI. Available at: https://www.ncbi.nlm.nih.gov/pubmed/15928695. Accessed: 5th Jan 2019

Mok H, Zhang M (2013) Superparamagnetic iron oxide nanoparticle-based delivery systems for biotherapeutics. Expert Opin Drug Deliv 10:73–87

Nahar S, Nayak AK, Ghosh A, Subudhi U, Maiti S (2017) Enhanced and synergistic downregulation of oncogenic miRNAs by self-assembled branched DNA. Nanoscale 10:195–202

Nielsen UB et al (2002) Therapeutic efficacy of anti-ErbB2 immunoliposomes targeted by a phage antibody selected for cellular endocytosis. Biochim Biophys Acta 1591:109–118

Oh JK, Park JM (2011) Iron oxide-based superparamagnetic polymeric nanomaterials: design, preparation, and biomedical application. Prog Polym Sci 36:168–189

Öztürk K, Esendağlı G, Gürbüz MU, Tülü M, Çalış S (2017) Effective targeting of gemcitabine to pancreatic cancer through PEG-cored Flt-1 antibody-conjugated dendrimers. Int J Pharm 517:157–167

Park K (2014) Controlled drug delivery systems: past forward and future back. J Control Release Off 190:3–8

Pathak Y, Benita S (2012a) Antibody-mediated drug delivery systems: concepts, technology, and applications. Wiley, Hoboken

Pathak Y, Benita S (2012b) Antibody-mediated drug delivery systems: concepts, technology, and applications. Wiley, Hoboken

Patri AK, Majoros IJ, Baker JR (2002) Dendritic polymer macromolecular carriers for drug delivery. Curr Opin Chem Biol 6:466–471

Prasad M et al (2018) Nanotherapeutics: an insight into healthcare and multi-dimensional applications in medical sector of the modern world. Biomed Pharmacother 97:1521–1537

Qian H et al (2017) Protecting microRNAs from RNase degradation with steric DNA nanostructures. Chem Sci 8:1062–1067

Qiao R, Yang C, Gao M (2009) Superparamagnetic iron oxide nanoparticles: from preparations to in vivo MRI applications. J Mater Chem 19:6274–6293

Ramakrishna D, Rao P (2011) Nanoparticles: is toxicity a concern? EJIFCC 22:92–101

Rao PR, Diwan PV (1997) Permeability studies of cellulose acetate free films for transdermal use: influence of plasticizers. Pharm ActaHelv 72:47–51

Singh R, Lillard JW (2009) Nanoparticle-based targeted drug delivery. Exp Mol Pathol 86:215–223

Smith BR et al (2008) Real-time intravital imaging of RGD-quantum dot binding to luminal endothelium in mouse tumor neovasculature. Nano Lett 8:2599–2606

Soundararajan S, Chen W, Spicer EK, Courtenay-Luck N, Fernandes DJ (2008) The nucleolin targeting aptamer AS1411 destabilizes Bcl-2 messenger RNA in human breast cancer cells. Cancer Res 68:2358–2365

Stacker SA, Achen MG, Jussila L, Baldwin ME, Alitalo K (2002) Lymphangiogenesis and cancer metastasis. Nat Rev Cancer 2:573–583

Sun C, Lee JSH, Zhang M (2008) Magnetic nanoparticles in MR imaging and drug delivery. Adv Drug Deliv Rev 60:1252–1265

Sun W et al (2015) Self-assembled DNA nanoclews for the efficient delivery of CRISPR-Cas9 for genome editing. Angew Chem Int Ed Eng 54:12029–12033

Susumu K et al (2007) Enhancing the stability and biological functionalities of quantum dots via compact multifunctional ligands. J Am Chem Soc 129:13987–13996

Tuerk C, Gold L (1990) Systematic evolution of ligands by exponential enrichment: RNA ligands to bacteriophage T4 DNA polymerase. Science 249:505–510

Turturro F (2007) Denileukindiftitox: a biotherapeutic paradigm shift in the treatment of lymphoid-derived disorders. Expert Rev Anticancer Ther 7:11–17

Varkouhi AK, Scholte M, Storm G, Haisma HJ (2011) Endosomal escape pathways for delivery of biologicals. J Control Release Off 151:220–228

Veiseh O, Gunn JW, Zhang M (2010) Design and fabrication of magnetic nanoparticles for targeted drug delivery and imaging. Adv Drug Deliv Rev 62:284–304

Wang AZ et al (2010) ChemoRad nanoparticles: a novel multifunctional nanoparticle platform for targeted delivery of concurrent chemoradiation. Nanomedicine (London) 5:361. https://doi.org/10.2217/nnm.10.6
Wanigasekara J, Witharana C (2016) Applications of nanotechnology in drug delivery and design-an insight. Curr Trends Biotechnol Pharm 10(1):78–91
Wen PY, Kesari S (2008) Malignant gliomas in adults. N Engl J Med 359:492–507
Wen J, Tao W, Hao S, Iyer SP, Zu Y (2016) A unique aptamer-drug conjugate for targeted therapy of multiple myeloma. Leukemia 30:987–991
Werner ME et al (2011) Folate-targeted nanoparticle delivery of chemo- and radiotherapeutics for the treatment of ovarian cancer peritoneal metastasis. Biomaterials 32:8548–8554
Wilczewska AZ, Niemirowicz K, Markiewicz KH, Car H (2012) Nanoparticles as drug delivery systems. Pharmacol Rep PR 64:1020–1037
Wu C et al (2013) Building a multifunctional aptamer-based DNA nanoassembly for targeted cancer therapy. J Am Chem Soc 135:18644–18650
Wu W, Wu Z, Yu T, Jiang C, Kim W-S (2015) Recent progress on magnetic iron oxide nanoparticles: synthesis, surface functional strategies and biomedical applications. Sci Technol Adv Mater 16:023501
Yang D et al (2016a) In vivo targeting of metastatic breast cancer via tumor vasculature-specific nano-graphene oxide. Biomaterials 104:361–371
Yang Y et al (2016b) Near-infrared light-activated cancer cell targeting and drug delivery with aptamer-modified nanostructures. Nano Res 9:139–148
Younes A et al (2008) Objective responses in a phase I dose-escalation study of SGN-35, a novel antibody-drug conjugate (ADC) targeting CD30, in patients with relapsed or refractory Hodgkin lymphoma. J Clin Oncol 26:8526–8526
Yun YH, Lee BK, Park K (2015) Controlled drug delivery: historical perspective for the next generation. J Control Release 219:2–7
Zaman MB, Baral TN, Zhang J, Whitfield D, Yu K (2009) Single-domain antibody functionalized CdSe/ZnS quantum dots for cellular imaging of cancer cells. J Phys Chem C 113:496–499
Zhang H et al (2009) Detection and downregulation of type I IGF receptor expression by antibody-conjugated quantum dots in breast cancer cells. Breast Cancer Res Treat 114:277–285
Zhang Q et al (2014) DNA origami as an in vivo drug delivery vehicle for cancer therapy. ACS Nano 8:6633–6643
Zhang H et al (2015) A controllable aptamer-based self-assembled DNA dendrimer for high affinity targeting, bioimaging and drug delivery. Sci Rep 5:10099
Zhao Y-X et al (2012) DNA origami delivery system for cancer therapy with tunable release properties. ACS Nano 6:8684–8691
Zheng J et al (2006) Cellular imaging and surface marker labeling of hematopoietic cells using quantum dot bioconjugates. Lab Hematol Off Publ Int Soc Lab Hematol 12:94–98

Recent Advances in the au NP Treatment Strategies of Lung Cancers

29

Parth Malik and Rakesh Kumar Ameta

Abstract

Of late, drug delivery using metal nanoparticles (NPs) having tunable geometric and physicochemical properties has been an intriguing domain of research. The foremost advantages of overwhelming scientific interest toward nanoparticle (NP)-mediated drug delivery encompass the size- and shape-specific response, capability of selective delivery to only affected cells, and, above all, the reduced patient sensitization via moderation of drug payloads. Among the several kinds of metal NPs, Au-based NPs are swiftly emerging as robust agents for delivering drugs to even most hard to reach body locations. The chief specialty of Au NPs is the diversity of their shapes and inert nature, making them much less toxic (than anticipated), thermally conducting chemical response which imparts them a synergistic chemical response along with the tagged drugs. A lot has been talked about the insufficient drug internalization once it has been delivered to treat cancers. The multifunctional Au NPs portray a benign solution in this regard via simultaneous surface tagging of diagnostic and drug delivery agents on their surface. Studies have shown a significant improvement in the drug internalization capability of Au nanorods (NRs) compared to their spherical counterparts, which has recently intensified investigations on NR-based anticancer therapies. Another interesting property of Au NPs has been their versatile and much easily available preparation methods that require less energy from outside and are also biocompatible. Among several methods, those involving chemical reduction have been the subject of interest, owing to a vast range of available reducing agents which could provide NPs of different features. With such insights, the present compilation summarizes the most recent attempts in Au NP-based lung cancer

P. Malik
School of Nano Sciences, Central University of Gujarat, Gandhinagar, Gujarat, India

R. K. Ameta (✉)
School of Chemical Sciences, Central University of Gujarat, Gandhinagar, Gujarat, India

S. Paul (ed.), *Biomedical Engineering and its Applications in Healthcare*,
https://doi.org/10.1007/978-981-13-3705-5_29

(LC) treatment strategies, focusing peculiarly on shape- and size-dependent differential impacts.

Keywords
Nanoparticles · Physicochemical Properties · Size and shape tenability · Multifunctional Au NPs · Biocompatible · Nanorods

29.1 Introduction

Occupying a formidable place in cancer-related deaths across the globe, cancer of lungs is characterized by the most diverse genotypic diversity, which comprise few repetitive mutations prevailing in high frequencies. Though advancements in diagnosis and treatment procedures have improved the curative aspects concerning LC in the past two decades, yet the pathogenesis dependence on a couple of abnormal gene expressions has proved an obstacle to contain this curbing menace. Histological and clinical distinctions differentiate the LCs into broadly two distinct types, namely, small cell lung cancer (SCLC) and non-small cell lung cancer (NSCLC). Cumulative studies have estimated NSCLC as more vulnerable compared to its SCLC counterpart, with nearly 85% of the cancer deaths witnessed across the globe. This distinction seems an obvious reason for a much tedious pathological background of NSCLC, having adenocarcinoma, squamous carcinoma, large-cell carcinoma, and bronchoalveolar cancer as its subtypes, distinguished on the basis of pathogenesis (Larsen and Minna 2011; Davidson et al. 2013; Cooper et al. 2013; Roh 2014). Consequently, the cure of each different form differs substantially, and this is why presence of more than one form (noticed in extreme cases) has often been deemed critical in terms of treatment.

Till date reports suggest nearly 85% of LCs to be caused by the carcinogens present in cigarette smoke, though up to 25% of similar incidences are also reported to infect the never smokers. Though falling in agreement with the belief of passive smoking being more dangerous, the sole reason behind LC occurrence in non-smokers is equally aided by their susceptible genotype and the exposure to harmful conditions and environment, along with the geography of living area. A startling observation complementing association of smoking or smoking history with greater LC vulnerability is that nearly 50% of newly recognized LCs in USA are noticed in yester smokers, who quit smoking, but the aftermaths of previous smoking habits still lasted in varying proportions. Never-smoking LC incidence is reportedly more common in women and East Asians, having an uncharacteristic commonness in younger age and rather frequently developing into the state of adenocarcinoma (Larsen and Minna 2011).

Conventional treatments involving elaborate therapy subjections and exposing the patient to elaborate immunological sensitization have often proven to be saturated in their containment efficacies, owing to which a progressive scientific alignment to newer, safer, and more efficient treatments is swiftly emerging. Among many such alternatives, targeting inorganic NPs, either as such or in combination with the conventional anticancer agents through targeted active or passive delivery approach, has been in the scientific limelight for the past few years. The NPs of inert metals harbor several unceremonious functional aspects, encompassing the shape- and size-selective responses and exploitable thermal, chemical, and optical properties that make them functionally flexible with the prevailing tumor state and regime (Jain et al. 2006; Jain 1987; Link and El-Sayed 2010; Murphy et al. 2010). The key benefits of using these entities to treat cancers range from much less drug dosage intake with an effective delivery to the cancerous cells only. Such attributes play a pivotal role in distinguishing the NP-delivered anticancer drugs from the conventional strategies, where loss of life has primarily been due to an excessive chemical stress of the medications and side effects of misfired chemotherapy trials. The NPs of inert metals like Au, Ag, and many others can be functionalized with several ways to enhance the localized cell-specific delivery of tagged anticancer agent, thereby making the treatment not only effective but reducing its cost and patient sensitization, both. To top all such benefits is the relative ease of the available methods through which such NPs can be prepared. One would wonder subjecting the dye or textile industry effluent streams to being cultured fungal and bacterial species and in return, not only getting their smooth disposal but the NPs of constituent deleterious agents, being formed at the extra or intracellular regions. Similarly, treatment of precursor salts with leaf or fruit extract of edible plants can fetch us NPs in varying yields and sizes, on the basis of reaction time and precursor agent stoichiometry variations (Gupta et al. 2019; Vijayan et al. 2017; Malik et al. 2014; Rajeshkumar 2016). A number of other green approaches, such as use of ultrasonication, microwave treatment, and emulsification of precursor and reducing agent mixtures for varying durations, can provide NPs in distinguished yields (Kalishwaralal et al. 2008; Roth 1996; Rizvi and Saleh 2018). So, there are perhaps numerous approaches which can provide us NPs as low as (Massion et al. 2002; Friend et al. 1986; Nobori et al. 1994; Steck et al. 1997; Weir et al. 2007; Amos et al. 2008; Hung et al. 2008; Thorgeirsson et al. 2008; Normanno et al. 2005; Hirsch et al. 2003; Franklin et al. 2002) nanometers or perhaps, in some cases, even still lower limits. The benefits of a lower size are the high kinetic energy (that prevents its arbitrary aggregation or sedimentation) apart from a maximized quantum confinement effect. The size- and shape-dependent chemical performance of the NPs are the clear-cut fascinations behind their remarkable photothermal properties which contribute in triggered drug release at the targeted site.

With such a viewpoint, this compilation focuses light on Au NP-treated LCs in the past one year, alongside a fundamental understanding of LC pathophysiology, role of its microenvironment in complicating its onset and typicality, and, at the last, the mechanisms and factors on which the drug delivery to cancer cells depend.

29.2 Genetic Diversity of Lung Cancer

Every cell harbors a peculiar optimum pH, temperature, and chemical environment for its sustainable growth ensuring its maximized benefit to the host. In this reference, it becomes highly imperative to understand that cancer cells are quite different from their normal counterparts. Two primitive features distinguishing the cancer cells from the normal ones are (i) a higher expression of cell-division regulating factors in cancer cells and (ii) the uncharacteristically lower or inhibited activity of cell death-inducing apoptotic factors. Nevertheless, the effect of changing cell death- and apoptosis-inducing genes is not the same on all cells, since a cancer existing for 2 years is likely to have much higher anti-apoptotic gene expression than the one which is just 6 months old. In similar ways, the signals and redox environment around the cancer cells is much different from the normal cells. This is so as unregulated growth potential of cancer cells consumes all the fed oxygen being consumed rather too fast and so it is likely that an advanced-stage tumor is likely to be oxygen starved at a given instant of time. This abnormal aspect of cancer cells alters its communication with the neighboring cells, in terms of growth progression. So, on this distinction, several genes are noticed to be wrongly expressed in the LCs (either more or less, depends on peculiar tumor adversity). Typically, expression of two distinct gene types is recognized vis-a-vis cancer cell development, namely, oncogenes (those which promote the manifestation of tumors) and the tumor suppressor genes (those which must be active to kill the tumor cells). A number of literature sources can be viewed for the fundamental distinction, pertaining to the expression patterns of these genes (Merlino et al. 1984; Lynch et al. 2004; Paez et al. 2004; Pao et al. 2004; Massion et al. 2002; Friend et al. 1986; Nobori et al. 1994; Steck et al. 1997; Weir et al. 2007). The oncogenes promote the growth of tumor cells, whereas the tumor suppressor genes prevent them, so it is equally implicit to understand the fundamental growth-stimulating or growth-inhibiting mechanisms through which oncogenes and tumor suppressor genes operate. Tables 29.1 and 29.2 comprise the set of overexpressed oncogenes and tumor suppressor genes in LC with their specified repercussions. A careful look at the signaling pathways involved in the activation of the relevant oncogenes and tumor suppressor genes (TSG) infers that there exists a very faint line of their clear-cut distinction in their respective oncogene and TSG activation. The functional expressions of oncogenes and TSG are interconnected to each other as well as the vicinity redox balance in a very delicate manner since even a random activation of any oncogene happens through a proportionate inactivation or suppressed functioning of a TSG. Furthermore, even existence of mutations contributing to a common condition is not discrete. So, the intricate dependence of these signaling mechanisms on one another along with the specific oxygen in the neighborhood of cancer cells prevents the existence of any particular signaling mechanisms in isolation. Among the several known genetic diversities complicating LC pathogenesis, the SNPs at 15q24-q25.1 loci have been exclusively correlated with the increased risk of nicotine addiction and consequent LC onset (Amos et al. 2008; Hung et al. 2008; Thorgeirsson et al. 2008). So, the diagnostic procedures often involve a testing of more than one kind

Table 29.1 Major oncogenes active in LC manifestation with their operating mechanism, treatment success, and prevalence in specific cancer regime (Lynch et al. 2004; Paez et al. 2004; Pao et al. 2004; Downward 2003; Karnoub and Weinberg 2008; Krystal et al. 1988; Richardson and Johnson 1993; Soda et al. 2007; West et al. 2004; Hay 2005; Kendall et al. 2007; Bass et al. 2009; Winslow et al. 2011)

Oncogene involved	Mechanism and till date treatment success	Prevalence
Epidermal growth factor receptor (EGFR)	Often operate in association with tyrosine kinases, activation of P13K/AKT and STAT3/STAT5 pathways, fundamental leads of erlotinib and gefitinib mode of action	Adenocarcinoma histology, women, never smokers, and East Asian ethnicity, (50–90)% incidence in NSCLCs
The RAS/RAF/MEK/MAPK pathway	Mutations in KRAS (80% in codon 12) operate independently of EFGR signaling, antisense oligonucleotides and oligodeoxyribonucleotides against RAS and RAF, unsuccessful trials in farnesyltransferase inhibitors	Primarily observed in lung adenocarcinomas of smokers, BRAF is direct effector of RAS, commonly mutated in (~70%) melanoma
MYC	Downstream effectors of the RAS/RAF/MEK/MAPK pathway, affects cell cycle, often requires co-expression of anti-apoptotic BCL2 proteins, apoptotic sensitization to mitochondrial apoptosis pathway	Most frequently activated in NSCLC
EML4-ALK fusion proteins	Fusion of PTK echinoderm microtubule-associated protein like-4 (EML4) with anaplastic lymphoma kinase (ALK), crizotinibentered a phase III clinical trial	Found exclusive of EGFR and KRAS mutations, occur predominantly in adenocarcinomas and never or light smokers
The PI3K/AKT/mTOR pathway	Through mutations in PI3K or PTEN as well as EGFR or KRAS, amplification of PIK3CA, PTEN loss, or activation of AKT	Significant efficacy in both NSCLC and SCLC cells with activated AKT signaling
SOX2 and NKX2–1	Critical roles in adenocarcinoma (via 14q13.3 amplification) and squamous cell carcinoma (3q26.33 amplification), SOX2 functions in a lineage dependent manner	Primary NSCLCs

of genetic aberrations, and, similarly, the therapeutic procedures also target more than one kind of signaling mechanism, since in advance cases, there is a substantial possibility that two or perhaps even more genes are overexpressed. A fitting scenario of such an eventuality can be noticed for the simultaneously aggravated prevalence of PI3K/AKT/mTOR pathway in SCLC as well as NSCLC subjects. Hence, the treatment procedures involving PI3K/AKT/mTOR pathway seem to be fairly similar in both cancer types. Such intricacies and perhaps unavoidable abnormal gene expressions make it essential to look for other abnormally expressed genes. Similarly, EGFR overactivity has been addressed much more

Table 29.2 Major tumor suppressor genes (TSG) active in LC manifestation with their operating mechanism, treatment success, and prevalence in specific cancer regime (Takahashi et al. 1989; Hollstein et al. 1991; Greenblatt et al. 1994; Harbour et al. 1988; Yokota et al. 1989; Ito et al. 2005; Kuroki et al. 2003; Feng et al. 2008; Hemminki et al. 1998; Sanchez-Cespedes et al. 2002; Carretero et al. 2004)

Tumor suppressor gene (TSG)	Mechanism with till date treatment success	Prevalence/preferential targeting
The p53 pathway	Induces the expression of cyclin-dependent kinase (CDK), inhibitors regulating cell cycle checkpoint signals, frequent observation of 17p13 hemizygous deletion is noted for p53 inactivating mutations	Substantial targeting in treating human lung cancers via blocking MDM2 expression, MDM2 ubiquitin ligase activity and targeting the p53-MDM2 interaction with small molecule inhibitors
The CDKN2A/ RB pathway	Halts the G1/S phase transition by binding to the transcription factor E2F1 and was the First tumor suppresser gene identified in lung cancer	Mutant RB protein is found in approximately 90% of SCLCs
Chromosome 3p TSGs	Operates through inhibition of protein tyrosine kinases such as EGFR, PDGFR, c-Abl, c-kit, and AKT176 as well as inhibition of MDM2-mediated degradation of p53177	Found in 96% of lung tumors and 78% of lung preneoplastic lesions
STK11 (LKB1)	Third most commonly mutated gene in lung adenocarcinoma, exerts itself through inhibition of the mTOR pathway via AMP-activated protein kinase	Prevalent in NSCLC, inactivating mutations are more common in tumors from male smokers and poorly differentiated adenocarcinomas

successfully compared to others, with only EML4-ALK fusion proteins abnormal expression able to reach the Phase III clinical trial. Significantly, the prevalence of former in NSCLC contrary to EML4-ALK fusion proteins hints at a further higher chance of meeting EGFR abnormal expression at a given instant of time (Normanno et al. 2005; Hirsch et al. 2003; Franklin et al. 2002; Herbst 2004). On the same lines, the differentiated activities of tumor suppressor genes prevail in different stages and types of LCs, with luckily uniquely specific operating mechanisms. The STK11 (LKB1) activation through inhibition of the mTOR pathway oncogene presents a logical link up to predict that a common genetic pathway switching on or off is responsible for simultaneous upstream and downstream functioning of a particular oncogene and its complementary tumor suppressor gene (Ding et al. 2008; Carretero et al. 2004; Matsumoto et al. 2007; Mahoney et al. 2009). So, collecting the database of such treatment feasibilities for a better attempt of ongoing and futuristic clinical trials presents an auspicious domain of research and investigation. After gaining a fair understanding of tumor growth-activating genetic pathways and controls, the signaling mechanisms regulating a particular gene activity are very minutely affected by the redox state of tumor-surrounding regions. In order to have a clear understanding of these biochemical fluctuations, the role of tumor microenvironment is highly fundamental to understand, through which novel routes of drug delivery to cancer cells could be consolidated much more reliably. So, the next section is devoted to the role of tumor microenvironment events in controlling the growth potential of tumor cells and its exploitation for accurate clinical trials. The prevalence of oxygen balance at a given time span can affect the structural efficacy of targeted drugs through their reduction or oxidation, based on which the trials differ at different stages of tumor development.

29.3 Consequences of a Peculiar Tumor Microenvironment

Two most critical characteristics of any cancer cell comprise its unstoppable growth and its spread to the nearby tissues or healthy cells. Perhaps the second feature has been found to be critical in limiting the effectiveness of curative tumor therapies despite their identification of right pathological aspects. The spread or manifestation of a particular cancerous fate is highly sensitively dependent on the cellular signals and events prevailing in its vicinity. The traversing to neighbors (frequently referred to as metastasis) proceeds via formation of new blood vessels, a process which needs a careful assessment to monitor the pathogenesis of each and every cancer cell. The process is recognized by the term "angiogenesis," which is pivotal for the expansion of a tumor from the microscopic state to the macroscopic and perhaps the clinically relevant stage. Several angiogenic proteins are known to aid in the formation of tumor cell native blood vessels, such as vascular endothelial growth factor (VEGF), platelet-derived growth factor (PDGF), fibroblast growth factor (FGF), interleukin-8 (IL-8), and angiopoietins 1 and 2. Among these, VEGF serves as a prime initiator of angiogenesis and is well-known to stimulate proliferation and migration and apoptosis inhibition and regulate the endothelial cell permeability (Korpanty et al. 2010). Such events are acutely affected by the redox status of tumor cell vicinity since intra- as well as intercellular communication proceeds via hormones that are fundamentally the chemical messengers. So, therapies to treat tumors at different stages in their progression depend critically on the functions performed by the tumor surrounding physiological region, also referred to as extracellular matrix (ECM). The ECM is the source of plentiful membranous proteins and important protein mechanism switches, such as proteoglycans, immune cells, bone marrow-derived cells, myofibroblasts, and fibroblasts, along with endothelial cells, innate cells, stromal cells, and several others.

Numerous studies have notified the compositional peculiarity of tumor microenvironment in NSCLC, having altered expression patterns of ECM proteins, collagens, and laminin-5. Evidences prevail regarding the involvement of excessive laminin-5 expression in the development of a tumor population having high phospho-EGFR/phospho-AKT expression (Chow et al. 2010). Studies also talk extensively about the involvement of abnormal cross-linking with ECM as a key feature in complicating the LC pathophysiology. To counter such claims, investigations have noticed the excessive expression level of lysyl oxidase (LOX), a Cu (II)-dependent amine-oxidase cross-linked to collagen, against the response to localized hypoxial environment. With regard to peculiar NSCLC pathobiology, the LOX expression has been estimated to be correlated with Src-phopshorylation and the expression of a pro-EMT transcription factor, Snail, together contributing in conferring nearly five-year survival rates (Wei et al. 2012). In fact, the LOX-mediated collagen cross-linkage is recurrent in several cancers other than NSCLC and operates through altered integrin-FAK signaling. So, screening normal biological functioning of phospho-EGFR/phospho-AKT and integrin-FAK signaling are assured recognizers of NSCLC invasion. The particular role of α5β4 integrin activities in the complicated pathogenesis of LCs could be traced in one of earlier compilations (Mukhopadhyay et al. 2013).

So, signaling events like these and many others promise a potential of monitoring the native functioning of ECM interactions with neighboring cells, to check the metastatic progress of a particular cancer. For example, many evidences of altered protease expression levels are also known to affect the ECM functioning in many ways. In this regard, NSCLC are known to secrete enhanced levels of matrix metalloproteinases (MMPs), MMP-1, 2, and 9 (Sauter et al. 2008). Investigations on examining LC genetic diversity have found a role of single nucleotide polymorphism (SPM) exaggerated activities of MMP-9 and MMP-2, as exclusive predictive tumor developing and survival factors, respectively (Gonzalez-Arriaga et al. 2012). Likewise, abnormally excessive activities of A-disintegrin and MMP-8 and 28 have been noted in cancer affected lung cells compared to the normal ones, as also those of TRAF4 in the NSCLC cells compared to normal lung fibroblasts (Kim et al. 2015, 2017). Concurrent investigations culminated at the unified observation of higher and abnormal VEGF expression in NSCLC as well as SCLC, making the VEGF-associated tumor cell signaling a substantial susceptible therapeutic target to contain the LC metastasis. Two salient mechanisms aimed at achieving that aim on blocking the VEGF binding with its extracellular receptors, via abnormal expression of VEGFR233 loci. The studies used VEGF-specific antibodies and recombinant fusion proteins (using TKIs), to forbid native VEGF binding to VEGFR233 receptor. With this success, the humanized monoclonal antibody (bevacizumab) (Mab) blocking the VEGF-A binding to its receptors, VEGFR1 and VEGFR1 gained approval for LC treatment. Still surprising fact is the failure of this Mab functioning in case of missing VEGF programmed expression pattern caused due to recurring SNPs (Watson et al. 2000; Heist et al. 2008). Thus, the tumor microenvironment acts as a crucial tracker of tumor cells via characteristic expression recognition of plentiful biomarkers in each developmental stage. The only hurdle in efficient exploitation of the unique biochemical variations in tumor vicinity is gaining the most accurate knowledge of receptor modulations caused by changing redox environment. Sometimes, even the polymorphisms are also caused due to sharp oxidative stress and oxygen starvation events near the tumor cells. Such challenges open several explorable gateways to further the research for arresting the troublesome LC physiology.

29.4 Specialty of NPs and NP Drug Delivery for Anticancer Viewpoint

Terminologies regarding consensus of a particle size limit fit for the nanoscale widely differ across the globe, and so far, there is no unanimity except the belief of an upper limit of 100 nanometers as the point below which nanoscale attributes express in significance (United States Department of Agriculture 2002; Nakache et al. 1999). The NPs derive their unique behaviors from the fundamental controls of their quantum mechanically distributed constitutional energy levels. Several studies and review articles outline the distinguished abilities of Au NPs in bettering the results of conventional tumor therapies, which comprise their shape- and size-dependent functional activities, abilities to be engineered for both active and passive delivery modes and inherent low stability (of specifically the Au NPs). Apart from these two qualities, the photothermal conducting nature of Au NPs serves as incentive in their selective tumor cell targeting and elimination. The easily controllable geometric features make it much easier to control and track the Au NP cellular distribution once they are inside the body, a prospect not so well observed with other elements in same group of periodic table. The major reason for using NPs to deliver drugs to the tumor cells is that the same drugs, once administered orally, are lost to a major extent before reaching their target cells (due to one-way proteolytic and tissue-specific varying degradations). Owing to this, conventional drug delivery incurs with the lame risk of consuming more drugs and concurrent manifestation of a higher chemical stress on the host body. With NPs, site-specific drug delivery along with much less drug quantities can be accomplished, that makes the overall drug delivery process much more effective, although commercial feasibility is yet to be gained for all diseases.

Figure 29.1 outlines the most practiced Au NP synthesis method, first documented by Turkevich and colleagues in 1951 (Turkevich et al. 1951). The method uses precursor salt and attains its chemical reduction using sodium citrate, with progressive studies reporting enormous modifications in the formed NP features, on the basis of temperature and reacting species stoichiometric variations. Sodium citrate performs the dual role of reducing and agglomeration stabilizing agent. Details of Turkevich method and its subsequent modifications to obtain different shapes and sizes (of Au NPs) could be traced in several other relevant literature sources (Hauser and Lynn 1940; Frens 1973; Freund and Spiro 1985; Takiyama 1958; Pei et al. 2004; Ojea-Jiménez et al. 2010). Owing to the non-feasibility of the subject concerned, here the discussion would be confined to size and shape dependent drug delivery attributes of Au NPs along with their active and passive administration to the tumor cells. Before looking at the specific efficiency in active and passive delivery modes, gaining understanding of the mechanisms of these drug delivery approaches would be a boost. So, further details of this section comprise these two aspects, with an intention to clarify the size- and shape-dependent drug delivery efficacy.

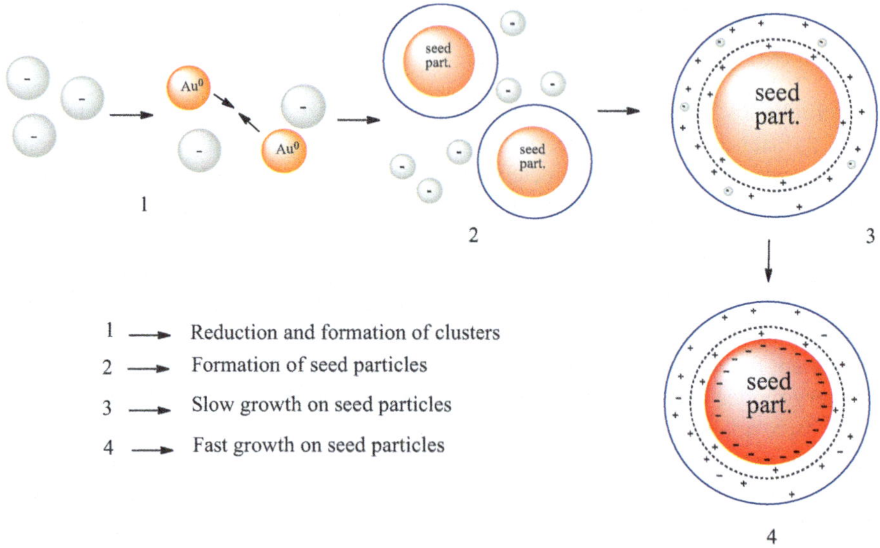

Fig. 29.1 Schematic depiction of Turkevich Method for Au NP preparation. Figure depicts a seed-based synthesis method during which growth is controlled via capping agent activity

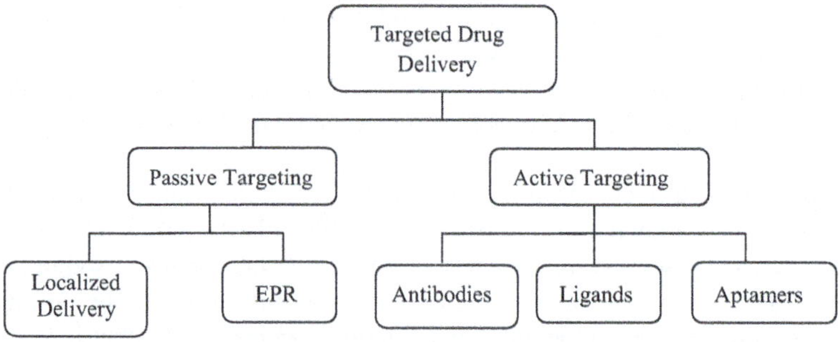

Fig. 29.2 The fundamental pillars of active and passive drug delivery routes, using NPs

29.4.1 Active and Passive Drug Delivery to Tumors

The tumor cells significantly differ from their normal counterparts, in terms of the biochemical activities of their microenvironment and their growth progression. The very first distinguishing aspect is the unrestricted growth of tumor cells, primarily caused by the absence of cell cycle and apoptosis-regulating factors. Along with this, the tumor cells harbor an unusual highly disorganized vasculature and congested extracellular environment in their near vicinity. Two distinct mechanisms (outlined in Fig. 29.2) are recognized for targeted drug delivery to tumors, quite similar to the analogy of transport across the cellular membranes. The more conventional and risk-prone mode is the active mode, characterized by the delivery of ligand encapsulated or conjugated drugs to the specifically overexpressed receptors on tumor cell surface. Once the ligand-carrying drug(s) meets the targeted receptor, it is engulfed and progressively internalized within the tumor cell. The second mechanism relies on the facilitated intracellular entry of drugs and is thereby known as the passive drug delivery method. The use of NPs in the active drug delivery mode is undertaken via binding the ligand containing encapsulated drug (in the NP or NPs). So, it becomes that for entry of drug inside the tumor cells, the ligand-receptor interaction should be maximum which would release the drug from the bound NPs on its surface or alternatively the NP-drug combine, which subsequently moves inside the cell. The ligand-cell-surface receptor binding is sensitively affected by the peculiar NP shape and the targeted cell-surface receptor nature, the peculiarities of which can be traced in Tables 29.3 and 29.4, respectively. The architecture of a NP has a highly sensitive association with the drug volume it can bind, interact, and accommodate so that the risk of undesired drug loading on healthy cells is minimized. Secondly, the targeted cell-surface receptor should also bear a close geometrical relationship with the specific ligand-drug conjugated assembly, so that only a particular cell-surface receptor is activated in response to bound drug. The third aspect is that no single drug is capable to function with same efficacy in different stages of tumor as different proteins are expressed in varying extents at different stages. So, for a wholesome containment consideration, a ligand should be capable of inducing multiple drug binding with no interference in the functioning of a particular drug. Quite often, it has been noticed that a mixture of drugs (in synergistic association) performs better than the single entity and also has a benefit to overcome the concurrent event of resistance.

Table 29.3 Shape diversity of Au NPs, with geometrically suited applications for varying morphology. The rod and shell morphology attract special interest due to their larger surface area (aiding in photothermal activity) and adsorption capacity (to carry sufficient drug quantities)

Characteristic shape/ morphology	Unique specialty	Positive aspects in drug delivery along with critical challenging prospect
Nanowires	Easier engineering of elongated surface	Enzyme immobilization suitability, challenge to remain stable in native form
Nanoshells	Excellent absorptive capacity	Peculiar suitability for engulfing destroyed tumor cells (via housing), may cause residual toxicity due to inefficient physiological elimination
Nanodots	3-dimensional quantum confined structures	Size-sensitive sensing behavior, suitable for fluorophore design, risk of being absorbed by living cells
Nanorods	Length prospect enables multiple functionalization	Elongated surface with high ARs results in better tumor internalization
Nanospheres	Hollow spheres for efficient absorption, conjugation and adsorption abilities	Can be used to develop multifunctional photo thermal on/off mechanisms, flexible geometry, using sideways reactions due to smaller size

AR aspect ratios

Table 29.4 Salient overexpressed cell-surface receptor proteins in lung cancer. These receptors are active members of active drug delivery approach and work on distinguished drug internalization mechanisms within the tumor cells (Carretero et al. 2004; Bhirde et al. 2009; Baker et al. 2002; Eren et al. 2008; Xie et al. 2016; Wu et al. 2018; Zhao et al. 2008; Sudimack and Lee 2000; Xie et al. 2014)

Sr. No.	Peculiar cell-surface receptor and examined cancer cell line	Impacts on cell growth and progression with most recent clinical application
1.	Epidermal growth factor receptor (EGFR), examined on A549 cell line	Cell-surface tyrosine kinase receptor responsible for cell growth, division, and proliferation; chitosan nanoparticle targeted at EGFR caused efficient silencing of mitotic checkpoint, Mad2, and cell-death onset
2.	CD44, exclusively expressed in NSLC, hyaluronic acid (HA) is the specific receptor, evaluated in A549 and SK-LU-1 cell lines	Successfully targeted for (30–60)% tumor growth inhibition via synergistic cisplatin and siRNA activities, induces macrophage repolarization and p53-apoptotis
3.	Transferrin, type II transmembrane glycoprotein, studied on H460 human lung cancers, expressed in 76% LADC and 93% SCLC	Limited therapeutic success, less use in lung cancer treatment, required for intercellular iron transport and growth regulation, specifically suited in treating blood-brain barrier (BBB) disorders
4.	Folate receptor alpha (FRA)	Overexpressed in 74% LADC and 13% SCLC, successfully targeted via polyspermine NPs conjugated with folic acid ligand, FA-conjugated chitosan graft-PEI used for small hairpin RNA delivery to cancer cells, induced stable sh-RNA condensation and protection from DNAse
5.	Integrins, evaluated on H1299 cell line	$\alpha5\beta1$ integrins present in all NSCLC, implicated successfully for improved diagnosis via PEG-grafted nanoparticle surface. Initiates EGFP gene silencing in human cell lung carcinoma (H1299), PLGA nanoparticle along with VEGF inhibition in a mouse model

LADC lung adenocarcinoma, *SCLC* squamous cell lung cancer, *PLGA* poly (lactic-co-glycolic acid), *VEGF* vascular endothelial growth factor

Erstwhile of active drug delivery approach, the passive mechanism relies on the gradients of physiological stimulus to propel the drug inside tumor cells. The facilitated transport is attained via concentration differences of cell-specific essentials, which could be bound with the carrier drug of interest. To systematically understand the facilitating stimulus regulating drug entry inside the tumor cells, a process termed as "enhanced permeation and retention (EPR)" must be regulated in proper manner. This mechanism takes into consideration the different factors responsible for drug internalization within the cancer cells (through the permeation strategy), and once the drug enters inside the tumor cell, it must spend a considerable time there so that its expression is adequate enough to destroy the tumor growth (this part is referred to as retention mode). In crux, this EPR mechanism also exploits the flawed tumor microenvironment, making extensive use of leaky vasculature and impaired lymphatic drainage of the tumor cells (Prabhakar et al. 2013; Matsumura and Maeda 1986). The mechanism holds substantial relevance for understanding the intracellular trafficking of larger NPs, since it is often difficult to retain the externally maintained lower sizes of drug-vehicle combinations within the physiological environment (Maeda 2001). The understanding of EPR considerably relies on the differences in tumor cell communication (inward and outward movements of the stimulus) in the tumor cells and normal cells. The most important factor contributing to significantly greater growth rate of tumour cells in the initial stage (of their development) is the elevated rate of angiogenesis, caused by the growth factors secreted from the gradually depleting normal cell population (Bates et al. 2002; Jain and Stylianopoulos 2010). The details of the EPR affecting physiological parameters can be looked for in a significant review by *Bertrand* et al., having its focus sheerly on the active and passive drug delivery to cancer cells (Bertrand et al. 2014).

The passive delivery of drugs to the tumors is generally less risky and addresses the challenge of cell-specific delivery with a greater caution. Furthermore, this mode also requires significant extent of interactions in the vicinity of cancer cells, where the conditions like prooxidant or antioxidant scenario of tumor microenvironment gain significance. This is because the carrier vehicle in the passive drug delivery should also ensure that the drug enters only inside the tumor cells and causes no damage to normal cells. For this purpose, the passive drug delivery mode should be capable of quicker drug internalization within the tumor cells along with the capability of slow and gradual release at the desired site only. Figure 29.3 distinguishes the action mechanism of active and passive drug delivery approaches where key difference is in the manner in which tumor cell is approached by the drug carrier. Here, it is important to mention that unlike the active mode, the passive mode does not restrict the carrier vehicle dimensions and can even internalize the NPs as large as >200 nm, although it is now well-known that the cross-sectional diameter of leaky tumor blood vessels is of the order of 600 nm (Bertrand et al. 2014). Therefore, the mechanism of passive transport substantially differs from active mode, where the carrier vehicle surface charge and physical nature (being coated with hydrophobic or hydrophilic monolayers) also have a significant consideration. In general, the positive-charged carrier vehicles are preferred since the cell membranes have negative polarities conferred by the phospholipid bilayers. Studies on passive delivery have noted higher tumor cell retention for the rod-shaped NPs due to their elongated surface and larger dimensions, although it seems ironical that a larger-sized foreign entity seems more likely to be caught or detected by immune cells (MPS tissues) as being a foreign entity (Chauhan et al. 2011). Perhaps, the scenario seems like after escaping the immune detection, entry, the elongated objects are not easily eliminated from within environment. Compared to this, the spherical shapes do escape the immune detection much easily but are also likely to initiate arbitrary or random interactions owing to their high kinetic energies. In all, a terminal stage cancer is more advisable to be treated using the active mode, since the risk in proceeding for a long-term relief is much preferable compared to longing for death. It is also

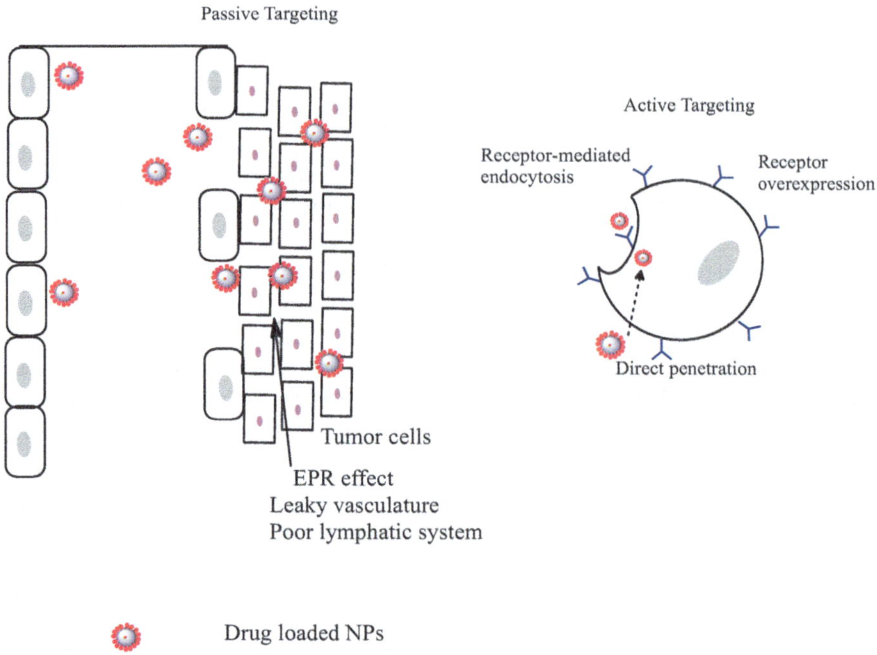

Fig. 29.3 An overview of active and passive drug delivery mechanisms, with former involving direct targeting of the ligand complimentary tumour cell-surface receptor, while the latter is facilitated approach encompassing drug internalization via changes in physicochemical properties of tumour microenvironment

fundamental to note here that conventional chemotherapies are active drug delivery systems, where even a minute risk of misfired or inapt drug delivery leaves the sufferer badly sensitized and more vulnerable to a host of chemical and physical stress abnormalities.

A detailed discussion on the NP factors affecting the success of a peculiar adopted approach is presented in the next section, while more variables affecting active and passive drug delivery can be looked in manifold extensive literature sources (Bertrand et al. 2014; Fahmy et al. 2005; Wakaskar 2017; Bazak et al. 2014, 2015; Sanna et al. 2014).

29.4.2 Key Nanoparticle Attributes Affecting the Drug Delivery Efficacy

The compositional make up of drug delivering NPs necessitates them as entities having inorganic interior (core) surrounded by the aggregation preventing organic monolayer on the surface. A smaller size confers a higher kinetic energy and thus, low sized NPs are more active and induce much more interactive interfaces compared to integrated larger counterparts. As far as particular size effects are concerned, it is imperative to note that even minor alterations in sizes can have serious effects on their properties. It is well known that lesser size confers a higher surface area to volume ratio to a NP, creating a pressure on the existing surface to accommodate a similar volume previously accommodated on a larger surface. Ultimately, a lower size leads to a dominant contribution of surface atoms compared to those in the bulk. Another important constraint regarding NP administration to the physiological environment is that these materials must not be recognized as foreign so that no sharp immune response is initiated against them. Therefore, prior to being delivered, it is ensured that NPs are surface coated with a biocompatible material not posing any toxicity or perhaps bearing a reasonable resemblance (opsonization) with the native biological environment. Owing to this kind of make-up, the prepared NPs have distinct roles of their bulk and surface atoms, with the former controlling their chemical interactions with the surroundings (here biological environment) and the physical properties being defined by the bulk atoms. The coated aggregation preventing monolayer on the surface serves as the barrier in between the NP bulk and surrounding stimulus. For the drug delivery purpose, it is a must that this monolayer enhances its aqueous solubility enabling an incorporation of biocompatible ligands. The most generally preferred ligands for this purpose are peptides, proteins, and DNA. Studies on size- and shape-dependent NP drug delivery illustrate a comparatively more critical role of their shapes in distinguishing their delivery efficacies and the studies on size-driven performance activity are not exclusive. Unlike passive mode, the active mode is risky and affected by a lot of closely related geometrical and physicochemical factors, such as shape, size, surface nature of carrier, etc. The risk in active mode is accompanied by the overall economic complications regarding the identification of the target receptor. Ideally, the receptor that is overexpressed in most of the tumors should be preferred since, otherwise, the event of misfired drug delivery could prove life threatening. The text ahead sheds light on major NP attributes affecting its drug delivery efficacy in the active mode, including size, shape, surface charge, and hydrophobicity.

Size The unanimous agreement prevailing regarding size is that larger NPs are likely to exhibit a faster sedimentation compared to their smaller counterparts. While smaller-sized particles can easily enter within the leaky blood vessels of tumor cells, there is a substantial possibility of their adverse reactions via entry to the normal cells too. Contrary to this, larger NPs can't move so freely within the physiological environment, due to which their distribution in the bloodstream remains fairly controlled. However, larger-sized NPs can prove tedious to be controlled once inside the live cells, and due to this, there is a substantial possibility that they may lead to undesired interactions. The episodes of larger NPs giving rise to inflammation-mediated toxicities via elevated oxidative stress in the local environment have been witnessed on a significant level, in mice and rat models. Relatively fewer studies have investigated the impact of sizes on the differences in drug delivery efficacy since two distinct studies (2011 and 2013) and observed comparable distribution for 30, 50, 70, and 100 nm Au NPs in the hyperpermeable murine colon adenocarcinoma (Cabral et al. 2011; Huo et al. 2013). However, in poorly permeable cancers, such as human pancreatic adenocarcinoma, only <70 nm NPs could accumulate efficiently, subsequently found to be dependent on tumor shrinking efficacy, tumor biology, and the relative endothelial permeability. Similarly, Wong et al. (2011) noticed the disintegration of 90 nm NPs into 10 nm quantum dots upon suffering degradation by the tumor proteinases, which subsequently improved their deeper intertumoral activities compared to the quantum dots loaded on similar-sized non-sliceable silica NPs. To make the NPs biocompatible, often polyethylene glycol (PEG) coating is administered on their surface, which imparts them extended stability and prevents the risk of any cross-reactivity (Cabral et al. 2011). In one such study, Perrault et al. investigated the cellular uptake tendencies of 20, 40, 60, 80, and 100 nm Au NPs conjugated with PEG, where it was noticed that half-life of smaller NPs (20 nm) was eight times (up to 51 hours) enhanced upon being coated with PEG. Subsequently, it was noted that larger-sized particles accumulated better within the tumors, and the results were in agreement with earlier observations of *Dreher and group*, arguing that small particles are not suitable for accumulation within tumor matrix due to their high diffusion rate (Perrault et al. 2009; Murakami et al. 1997; Dreher et al. 2006). Some other studies also distinguished the NP size-dependent cellular internalization abilities, with smaller sizes (up to 50 nm) able to travel deep within the tumor cells while those having >100 nm size remained along the cell surface. Similarly, larger-sized NPs were often seen to be fairly distributed within blood, liver, lung, spleen, kidney, brain, heart, and stomach, while the larger ones remained confined to the blood, liver, and spleen. These observations make it evident that smaller particles gain easier cell and tissue entry and escape immune detection much easily compared to the larger ones, which could otherwise manifest in the local toxicity through elevating the inflammatory responses (Dreher et al. 2006). With few reports on NP size dependent LC-specific treatment, the best generalization could be made through considering size effects on other cancers with tumour specific biochemical environment along with its age playing decisive roles.

Shape Figure 29.4 describes the shape diversity of Au NPs, with the characteristic specialty of the peculiar morphology illustrated in Table 29.2. Presently, the rod and spherical morphologies are in active consideration to deliver the drugs within the cancer cells, and the performance has been found to be very sensitively dependent on the aspect ratios of a particular design. For example, despite having nearly 10 to 20 times higher dimensions than glomerular filtration upper limit, the single-walled carbon nanotubes were found to be efficiently eliminated by the kidneys, making it logical to believe that rod shapes have higher tumor internalization ability than the spherical ones (Ruggiero et al. 2010). These beliefs are in strong agreement with the experimental findings, reporting nearly 4 times faster internalization of rod shape to the deeper regions of the tumor and concurrently interacting with nearly 1.7 times higher surrounding volume (Chauhan et al. 2011). Similarly, results of another interesting study offer valuable leads in this regard, where spherical and rod-shaped Au NPs were intravenously administered to mice models carrying

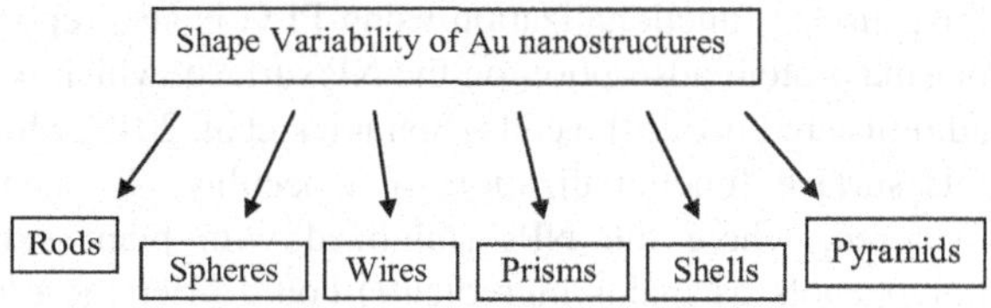

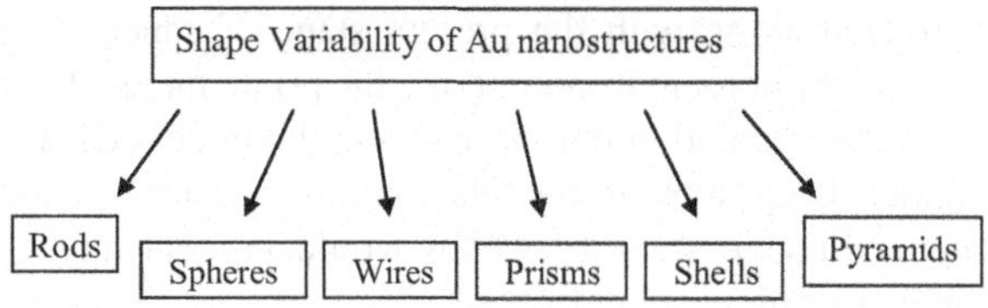

Fig. 29.4 The different shapes of Au nanostructures employed for drug delivery. Each of the shapes has its own specialty and therefore harbors particular reasons for incidental preference

ovarian tumors. It was noted that rod-shaped particles possessed longer blood circulation times, had lower uptake by liver, and were tumor internalized to a higher extent than spherical NPs. The cellular uptake studies also noted a much lower detection of rod-shaped NPs by the macrophages, making the rod-shaped particles superior drug delivery agents (Arnida et al. 2011).

Surface Charge and Hydrophobicity Binding a ligand to NPs in their native state is not a simple task and therefore requires the NPs to be in interacting proximity toward the ligand. Fundamentally and highly practically, the ligand-NP interactions must be stabilized through moderate interactive forces, since too strong interactions would hinder the release of drug from the bound NPs via preventing the NPs to act freely. It must be noted that the sole purpose of ligand is to bring the NP carrying drug near the cancer cell required position, so that the release of drug happens at the right place and in right manner. So, it assumes paramount important that ligand-NP binding should not cause any deviation in the native structure of ligand which could prove a hurdle in its recognition of complementary cell-surface receptor. To ensure that ligand-NP interactions remain mild and do not adverse the molecular structures of each, usually a molecule having distributed hydrophilic and hydrophobic chemical constitution is intervened between the two. Till date, the most common spacer of such type is the polyethylene glycol (PEG) although synthesis of ligand-NP assemblies having PEG linkages has not been easy. More importantly, the redox sensitive environment in tumour vicinity affects the ligand-NP binding in several ways, necessitating an in vitro stability evaluation of stable association. Ideally, the binding should be moderate but must not be weaker enough which loses its therapeutic essence amidst the pH, temperature and oxidative effects of specific tissues. Several studies agree with these possibilities, where NPs entirely without steric stabilization and PEG surface functionalization have both lost their specific activities to bind the requisite target receptor. For instance, silica NPs functionalized with transferrin without any steric stabilization on their surface, when administered against transferrin receptor, lost their specificity toward transferrin receptors, due to adsorption of plasma proteins on their surface (Salvati et al. 2013). Contrary to this, surface functionalization using PEG is also reported to delay the opsonin and plasma protein adsorption on the NP surface, which is dire essential before they are administered inside the cells (Stefanick et al. 2013; Mori et al. 1991). So, optimizing this surface functionalization of a peculiar NP seems an interesting area to be explored, where the NPs stabilized with plant extracts (carrying amphipathic polyphenols as major ingredients) could offer more advantages from the in vivo handling point of view. Similarly, the specific cancer cell to which the NP is to be delivered along with the present stage of cancer (which would critically affect a particular microenvironment) are the prime factors for which the delivered NP would be in the circulation before reaching the target cell. In this reference, computational studies using docking models could be a much-needed advance to design NPs with higher in vivo stability which would prevent any cross-reactivity and ensure its target-specific drug delivery.

29.5 Past One Year Studies on au NP-Treated Lung Cancers

This section describes the past one year salient attempts to treat LC using varied shapes and sizes of Au NPs. Discussion on the past one year studies is presented with an intent to gain knowledge and awareness about the latest developments. The vulnerable pathophysiology of LC make it further essential to know about the latest status of completed investigations so that a track record of any new emergent marker is aptly taken into consideration. The most recent study portrays the near infrared (NIR) absorption of aptamers modified Au nanoshells (NS) to target Mucin 1, an overexpressed transmembrane glycoprotein in LC. The scientists examined the effect of aptamer-stabilized hollow Au NS (targeted against Mucin-1) toward the spheroids cultured in a microfluidic system that resembled an early-stage nonvascular tumor, in terms of morphology, concentration differences of oxygen, cellular interactions, and cell growth kinetics. Evaluating the toxicity effect of NIR-activated Au NS on the normal (MRC-5, MCF-10A) and tumor cells, the scientists observed its enhanced accumulation within the tumor cells through EPR mechanism. This conclusion was drawn on the basis of reduced size and viability of tumor cells compared to the normal cells upon being administered with aptamer-stabilized Au NS. Control studies confirmed this suppression as significant only in the presence of NIR, making it feasible to be called as photothermally aided Au NS anticancer effect (Kalinowska et al. 2019). Another potential investigation based on nearly same theme examined the dose optimization effects of Au NPs in the distinct irradiation sources, Iridium (^{192}Ir) based brachytherapy and Xoft Axxent based Electronic Brachytherapy (eBx). To study the effect of Au NP inclusion on their moderated durations, the scientists administered 1 mM Au NPs (of 15 nm) to the lung (A549) and prostate (DU145) cancer cell lines. Analyzing the dose enhancement factors of Ir and eBx brachytherapy, the scientists observed the former to be enhanced by 1.54 compared to 2.06 for the latter in A549 cancer cells, indicating a stronger dose enhancement in the eBx-based brachytherapy. These differences are due to inherent lower average energy of eBx brachytherapy (closer to optimal energy required for drug enhancement) compared to Ir brachytherapy. In general, Ir brachytherapy is employed for LC treatment, while the results of this investigation project eBx brachytherapy as a better treatment option due to its higher dose enhancement conferred via Au NPs inclusion. So, clinicians and cancer patients, getting Ir brachytherapy, can heave a sigh of relief since similar anticancer efficacy could be achieved with much lower eBx brachytherapy (Shahhoseini et al. 2018). In another versatile effort, *Malekmohammadi* et al. immobilized Au NPs on folic acid decorated mesoporous networks of silica-coated reduced graphene oxide nanosheets. This study also manifests a breakthrough attempt to the combinatorial advancing of reduced graphene oxide (rGO) anticancer attributes. The scientists characterized the prepared GO nanosheets (having Au NPs attached to silica coating) for their physical and chemical properties using spectroscopy, electron microscopy, and surface area measurement, which found the prepared nanocomposite (NC) to have good biocompatibility, biodegradability, and high surface area aiding in adequate drug-loading abilities. The scientists studied drug delivery efficacy of prepared nanocomposite via its curcumin-delivering potential to A549 human LC and MCF-7 breast cancer cell lines, where free curcumin, NC without curcumin, and curcumin-loaded NCs without folate conjugation were examined as controls. The analysis revealed a highest cytotoxicity for folate-conjugated curcumin-loaded NC due to higher cellular uptake of folate-conjugated drug vehicle, confirmed via fluorescence microscopy and variable staining techniques. Interestingly, folate conjugation was found to play decisive roles in conferring suitable curcumin loading capacity, pH-controlled release, and photothermal and active targeting abilities, which was held in strong agreement by the better results in the breast cancer studies, owing to their larger folate receptor surface expressions. So, this study explained the utility of Au NPs in targeted toxicity of curcumin with aid from another nanocarrier (silica-functionalized rGO nanosheets) in the lung and breast cancer treatment (Malekmohammadi et al. 2018). The structural versatility and biocompatibility potential of rGO has been on the exclusive scientific radar (since molecular interactions through higher number of binding loci). In this reference, *Kadiyala* et al. made use of Syzygium cumini seed extract to achieve the simultaneous chemical reduction of graphene and Au NP precursor, $HAuCl_4$ (Kadiyala et al. 2018). Following the reduction process, the structural and morphological characterizations predicted a stable formation of Au NP-rGO NC. The scientists also noted significant anticancer potential of the complex by evaluating its activity on A549 (lung) and HCT116 (colorectal) cancer cell lines, where stronger activity against A549 cells was noted. The anticancer activity was ascertained to be expressed via induction of reactive oxygen species (ROS). Such studies offer green solution via efficient usage of bioactive polyphenols in Syzygium cumini seeds, replacing the need of costly chemical reducing agents, where additional risks of nonspecific and uncontrolled toxicity may also crop up.

Another splendid effort from Spain illustrated the application of epitaxially grown Au NRs stabilized with poly(sodium-4-styrenesulfonate) (PSS), poly-L-lysine hydrobromide (PLL), and hyaluronic acid (HA) to deliver doxorubicin into the cancer cells. Though the study is not exclusively conducted on LC cells, the outermost layer of fabricated nanorod (NR) assembly carried HA which is a specific ligand of the overexpressed CD44 cell-surface receptor on the LCs. One important aspect of this study is the combinatorial and controlled drug delivery efficacy via exploiting photothermal properties of elongated surface of rod-shaped NPs. The adjacent layers of polymers incorporated the benefit of encapsulating variable drug payloads that aided in sustained drug release and enabled a smoother drug delivery than inducing sudden abrupt toxicity (Alvarez et al. 2018). An attempt nearly similar to this made use of biocompatible HA in achieving paclitaxel targeted delivery via coating on bovine serum albumin (BSA) stabilized Au nanocluster (NC). The scientists examined the delivery efficacy of paclitaxel in 4 T1 and A549 LC cell lines and observed a better performance in the former due to low HA-specific CD44 expression in A549 cells. The tumor internalization studies noted a higher cellular residence of the BSA-stabilized HA-coated Au NC even for 200 nm size, a surprising observation pertaining to the internalization of size as high as 200 nm (Liu et al. 2018). Here, the use of EPR along with structural compatibility of HA (having ubiquitous –OH groups) played a crucial role in cellular internalization of even 200 nm carrier vehicle through stabilized hydrophilic-hydrophobic force gradients. This and the last study offer significant leads to still investigate the most stable interaction mechanism through in vitro computational docking so that the cost and time expenditure in implementing experimental trials could be mitigated.

Another worthwhile study by *Liaskoni* et al. has used mercaptooctanoic acid (MOA) and folic acid (FA) functionalized Au NPs loaded with paclitaxel nanocarrier, encapsulating pH-sensitive poly(2-vinylpyridine)-b-poly(ethylene oxide) (P2VP-PEO) vesicles (Liaskoni et al. 2018). The study aimed to achieve a long-term stable paclitaxel action to treat LC via encapsulation within <200-nm-sized nanovesicles, evaluated for intracellular entry within A549 cells through EPR mechanism. The vesicle preserved paclitaxel spilling in extracellular locations and enabled a favorable release profile in acidic environment. The flow cytometry and confocal microscopy examination revealed a time-dependent paclitaxel intracellular entry alongside reducing its localized toxicity, compared to its unaided delivery. In one of the most rigorous 2018 effort, *Miller and Frieboes* have put forward the response of tumor cells to the targeted drugs, through a one-to-one computational modeling-based tumor microenvironment screening. The scientists have used implicit computational modeling to monitor the variances of oxygen and nutritive transport, NP intracellular trafficking, and the drug internalization, where heterogenicity of tumor tissue vicinity has been noted as one of the most significant factors affecting the movement of drugs and nutrients into the leaky vasculature of tumor cells. The study found a nonlinear relationship of NP tumor cell penetration with the microenvironment heterogeneity, providing highly authentic leads regarding the drug strength and its anticancer efficacy. The scientists have argued that considering tumor heterogeneity as a function of vascular density, a drug efficacy pertaining to higher drug strength is highest for the tumors that exhibit least heterogeneity. Similarly, it was observed that for weaker drug strengths, a drug operates best against tumors having high vascular heterogeneity. The studies have taken special account of LC, recalling the attempts to modulate the vascular density for attaining an improved response, in course of NSCLC treatment using small molecule tyrosine kinase inhibitors or monoclonal antibodies targeted at alterable

VEGF expression. Likewise, the scientists argue for attempting angiogenesis manipulation in case tumor vascular density is inadequate to effectuate a desired anticancer response of nanotherapy (Miller & Frieboes 2018). So, this study provides multifold explorable inputs to regulate the drug delivery efficacy using nanocarriers. Although we have covered extensive studies aimed toward using Au NPs LC treatment, few more could be traced in one of our recent contributions (Malik and Mukherjee 2018).

29.6 Cautions Regarding Opportunistic Nanoparticle Toxicity

With an atomic number of 79, >1000 K melting and boiling points, the physical stability of Au in the native state seems reasonably justifiable. The outermost electron of Au resides in 6 s subshell, too far from its nucleus. But after its loss, the shells are completely filled till 4f, and studies predict the existence of these electrons at an energy which mostly remains unmatchable with other substances due to relativistic effects (Schwerdtfeger 1989; Yatsimirskii 1995). In fact, a high ground state chemical stability is the reason behind the existence of highly diverse shapes of Au at the nanoscale. This is further supported by its preferential choice to deliver the drugs compared to other metals; however, merely these aspects do not define the risks associated with Au NP drug delivery. The pacing of clinical trials for successful launch of a peculiar biomarker targeted anticancer drug is a significant indication of the associated constraints. The key issues which even now remain unanswered comprise the aftermaths of the situations when NPs get entrapped within liver, kidney, and spleen in case these are not efficiently internalized with the cancer cells. Though significant progress has been made in improving the surface protection mechanisms, and biocompatible materials limiting the cross-reactivity are being swiftly evaluated, the results are slow in deliverance and therefore mandate a cautious consideration. This is so as a lot of in vitro successful evaluations get collapsed in the in vivo environment. So, strategies to improve pH and temperature stability along with the optimum fittings of size, shape, and surface charge need a thorough attention so that either the desired site drug delivery is enhanced or, even if it is not met, there should be any undue chemical stress to eliminate the NPs through normal biological mechanisms. Studies reporting oxidative stress like events in few animal models do catch the attention in this regard (Pourali et al. 2017; Lee et al. 2013; Lasagna-Reeves et al. 2010). Nevertheless, better controls and more realistic trials (focused on having higher structural resemblance of ligand and receptor) are in active consideration to address such concerns. These concerns remain substantial in the actively targeted therapies, and studies have elucidated that bettering the MPS and macrophage escape can remarkably enhance the tumor internalization of the administered drugs (Farokhzad and Langer 2009; Qiu et al. 2010; Wilhelm et al. 2016). It is important to note here that most of the drugs operate by enhancing the oxygenated conditions (prooxidant hypothesis); the only challenge is to attain such deliverance at the requisite site.

29.7 Conclusions

Containment of LCs via Au NP drug delivery resolves several challenges of conventional drug delivery, especially multidrug resistance and systemic toxicity. The potentialities of shape- and size-driven physicochemical responses bestow Au NPs as reliable need-based therapeutic agents which not only deliver drugs effectively but also aid in the earlier diagnosis of cancers. The factors in progression aim at bettering the ligand-receptor binding activities via non-covalent and non-electrostatic linkages as well as bettering the cellular internalization facilitating externally modulated EPR stimulus. A primitive area of concern is the improvement in correct identification of target so that there is no adverse activity of delivered drugs. Equally essential is the need to effectively eliminate the carrier vehicle and to modulate the carrier design with the rationale to have minimized risk of cross-reactivity. In a nut shell, Au NPs could pave way for designing effective theranostic agents in singular as well as combinatorial mode where the concentration affects streamlining that needs to be tapped. The results of latest therapeutic models need to be compiled as prompt database so that every subsequent trial attempts to monitor a new therapeutic linkage with the previously successful attempts.

Acknowledgments Authors would like to thank Central University of Gujarat, Gandhinagar, for all the infrastructural and round the clock internet facility.

References

Alvarez EV, Cambón A, Pardo A, Mosquera VX, Mosquera AB, Topete A, Barbosa S, Taboada P, Mosquera V (2018) Gold nanorod-based nanohybrids for combinatorial therapeutics. ACS Omega 3:12633–12647

Amos CI, Wu X, Broderick P, Gorlov IP, Gu J, Eisen T, Dong Q, Zhang Q, Gu X, Vijayakrishnan J, Sullivan K, Matakidou A, Wang Y, Mills G, Doheny K, Tsai YY, Chen WV, Shete S, Spitz MR, Houlston RS (2008) Genome-wide association scan of tag SNPs identifies a susceptibility locus for lung cancer at 15q25.1. Nat Genet 40:616–622

Arnida, Janát-Amsbury MM, Ray A, Peterson CM, Ghandehari H (2011) Geometry and surface characteristics of gold nanoparticles influence their biodistribution and uptake by macrophages. Eur J Pharm Biopharm 77:417–423

C.H. Baker, D. Kedar, M.F. McCarty, R. Tsan R, K.L. Weber, C.D. Bucana and I.J. Fidler (2002) Blockade of epidermal growth factor receptor signaling on tumor cells and tumor-associated endothelial cells for therapy of human carcinomas. Am J Pathol 161:929–938

Bass AJ, Watanabe H, Mermel CH, Yu S, Perner S, Verhaak RG, Kim SY, Wardwell L, Tamayo P, Gat-Viks I, Ramos AH, Woo MS, Weir BA, Getz G, Beroukhim R, O'Kelly M, Dutt A, Rozenblatt-Rosen O, Dziunycz P, Komisarof J, Chirieac LR, Lafargue CJ, Scheble V, Wilbertz T, Ma C, Rao S, Nakagawa H, Stairs DB, Lin L, Giordano TJ, Wagner P, Minna JD, Gazdar AF, Zhu CQ, Brose MS, Cecconello I, Ribeiro U Jr, Marie SK, Dahl O, Shivdasani RA, Tsao MS, Rubin MA, Wong KK, Regev A, Hahn WC, Beer DG, Rustgi AK, Meyerson M (2009) SOX2 is an amplified lineage-survival oncogene in lung and esophageal squamous cell carcinomas. Nat Genet 41:1238–1242

Bates DO, Hillman NJ, Williams B, Neal CR, Pocock TM (2002) Regulation of microvascular permeability by vascular endothelial growth factors. J Anat 200:581–597

Bazak R, Houri M, Elachy S, Hussein W, Reefat T (2014) Passive targeting of nanoparticles to cancer: A comprehensive review of the literature. Mol Clin Oncol 2:904–908

Bazak R, Houri M, Alchy SE, Kamel S, Refaat T (2015) Cancer active targeting by nanoparticles: a comprehensive review of literature. J Cancer Res Clin Oncol 141:769–784

Bertrand N, Wu J, Xu X, Kamaly N, Farokhzad OC (2014) Cancer nanotechnology: the impact of passive and active targeting in the era of modern cancer biology. Adv Drug Deliv Rev 66:2–25

Bhirde AA, Patel V, Gavard J, Zhang G, Sousa AA, Masedunskas A, Leapman RD, Weigert R, Gutkind JS, Rusling JF (2009) Targeted killing of cancer cells in vivo and in vitro with EGF-directed carbon nanotubebased drug delivery ACS Nano 3:307–316

Cabral H, Matsumoto Y, Mizuno K, Chen Q, Murakami M, Kimura M, Terada Y, Kano MR, Miyazono K, Uesaka M, Nishiyama N, Kataoka K (2011) Accumulation of sub-100 nm polymeric micelles in poorly permeable tumours depends on size. Nat Nanotechnol 6:815–823

Carretero J, Medina PP, Pio R, Montuenga LM, Sanchez-Cespedes M (2004) Novel and natural knockout lung cancer cell lines for the LKB1/STK11 tumor suppressor gene. Oncogene 23:4037–4040

Chauhan VP, Popović Z, Chen O, Cui J, Fukumura D, Bawendi MG, Jain RK (2011) Fluorescent nanorods and nanospheres for real-time in vivo probing of nanoparticle shape-dependent tumor penetration. Angew Chem Int Ed 50:11417–11420

Chow G, Tauler J, Mulshine JL (2010) Cytokines and growth factors stimulate hyaluronan production: role of hyaluronan in epithelial to mesenchymal-like transition in non-small cell lung cancer. J Biomed Biotechnol 2010:485468

Cooper WA, Lam DCL, O'Toole SA, Minna JD (2013) Molecular biology of lung cancer. J Thorac Dis 5:S479–S490

Davidson MR, Gazdar AF, Clarke BE (2013) The pivotal role of pathology in the management of lung cancer. J Thorac Dis 5:S463–S478

Ding L, Getz G, Wheeler DA, Mardis ER, McLellan MD, Cibulskis K, Sougnez C, Greulich H, Muzny DM, Morgan MB, Fulton L, Fulton RS, Zhang Q, Wendl MC, Lawrence MS, Larson DE, Chen K, Dooling DJ, Sabo A, Hawes AC, Shen H, Jhangiani SN, Lewis LR, Hall O, Zhu Y, Mathew T, Ren Y, Yao J, Scherer SE, Clerc K, Metcalf GA, Ng B, Milosavljevic A, Gonzalez-Garay ML, Osborne JR, Meyer R, Shi X, Tang Y, Koboldt DC, Lin L, Abbott R, Miner TL, Pohl C, Fewell G, Haipek C, Schmidt H, Dunford-Shore BH, Kraja A, Crosby SD, Sawyer CS, Vickery T, Sander S, Robinson J, Winckler W, Baldwin J, Chirieac LR, Dutt A, Fennell T, Hanna M, Johnson BE, Onofrio RC, Thomas RK, Tonon G,Weir BA, Zhao X, Ziaugra L, Zody MC, Giordano T, Orringer MB, Roth JA, Spitz MR, Wistuba II, Ozenberger B, Good PJ, AC ACC, Beer DG, Watson MA, Ladanyi M, Broderick S, Yoshizawa A, Travis WD, Pao W, Province MA, Weinstock GM, Varmus HE, Gabriel SB, Lander ES, Gibbs RA, Meyerson M, Wilson RK (2008) Somatic mutations affect key pathways in lung adenocarcinoma. Nature 455:1069–1075

Downward J (2003) Targeting RAS signalling pathways in cancer therapy. Nat Rev Cancer 3:11–22

Dreher MR, Liu W, Michelich CR, Dewhirst MW, Yuan F, Chilkoti A (2006) Tumor vascular permeability, accumulation, and penetration of macromolecular drug carriers. J Nat Cancer Inst 98:335–344

Eren B, Sar M, Oz B, Dincbas FH (2008) MMP-2, TIMP-2 and CD44v6 Expression in Non-small-cell lung carcinomas. Ann Acad Med Singap 37:32–39

Fahmy TM, Samstein RM, Harness CC, Mark Saltzman W (2005) Surface modification of biodegradable polyesters with fatty acid conjugates for improved drug targeting. Biomaterials 26:5727–5736

Farokhzad OC, Langer R (2009) Impact of nanotechnology on drug delivery. ACS Nano 3:16–20

Feng Q, Hawes SE, Stern JE, Wiens L, Lu H, Dong ZM, Jordan CD, Kiviat NB, Vesselle H (2008) DNA methylation in tumor and matched normal tissues from non-small cell lung cancer patients. Cancer Epidemiol Biomark Prev 17:645–654

Franklin WA, Veve R, Hirsch FR, Helfrich BA, Bunn PA Jr (2002) Epidermal growth factor receptor family in lung cancer and premalignancy. Semin Oncol Suppl 4:3–14

Frens G (1973) Controlled nucleation for the regulation of the particle size in monodisperse gold suspensions. Nat-Phys Sci 241:20–22

Freund PL, Spiro M (1985) Colloidal catalysis: the effect of sol size and concentration. J Phys Chem 89:1074–1077
Friend SH, Bernards R, Rogelj S, Weinberg RA, Rapaport JM, Albert DM, Dryja TP (1986) A human DNA segment with properties of the gene that predisposes to retinoblastoma and osteosarcoma. Nature 323:643–646
Gonzalez-Arriaga P, Pascual T, Garcia-Alvarez A, Fernandez-Somoano A, Lopez-Cima MF, Tardon A (2012) Genetic polymorphisms in MMP 2, 9 and 3genes modify lung cancer risk and survival. BMC Cancer 12:121
Greenblatt MS, Bennett WP, Hollstein M, Harris CC (1994) Mutations in the p53 tumor suppressor gene: clues to cancer etiology and molecular pathogenesis. Cancer Res 54:4855–4878
Gupta R, Malik P, Das N, Singh M (2019) Antioxidant and physicochemical study of Psidium guajava prepared zinc oxide nanoparticles. J Mol Liq 275:749–767
Harbour JW, Lai SL, Whang-Peng J, Gazdar AF, Minna JD, Kaye FJ (1988) Abnormalities in structure and expression of the human retinoblastoma gene in SCLC. Science 241:353–357
Hauser EA, Lynn JE (1940) Experiments in colloid chemistry. New York, McGraw-Hill
Hay N (2005) The Akt-mTOR tango and its relevance to cancer. Cancer Cell 8:179–183
Heist RS, Fidias P, Huberman M, Ardman B, Sequist LV, Temel JS, Lynch TJ (2008) A phase II study of oxaliplatin, pemetrexed, and bevacizumab in previously treated advanced non-small cell lung cancer. J Thorac Oncol 3:1153–1158
Hemminki A, Markie D, Tomlinson I, Avizienyte E, Roth S, Loukola A, Bignell G, Warren W, Aminoff M, Höglund P, Järvinen H, Kristo P, Pelin K, Ridanpää M, Salovaara R, Toro T, Bodmer W, Olschwang S, Olsen AS, Stratton MR, de la Chapelle A, Aaltonen LA (1998) A serine/threonine kinase gene defective in Peutz-Jeghers syndrome. Nature 391:184–187
Herbst RS (2004) Review of epidermal growth factor receptor biology. Int J Radiat Oncol Biol Phys 59(Suppl):21–26
Hirsch FR, Varella-Garcia M, Bunn PA Jr, Di Maria MV, Veve R, Bremmes RM, Barón AE, Zeng C, Franklin WA (2003) Epidermal growth factor receptor in non-small-cell lung carcinomas: correlation between gene copy number and protein expression and impact on prognosis. J Clin Oncol 21:3798–3807
Hollstein M, Sidransky D, Vogelstein B, Harris CC (1991) p53 mutations in human cancers. Science 253:49–53
Hung RJ, McKay JD, Gaborieau V, Boffetta P, Hashibe M, Zaridze D, Mukeria A, Szeszenia-Dabrowska N, Lissowska J, Rudnai P, Fabianova E, Mates D, Bencko V, Foretova L, Janout V, Chen C, Goodman G, Field JK, Liloglou T, Xinarianos G, Cassidy A, McLaughlin J, Liu G, Narod S, Krokan HE, Skorpen F, Elvestad MB, Hveem K, Vatten L, Linseisen J, Clavel-Chapelon F, Vineis P, Bueno-de-Mesquita HB, Lund E, Martinez C, Bingham S, Rasmuson T, Hainaut P, Riboli E, Ahrens W, Benhamou S, Lagiou P, Trichopoulos D, Holcátová I, Merletti F, Kjaerheim K, Agudo A, Macfarlane G, Talamini R, Simonato L, Lowry R, Conwayl DI, Znaor A, Healy C, Zelenika D, Boland A, Delepine M, Foglio M, Lechner D, Matsuda F, Blanche H, Gut I, Heath S, Lathrop M, Brennan P (2008) A susceptibility locus for lung cancer maps to nicotinic acetylcholine receptor subunit genes on 15q25. Nature 452:633–637
Huo S, Ma H, Huang K, Liu J, Wei T, Jin S, Zhang J, He S, Liang XJ (2013) Superior penetration and retention behavior of 50 nm gold nanoparticles in tumors. Cancer Res 73:319–330
Ito M, Ito G, Kondo M, Uchiyama M, Fukui T, Mori S, Yoshioka H, Ueda Y, Shimokata K, Sekido Y (2005) Frequent inactivation of RASSF1A, BLU, and SEMA3B on 3p21.3 by promoter hypermethylation and allele loss in non-small cell lung cancer. Cancer Lett 225:131–139
Jain RK (1987) Transport of molecules across tumor vasculature. Cancer Metastasis Rev 6:559–593
Jain RK, Stylianopoulos T (2010) Delivering nanomedicine to solid tumors. Nat Rev Clin Oncol 7:653–664
Jain PK, Lee KS, El-Sayed IH, El-Sayed MA (2006) Calculated absorption and scattering properties of gold nanoparticles of different size, shape, and composition: applications in biological imaging and biomedicine. J Phys Chem B 110:7238–7248

Kadiyala NK, Mandal BK, Ranjan S, Dasgupta N (2018) Bioinspired gold nanoparticles decorated reduced graphene oxide nanocomposite using Syzygium cumini seed extract: Evaluation of its biological applications. Mater Sci Eng C Mater Biol Appl 93:191–205

Kalinowska D, Grabowska-Jadach I, Liwinska M, Drozd M, Pietrzak M, Dybko A, Brzozka Z (2019) Studies on effectiveness of PTT on 3D tumor model under microfluidic conditions using aptamer-modified nanoshells. Biosens Bioelectron 126:214–221

Kalishwaralal K, Deepak V, Ramkumarpandian S (2008) Biosynthesis of silver nanocrystals by Bacillus licheniformis. Colloids Surf B: Biointerfaces 65:150–153

Karnoub AE, Weinberg RA (2008) Ras oncogenes: split personalities. Nat Rev Mol Cell Biol 9:517–531

Kendall J, Liu Q, Bakleh A, Krasnitz A, Nguyen KC, Lakshmi B, Gerald WL, Powers S, Mu D (2007) Oncogenic cooperation and coamplification of developmental transcription factor genes in lung cancer. Proc Natl Acad Sci USA 104:16663–16668

Kim W, Youn H, Kang C, Youn B (2015) Inflammation-induced radioresistance is mediated by ROS-dependent inactivation of protein phosphatase 1 in non-small cell lung cancer cells. Apoptosis 20:1242–1252

Kim E, Kim W, Lee S, Chun J, Kang JH, Park G, Han IJ, Yang HJ, Youn HS, Youn BH (2017) TRAF4 promotes lung cancer aggressiveness by modulating tumor microenvironment in normal fibroblasts. Sci Rep 7:8923

Korpanty G, Smyth E, Sullivan LA, Brekken RA, Carney DN (2010) Antiangiogenic therapy in lung cancer: focus on vascular endothelial growth factor pathway. Exp Biol Med 235:3–9

Krystal G, Birrer M, Way J, Nau M, Sausville E, Thompson C, Minna J, Battey J (1988) Multiple mechanisms for transcriptional regulation of the myc gene family in small-cell lung cancer. Mol Cell Biol 8:3373–3381

Kuroki T, Trapasso F, Yendamuri S, Matsuyama A, Alder H, Williams NN, Kaiser LR, Croce CM (2003) Allelic loss on chromosome 3p21.3 and promoter hypermethylation of semaphoring 3B in non-small cell lung cancer. Cancer Res 63:3352–3355

Larsen JE, Minna JD (2011) Molecular biology of lung cancer: clinical implications. Clin Chest Med 32:703–740

Lasagna-Reeves C, Gonzalez-Romero D, Barria MA, Olmedo I, Clos A, Sadagopa Ramanujam VM, Urayama A, Vergara L, Kogan MJ, Soto C (2010) Bioaccumulation and toxicity of gold nanoparticles after repeated administration in mice. Biochem Biophys Res Commun 393:649–655

Lee JH, Kim YS, Song KS, Ryu HR, Sung JH, Park JD, Park HM, Song NW, Shin BS, Marshak D, Ahn K, Lee JE, Yu IJ (2013) Biopersistence of silver nanoparticles in tissues from Sprague-Dawley rats. Part Fibre Toxicol 10:36

Liaskoni A, Angelopoulou A, Voulgari E, Popescu MT, Tsitsilianis C, Avgoustakis K (2018) Paclitaxel controlled delivery using a pH-responsive functional-AuNP/block-copolymer vesicular nanocarrier composite system. Eur J Pharm Sci 117:177–186

Link S, El-Sayed MA (2010) Shape and size dependence of radiative, non-radiative and photothermal properties of gold nanocrystals. Int Rev Phys Chem 19:409–453

Liu R, Xiao W, Hu C, Xie R, Gao H (2018) Theranostic size-reducible and no donor conjugated gold nanocluster fabricated hyaluronic acid nanoparticle with optimal size for combinational treatment of breast cancer and lung metastasis. J Control Release 278:127–139

Lynch TJ, Bell DW, Sordella R, Gurubhagavatula S, Okimoto RA, Brannigan BW, Harris PL, Haserlat SM, Supko JG, Haluska FG, Louis DN, Christiani DC, Settleman J, Haber DA (2004) Activating mutations in the epidermal growth factor receptor underlying responsiveness of non-small-cell lung cancer to gefitinib. N Engl J Med 350:2129–2139

Maeda H (2001) SMANCS and polymer-conjugated macromolecular drugs: advantages in cancer chemotherapy. Adv Drug Deliv Rev 46:169–185

Mahoney CL, Choudhury B, Davies H, Edkins S, Greenman C, van Haaften G, Mironenko T, Santarius T, Stevens C, Stratton MR, Futreal PA (2009) LKB1/KRAS mutant lung cancers constitute a genetic subset of NSCLC with increased sensitivity to MAPK and mTOR signalling inhibition. Br J Cancer 100:370–375

Malekmohammadi S, Hadadzadeh H, Farrokhpour H, Amirghofran Z (2018) Immobilization of gold nanoparticles on folate-conjugated dendritic mesoporous silica-coated reduced graphene oxide nanosheets: a new nanoplatform for curcumin pH-controlled and targeted delivery. Soft Matter 14:2400. https://doi.org/10.1039/C7SM02248D

Malik P, Mukherjee TK (2018) Recent advances in gold and silver nanoparticle based therapies for lung and breast cancers. Int J Pharm 553:483–509

Malik P, Shankar R, Malik V (2014) Green chemistry based benign routes for nanoparticle synthesis. J Nanoparticles 2014:14. Article ID 302429

Massion PP, Kuo WL, Stokoe D, Olshen AB, Treseler PA, Chin K, Chen C, Polikoff D, Jain AN, Pinkel D, Albertson DG, Jablons DM, Gray JW (2002) Genomic copy number analysis of non-small cell lung cancer using array comparative genomic hybridization: implications of the phosphatidylinositol 3-kinase pathway. Cancer Res 62:3636–3640

Matsumoto S, Iwakawa R, Takahashi K, Kohno T, Nakanishi Y, Matsuno Y, Suzuki K, Nakamoto M, Shimizu E, Minna JD, Yokota J (2007) Prevalence and specificity of LKB1 genetic alterations in lung cancers. Oncogene 26:5911–5918

Matsumura Y, Maeda H (1986) A new concept for macromolecular therapeutics in cancer chemotherapy: mechanism of tumoritropic accumulation of proteins and the antitumor agent smancs. Cancer Res 46:6387–6392

Merlino GT, Xu YH, Ishii S, Clark AJ, Semba K, Toyoshima K, Yamamoto T, Pastan I (1984) Amplification and enhanced expression of the epidermal growth factor receptor gene in A431 human carcinoma cells. Science 224:417–419

Miller HA, Frieboes HB (2018) Evaluation of drug-loaded gold nanoparticle cytotoxicity as a function of tumor vasculature-induced tissue heterogeneity. Ann Biomed Eng 47:257. https://doi.org/10.1007/s10439-018-02146-4

Mori A, Klibanov AM, Huang TL (1991) Influence of the steric barrier activity of amphipathic poly(ethyleneglycol) and ganglioside GM1 on the circulation time of liposomes and on the target binding of immunoliposomes in vivo. FEBS Lett 284:263–266

Mukhopadhyay S, Malik P, Arora SK, Mukherjee TK (2013) Role of β1 integrins in the complication and drug resistance against lung cancer: targeting β1 integrins to eradicate lung cancer. In: Bonavida B (ed) Molecular mechanisms of tumor cell resistance to chemotherapy, resistance to targeted anti-Cancer therapeutics, Vol 1. https://doi.org/10.1007/978-1-4614-7070-0-5

Murakami Y, Tabata Y, Ikada Y (1997) Tumor accumulation of poly(ethylene glycol) with different molecular weights after intravenous injection. Drug Deliv 4:23–31

Murphy CJ, Thompson LB, Alkilany AM, Sisco PN, Boulos SP, Sivapalan ST, Yang JA, Chernak DJ, Huang J (2010) The many faces of gold nanorods. J Phys Chem Lett 1:2867–2875

Nakache E, Poulain N, Candau F, Orecchioni AM, Irache JM (1999) In: Nalwa HS (ed) Handbook of nanostructured materials and nanotechnology. Academic, New York

Nobori T, Miura K, Wu DJ, Lois A, Takabayashi K, Carson DA (1994) Deletions of the cyclin-dependent kinase-4 inhibitor gene in multiple human cancers. Nature 368:753–756

Normanno N, Bianco C, Strizzi L, Mancino M, Maiello MR, De Luca A, Caponigro F, Salomon DS (2005) The ErbB receptors and their ligands in cancer: an overview. Curr Drug Targets 6:243–257

Ojea-Jiménez I, Romero FM, Bastius NG, Puntes V (2010) Small gold nanoparticles synthesized with sodium citrate and heavy water: insights into the reaction mechanism. J Phys Chem C 114:1800–1804

Paez JG, Janne PA, Lee JC, Tracy S, Greulich H, Gabriel S, Herman P, Kaye FJ, Lindeman N, Boggon TJ, Naoki K, Sasaki H, Fujii Y, Eck MJ, Sellers WR, Johnson BE, Meyerson M (2004) EGFR mutations in lung cancer: correlation with clinical response to gefitinib therapy. Science 304:1497–1500

Pao W, Miller V, Zakowski M, Doherty J, Politi K, Sarkaria I, Singh B, Heelan R, Rusch V, Fulton L, Mardis E, Kupfer D, Wilson R, Kris M, Varmus H (2004) EGF receptor gene mutations are

common in lung cancers from "never smokers" and are associated with sensitivity of tumors to gefitinib and erlotinib. Proc Natl Acad Sci 101:13306–13311
Pei L, Mori K, Adachi M (2004) Formation process of two-dimensional networked gold nanowires by citrate reduction of AuCl4- and the shape stabilization. Langmuir 20:7837–7843
Perrault SD, Walkey C, Jennings T, Fischer HC, Chan WC (2009) Mediating tumor targeting efficiency of nanoparticles through design. Nano Lett 9:1909–1915
Pourali P, Badiee SH, Manafi S, Noorani T, Rezaei TA, Yahyaei B (2017) Biosynthesis of gold nanoparticles by two bacterial and fungal strains, Bacillus cereus and Fusarium oxysporum, and assessment and comparison of their nanotoxicity in vitro by direct and indirect assays. Electron J Biotechnol 29:86–93
Prabhakar U, Blakey DC, Maeda H, Jain RK, Sevick-Muraca EM, Zamboni W, Farokhzad OC, Barry ST, Gabizon A, Grodzinski P, Blakey DC (2013) Challenges and key considerations of the enhanced permeability and retention effect for nanomedicine drug delivery in oncology. Cancer Res 73:2412–2417
Qiu Y, Liu Y, Wang LM, Xu LG, Bai R, Ji Y, Wu X, Zhao Y, Li Y, Chen C (2010) Surface chemistry and aspect ratio mediated cellular uptake of Au nanorods. Biomaterials 31:7606–7619
Rajeshkumar S (2016) Anticancer activity of eco-friendly gold nanoparticles against lung and liver cancer cells. J Genet Eng Biotechnol 14:195–202
Richardson GE, Johnson BE (1993) The biology of lung cancer. Semin Oncol 20:105–127
Rizvi SAA, Saleh AM (2018) Applications of nanoparticle systems in drug delivery technology. Saudi Pharm J 26:64–70
Roh MS (2014) Molecular pathology of lung cancer: current status and future directions. Tuberc Respir Dis 77:49–54
Roth J (1996) The silver anniversary of gold: 25 years of the colloidal gold marker system for immunocytochemistry and histochemistry. Histochem Cell Biol 106:1–8
Ruggiero A, Villa CH, Bander E, Rey DA, Bergkvist M, Batt CA, Manova-Todorova K, Deen WM, Scheinberg DA, MR MD (2010) Paradoxical glomerular filtration of carbon nanotubes. Proc Natl Acad Sci USA 107:12369–12374
Salvati A, Pitek AS, Monopoli MP, Prapainop K, Bombelli FB, Hristov DR, Kelly PM, Aberg C, Mahon E, Dawson KA (2013) Transferrin-functionalized nanoparticles lose their targeting capabilities when a biomolecule corona adsorbs on the surface. Nat Nanotechnol 8:137–143
Sanchez-Cespedes M, Parrella P, Esteller M, Nomoto S, Trink B, Engles JM, Westra WH, Herman JG, Sidransky D (2002) Inactivation of LKB1/STK11 is a common event in adenocarcinomas of the lung. Cancer Res 62:3659–3662
Sanna V, Pala N, Sechi M (2014) Targeted therapy using nanotechnology: focus on cancer. Int J Nanomed 9:467–483
Sauter W, Rosenberger A, Beckmann L, Kropp S, Mittelstrass K, Timofeeva M, Wölke G, Steinwachs A, Scheiner D, Meese E, Sybrecht G, Kronenberg F, Dienemann H, LUCY-Consortium, Chang-Claude J, Illig T, Wichmann HE, Bickeböller H, Risch A (2008) Matrix metalloproteinase 1 (MMP1) is associated with early-onset lung cancer. Cancer Epidemiol Biomark Prev 17:1127–1135
Schwerdtfeger P (1989) Relativistic effects in gold chemistry. 2. The stability of complex halides of gold (III). J Am Chem Soc 111:7261–7262
Shahhoseini E, Ramachandran P, Patterson WR, Geso M (2018) Determination of dose enhancement caused by AuNPs with Xoft® Axxent® Electronic (eBx™) and conventional brachytherapy: in vitro study. Int J Nanomed 13:5733–5741
Soda M, Choi YL, Enomoto M, Takada S, Yamashita Y, Ishikawa S, Fujiwara S, Watanabe H, mKurashina K, Hatanaka H, Bando M, Ohno S, Ishikawa Y, Aburatani H, Niki T, Sohara Y, Sugiyama Y, Mano H (2007) Identification of the transforming EML4-ALK fusion gene in non-small-cell lung cancer. Nature 448:561–566
Steck PA, Pershouse MA, Jasser SA, Yung WK, Lin H, Ligon AH, Langford LA, Baumgard ML, Hattier T, Davis T, Frye C, Hu R, Swedlund B, Teng DH, Tavtigian SV (1997) Identification of a candidate tumour suppressor gene, MMAC1, at chromosome 10q23.3 that is mutated in multiple advanced cancers. Nat Genet 15:356–362

Stefanick JF, Ashley JD, Kiziltepe T, Bilgicer B (2013) A systematic analysis of peptide linker length and liposomal polyethylene glycol coating on cellular uptake of peptide-targeted liposomes. ACS Nano 7:2935–2947

Sudimack J, Lee RJ (2000) Targeted drug delivery via the folate receptor. Adv Drug Deliv Rev 41:147–162

Takahashi T, Nau MM, Chiba I, Birrer MJ, Rosenberg RK, Vinocour M, Levitt M, Pass H, Gazdar AF, Minna JD (1989) p53: a frequent target for genetic abnormalities in lung cancer. Science 246:491–494

Takiyama K (1958) Formation and aging of precipitates. viii. formation of monodisperse particles (1) gold sol particles by sodium citrate method. Bull Chem Soc Jpn 31:944–950

Thorgeirsson TE, Geller F, Sulem P, Rafnar T, Wiste A, Magnusson KP, Manolescu A, Thorleifsson G, Stefansson H, Ingason A, Stacey SN, Bergthorsson JT, Thorlacius S, Gudmundsson J, Jonsson T, Jakobsdottir M, Saemundsdottir J, Olafsdottir O, Gudmundsson LJ, Bjornsdottir G, Kristjansson K, Skuladottir H, Isaksson HJ, Gudbjartsson T, Jones GT, Mueller T, Gottsäter A, Flex A, Aben KKH, de Vegt F, Mulders PFA, Isla D, Vidal MJ, Asin L, Saez B, Murillo L, Blondal T, Kolbeinsson H, Stefansson JG, Hansdottir I, Runarsdottir V, Pola R, Lindblad B, van Rij AM, Dieplinger B, Haltmayer M, Mayordomo JI, Kiemeney LA, Matthiasson SE, Oskarsson H, Tyrfingsson T, Gudbjartsson DF, Gulcher JR, Jonsson S, Thorsteinsdottir U, Kong A, Stefansson K (2008) A variant associated with nicotine dependence, lung cancer and peripheral arterial disease. Nature 452:638–642

Turkevich J, Stevenson PC, Hillier J (1951) A study of the nucleation and growth processes in the synthesis of colloidal gold. Discuss Faraday Soc 11:55–75

United States Department of Agriculture (2002) Nanoscale science and engineering for agriculture and food systems; Report submitted to Cooperative State Research, Education and Extension Service. United States Department of Agriculture, National Planning Workshop, Washington, DC

Vijayan R, Joseph S, Mathew B (2017) Indigofera tinctoria leaf extract mediated green synthesis of silver and gold nanoparticles and assessment of their anticancer, antimicrobial, antioxidant and catalytic properties. Artif Cells Nanomed Biotechnol 46:861–871

Wakaskar RR (2017) Passive and active targeting in tumor microenvironment. Int J Drug Dev Res 9:37–41

Watson CJ, Webb NJ, Bottomley MJ, Brenchley PE (2000) Identification of polymorphisms within the vascular endothelial growth factor (VEGF) gene: correlation with variation in VEGF protein production. Cytokine 12:1232–1235

Wei L, Song XR, Sun JJ, Wang XW, Xie L, Lv LY (2012) Lysyl oxidase may play a critical role in hypoxia-induced NSCLC cells invasion and migration. Cancer Biother Radiopharm 27:672–677

Weir BA, Woo MS, Getz G, Perner S, Ding L, Beroukhim R, Lin WM, Province MA, Kraja A, Johnson LA, Shah K, Sato M, Thomas RK, Barletta JA, Borecki IB, Broderick S, Chang AC, Chiang DY, Chirieac LR, Cho J, Fujii Y, Gazdar AF, Giordano T, Greulich H, Hanna M, Johnson BE, Kris MG, Lash A, Lin L, Lindeman N, Mardis ER, McPherson JD, Minna JD, Morgan MB, Nadel M, Orringer MB, Osborne JR, Ozenberger B, Ramos AH, Robinson J, Roth JA, Rusch V, Sasaki H, Shepherd F, Sougnez C, Spitz MR, Tsao MS, Twomey D, Verhaak RG, Weinstock GM, Wheeler DA, Winckler W, Yoshizawa A, Yu S, Zakowski MF, Zhang Q, Beer DG, Wistuba II, Watson MA, Garraway LA, Ladanyi M, Travis WD, Pao W, Rubin MA, Gabriel SB, Gibbs RA, Varmus HE, Wilson RK, Lander ES, Meyerson M (2007) Characterizing the cancer genome in lung adenocarcinoma. Nature 450:893–898

West KA, Linnoila IR, Belinsky SA, Harris CC, Dennis PA (2004) Tobacco carcinogen-induced cellular transformation increases activation of the phosphatidylinositol 3'-kinase/Akt pathway in vitro and in vivo. Cancer Res 64:446–451

Wilhelm S, Tavares AJ, Dai Q, Ohta S, Audet J, Dvorak HF, Chan WCW (2016) Analysis of nanoparticle delivery to tumours. Nat Rev 1:1–12

Winslow MM, Dayton TL, Verhaak RG, Kim-Kiselak C, Snyder EL, Feldser DM, Hubbard DD, DuPage MJ, Whittaker CA, Hoersch S, Yoon S, Crowley D, Bronson RT, Chiang DY,

Meyerson M, Jacks T (2011) Suppression of lung adenocarcinoma progression by Nkx2-1. Nature 473:101–104

Wong C, Stylianopoulos T, Cui J, Martin J, Chauhan VP, Jiang W, Popović Z, RK RKJ, Bawendi MG, Fukumura D (2011) Multistage nanoparticle delivery system for deep penetration into tumor tissue. Proc Natl Acad Sci 108:2426–2431

Wu Y, Xu J, Chen J, Zou M, Rusidanmu A, Yang R (2018) Blocking transferrin receptor inhibits the growth of lung adenocarcinoma cells in vitro. Thorac Cancer 9:253–261

Xie H, Hu J, Pan H, Lou Y, Lv P, Chen Y (2014) Adenovirus vector-mediated FAM176A overexpression induces cell death in human H1299 non-small cell lung cancer cells. BMB Rep 47:104–109

Xie Y, Killinger B, Moszczynska A, Merkel OM (2016) Targeted delivery of sirna to transferrin receptor overexpressing tumor cells via peptide modified polyethylenimine. Molecules 21: E1334

Yatsimirskii KB (1995) Relativistic effects in chemistry. Theor Exp Chem 31:153–168

Yokota J, Mori N, Akiyama T, Shimosato Y, Sugimura T, Terada M (1989) Multiple genetic alterations in small-cell lung carcinoma. Princess Takamatsu Symp 20:43–48

Zhao X, Li H, Lee RJ (2008) Targeted drug delivery via folate receptors. Exp Opin Drug Del 5:309–319

Part XII

Medical Ethics and Policies

Medical Ethics and Policies Related to Biomedical Equipment

30

Ramkrishna Mondal

Abstract

The relationship between a doctor and a patient is rightly stated by Charaka: 'No other gift is greater than the gift of life'. Sometime patient may not believe his sons or even parents, but he believe his physician. It is the physician's duty to care for the patient as his own relative. Rapid changes in medical science and technology have outspread numerous complicated social, ethical and legal issues along with medical. To establish the faith and trust, ethics is a must. Every doctor has some ethical duties to be carried out for his patient, colleagues and society in large. In this chapter, various ethical issues in different situations like poor patient, female patient and clinical research were discussed in details. Though technology has advanced a lot, still it is a sad reality that physician's focus has been gradually shifting towards cell or even DNA level instead of the patient as a whole.

Keywords

Ethics · Oath · Medical ethics · Geneva Declaration

30.1 Introduction

'The test of a man's real character is what he would do if knew he would never be found out'. – Thomas Macaulay.

'Ethics: A set of rules laid out by professionals to show the way they would like to act if it was profitable'. – Frank Dane.

Ethics has been defined in various ways. One of the definitions is as follows:

R. Mondal (✉)
DMS, IQ City Medical College, Durgapur, West Bengal, India

S. Paul (ed.), *Biomedical Engineering and its Applications in Healthcare*,
https://doi.org/10.1007/978-981-13-3705-5_30

It is a science that is concerned with the nature of moral obligations distinguishing right from wrong and the reason of the same.

30.1.1 Ethics and Moral

'*Ethics*' a Greek origin word meaning custom, habits, whereas.

'*Moral*' is a Latin-origin word 'Mos' or 'Moris' meaning manners.

30.2 Medicine as Profession

The Supreme Court of India in a case of *Shantha VP* vs *IMA* made clear distinction between an occupation and a profession.

An occupation is considered a profession if it fulfils the following criteria:

- The occupation should involve intellectual work or skilled controlled manual work.
- The practitioners should have an association to control their activities.
- They should adhere to some ethical principles.
- They should have a high status in the society.

30.3 Code of Ethics in Medical Practice

Ethical codes have evolved over time. Over 4000 years ago, the King of Hammurabi of Babylonia codified the laws of human behaviour. These codes included penalties for physicians and surgeons who did not cure.

Today we have codes of Hippocrates, Charaka and Susruta.

30.3.1 The Declaration of Geneva

Due to the atrocities of World War II, the World Medical Association was formed at the instigation of British Medical Council, and it restarted the Hippocratic oath in a new manner, known as the Declaration of Geneva, in 1948. Which was approved by the General Assembly of World Medical Association in 1949, known as The International Code of Medical Ethics. The code of conduct subsequently amended at Sydney in 1968 and again in 1983 in Venice.

30.4 The International Medical Ethics and Medical Profession

The Geneva Declaration formulated following duties for the doctors.

(A) General duties of doctors:
  (a) Always maintain highest standard of professional conduct.
  (b) Always practice the profession and exercise free and independent judgment uninfluenced by motive of profit.
  (c) Deals honestly with patients and fellow colleagues and try to expose incompetent or fraud physicians.
(B) Duties of the doctor to the sick:
  (a) Preservation of human life is an obligation on him.
  (b) He has to give full loyalty to his patients and all benefits of resources at his command.
  (c) If he feels inadequacy in examination or treatment of the patient, he must refer or summon other physician who has the ability or who can treat.
  (d) He will maintain absolute secrecy and respect the confidence entrusted in him. Such knowledge as privileged communication and absolute confidentiality has to be maintained even after the death of the patient.
  (e) He shall dedicatedly provide his competent medical services in full technical and moral independence with kind-heartedness and admiration for human dignity.
  (f) He shall respect the patients' rights and also respect colleagues and other health professionals.
  (g) Always extend emergency treatment as a philanthropic duty when necessary, except when he assured that patient will get it from other doctors.
(C) Duties of doctors towards each other:
  (a) Treat his colleagues as he would like himself to be treated.
  (b) Will not attract the patients from his colleagues.
  (c) Follow the doctrines of ethics of his law of land.

Practices considered as unethical conduct:

1. Any self-advertisement except permissible under the law of land.
2. Working in collaboration with form of medical science in which he does not have professional independence.
3. Receiving extra remuneration or money.
4. Paying or receiving any fees or any consideration for referral.
5. Any evidence or act which weaken the patient.
6. Testing or certifying which he has no information about.
7. Use great caution in making claims about new techniques which he has not been satisfactorily substantiated.

30.5 Indian Code of Medical Ethics

Permitted by the Central Government U/S 33 (m) of the Indian Medical Council Act 1956, vide letter no F-17-4/64-MPT dated 23.10.70.

At the time of registration, every doctor was given a copy of the declaration by the Registrar of the Medical Council.

Declaration:

1. I solemnly pledge myself to devote my life to service of humanity.
2. I will not use my medical knowledge opposing to the laws of humanity, even under threat.
3. I will maintain the utmost respect for human life from the time of conception.
4. I will not permit considerations of religion, nationality, race, party, politics or social standing to intervene between my duty and my patient.
5. I will practice my profession with integrity and dignity.
6. The health of my patient will be my first consideration.
7. I will respect the secrets which are confided in me.
8. I will give my teachers the respect and gratitude which is their due.
9. I will sustain, by all means in my power, the honour and moral traditions of medical profession.
10. My colleagues will be my brothers.

I make the above undertakings solemnly, liberally and upon my honour.

30.6 Professional Secrecy

Secrecy is an important component of medical ethics. Whatever a physician learned during the treatment of the patient is to be treated as confidential, but the physician has responsibility to society too.

In such cases, disclosure of communications is called privileged communication. It's of two types:

1. Absolute: any statement made in court of law, Parliament, etc.
2. Qualified: any statement made by doctor beside above in the following situation:
   (a) Statement made in good faith without malice anyone.
   (b) Only relevant information and not the whole without any identity.

30.7 Ethical Issues and Poor Patients

- Every patient should be provided healthcare according to his need and not by status of poor or rich.
- Both poor and rich should get similar treatment for similar diseases.

- Privacy of poor patient too should be respected.
- Poor should not be lured to trade organ.
- Poor should not be exploited sexually.

30.8 Ethical Issues and Female Patients

- All men and women should be treated equally.
- Privacy should be ensured, and proper consent should be taken before examining female patient in the presence of female attendant.
- Should not be abuse his position for sexual motives.
- Regarding contraception all risk should be explained and keep the matter confidential.
- Regarding Medical Termination of Pregnancy (MTP) act according to MTP Act and keep it confidential.
- Judicious use of drugs in pregnant and nursing mother.
- Proper care should be taken for victims of rape.
- Regarding sex determination, act according to the Pre-conception and Pre-Natal Diagnostic Techniques (PCPNDT) Act.

30.9 Ethical Issues in Clinical Research in Human Being

Three major ethical issues are:

1. Is it ethical to have a control group?
2. Is it ethical to give placebo to some patients?
3. Last but not least whether patient consent was taken?

Human rights in research:

- Right of privacy.
- Right of self-determination to give informed consent.
- Right of conservation of personal resources.
- Right to freedom from arbitrary harm.
- Right to minors and incompetent persons.

Informed consent should include the followings six basic elements:

- A non-discriminatory description of the procedure.
- A depiction of discomforts and expected risks.
- A description of any expected benefits.

- Confession of all appropriate alternative procedures if any available.
- A proposition to answer all questions regarding the procedure.
- An education that participant is free to withdraw the consent at any time and discontinue the procedure.

30.10 Ethical Codes and Declarations Relevant to Health Professionals

Physicians
1. Declaration of Geneva.
2. International code of Medical Ethics.
3. Declaration of Tokyo.
4. Declaration of Helsinki.
5. Declaration of Malta on Hunger Strikes.

Psychiatrists
1. Declaration of Madrid.

Nurses
1. Nurses and Human Rights.

Psychologists
1. The Oath of Athens.

30.11 Conclusion

Till today, there were no documented evidence of ethics and policies available specific for biomedical engineers, but the roles played by them are of immense importance. Medical Professionals are still responsible for the use of new modern equipments in the treatment of patients.

Medical professional should use the modern biomedical equipment with great care and compassion. And it is the responsibility of the biomedical engineers or researcher to give accurate and updated knowledge to medical professionals with due consideration of Medical Ethics.

Few other organizations like FDA, HIPPA, WHO take part in the overall aspect of biomedical equipment issues, which is a good indicator towards better healthcare.

The manufacturer's authorised representative in the EU is Springer Nature Customer Service Centre GmbH, Europaplatz 3, 69115 Heidelberg, Germany. If you have any concerns regarding our products, please contact ProductSafety@springernature.com

Printed and bound by CPI Group (UK) Ltd, Croydon, CR0 4YY
22/09/2026
02221048-0017